The nutrition of a vegetative fungal colony can be viewed as a web of interconnected processes. In this volume, the author provides a mechanistic basis to the subject, focusing on processes at the plasma membrane, considering the modulating effects of the fungal wall and describing the fate of nutrients entering the fungus. The emphasis is physiological, but biochemical and molecular biological information has been drawn upon as appropriate to reflect the power of the multi-faceted approach and encourage such study further. A comprehensive review of what is known for the more commonly studied fungal species is complemented by information on other fungi to provide an indication of the diversity of nutritional processes that exist in the Fungal Kingdom.

T0296824

THE PHYSIOLOGY OF FUNGAL NUTRITION

THE PHYSIOLOGY
OF FUNGAL NUTRITION

D. H. JENNINGS
Emeritus Professor of Botany, University of Liverpool

CAMBRIDGE
UNIVERSITY PRESS

CAMBRIDGE UNIVERSITY PRESS
Cambridge, New York, Melbourne, Madrid, Cape Town, Singapore, São Paulo

Cambridge University Press
The Edinburgh Building, Cambridge CB2 8RU, UK

Published in the United States of America by Cambridge University Press, New York

www.cambridge.org
Information on this title: www.cambridge.org/9780521355247

First published 1995
Reprinted 1996
This digitally printed version 2007

A catalogue record for this publication is available from the British Library

Library of Congress Cataloguing in Publication data

Jennings, D. H. (David Harry), 1932–
   The physiology of fungal nutrition / D. H. Jennings.
      p.   cm.
   Includes bibliographical references and index.
   ISBN 0 521 35524 9
   1. Fungi – Physiology.   2. Fungi – Nutrition.   I. Title.
QK601.J46   1995
589.2´0413 – dc20   93-45578CIP

ISBN 978-0-521-35524-7 hardback
ISBN 978-0-521-03816-4 paperback

Contents

Introduction   *page* xiii
1 *Primary active transport*   1
   Introduction   1
   The plasma membrane H^+-ATPase of *Neurospora crassa*   4
   Fungi in general   13
   Plasma membrane redox systems   19
2 *The relationship between membrane transport and growth*   25
   Introduction   25
   The affinity constant   26
   Transport flux   31
   Hyphal extension   33
   Enzyme secretion   39
   Spatial aspects of nutrient acquisition in a colony   41
3 *Walls and membranes*   44
   Introduction   44
   The wall   44
      Introduction   44
      Saccharomyces cerevisiae and other yeasts   45
      Other fungi   55
      Hydrophobicity of walls   56
   The plasma membrane   58
      Introduction   58
      Saccharomyces cerevisiae and other yeasts   59
4 *The vacuolar compartment (vacuole)*   66
   Introduction   66
   Solute concentrations inside the vacuolar compartment   67
   Isolated vacuoles   74
   The tonoplast ATPase   75

The tonoplast pyrophosphatase   79
Transport systems at the tonoplast   81
5   *Carbon*   87
A. Monosaccharide utilisation   87
General features of glucose utilisation in fungi other than
   yeasts   87
Monosaccharide utilisation by yeasts   95
   Glucose utilisation by *Saccharomyces cerevisiae*   96
   Monosaccharide utilisation by other yeasts   101
Anaerobic fungi   108
Ethanol tolerance   109
Monosaccharide transport   117
   Saccharomyces cerevisiae   117
   Kluyveromyces marxianus (Saccharomyces fragilis)   131
   Rhodotorula glutinis   133
   Neurospora crassa   137
   Concluding remarks   139
B. Disaccharide utilisation   140
General   140
Disaccharide transport   145
C. Carbon polymer utilisation   148
Introduction   148
Breakdown of cellulose by *Trichoderma* species   149
Other cellulolytic systems   158
   Rumen fungi   158
   Cellulose degradation by brown-rot fungi   158
Wood degradation by white-rot fungi   160
   Introduction   160
   Studies with *Phanaerochaete chrysosporium*   165
   Selective delignification and resistance to delignification   183
D. One-carbon compounds   185
E. Hydrocarbons   190
6   *Nitrogen*   195
Utilisation of nitrogen compounds   195
Absorption of inorganic nitrogen   198
Assimilation of nitrate   200
Uptake of organic nitrogen compounds (except peptides) into
   Saccharomyces cerevisiae   205
Uptake of amino acids   213
   Achlya   213

Aspergillus nidulans 215
Neurospora crassa 217
Penicillium 221
Peptide utilisation 223
Candida albicans 223
Neurospora crassa 225
Saccharomyces cerevisiae 228
Regulation of nitrogen utilisation 229
Introduction 229
Aspergillus nidulans, Neurospora crassa and *Saccharomyces cerevisiae* 230
Other fungi 239
Ectomycorrhiza of forest trees 242
Protein utilisation 248
Nitrogen storage 250

7 *Phosphorus* 251
Introduction 251
Phosphorus solubilisation 251
Inorganic phosphate 251
Organic phosphates 253
Orthophosphate transport 255
Neurospora crassa 256
Saccharomyces cerevisiae 261
Other fungi 267
Phosphorus compartmentation in the protoplasm 269
Mycorrhiza 282
Orthophosphate uptake 282
Sources of phosphorus other than orthophosphate 285

8 *Sulphur* 288
Introduction 288
Sulphate transport 289
Neurospora crassa, Aspergillus and *Penicillium* spp. 289
Saccharomyces cerevisiae 292
Assimilation of sulphur 294
Storage of sulphur 296
Inorganic sulphur oxidation 298
Tolerance to sulphur dioxide 300

9 *Growth factors* 303
General 303

Membrane transport   308
  Thiamine   308
  Biotin   311
  Vitamin B_6 312
  Riboflavin   313
  Inositol and choline   314
10 *Potassium and other alkali metal cations*   317
  Introduction   317
  Monovalent cation transport   320
    Saccharomyces cerevisiae   320
    Neurospora crassa   332
11 *Multivalent metals (required or toxic)*   336
  Introduction   336
  Transport of multivalent metals (other than iron) across
    the plasma membrane   336
    Specific systems   336
    The question of the presence of less specific systems with
      special reference to *Saccharomyces cerevisiae*   340
    Other fungi   345
  Calcium   347
  Compounds binding multivalent cations   359
    Metallothioneins   360
  Membrane processes in relation to tolerance to metals   365
  Iron   367
  Mycorrhiza   372
12 *Organic acids*   377
  Introduction   377
  Transport   377
  Metabolism of organic acids   382
  Synthesis of organic acids   384
  The role of organic acids in charge balance and control of
    cytoplasmic pH in fungi   392
  Carbon dioxide fixation and oligotrophy   395
13 *Water relations and salinity*   398
  Introduction   398
  Growth   398
    The general picture   398
    Turgor   400
    Some theoretical considerations   403
    Conditions far from equilibrium   411

Contents

Generation of the osmotic (solute) potential 415
Tolerance to low water potential 421
Internal osmotic potential as a function of the external
   water potential 424
  Some theoretical considerations 424
  Experimental observations 426
Cell wall and cell volume changes – a comment 433
The special case of sodium chloride 435
  The need to exclude sodium 435
  Exclusion of sodium 438
  Adaptation for growth in the presence of high
    concentrations of sodium chloride 440
  Genetic studies 442
Compatible solutes 443
14 *Nutrient movement within the colony* 447
Introduction 447
Mechanism of translocation 449
  Mould mycelium 450
  Vesicular–arbuscular mycorrhiza 454
  Serpula lacrymans 456
  Basidiomycetes forming linear aggregates 460
  Basidiomycete sporophores 466
Concluding remarks 467
Literature cited 468
Index 595

Introduction

As far as I am aware, there has only been one other book about fungal nutrition. This contrasts very markedly with the large number of books devoted to the nutrition of higher plants. Here, we have a coherent field of study, well established through the need to understand how plant yield might be increased through the provision of inorganic nutrients. The economic benefits of an effective fertiliser regime for an agricultural or horticultural crop have been an important driving force for establishing plant nutrition as an identifiable discipline in plant physiology. But this identity for plant nutrition is aided in general germs by the fact that the higher plant has a specific organ, the root, for absorbing nutrients and the effectiveness of that organ can be determined by observable responses in other parts of the plant, such as change in shape and colour.

Fungal nutrition is clearly about what kinds and amounts of nutrients will support growth or bring about differentiation, whether it be secondary metabolism on reproductive structures. However, to focus on fungal nutrition thus described would be to produce recipes and little else. This book is about the nutrition of fungi as a web of processes, attempting to provide a mechanistic basis for the subject. For the most part fungi do not produce specialised organs of nutrition, nor, because of their great physiological plasticity, which is much greater than that of higher plants, can one observe in such a clear manner the consequences of changed nutrient conditions for the vegetative colony. Usually, any observable response, apart from a change in growth rate, is differentiation.

It should be apparent that nutrition of the vegetative colony is not easily defined as a topic. The fungus responds to a change in nutrient conditions homeostatically, with changes both qualitatively and quantitatively in a plethora of processes. My focus in this book is on how the fungus interfaces with the external medium, both in terms of acquiring

xiii

nutrients from it and of avoiding the deleterious effects of constituents within it that might be toxic to the fungus. So the emphasis of the volume is for the most part on processes at the plasma membrane as well as a consideration of the wall as a structure modulating the interactions between the external medium and the protoplasm of the fungus. Inevitably one cannot isolate processes occurring at the plasma membrane from other processes occurring within the protoplasm. Here, my decisions as to what or what not to include have been to a certain extent arbitrary. But as a rough rule, I have indicated how a nutrient once entering a fungus interacts with primary metabolism or might have its concentration regulated by internal processes.

In this volume, the external medium is essentially the non-living medium/substratum inhabited for the most part by saprotrophic fungi. I have resisted the temptation to consider parasitic fungi because of the complexity of the interactions between fungus and host but also because my assessment of present knowledge is that our *physiological* understanding of what may be occurring is somewhat limited. To introduce too much speculation would run counter to the general philosophy underlying the text, namely that the emphasis should be on the facts about any process under consideration. Although the book is very much concerned with saprotrophic fungi, I have included, where appropriate, information about parasitic fungi when they are grown in culture and mycorrhizal fungi when they are behaving saprotrophically.

One aim in writing this book has been to indicate for those processes coming within its preview the diversity that exists within the Fungal Kingdom. Mycological research is dominated by studies on *Saccharomyces cerevisiae* and to a lesser extent on *Aspergillus* sp., *Neurospora crassa*, *Penicillium* and *Schizosaccharomyces pombe*. One must not ignore the very important work on these fungi and I have attempted to give adequate coverage of the information relevant to this volume. However, wherever possible, I have tried to give a sensible picture of what is known for other fungi for three reasons. First, I believe it is important that we do not, without good reason, try to extrapolate from what we know about processes occurring in the above-named species to equivalent processes occurring in species in other parts of the Fungal Kingdom. Second, by referring to the little that is known for much less-studied species I hope to encourage more research on those species. Third, I am very anxious that we avoid more positively the tendency to see the fungi as a physiologically homogeneous group of organisms. Though species possess similar physiological features, there are nevertheless many striking differences. A stumbling block often to the appreciation of those differences is our ignorance of what a fungus might be doing in its natural environment. The need to attempt to draw together physiology and ecology of fungi is a theme underlying the volume.

Essentially this book has been written for those involved in fungal physiology, as teachers or students, but particularly those are active in research. For that reason I have concentrated on known facts and eschewed speculation, unless it points up a clear line of research. I have almost punctiliously avoided referring to what might be occurring in other eukaryotic organisms to avoid clouding the readers' judgement as to what might be happening in fungi when present knowledge seems inadequate. While there is a considerable amount of information in the volume, I cannot claim that it is encyclopaedic. Nevertheless, in spite of the rate of advance of the subject, I hope the volume will be a useful source of information for some period for those with interests in fungal nutrition. For those areas where coverage has been to an extent superficial, I hope the literature to which I have referred will allow the reader to probe deeper.

Finally, though the major emphasis of the volume is physiological, biochemical and molecular biological information has been drawn upon wherever appropriate. Not one of those approaches can give more than a partial picture of what might be taking place in a particular process. I stress this, because today's emphasis on molecular biology, the undoubted power of which must be acknowledged, has tended to lead to much less reliance on other approaches, particularly the physiological. Wherever possible, I have tried to show the success of the multi-faceted approach to the study of fungal nutrition. If this volume encourages future research also to be more multi-faceted then I shall be pleased.

On a more personal note. I wish to thank Karen McGowan for her great skill in transferring my poor handwriting onto computer disc. Also, I am particularly grateful to Sandi Irvine of Cambridge University Press for her meticulous editing of the manuscript. Any mistakes that are still present are my responsibility! But an especial word of thanks is due to all those many fellow scientists who over the years so kindly responded to my many requests for reprints. Without them, my job in preparing the book would have been immeasurably harder.

David Jennings

1

Primary active transport

Introduction

A living cell must do work to maintain the composition of its internal medium different from that of the external. If that cell is growing, work has also to be done to generate the small but necessary osmotic gradient for the inwardly directed influx of water to take place. In a cell with a wall, such as a fungus, the osmotic gradient is much more significant, in order to maintain the necessary internal hydrostatic pressure for turgor (see Chapter 13). Two important loci for the above work are the plasma and the vacuolar membranes. When work is carried out in moving solutes across these two or indeed any other membranes, we can speak of active transport, i.e. the movement of solutes against their electrochemical potential gradient. Of course, there may also be diffusion of solutes across the membrane, albeit almost always, if the solute is polar, aided by the presence of carrier proteins that overcome the activation energy required by such a solute to enter through the lipid portion of the membrane. When diffusion occurs, it must be down the electrochemical potential gradient. However, it needs to be remembered that if the affinity of such a carrier for the solute on one side of the membrane were to be much higher than on the other or the mobility of solute transfer were to be faster in one direction than in the other – both of which changes would depend on the cell doing work – then active transport would occur (Jennings, 1974).

Active transport may be classified as primary or secondary. Primary active transport or translocation is brought about by reactions that involve the exchange of primary bonds between different chemical groups or the donation or acceptance of electrons. These reactions lead to the translocation of a chemical group or solute across the membrane. Thus, chemical energy is used to generate a vectorial process. Secondary transport or translocation does not involve primary bond exchange between different chemical groups or donation or acceptance of electrons. Thus, the involvement of proteins in secondary active transport is different from that in primary active transport. In the latter process, the protein is acting like an enzyme, catalysing a vectorial rather than a scalar process. Proteins involved in secondary active transport do not function like the classical concept of an enzyme. It is for this reason that they are called 'porters' (Mitchell, 1967). A consequence of the mode of action of porters is that they are able to catalyse vectorial processes in a reversible manner.

The reactions bringing about primary active transport can be described as chemiosmotic (Mitchell, 1979). It is these reactions that ultimately drive secondary active transport or purely osmotic reactions (Figure 1.1). These latter reactions can be described either as uniport, namely the bringing about of exchange diffusion (Ussing, 1947), or sym- or antiport, in which there are coupled flows (Figure 1.2). A uniport will lead only to accumulation of a solute if it is charged and there is an appropriate electrical potential difference across the membrane to provide the driving force for such accumulation. Where there are coupled flows, the flow of one solute, i.e. either protons or sodium (Figure 1.1), down its electrochemical potential gradient can bring about the movement of

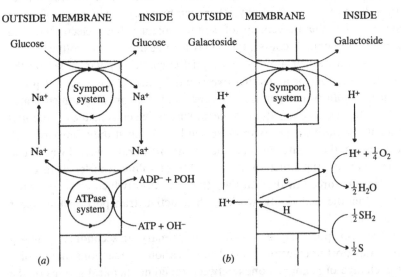

Figure 1.1. Coupling in two well-studied transport systems: (*a*) between the sodium-transporting ATPase and the sodium/glucose symporter in mammalian intestinal mucosa; (*b*) between a proton-motive respiratory chain system and an H^+/galactoside symporter in *Escherichia coli*. S, sulphur. (From Mitchell, 1979.)

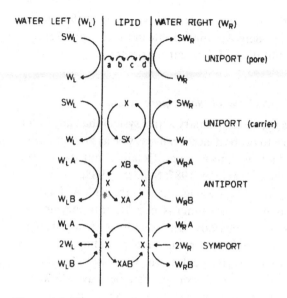

Figure 1.2. Diagrams of porter-catalysed translocation of solutes (S, A and B) across a lipid membrane between aqueous phases in which S, A and B exist as hydrates SW, WA and WB. Left and right aqueous phases are denoted by suffixes L and R. a–d and X represent a chemically specific pore and many carriers. (From Mitchell, 1979.)

another solute against its electrochemical potential gradient. Thus, there is active transport of one solute across the membrane brought about by the free energy in the gradient of the other solute across the membrane. The latter gradient of course is maintained by primary active transport. In considering secondary translocation, movement of charge is important, as well as any ensuing changes in concentrations on either side of the membrane. Figure 6.3 (p. 209) shows possible movements of charge that can occur as a result of secondary translocation.

As is discussed more fully below, primary active transport at the plasma membrane in fungi is predominantly the extrusion of protons using the free energy of hydrolysis of ATP. The evidence for any other primary active transport process is tenuous, a matter that also is considered below. As far as we know, therefore, fungi interact with the external medium almost entirely through a proton economy. There are indications that sodium can accompany solutes such as phosphate (see Chapter 7) in translocation processes but essentially it is protons that cotransport with other solutes requiring entry into or expulsion from a fungus. Secondary active transport is not considered as a particular topic. There is now such a plethora of secondary active transport processes occurring at the plasma membrane that they are considered as appropriate in the relevant sections of the text dealing with utilisation of different nutrients.

The plasma membrane H^+-ATPase of *Neurospora crassa*

We know more about this particular primary active system than any other because it has been possible to probe it electrically, as well as biochemically and by use of molecular biological techniques. Details about the system have been reviewed elsewhere (Slayman, 1987; Sanders, 1988).

The reason there is greater knowledge about the H^+-ATPase of *Neurospora crassa* than that of other fungi is because this enzyme has been probed more extensively electrically. One cannot stress too strongly the importance of electrical studies for our understanding of active transport of charged atoms or molecules. In the case of primary active transport in *N. crassa*, electrophysiological studies have allowed a kinetic description of the process of proton extrusion, taking into account not only the characteristics of the ATPase *in vitro* and in membrane vesicles but also the membrane electrical field in which the enzyme resides.

The success of electrophysiological studies with *N. crassa* is due to the fact that, within mycelium growing on cellophane overlaying agar, there are hyphae 10–20 μm in diameter that are found 7–9 mm behind the growing margin (Slayman, 1965). The diameter is sufficiently large for insertion of electrodes for measuring the voltage and for either injecting current or measuring pH.

The functioning H^+-ATPase leads to the extrusion of protons electrogenically from the hyphae. Thus, we can speak of the enzyme acting physiologically as a proton pump. If the pump is non-functional, the membrane potential difference is *c.* -25 mV, inside negative (Slayman, 1970; Slayman, Long & Lu, 1973). When the pump is operational, the potential difference under normal physiological conditions is *c.* -270 mV, thus something like -200 mV of potential is generated by pump activity. When a mycelium is in 100 mM potassium chloride in a calcium-free medium, the potential difference is as low as -60 mV (Slayman, 1965); on the other hand, in the absence of chloride (which tends to depolarise the membrane) the potential difference can be as great as -300 mV (Blatt & Slayman, 1983).

The evidence that ATP is the substrate for the pump comes from the equivalence of the decay of the potential with the decay of ATP in the mycelium upon the addition of sodium azide. Both decay in an exponential manner with maximal rate constants of $0.18\,s^{-1}$ and half-times of $3.7\,s$. One proton is pumped out of the hyphae for each molecule of ATP hydrolysed (Warncke & Slayman, 1980). This stoichiometry has been confirmed with studies on membrane vesicles (Perlin *et al.*, 1986; see below). With this stoichiometry and knowing the average rate of ATP synthesis (Slayman, 1973), it has been possible to calculate the proportion of ATP so synthesised that is utilised in pump activity (Gradmann *et al.*, 1978). The calculation shows that 38%–52% of the total ATP production is consumed by the pump.

The dissection of pump activity by current–voltage analysis has led to the production of a four-step kinetic model of pump operation (Slayman & Sanders, 1984; Sanders, 1988; Figure 1.3). The fast steps of the process of proton extrusion are the transmembrane movement of the proton and its dissociation from the carrier; the slow steps are the binding of protons and of ATP. Carrier cycling in the direction of efflux is limited by binding of protons on the cytoplasmic side; the model gives a pK_a for the internal binding site of 5.4 (cytoplasmic pH 7.2) and for the external site of 2.9 (external pH *c.* 5.5). This is important physiologically, because it means that the external pH must drop very considerably before there is an effect on pump activity. Thus, the pump can respond to changes in the internal pH without being very much influenced by the external pH (Sanders, 1988).

Under normal conditions of culture (pH 5.8) and with a potential

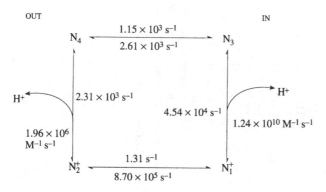

Figure 1.3. Four-step reaction kinetic model for the plasma membrane proton pump of *Neurospora crassa* derived from pH-dependent behaviour of the pump current–voltage relationships. Selective manipulation of the external and internal pH enables specific identification of the H^+-binding reactions. Those reactions that are voltage insensitive and thus not identifiable are lumped in the transition between carrier states 3 and 4. Values for the reaction constants from Slayman & Sanders (1984) are shown. (From Sanders, 1988.)

Table 1.1. *Physical and chemical properties of the different* H^+-*ATPases in Neurospora crassa. Citations in each column indicate references where the relevant information can be located*

	Plasma membrane	Mitochondrial membrane	Vacuolar membrane
Subunit masses (kDa)	104	59, 56, 36, 22, 21, 19, 16, 15, 12, 8	70, 62, 15
pH optimum	6.7	8.3	7.5
K_m (ATP) (mM)	1.8	0.3	0.2
Nucleotide specificity	ATP	ATP>GTP ITP>UTP>CTP	ATP>GTP ITP>UTP>CTP
Divalent cation specificity	Mg, Co>Mn	Mn>Mg>Co	Mg, Mn>Co
Inhibitors	DCCD, vanadate	DCCD, oligomycin, azide	DCCD, SCN$^-$, NO$_3^-$
References	Bowman & Slayman (1977) Scarborough (1977) Bowman et al. (1978) Bowman, Blasco & Slayman (1981) E. J. Bowman et al. (1981) Scarborough & Addison (1984)	Jackl & Sebald (1975) Mainzer & Slayman (1978) Sebald & Hoppe (1981) E. J. Bowman (1983)	Bowman & Bowman (1982) Bowman et al. (1986)

DCCD, N,N'-dicyclohexylcarbodiimide.
From Slayman (1987).

difference across the plasma membrane of $-200\,\text{mV}$, the estimated current carried by the pump is 0.123–$0.25\,\text{A}\,\text{m}^{-2}$ (Gradmann *et al.*, 1978; Sanders, Hansen & Slayman, 1981; Gradmann, Hansen & Slayman, 1982; Sanders, 1988). This current is equivalent to a proton flux of 1.2–$2.6\,\mu\text{mol}\,\text{m}^{-2}\,\text{s}^{-1}$ (Slayman & Slayman, 1968). The net loss (net efflux) of hydrogen ions from the hypha under the same conditions is $0.04\,\mu\text{mol}\,\text{m}^{-2}\,\text{s}^{-1}$ (Slayman & Slayman, 1968). This difference is highly likely to be due to the passive influx of protons.

There is now an extensive biochemical and molecular description of the proton pump. The major problem in the isolation of the plasma membrane H^+-ATPase is contamination with those ATPases that reside in other membranes, i.e. mitochondrial and vacuolar. Plasma membranes of *N. crassa* have been prepared in two ways. One way has been to use the cell wall-less mutant (*sl*) (Scarborough, 1975, 1988; Smith & Scarborough, 1984). The membranes are first stabilised against fragmentation and vesiculation with concanavalin A. Membranes thus treated can be purified by low-speed centrifugation and converted to vesicles by removal of the concanavalin A. Of course, the lectin acts not only as a stabilising agent but also as a marker for the plasma membrane. The other method of preparation involves the removal of hyphal walls by snail digestive enzyme, followed by gentle lysis (Bowman, Bowman & Slayman, 1981; Bowman & Bowman, 1986). The plasma membrane vesicles so formed can be separated by low speed or density gradient centrifugation.

The characteristics of the ATPases found in *N. crassa* are given in Table 1.1. It is not appropriate to go into further details than are presented there. However, it is necessary to highlight the different inhibitor sensitivities and other properties of the three ATPases, noting particularly the sensitivity of that of the plasma membrane to vanadate, the high specificity for ATP and the much lower pH optimum compared with that of the mitochondrial ATPase. While such features characterise the plasma membrane ATPase, they do not characterise it as a proton pump. For that to be possible, it has been necessary to study the functioning of the enzyme in plasma membrane vesicles.

The use of such vesicles has been pioneered by Scarborough (1975, 1976, 1980). Vesicles prepared as above contain a significant population of the everted type, namely those where the morphologically inner surface is exposed to the experimental solution. ATP hydrolysis therefore takes place when the compound comes into contact with such vesicles. Fluorescent compounds placed inside the vesicles during their preparation

(Scarborough, 1980) or radiolabelled or fluorescent compounds that distribute across the membrane according to either the pH gradient or the potential difference across it (Scarborough, 1976; Perlin *et al.*, 1984) can be used to determine the pH inside the vesicles and potential difference across their membranes. The temporal changes of internal pH and membrane potential on addition of ATP to the vesicles and the effects of inhibitors such as vanadate on such changes is in keeping with the ATPase acting as an electrogenic pump (Figure 1.4). Calibration of the potential change has been said to give a value of 120 mV (Slayman, 1987). This is much lower than what is observed for intact hyphae (see above); the difference is probably due to the leakiness of the vesicle membrane to protons. Vesicle preparations have been used to determine the apparent

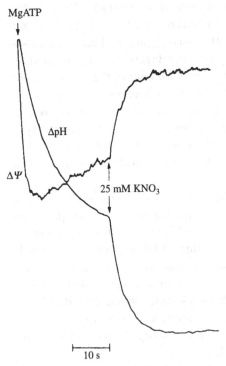

Figure 1.4. Superimposed demonstrations of the creation of a pH difference and a membrane potential in everted vesicles of the plasma membrane of *Neurospora crassa*. ΔpH, pH difference monitored via quenching of acridine orange fluorescence. ΔΨ, membrane potential (inside positive) monitored via quenching of oxonol V fluorescence. Quenching in both instances was initiated with 1.5 mM MgATP. Note that addition of the permeant nitrate ion at 25 mM reverses oxonol quenching and enhances acridine orange quenching. (From Perlin 1984.)

stoichiometry of H^+ translocated per ATP split by the ATPase; values close to unity were obtained, in keeping with values obtained by electrophysiological measurements (Perlin *et al.*, 1986).

A picture of the molecular structure of the H^+-ATPase in relation to its functioning is now starting to emerge. The structural gene for the enzyme has been cloned and sequenced (Addison, 1986; Hager *et al.*, 1986). The single polypeptide coded by the gene contains 920 amino acid residues and has a molecular mass of 100 kDa, confirming the similar value (105 kDa) found by gel electrophoresis studies (Dame & Scarborough, 1980). A hydropathy plot has allowed some idea of the transmembrane orientation of the polypeptide. The relationship of the structure of the polypeptide to function has been probed further either by incubation of isolated plasma membrane vesicles with trypsin (Dame & Scarborough, 1980; Addison & Scarborough, 1982; Hennessey & Scarborough, 1990; Scarborough & Hennessey, 1990; Hennessey & Scarborough, 1991; Figure 1.5) or by site-specific agents such as *N*-ethylmaleimide (for sulphydryl groups) (Brooker & Slayman, 1982, 1983; Davenport & Slayman, 1988; Chang & Slayman, 1990), phenylglyoxyl or 2,3-butanedione (for arginine) (Kasher *et al.*, 1986), *N*-(ethoxycarbonyl)-2-ethoxy-1,2-dihydroquinoline for carboxyl groups (Addison & Scarborough, 1986; Table 1.2). As a result of such studies we are obtaining a clearer view of the residues that are essential for catalytic activity. In particular, we now have a view of the transmembrane topography of the polypeptide and of those portions associated with the cytoplasm (Rao, Hennessey & Scarborough, 1991; Figure 1.6).

An important question for which there is an uncertain answer concerns the number of polypeptides constituting the functional proton

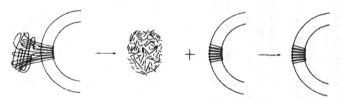

Figure 1.5. A diagram of the experimental approach to defining the membrane-embedded regions of the H^+-ATPase of *Neurospora crassa*. Vesicles with the enzyme molecules oriented predominantly with their cytoplasmic surface facing outwards are treated with trypsin. That part of the polypeptide molecule outside the membrane is cleaved, leaving that inside the membrane untouched. The peptides released by trypsin and those remaining in the membrane can be analysed separately. (From Rao *et al.*, 1991.)

Table 1.2. *Use of group-specific reagents to define functionally important residues of the plasma membrane H^+-ATPase of Neurospora crassa*

Residue	Inhibitor	Protection by ATP	Reference
Lys474	Fluorescein isothiocyanate	Yes	Pardo & Slayman (1988)
Cys532	N-ethylmaleimide	Yes	Pardo & Slayman (1989)
Cys545	N-ethylmaleimide	No	Pardo & Slayman (1989)
Arg	Phenylglyoxal, butanedione	Yes	Di Pietro & Goffeau (1985)
			Kasher et al. (1986)
Glu129	N,N'-dicyclohexylcarbodiimide	No	Sussman et al. (1987)
Glu/Asp	N-ethoxycarbonyl-2-ethoxy-1,2,-dihydroquinoline	No[a]	Addison & Scarborough (1986)

[a] Protection is seen in the presence of MgATP + vanadate (Addison & Scarborough, 1986). From Nakamoto & Slayman (1989).

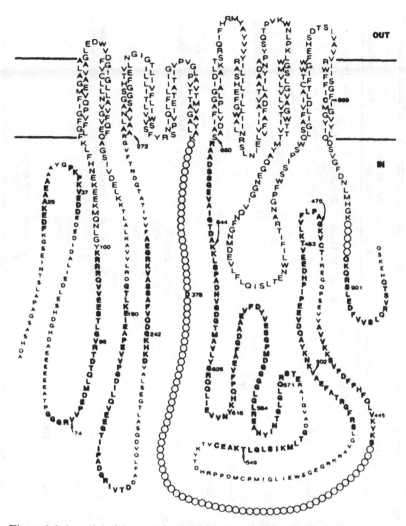

Figure 1.6. A model of the transmembrane topography of the complete H^+-ATPase molecule. The different fonts used for the letters designating the various amino acid residues represent information obtained from different investigations. Single-letter code is used for amino acids. (From Rao *et al.*, 1991.)

pump. Studies using radiation inactivation suggest two polypeptides (Bowman, Beranski & Jung, 1985). On the other hand, studies on the incorporation of ATPase into liposomes capable of generating a transmembrane proton potential difference indicated that only one polypeptide was incorporated into the liposome membrane (Goormaghtigh, Chadwick & Scarborough, 1986).

Table 1.3. *Evidence for the presence in named fungi of an H^+-ATPase acting in the plasma membrane as a proton extrusion pump. Citations in each column indicate the references where the relevant information can be located*

Organism	Isolation of enzyme	Proton pumping in vesicles	Electrophysiology	Molecular mass	Amino acid sequence	Review of literature
Achlya bisexualis			Kropf (1986)			
Candida albicans	Hubbard et al. (1986)			97, 398 Da (Monk et al., 1991)	Monk et al. (1991)	Gow (1989a)
Candida tropicalis	Blasco, Chapuis & Giordani (1981)	Blasco & Gidrol (1982)				
Debaryomyces hansenii	Comerford, Spencer-Phillips & Jennings (1985)					Clipson & Jennings (1992)
Dendryphiella salina	Garrill & Jennings (1991)		Davies, Brownlee & Jennings (1990a, b)			
Metchnikowia reukaufii	Gläser & Höfer (1986)	Gläser & Höfer (1987)				Clipson & Jennings (1992)
Phytophthora megasperma f. sp. *glycinea*	Giannini, Holt & Briskin (1988)	Giannini et al. (1988)				
Saccharomyces cerevisiae	Serrano (1978, 1988b); Goffeau & Dufour (1988)	Malpartida & Serrano (1981a, b); Serrano (1988b)		105 kDa (Malpartida & Serrano, 1980); 99, 533 Da (Serrano, Keilland-Brandt & Fink, 1986)	Serrano et al. (1986)	Serrano (1988a, b)
Schizosaccharomyces pombe	Delhez et al. (1977); Dufour, Amory & Goffeau (1988)	Villalobo, Boutry & Goffeau (1981); Mair & Höfer (1988)		105 kDa (Dufour & Goffeau, 1978); 99, 769 Da (Ghislain, Schlesser & Goffeau, 1987)	Ghislain et al. (1987)	Goffeau, Coddington & Schlesser (1989)
Thraustochytrium aureum	Garrill, Clipson & Jennings (1992)	Garrill et al. (1992)				
Zygosaccharomyces rouxii	Watanabe et al. (1991)			100 kDa (Watanabe et al. 1991)	Watanabe et al. (1991)	

Fungi in general

Our knowledge about the plasma membrane H^+-ATPase of other fungi is summarised in Table 1.3. There are also reviews by Goffeau & Slayman (1981), Serrano (1984, 1985, 1988*a*, 1989), and Goffeau & Green (1990) that focus on the enzyme from *N. crassa*, *Saccharomyces cerevisiae* and *Schizosaccharomyces pombe*. While none of the information for other fungi is quite as extensive as that for *N. crassa*, there is sufficient, particularly with respect to the amino acid sequence, for *S. cerevisiae* and *Sch. pombe* to conclude that the enzyme must function in a manner similar to that of *N. crassa* in these two yeasts. As an example of the extent of the similarity, the ATPase of *S. cerevisiae* has 74% sequence identity with that of *N. crassa*, with only 238 differences in the amino acid sequences, the total number of residues being 918 and 920, respectively. Of these differences, 105 are conservative substitutions of related amino acids. Most differences occur in the N-terminal and C-terminal regions, with 74 differences in the first 100 and 42 in the last 100 residudes (Serrano, 1988*a*). Figure 1.7 shows the transmembrane structure of H^+-ATPases as proposed

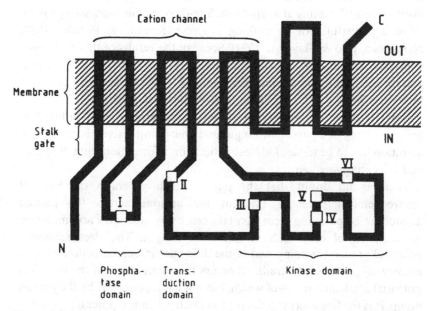

Figure 1.7. Diagram of the proposed transmembrane structure and functional domains of H^+-ATPases. I to VI are the conserved motifs described in Table 1.4. (From Serrano, 1989: reproduced, with permission, from the *Annual Review of Plant Physiology and Plant Molecular Biology*, **49**, © 1989 by Annual Reviews Inc.)

Table 1.4. *Conserved motifs of ATPases acting as ion pumps and the proposed functions of these motifs (see Figure 1.7)*

Motif	Proposed functions
I. TGES	Phosphatase activity (E)
II. D(K,R)TGT(L,I)T	Phosphorylation and transduction (D)
III. KGAP (only eukaroytes)	ATP binding and/or kinase activity (K)
IV. DPPR (only eukaryotes)	ATP binding (D)
V. M(L,I,V)TGD	ATP binding (D)
VI. GDGXND(A,S)P(A,S)LK (K only in eukaryotes)	ATP binding (two D and K)

X, any amino acid. Single-letter code for amino acids is used.
From Serrano (1989). Reproduced, with permission from the *Annual Review of Plant Physiology and Plant Molecular Biology* **49**, © 1989 by Annual Reviews Inc.

by Serrano (1989), showing the location of conserved motifs and postulated functional domains (Table 1.4). Further combined genetic and biochemical studies are needed before we can be certain about how various parts of the polypeptide function with the membrane. However, information is likely to be forthcoming at a rapid rate. Site-directed mutagenesis is proving to be a powerful tool for making progress (Serrano & Portillo, 1990; Nakamoto, Rao & Slayman, 1991). Some of the results of this mutagenic approach are discussed below. Other very recent information about mutations affecting the plasma membrane H^+-ATPase has been provided by Perlin, Brown & Haber (1988) and Perlin *et al.* (1989).

When it comes to considering the role of the proton pump in terms of the physiology of a particular fungus and how pump activity is controlled, attention has to be focused almost entirely on information from *N. crassa* and *S. cerevisiae*.

There is no doubt that the pump, which generates the bulk of electrochemical potential gradient for protons across the plasma membrane, by pumping them out of the cell, brings about the accumulation or extrusion of other solutes inside the fungus. The electrochemical potential gradient is composed of the difference in concentration (strictly activity) of protons (pH gradient) across the membrane and the electrical potential gradient – much of which, but not all, is generated by the proton pump. It is the free energy in the proton electrochemical potential gradient that either drives the accumulation of another solute, symported by a carrier with protons moving down the gradient, or drives the extrusion of a solute, in this case antiported with the protons via the same carrier.

In terms of theory (Sanders, 1988), the general reaction for symport at the plasma membrane can be written as

$$nH_o^+ + S_o = nH_c^+ + S_c$$

in which the subscripts o and c refer to the external medium and cytoplasm, respectively, n is the number of H^+ transported per molecule of solute, S. The change in Gibbs' free energy (G) for this reaction (left to right) is given as

$$\Delta G = n\Delta\bar{\mu}_{H^+} + \Delta\bar{\mu}_s$$

where $\Delta\bar{\mu}_{H^+}$ and $\Delta\bar{\mu}$ are respectively the changes of electrochemical potential for H^+ and S. The equation can be expanded and the result expressed in terms of the components of the respective electrochemical gradients as

$$\Delta G = (n+z)F\Delta\Psi + RT \ln\left[\frac{[H^+]_c^n[S]_c}{[H^+]_o^n[S]_o}\right]$$

where z is the valence of S, F is Faraday's constant, $\Delta\Psi$ is the electrical potential difference across the membrane, R is the gas constant and T is the absolute temperature.

Thus, the value of n has a profound influence on the thermodynamic poise of the reaction. The effect on the maximum attainable solute accumulation ratio can be found by setting the previous equation to zero and rearranging, which yields

$$\frac{[S]_c}{[S]_o} = \frac{[H^+]_o^n}{[H^+]_c^n}\exp\left(\frac{-(n+z)F\Delta\Psi}{RT}\right)$$

For a typical value of $\Delta\Psi = -150\,mV$ and for $[H^+]_o/[H^+]_c = 100$, a unit increase in n generates an increase in the equilibrium solute accumulation ratio of 36 000-fold. Evaluation of n is therefore essential if the thermodynamics of transport are to be understood.

The stoichiometry of cotransport systems has been studied on a considerable number of occasions (see Chapters 5, 6 and 7). Very often, the stoichiometry has been determined by change in the external pH. But if the symport carries current, this will generate other ion fluxes across the membrane. If protons constitute one such flux, in particular proton extrusion, n will be underestimated (Sanders, 1988). Two procedures can be used to obviate this problem. One is to use 'de-energised' cells. Seaston, Inkson & Eddy (1973), studying amino acid and maltose transport in

Saccharomyces carlsbergensis and *S. cerevisiae*, used washed cells treated with 2-deoxy-D-glucose and antimycin to inhibit energy metabolism. The other method, more rigorous, is to measure the current through the cotransport system directly via a voltage–clamp experiment. The current is determined in the absence of the solute and then again as soon as possible after the solute has been presented to the fungus (Slayman & Pall, 1983; Sanders, 1988).

That the plasma membrane H^+-ATPase pumps protons from one side of the membrane to other on the outside of the protoplast suggests that the pump has an important role in controlling cytoplasmic pH. The cytoplasmic pH of fungi can be measured by a number of means, such as ^{31}P nuclear magnetic resonance (NMR) (Gillies *et al.*, 1981), ^{15}N NMR (Legerton *et al.*, 1983), glass microelectrodes (Sanders *et al.*, 1981) – though neutral carrier-based hydrogen ion selective electrodes (Ammann *et al.*, 1981) are to be preferred – and fluorescent probes (Davies, Brownlee & Jennings, 1990*b*). That such measurements are possible means that one can probe the various processes for maintaining the cytoplasmic pH at the appropriate value.

Before discussing the evidence for the role of the plasma membrane H^+-ATPase in controlling cytoplasmic pH, it is necessary to point out that the activity of protons in the cytoplasm (or indeed any other compartment in the protoplasm) is determined by the following relationship.

$$H^+ - OH^- + C^+ - A^- + WBH^+ - WA^- = 0$$

where C^+ and A^- are the sums of the concentration of strong cations and of strong anions, respectively, and WBH^+ and WA^- the sums, respectively, of the concentrations of the ionised forms of weak bases and of weak acids (Guern *et al.*, 1991). It is easy to see that H^+ (and OH^-) concentrations are very dependent on the concentrations of other ions, particularly the major strong ions if their concentration in the external medium fluctuates significantly.

It follows from what has just been stated that, for the pump to change the cytoplasmic pH, there will need to be the flux of strong cations in the reverse direction or strong anions in the same direction. Without such fluxes there will be limited pH change; there will be a large electrogenic effect, which will lead to an enhanced backward passive flux of protons into the fungus. All features just mentioned are exemplified in the response of *N. crassa* to enhanced acid load brought about by exposing the mycelium to butyric acid (Sanders *et al.*, 1981). Electrophysiological experiments show that the pump appears to respond directly to the acid load. There is an increase in membrane conductance shortly after this due to a change of permeability to one or more ions (organic anions?). It is the increased conductance that stabilises the cytoplasmic pH, albeit at a slightly more acidic value than at the start of the experiment.

However, Sanders & Slayman (1982) showed that cytoplasmic acidification produced by respiratory blockade is unaffected by the size or direction of the proton gradient across the membrane. Furthermore, blocking pump activity by vanadate appears to have little effect on the cytoplasmic pH. A fall in pH caused by cyanide is corrected when cyanide is removed, even when the pump is inhibited. It was concluded that metabolism is equally important in controlling cytoplasmic pH. This suggests that there are elements other than the proton pump involved in pH control. They may be, amongst others, pump activity at vacuolar membranes (see Chapter 4), plasma membrane redox systems (see below) and a biochemical pH-stat (see Chapter 12).

However, there is other evidence from *S. cerevisiae* that, while the regulation of cytoplasmic pH is due to a number of processes occurring in a fungus, the activity of the H^+-ATPase is nevertheless an important element within the complex of processes. First, Eraso & Gancedo (1987) have observed that, when cells are grown on media with a pH lower than 4, there is a two- to three-fold increase in plasma membrane ATPase activity in late logarithmic or stationary cells. Equally, Vallejo & Serrano (1989) introduced two mutations containing insertions and deletions in the promoter in the plasma membrane H^+-ATPase (*PMA1*) gene. The resulting strains had 15% and 23%, respectively, of the wild-type ATPase activity. There were accordingly decreased rates of proton efflux and proton-associated transport of other solutes. While the growth of the mutants and intracellular pH (measured by distribution of ^{14}C-labelled propionic acid) had the same value (7.3) as the wild-type at pH 6, when the mutant strains were grown at pH 4, their growth rates were less than that of the wild-type and the intracellular pH values were lower. The strain with the least active ATPase had the most reduced growth rate and the lowest cytoplasmic pH.

Another key issue about the functioning of the H^+-ATPase concerns the extent to which activity of the enzyme functioning as a proton pump is governed solely by the concentration of ATP in the cytoplasm. One part of the evidence comes from studies on the so-called *poky f* strain of *N. crassa*, which has a primary defect in the mitoribosomes such that it is depleted of cytochromes *a* and *b* but contains high levels of the alternative oxidase (Lambowitz *et al.*, 1972; La Polla & Lambowitz, 1977). In the wild-type, as indicated on pp. 4–5, there is exponential decay of the ATP. In *poky f* strain, there is an exponential decay in the presence of 1 mM cyanide but, when the mycelial concentration reaches about 40% of the initial value, there is recovery to about 70% (Gradmann & Slayman, 1975). Whereas in the wild-type the membrane potential depolarises exponentially, in *poky f* there is depolarisation (30–100 mV) followed by slight hyperpolarisation, then two to four damped cycles, which indicates continued regulation of the pump. Current–voltage characteristics of the pump in *poky f* show that, in the presence of cyanide, there is a change in the proton stoichiometry of the pump from one to two protons per ATP molecule (Warncke & Slayman, 1980). It seems that the mechanisms that result in the maintenance of a depolarised membrane and relatively high levels of ATP in the mycelium and the changed stoichiometry of the pump in the presence of cyanide are the same. The data from electrophysiological studies suggests that cyclic AMP (cAMP) plays a role. There is evidence from wild-type mycelium that membrane potential changes do not always track changes in intrahyphal concentrations of ATP, implying that here too there must be other kinds of metabolic control than via ATP concentration (Gradmann & Slayman, 1975) and one presumes here, as with *poky f*, that cAMP must be involved.

Another view of this matter of the regulation of plasma membrane H^+-ATPase activity comes from studies on *S. cerevisiae*. There are two important observations. First, that of Portillo & Serrano (1989), who produced by site-directed mutagenesis a series of mutants of some residues essential for the phosphorylation and hydrolytic steps of the enzyme catalytic cycle. In a medium at pH 4 it was shown that the relationship between growth of each mutant was a strictly linear function of its relative (to wild-type) ATPase activity. Second, Serrano (1983) showed that ATPase activity increased several-fold during glucose metabolism.

There have been suggestions (Serrano, 1989) that this relationship between H^+-ATPase activity and growth and glucose metabolism is mediated via cAMP. It now seems that the process in mind, namely the Ras-adenylate cyclase signalling pathway, operates only in the transition from the derepressed state (growth of cells of *S. cerevisiae* on respiratory carbon sources) to the repressed state (growth on fermentative carbon sources) (Thevelein, 1991). However, Ulaszewski, Hilger & Goffeau (1989) have demonstrated that a thermosensitive mutant defective in adenylate cyclase activity, when shifted to a restrictive temperature, results in a 50% reduction of plasma membrane H^+-ATPase activity within 1 h. The data indicated that cAMP acts as a positive effector of the H^+-ATPase of the plasma membrane but a negative effector of that on the mitochondrion. How cAMP exerts its effects is not known. It does not seem to be through the action of a cAMP-dependent protein kinase, since the H^+-ATPase of *S. cerevisiae* does not contain any potential phosphorylation sites (clusters of arginine residues followed by serine) recognised by such kinases. Nevertheless there is substantial evidence that the H^+-ATPase can be phosphorylated *in vivo* (McDonough & Mahler, 1982; Portillo & Mazón, 1986), possibly by a protein serine kinase in the plasma membrane (Yanagita *et al.*, 1987; Kolarov *et al.*, 1988), though there are other possibilities (Chang & Slayman, 1991). Chang & Slayman (1991) have shown that, with growth on glucose, there is phosphorylation at sites on the enzyme that are unique to its location in the plasma membrane (there are other sites that are phosphorylated before the enzyme arrives at the membrane). There is dephosphorylation when glucose is removed and rapid reversal (2 min) on readdition of glucose. It is likely that glucose is exerting its effect also via the regulatory domain at the C-terminal end of the H^+-ATPase polypeptide (Serrano & Portillo, 1990).

Plasma membrane redox systems

As far as is known there are no other primary active transport systems present in the plasma membranes of fungi, but this statement cannot be properly sustained without studies on many more fungi. However, the known presence of plasma membrane oxidoreductases in *S. cerevisiae* (Crane *et al.*, 1982) suggests that there can be other biochemical processes occurring at that membrane that are relevant to transport processes occurring therein. Any speculation about them can really only be made on the basis of what we know from the studies on the plasma membranes of flowering plants (Crane, 1989; Crane & Barr, 1989).

The essential feature of the plasma membrane oxidoreductase system is the NADH and NADPH:cytochrome *c* reductase (NCR) activity. In flowering plants, oxygen, semidehydroascorbate and nitrate have been identified as probable natural electrons acceptors. Also in flowering plants the NCR activity is only part of a battery of redox activity in the plasma membrane (Figure 1.8). The NCR activity is thought to have an important role in signal transduction mediated by indole-3-acetic acid (auxin), a key response being the extrusion of protons and acidification of the wall leading to cell expansion. The system can be represented diagrammatically

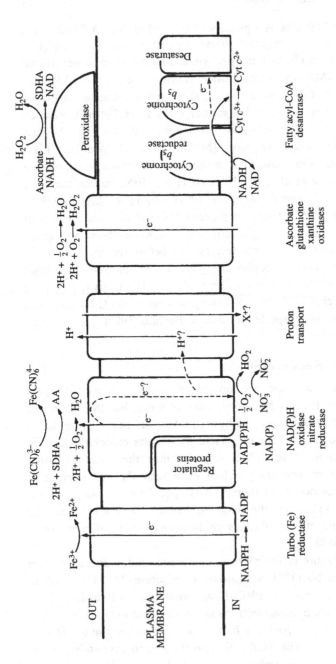

Figure 1.8. Oxidoreductase enzymes intrinsic to the plasma membrane of plant cells. (From Crane, F. L. (1989), in *Second Messengers in Plant Growth and Development*, ed. W. F. Boss & D. F. Morré, copyright © 1989 Alan R. Liss, reprinted by permission of John Wiley & Sons, Inc.)

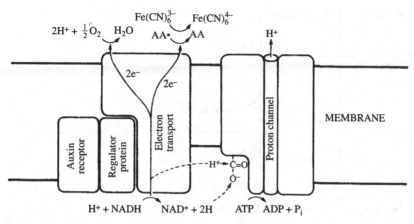

Figure 1.9. Possible components and arrangement of the trans-plasma membrane NADH oxidase, which is activated by indole-3-acetic acid (auxin) and can activate the excretion of protons from the cytoplasm through an associated proton channel. Ascorbate radical (AA•) in the wall may act as an alternative electron acceptor. Ferricyanide can act also as an alternative electron acceptor. The auxin receptor may be facing inside or outside. (From Crane, F. L. (1989), in *Second Messengers in Plant Growth and Development*, ed. W. F. Boss & D. F. Morré, copyright © 1989 Alan R. Liss, reprinted by permission of John Wiley & Sons, Inc.)

(Figure 1.9); however, the actual source of the protons moving into the wall is not clear. They are produced either via oxidation of a substrate involving NADH or NADPH or from other metabolic processes in the cytoplasm. It has been suggested that the NCR system is involved in intracellular pH regulation (Crane, 1989; Crane & Barr, 1989; Guern *et al.*, 1991). Although the presence of ferricyanide at the surface of a flowering plant cell has been shown to reduce the cytoplasmic pH (evaluated by ^{32}P NMR (Guern *et al.*, 1991), there is as yet no firm evidence for a regulatory role under normal conditions.

In *S. cerevisiae*, both respiratory-competent and respiratory-deficient (petite strain) cells are able to reduce ferricyanide (Crane *et al.*, 1982). The reduction is stimulated by ethanol and inhibited by the alcohol dehydrogenase inhibitor pyrazole. With reduction of ferricyanide there is an increased loss of protons from the cells. Presumably NADH is the internal reductant regenerated by the oxidation of ethanol, the protons being released in the same manner as in flowering plant cells.

Of course, the presence of a plasma membrane redox system in *S. cerevisiae* had been proposed at an earlier date by E. J. Conway and hypothesised to drive K^+–H^+ and K^+–Na^+ exchange (Figure 1.10;

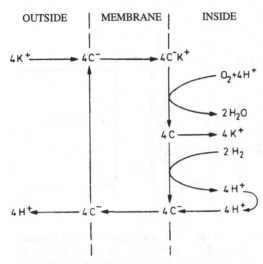

OUTSIDE | MEMBRANE | INSIDE

Figure 1.10. A diagrammatic representation of the redox theory of Conway as applied to yeast. (From Jennings, 1976.)

Jennings, 1963, 1976). Conway & Kernan (1955) studied the effect of dyes with redox potentials from $+290$ to -160 mV at $100\,\mu$M concentration. Without any dye, the redox potential was about $+180$ mV. When it was raised above this value by the addition of the appropriate dye, there was increased K^+-H^+ exchange; the reverse occurred when the potential was reduced (Figure 1.11). Ryan (1967) studied specifically the reduction of $125\,\mu$M methylene blue by cells of *S. cerevisiae* in the presence of ethanol. There was rapid reduction at the surface of the cells (the dye does not penetrate). The rate of reduction is strongly inhibited by potassium ions in the medium. Other alkali cations also have an inhibitory effect and significantly the relative concentrations at 50% inhibition were found to be similar to the reciprocals of transport affinities of the same cations for the alkali cation carrier (Table 1.5). In view of what has been described above, it is interesting to draw attention to the interpretation of these results. It was suggested (Ryan, 1967) that the alcohol dehydrogenase reduces the K^+ carrier and in the absence of external K^+ passes both electrons and protons to methylene blue. Alkali cations, particularly K^+, transported by the carrier compete with methylene blue for the carrier when it is in the reduced form.

The hypothesis can probably be ignored in terms of alkali cation transport (see Chapter 10) but the findings are certainly significant for an understanding of the role of plasma membrane redox systems. Clearly the

Table 1.5. *Effect of various cations on the rate of reduction of methylene blue by* Saccharomyces cerevisiae *under anaerobic conditions*

Cation	Concentration at 50% inhibition (mM)	Relative concentration at 50% inhibition	Reciprocals of transport affinities
K^+	0.4	1	1
Rb^+	1.1	2.8	2.4
Cs^+	3.9	9.8	14.0
Na^+	8.0	20.0	26.0
Li^+	13.6	34.0	200.0

From Ryan (1967). Reciprocals of transport affinities are calculated from data of Conway & Duggan (1958).

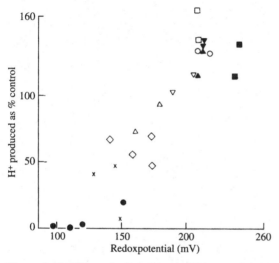

Figure 1.11. Effect of redox dyes on H^+ secretion by yeast fermenting in 40 mM K^+–H^+-succinate at pH 4.5. Glucose was added after expulsion of oxygen. The following dyes were used (at 100 μM): (●) Nile blue; (×) Safranin; (◇) neutral red; (△) benzyl-viologen; (▽) phenol-1-indo-2,6-dichlorophenol; (▲) 1-naphthol-2-sodium sulphonate indophenol; (○) O-cresol-indo-2,6-dichlorophenol; (▼) phenol-indo-2,6-dibromophenol; (■) O-chloroindo-2,6-dichlorophenol; (□) O-chlorophenol indophenol. (From Conway & Kernan, 1955.)

role of such a system must be as part of a (or indeed one such) system involved in proton extrusion. There must be further studies on plasma membrane redox systems in fungi, not only in *S. cerevisiae*. In particular, there is a need to establish whether such systems are involved in the regulation of cytoplasmic pH. In moulds, it is possible that plasma membrane redox systems are involved in the oxidation and reduction of polyols. Holligan (1970) and Holligan & Jennings (1972c) have shown that in *Dendryphiella salina* internal polyol concentrations are affected by the presence of methylene blue in the external medium. A related observation may be that of Dijkema *et al.* (1986), who showed that polyol levels in *Aspergillus nidulans* are influenced by the pH of the medium. Jennings (1984a) has suggested that polyols may be involved in control of cytoplasmic pH through being sinks for, and sources of, protons in redox reactions. That suggestion, based on intuition rather than experimental evidence, could have veracity if polyol metabolism were to be linked with the plasma membrane redox system.

2

The relationship between membrane transport and growth

Introduction

It is a truism to say that for a heterotrophic organism the rate of nutrient uptake governs the rate of growth. For a fungus, there is a need to supply not only the nutrients from which the polymeric materials of the organism are built but also the inorganic ions that act as cofactors and the nutrients that either directly or indirectly help to generate the internal solute potential. Thus, the rate of functioning of those processes at the plasma membrane will be an important determinant of the rate of growth. At its simplest, the processes at the plasma membrane or indeed any other membrane can be described in terms of two characteristics, namely: (a) the affinity constant, K_T, of the transport system (equivalent to the Michaelis constant, K_m, but not in a strictly formal manner); and (b) the flux, the number of moles of solute transported per unit surface area of membrane per unit time ($mol\,cm^{-2}\,s^{-1}$). Assessment of the affinity constant makes assumptions, albeit very rudimentary, about the mechanism of transport, i.e. that there is a transporting molecule to which the solute first binds before crossing the membrane. This is true for many solutes but is not necessarily obligatory. The determination of the flux of a solute, however, makes no presuppositions about the mechanism of transport. The net flux is the difference between influx and efflux, and the flux determined for one of these directions is the sum of those processes acting in that direction.

The values for the flux of a particular nutrient into a fungus can be used to determine the competitive ability of the fungus for that nutrient. Clearly the species of a group of fungi that has the highest flux will be able to accumulate more of the nutrient than its competitors. Thus, in terms of the ecology of fungi, flux values are singularly important (Jennings, 1990a). Nevertheless true flux values, i.e. those based on

25

membrane surface area (and not on biomass), are not easy to determine. So not surprisingly there are few flux values reported in the literature. I discuss this matter in more detail below.

The affinity constant

Here I focus on K_T, the affinity of the uptake process(es) for a particular nutrient. I do so in the simplest terms for two reasons. First, experimental data relating an affinity constant to growth rate are very limited and, second, formulation of anything more than simple models relating transport to growth is a mathematically impossible task, as pointed out by Lievense & Lim (1982). These authors stated that

the challenge in mathematical modelling lies, therefore, in the task of forming a simplified concept of a hopelessly complex system and transforming this concept into a mathematical form which is tractable. Given a tractable model, its utility must be judged on its ability to quantitatively correlate data on the behaviour of the microbial system and the degree to which the model lends insight to the nature of the system.

I subscribe to this view but, at present, even at a relatively unsophisticated level, there is nothing more than the very simple model, namely the Monod equation, to relate transport to growth in a useful way. Nevertheless application of the Monod equation can be a valuable exercise.

The Monod equation relates growth of a microorganism to the external concentration of that nutrient that is limiting growth, namely

$$\mu = \mu_m s/(s + K_S)$$

where μ is the specific growth rate, μ_m the maximum specific growth rate, s the nutrient concentration in the external medium and K_S the saturation constant. The question is whether K_S is the same as K_T, namely the affinity constant of the transport system for the solute under consideration. van Uden (1969) pointed out that the Monod equation is a special case in which transport is completely unidirectional. A significant efflux demands that the kinetics be altered.

The K_T of the transport system quantifies only the affinity of the system for a nutrient. The constant says nothing about the number of transport sites. It is that number which determines, at any one concentration of nutrient in the external medium, the flux of the nutrient into the organism. When different organisms are compared, the affinity constant is an important characteristic.

Returning to the Monod equation, a comparison between the saturation constant K_S and the affinity constant of the transport system K_T for the nutrient in question should be between (a) K_S, determined from the application of the Monod equation to cells or mycelium growing under steady-state conditions with the particular nutrient at limiting concentration, and (b) K_T, determined for those cells or mycelium abstracted from the fermenter but exposed only to the nutrient and with the pH and solute potential of the external solution being the same as that of the nutrient medium. The solute potential may have to be adjusted with a non-penetrating solute. If, however, the specific growth rate is low compared with the time taken for the determination of initial uptake rates, then the external solution can have the same composition as the medium in the fermenter.

The first study to examine the equivalence of K_S and K_T is that by van Uden (1967), who obtained for cells of *Saccharomyces cerevisiae* growing under glucose limitation a value for K_S of 7.2×10^{-4} M and a K_T for glucose transport of 5.6×10^{-4} M. The two appear to be equivalent. However, a study undertaken by Postma, Scheffers & van Dijken (1989b) extends our understanding. First, it was found that, for all dilution rates in the fermenter, glucose uptake exhibits biphasic kinetics, indicating the presence of two transport processes having K_T values of 1 and 20 mM (all of which is in keeping with what we now know about glucose transport in *S. cerevisiae*, see Chapter 5). Because there are two transport processes, the rate of functioning (v) of each for any one *in situ* (residual) glucose concentration in the fermenter (s) could be calculated from the Michaelis-type equation:

$$v = V_{max}s/(s + K_T)$$

where V_{max} is the maximum rate of transport.

If transport were limiting the growth, the sum of the rates of glucose transport by the two systems should equal the rate of glucose consumption in the fermenter as calculated from cell yield and dilution rate. Figure 2.1 shows that there is agreement at the lower dilution (growth) rates, which incidentally cover the range used by van Uden (1967). At higher dilution rates, there is divergence between the two sets of values. The reason for this divergence is not clear.

While there is divergence, there is sufficient similarity between the values for glucose consumption to consider the data further. First, at dilution rates greater than $0.38\ h^{-1}$, glucose consumption calculated either way increases markedly with increase in dilution rate. That appears to be due

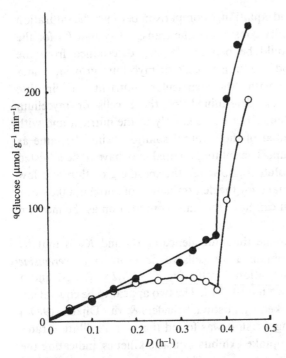

Figure 2.1. *In situ* uptake rate of glucose (O) by cells of *Saccharomyces cerevisiae* growing in a fermenter at different dilution rates, as calculated from the residual substrate and the kinetic constants of the two glucose carriers at these dilution rates. Also plotted is the rate of glucose consumption (●) by the culture as calculated (q) from cell yield and the dilution rate. (From Postma, E., *et al*. (1989b), *Yeast*, copyright © 1989, reproduced by permission of John Wiley and Sons Limited.)

to the onset of fermentation at dilution rates above $0.38 \, h^{-1}$ (Postma *et al*., 1989b, c). It appears that the onset of fermentation is accompanied by a reduced yield of biomass per mole of glucose. But what is interesting is that under these conditions the residual glucose concentration in the fermenter vessel is dependent on the reservoir glucose and not independent of it, as one would expect from fermenter theory (Herbert, Elsworth & Telling, 1956), i.e. above the dilution rate of $0.39 \, h^{-1}$ Monod kinetics are not observed. It is argued that this divergence from the expected kinetics is due to the effect of the extracellular products of fermentation (ethanol, acetaldehyde, acetate and pyruvate) on glucose consumption.

Below a dilution rate of $0.39 \, h^{-1}$ when the residual glucose concentration was independent of the dilution rate, the organism follows Monod kinetics.

Nevertheless as one can see from Figure 2.1, glucose transport – certainly at the lower dilution rates – is proportional to dilution rate. The capacity for transport, i.e. the transport flux, increases with dilution rate or more importantly with growth. Thus the number of carriers increases with the growth rate.

In terms of the functioning of the two carriers, below a dilution rate of $0.39\,h^{-1}$ there is very little contribution from that with the low affinity, since the residual substrate concentration ($c.\ 100\,\mu M$) is nearly 200 times lower than the K_T. Above a dilution rate of $0.39\,h^{-1}$, the low affinity carrier becomes more important. There is an increased residual glucose concentration but, in any case, the amount of the high activity system decreases.

The work on *S. cerevisiae* has been complemented by similar studies on *Candida utilis* (Postma, Scheffers & van Dijken 1988). As is discussed in Chapter 5, transport in the former yeast appears to be by facilitated diffusion, though there is some slight uncertainty about this. What is certain is that transport is not via a proton symport, i.e. that the free energy of the proton electrochemical gradient across the plasma membrane drives glucose transport. On the other hand, there is no doubt that glucose transport in *C. utilis* is via a proton symport (see Chapter 5). So comparison of the two yeasts is instructive.

When *C. utilis* is grown in continuous culture under glucose limitation, at all dilution rates the residual glucose concentration is independent of the reservoir concentration (Postma *et al.*, 1988). Cells of *C. utilis* do not exhibit fermentation under aerobic conditions (they are said to be Crabtree negative, as opposed to *S. cerevisiae*, which is Crabtree positive; see Chapter 5), thus confirming the supposition that fermentation products affect the rate of glucose utilisation by *S. cerevisiae*. Three different systems are apparent, with K_T values of $25\,\mu M$, $190\,\mu M$ and $2000\,\mu M$. Figure 2.2 shows how the three transport systems contribute to the supply of glucose to the cells at different specific growth rates. At the lowest specific growth rate studied only the high affinity system ($25\,\mu M$) is functional, while at the highest specific growth rate (close to the maximum) only the low affinity system ($2000\,\mu M$) is functional. As with *S. cerevisiae*, the rate of glucose uptake in $\mu mol\ (g\ dry\ weight)^{-1}\,h^{-1}$ increases with dilution rate up to $c.\ 0.35\,h^{-1}$, with no further increase at higher rates. It would seem that, as in *S. cerevisiae*, the number of transport sites increases with growth rate.

When cells of *C. utilis* are grown under nitrogen (as NH_4^+) rather than under glucose limitation, the $25\,\mu M$ system is repressed but the $2000\,\mu M$

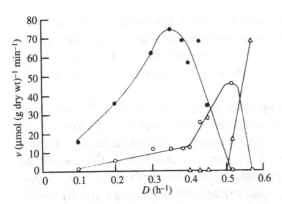

Figure 2.2. *In situ* uptake rates of glucose by cells of *Candida utilis* growing in a fermenter, at different dilution rates for the individual glucose transport systems, calculated from the free glucose concentration in the fermenter and from kinetic characteristics of the systems: (●) 25 μM; (○) 190 μM; (△) 2000 μM. (From Postma *et al.*, 1988.)

system is present. Since the residual glucose concentration under nitrogen limitation is 2.5 mM rather than the 3 μM under glucose limitation, it can be concluded that the high affinity system is repressed.

I have dwelt on these studies on *S. cerevisiae* and *C. utilis* for two reasons. First, the studies indicate that one can equate K_S and K_T only under relatively special circumstances. Second, as indicated earlier, the consideration of the relationship between growth of a fungus and transport of a nutrient is probably inappropriate at the present time. The above studies on the relationship between glucose transport and growth show the situation to be complex, with homeostatic mechanisms within the cell controlling transport either through the number of transport systems present or through the type of system that is functional.

If attention were to turn to the relationship between growth and uptake of a nutrient, such as potassium, that is not metabolised, it is likely that interpretation of the findings in terms of some sort of growth model will be more difficult. Although no such study has been carried out, there is sufficient information on the growth of *C. utilis* under potassium limitation (see Chapter 10) to show that the biochemical effects of limiting potassium can be complex.

The message from all this is that there is a need for more information about how the transport processes in cells respond under steady-state conditions to a limiting nutrient. The extent of constancy of the number of transport sites is obviously one matter needing further investigation.

Another is the effect of having one nutrient at limiting concentration on the transport of other nutrients. A list of other items of required information may be readily compiled. Also required is information on filamentous fungi. Growth of such fungi under steady-state conditions has been difficult because of non-uniform growth; but it may be possible now to rectify the deficiencies in the satisfactory growth of filamentous fungi in fermenters by using more tractable species, of which *Fusarium graminearum* is a good example (Trinci, 1992).

Transport flux

The use of steady-state conditions provides valuable insights into the relationsip between membrane transport and growth. But a fungus growing in the natural environment is exposed to a fluctuating environment due either to the heterogeneous nature of the substratum or to the metabolic activities of the fungus itself, such as extrusion of protons. Further, in the natural environment a fungus needs to capture nutrients in competition with other microbes. As pointed out above, it is the flux of the nutrient into the fungus that governs its ability to acquire nutrients and the greater the influx of a nutrient the more competitive will the fungus be when faced with limited supplies of the nutrient. Thus, the higher flux of phosphorus into beech mycorrhizal roots compared to that into non-mycorrhizal roots is believed to underpin the better growth of mycorrhizal seedlings compared with those which are non-mycorrhizal (see Chapter 7).

Although there is this striking example, there are few flux data of any kind for fungi. This is because virtually all studies are in terms of dry matter rather than surface area. Even those values that have been obtained are for the most part only assessments because the surface area has not been measured directly but calculated from fresh and dry weights and mean hyphal diameters (Slayman & Slayman, 1968; Jennings & Aynsley, 1971). There is a need for flux values calculated on the basis of surface area.

The determination of the surface area of the plasma membrane through which the flux is occurring is now relatively simple, with stereology as an established discipline (Clipson, Jennings & Smith, 1989). As far as the determination of the amount of nutrient flowing into the mycelium is concerned, the measurements need to be made with young mycelium. If one considers liquid culture, with increasing pellet size there will be increasing uncertainty about (a) the physiological state throughout the mycelium of the plasma membrane as well as the protoplasm that it bounds and (b) the extent to which the concentration of the nutrient in the medium is anywhere near uniform external to the plasma membrane. As the pellet becomes larger, declining oxygen tension within the pellet will affect directly the rate of transport but also other metabolic characteristics, while a non-uniform concentration will confound the determination of the nutrient flux (Trinci, 1970) (see Chapter 14).

Of course, information about flux of a particular nutrient is of little value for interpreting what may be occurring in the natural environment without knowledge of the concentration of the nutrient and also the pH in the locale where the fungus is growing. Knowledge about the latter is crucial for those nutrients whose flux is driven by the proton electrochemical gradient across the plasma membrane. Some of the ways in which this information might be garnered have been discussed by Jennings (1990a). However, it is not difficult to see that the interpretive jump from laboratory culture to the natural environment at the present time is almost impossible. Not only is there the problem of determining nutrient concentrations but also morphology in nature might be different from that in the laboratory. This can certainly be so for yeasts in nature when on a solid substrate. Then the cells take up a colonial disposition, though the kinetics for growth of such a colony are well established (Pirt, 1967). There is also the issue in filamentous fungi of the extent to which growth can be supported by nutrients translocated from older parts of the mycelium to the extending hyphal apices (see Chapter 14). The nutrients transported may not only be those absorbed from the external medium but may equally be those arising from the breakdown of polymers in protoplasm that has become aged.

What are needed now are experiments in which fungi are grown in situations intermediate between stirred liquid culture and the natural environment. Such model or artificial systems mimic important features of the habitat in which a fungus may be found (Jennings, 1987a). In terms of what is being discussed here, any procedure used must be such as to allow the investigator to have control over the concentration of nutrients available to the fungus being studied. The system used will depend on the fungus and the natural environment in which it grows normally.

Finally, though I have stressed the importance of flux values for comparative purposes, if the surface area through which transport is occurring is the same for two fungi, effective comparison can be via the affinity constant K_T. This is exemplified by the study of Postma et al. (1989a), who were concerned with competition between the yeasts *S. cerevisiae* and *Candida utilis* for glucose in glucose-limited chemostat culture. Under aerobic conditions, *C. utilis* always dominated over *S. cerevisiae*; under anaerobic conditions, the reverse was true. It would seem that under aerobic conditions the affinity of the glucose transport system in *C. utilis* ($K_T = 25\,\mu M$) ensures that this yeast has a higher influx of glucose than does *S. cerevisiae* ($K_T = 220\,\mu M$). That this is so is because at limiting glucose concentrations ($4\,\mu M$ in mixed cultures) $K_T > s$ and thus K_T has an influential effect on the flux (v) as predicted by the Michaelis-type (Monod) equation (i.e. $\mu = \mu_{max} \times s/(K_S + s)$, where μ and μ_{max} are the specific growth rate and maximum specific growth rate, respectively). It would seem that *S. cerevisiae* is better adapted than *C. utilis* to anaerobic conditions, possibly at the level of glucose transport, since, unlike in *C. utilis*, the process in *S. cerevisiae* does not require the input of metabolic energy. However, there could be other reasons for the better growth of *S. cerevisiae* under anaerobic conditions.

Hyphal extension

The growth of hyphae by apical extension is accompanied by the presence of electric currents. These transhyphal currents are measured in the external medium by a vibrating microelectrode called the vibrating probe. This microelectrode is made of glass or platinum/iridium wire, on the end of which is electroplated a ball of platinum. The electrode is vibrated between 200 and 800 Hz over 10–50 μm; the voltage is measured between the extremities of the vibration. By use of a modified form of Ohm's Law (Jaffe & Nuccitelli, 1974) the current can be calculated, the reciprocal of conductivity of the medium giving the resistivity. The direction of current is defined with respect to the net movement of positive charge. A net flow into the hyphae must therefore either be a net influx of cations or a net efflux of anions. The sensitivity of the electrode and amplification system can be as small as $0.5\,\mu A\ cm^{-2}$ and $1\ nV\ m^{-1}$. In the original development the probe was vibrated in one plane (Jaffe & Nuccitelli, 1974; Dorn & Weisenseel, 1982; Nawata, 1984; Scheffey, 1988). It is now possible to vibrate the probe in two dimensions so that current in two axial planes can be measured simultaneously (Freeman *et al.*, 1985; Nuccitelli, 1986; Scheffey, 1988).

The ionic composition of the current in fungi has been assessed up to the present almost exclusively by changing the composition of the external medium. In the future, the use of ion-sensitive vibrating electrodes will

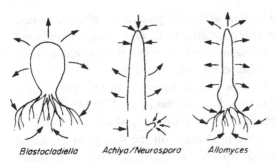

Blastocladiella   *Achlya/Neurospora*   *Allomyces*

Figure 2.3. Circulating ion currents in the fungi described in the text. The arrows indicate the movement of positive electrical current. The direction of growth of the hyphae of *Achlya*, *Neurospora* and *Allomyces* is towards the upper part of the figure. (From Gow, 1989*a*.)

facilitate such an assessment (Gow, 1989*a*). Changes of pH round the hyphae also give clues about the extent of involvement of protons in the current.

The essential feature of the current in fungi is that there is a circuit that enters the hypha at one point and comes out at another point a significant distance away. Thus, current is carried within the hypha itself. Figure 2.3 shows the general pattern of current flow in the four species that have been most studied to date. *Blastocladiella emersonii* consists of a sphaeroid-shape thallus to which are attached rhizoids. The thallus eventually differentiates into a sporangium. When the zygote of *Allomyces macrogymus* germinates it also produces rhizoids but there is also the development of apically growing hyphae. *Achlya* sp. and *Neurospora crassa* are essentially similar in the manner of their vegetative growth, though it needs to be remembered that the two fungi are representatives of very different parts of the Fungal Kingdom. It has been argued that the Oomycetes to which *Achlya* belongs are not true fungi. The findings for *N. crassa* can be taken to represent those for a number of other true fungi that have been investigated, namely *Basidiobolus variarum*, *Mucor mucedo*, *Aspergillus nidulans*, *Coprinus cinereus* and *Schizophyllum commune* (Gow, 1984, 1989*b*) and *Trichoderm harzianum* (Horwitz *et al.*, 1984).

In all the fungi investigated, the most likely candidates for the ions that carry the current are protons. The evidence, which is not equally strong in the four particular cases to be considered, is as follows:

(1) *Blastocladiella emersonii*: Vibrating probe studies show that positive current enters the rhizoids and leaves via the thallus (Stump *et al.*, 1980). When the organism is grown in culture in the

presence of glass fibres coated with insoluble inhibitors or ion exchange resin to which inhibitors are absorbed, the orientation is altered by proton conductors (Harold & Harold, 1980). Potassium and calcium ionophores have no effect. When microelectrodes are inserted into the thallus, the membrane potential behaves like a potassium diffusion potential (van Brunt, Caldwell & Harold, 1982). Potassium ions diffuse out of the cell by a pathway that can carry rubidium and barium ions but not lithium, sodium, caesium, magnesium and calcium ions, though the last on the list is required for the maintenance of the potential difference. Anions have no effect on the potential. With potassium-starved cells, potassium is absorbed by the thalli, with an expulsion of protons in a precisely correlated manner. There is a pH gradient across the membrane that is abolished by inhibitors of respiratory ATP generation. Potassium influx can be inhibited by proton-coupling uncouplers or by inhibitors of respiratory ATP production. These observations indicate that potassium uptake is by some active process in which potassium ions are exchanged for protons. Thus, the most likely total picture is that protons enter by the rhizoids and leave by the thallus. However, there are strong arguments for the outward limb of the current being carried by potassium ions and not protons.

(2) *Allomyces macrogynus*: Unlike all other fungi, there is an outward flow of positive electric current from the apex ($0.16 \, \mu A \, cm^{-2}$) with inward current at the rhizoids ($0.55 \, \mu A \, cm^{-2}$). There is no particular evidence that protons carry the current. Nevertheless, a range of experimental evidence rules out calcium (Youatt, Gow & Gooday, 1988).

(3) *Achlya* and *Neurospora crassa*: These two species, though taxonomically far apart, behave similarly with respect to their circulating ion currents, so are treated together. Of the two species, *Achlya* is the much better studied. In *Achlya bisexualis*, it has been shown that both inward and outward currents are carried by protons. Those that enter at the apex can do so in cotransport with amino acids, especially, but not exclusively, methionine (Gow, Kropf & Harold, 1984; Kropf et al., 1984; Harold, Kropf & Caldwell, 1985a; Harold et al., 1985b; Kropf, 1986; Schreurs & Harold, 1988). The outward current is carried by protons via an H^+-ATPase (the evidence for which is presented in Chapter 1). Evidence for protons comes from experiments using different media but particularly from studies relating pH and current density profiles along a hypha (Figure 2.4). In passing, one should note that the alkalinity at the tip may be a function not of hydrogen depletion due to proton uptake here but of release of

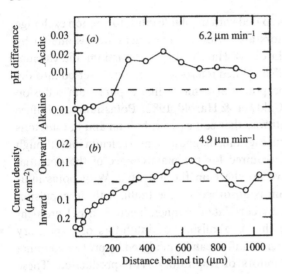

Figure 2.4. Comparison of (*a*) the pH and (*b*) the electrical current profiles of the same hypha of *Achlya bisexualis*. Both profiles were recorded within a period of 45 min. (From Gow *et al.*, 1984.)

ammonia that is known to occur during the growth of *A. bisexualis* (Schreurs & Harold, 1988).

Though the above gives the overall picture of the pattern of circulating currents in *Achlya*, there is an underlying complexity that is not completely understood. Armbruster & Weisenseel (1983) found the current was spikey not steady. Also, the current at the apices of hyphae that are branching can turn transiently outward (Kropf *et al.*, 1983). There is no change in growth rate with change in direction of the current. This change in current direction takes place when branches are produced. An inwardly directed current precedes the production of a branch and the location of such a current predicts the site where the branch will form (Figure 2.3). Finally, Schreurs & Harold (1988) have shown that hyphae of *A. bisexualis* have a normal diameter and extension rate but almost no electric current in a medium lacking amino acids but supplemented with thioglycolate and urea.

For *N. crassa*, the evidence for protons being the principal ions responsible for the transcellular current comes from ion substitution experiments (McGillivray & Gow, 1987; Takeuchi *et al.*, 1988). While the current is carried principally by protons it requires the presence of phosphate and glucose in the growth medium. Potassium ions may be required, though the position is not clear from ion substitution experiments. However, the changes that have been shown to occur in the absence of potassium can be mimicked by the potassium-channel blocking agent 3,4-diaminopyridine, although this compound does not inhibit potassium uptake (Takeuchi *et al.*, 1988). Calcium ions are required for hyphal extension but not for current generation.

It should be noted that, although current generation requires the presence of phosphate and glucose, phosphate itself cannot carry the current, since to do so the ion would have to move out of the hyphal apex; it seems unlikely that a glucose/proton symport would be functioning because it is repressed by the high concentrations of glucose used in the experiments (see Chapter 5). However, a phosphate/proton symport of stoichiometry $H_2PO_4^-:H^+$ of 1:2 could generate current in the observed direction. Potassium ions will be required for charge balance (see Chapter 7). The lack of clear evidence for the role of potassium might be confounded by the use of sodium in the ion substitution experiments, i.e. sodium may replace potassium in some of its functions.

In keeping with what has been discussed immediately above, Turian (1979*a*) has demonstrated cytochemically that, within extending hyphae of *N. crassa*, the lowest pH and highest concentration of phosphate are found in the immediate subapex. However, Turian (1979*b*, 1980) believed that protons are *ejected* into the medium at the hyphal apex generated by the mitochondria underneath the apex.

While it is assumed that the protons carrying the current leave the hyphae in the distal region via the H^+-ATPase, there is no direct evidence in *N. crassa* that this is so. However, Galpin & Jennings (1975) carried out cytochemical studies, on hyphae of the marine hyphomycete *Dendryphiella salina*, that indicated that ATPase activity was not present in the apical region.

It is possible to isolate, from a filamentous fungus, samples of protoplasts that represent different regions of hyphal cytoplasm (Isaac, Ryder & Peberdy, 1978; Isaac, Briarty & Peberdy, 1979). With lysis of hyphal walls by enzymes, protoplasts are released first from the apex and then progressively from older regions of the hyphae. The postulated absence of ATPase in the hyphal apex of *D. salina* is supported by an inability to detect H^+-ATPase in protoplasts after 1 h of exposure of hyphae to wall lytic enzymes, though activity was readily detectable in protoplasts isolated after 3 h of exposure (Garrill & Jennings, 1991). In *Aspergillus nidulans*, apical protoplasts have been shown to possess a lower $K^+:Na^+$ ratio than those from older regions (Isaac, Gokhale & Wyatt,

1986). This is in keeping with X-ray microanalytical studies on hyphae of
D. salina, which show a K^+:Na^+ of 0.91 at the apex and 3.30 in the region
1–50 μm behind the apex (Galpin *et al.*, 1978; Jennings, 1979). Thus, there
is evidence from other than electrical current studies that the hyphal apex
in fungi related to *N. crassa* differs in its transport properties from those
of the older regions. In the oomycete *Saprolegnia ferax* there is even more
direct evidence that this is so from patch-clamp studies. By developing a
technique in which protoplasts are produced in distinct linear arrays,
Garrill, Lew & Heath (1992) were able to show that the hyphal tip, but
rarely distal protoplasts, contains two stretch-activated ion channels, one
permeable to both calcium and potassium and the other to magnesium.

The bulk of the initial studies on circulating ion currents in fungi related
to a concern that these currents might be fundamental to the polar
growth of fungal hyphae (Harold *et al.*, 1985b; Gow, 1987; Harold, Caldwell
& Schreurs, 1987). The consensus now is that the data thus far do not
allow an unequivocal statement about the relationship between ion
currents and cell polarity (Gow, 1989a). As Gow (1989b) has
pointed out: (a) in those fungi that have inward apical currents, outward
currents are found occasionally and transiently in hyphae that extend at
normal rates; (b) there is no good correlation between either the extension
rate or the length of the peripheral growth zone (Trinci, 1971) and the
current density; and (c) in *Allomyces macrogynus* in particular there can be
reverse development of a hypha, i.e. it widens backwards after growth
under low oxygen, yet the current direction remains unchanged.
Nevertheless, as Gow (1989b) has pointed out, it may not be net flow of
ions that relates to the direction of growth but the flow of protons. When,
for instance, there is a change in direction of current without a change in
the rate of hyphal extension, it may be the result of a new flow of other
ions, with the flow of protons remaining constant.

Whatever the role of transhyphal currents in the process of hyphal
development, there is little doubt that the study of these currents has led
to important findings about the spatial aspects of nutrient acquisition by
fungal hyphae. We now know therefore that amino acids can be absorbed
at the hyphal apex of *Achlya* and that it is likely that phosphorus can be
absorbed in similar manner in *N. crassa*. In the case of *Achlya*, the presence
of amino acid/proton symport systems in the membrane of the hyphal
apex provides part of the molecular basis for the ability of hyphae of
this fungus and other members of the Mastigomycotina to show
positive chemotropism towards amino acids, especially methionine
(Musgrave *et al.*, 1977; Manavathu & Thomas, 1985). At the very least,

one presumes that, if there is an array of transport proteins for an amino acid that is evenly distributed within the membrane over the apex, then such an array can detect the location of sources of amino acids as a consequence of those transport proteins nearest the source being exposed to higher concentrations. How the sensory information is translated into a change in direction can only be surmised at present.

In *Blastocladiella emersonii* the rhizoids are known to be important in the acquisition of nutrients. Kropf & Harold (1982) showed that rhizoids are capable of absorbing a variety of nutrients and that specifically phosphate ions and amino acids appeared to be taken up preferentially by the rhizoids compared to the thallus. The rhizoids orientate strongly to gradients of phosphate and a defined mixture of amino acids – though not to glutamic acid, glucose, potassium or calcium (Harold & Harold, 1980). The rhizoids (but not the hyphae) of *Allomyces macrogynus* are chemotropic, suggesting that these structures are significant in nutrient acquisition (Gow, 1989a).

Enzyme secretion

The success of fungi as heterotrophic organisms lies in their ability not only to scavenge nutrients and invade new substrata but also to break down insoluble material into soluble components to provide part of their diet. There are numerous examples of this ability throughout this volume. Enzymes are secreted to bring about the solubilisation of nutrient substrates. However, the exact location for enzyme secretion has for years never been clear. The enzymes that are secreted have a size larger than the pore size of hyphal walls determined with isolated cell wall fractions (Trevithick & Metzenberg, 1966) or by molecular-exclusion using living hyphae (Money, 1990a). The paradox was highlighted by Chang & Trevithick (1974), who pointed out that ribonuclease, the smallest exoenzyme secreted by *Neurospora crassa* has a molecular mass of 13.7 kDa yet isolated walls do not permit penetration of polymers with molecular masses greater than 4.75 kDa. These authors proposed that the passage of exoenzymes is likely to occur more easily through the more plastic and porous nascent cell wall in the hyphal apex (Figure 2.5). It is implicit in the hypothesis that the porosity of the hyphal apex does not represent that of the wall of the bulk of the mycelium. Thus, as the apical wall is transformed into the lateral wall the structure becomes less porous, causing some exoenzymes to be trapped in transit and thus become bound to the wall.

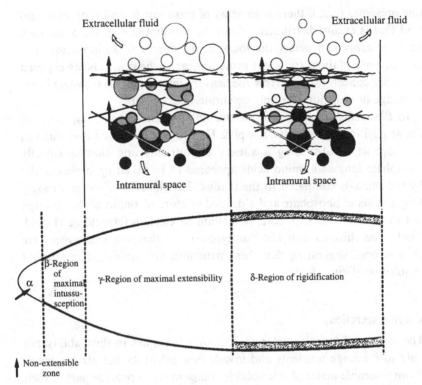

Figure 2.5. A diagrammatic view of the secretion of enzymes by a fungal hypha according to the hypothesis of Chang & Trevithick (1974).

Chang & Trevithick (1974) produced the following evidence for their hypothesis:

(1) During germ-tube outgrowth from spores and during early log phase, when hyphal apices form a greater proportion of the total mycelial surface area than in older cultures, there is a decrease in molecular sieving of secreted exoenzymes. Thus, at these early stages of growth, the proportion of the light (c. 51.5 kDa) invertase in the medium compared with the heavy (c. 210 kDa) is very much smaller than at the later stages (Table 2.1).

(2) Under a particular growth regime, a group of three exoenzymes of high molecular weight (aryl-β-glucosidase, invertase and trehalase) were shown to have a significantly higher percentage of their total activity in the wall than a group of three exoenzymes of low molecular weight (amylase, acid protease, ribonuclease).

Table 2.1. *Light invertase as a percentage of the total enzyme in culture filtrates during vegetative growth of* Neurospora crassa *in stirred liquid culture*

Hours of growth	Stage of growth	% light invertase in culture filtrate
0–2	Early germination	16.1
2–4	Mid-germination	18.7
4–6	Late germination	5.8
6–8	Early log phase	12.9
24–36	Early stationary phase	43.3

More direct evidence for the hypothesis has come from studies on protein secretion in *Aspergillus niger* by Wösten *et al.* (1991). The mycelium was grown in a thin agarose layer sandwiched between two perforated polycarbonate membranes to allow free diffusion of gases and nutrients and water (from nutrient agar on which the polycarbonate sandwich was placed), and also the free diffusion of secreted protein. The colonies grew essentially in a two-dimensional manner, allowing microscopical examination of individual hyphae. ^{35}S was fed to the mycelium, and protein secretion located by the distribution of radioactivity in the medium. The location of amylase was visualised through immunogold labelling. Though there were some technical difficulties in using these procedures, the bulk of the evidence is in keeping with the hypothesis of Chang & Trevithick (1974). Thus, in 4-day-old cultures both growth and protein secretion were localised at the periphery of the colony. Glucoamylase was associated at all stages of growth with leading hyphae, all of which were almost certainly growing as a result of their ability to incorporate *N*-acetylglucosamine into the wall. With 5-day-old cultures, protein secretion occurred in more central parts of the mycelium. Here, conidiophore differentiation was observed and there was some evidence that also at this location there could be secretion of idiophasic enzymes by growing branches.

Spatial aspects of nutrient acquisition in a colony

Extension of a hypha from a germinated spore is an exponential function of time, until such time as the protoplast has doubled. A branch is then formed and further branches continue to be formed after protoplasmic doubling – the so-called duplication cycle (Trinci, 1978). After formation of the first branch, leading hyphae extend in a linear manner. Each hypha is dependent on a supply of materials from further back. The evidence is that these materials are wall precursors and enzymes that are brought to the tip in vesicles probably moved by contractile elements in the cytoplasm (Heath, 1990). The length of hypha bringing about the supply constitutes the peripheral growth zone (Trinci, 1971).

As the circumference of the colony increases, the branches fill the space generated by the divergence of individual leading hyphae. The question then is how does a branch take up its position on the substratum approximately equidistant between the two neighbouring hyphae? Whatever the mechanism, the ability of the colony to arrange its hyphae in an even manner over a homogeneous substratum makes for the most effective utilisation of the nutrients in that substratum. Indeed the position taken by the branch is by definition the one at which the concentration of nutrients (including oxygen dissolved in the substratum, particularly so in the case of a substratum like agar) is at a maximum relative to the two leading hyphae bounding the space and before the branch moves into it. In an analogous manner, the same position will be where the two leading hyphae have had minimum effect on pH as the result of metabolic activity, while compounds released by the same hyphae that might affect growth of the branch are at minimum concentration. Thus, it seems highly likely that the growing tip of the developing branch must be able to sense the concentration of solutes within the agar such that the direction of growth is along the line of maximum or minimum concentration of the particular solute(s) being sensed.

There have been many studies on chemotropism in fungi, most of which have been concerned with the orientation of spore germ tubes. It has been concluded generally from such studies that only the Mastigomycotina show a tropic response to nutrients such as amino acids (though there is recent evidence (Horan & Chilvers, 1990) suggesting ectomycorrhizal fungi might be considered an exception). For the higher fungi, it has been argued that, because leading hyphae grow approximately parallel to each other and are further apart on 'nutrient-poor' than on 'nutrient-rich' media, hyphae react in a negative chemotropic manner to compounds produced by individual hyphae. In fact, many of the conditions used in the past by experimenters have been such that the direction of growth could well have been affected by the oxygen gradient in the medium. It seems now that many fungi could well be positively chemotropic to oxygen. Robinson (1973), who presented much of the evidence for such a possibility, believes that, in some way, differential respiratory activity on two sides of a hypha leads to bending towards a higher oxygen concentration.

The molecular basis of chemotropism in the Mastigomycotina has been alluded to already in the previous section. There the amino acid proton symport was suggested to be the sensing system. Conceivably in fungi other than the Mastigomycotina it is protons not oxygen that are the important determinant of chemotropic growth (Jennings, 1991a). Thus, growth along a gradient of oxygen concentration generated by metabolic activity might equally be growth away from an increased concentration of hydrogen ions (decreased pH) in the medium concomitant upon that same metabolic activity. The virtue of protons as against oxygen as a key element is that protons essentially can only cross the plasma membrane via a transport protein (porter). Thus there will be specific sensing agents on the plasma membrane, arguably the most appropriate locus for such agents (Jennings, 1974). If a difference in concentration of oxygen between one side of a hypha and the other is to be sensed, it is necessary that there should be no turbulence within the cytoplasm to disturb diffusion of gas to the mitochondrion. There is no doubt from microscopic observations that this is not so, there being much movement of the cytoplasm during growth. In addition to discussing chemotropism in terms of concentrations of oxygen and hydrogen ions in the medium, it has to be said that as yet there is no firm evidence to rule out inorganic ions, such as potassium and phosphate, as being important determinants of chemotropism.

Other aspects of the spatial aspects of nutrient acquisition are discussed in Chapter 14.

3
Walls and membranes

Introduction

The cell wall, plasma membrane and tonoplast are major barriers to the movement of solutes into and out of fungi. The wall is very much less of a barrier than the two membranes. Nevertheless it cannot be considered solely as a porous structure that allows through it only molecules below a certain size. The wall has other properties that influence the nutritional behaviour of fungi. These properties as well as the porosity of the wall are addressed below. The consideration of membrane focuses on the somewhat scattered information about the lipid part of the membrane; the various transport systems are dealt with in other chapters.

The wall

Introduction

Our knowledge of the polysaccharide composition of the walls in both filamentous fungi and yeasts covers a wide spectrum of taxonomic groupings and for many species there is much detailed information (for reviews, see Farčaš, 1979, 1990; Gooday & Trinci, 1980; Arnold, 1981a; Wessels & Sietsma, 1981; Bartnicki-Garcia & Lipmann, 1982; Bartnicki-Garcia, 1987; Fleet, 1991). On the other hand, we know much less about the other components of the wall. Enzymes are present (see Chapter 2) and polyphosphates can be present (see Chapter 7). Here I concentrate on what we know about the physicochemical properties of the wall of *Saccharomyces cerevisiae* and other yeasts, drawing attention afterwards to some salient features of the walls of other fungi.

The relative volume of the fungal wall is rarely considered. If one can hazard a guess, the volume of an actively growing fungal cell or hypha is probably thought to be no more than 15% of the total volume. An estimate of the volume is rarely made because the need for the value is not perceived. More likely than not, the wall is assumed (a) to influence solute movement into the protoplast only through an ability to keep out molecules above a certain size, and (b) to influence through its charged polymers the rate of movement of charged solutes across the wall. But as is pointed out in Chapter 13, the wall may be important in maintaining the environment outside the plasma membrane at a pH lower than that of the bulk medium. In this way the proton-motive force across the membrane is enhanced. Under such circumstances, the larger the volume of the wall the more effective is the maintenance of that lower pH.

The effective volume of the wall may be increased by the secretion of mucilage to form a layer over the wall. Marine fungi have been shown to secrete mucilage under a variety of circumstances (Jones, 1994). One purpose is clearly to increase adhesion to the substratum on which the fungus is growing. It is possible that mucilage may increase the probability that the environment close to the plasma membrane is less affected by changes in the external medium. Mucilage appears to be important in maintaining the stability of extracellular enzymes, as is discussed in Chapter 5 with respect to those involved in lignin degradation.

Saccharomyces cerevisiae and other yeasts

The walls of yeasts are by far the best studied of all fungal walls in terms of their physicochemical properties, with the greater part of the information coming from studies on cells of *S. cerevisiae*.

de Nobel & Barnett (1991) have critically reviewed what is known about the porosity of yeast cell walls. These authors point out the discrepancy between the first groups of determinations of wall porosity as exemplified by those of Gerhardt & Judge (1964) and Scherrer, Louden & Gerhardt (1974) and those made more recently. Gerhardt & Judge used fragmented cell walls, estimating the penetration of solutes into the wall material by assessing the degree of dilution of a solute in a known volume of solution when in contact with a known volume of packed wall material. Scherrer *et al.* (1974) made studies on intact cells. The study with isolated cell walls gave a value of 4500 Da for the mass of the molecules (referred to here by the symbol M_n to allow for the fact that in this particular investigation polydisperse polymer samples were used, requiring the use of a number-average molecular weight) capable of penetrating the wall, while the other study with intact cells and polyglycols gave an M_n value of 740 Da. The difference between the two values was ascribed by Scherrer

et al. (1974) to the fact that in the intact cell the outer layer of the wall
has smaller pores than the inner layer. Electron microscopic evidence was
brought forward in support of this proposal and Zlotnick *et al.* (1984)
have provided confirmatory evidence from enzyme studies on intact cells.
Whereas in intact cells the pores in the outer layer would result in a low
value of M_n, with fragmented wall material, there would not be this
screening effect of the outer layer. Hence a higher value of M_n would be
possible. The value of M_n for intact cells was given some support by the
study of Arnold & Lacy (1977) on the accessibility of substrates and
products of the action of β-fructofuranosidase and acid phosphatase in
the cell wall.

It was, however, known from the studies by Ottolenghi (1967) and
Schlenk & Dainko (1965) and Schlenk (1970) that proteins of a molecular
mass of at least 65 kDa seemed to be able to penetrate the wall and cause
the plasma membrane to break down. Scherrer *et al.* (1974) argued that
relatively large pores occur in the cell wall only after the addition of the
active protein but not before. This argues that the protein has an effect
on the wall analogous to that on the plasma membrane. If this were not
so, it seems hard to explain why invertase and acid phosphatase are
retained in the wall.

Since 1974, it has become increasingly clear, however, that, while the
observations reported above are internally consistent, they are in conflict
with a whole range of other data concerning the porosity of the wall of
yeast cells to large molecules. These other data are of two kinds:

(1) Wall porosity measurements: These are based on the ability of
polycations to damage the plasma membrane (see Chapter 4). The
technique developed by de Nobel *et al.* (1990*a*) is the simplest and
depends on the relative ability of polycations to penetrate the wall.
de Nobel *et al.* (1990*a*) used poly-L-lysine and DEAE-dextran. The
former lacks a tertiary structure and is capable of reptation – the
ability to 'squirm along' a narrow passage – thus penetration is not
related to the mass of the molecule. On the other hand, DEAE-dextran
does penetrate porous structures to an extent determined by its
molecular mass. It is easy to see that the ratio of leakage of
protoplasm-located material brought about by poly-L-lysine and
DEAE-dextran can be used as a relative measure of wall porosity. de
Nobel *et al.* (1990*a*) found that DEAE-dextrans of a molecular mass
of 380 kDa were capable of penetrating the wall of *Kluyveromyces
lactis* and *Schizosaccharomyces pombe* as well as *Saccharomyces
cerevisiae*.

Table 3.1. *Homologous compounds secreted by* Saccharomyces cerevisiae

Compound	Glycosylated	Approx. $M_r{}^a$	Location outside plasmalemma[b]
Asparaginase II	+	800 000	w
Invertase	+	270 000 (d)	w, m
	+	700 000 (o, m)	w
Acid phosphatase	+	252 000 (d)	w
	+	560 000 (o)	w
Endochitinase	+	115 000	w
KT 28 receptor	+	185 000	w
'Structural' wall protein	+	180 000 (m)	w
α-Agglutinin	+	160 000	w
	+	225 000	w
a-Agglutinin	+	22 000	w
Lysophospholipase	+	200 000	w, m
		280 000	
α-Galactosidase	+	300 000	w, m
		550 000 (o?, m)	w
D-Galactose specific lectin	+	320 000	m
Glucoamylase	+	250 000	m
'Barrier' proteinase	+	> 200 000	m
Glucan 1,3-β-glucosidase	+	56 000	m
	+	93 000	m,
IIIA	+	68 000	w, m
	+	29 000	w
Glucan *endo*-1,3-β-glucosidase			
I	+	218 000	w, m
IIIB	+	190 500	w
Serine carboxypeptidase (carboxypeptidase Y)	+	61 000	m
α-Factor mating pheromone	−	1500	m
a-Factor	−	1400	m
Killer toxin KT 28	+	16 000	m
K1	−	11 470	m

[a] (d) Dimeric form; (o) oligomeric form; (m) *mmn9* mutant.
[b] w, Cell wall or periplasmic space; m, medium.
From de Nobel & Barnett (1991). From *Yeast* 7, 313–24, reproduced by permission of John Wiley and Sons Limited.

(2) There is now a large body of information about the secretion of both homologous and heterologous proteins from *S. cerevisiae* (Tables 3.1 and 3.2). Table 3.3 presents information about some analogous compounds secreted by other yeasts. Inspection of the information indicates that walls can be porous to molecules with a relative

Table 3.2. *Some heterologous compounds secreted by* Saccharomyces cerevisiae

Compound	Glycosylated[a]	Approx. M_r	Location[b] i	w	m
Viral haemagglutinin	+	>250 000			+
Fungal cellobiohydrolase I	+	200 000	−	20	80
Aspergillus glucoamylase	+	100 000			90
Mouse IgG heavy chain	+	70 000			15
Mouse IgG light chain	−	28 000			40
Human serum albumin	−	67 000			+
Calf prochymosin	+	60 000	90	−	10
Bacillus α-amylase	+	60 000	25	−	75
Human α-amylase		55 000	30	25	45
Mouse α-amylase		55 000			70
Wheat α-amylase	−	42 000		−	+
Human interleukin 1β	+	22 000	−	−	+
Interferon-Con1	+	20 000	5	90	5
Human interferon		20 000	60	20	20
Human lysozyme		14 700			67
Human ANP		7400			+
Epidermal growth factor		5500	5	−	90
Human calcitonin		3500	−	−	95
Human β-endorphin		3000	−	−	95

IgG, immunoglubulin; ANP, atrial natriuretic peptide.
[a] −, not glycosylated; +, glycosylated.
[b] Percentage of heterologous protein found intracellularly (i), in the cell wall (w), or in the suspending medium (m); −, not present; +, present but not quantified.
From de Nobel & Barnett (1991). From *Yeast* 7, 313–24, reproduced by permission of John Wiley and Sons Limited.

molecular mass of 200 000–700 000 and smaller. Molecules with greater molecular masses are retained by the wall (the ability of amylose with a mass of 2000 kDa to enter the medium from cells of *Cryptococcus laurentii* (Table 3.3) clearly needs reinvestigation).

How then is the conflict between these more recent observations indicating that cell walls are porous to molecules with an M_n of c. 300 000 or more and the earlier ones indicating that only molecules of M_n of 760 could penetrate the walls to be resolved? The explanation appears to lie in the nature of the cells used for the various studies. The earlier studies used stationary phase cells; the later studies used growing cells. It is suggested by de Nobel & Barnett (1991) that there is in cells of yeasts a situation akin to that in hyphal apices (see Chapter 2), i.e. that a plastic

Table 3.3. *Some compounds secreted by yeasts other than* Saccharomyces cerevisiae

Yeast	Compound	Approx. M_r	Location outside plasmalemma
Candida albicans	Protease (mannoprotein)	44 000	Medium
	N-acetyl-β-glucosaminidase	66 000	Cell wall and medium
	Exo-1,3-β-glucanase	?	Medium
	Epitope (glycoprotein)	Up to 205 000	Medium
Candida glabrata	Killer toxin (glycoprotein)	16 000	Medium
Cryptococcus laurentii	Amylose	2 000 000	Medium
Filobasidiella neoformans var. *bacillispora*	Capsular polysaccharide	700 000	Capsule
Hanseniaspora uvarum	Killer protein	18 000	Medium
Kluyveromyces lactis	Killer polypeptide	100 000	Medium
Kluyveromyces marxianus	Inulinase	165 000	Medium
		335 000	Cell wall
Schwanniomyces occidentalis	α-Amylase (polypeptide)	50 000	Medium
Williopsis mrakii	Killer toxin (polypeptide)	8900	Medium
Yarrowia lipolytica	Protease	30 000	Medium

From de Nobel & Barnett (1991). From *Yeast* 7, 313–24, reproduced by permission of John Wiley and Sons Limited.

expanding wall is more porous than one that is not expanding. Consistent with this view are the results of a study by de Nobel *et al.* (1991) of cell-cyclic variations in the porosity of the cell wall of *S. cerevisiae*. Fractionation of an asynchronous culture into broadly different stages of the growth cycle showed that cell wall porosity increased with enrichment of the fractions with budded cells (Figure 3.1). In synchronous cultures, porosity increased sharply shortly before buds became visible, with maximum porosity during the initial stages of bud growth (which is what might be expected if the analogy with the extending hyphal apex holds). Thereafter there was a decrease in porosity, with the lowest values during bud abscission and in unbudded cells. Interestingly glucanase-extractable mannoproteins showed cyclic oscillations, with maximum incorporation

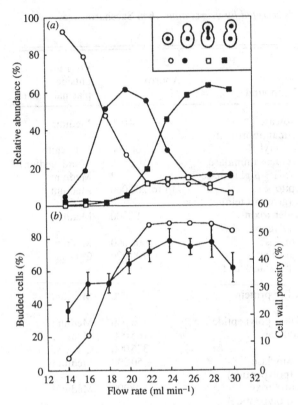

Figure 3.1. Cell wall porosity in successive age fractions, as obtained by centrifugal elutriation, of an asynchronous population of cells. (*a*) The relative abundance of different cell stages (see inset) was determined after nuclear staining of the cells in the successive subpopulations. (*b*) The relative wall porosity (●) of the cells was determined and compared with the total amount of budded cells (○) in these subpopulations. The results presented are the mean ±S.E.M. of five independent experiments (de Nobel, J. G., *et al.* (1991), *Yeast*, copyright © 1991, reproduced by permission of John Wiley and Sons Limited.)

after nuclear migration (just before bud abscission). This later observation confirms the view that mannoproteins are important in governing the extent of cell wall porosity (Zlotnick *et al.*, 1984; de Nobel *et al.*, 1990*b*).

In terms of retention of proteins by the wall, factors other than wall porosity could also be important. Ionic groups appear to be significant in restraining proteins from moving from the wall. Some evidence for this is the fact that EDTA, while increasing the porosity of walls of cells of *S. cerevisiae*, does not bring about the expected release of invertase (de Nobel *et al.*, 1989). The removal of divalent cations is thought to expose negative charges, which interact with positive charges on the invertase molecule. On the other hand, glycosylation of the proteins does not appear to be important in retention except in terms of increasing the size of the protein.

It is well known since the classic studies of Rothstein and his colleagues (e.g. Rothstein, 1954) that yeast cell walls contain negative charges due to phosphate and carboxyl groups. Of the phosphate groups, which are polyphosphate, 4.6×10^7 per cell are involved in some way in glucose fermentation (see Chapter 5). The chemical nature of the phosphate groups was elucidated by a study of the competition for uranyl ions between yeast cells and solutes in the medium (Rothstein, Frenkel & Larrabee, 1948; Figure 3.2). The presence of carboxyl groups was inferred from the fact that at higher concentration of uranyl and other bivalent cations such as Mn(II) an additional species of binding site becomes apparent (Rothstein & Hayes, 1956). Perhaps the best evidence that the groups of lower affinity are carboxyl is the fact that wall-bound invertase is inhibited by high concentrations of uranyl ions (Demis, Rothstein & Meier, 1954). Invertase is known to possess a high proportion of carboxyl groups.

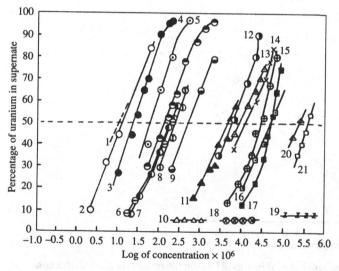

Figure 3.2. The relationship between yeast concentration and initial uranyl nitrate concentration at a fixed inhibition. 1, yeast; 2, metaphosphate polymer; 3, hexametaphosphate; 4, deoxyribonucleic acid; 5, pyrophosphate; 6, triphosphate; 7, ATP; 8, metaphosphate; 9, nucleic acid (technical); 10, adenylic acid; 11, egg albumin; 12, serum albumin; 13, citrate; 14, hexose bisphosphate; 15, orthophosphate; 16, maleate; 17, glycerophosphate; 18, glucose 1-phosphate; 19, glucose; 20, acetate; 21, fructose. (From Rothstein, 1954.)

   These negatively charged groups affect the electrophoretic properties
of yeast cells (Eddy & Rudin, 1958a). Mobility of cells of strains of
S. cerevisiae and *S. carlsbergensis* in an electric field can be described in
terms of two types: one in which the mobility is nearly independent of
pH and the other in which the mobility varies continuously between pH 3
and pH 6, with an alteration in charge from positive to negative at pH 4.
This latter type of mobility is thought to be due to the proteins in the
wall. Alkaline solutions remove the charged groups governing the
pH-sensitive mobility. The groups contributing to the mobility that is
independent of pH are believed to be phosphate, since the necessary
charged groups disappear when cells are grown in phosphate-deficient
media and there is a good correlation between phosphorus content and
mobility (Figure 3.3).

   Electrophoretic mobility is a function of the electrical double layer
round the cell, properties of which are defined by the Gouy–Chapman
equation, one being the surface charge density (Theuvenet &
Borst-Pauwels, 1976a, b). Though one might expect from this that the
charged groups in the wall would affect the rate of ion transport, studies

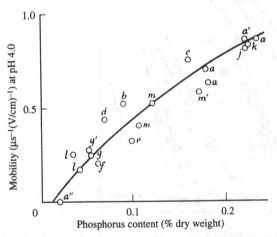

Figure 3.3. Dependence of mobility at pH 4.0 on the respective phosphorus contents
of the cell walls of 11 yeasts. Each point refers to a different cell wall preparation,
certain yeasts being represented more than once. Yeasts are designated by the
letters a–g and j–m and were normally harvested from mature standing cultures
in the malt-extract medium (72 h old). However, a prime attached to the letter
means that the yeast in question was from a young culture (18 h old); the double
prime (*a″*) refers to a culture of yeast *a* that was starved of phosphate. (From Eddy
& Rudin, 1958a.)

on plasmolysed cells and on protoplasts indicate that it is the surface charges on the plasma membrane that have a decisive effect on the characteristics of monovalent cation transport.

The wall plays a crucial role in yeast flocculation. This is the phenomenon 'wherein yeast cells adhere in clumps and sediment rapidly from the medium in which they are suspended' (Stewart *et al.*, 1976). While the definition also covers mating aggregates, the two types can be distinguished by flocs being dispersed by EDTA and reflocculation occurring in the presence of calcium ions. Flocculation is important in applied fungal nutrition because in brewing, if flocculation occurs too early, there will be a hung fermentation containing residual sugar. Flocculation at the correct time in a fermentation is, however, a most cost-effective way of removing cells and producing a clear liquid. While not wanting to dwell long on the process of flocculation – the topic has been well reviewed by Calleja (1987) and Stratford (1992*a*, *b*), and Gregory (1989) has provided an excellent introduction to the physics of the phenomenon – I discuss briefly the process of flocculation in *S. cerevisiae* because ions have been implicated.

Calleja (1987) pointed out that the enhancing effect of inorganic salts on flocculation has been known since the turn of the century. Dispersed in distilled water, cells reflocculate when suspended in salt solutions such as chlorides and sulphates of calcium, magnesium, sodium and potassium. The literature on the subject contains many conflicting reports on which ion(s) are the most effective. With time, it became clear that calcium seemed particularly important for flocculation. This led to the calcium-bridging hypothesis of Harris (1959) and Mill (1964), which suggested that calcium ions link cells via their interaction with the surface carboxyl groups, stability being brought about by hydrogen bonding between hydroxyl groups of the wall carbohydrates (Figure 3.4). The involvement of carboxyl groups was proposed on the basis of the effect of pH on flocculation. The involvement of such groups is supported by the irreversible inhibition of flocculation by the esterifying agent 1,2-epoxypropane (Mill, 1964; Jayatissa & Rose, 1976) and by the fact that onset of flocculation in the stationary phase of growth is accompanied by an increase in the density of wall carboxyl groups as determined by electrophoretic mobility measurements (Beavan *et al.*, 1979). On the basis of what has been concluded from wall porosity determinations, this increase in carboxyl groups could be due to increased retention of secreted proteins within the wall. This is consistent with the observed loss of flocculation brought about by proteases and protein-denaturing agents

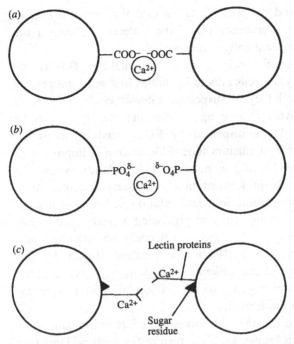

Figure 3.4. Suggested mechanisms of yeast flocculation. (*a*) Divalent calcium ions form bridges between surface carboxyl groups on different cells. These are supported by hydrogen bonding between mannan hydroxyl groups and hydrogen moieties. (*b*) Phosphate groups of wall mannan form the attachment groups for calcium bridging. (*c*) Surface proteins with lectin properties specifically bind to sugar residues in the wall mannan of neighbouring cells. Calcium ions are required to maintain the lectins in active conformations. (From Stratford, 1992*a*.)

(Eddy & Rudin, 1958*b*; Nishihara, Toraya & Fukui, 1977, 1982; Nishihara & Toraya, 1987). There is strong evidence against phosphodiester groups being involved (Jayatissa & Rose, 1976; Beavan *et al.*, 1979).

Stratford (1992*a*) has pointed out that, though there is little direct evidence against the calcium-bridging hypothesis, there are two facts that militate against its acceptance. First, calcium ions are not divalent, as can be the case with calcium metal, but are dipositive. Thus, unlike calcium metal, calcium ions are unlikely to form bonds between two structures. Second, sugars, specifically mannose, can inhibit flocculation (Taylor & Orton, 1978; Lipke & Hull-Pillsbury, 1984), though it has been argued by Calleja (1984) that this is not a direct effect but indirect through metabolism of the sugars (but see below).

While the calcium-bridging hypothesis does not appear tenable, the question of the involvement of calcium still remains. The matter has been complicated by reports that flocculation could be brought about by magnesium ions and in some strains by low concentrations of potassium or sodium ions (Stewart & Goring, 1971). We are fortunate that a study by Stratford (1989) helps to clarify the situation. He has shown the following. Flocculation does require the presence of multi-charged ions. Calcium can fulfil this role over a broad pH range. At near neutral pH, magnesium and a variety of transition elements can induce flocculation. EDTA and EGTA inhibit flocculation, which can only be reversed by calcium ions; magnesium and the transition elements do not possess this capability. Sodium inhibits calcium-induced flocculation. Studies with ^{45}Ca showed that other ions cause flocculation by release of cellular calcium ions, which can then exert their effect at the wall surface. The effectiveness of neutral pH in bringing about flocculation with magnesium is thought to be due to the greater efflux of calcium at this pH. Sodium has its effect by competitively inhibiting calcium at the wall binding sites.

The evidence now available supports the view that specific surface proteins of flocculent cells bind mannose residues intrinsic to mannan in other yeast cell walls, thus forming bonds between cells (see Figure 3.4, above). The idea of protein–carbohydrate bonds was first proposed by Taylor & Orton (1978) but much more fully developed by Miki *et al.* (1982*a*) and Miki, Poon & Seligy (1982*b*) as the 'lectin-like' hypothesis. Stratford (1992*a*) has summarised the evidence for the hypothesis as follows:

(1) The putative involvement of proteins in flocculation becomes explicable.
(2) The effect of calcium is consistent with lectin behaviour.
(3) Receptors found in flocculent and non-flocculent cells are not proteins but are sensitive to blocking with concanavalin A (Miki *et al.*, 1982*a*; Nishihara & Toraya, 1987), which indicates α-gluco- or mannopyranose carbohydrate groups.
(4) Specific sugar inhibition with mannose.

Details of the 'lectin-like' hypothesis were discussed by Stratford (1992*a*), and the genetic basis of flocculation was considered in detail by Stratford (1992*b*).

Other fungi

There is a regretable paucity of information about the ionic characteristics of the walls of fungi other than *S. cerevisiae*. We know from studies on

the uptake of multivalent metal cations that they bind readily to the
hyphal surface (see Chapter 11). There is much other information in the
literature about ions binding to fungal walls but little of coherence.
Certainly, there is no information about those ligands involved in ion
binding. The studies with *S. cerevisiae* point the way forward.

Nevertheless there are some intriguing observations reported, and I
record a few, essentially without comment.

(1) Mycelium of *Coprinus cinereus* binds sugars to a small but significant
    extent (Moore & Devadatham, 1979).
(2) Mycelium of filamentous fungi can adsorb particulate matter such as
    clays, calcium silicate, elemental sulphur, lead sulphide and zinc dust
    onto their surface from water suspension (Wainwright, Grayston &
    De Jong, 1986). The ability of *Mucor flavus* to absorb particulates was
    decreased in the presence of Cu(II) and Hg(II) ions but stimulated by
    magnesium ions (Singleton, Wainwright & Edyvean, 1990). Magnetite
    can be adsorbed such that the mycelium becomes susceptible to a
    magnetic field (Wainwright, Singleton & Edyvean, 1990).
(3) Walls of hyphae of *Dendryphiella salina* bind potassium selectively
    rather than sodium (Clipson, Hajibagheri & Jennings, 1990).
(4) For higher plants, the cation exchange capacity (CEC) has proved to
    be a useful procedure for comparing the ability of roots of higher
    plants to bind cations (Crooke, 1964). Pulverised dry material is used. It
    is suspended in acid for a standard time, washed and titrated with
    standard alkali to neutrality. McKnight, McKnight & Harper (1990)
    have determined the CEC of stipe tissue from sporophores of 18
    basidiomycete species. As one can see from Figure 3.5 there can be
    considerable differences between species. That in itself is interesting
    but there is another point: dried mycelium as used in this investigation
    would be one way of probing further the ion binding capability of
    walls of mycelial fungi.

Hydrophobicity of walls

Over the last decade, it has become clear that cell walls of fungi are
capable of being involved, to a greater or lesser extent, in hydrophobic
interactions with other surfaces, indeed other walls. This interest in
hydrophobic interactions arises from the realisation that hydrophobic
interactions play a role in many, if not most, microbial adhesion
phenomena (Doyle & Rosenberg, 1990). The surface hydrophobicity is a

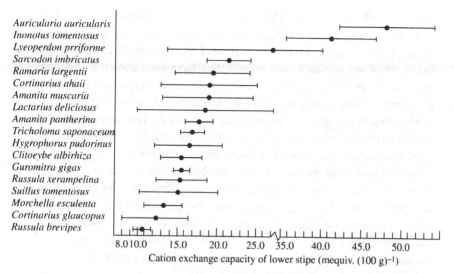

Figure 3.5. Average cation exchange capacity of lower stipe tissue from sporophores of 18 species of basidiomycete fungi. Standard deviations (horizontal bars) were based on three samples. (From McKnight, K. B. McKnight, K. H. & Harper, K. T. (1990), reprinted by permission from *Mycologia* **82**, 91–8, Copyright 1990 The New York Botanical Garden.)

factor in the adhesion to soft tissues of animal hosts, submerged surfaces of aquatic organisms as well as medical implants and prostheses. Since this is so, it is probable that, when considering the ecology of a fungus in nutritional terms, knowledge of hydrophobic interactions is important in assessing the ability to utilise particular nutrient substrata. Certainly our knowledge of human pathogenic fungi is being aided by such knowledge (Hazen, 1990). But equally such knowledge is important for the industrial use of fungi, since adhesion to surfaces within the fermentation plant and to others of industrial importance can have serious economic consequences (Mozes & Rouxhet, 1990). Flotation of yeast and other microbial cells by rising bubbles in a fermentation process is a particular aspect of the same general phenomenon. Hydrophobic interactions are certainly important in the utilisation of alkanes and other hydrocarbon-containing substrates (see Chapter 5).

A very important discovery has been the identification of the genes specifically expressed in the fruiting dikaryon of *Schizophyllum commune* (Wessels, 1991). Three gene products (those of *Sc1*, *Sc3* and *Sc4*) are particularly abundant. There is a remarkable similarity between the general structures and coding sequences of the three genes (Dons *et al.*, 1984*a*, *b*;

Schuren & Wessels, 1990). The three genes are evolutionarily related and code for small similar hydrophobic proteins, each with eight cysteine residues conserved in the same positions. Putative signal peptides suggest the proteins are secreted, and indeed they have been shown to be major components of the wall and the medium in which the mycelium is growing. It is highly likely that these proteins are involved in the aggregation of hyphae leading to the formation of the sporophore. One can surmise that similar proteins are involved in the formation of other aerial (and linear structures) whose physiology is discussed in Chapter 14.

Determination of the surface hydrophobicity in fungi is not easy because of either variation in cell size in the case of yeast or the considerable complexity of the mycelial system (when compared with other microorganisms) of filamentous fungi (Hazen, 1990). For these reasons, the information about fungal surface hydrophobicity is considerably fragmented. The reader should consult the references cited above for an overview of what is known to date.

The plasma membrane

Introduction

It is my intention to consider here only one feature of the plasma membrane, namely the relationship between lipid composition and function. The characteristics of transport across the membrane are considered throughout the text. Such matters as the involvement of the membrane in wall synthesis and protein secretion or it being the locus of light reception are not considered to be within the compass of this volume.

There is a considerable body of information on lipids in fungi that has been well reviewed by Weete (1980) and Losel (1988) and the lipids of yeasts by Rattray (1988). These reviews contain information of relevance to any consideration of the lipids in the plasma membrane. Apart from a review of Rose (1976), there is no good recent review of the plasma membrane of filamentous fungi focusing on the role of the lipids; the situation with respect to yeasts is more satisfactory, with comprehensive reviews by Rank & Robertson (1983), Prasad (1985), Prasad & Rose (1986), Henschke & Rose (1991).

The publications mentioned in the last paragraph should be consulted for what is known about the details of lipid composition of the plasma membrane. Here, only two items of information need be singled out. First, the two cultivatable genera of the Peronosporales, which are mainly obligate plant pathogens, namely *Pythium* and *Perenospora*, are unusual in fungi in that when they are grown on media without sterols their membranes do not contain such compounds (Elliott *et al.*, 1964; Elliott, Hendrie & Knights, 1966). Sterols are, however, required for sexual reproduction (Elliott, 1977, 1979). Second, although this chapter focuses on the plasma membrane, the lipid composition of other structures in a fungus are also important nutritionally. In this context, one is thinking particularly about the tonoplast, about which virtually nothing is known in terms of membrane function in relation to lipid composition. Major differences in the properties of the tonoplast *vis à vis* the plasma membrane are considered in Chapter 4. But it needs also to be kept in mind that lipid can be found in fungal walls (Losel, 1988). Much of the evidence comes from a strong osmiophilic reaction to stains for lipid but there are also analytical data. The amount can vary from 1–2% to 19% of wall dry weight, with walls of *Penicillium charlesii* containing as much as 37.5% (Bulman & Chittenden, 1976; Losel, 1988; Fleet, 1991). The role of this lipid is unknown but conceivably it might be important in reducing water loss.

Saccharomyces cerevisiae and other yeasts

There is now a significant body of information for *S. cerevisiae* and other yeasts on the effect of altering the lipid composition of membranes in the cells of the yeast on its physiology. Table 3.4 lists some of the physiological functions that can be altered, the classification of the information being by the method used to alter lipid composition. One should note that not all the methods used yield unambiguous information. Thus, changes in lipid composition brought about by altering growth conditions (except under an anaerobic environment) usually involve several classes of lipid. However, growth under anaerobic conditions provides a means of making more specific changes. Under these conditions, the yeast requires a sterol and unsaturated fatty acid for growth (Andreasen & Stier, 1953, 1954). However, the requirements for these two types of compound are relatively nonspecific. Therefore the growth medium can be supplemented with a specific sterol and/or fatty acid of choice. There appears to be no alteration to the content and composition of other types of lipid in the cells (Hossack & Rose, 1976).

There is clear evidence that altering the plasma membrane lipid composition in the way just described alters the transport characteristics for particular solutes. Table 3.5 gives kinetic data for the transport of four amino acids into cells of *S. cerevisiae* in which the plasma membrane has

Table 3.4. *Procedures used for altering lipid composition, with consequences for the physiology of yeasts*

Conditions	Alteration in lipid composition	Affected structure or function	References
A. By varying the environment			
1. Anaerobically grown *Saccharomyces cerevisiae* supplemented with ergosterol oleic or linoleic acids	Enriched with oleyl or linoleyl residues	Solute transport	See text
2. Growth of psychrophilic, mesophilic and thermophilic yeasts	Fatty acid composition	Solute transport, membrane fluidity	Watson (1980)
3. Supplementation of growth medium with choline or ethanolamine	Enriched with phosphatidylcholine or phosphatidylethanolamine	Solute transport, osmotic fragility of plasma membrane	Hossack, Sharpe & Rose (1977); Trivedi *et al.* (1982); Trivedi, Singhal & Prasad (1983)
4. Growth of *Zygosaccharomyces rouxii* in the presence of sodium chloride	Fatty acid composition	Membrane fluidity; tolerance to sodium chloride	Hosono (1992); Watanabe & Takakuwa (1984, 1987, 1988)
B. By using specific inhibitors/drugs to affect lipid synthesis			
1. 2,3-Imino-squalene	Complete blockage of ergosterol synthesis in anaerobic *S. cerevisiae*	Stereochemical specificity of sterols	Pinto & Nes (1983)

2. Hydroxylamine hydrochloride	Blocks phosphatidylserine decarboxylase and results in phosphatidylserine accumulation in *S. cerevisiae* and *Candida albicans*	Solute transport	Trivedi *et al.* (1983)
3. Hydroquinone and ascorbic acid	Respectively increase or decrease in ergosterol levels	Solute transport, polyene sensitivity	Singh *et al.* (1979*a, b*)
4. Cerulenin	Fatty acid synthesis	Solute transport	Otoguro *et al.* (1981)
5. Polyene antibiotics	Membrane sterols	Amino acid transport and potassium release; phosphate uptake	Singh *et al.* (1979*a*) Berdicevsky & Grossowicz (1977)

C. By using mutants defective in lipid synthesis
See text for a discussion of this matter

D. By using phospholipases

1. Phosphoinositidase C	Phospholipid composition	Kynurenine hydroxylase, conformational changes in protein	Esfahani & Devlin (1982); Esfahani *et al.* (1979)

Table 3.5. *Effect of phospholipid fatty-acyl unsaturation on kinetic constants for transport of selected amino acids into strains of Saccharomyces cerevisiae*

			Constant for cells with membranes enriched in				
			Oleyl residues		Linoleyl residues		
Strain	Amino acid	Affinity of uptake system	K_T (M)	V_{max} (nmol (mg dry wt)$^{-1}$ min^{-1})	K_T (M)	V_{max} (nmol (mg dry wt)$^{-1}$ min^{-1})	Reference
NCYC366	L-Arginine	High	1×10^{-6}	2.0	1.5×10^{-5}	4.5	Keenan & Rose (1979); Keenan, Rose & Silverman (1982)
NCYC366	L-Lysine	Low	2×10^{-2}	35.0	9.6×10^{-2}	218	Keenan & Rose (1979); Keenan et al. (1982)
		High	6.5×10^{-6}	4.2	2.5×10^{-7}	0.5	
Y185	L-Threonine	Low	2.8×10^{-4}	1.9	1.1×10^{-4}	3.1	Calderbank, Keenan & Rose (1984a)
		—	2.8×10^{-4}	53.4	2.1×10^{-4}	43.9	
Y185	L-Histidine	—	3.71×10^{-6}	1.42	2.55×10^{-6}	1.48	Calderbank et al. (1984b)

From Henschke & Rose (1991).

been enriched with oleyl or linoleyl residues (Henschke & Rose, 1991). The characteristics of transport of L-arginine, L-lysine, L-threonine and L-histidine are all affected, but particularly those of the first two amino acids. The characteristics of both high and low affinity systems can be altered. The kinetics of transport of L-asparagine, L-aspartic acid, L-glutamic acid, L-glutamine, L-isoleucine, L-leucine, L-methionine, L-serine, D-glucose, phosphate and calcium have been shown not to be affected by such changes.

Another facet of the effect of plasma membrane lipid composition on transport has been opened up by the studies of Calderbank, Keenan & Rose (1984b). They showed that cells enriched with linoleyl residues more rapidly acquired activity of the general amino acid transport system after derepression than did cells enriched with oleyl residues. When the cells were in the derepression buffer before the transport studies, there was no difference in the size of the amino acid-N pool, the concentration of L-alanine in that pool or the rates of protein synthesis and glucose fermentation. These latter findings support the view that the effect on expression of the general amino acid permease is on the protein when it is within the membrane.

There are other reports in which transport of solutes is specifically affected by changes in plasma membrane composition. Singh, Jayakumar & Prasad (1979a) showed that uptake of several (but not all) amino acids and efflux of potassium in cells of *S. cerevisiae* are altered by changes in the ergosterol content, while growth of cells of *Candida albicans* in alkanes of different chain lengths brings about reduced uptake of some amino acids but not others.

It is not easy to interpret these findings. Thus, it is not clear whether transport processes at the plasma membrane or at the tonoplast or at both membranes are affected. Also most interpretations are in terms of changes in the environment in which the transport protein functions. One cannot say whether such interpretations are correct without a great deal more work. While lipid modulation of transport processes is thought to be important, a survey by Cooke & Burden (1990) of what is known about lipid modulation of plasma membrane ATPases in plants, animals and fungi indicated that the matter is controversial in terms of the processes involved. There may be effects through protein–lipid binding or there may be an effect due to lipid fluidity. Other approaches are certainly needed if we are to establish the extent to which lipid composition of the plasma membrane affects its functioning. One obvious approach would be studies on the functioning of transport proteins in artificial lipid membranes of known composition. There is also the genetic approach, of which there is one very good example, detailed below.

Höfer *et al.* (1982) showed that a nystatin-resistant strain of *Rhodotorula gracilis* contains very much less ergosterol compared with the wild-type. The mutant cells have a decreased capacity for xylose transport, though there is no effect on the magnitude of the proton electrochemical potential gradient (Höfer, Huh & Künemund, 1983; Künemund & Höfer, 1983) (see Chapter 5). Nystatin has been shown to have its effect in cells of *R. gracilis* only at the plasma membrane (von Hedenström & Höfer, 1979). So it is likely that this membrane is the one affected by the mutation. It appears that the mutant cells have a lower passive proton and potassium ion permeability than do the wild-type cells. Thus, unlike the wild-type cells, the membrane potential of mutant cells is virtually independent of medium pH (down to 4.5) and of extracellular potassium concentration. The governing process appears to be the passive flux of protons, since the uncoupler carbonyl *p*-trifluoromethoxyphenyl hydrazone (CCCP), which is a specific ionophore for protons, increases the flux of both protons and potassium, the efflux of the latter always being high enough to compensate for the CCCP-induced influx of protons. The decreased sugar uptake therefore appears to be due to the decrease in the velocity of charge compensation necessary for the effective functioning of the proton/sugar symport. These studies need to be followed up by other studies on the lipid part of the plasma membrane.

Nystatin-resistant mutants with altered sterol composition and different sterol content have been isolated from *Zygosaccharomyces rouxii* by Ushio, Sawatani & Nakata (1991). The mutants possessed properties different from those of the wild-type. In particular two mutants with altered sterol composition grew much less well in the presence of 1.47 M sodium chloride, leaking much more glycerol into the medium than did wild-type cells (see Chapter 13). The process of leakage has not been characterised.

These studies on the nystatin-resistant mutants of *S. cerevisiae* and *Z. rouxii* support the view that an approach via the isolation of mutants defective in the lipid portion of the plasma membrane could be a potent way of studying its function. Indeed, now there are a considerable number of lipid mutants of *S. cerevisiae* whose transport properties need investigation (Henry, 1982). Nevertheless, a word of caution is needed. Henry (1982) has pointed out, from a study of many of the mutants, that the mitochondrial membrane appears to be the most sensitive to changes in lipid composition. Thus, changes in the transport properties of lipid mutants might be due to the energy-generating capacity of the cells. Again, as Henry (1982) has pointed out, when cells of *S. cerevisiae* are grown fermentatively or anaerobically, the cells will accept, as we have seen above, a wide range of sterols and fatty acids as substitutes for the naturally occurring ergosterol and 16- to 18-carbon fatty acids. Also glucose-grown cells with a wide range of phospholipid compositions can be generated very readily by altering, in a relatively simple manner, the characteristics of the growth regime. That such considerable changes can be made to the membrane lipids of cells in which mitochondrial function is not required indicates that the plasma membrane (and also the tonoplast) can be relatively flexible in their lipid composition, without loss of function. Thus, one must expect that changes to the characteristics of transport processes at the plasma membrane or tonoplast in lipid mutants are likely to be small and subtle. This is what has been observed for the effects of altering the lipid composition of yeast cells on amino acid transport, as described above.

There is a strong argument for turning to filamentous fungi for mutants with altered plasma membrane lipid composition. Given that the hyphae are of sufficiently wide diameter, not only can the kinetic properties of transport systems be investigated but so too can the electrical properties of the membrane in which the systems reside. *Neurospora crassa* is the obvious organism to use (see Chapter 1) and the value of using this species for the indicated studies has been well demonstrated (Friedman, 1977; Friedman & Glick, 1980).

4

The vacuolar compartment (vacuole)

Introduction

The term 'vacuolar compartment' is used advisedly. In the green plant cell, one can speak of the vacuole as a distinct entity, since it can routinely occupy c. 90% of the volume of the cell. The size of the vacuole can be such – as is the case in large algal coenocytes such as *Chara* and *Nitella* – that the vacuolar sap can be withdrawn readily for analysis (Hope & Walker, 1975). Further, the large size of the vacuole relative to the cytoplasm in green plant cells means that the vacuole makes a significant contribution to the cell osmotic potential. Study of the flux of radiotracers into and out of the above coenocytes has allowed the fluxes of the major ions across the vacuolar membrane (tonoplast) to be determined. Since the cytoplasmic and vacuolar concentrations of these ions can be determined readily and the potential difference across the tonoplast determined without too much difficulty, by insertion of an electrode into the vacuole, the driving force on any particular major ion can be determined. The literature on green plant vacuoles has been reviewed relatively recently (Marin, 1986; Raven, 1987).

In fungi, it is not often that there is a single vacuole. Certainly, there does not appear to be a single vacuole in actively growing hyphae, indeed on the basis of the different responses of vacuoles to stains there may be more than one type of vacuole (Clipson *et al.*, 1989). In fact, it has to be admitted that the cytological aspects of vacuolar organisation in hyphae have not been studied with the desired degree of rigour. One is unlikely to know whether the vacuoles that can be observed in microscope section are discrete entities or are interconnected.

In *Saccharomyces cerevisiae*, depending on cultural and cellular conditions, vacuoles might be dispersed as several smaller units in the cytoplasm or may fuse to form a single large unit (Robinow & Johnson,

1991). The vacuolar volume changes along with cell volume during the cell cycle (Wiemken, Matile & Moor, 1970). All the above observations suggest caution in thinking about a specific organelle. As Schwencke (1991) has pointed out, while 'vacuole' is a useful descriptive and operational term, the organelle is in fact an integral part of a dynamic endomembrane system that is involved in the flow of glycoproteins designed either for secretion or for the vacuole itself. Endoplasmic reticulum, Golgi apparatus and other vesicles are part of this system. It is not clear how large vacuoles relate to small vacuoles or, probably better, vesicles. However, vacuoles can be isolated from both yeasts and filamentous fungi and it is also possible using cytochemical and other procedures to say something about the properties of vacuoles *in situ*. This chapter is devoted to a consideration of what we now know about the vacuole, particularly as it relates to the nutrition of fungi. Thus, the vacuole of *S. cerevisiae*, and indeed those of other fungi, contains a variety of enzymes (Table 4.1). I do not consider here the synthesis and processing of vacuolar enzymes or the relation of the vacuole to other parts of the endomembrane systems and their role in exo- and endocytosis. The subject is reviewed elsewhere (Schekman, 1985; Rothblatt & Schekman, 1989; Rothman *et al.*, 1989; Klionsky, Herman & Emr, 1990; Schwencke, 1991). Attention in this chapter is focused on those soluble ions and molecules found in the vacuole and solute transfer across the tonoplast. Emphasis is on what is known about vacuoles in *S. cerevisiae*, about which we know most, but important information has also been obtained from *Neurospora crassa*.

Solute concentrations inside the vacuolar compartment

We have significant information about the concentrations of solutes within the vacuoles of five fungi, *Candida utilis*, *Dendryphiella salina*, *N. crassa*, *Saccharomyces carlsbergensis* and *S. cerevisiae*. The information has been obtained by three different procedures:

(1) Differential extraction by such agents as cytochrome *c*, DEAE-dextran or cupric ions. These agents make the plasma membrane permeable to ions and small molecules without affecting the tonoplast (see p. 75). This procedure has been used with yeasts; determination of the cell content of a solute before and after treatment gives the vacuolar content. Table 4.2 provides information, obtained by Wiemken & Nurse (1973) by permeabilising the plasma membrane of cells of *C. utilis*, on the percentage contribution of particular amino

Table 4.1. *Enzyme activities in vacuoles isolated from* Saccharomyces cerevisiae

Enzymes	Relative activity[a]
Intravacuolar hydrolases	
Proteinase A (acidic endoproteinase)	130
Proteinase B (serine-type, neutral endoproteinase)	143.7
Carboxypeptidase Y (serine-type, exoproteinase)	144.0
Aminopeptidase I (metallo, exoproteinase)	202.0
Endopyrophosphatase	73.7
Exopyrophosphatase	65.8
Ribonuclease	135.0
Trehalase	33.6
Extracellular hydrolases located also in vacuoles	
Acid phosphatase (derepressed cells)	15.0
Chitinase	14.0
Exo-β-1,3-glucanase	27.0
Invertase	20.1
p-Nitrophenylacetate esterase	20.5
Tonoplast-bound hydrolases	
ATPase (Mg^{2+}) insensitive to sodium azide: pH 7.0	41.4
α-Mannosidase	109.0
X-Prolyldipeptidyl aminopeptidase	65.0
Alkaline phosphatase (Mg^{2+})	85.0
Oxidoreductases probably localised in tonoplasts	
NADH:cytochrome-*c* oxidoreductase	ND
NADH:dichlorophenol–indophenol oxidoreductase	0.638
Cytosol hydrolases absent from vacuoles	
α-Glucosidase	<0.01
α-Glycerophosphatase, pH 7.0	7.7
Aminopeptidase II	<0.05
Alcohol dehydrogenase	<0.01
Glucose-6-phosphate dehydrogenase	<0.01
Mitochondrial enzymes absent from vacuoles	
Citrate synthase	ND
Cytochrome-*c* oxidase	<0.01
L-Lactic dehydrogenase	ND
Succinate dehydrogenase	<0.01
Succinate–cytochrome-*c* oxidoreductase	<0.01

ND, activity not detected.
[a] Ratio of specific activity in vacuoles and specific activity in total spheroplast homogenate.
From Schwencke (1991).

Table 4.2. *The proportion of soluble amino acids in cytoplasmic and vacuolar pools of cells of* Candida utilis *grown on different carbon and nitrogen sources*

N and C source	Pool	Extracted amino acids (α-amino nitrogen as % of total cellular pool)	Extracted amino acids as % of total cytoplasmic and total vacuolar pools					
			Basic amino acids	Lys	Orn	Arg	Acidic neutral amino acids	Glu
NH_4^+, glucose	Cytoplasmic	17	9.9	1.8	2.4	4.0	90.1	46.0
	vacuolar	83	19.7	4.1	1.9	12.0	80.3	24.0
Arg, glycerol	Cytoplasmic	9	23.8	1.1	5.0	17.0	76.2	35.0
	vacuolar	91	66.2	3.5	13.0	49.0	33.8	12.0

From Wiemken & Nurse (1973).

Table 4.3. *Distribution of potassium, sodium and chloride between cytoplasm (c) and vacuole (v) in different fungi determined by the three procedures referred to in the text*

Organism (units) Reference	Total monovalent cation c	v	Potassium c	v	Sodium c	v	Potassium: Sodium c	v	Chloride c	v
Candida utilis[a] (μmol (g protein)$^{-1}$)	2718	2020	2123	1648	595	972	3.6	1.1		
Huber-Walchli & Wiemken (1979)										
Dendryphiella salina[b] (mM)	185	103	164	89	21	14	7.8	6.4	33	24
	208	185	112	62	96	123	1.2	0.5	121	101
	283	285	110	86	173	199	0.6	0.4	199	173
Clipson et al. (1990)										
Neurospora crassa[c] (μmol ml^{-1})				0.7		1.5				
Cramer & Davis (1984)										
Saccharomyces carlsbergensis[d] (mM)			60	470						
Okorokov, Lichko & Kulaev (1980)										
Saccharomyces cerevisiae[e] (mM (kg fresh weight compartment)$^{-1}$)	180–190	75–90								
Roomans & Seveus (1976)										

[a] Differential extraction of soluble pools using DEAE-dextran.
[b] X-ray microanalysis of sections of material prepared by freeze substitution of mycelium. First row of results, mycelium grown in medium low in sodium; second and third rows, medium containing 100 mM and 500 mM sodium, respectively.
[c] Isolated vacuoles.
[d] Differential extraction using cytochrome c.
[e] X-ray microanalysis of sections of rapidly frozen cells loaded with rubidium and caesium chloride but containing potassium.

acids to the cytoplasmic and vacuolar pools. As can be seen, basic amino acids particularly arginine are accumulated in the vacuole; acidic amino acids tend to be excluded (Kitamoto *et al.*, 1988; see below, and Chapter 6).

The investigator using this procedure needs to keep in mind that there is as yet no confirmatory evidence (a) that the presence of the wall does not have an effect on the outflow of charged substances through its capacity to act as an ion exchanger and (b) that what is called the vacuolar compartment is *de facto* the vacuole and not other compartments within the fungus. It should be noted that values for concentration are based on an estimate of the vacuolar volume based on light microscopical observations. Also the procedure does not appear yet to have been used with filamentous fungi.

(2) The procedure for the isolation of vacuoles is described in the next section. It is assumed that there is little leakage of solutes during vacuolar isolation. As is indicated below, this need not be the case.

(3) X-ray microanalysis is free from the problems associated with the above two procedures. The use of X-ray microanalysis is well proved for green plant tissue (Harvey *et al.*, 1981). Preparation of material for analysis immobilises the solutes throughout the tissue. So loss from a compartment is reduced to a minimum. Further, a concentration value is obtained almost directly. Unfortunately, the method is applicable only to elements and thus is of real value only for metal cations. Also it has to be assumed that they are free in solution, which may not be the case.

As yet there has been no direct comparison of the three procedures. Table 4.3 provides data obtained, for five different fungi, for potassium, sodium and chloride concentrations in the cytoplasm and vacuole. The values for potassium and sodium allow a comparison between the fungi, albeit superficial. At this stage, one must question the very low concentrations of the two cations in isolated vacuoles of *N. crassa*, suggesting loss of the cations during the isolation procedure. On the other hand, the very high concentration of potassium in the vacuoles of *S. carlsbergensis* suggests influx of the ion into the vacuole during differential extraction.

As well as the above three procedures, there are four other procedures that, though indirect, can provide relevant information on the solute composition of vacuoles.

(1) Radiotracer compartmental analysis, a procedure used often to great

effect by green plant physiologists (Walker & Pitman, 1976). Success in obtaining values for vacuolar concentration demands that the permeability of the vacuolar membrane is such that the cells or mycelium behave as a three-compartment system (wall, cytoplasm and vacuole), that the compartments are in series, and that the volume of the cytoplasm and vacuole are known. There have been three studies of tracer efflux of ^{42}K from fungal mycelium, one using *N. crassa* (Slayman & Tatum, 1965), the other two using *D. salina* (Jennings & Aynsley, 1971; Clipson & Jennings, 1990); all three have found that mycelium behaves as a two-compartment system, with no evidence of an additional slower flux of radioactivity from the protoplasmic compartment.

(2) It is possible to obtain a reasonable idea of the content of certain metabolites in the vacuole by feeding these same metabolites labelled with radioisotopes of high specific activity. The extent of dilution of the label when it appears in some end product, e.g. protein in the case of amino acids, is then determined. Essentially, the extent of dilution provides an assessment of the cytoplasmic concentration. From the value so obtained, the vacuolar concentration can be inferred. This procedure has been used very effectively in the estimation of vacuolar concentration of arginine and ornithine in *N. crassa* (for a summary of these studies, see Davis, 1986).

(3) For those elements that have a nucleus with a nuclear magnetic moment – that is, any nucleus with a non-zero nuclear spin quantum number – it may be possible to obtain vacuolar concentrations non-invasively by nuclear magnetic resonance (NMR). Since most elements have isotopes with this type of quantum number, NMR can be used to determine their molecular environment. The potential use of NMR has been well reviewed by Loughman & Ratcliffe (1984) for plant tissues. One need hardly mention the value of NMR for the study of metabolic reactions *in vivo*. Then it is customary to assume that the concentration of a particular compound is the concentration involved in the metabolic flux under consideration. However, there may be compartmentation with the ion or molecule distributed between cytoplasm and vacuole. With certain molecules it has been possible to detect and quantify the extent of that compartmentation as a consequence of the environments of cytoplasm and vacuole having differential effects on the resonances observed. An important element of the two environments is pH, which can exert its effect through

chemical shifts. The extent of the shifts can, however, depend on the ionic strength as well as on the pH. Further, the NMR signal of basic amino acids is altered by their binding to polyphosphate.

Although the exact cause of distinct resonance peaks for a molecule in the cytoplasm and in the vacuole within certain limits can only be inferred, nevertheless the presence of distinct peaks allows the concentration of the molecule in the cytoplasm or vacuole to be estimated. This has been done for orthophosphate in the yeasts *Brettanomyces intermedius* and *C. utilis* during substrate utilisation under, in the case of the former, aerobic conditions, and, in the case of the latter, aerobic and anaerobic conditions (Nicolay *et al.*, 1983). Legerton *et al.* (1983) confirmed with NMR that proline and alanine are cytoplasmic in *N. crassa*, whereas histidine and arginine are located primarily in the vacuole. As well, with careful interpretation of the data, NMR can be used to determine the hydrogen ion concentration within the cytoplasm and vacuole, as considered below.

(4) Reference is made below to the use of fluorescence imaging using fura-2 to determine the concentration of calcium in the cytoplasm of *S. cerevisiae* (Iida, Yagawa & Anraku, 1990). Fluorescence imaging has been used to determine the bulk cytoplasmic pH within hyphae of *D. salina* (Davies *et al.*, 1990*b*); some of the background to the procedure is provided by Jennings & Garrill (1994) as well as information on other uses of fluorescent probes in marine fungi. It should be noted that the studies of Iida *et al.* (1990) could not give a value for the calcium concentration in the vacuole; interpretation of the fluorescence image depended on more knowledge of how the environment in the vacuole might affect the fluorescence. There is also the problem of accumulation of fura-2 itself in the vacuole. Nevertheless, the study of short-term loading (3 min) gave an indication of the cytoplasm calcium concentration (see p. 351), which allowed conclusions to be drawn about how the tonoplast was functioning with respect to calcium. It should be noted that fura-2 does not by itself readily penetrate cells of *S. cerevisiae* (Iida *et al.*, 1990); the compound has to be made to penetrate by electroporation, i.e. the production of pores by the generation round the cells of a large but short-term electric field.

Halachmi & Eilam (1989) have used the calcium-sensitive fluorescent dye indo-1 to measure calcium in the cytoplasm of cells of *S. cerevisiae*. This dye seems to penetrate the plasma membrane fairly readily but not the tonoplast.

Table 4.4. *Comparative information on the gross chemical composition of the*
plasma membrane and of the tonoplast of Saccharomyces cerevisiae

	Plasma membrane	Tonoplast
Protein:total lipid	2.1 ± 0.125	0.66 ± 0.067
Phospholipid (% of total lipid)	5.4 ± 0.71	40.0 ± 4.87
Neutral lipid (% of total lipid)	94.6	60.0
Ergosterol (% in neutral lipid fraction)	25.0 ± 0.61	6.05 ± 0.74
Sterol:phospholipid		
By weight	4.6 ± 0.53	0.16 ± 0.015
Molar	8.9	2.9

From Kramer *et al.* (1978).

Isolated vacuoles

Successful isolation of vacuoles from fungi depends on the method of
opening up the cell so that the vacuoles can be released. It is necessary
first to produce protoplasts. Intact vacuoles can be released by procedures
that gently disrupt the plasma membrane. Successful procedures are based
on utilising the different properties of the plasma membrane and tonoplast.
Separating vacuoles from most other constituents of the cell is relatively
easy, due to the lower buoyant density of the former. Procedures used
with *S. cerevisiae, C. utilis* and *Candida tropicalis* were summarised by
Lloyd & Cartledge (1991), whereas procedures of a similar kind for the
isolation and purification of vacuoles from *N. crassa* have been described
by Cramer, Vaughn & Davis (1980), Vaughn & Davis (1981) and Bowman
& Bowman (1982).

The tonoplast is very much more elastic than the plasma membrane
(Matile & Wiemken, 1967); this greater elasticity probably allows the
tonoplast to remain intact when protoplasts are burst under hypoosmotic
conditions. The difference in elasticity of the two membranes is almost
certainly a function of their different chemical composition (Kramer
et al., 1978; Table 4.4). The low sterol content can account for the resistance
of the tonoplast to nystatin (Schwencke & de Robichon-Szulmajster, 1976).
Also the tonoplast has a lower carbohydrate content than does the plasma
membrane (Niedermeyer, 1976). Thus, whereas concanavalin A binds to
the outer surface of the plasma membrane it does not bind to the outer
surface of isolated vacuoles. This difference is not due to the different
relationship to the cytoplasm of the exposed faces of the membranes in
the two studies, though there can be a small amount of binding to the

inner side of the tonoplast when it is made permeable to the lectin (Boller, Dürr & Wiemken, 1976).

Also, in comparison with the plasma membrane, the tonoplast is about ten times more resistant to polycationic molecules such as cytochrome *c*, polylysine and DEAE-dextran (Svihla, Dainko & Schlenk, 1963, 1969; Dürr, Boller & Wiemken, 1975). The greater sensitivity of the plasma membrane to these compounds has allowed the determination of the cytoplasmic and vacuolar concentrations of solutes as described above.

Finally, on one hand, vacuoles are very sensitive to pH values less than 6.5, high concentrations of potassium chloride and relatively low concentrations (5 mM) of calcium or magnesium; on the other hand, protoplasts are stable under these conditions (Schwencke, 1991). Vacuoles are more stable than protoplasts in weakly alkaline (pH 7.5–8.0) conditions.

The tonoplast ATPase

We know from information obtained from three fungi, *S. cerevisiae* (Anraku *et al.*, 1989), *S. carlsbergensis* (Lichko & Okorokov, 1985) and *N. crassa* (Bowman & Bowman, 1986), that the tonoplast possesses an ATPase that can be differentiated from plasma membrane and mitochondrial F_0F_1 ATPases on the basis of pH optimum, subunit composition and structure, and sensitivity to inhibitors (as indicated in Table 1.1, p. 6). Essentially, the ATPase in the tonoplast possesses: (a) a pH optimum of approximately 7.0–7.5; (b) a molecular mass of around 400–500 kDa; (c) at least three types of subunit (see below); (d) an insensitivity to vanadate, azide and oligomycin; and (e) the property of being inhibited by nitrate, thiocyanate and bafilomycin. Though information is limited to three fungi, it is almost certain that the ATPase will be present in all other fungi, since a family of similar ATPases is known to be distributed in a wide variety of eukaryotic endomembrane systems that possess a lysosomal function (Sze, 1985; Rudnick, 1986; Anraku *et al.*, 1989).

The ATPase appears to be firmly embedded in the tonoplast. Use of detergent is necessary to solubilise the enzyme. Loss of activity occurs, probably because of delipidation. For all three fungi, the purified enzyme has been shown to possess two large polypeptides and one smaller one. Table 4.5 gives information about the values for the molecular masses that have been obtained, the designations used and the genes, where identified, for the three polypeptides as isolated from *S. cerevisiae*, *S. carlsbergensis* and *N. crassa*.

Table 4.5. *Values for the molecular mass of the subunits, and where appropriate their designation and the genes coding for them, of the vacuolar ATPase from Saccharomyces cerevisiae, S. carlsbergensis and Neurospora crassa*

Organism	Molecular mass (kDa)	Designation	Gene	Reference
Saccharomyces cerevisiae	118.6[a]	Subunit a	*VMA1*[b]	Hirata et al. (1990)
	69[a]			Kane, Yamashiro & Stevens (1989)
		Subunit A		Nelson, Mandiyan & Nelson (1989)
	89	Subunit a		Uchida et al. (1985)
	67	Subunit a		Hirata, Ohsumi & Anraku (1989)
	57[a]	Subunit B	*VMA2*[b]	Anraku et al. (1989)
				Nelson et al. (1989)
				Rothman et al. (1989)
	64	Subunit b		Uchida et al. (1985)
	60			Kane et al. (1989)
	16[a]		*VMA3*[b]	Nelson & Nelson (1989); Umemoto et al. (1990)
	19.5	Subunit c		Uchida et al. (1985)
	20			Hirata et al. (1989)
	17			Kane et al. (1989)
Saccharomyces carlsbergensis	75			Lichko & Okarokov (1985)
	62			
	9			
Neurospora crassa	67[a]		*vma-1*	Bowman et al. (1988)
	70			Bowman et al. (1986)
	57[a]		*vma-2*	Bowman et al. (1988)
	60			Bowman et al. (1986)
	16			Bowman et al. 1986)

[a] Calculated from amino acid composition obtained from genomic studies.
[b] Designation of Klionsky et al. (1990).

The larger of the two large polypeptides is believed to contain the catalytic site for ATP hydrolysis. It has been shown for both *S. cerevisiae* and *N. crassa* that the polypeptide can bind radioactive ATP analogues and inhibitors of ATPase activity (Bowman *et al.*, 1986; Uchida, Ohsumi & Anraku, 1988). The amino acid sequence of the polypeptide from *N. crassa* contains a putative nucleotide binding region (Bowman, Tenney & Bowman, 1988); the sequence information also shows that there is considerable degree of identity between this subunit of the vacuolar ATPase and both the α and β subunits of F_0F_1 ATPases. The gene (*VMA1*) encoding the polypeptide from *S. cerevisiae* has been isolated (Hirata *et al.*, 1990). The molecular mass of the polypeptide is 118.6 kDa, which is very much greater than the M_r determined by polyacrylamide gel electrophoresis (67 000). Comparison with the amino acid sequence of the equivalent polypeptide from *N. crassa* indicates an insert of 454 amino acid residues. It is argued that this insert is excised during biogenesis of the functional polypeptide. The *VMA1* gene has been shown to be the same as *TFP1* (Shih *et al.*, 1988), whose dominant mutant allele confers, amongst other properties, calcium-sensitive growth, which suggests that the H^+-ATPase of the tonoplast participates in maintenance of cytoplasmic calcium ion homeostasis.

The smaller of the two larger polypeptides of *N. crassa* shows the same degree of similarity to the α and β subunits of the F_0F_1 ATPase (of *Escherichia coli*) as it does to the larger polypeptide (Bowman *et al.*, 1988). There are similar findings for the equivalent polypeptide of *S. cerevisiae* (Anraku *et al.*, 1989). The function of this smaller polypeptide is not clear.

The smallest polypeptide binds *N,N'*-dicyclohexyl-carbodiimide (DCCD), which blocks proton flow across the membrane; the polypeptide is therefore thought to be the proton channel (Lichko & Okorokov, 1985; Uchida *et al.*, 1985; Bowman *et al.*, 1986). The gene encoding the polypeptide has been sequenced (Nelson & Nelson, 1989; Umemoto *et al.*, 1990). The polypeptide consists of 160 amino acid residues, forming a highly hydrophobic protein with four transmembrane segments. Haploid mutants have been constructed from the disrupted *VMA3* gene (Umemoto *et al.*, 1990). Although cells can grow in a nutrient-enriched medium, the vacuolar ATPase activity has been lost and isolated vacuoles do not acidify in the presence of ATP. The polypeptide encoded by *VMA3* appears to be essential for the total assembly of the tonoplast ATPase.

The nature of the tonoplast ATPase in *N. crassa* has been probed further by an investigation of the effects of nitrate. Bowman *et al.* (1989) showed that inhibition of enzyme activity by nitrate was associated with the loss of polypeptides (six in number, of molecular mass 67, 57, 51, 48, 30 and 16 kDa). Nitrate had this effect only when ADP, ATP or ITP were present. N-ethylmaleimide, which has been shown to bind to the 67 kDa polypeptide at the adenylate-protectable site (Bowman *et al.*, 1986), prevents release of the polypeptides; DCCD, which binds to the small hydrophobic polypeptide, does not do so. Examination of the vacuolar membrane preparations by electron microscopy using negative staining showed the presence of 'ball and stalk' structures similar in size, but somewhat different in shape, to similar structures on mitochondrial membranes. Nitrate removes these structures from vacuolar membranes but not from mitochondrial membranes.

These studies with nitrate indicate a more complex organisation within the functioning ATPase than has been indicated hitherto. This should be expected in view of the probable structure of the vacuolar ATPase from chromaffin granules (Wang *et al.*, 1988; Nelson, 1989; Figure 4.1); it is known that the presently accepted polypeptides consituting the enzyme of *S. cerevisiae* show very considerable sequence identity, particularly in

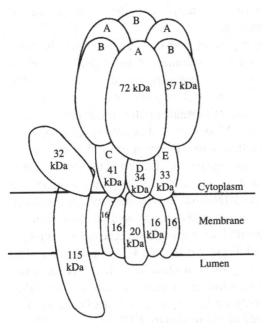

Figure 4.1. A model for the subunit structure of vacuolar ATPase from chromaffin granules. The Ac39 is marked as 32 kDa according to its calculated molecular mass (Wang *et al.*, 1988). Six copies of the 16 kDa proteolipid and a few copies of 32 and 115 kDa subunits per enzyme are present. (From Nelson, 1989.)

the proteolipid subunits, with their counterparts from the chromaffin granule enzyme (Nelson & Nelson, 1989). Other evidence that there is this complexity is coming forward. It has been shown that there is a 26.6 kDa polypeptide in *S. cerevisiae* encoded by the gene *VMA4*, which is probably also part of the ATPase (Foury, 1990). Null *vma4* haploid mutants, while viable, do not accumulate in their vacuole the red pigment of *ade2* strains and do not possess bafilomycin-sensitive ATPase activity.

Finally, it needs to be mentioned that the ability of the H^+-ATPase in the tonoplast membrane to acidify the vacuolar compartment is important for protein sorting with respect to endo- and exocytosis (Mellman, Fuchs & Helenius, 1986; Klionsky *et al.*, 1990; Yamashiro *et al.*, 1990).

The tonoplast pyrophosphatase

The tonoplast of green plant cells possesses not only an H^+-ATPase but also a pyrophosphatase (Rea & Poole, 1985, 1986) that generates a pH gradient (Chanson *et al.*, 1985; Rea & Poole, 1985) and a potential difference (Wang *et al.*, 1986) across the membrane. The presence of a pyrophosphatase in the protoplasm of *S. cerevisiae* is well authenticated (Kasho & Avaeva, 1978); it constitutes 90% of total pyrophosphatase activity of the cell. A small fraction of that pyrophosphatase has been shown to be associated with vacuoles (Lichko & Okorokov, 1991). Some of that activity is membrane-bound; the other part is soluble. According to the observations of Lichko & Okorokov (1991) the membrane-bound enzyme is a proton pump, a pyrophosphate-generated pH gradient across the tonoplast of isolated vacuoles can be observed and pyrophosphate hydrolysis by membrane preparations is not affected by the presence of inhibitors of ATPase and phosphatase activity. There was no detectable effect of pyrophosphate on the membrane potential but reasons can be advanced for the lack of any observed effect (Lichko & Okorokov, 1991). The relative molecular mass of this enzyme is 120 000 and appears to consist of three subunits ($M_r = 41 000$ each). The soluble vacuolar pyrophosphatase has an $M_r = 82 000$ and appears to have three subunits of $M_r = 28 000$. The bulk of the pyrophosphatase in the protoplasm has a relative molecular mass of 71 000 and consists of two polypeptides each of $M_r = 35 000$ (Kasho & Avaeva, 1978). The relationship between the three enzymes is not clear.

More studies are needed to establish clearly the role of the pyrophosphatase in generating a proton-motive force across the tonoplast. The possible relationship between two proton pumps located in the

Table 4.6. *Solutes known to be located in fungal vacuoles and not referred to in the text*

Compound	Fungus	Method of demonstration	Reference
Putrecene	*Neurospora crassa*	Isolated vacuoles	Davis & Ristow (1991)
Cobolt	*Saccharomyces cerevisiae*	DEAE-dextran fractionation	White & Gadd (1986)
Iron	*Saccharomyces cerevisiae*	Isolated vacuoles	Raguzzi, Lesuisse & Crichton (1988)
Manganese	*Saccharomyces carlsbergensis*	Isolated vacuoles	Okorokov et al. (1977)
Zinc	*Saccharomyces cerevisiae*	Fluorescent staining	Bilinski & Miller (1983)
		DEAE-dextran fractionation	White & Gadd (1987a)

tonoplast – an H^+-ATPase and an H^+-translocating pyrophosphatase – has been discussed by Rea & Sanders (1987). While it is known that the ATPase is able to generate an electrochemical potential difference of protons (180 mV) across the tonoplast of isolated vacuoles of *S. cerevisiae* (inside positive, the potential difference determined by flow dialysis) (Kakinuma, Ohsumi & Anraku, 1981), one cannot be sure how far this relates to the situation *in vivo*, in which amongst other factors a proton-translocating pyrophosphatase may be functioning. In *Neurospora crassa*, however, for which there is a value for the potential difference *in vivo* across the tonoplast (determined by microelectrode), the electrochemical potential difference for protons is around 100–110 mV (Bowman & Bowman, 1986). Thus, although one cannot be totally certain of the extent to which a pyrophosphatase is contributing to the proton gradient, there is an approximate idea of the possible driving force on solutes being secondarily actively transported across the tonoplast.

Transport systems at the tonoplast

The diversity of solutes in the vacuole must mean that there is an almost equivalent diversity of transport systems in the tonoplast membrane. Whether all such systems are present at all times is questionable. As with the plasma membrane, some systems are constitutive, whereas there are likely to be others that are inducible. It was mentioned above that amino acids and potassium, sodium and chloride ions are able to move across the tonoplast. To these solutes we must add phosphate (see Chapter 7) and magnesium (Chapters 7 and 11). Table 4.6 lists other solutes also known to have a vacuolar location.

I have indicated already that there is accumulation of basic amino acids in the vacuole. Some, such as arginine and ornithine, are at concentrations in the vacuole at least 50 times that in the cytoplasm (Weiss, 1973), though it has been stated that a 1000-fold concentration difference is possible (Bowman & Bowman, 1986). It seems probable that the proton electrochemical potential gradient, if it drives secondary active transport, is not sufficient to generate the total concentration gradient of these amino acids across the tonoplast. The shortfall must be brought about by the negative charges of polyphosphate. These could contribute to the accumulation of other positively charged species including inorganic ions such as magnesium. It must not be assumed therefore that any perceived accumulation of a positively charged species is necessarily driven by the proton electrochemical potential gradient.

In terms of reasonably well-authenticated transport systems at the tonoplast, the following have been shown to be present. They are described under the heading of the major solute(s) transported.

S-adenosyl-L-methionine in *Candida utilis*

Studies by Schlenk and colleagues (Svihla & Schlenk, 1960; Svihla *et al.*, 1969; Schlenk, Dainko & Svihla, 1970) showed that this amino acid is concentrated in the vacuole of cells of *C. utilis*. Schwencke & de Robichon-Szulmajster (1976) showed that isolated vacuoles were also effective in the accumulation of the amino acid. The process is insensitive to the presence of glucose, phosphoenolpyruvic acid, ATP and other triphosphorylated nucleotides. Transport is optimal at pH 7.4 and is insensitive to azide, 2,4-dinitrophenol and nystatin. The K_T was found to be 68 μM and transport competitively inhibited by S-adenosyl-L-homocysteine and S-adenosyl-L-methionine. The latter, when inside the vacuole, does not leak out again. It is believed that the amino acid penetrates by facilitated diffusion, accumulation being due to binding to polyphosphate.

Arginine in *Neurospora crassa*

Mycelium growing in the presence of ammonia* accumulates arginine and ornithine in the vacuoles (Weiss, 1973). Vacuolar membrane vesicles can accumulate unmodified arginine 14-fold over the bathing medium (Zerez *et al.*, 1986). The process is dependent on ATP and magnesium. The K_T for transport is 0.4 mM and, although transport is not inhibited by glutamine, ornithine or lysine, there is inhibition with nitrate and uncouplers. The pH optimum for transport is 7.8 (that for the tonoplast H^+-ATPase is 7.5). These data suggest that argine transport is driven by the proton electrochemical potential gradient across the tonoplast.

In intact mycelia, the accumulation of arginine in the vacuole (relative to that in the cytoplasm) appears to be much greater than has been achieved with isolated vacuoles (Subramanian, Weiss & Davis, 1973). The binding by polyphosphate could be important in the process of accumulation of arginine. However, it has been shown that the amino acid content of the vacuolar pool is almost wholly independent of its polyphosphate content when the latter is varied by starvation of the mycelium of phosphate followed by readdition of the ion to the medium (Cramer *et al.*, 1980). Also vacuoles isolated from mycelium depleted of

---

* Throughout the text, I have used ammonia in the generic sense, since often it is not clear whether or not ammonia or ammonium is the species under consideration.

phosphate retain their amino acids (Cramer & Davis, 1984). In any case, half the polyphosphate of normal vacuoles is associated with spermidine and magnesium and calcium. Nevertheless the presence of polyphosphate appears to be not without effect on the activity of the basic amino acids in the vacuole, since mycelium that has been grown on arginine and starved of phosphate contains very large vacuoles, suggesting that under these conditions the amino acids generate a much lower osmotic potential.

More recently, there has been progress towards identification of the arginine carrier in the tonoplast (Paek & Weiss, 1989). Radioactive *N*-α-*p*-nitrobenzyloxycarbonylarginyldiazomethane (synthesised from L-[2,3-^{3}H]arginine) which inhibits arginine transport, can be used to label the carrier. The protein has a molecular mass of 40 kDa.

Basic amino acids in *Saccharomyces cerevisiae*

Ohsumi & Anraku (1981) showed that arginine transport into vacuole membrane vesicles depended on ATP, which was shown to generate a proton gradient across the membrane. Transport of arginine was blocked completely with nigericin. It is likely that arginine transport is a proton/amino acid antiport. Later, it was shown by Sato, Ohsumi & Anraku (1984a), on the basis of blockage of transport by the protonphore 3,5-di-*tert*-butyl-4-hydroxybenzilidenematononitrile (SF 6847) and competition studies, that there appear to be at least seven independent proton/amino acid antiports in the vacuolar membrane (Table 4.7). Interestingly, the transport systems have affinities for their substrates of the order of 10^{-3} M, one to two orders of magnitude less than amino acid transport systems at the plasma membrane, which may be related to the need to control amino acid concentrations in the cytoplasm rather than to the transport system acting as a mechanism for concentrating amino acids in the vacuole. The pH optimum for transport is in the range 6.3–7.0, except for lysine (pH 7.9) and arginine (pH 7.5).

Subsequently, Sato *et al.* (1984b) have argued for the presence of a third arginine transport system. When vacuolar vesicles were preloaded with histidine in the absence of ATP, and the reaction mixture containing c. 50 μM [^{14}C]arginine instead of the original 10 mM histidine, was diluted 50-fold, there was a transient influx of arginine and downhill efflux of histidine. Interestingly there was inhibition of histidine uptake by lysine, the inhibition appearing to be competitive (Sato *et al.*, 1984a). However, histidine had no effect on lysine uptake. Other studies by Ohsumi, Kitamoto & Anraku (1988), using copper(II)-treated cells (Kitamoto *et al.*, 1988) to assess processes occurring at the tonoplast, have shown

Table 4.7. *A summary of the properties of seven proton/amino acid antiport transport systems located in the tonoplast of cells of* Saccharomyces cerevisiae

System	Substrate	K (mM)	Inhibitor
Arg	Arginine	0.4	D-Arginine
Arg-Lys	Lysine	0.56	N-ε-Methyl-L-lysine
	Arginine	1.5	D-Lysine
	Ornithine	2.5^a	
	Citrulline	2.5^a	
His	Histidine	1.2	L-Lysine
			N-ε-Methyl-L-lysine
			D-Histidine
			D-Lysine
Phe-Trp	Phenylalanine	1.0	
	Tryptophan	3.0	
Tyr	Tyrosine	5.0	
Gln-Asn	Glutamine	8.8	
	Asparagine	3.0^b	
Ile-Leu	Isoleucine	5.0	
	Leucine	30	

^{a} The K_i value for lysine uptake instead of the K_T value was determined, as no labelled compound was available.
^{b} The K_i value for glutamine uptake instead of the K_T value was determined, as the activity of asparagine uptake is weak.
From Sato *et al.* (1984a).

that arginine can exchange across the membrane in a 1:1 manner with potassium. It is not clear how this relates to what might be occurring in the intact cell.

Calcium

From DEAE-dextran differential extraction, it would seem that in *S. cerevisiae* most of the calcium is located in the vacuole (Eilam, Lavi & Grossowicz, 1985). Cytoplasmic calcium concentration was found to be maintained at 5.8×10^{-6}–23×10^{-6} M. On the other hand, the vacuolar (free) concentration rose from 2.25×10^{-4} M to 3.4×10^{-3} M. Cells permeabilised by DEAE-dextran can still take up calcium, requiring ATP not initially but after 2 h. These latter findings are consistent with those of Ohsumi & Anraku (1983). They showed that vacuolar membrane vesicles can accumulate calcium in the presence of ATP. The K_T for transport is 0.1 mM, with an optimum pH of 6.4. Transport is inhibited by protonophore uncouplers and inhibitors of the tonoplast ATPase. The

ATP-dependent formation of a pH gradient across the vesicle membrane is dissipated by the presence of calcium in the bathing medium. These findings indicate that calcium enters the vacuole by a calcium/proton antiport. The system can also transport strontium.

The role of the tonoplast ATPase in bringing about accumulation of calcium in vacuoles is confirmed by studies on the calcium-sensitive *cls* mutants (Ohya, Ohsumi & Anraku, 1986). These cells die in the presence of excess calcium chloride. Other mutants containing a defect in the tonoplast H^+-ATPase brought about by the disruption of the *VMA1* and *VMA3* genes show kinetics of cell death similar to that of a *cls Pet⁻* mutant (*Pet⁻* cosegregates with *cls*) (Hirata *et al.*, 1990; Umemoto *et al.*, 1990). Ohya *et al.* (1991) have shown that *cls Pet⁻* cells have a cytoplasmic calcium concentration (determined with the calcium-specific probe fura-2) of 9×10^{-4} M compared with 15×10^{-8} M in the wild-type, indicating reduced ability of the vacuole to accumulate calcium.

There are strong indications, from the studies of Miller, Vogg & Sanders (1990) on changes in the cytoplasmic concentration of calcium (determined by a calcium-sensitive electrode) in hyphae of *N. crassa*, that in this fungus also vacuoles may act as sinks of calcium. Support for this view comes from studies by Cornelius & Nakashima (1987) of isolated vacuoles of mutants that will grow well only in medium with low calcium. The vacuoles show a much lower rate of uptake of the element than into those of wild-type mycelium. Calcium contents of such vacuoles were found to be in keeping with the observed rates of uptake.

In situ localisation of calcium in stage 1 (Castle, 1942) sporangiophores of *Phycomyces blakesleeanus*, at the electron microscope level, has been carried out by Morales & Ruiz-Herrera (1989) using potassium pyroantimonate. While there was some accumulation of the element in the mitochondria, endoplasmic reticulum and 'Golgi-like' bodies, there was marked accumulation in the vacuoles.

Ions in *Saccharomyces carlsbergensis*

The main mechanism for transport established from studies on isolated vacuoles using pH and membrane potential probes appears to be a proton/ion antiport (Okorokov *et al.*, 1985). The K_T values of the transport systems for calcium, magnesium, manganese, zinc and phosphate are 0.06, 0.3, 0.8, 0.055, 0.17 and 1.5 mM, respectively – all values close to the putative concentrations of these ions in the cytoplasm. The stoichiometry of transport for calcium to protons appears to be 1:2, and for zinc to protons 1:1. Potassium and arginine are thought to be transported in a manner similar to that of the above ions.

Organic acid anions and guanosine in *Saccharomyces pastorianus*

Isolated vacuoles accumulate citrate, α-oxoglutarate, malate and guanosine (Kulakovskaya, Matys & Okorokov, 1991). The accumulation is ATP dependent and is inhibited by protonophores. Magnesium increases the proton gradient across the membrane and increases the uptake of all compounds; calcium hyperpolarises the membrane and decreases the proton gradient, inhibits guanosine but stimulates citrate uptake. Kulakovskaya *et al.* concluded that guanosine moves across the membrane via a proton/guanosine antiporter, whereas citrate, malate and α-oxoglutarate are transported by uniporters driven by the membrane potential.

Monovalent cations in *Saccharomyces cerevisiae*

Using artificial planer bilayer membranes containing vacuolar membranes, Wada *et al.* (1987) demonstrated the presence of a membrane-potential-dependent monovalent cation channel. The channel appears to require calcium on the cytoplasmic face for activity, though the effective concentration was found to be unexpectedly unphysiological (1 mM).

5
Carbon

A. MONOSACCHARIDE UTILISATION

General features of glucose utilisation in fungi other than yeasts

Almost invariably glucose is one of the sugars most readily utilised by nearly all fungi. This view is supported by the information in Table 5.1, albeit for a minute selection of fungi, although of widespread taxonomic distribution. However, the dominance of growth media based on glucose or glucan as the carbon source adds confidence to the view that glucose is a very important source of carbon and energy for growth. In all probability, almost every fungus metabolises glucose to pyruvate in a manner that is the same or extremely similar to that in *Saccharomyces cerevisiae*, i.e. by glycolysis or the pentose phosphate pathway/hexose monophosphate shunt. The evidence that fungi utilise glucose in such a manner has been well reviewed by Cochrane (1963), Gottlieb (1963), Blumenthal (1965, 1975) and McCullough *et al.* (1986). The evidence comes from a variety of approaches but it has to be stressed that it is only for a relatively few fungi that compelling evidence has been accumulated using a majority of the approaches. The effectiveness of any one approach is well discussed with respect to fungi by Blumenthal (1965), but the reader should also consult ap Rees (1974, 1980*a*, *b*) for a critique of the same methodology as applied to flowering-plant tissues, in particular with respect to assessing the extent to which the metabolic flux of glucose to pyruvate is via glycolysis or the pentose phosphate pathway. Briefly the approaches used have been:

(1) Stoichiometric conversion of glucose to ethanol or other products derived from pyruvate. This has been demonstrated with respect to the production of ethanol for *Fusarium lini* (Cochrane, 1956; Heath, Nasser & Koffler, 1956).

Table 5.1. *Relative carbon nutrition of nine species of fungi from different parts of the Fungal Kingdom. Growth on glucose taken as 100*

Carbon source	Chytridium sp. (Crasemann, 1954)	Allomyces javanicus (Ingraham & Emerson, 1954)	Blastocladiella pringsheimii (Cantino, 1949)	Saprolegnia delica (Bhargava, 1945)	Phytophthora cactorum (Mehotra, 1951)	Coprinus sp. (Johnson & Jones, 1941)	Stereum gauspatum (Herrick, 1940)	Fusarium oxysporum f. nicotianae (Wolf, 1955)	Penicillium digitatum (Fergus, 1952)
Fructose	101	0	88	102	50	107	104	104	108
Mannose	101	109	97	0	38	127	104	109	108
Galactose	99	0	7	0	63	76	89	78	99
Sorbose	8		7					59	
Xylose	7	0	6	0	42	52	123	135	102
L-Arabinose	4		7	0	55	56	43	68	95
Rhamnose	4				58		41	39	65
Mannitol				0	63	111		41	
Glycerol	17	0	39	14	102	105	24	48	21
Maltose	10	103	101	107	125	84	160	138	92
Sucrose	9	0	99	14	133		158	113	101
Cellobiose	100						112	108	
Glycogen	94			93	125	112	160	59	
Starch	50		95	100	83	77	70		2
Inulin	3								
Cellulose	3		6						
Acetate	0	0	5						
Succinate	0	0	9						2
Citrate	0	0	6						28

(2) Use of inhibitors known to inhibit specific steps of the pathway.
(3) Isolation of relevant enzymes.
(4) Use of mutants. The relevant ones are those that cannot grow on hexoses but can grow on acetate, alanine or butyrate (Payton & Roberts, 1976).
(5) Enzyme changes when mycelium growing on a non-hexose substrate is transferred to a growth medium in which the carbon source is a hexose.

With respect to (4) and (5), it must be realised that only certain enzymes are involved, namely phosphofructokinase and pyruvate kinase. In the case of mutants, loss of other enzymes would not allow gluconeogenesis.

(6) Use of glucose specifically labelled with ^{14}C. Glucose so labelled can be used for two purposes. First, location of the label in the product of fermentation, whether it be ethanol or lactate (see below), can be compared with what might be expected from a knowledge of their flow through the pathway of glycolysis of each of the original carbon atoms in glucose. Second, when glucose is dissimilated by the pentose phosphate pathway, carbon dioxide is released in the first steps of the pathway, but when there is glycolytic breakdown of glucose carbon dioxide is released only at the end of that pathway. This means that the determination of the specific activity of carbon dioxide lost over time from 1-^{14}C- or 6-^{14}C-labelled glucose allows estimates of the extent to which glucose is dissimilated to pyruvate via glycolysis or the pentose phosphate pathway. There is a problem of recycling of label. For instance, it is now clear that, for those fungi that contain mannitol, earlier estimates of the proportion of carbon from glucose channelled through each pathway must be viewed with caution. This is because when either 1-^{14}C- or 6-^{14}C-labelled glucose enters the fungus, the early conversion of the sugar to the hexitol means randomisation of radioactivity between C-1 and C-6 because mannitol is symmetrical. Conversion of mannitol back to hexose leads to dilution of the specific activity of either 1-^{14}C or 6-^{14}C atoms, depending on which was initially labelled in the hexose molecules fed to the fungus (Holligan & Jennings, 1972*c*, *d*).
(7) ^{13}C NMR studies. Natural abundance high-resolution ^{13}C NMR has been used to probe the metabolism of carbon sources by *Aspergillus nidulans* via glycolysis and gluconeogenesis (Dijkema, Kester & Visser, 1985; Dijkema *et al.*, 1986; Dijkema & Visser, 1987). The potential of

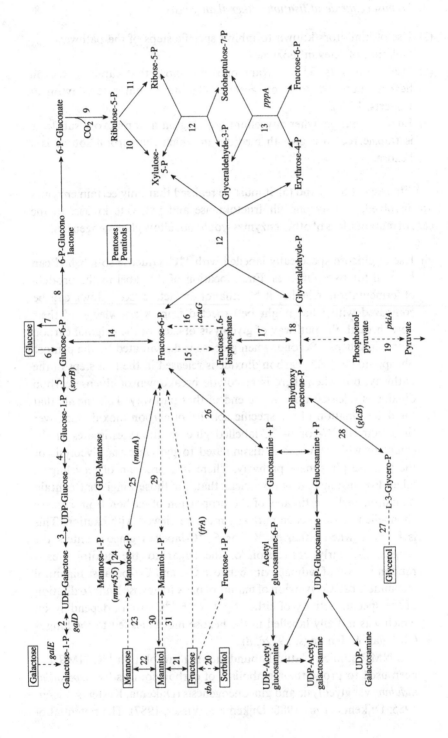

this non-destructive method of examination of primary metabolism in filamentous fungi has yet to be realised.

In spite of confidence that glycolysis and the pentose phosphate pathway are found in virtually all fungi, apart from *S. cerevisiae* (see below), knowledge of the two pathways for any one particular fungus is not as extensive as one would hope. Nevertheless for *Aspergillus* there is a reasonably coherent picture of the metabolism of glucose via these two pathways as well as their relationship to the production of polyols, mannitol, arabitol and erythritol and to the metabolism of galactose (Figure 5.1).

We need to keep uncertainty about the extent of glycolysis and the pentose phosphate pathway in the Fungal Kingdom in perspective. There are very few fungi that do not utilise glucose. Examples to date are two members of the Leptomitales, *Araispora* and the sewage fungus *Leptomitus lacteus* (Schade, 1940; Gleason, 1968), though there appear to be isolates of the latter fungus that can grow on glucose (Zehender & Böck, 1964). *Leptomitus lacteus* is nutritionally variable in a number of respects (Gleason, 1968). *Araispora* grows on D(−)- and L(+)-lactate and on succinate, acetate and pyruvate but not on D-alanine, L-glutamate and L-leucine, while the strain of *L. lacteus* that cannot utilise sugars grows only on L(+)-lactate, succinate, alanine, glutamate and leucine (Gleason, 1968). In addition, *Ostracoblabe implexa* a marine phycomycete of uncertain affinities, found on oyster shells, grows only on sodium malate out of a large range of sugars and organic acids examined as carbon sources (Alderman & Jones, 1971). All three fungi deserve further investigation.

In terms of anerobic fermentation of glucose, a number of lower fungi

---

Figure 5.1. Carbohydrate metabolism in *Aspergillus nidulans*. Continuous lines, constitutive enzymes; and broken lines, enzymes whose synthesis is regulated. The enzymes indicated are: 1, galactokinase; 2, UDP-galactose pyrophosphorylase; 3, UDP-glucose epimerase; 4, UDP-glucose pyrophosphorylase; 5, phosphoglucomutase; 6, hexokinase; 7, glucose-6-phosphatase; 8, glucose-6-phosphate dehydrogenase; 9, 6-phosphogluconate dehydrogenase; 10, phosphoketopentoepimerase; 11, phosphoriboisomerase; 12, transketolase; 13, transaldolase; 14, phosphoglucoisomerase; 15, phosphofructokinase; 16, fructose-1,6-bisphosphatase; 17, aldolase; 18, triose phosphate isomerase; 19, pyruvate kinase; 20, sorbitol dehydrogenase; 21, mannitol dehydrogenase; 22, mannose dehydrogenase; 23, mannokinase; 24, phosphomannose mutase; 25, phosphomannose isomerase; 26, L-glutamine: D-fructose-6-phosphate amidotransferase; 27, glycerol kinase; 28, glycerol phosphate dehydrogenase; 29, mannitol-1-phosphate dehydrogenase; 30, mannitol-1-phosphate phosphatase; 31, mannitol kinase. (From McCullough, Paton & Roberts, 1977.)

are known to produce lactic acid (Gleason & Price, 1969; Emerson & Natvig, 1981). Those known to be capable of so doing are: the Oomycetes (members of the Leptomitales) *Aqualinderella, Mindeniella, Rhipidium* and *Sapromyces* and (member of the Peronosporales) *Pythiogeton*; Chytridiomycetes *Blastocladia, Blastocladiella, Chytridium* and *Macrochytridium*; and Zygomycetes *Mucor* and *Rhizopus*. The first two groups of fungi produce only D($-$)-lactate, whereas the zygomycetes produce both D($-$)- and L($+$)-lactate.

The first two groups are capable of living in habitats containing very low concentrations of oxygen (Emerson & Natvig, 1981). At one stage, it was thought that one of the most extreme species *Aqualinderella fermentans* had no metabolism involving molecular oxygen (Emerson & Held, 1969; Held *et al.*, 1969). Certainly fermentation is insensitive to oxygen, and growth is reduced under aerobic conditions. Further, the organism lacks recognisable cristate mitochondria and cytochromes. The better growth under anaerobic conditions contrasts markedly with the majority of fungi, whose dry matter production is reduced to at least 60% and often much less of that in aerobic conditions by anaerobiosis (Curtis, 1969; Tabak & Cooke, 1968). *Aqualinderella fermentans* is said to be an obligate fermenter; *Blastocladia* and *Pythiogeton* also come into this category. Later studies (Natvig, 1982; Natvig & Gleason, 1983) have shown that the three organisms are capable of significant oxygen uptake, though by an oxidase that is completely insensitive to cyanide but sensitive to salicylhydroxamic acid (SHAM); the oxidase is thus not a cytochrome. In view of the sensitivity of growth of many of the species listed above to oxygen, it is interesting to find that those whose growth was so sensitive possessed lower specific activities of superoxide dismutase. True anaerobic fungi are dealt with in a later section.

Higher filamentous fungi appear to ferment glucose only to ethanol. The zygomycete *Rhizopus oryzae* is somewhat intermediate, forming both ethanol and lactic acid, but, unlike its relatives above, the D($+$) isomer is formed (Gibbs & Gastel, 1953; Margulies & Vishniac, 1961; Obayashi *et al.*, 1966). But the statement about the higher filamentous fungi is simplistic. We know all too little about the fermentative metabolism of such fungi. The diversity of fermentative behaviour in yeasts (see below) indicates by comparison why further study of fermentation in higher fungi is required urgently.

There is little doubt that the pentose phosphate pathway is widely distributed in fungi. But the evidence is less compelling than that for the presence of glycolysis. Nevertheless, the role of the pathway in generating the reducing form of NADPH in the extramitochondrial cytoplasm (Lehninger, 1970) would indicate the probable ubiquity of the pathway to accompany the glycolytic breakdown of glucose. Indeed, in the obligately aerobic yeasts of the genus *Rhodotorula*, which appear to lack phosphofructokinase, carbon flow is initially through the pentose phosphate pathway (Brady & Chambliss, 1967; Höfer, Betz & Becker, 1970).

There have been a considerable number of assessments of the contribution of the pentose phosphate pathway to the dissimilation of glucose. The accuracy of many of the estimates must be in doubt because of the unknown amount of possible recycling of label which might have been occurring, in the manner indicated above and also in other ways which do not necessarily lead to randomisation but lead to a reduction in specific activity. Nevertheless the small amount of information available does indicate the importance of the pathway in generating NADPH, in particular for the utilisation of highly oxidised nitrogen, i.e. nitrate. An increased flux of carbon through the pathway has been demonstrated when nitrate is supplied to *Dendryphiella salina* as compared with ammonium (Holligan & Jennings, 1972c). In *Aspergillus nidulans*, there is an increased level of the pathway enzymes glucose-6-phosphate dehydrogenase, 6-phosphogluconate dehydrogenase, transketolase and transaldolase, as well as phosphoglucoisomerase, when the fungus is grown with nitrate as opposed to urea as the nitrogen source (Hankinson & Cove, 1974). However, the level of mannitol-1-phosphate dehydrogenase is reduced under the same conditions (Hankinson & Cove, 1975). This can be accounted for by the need to reduce competition for fructose 6-phosphate between that enzyme and phosphoglucoisomerase so that the sugar moiety can be recycled through the pentose phosphate pathway. Reduction in the flux of fructose 6-phosphate to mannitol phosphate could also mean that one source of demand on NADH is much reduced when the possible supply from glycolysis is also reduced.

This reference to mannitol metabolism is a useful reminder of the importance of polyol metabolism in fungi. It is not appropriate here to give details of the pathways leading to the synthesis and degradation of polyols in filamentous fungi; there are reviews elsewhere (Jennings, 1984a; Jennings & Burke, 1990). Nevertheless the levels of individual polyols are determined by the nature of the external medium such as the carbon or nitrogen source (Dijkema *et al.*, 1985; Holligan & Jennings, 1972b; Pfyffer & Rast, 1988, 1989; Figure 5.2) and pH (Dijkema *et al.*, 1986), though neither of these influences has been investigated in anything other than

94 *Carbon*

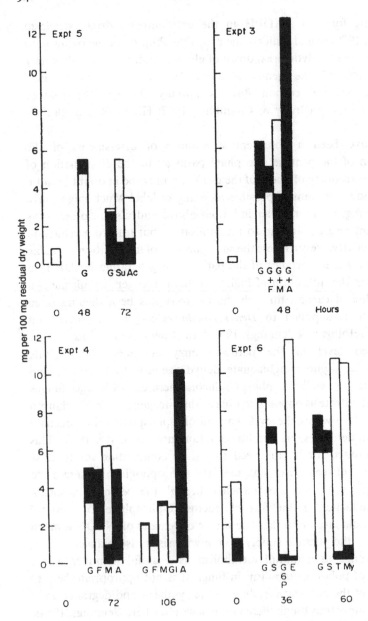

Figure 5.2. The level of mannitol (open columns) and arabitol (filled columns) within mycelium, initially starved, of *Dendryphiella salina* at various times after regrowth in media with nitrate as the nitrogen source and various carbon sources. A, arabitol; Ac, acetate, E, erythritol; F, fructose, G, glucose; G6P, glucose 6-phosphate, M, mannitol; My, *myo*-inositol; S, sucrose; T, trehalose. (From Holligan & Jennings, 1972*b*.)

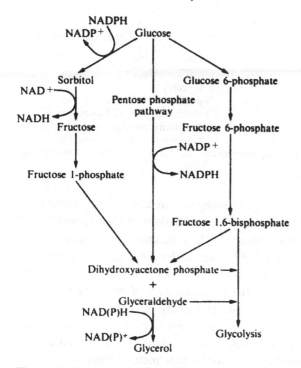

Figure 5.3. The sorbitol (glucitol) bypass as it might apply to fungal metabolism. (From Jennings, 1984*a*.)

a sketchy manner. There is, however, a significant body of information about the interrelationships between polyol concentrations within a fungus, both filamentous and yeast, and the osmotic potential of the medium; this is discussed in Chapter 13. Though it is not my intention to dwell on the details of polyol metabolism, it is important to point out here that there could be a route, other than glycolysis and the pentose phosphate pathway, to triose phosphate in fungi via the sorbitol (glucitol) bypass proposed for animal cells by Jeffery & Jornvall (1983) (Figure 5.3). Finally, we now have a comprehensive view of the taxonomic distribution of polyols in fungi (Pfyffer, Pfyffer & Rast, 1986), the details of which are of great value in helping to anticipate the patterns of glucose metabolism in insufficiently studied genera (Table 5.2).

Monosaccharide utilisation by yeasts

The need to highlight monosaccharide utilisation in yeasts is a consequence not only of the great attention paid to glucose metabolism in *S. cerevisiae*

Table 5.2. *Polyol distribution pattern in fungi*

Polyols present	Taxonomic group
None	Oomycetes
One or more polyols	Zygomycetes, Hemiascomycetes
Mannitol and other polyols	Chytridiomycetes, Euascomycetes, Basidiomycetes and Deuteromycetes (except some imperfect yeasts)

From Pfyffer *et al.* (1986).

but also because of the extensive knowledge about carbohydrate utilisation by yeasts in general. This knowledge has its basis in the fundamental role of nutritional tests in yeast taxonomy. Thus, we know much about which carbohydrates can or cannot be used under aerobic and anaerobic conditions (Barnett, Payne & Yarrow, 1983; Kreger-van Rij, 1984). It is this knowledge, together with the potential ease of handling of yeasts industrially, that has led to the intensive study of pentose utilisation in yeasts in the biotechnological attack on the problem of lignocellulose and pentosan degradation to ethanol. Such study has led to important information about the control of carbohydrate metabolism in yeasts.

Glucose utilisation by *Saccharomyces cerevisiae*

Saccharomyces cerevisiae is the best-known organism for producing ethanol, which the yeast achieves by fermentation of glucose under both aerobic and anaerobic conditions. Under aerobic conditions, only a small proportion (*c.* 3%) of the glucose consumed is respired, though 34% of the ATP could be produced by this route. With galactose, 29% of the sugar consumed is respired, with potentially 88% of the ATP being produced by respiration (Lagunas, 1986).

As one can see from the above, the yeast is a facultative anaerobe. However, it is not a model one. Facultative anaerobes are usually defined (Lagunas, 1986) as having the following three characteristics:

(1) They are able to grow aerobically or anaerobically using oxygen in respiration or organic compounds in fermentation as final electron acceptors.

(2) They use oxygen preferentially if it is available as the final electron acceptor, because of the greater energy yield from respiration than from fermentation.

Table 5.3. *Fermentation and respiration rate of sugars by cells of* Saccharomyces cerevisiae

Sugar in the medium	Hexose consumption rate (mmol (g protein)$^{-1}$ h^{-1})		
	Aerobiosis		Anaerobiosis
	Fermentation	Respiration	Fermentation
Glucose	39	1.1	42
Fructose	37	1.0	41
Maltose	28	1.4	26
Galactose	11	4.5	17

From Lagunas (1986). From *Yeast* 2, 221–8, reproduced by permission of John Wiley and Sons Limited.

(3) The lower rate of breakdown of sugars in aerobiosis than in anaerobiosis is a consequence of the greater yield of ATP from the former process. This can be considered as the inhibition of fermentation by respiration, the so-called 'Pasteur effect'.

Saccharomyces cerevisiae does not fit these criteria. Furthermore, there are other features of the organism that confirm that the yeast cannot be considered as a true facultative anaerobe in terms of glucose consumption.

(1) Anaerobic growth requires nutrient conditions different from those for aerobic growth. The synthesis of ergosterol and non-saturated fatty acids requires molecular oxygen (Hunter & Rose, 1970). Therefore these compounds must be present in the medium when *S. cerevisiae* is growing anaerobically.

(2) Sugars repress respiratory enzymes (Slonimski, 1953; Utter, Duell & Bernofsky, 1968; Polakis, Bartley & Meek, 1965) so that cells show a strong decrease in their ability to respire. Only when the sugars are exhausted or present in very low concentration are the respiratory enzymes derepressed.

(3) Experiments with batch cultures have shown that glucose, fructose and maltose are consumed at similar rates in anaerobic and aerobic conditions (Table 5.3). The Pasteur effect does not occur with these sugars. The data presented in Table 5.4 also show that even the theoretical benefits of aerobic conditions in terms of ATP production outlined above are not seen. Either the amount of ATP generated is lower than the maximum theoretical amount (galactose) or there is a greater expenditure of ATP under aerobic than under anaerobic conditions.

Table 5.4. *Calculated rate of ATP production for cells of* Saccharomyces cerevisiae *showing the theoretical benefit of aerobiosis versus anaerobiosis*

	ATP production rate (mmol $(g\,protein)^{-1}\,h^{-1}$)				
	Aerobiosis			Anaerobiosis	
Sugar in the medium	Fermentation	Respiration	Total	Total	Theoretical benefit[a]
Glucose	78	40	118	84	1.4
Maltose	56	51	107	53	2.0
Galactose	22	162	184	34	5.4

For the calculation of ATP in fermentation, 2 mol of ATP per mol of sugar fermented has been assumed. ATP:O = 3 is assumed in respiration.
[a] Theoretical benefit is expressed as the ratio between total ATP formed in aerobiosis and that in anaerobiosis.
From Lagunas (1986). From *Yeast* **2**, 221–8, reproduced by permission of John Wiley and Sons Limited.

The Pasteur effect can, however, be demonstrated in resting cells supplied with glucose in the absence of a nitrogen source (Lagunas *et al.*, 1982; Lagunas & Gancedo, 1983) and also in cells grown in a chemostat with very low glucose concentrations (Rogers & Stewart, 1974). Under aerobic conditions, the proportion of sugar respired by resting cells becomes as high as 100%. This proportion is a consequence of the irreversible inactivation of sugar transport systems as soon as resting conditions commence (Busturia & Lagunas, 1985, 1986). Thus, entry into the cells occurs only very slowly. This results in a drop in the rate of fermentation, with no change in the rate of respiration. Consequently the Pasteur effect makes its appearance (Table 5.5). The reason for the drop in the rate of fermentation with reduced sugar uptake can be explained in terms of the enzyme competition theory of Holzer (1961). Decreased sugar uptake results in a decreased concentration of pyruvate. The affinity of pyruvate oxidoreductase (route to respiration) for pyruvate is very much higher than that of pyruvate decarboxylase (route to ethanol). It is not difficult to see that with reduced sugar uptake there is likely to be less ethanol formed. In chemostat culture, when there is a very low sugar concentration there is a low rate of sugar uptake by the cells, with a decrease in fermentation in the manner just described. The corollary of these observations is that, in order to maximise the yield of biomass of

Table 5.5. *The appearance of the Pasteur effect in resting cells of* Saccharomyces cerevisiae

Sugar in the medium	NH_4^+ starvation (h)	Respiration (% of sugar consumed)	Pasteur effect[a]
Glucose	0	5	1.0
	4	10	1.3
	24	25	1.8
Maltose	0	10	1.1
	3	28	1.4
	7	100	1.8
Galactose	0	23	1.2
	3	47	1.8
	8	100	2.2

[a] Pasteur effect is expressed as the ratio between the rate of sugar consumption in anaerobiosis and that in aerobiosis. (A value close to 1 means that no Pasteur effect occurs.)
From Lagunas (1986). From *Yeast* **2**, 221–8, reproduced by permission of John Wiley and Sons Limited.

S. cerevisiae, good aeration and low sugar supply are required such that fermentation is suppressed.

The ability of glucose to repress respiratory enzymes in *S. cerevisiae* is part of a general process by which a number of enzymes, mainly of the oxidative part of catabolism, are repressed (Figure 5.4). As a consequence of such repression, respiration is prevented and ethanol formation now becomes an alternative pathway for energy generation and the reoxidation of reducing equivalents. It is also possible that, when a fermentable substrate is introduced into the cells, there are enzymes such as those for gluconeogenesis (see Figure 5.4) that are already at high level. As with the sugar transport systems referred to above, these enzymes are also inactivated. The process has been termed 'catabolite inactivation' by Holzer (1976). Thevelein (1984) has pointed out that catabolite activation can also occur.

The process of inactivation needs to be clearly distinguished from repression. The relevant enzyme loses its activity when inactivation occurs; with repression, activity is lost by dilution with growth because, although enzyme molecules are still present, they are no longer being synthesised (Cooper & Sumadra, 1983). This is not always made clear in many investigations. In this volume, unless the evidence for inactivation is

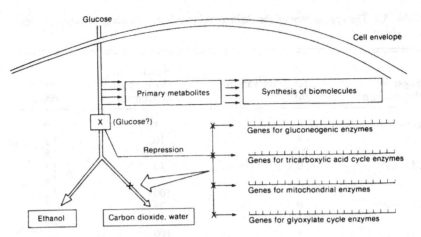

Figure 5.4. Diagram showing glucose repression as a regulatory mechanism to explain aerobic ethanol formation in yeasts. In the early stages a catabolite ($\times$) is formed that triggers the repression of several enzymes, mainly of the oxidative part of metabolism. In consequence, respiration is cut off and ethanol formation serves as the alternative pathway for energy generation and reoxidation of reducing equivalents. (From Käppeli, 1986*a*.)

available, it is implied that only repression is occurring. With respect to the process of inactivation, it is believed that proteolysis is probably involved (Achstetter & Wolf, 1985), though vacuolar proteinases can be ruled out on the basis of genetic evidence. We need to note that in the case of one gluconeogenic enzyme, fructose 1,6-bisphosphatase, activation is brought about by cyclic AMP phosphorylation (Müller & Holzer, 1981; Tortora *et al.*, 1981; Mazon, Gancedo & Gancedo, 1982; Purwin, Leidig & Holzer, 1982).

As well as the regulation of processes shown in Figure 5.4, there is also regulation of those transport systems and enzymes concerned with the metabolism of those compounds other than glucose that are capable of supplying the cell with carbon skeletons and energy for growth. Both repression and inactivation may be operating. On the addition of glucose to maltose-fermenting cells there is therefore repression of the synthesis of the maltose permease but also inactivation, as referred to above (see p. 147). The amino acid sequence of the permease has revealed regions that are believed to be targets for proteolysis (Cheng & Michels, 1989).

Enzyme repression and inactivation, as consequences of the presence of glucose as the substrate for fungal growth, are almost certainly widespread phenomena amongst fungi, as is indicated by the several mentions of glucose repression in this volume. Nevertheless, virtually all the research on the phenomenon has been on *S. cerevisiae*.

There is a very considerable literature on carbohydrate metabolism in *S. cerevisiae*, much of it not directly relevant to the general theme of this book. For good overviews of the subject, including the complex internal events involved in glucose repression, the reader should consult: Fiechter, Fuhrmann & Käppeli (1981); Fraenkel (1982); Entian (1986); Gancedo (1986, 1992); Käppeli (1986a); Käppeli & Sonnleitner (1986); and Wills (1990). The last review provides an excellent molecular biological view of sugar and ethanol metabolism. For information about the genes controlling glucose repression, the reader should consult the article by Rose, Albig & Entian (1991).

Monosaccharide utilisation by other yeasts

Yeasts may be classified on the basis of their ability to ferment glucose. Thus species can be:

(1) Non-fermentative.
(2) Obligately fermentative.
(3) Facultatively fermentative.

The extent to which yeasts are unable to ferment glucose is not entirely clear. Approximately a third of 439 yeast species listed by Barnett *et al.* (1979) are non-fermentative (Barnett, 1981), but many of the data are based on a test that depends on visible carbon dioxide production in an inverted tube inserted within the tube containing the medium and organism. There is a degree of uncertainty about the nature of the results obtained with this test (Scheffers, 1987). If the rate of production of carbon dioxide does not exceed the rate of diffusion of the gas into the atmosphere at the top of the tube no gas is collected in the inverted tube. Also, only gas produced at the bottom of the tube, and then only a small quantity, may find its way into the inverted tube. Finally, if oxygen is required for fermentation, carbon dioxide will be produced only near the air/medium interface and not trapped.

The classification above may be amplified by the following effects:

(1) Pasteur effect – the greater rate of fermentation under anaerobic than under aerobic conditions.
(2) Custers effect – the inhibition of fermentation by lack of oxygen (Scheffers, 1966).

(3) Crabtree effect – the aerobic fermentation of glucose despite sufficient oxygen (De Deken, 1966).

(4) Kluyver effect – some yeasts are able to ferment glucose anaerobically but not other sugars, though these same sugars may be assimilated aerobically (Kluyver & Custers, 1940).

It must be stressed that, even in a single organism, the presence of one of the above phenomena depends on the cultural conditions used – as has been shown above with respect to the Pasteur effect in *S. cerevisiae*. Having said that, I now turn to individual species.

Candida pintolopesii, which is a naturally occurring respiratory-deficient yeast, is obligately fermentative. *Saccharomyces cerevisiae* is facultatively anaerobic and, in view of what has been described above, clearly Crabtree positive. *Candida utilis* is essentially Crabtree negative, producing no ethanol under aerobic conditions, the indications being that fermentation is impeded at the level of redox reactions (van Dijken & Scheffers, 1986). A study of glucose transport in four Crabtree-positive and four Crabtree-negative yeasts indicates that the former possess a low affinity facilitated diffusion system for glucose while the latter have a high affinity proton symport (van Urk *et al.*, 1989). The reader should consult the section on glucose transport in *S. cerevisiae* (p. 117) as an example of a facilitated diffusion system; that on glucose transport in *Rhodotorula glutinis* (though not used in the above study) is the best-studied example of a glucose/proton symport. It is believed by van Urk *et al.* (1989) that facilitated diffusion allows unrestricted flow of glucose into the yeast, hence aerobic fermentation, whereas a proton symport does not allow this to happen.

Brettanomyces spp. occupy a position intermediate between those that are Crabtree positive and those that are Crabtree negative (Wijsman *et al.*, 1984; van Dijken & Scheffers, 1986). When *B. intermedius* is grown aerobically on glucose in batch culture, the first phase of growth is the fermentative production of equal amounts of ethanol and acetate. When glucose is depleted, growth continues, ethanol being converted to acetic acid. Then, when ethanol is completely utilised, growth stops. Growth resumes, however, after a long lag period, when it is supported by the utilisation of acetate. *Brettanomyces intermedius* has a proton symport for glucose uptake (van Urk *et al.*, 1989), which suggests that one needs to be cautious about relating the type of transport of glucose to its subsequent metabolism, as indicated above. Nevertheless, there is sufficient information provided by van Urk *et al.* (1989) to indicate that *B.*

intermedius is only weakly Crabtree positive. This species would repay more intensive investigation.

When cells of *B. intermedius* are subjected to anaerobiosis, both growth and glucose utilisation cease. Here is an example of the Custers effect. After 7–8 h, growth resumes. The lag phase can be shortened by the presence of hydrogen acceptors such as oxygen or acetoin. The effect appears to be due to an unfavourable redox balance brought about by the formation of acetic acid, which is associated with an overproduction of NADH. Thus:

$$C_6H_{12}O_6 + 4NAD^+ + 2H_2O \rightarrow 2CH_3COOH + 2CO_2 + 4NADH + 4H^+$$

This NADH can be reoxidised only by the production of reduced products, which is very slow under anaerobic conditions. Exogenous hydrogen acceptors such as oxygen or acetoin rectify readily the redox imbalance. Thus:

$$CH_3CHOHCOCH_3 + H^+ \rightarrow CH_3CHOHCHOHCH_3 + NAD^+$$

The transient inhibition of anaerobic fermentation as described above is characteristic also of the genera *Dekkera* and *Eeniella* (Scheffers, 1961; Wikén, Scheffers & Verhaar, 1961; Scheffers & Wikén, 1969).

A key criterion for the anaerobic dissimilation of glucose is the ability of the organism to regenerate NADP from NADPH. This ability is dependent upon the carbon and nitrogen sources for growth. It has been pointed out above with reference to filamentous fungi that the greater requirement for NADPH in the utilisation of nitrate is associated with a greater flux of carbon through the pentose phosphate pathway, the most important source of NADPH in the cell. Given the absence of a mannitol cycle, which can also generate NADPH (Hult & Gatenbeck, 1978; Jennings & Burke, 1990), the only other possible source, namely $NADP^+$: isocitrate dehydrogenase, cannot produce sufficient NADPH (Bruinenberg, van Dijken & Scheffers, 1983a), and transhydrogenation, viz:

$$NADH + NADP^+ \rightleftharpoons NAD^+ + NADPH$$

does not appear to take place to any significant extent in the yeasts *C. utilis* (Bruinenberg *et al.*, 1983b) and *S. cerevisiae* (Lagunas & Gancedo, 1983).

The importance of the pentose phosphate pathway in generating NADPH has been demonstrated in a theoretical analysis (Bruinenberg *et al.*, 1983a). As with *Aspergillus nidulans*, it has been shown that growth of *C. utilis* on nitrate in a chemostat leads to elevated activity of the

pentose phosphate pathway enzymes when compared with ammonia as
the nitrogen source (Bruinenberg *et al.*, 1983*b*). This increase in activity
is independent of the carbon source whether it be glucose, gluconate,
xylose, acetate or ethanol.

Under aerobic conditions, excess NADH or NADPH can be removed
by mitochondrial respiration. Under anaerobic conditions, there must be
oxidative reactions for the two reduced coenzymes that balance those
reactions leading to the formation of NADH or NADPH. This need for
such a balance underpins any consideration of the utilisation of pentose
sugars by yeasts.

Of the 439 species of yeast listed by Barnett *et al.* (1979), 291 can ferment
glucose facultatively, 317 can grow on xylose, while 193 share both
properties (van Dijken & Scheffers, 1986). But of these 193, only a very
few can ferment xylose anaerobically, those not being able to do so
exhibiting the Kluyver effect. *Bretanomyces naardenensis*, *Candida
shehatae*, *Candida tenuis*, *Pachysolen tannophilus*, *Pichia segobiensis* and
Pichia stipitis, presumptive telomorph of *C. shehatae*, are amongst those
few that are capable of fermenting xylose (Toivola *et al.*, 1984). The ability
to ferment this and other pentose sugars is important because of their
presence in hemicelluloses, a potentially important substrate for the
microbiological production of ethanol.

The block on the fermentation of xylose by those yeasts that exhibit
the Kluyver effect occurs in the initial stages of metabolism (Figure 5.5).
In *Candida utilis*, xylose is first converted to xylitol via the activity of an
$NADP^+$-requiring reductase and the xylitol is then oxidised to xylulose
via the activity of an NADH-requiring oxidoreductase. The yeast does
not contain xylose isomerase (converting xylose to xylulose directly) nor,

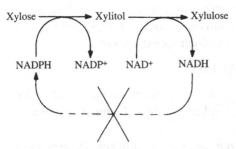

Figure 5.5. Initial reaction of xylose metabolism in yeasts. Due to the absence of
transhydrogenase activity, the NADH produced in the conversion of xylitol to
xylulose cannot be used for the NADPH-linked reduction of D-xylose to xylitol.
(From van Dijken & Scheffers, 1986.)

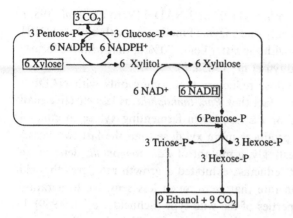

Figure 5.6. Schematic representation of xylose metabolism via NADPH-linked xylose reductase, assuming ethanol fermentation is taking place. The scheme makes clear that in this case there is a discrepancy between the production and consumption of NADH in the overall conversion of xylose to ethanol. (From Bruinenberg *et al.*, 1983.)

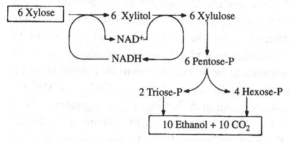

Figure 5.7. Schematic representation of xylose metabolism via NADH-linked redox balance. Formation of ethanol can proceed with a closed redox loop. (From Bruinenberg *et al.*, 1984.)

as has been pointed out above, is it capable of transhydrogenation. Therefore metabolism of xylose under anaerobic conditions would lead to an overproduction of NADH (Figure 5.6) (Bruinenberg *et al.*, 1983).

In those yeasts that are xylose-fermenting, the unbalance of the $NAD^+/NADH$ redox system is circumvented by the presence of enzyme activity capable of reducing xylose to xylitol in the presence of NADH (Bruinenberg *et al.*, 1984). This means of course that the NADH produced by the oxidation of xylitol can be used to reduce xylose to xylitol, thus generating a closed redox loop (Figure 5.7). In *Pachysolen tannophilus* not only is there a xylose reductase specific for NADPH but there is also a form that is active with both NADPH and NADH (Verduyn *et al.*, 1985). In *Pichia stipitis* there is only one enzyme functioning with both NADPH and NADH, the activity with the latter being 70% of that with the former reduced coenzyme (Verduyn *et al.*, 1985a, b).

The presence of the xylose reductase functioning only with NADPH could well account for the fact that *Pac. tannophilus* is less effective than either *Candida shehatae* or *Pic. stipitis* in fermenting xylose to ethanol and that the first yeast produces more xylulose than the last two yeasts (Schneider, 1989). Interestingly, a mutant of *Pac. tannophilis* deficient in the NADPH-dependent reductase exhibited a growth rate, growth yield and xylose consumption rate that were much less sensitive to aeration rate than the same properties of the wild-type (Schneider *et al.*, 1989). It has been suggested from these studies that the NADPH-dependent xylose reductase is linked to oxygen utilisation and that it plays an important role in growth on xylose under aerobic conditions. But the properties of the mutant are also relevant to the arguments put forward above, namely that the effective fermentation of xylose to ethanol depends on the redox balance. It was found that the mutant, though growing more slowly than the wild-type, produced much less xylitol under both aerobic and anaerobic conditions and more ethanol under the latter. That said, the extent to which there is effective fermentation under anaerobic conditions must be questioned, since there appears to be an enhancement by limited aeration (du Preez, Prior & Monteiro, 1984; Slininger *et al.*, 1985; Delgenes, Moletta & Navarro, 1986; Sreenath, Chapman & Jeffries, 1986) suggesting that a degree of imbalance of the $NAD^+/NADH$ redox system nevertheless occurs (du Preez, van Driessel & Prior, 1989). In *Pac. tannophilus* the degree of aeration modulates the ratio of NADH- to NADP-linked xylose reductase activity (Ligthelm, Prior & du Preez, 1988a, b). Anaerobic conditions can also reduce enzyme levels through protein degradation, as appears to be the case with *Pac. tannophilus* (Neirinck, Maleska & Schneider, 1982; du Preez *et al.*, 1989). However, there is not a clear picture for all yeasts.

Though xylose-fermenting yeasts seemingly possess biotechnological potential in degrading a major component of plant hemicelluloses, in actuality the use of these yeasts is made difficult by the presence of hexose, particularly glucose, in the material being degraded. First, xylose utilisation is repressed by glucose (Bicho *et al.*, 1988; Delgenes *et al.*, 1986; Panchal *et al.*, 1988). Second, the fermentation of hexose produces ethanol, which may inhibit the utilisation of xylose (Webb & Lee, 1990); pentose-fermenting yeasts have a low ethanol tolerance (du Preez, Bosch & Prior,

1987; and see below). Third, not only are there these inhibitory effects but the sequential utilisation of hexose then xylose either increases the time of fermentation or makes for a more complicated fermentation process (Grootjen *et al.*, 1991*a, b*). The biotechnological fermentation of xylose to ethanol has been well discussed by Prior, Kilian & du Preez (1989) and Schneider (1989).

Finally, in this consideration of the utilisation of monosaccharides by yeasts, I turn to another kind of Kluyver effect. It concerns the fact that, of the yeasts that can ferment glucose anaerobically, over 40% of species can utilise galactose and certain glycosides aerobically but cannot ferment them (Sims & Barnett, 1978). These yeasts absorb the glycosides, which are hydrolysed internally. Initially, it was thought the Kluyver effect for these substrates had two possible causes:

(1) Under anaerobic conditions the glycosidases are inactivated (Kluyver & Custers, 1940).
(2) Transport of glycosides into the cell depends upon the presence of a proton symport (see pp. 145–8) that becomes inoperative under anaerobic conditions because of insufficient ATP to maintain the necessary proton-motive force to drive glycoside accumulation (Sims & Barnett, 1978; Barnett & Sims, 1982; Schultz & Höfer, 1986).

The first cause can almost certainly be ruled out (Sims & Barnett, 1978, 1991; Sims, Stålbrand & Barnett, 1991). Transport is certainly not blocked under anaerobic conditions; those yeasts studied change from active transport to facilitated diffusion (Barnett & Sims, 1982). Most recent studies suggest that there is a suite of effects that leads to an inability of the yeast to ferment the glycoside (Sims *et al.*, 1991).

(1) The change from active transport to facilitated diffusion could impose a rate limitation on the supply of the glucoside to the cytosol.
(2) The internal glucosidases have much lower affinities than hexokinases, which again could limit the carbon flux into glycolysis.
(3) When yeasts showing the Kluyver effect are grown on a glycoside, there is a much lower specific activity of pyruvate decarboxylase. This again lowers the flux of carbon through glycolysis.

Thus, the Kluyver effect seems to be brought about by the interplay of several factors.

Anaerobic fungi

Anaerobic chytridomycete fungi are found in the rumen of sheep and cattle and other large herbivores. They were first discovered by Orpin (1976) in the rumen of sheep and identified on the basis of their morphology, life cycle and presence of chitin in the walls of the vegetative stage (Orpin, 1977*b*). There is now a considerable amount known about these fungi. The reader is referred to reviews by Bauchop (1989), Theodorou, Lowe & Trinci (1992), Orpin (1993).

Rumen fungi are obligate anaerobes and fail to grow in the presence of oxygen. Ultrastructural studies show that these organisms do not contain mitochondria but do contain hydrogenosomes (Yarlett *et al.*, 1986). Glucose is the major substrate for growth of these fungi, derived in the gut from the breakdown of cellulose (see p. 158). The primary pathways for the metabolism of glucose in *Neocallimastix frontalis* have now been established (Yarlett *et al.*, 1986; O'Fallon, Wright & Calza, 1991). There is strict anaerobic integration of the metabolism. Glycolysis is coupled to malate dehydrogenase, malic enzyme and the hydrogenosome (Figure 5.8). The coupling with the hydrogenosome is via NADH and NADPH, which are oxidised by ferredoxin, with the generation of hydrogen. The other

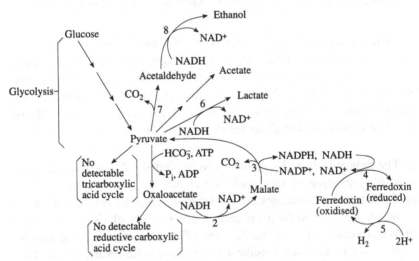

Figure 5.8. Scheme for the conversion of glucose to acetate, ethanol and lactate by *Neocallimastix frontalis*. Enzymes: 1, pyruvate carboxylase; 2, malate dehydrogenase; 3, 'malic' enzyme; 4, NADH (NADPH):ferredoxin oxidoreductases; 5, hydrogenase; 6, lactate dehydrogenase; 7, pyruvate decarboxylase; 8, alcohol dehydrogenase. (From O'Fallon *et al.*, 1991.)

fermentation products are acetate, ethanol and lactate. All the ATP is synthesised via glycolysis.

Ethanol tolerance

The accumulation of a metabolic end-product in the medium surrounding a fungus can represent a chemical stress to the organism. This may well happen if the organism is growing in very limited volume but with an adequate supply of carbon and other relevant nutrients. Equally it may also occur when a fungus is being used to produce a product of economic value. High yield of the product can enhance the economic return on the fermentation. However, the product as it increases in concentration may start to have untoward effects on the fungus such that it becomes more unstable. At the very least there can be both metabolic inhibition and cell deactivation, i.e. the loss of the ability to replicate, while the cell still remains viable and metabolically active. A major consequence will be a reduction in growth rate and probable changes in metabolic end-products. Jones (1990) considered deactivation in relation to yeast ethanol fermentation but he pointed out the generality of the concept.

Here, too, I focus on the effects of ethanol on fungi, and for three reasons: first, as made clear above, ethanol is the major end-product of fermentation for most fungi; second, as a specific example of the more general situation referred to above; and third, because understanding the effects of ethanol on fungi and their ability to tolerate the compound involves membrane processes that are a recurring theme in fungal nutrition. The matter of ethanol tolerance in yeasts has been reviewed as a whole (Ingram & Buttke, 1984; Casey & Ingledew, 1986; D'Amore *et al.*, 1990; Mishra, 1993) or in terms of specific aspects: intracellular ethanol accumulation (Jones, 1988), the role of the membrane in tolerance (Ingram, 1986; Jones & Greenfield, 1987), and lipids as modulators of tolerance (Mishra & Kaur, 1991).

There is almost universal agreement that ethanol inhibits growth of and fermentation by species of *Saccharomyces* in a non-competitive manner, i.e. ethanol affects the maximum specific rates of the two processes but not substrate affinity (Casey & Ingledew, 1986). *In vitro* studies by Millar *et al.* (1982) on the effect of ethanol on glycolytic enzymes have established the ethanol concentrations required to cause various percentage denaturation levels of the proteins (Table 5.6). There is very little effect on the enzymes below 13% (w/v) ethanol. At least 17% ethanol is required to give 50% denaturation. It is generally accepted that most yeast cells

Table 5.6. *The percentage (w/v) concentration of ethanol required to cause 10%, 50% and 90% denaturation of various enzymes in 30 min*

Enzyme	Loss (%)		
	10	50	90
Hexokinase	16	19	25
Phosphoglucose isomerase	22	35	>40
Phosphofructokinase	14	19	22
Fructose 1,6-bis P aldolase	15	18	20
Triose phosphate isomerase	25	>35	>40
Glyceraldehyde-P dehydrogenase	13	17	21
Phosphoglycerate kinase	16	19	21
Phosphoglycerate mutase	20	35	>40
Enolase	12	19	28
Pyruvate kinase	18	21	27
Pyruvate decarboxylase	14	17	19
Alcohol dehydrogenase	25	>35	>40

P, phosphate.
From Millar *et al.* (1982).

are tolerant to concentrations around 12%–15% ethanol, while in the case of saké fermentations the yeasts used are able to produce 20% ethanol.

If yeast cells concentrate ethanol as a consequence of the rate of production being greater than the rate of release into the medium, then it is conceivable that the effect of the alcohol on growth might be through the inhibition of glycolysis. However, there must be a strong question mark against those results which show that ethanol accumulates in cells. The matter has been discussed at length by Casey & Ingledew (1986) and Jones (1988). There is little doubt now that, where ethanol accumulation has been purported to occur, there have been defects in the methods used. Two major defects have been a failure to realise (a) that ethanol production can continue during centrifugation when the cells are separated from the medium and (b) that the volume of the yeast cell determined under one set of experimental conditions is the same under another set.

The most detailed study on ethanol accumulation has been that of Guijarro & Lagunas (1984). On the basis of the measured distribution of $[1\text{-}^{14}C]$ethanol between medium and cells, these investigators concluded that the solute is passively distributed across the plasma membrane and that accumulation does not occur. The evidence is as follows: (a) uptake obeys first-order kinetics, i.e. there is no saturation of the rate of uptake as a function of ethanol concentration, which would occur if a carrier were involved; (b) equilibrium across the membrane is rapid, being achieved within 5 s; (c) uptake of ethanol cannot be altered by structural analogues such as acetaldehyde, methanol, propanol and ethylene glycol, treatment with cycloheximide or varying the pH of the medium. Furthermore, these same workers showed that there was no effect of osmotic pressure in the medium on the efflux of ethanol and, in media of varying glucose concentration, the ethanol concentration in cells was found to be always the same as that of the medium. Dombek & Ingram (1986*a*), also using a procedure avoiding many of the technical problems of previous studies, measured the intracellular concentrations of ethanol in fermenting cells of *S. cerevisiae*. The results indicated that these concentrations are less than or equal to those in the external medium and thus the results are similar to those of Guijarro & Lagunas (1984).

Both Casey & Ingledew (1986) and Jones (1988) pointed out that, if accumulation of ethanol does occur, it would mean that the permeability of the yeast cells to the solute would have to be much lower than for other eukaryotic cells, plant and animal, namely of the order of 0.4×10^{-4} to 2×10^{-4} cm s^{-1}.

One of the reasons for the interest in the possibility of ethanol accumulation has been the fact that externally added ethanol is much less toxic than ethanol produced internally by fermenting yeast cells (Nagodawithana & Steinkrauss, 1976). There is interest in the idea that as well as growth inhibition by ethanol per se there is also synergy with other inhibitory metabolic by-products (Maiorella, Bland & Wilke, 1983; Lafon-Lafourcade, Geneix & Ribéreau-Gayon, 1984). Jones (1988) has drawn attention to the fact that there are at least 22 secondary alcohols, 12 aldehydes, 12 α-keto acids, 28 esters, 22 free acids (fatty acids) and 13 other carbonyl and phenolic compounds all produced during fermentation.

One can safely discount the possibility that reduced growth brought about by ethanol is via an effect on enzymes, since the necessary concentrations of ethanol inside the cells are not achieved under those conditions when growth is inhibited. That conclusion is endorsed by the study of ethanol-sensitive mutants of *S. cerevisiae* by Aguilera & Benitez (1986), who showed that, out of 21 monogenic mutants, 20 showed no alternation in their glycolytic pathway (or in their lipid biosynthetic pathway). Mishra (1993) has listed the major physiological and biochemical processes in *S. cerevisiae* and other yeasts that are known to be affected by ethanol. Apart from inducing stress proteins, which is considered below, and morphological transitions in the dimorphic yeasts *C. albicans* and

C. tropicalis, which are not considered, all the other effects are on membranes and membrane processes.

Evidence has now accumulated to show that a major if not *the* major locus for the action of ethanol on yeast cells is the plasma membrane. The evidence is as follows:

(1) Ethanol and related alkanols inhibit in a non-competitive manner the uptake of glucose, maltose, ammonia and amino acids (via the general amino acid permease). V_{max} is decreased by ethanol; K_T is unaffected (Leão & van Uden, 1982, 1983, 1984a). It has been shown also that potassium loss accompanying amino acid uptake in *S. cerevisiae* (see Chapter 6) is inhibited by ethanol in a similar manner. Leão & van Uden (1983) found that the minimum concentration of alkanol (ethanol, isopropanol, propanol and butanol) inhibiting ammonium uptake and the concentration that halved the maximal initial uptake rate of methylammonium were inversely related to the lipid/buffer partition coefficients of the alkanols (Figure 5.9). Phosphate uptake is also inhibited by ethanol (Thomas & Rose, 1979).

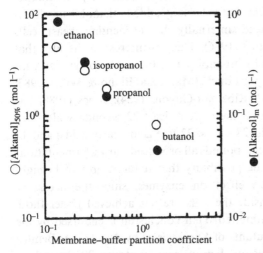

Figure 5.9. Relations between the minimum inhibitory concentrations of alkanols (filled symbols) and the concentrations of alkanols that halved the maximum initial uptake rates of methylammonium by *Saccharomyces cerevisiae* (open symbols) and the membrane–buffer partition coefficients of the alkanols. The values for the coefficients are from Seeman (1972). (From Leão, C. & van Uden, N. (1983), *Biotechnology and Bioengineering*, copyright © 1983, reprinted by permission of John Wiley & Sons, Inc.)

(2) Ethanol (and the other alkanols referred to in (1)) enhance the passive influx of protons into de-energised cells of *S. cerevisiae* (Leão & van Uden, 1984*b*). The influx follows first-order kinetics, with a rate constant that increases exponentially with alkanol concentration. The exponential enhancement constants increase with lipid solubility of the alkanol. This indicates that hydrophobic membrane regions are a target site, though not the only one, since increased protonation of the membrane as the external pH is decreased makes the membrane less sensitive to alkanols. In keeping with these observations Cartwright *et al.* (1986) found that ethanol dissipates the proton-motive force across the membrane of *S. cerevisiae*. It is interesting in this respect that extracellular acidification has been suggested as a possible means of assessing ethanol tolerance in yeast. Jimenez & van Uden (1985) showed that a number of yeast strains fitted a reciprocal relation between the concentration of ethanol that reduces the specific growth rate by 50% and the exponential constant (see above) of ethanol-enhanced proton diffusion out of the cells. However, since that publication, other reports have shown that the situation is much more complicated. There appear to be yeast species, such as *C. shehatae, C. utilis, Candida wickerhamii, Hansenula polymorpha* and *Saccharomyces unisporus*, in which proton diffusion is not or is hardly affected by ethanol at concentrations that strongly inhibit growth (Kilian, du Preez & Gericke, 1989; Prell, Páca & Sigler, 1992).

(3) Ethanol stimulates the leakage of solutes other than protons from cells of *S. cerevisiae*. Efflux of amino acids and compounds that absorb light at 260 nm are lost, with kinetics that are first-order (Salgueiro, Sá-Correia & Novais, 1988). Significantly, the addition to a fermentation medium of the intracellular material resulting from ethanol treatment was found to enhance alcohol fermentation by both *S. cerevisiae* and *Saccharomyces bayanus*.

(4) There are now numerous instances (Table 5.7) where addition of lipids to the growth medium has been shown to enhance ethanol tolerance (Mishra & Kaur, 1991; Mishra, 1993). There is a problem (discussed in Chapter 3) of relating gross changes in membrane composition to membrane function. Interpreting the observations on changes in plasma membrane composition in relation to ethanol tolerance have proved equally difficult. Jones & Greenfield (1987) have produced a valuable commentary on the relation between membrane fluidity and ethanol tolerance.

(5) Ethanol at concentrations (3–10% (v/v)) that reduce the specific growth

Table 5.7. *Details of lipids and lipid-containing material that have been used to modulate ethanol tolerance of yeasts*

Lipid component	Yeast strain	Ethanol tolerance	References
Proteolipid (*Aspergillus oryzae*)	*Saccharomyces sake*	Increased growth, fermentative activity and ethanol endurability	Hayashida, Feng & Hongo (1974); Hayashida et al. (1976); Hayashida & Ohta (1978)
Crude egg yolk PC	*S. sake*	Increased growth and ethanol endurability	Hayashida & Ohta (1980)
Purified PC	*S. sake*	Increased growth and fermentative activity	Hayashida & Ohta (1980)
Protein–phospholipid complex	*S. cerevisiae*	Increased ethanol production	Jin, Chiang & Wang (1981)
PC albumin	*S. sake*	Increased ethanol production	Hayashida et al. (1974)
PC, palmitic acid and cholesterol	*S. cerevisiae*	Increased growth	Ghareib, Youssef & Khalil (1988)
PS	*S. cerevisiae*	Increased alanine uptake, H^+ efflux and fermentative activity	Mishra & Prasad (1988)
Ergosterol	Grape must	Increased growth and viability but restricted fermentative activity	Larue, Lafon-Lafourcade & Ribereau-Gayon (1980)
Ergosteryl-oleate or ergosterol + oleic acid	*S. sake*	Increased growth, alcohol endurability and no enhancement in fermentative activity	Hayashida & Ohta (1980)

Ergosteryl-oleate + egg yolk PC	*S. sake*	Increased growth, alcohol endurability and fermentative activity	Hayashida & Ohta (1980)
Ergosterol + oleic and linoleic acid	*Kluyveromyces fragilis*	Increased growth rate, biomass production	Janssens et al. (1983)
Ergosterol or oleic acid	*S. sake*	No effect on alcohol endurability and fermentative activity	Hayashida & Ohta (1980)
Yeast hull (mixture of sterols + UFAs)	Wine yeast	Enhanced growth rate and fermentation activity	Munoz & Ingledew (1989)
Tween 80, ergosterol + albumin	*S. sake*	Increased fermentative activity and reduced fermentation time	Ohta & Hayashida (1983)
Ergosterol or campesterol + linoleic acid	*S. cerevisiae*	Increased viability and nutrient uptake	Thomas, Hossack & Rose (1978)
Linolenic acid	*S. cerevisiae*	Increased viability and nutrient uptake	Thomas & Rose (1979)
Linoleic acid or Tween 80	*S. uvarum*	Increased ethanol production	Panchal & Stewart (1981)
Oleic acid or linoleic acid or linolenic acid	*S. cerevisiae*	Sequential increase in alanine uptake, H^+ efflux and fermentative activity	Mishra & Prasad (1989)

PC, phosphatidylcholine; PS, phosphatidylserine; UFAs, unsaturated fatty acids.
From Mishra & Kaur (1991).

rate of cultures of *S. cerevisiae* have been shown to lead to the extraction of the plasma membrane H^+-ATPase with a higher specific activity. This increase decays when the cells are transferred to an ethanol-free medium (Rosa & Sá-Correia, 1991). Cycloheximide appears to inhibit the activation *in vivo*. Ethanol, but at relatively high concentration ($K_i \approx 2$ M; 9.2% (v/v)), inhibits plasma membrane activity *in vitro* (Cartwright, Veazey & Rose, 1987); at low concentrations of ethanol, the activity may be increased (Petrov & Okorokov, 1990). It is not clear how these studies *in vitro* relate to the findings *in vivo*. There is some indication from a comparative study on a yeast less tolerant to ethanol, *Kluveromyces marxianus*, that the specific activity of ATPase in cells growing in the presence of ethanol is related to the ethanol tolerance of the yeast (Rosa & Sá-Correia, 1992). These findings need to be explored further.

While ethanol clearly exerts a major effect on membrane functioning, it cannot be assumed as yet that enhanced tolerance resides necessarily in an altered membrane structure. Also, because of the obvious facility in their study, all the membrane properties found so far to be altered by ethanol have been those of the plasma membrane. We know nothing therefore of the effect of ethanol on the tonoplast, yet we know it has a composition different from that of the plasma membrane (see Chapter 4). In establishing the mechanism of tolerance, there will be a need to consider not only the facts referred to above but also others that at present are less readily associated with any particular processes.

(1) Calcium and magnesium appear to enhance ethanol tolerance in *Saccharomyces* spp. (Dombek & Ingram, 1986b; Nabais *et al.*, 1988). It is not clear the extent to which one is dealing with an effect of bivalent cations per se rather than that relating to action of the specific cations. Nor is it clear whether the target of action is a membrane or some part of metabolism, i.e. the reduction in the amount of those toxic products of fermentation other than ethanol. However, it should be noted that Petrov & Okorokov (1990) have found that in plasma membrane vesicles from *S. carlsbergensis* an increase in proton and anion permeability by ethanol is partially prevented by magnesium.

(2) Ethanol stimulates the production of heat shock proteins in *C. albicans* (Del Castillo Agudo, 1985), *N. crassa* (Roychowdhury & Kapoor, 1988) and *S. cerevisiae* (Plesset, Palm & McLaughlin, 1982; Watson & Cavicchioli, 1983). It is not appropriate here to discuss the putative role of such stress proteins in fungi; the reader should consult

Plesofsky-Vig & Brambl (1993) for the most recent review on this topic. The key observation is that when cells of *S. cerevisiae* are subjected to heat shock there is not only increased thermotolerance but also increased tolerance to ethanol (Watson & Cavicchioli, 1983).

It is likely therefore that a mosaic of internal factors is necessary for ethanol tolerance. Further physiological studies are required but there is also a great need for a better genetic understanding of ethanol tolerance (Mishra, 1993). Progress here may be hindered by the fact that it is rather difficult to isolate ethanol-tolerant mutants by conventional procedures. The most effective method is one of relatively long-term use of continuous culture (Brown & Oliver, 1983).

Monosaccharide transport

Much is known about monosaccharide transport in a range of fungi. The situation is complex in terms of: (a) whether or not active transport occurs; (b) if it does occur, the type of process involved; and (c) how transport is regulated. It is also clear that it is not always possible to argue from what occurs in one fungus to what might happen in another. It is for this reason that consideration of monosaccharide transport is presented here in terms of information about four particular species: those that have been studied most. Three are yeasts (*S. cerevisiae*, *K. marxianus*, *R. glutinis*); the fourth is the filamentous *N. crassa*. Information about other fungi is presented in summary form in Table 5.8.

Saccharomyces cerevisiae

Baker's yeast possesses a constitutive monosaccharide transport system with the highest affinity being for D-glucose and 2-deoxy-D-glucose (Cirillo, 1962*a*,*b*, 1968*a*; Kotyk, 1967; Heredia, Sols & de la Fuente, 1968; Table 5.9). Analysis of results of inhibition studies using L-sorbose or D-xylose as the transport sugar indicates that the system has rather broad specificity for pyranoses. Single changes at each of the five carbon atoms of D-glucose (but not those of 2-deoxy-D-glucose) result in decreases in affinity for the transport system. The largest decreases are brought about by methylation or glucosylation of the anomeric hydroxyl group. The relatively high affinity for fructose is surprising but this may be because of lack of knowledge of the molecular configuration of fructose in solution.

Growth of baker's yeast on D-galactose results in the induction of a

Table 5.8. Summary of those studies concerned with sugar transport in fungi

Sugar	K_T (mM)	Proton symport	Glucose repression	Special features	References
Achlya sp.					
2-Deoxy-D-glucose	c. 0.01	?	?	Only D-galactose and D-xylose competitively inhibit transport	Goh & Lé John (1978)
Aspergillus nidulans					
D-Glucose		?	?	Separate systems for the two sugars, although each sugar can be transported by each system	Mark & Romano (1971)
D-Galactose					
D-Fructose		?	Yes	Declines with age of culture	Kurtz (1980)
Candida intermedia					
D-Glucose	0.16	Yes	Yes	Accumulation of L-sorbose by this system, transport of sorbitol in ethanol-grown cells	Loureiro-Dias (1987)
	2.0	No	No	L-Sorbose transported but not accumulated	
Sorbitol		Yes	Yes	Induced by growth on sorbitol	
Candida shehatae					
D-Glucose	0.12	Yes	Yes	Transports also D-mannose	Lucas & van Uden (1986)
	2.0	No	No	Transports also D-mannose and D-xylose	
D-Xylose	1.0	Yes	Yes	Transports also D-galactose	
L-Arabinose		Yes	Yes		

Candida utilis					
D-Glucose	0.015	Yes	Yes	Repression above 0.7 mM glucose	Peinado, Cameira-dos-Santos & Loureiro-Dias (1989)
	?	No	No	Facilitated diffusion in repressed cells; complicated kinetics depending on the glucose concentration	
Candida wickerhamii					
D-Glucose	0.18	Yes	Yes	Transports also 2DG, 6DG and D-xylose: non-competitively inhibited by cellobiose	Spencer-Martins & van Uden (1985a, b, 1986)
	1.7	No	No	Competitively inhibited by cellobiose	
Cladosporium resinae					
3-O-methylglucose	0.42	?	No	Accumulation of 3O MG; starvation does not result in appearance of low K_T system; inhibition of uptake by azide, cyanide and DNP	Teh (1975)
Coprinus cinereus					
D-Glucose	0.027	?	Yes	Accumulation; transports also 2DG, 3O MG, D-glucosamine, D-fructose and L-sorbose; inhibited by cyanide, DNP and N-ethylmaleimide; cycloheximide leads to decay of system but has no effect on derepression	Moore & Devadathan (1979); Taj-Aldeen & Moore (1982)
	3.3	?	No	Accumulation; transports 2DG and 3O MG; pH changes affinity for 2DG	
				ftr cistron is the structural gene for *both* systems	

Table 5.8. *Continued*

Sugar	K_T (mM)	Proton symport	Glucose repression	Special features	References
Dendryphiella salina					
D-Glucose	?	Yes	Probably	Evidence from electrophysiological studies	Davies *et al.* (1990*a*)
Endomyces magnusii					
D-Glucose	0.1	?	Yes	Derepression inhibited by cyclo-heximide	Ehwald, Mavrina & Wilken (1981)
	10	Unlikely	No	Facilitated diffusion	
Kluyveromyces lactis					
2-Deoxy-D-glucose	1.7		Yes	Glucose-grown cells; concentration inside lower than outside; encoded by *RAG1* gene; protein high degree of homology with that encoded by *HXT2* of *Saccharomyces cerevisiae*	Royt & MacQuillan (1976); Royt (1981, 1983); Royt & Tombes (1982); Chen, Wésolowski-Louvel & Fuhukara (1992); Wésolowski-Louvel *et al.* (1992)
	3.4			Succinate-grown cells; rate of uptake slower (and also phosphorylation of 2DG) than glucose-grown cells	
6-Deoxy-D-glucose				Facilitated diffusion	
Kluyveromyces marxianus				See text	
Metschnikowia reukaufii					
D-Glucose	0.9	Yes	No	Accumulation; also accumulates 3O MG and D-xylose	Aldermann & Höfer (1981)

					Reference
Neocosmospora vasinfecta					Budd (1976)
D-Glucose	0.02	Probably	Yes		
	2.0	?	No		
Neurospora crassa				See text	
Penicillium notatum					Pitt & Barnes (1987)
D-Glucose	0.1	?	?	Accumulation at low but not at high concentrations; inhibited by azide	
Phanerochaete chrysosporium					Greene & Gould (1984*a*)
D-Glucose	?	Yes	?	Membrane vesicles used for study	
Phytophthora palmivora					Sheard & Farrar (1987)
D-Glucose	2.1	Possibly	?	Inhibited by DNP	
	217	?	?	Not inhibited by DNP	
Pichia heedii					Does & Bisson (1990)
D-Xylose	0.1	Yes	No	Glucose non-competitive inhibitor; reduced by elevated xylose concentrations	
	4.5	Yes	No	Induced by elevated xylose concentrations. There may be more than two transport systems	
Pichia ohmeri					Verma, Spencer-Martins & van Uden (1987)
D-Glucose		Yes	Yes	Formation prevented by cycloheximide	
		No	No	Facilitated diffusion; cycloheximide has no effect on conversion from high affinity system	

Table 5.8. *Continued*

Sugar	K_T (mM)	Proton symport	Glucose repression	Special features	References
Pichia stipitis					
D-Xylose	0.04–0.4	Yes	No	Competitive inhibition by glucose	Kilian & van Uden (1988)
	1.9–3.7	Yes	No	Shared with glucose	
				Neither system greatly affected by growth of cells in glucose or xylose, or by short starvation. Growth in glucose slightly increases K_T of both systems	
D-Glucose	0.26–0.73	Yes	No	No low affinity system	
Pilobolus longipes					
6-Deoxy-D-glucose	14	No	No	Dormant spores	Bourret (1985)
	40	No	No	Activated spores: uptake inhibited by glucose, which brings about counterflow; DNP no effect; internal concentration lower than external	
				Activation of spores brought about by glucose or 6DG. V_{max} of activated spores higher than dormant	
Rhodotorula glutinis				See text	
Saccharomyces cerevisiae				See text	

Schizosaccharomyces pombe					
D-Glucose	3	Yes	No?	System transports only glucose, 2DG and glucosamine in both aerobic and anaerobic conditions but under the latter glucose must be present to provide energy; 2DG and glucosamine accumulated, glucose metabolised	Höfer & Nassar (1987)
6-Deoxy-D-glucose	45	No	No?	Facilitated diffusion	
Torulopsis candida					
L-Sorbose	0.3	?	?	Transports glucose, galactose and xylose	Háskovec & Kotyk (1973)
Mannitol	0.6	?	?	Transports D- and L-arabitol, ribitol and xylitol	
Erythritol	0.2	?	?	Transports D-ribose Also low affinity system for mono-saccharides and polyols	
Ustilago maydis					
3-O-methylglucose	0.08	?	?		
	1.3	?	?		Miller & Harun (1978)

2DG, 2-deoxy-D-glucose; 6DG, 6-deoxy-D-glucose; 3O MG, 3-O-methylglucose; DNP, 2,4-dinitrophenol.

Table 5.9. *Comparative affinities of sugars as substrates for the glucose transport system of* Saccharomyces cerevisiae

Sugar	Relative affinity
D-Glucose	1
2-Deoxy-D-glucose	1
D-Fructose	0.2
D-Galactose	0.13
D-Mannose	0.1
(1→5)-Anhydro-D-glucitol	0.1
3-*O*-Methyl-D-glucose	0.02
D-Xylose	0.13
D-Fucose	0.02
D-Arabinose	0.02
L-Sorbose	0.005
L-Glucose	0.0025
α-Methyl-D-glucoside	0.0025

From Cirillo (1968*a*).

transport system that has a higher specificity for hexose and its non-metabolisable analogues than does the constitutive system (Cirillo, 1968*b*; Kotyk & Håskovec, 1968). Subsequent studies (Kuo, Christensen & Cirillo, 1970; Kuo & Cirillo, 1970) showed that D-fucose and L-arabinose, which are not metabolised, are transported by the system by facilitated diffusion. Galactose is also transported in like manner in cells that are galactokinaseless; this is germane to considerations below, since it indicates that transport does not require phosphorylation of the sugar to take place. Equally important is the finding that facilitated diffusion occurs at a rate four to ten times greater than the rate of galactose metabolism. Phosphorylation of galactose takes places in cells that are uridine diphosphate, galactose-1-phosphate:uridylyl-transferase negative, but addition of [1-^{14}C]galactose 1-phosphate results in a higher rate of incorporation of radioactivity into the free-sugar pool than the galactose 1-phosphate pool. Again, this is in keeping with transport of the free sugar into the cell followed by phosphorylation therein.

Kinetic analysis of glucose transport into plasma membrane vesicles of baker's yeast shows unambiguously that the process is by facilitated diffusion (Christensen & Cirillo, 1972; Fuhrmann, Boehm & Theuvenet, 1976; Franzusoff & Cirillo, 1983*a*, *b*; Fuhrmann *et al.*, 1989). A very interesting aspect of the study of Fuhrmann *et al.* (1989) was the demonstration that the K_T for transport was dependent upon conditions of growth and the genetic constitution of the cells used in the preparation of the membranes. Glucose-repressed cells had a K_T of 8.6 mM, galactose-derepressed cells a K_T of 2.9 mM and a triple kinaseless mutant one of 1.8 mM.

The kinetics of uptake of glucose, fructose and 6-deoxy-D-glucose in wild-type cells have been shown to be biphasic (Bisson & Fraenkel, 1983a, b; Lang & Cirillo, 1987). It has been argued that there are two components, one with a higher affinity (lower K_T) than the other (Figure 5.10). The presence of the high affinity system is dependent on the presence in the cell of the three kinases responsible for the phosphorylation of glucose (hexokinase 1, hexokinase 2 and glucokinase). Mutants lacking these kinases have the simple kinetics of the low affinity system. 6-Deoxy-D-glucose, which is a structural homomorph of D-glucose, lacks the hydroxyl group at carbon-6 and thus cannot be phosphorylated. The kinases are not therefore needed for sugar phosphorylation to be a step in the transport process. This is confirmed by studies on liposomes in which glucose transport activity has been reconstituted. (Ongjoco, Szkutnicka & Cirillo, 1987). In these vesicles, the K_T for glucose transport (c. 8 mM) is that for kinase-independent low affinity transport in whole cells. Trapping of hexokinase and ATP inside the vesicles has no effect on the K_T.

In all probability, the kinetic interpretation of the changes referred to is incorrect. Fuhrmann & Völker (1992) give a critique of the kinetic analysis used; essentially they assume that over the whole range of medium glucose concentrations only one carrier is operating (this will not necessarily be so for the transport of all solutes, as is well demonstrated in Chapter 7 in the consideration of phosphate transport in *Neurospora crassa*). A more critical analysis by Fuhrmann & Völker (1992) indicates that glucose transport in *S. cerevisiae* can be adequately described kinetically by facilitated diffusion via a single carrier alongside simple diffusion. However, even with only a single carrier involved, changes in the genotype of the cell can still bring about changes in the kinetics of glucose transport.

The role of the kinases is unclear. Hexokinases 1 and 2 have been shown to be phosphoproteins (Vojtek & Fraenkel, 1990). Genetic and biochemical analyses showed that the hexokinases, particularly hexokinase 2, are responsible for glucose repression (Rose *et al.*, 1991), though the mechanism by which this operates is unclear.

There is a plethora of evidence that the characteristics of glucose and galactose transport (in induced cells) are affected by the metabolic state

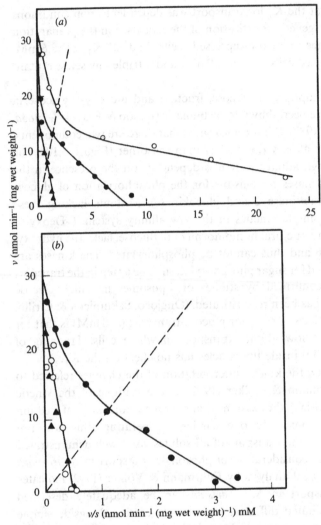

Figure 5.10. Hofstee (1952) plots of sugar uptake by wild-type and mutant strains of *Saccharomyces cerevisiae* over a 5 s period. The broken line represents a substrate concentration of 10 mM. (*a*) Glucose uptake; (*b*) fructose uptake: (●) wild-type; (○) strain without hexokinase 1 and 2; (▲) strain without both hexokinases and glucokinase. Note that, when there is no enzyme to phosphorylate the hexose, there is a linear relationship between v and v/s. (From Bisson & Fraenkel, 1983*b*.)

of the cell (van Steveninck & Rothstein, 1965; Wilkins & Cirillo, 1965; Spoerl, Williams & Benedict, 1973; Serrano & de la Fuente, 1974). Thus the K_T and V_{max} for glucose transport in iodoacetate-poisoned cells is higher than those in control cells. The latter cells show a much greater sensitivity of glucose transport to uranyl (UO_2^{2+}) ions than do iodoacetate-poisoned cells. When glucose uptake takes place in normal cells, cobalt (Co^{2+}) ions bound to the cell surface are lost from it. They rebind when all the glucose has been absorbed. Cobalt binding is reduced by 25% by iodoacetate. L-Sorbose has no effect on cobalt binding, nor are the kinetics of the transport of this sugar affected by iodoacetate and this is the same for D-xylose. Neither sugar is metabolised by glucose-grown cells.

Genetic studies give additional perspective on the transport of glucose. First *snf3* (sucrose non-fermenting) mutants do not show biphasic kinetics for glucose uptake (Bisson *et al.*, 1987). Growth of such mutants is much impaired at low glucose concentrations. Interestingly the *SNF3* gene has been shown to encode a protein that has a significant degree of similarity to the mammalian glucose transporter, both in terms of sequence and probable membrane-spanning regions (Celenza, Marshall-Carlson & Carlson, 1988). Also, a gene has been identified and sequenced that is able to complement the defect in glucose transport (Kruckeberg & Bisson, 1990). This gene has been designated *HXT2* (hexose transporter). It has been shown to have a high sequence and structural similarity to a large family of sugar transport proteins, including that encoded by *SNF3*. It is most similar to the galactose transporter of baker's yeast, which is encoded by *GAL2* (Szkutnicka *et al.*, 1989). These two proteins are identical at 65% of their residues and at 72% in regions predicted to be transmembrane domains. Unlike *snf3* mutants, *hxt2* null mutants are only partially defective in growth on low glucose. HXT2 protein has two consensus cyclic AMP-dependent protein kinase phosphorylation sites. Wild-type levels of high affinity glucose transport require the products of both *SNF3* and *HXT2* genes.

More recently, two novel genes affecting hexose transport have been identified (Lewis & Bisson, 1991). The gene *HXT1* was found to encode a protein very similar to HXT2, being 66% identical. Disruption of the *HXT1* gene resulted in the loss of a portion of high affinity glucose and mannose transport but not fructose, suggesting the protein might be specific for aldohexoses. There was a greater effect on mannose transport, compared with that of glucose, but that could be due to the presence of other glucose transport proteins. The other gene, *ORF2*, was capable of suppressing the *snf3* null mutant phenotype by restoring high affinity glucose transport and increased low affinity transport. *ORF2* has been partially sequenced and no similarity to any previously described transport gene has been found.

It is clear from the genetic studies that there is considerable complexity at the molecular level. There is some evidence that the transporter encoded by *HXT1* is operative mainly during lag phase and early exponential growth (Lewis & Bisson, 1991). Elucidation of the roles of these transport proteins requires extensive physiological studies. It is likely that the situation will not be simple. Experiments by Johnston & Barford (1991) with *S. cerevisiae* in batch culture with mixtures of glucose and fructose in the medium have shown that there are two mechanisms of uptake: approximately equal uptake of both sugars and the preferential uptake of glucose to fructose (Orlowski & Barford, 1987). Within a batch, there can be a switch from one system to the other that does not appear to involve protein synthesis. How far this observation relates to those above is completely uncertain; however, it indicates the complexity of hexose transport in *S. cerevisiae*, of which the presence of several transport proteins is only one aspect.

Readers should consult the review by Bisson *et al.* (1993) for the molecular biology of the transport of glucose (and other sugars) into *S. cerevisiae* to obtain a picture of the molecular complexity of the process. These authors have pointed out that there is a question about the actual role of the several transport proteins in glucose catabolism. Some appear to be expressed at a low level, and point mutations at the relevant genes may confer pleiotropic effects. It is argued that the proteins are not acting solely in a transport role but also sensing the external concentration of glucose. The evidence for this is two-fold. First, there is indirect support for the idea from studies on mammalian cells, which have provided good evidence for the sensing role of glucose transporters. Second, there is extensive information about glucose repression in *S. cerevisiae* (Thevelein, 1991), which indicates the need for a sensing system; but it has to be admitted that there no direct evidence that the glucose transport proteins comprise the system. Bisson *et al.* rightly point out that another reason for the several glucose transporters could be the greater flexibility that might be provided for dealing with a range of environmental conditions, of which glucose concentration is likely to be an important variable.

Cells of a glucose isomeraseless mutant (Perea & Gancedo, 1978), when incubated with glucose, show a rise in cellular glucose 6-phosphate concentration to 15 mM but the transport of xylose into them is inhibited by only 20%. Thus, contrary to an earlier suggestion (Azam & Kotyk, 1969), it can be concluded that glucose 6-phosphate is not a regulatory metabolite of monosaccharide transport.

Other studies indicate that glucose (and galactose) transport are subject to catabolite repression and inactivation. The high affinity glucose transport is low during growth on high concentrations (100 mM) of glucose but high on low or normal concentrations of galactose, or on lactate plus glycerol or ethanol (Bisson & Fraenkel, 1984). Shift from high to low glucose leads to a 10-fold increase in the putative high affinity uptake (see above), an increase inhibited by 2,4-dinitrophenol and cycloheximide. The reverse can occur such that there is loss of high affinity uptake (but not sensitive to cycloheximide). Essentially similar findings were made by Busturia & Lagunas (1986). Ramos, Szkutnicka & Cirillo (1988) have shown that the low affinity transport process is constitutive. Its activity is inhibited in proportion to the extent of the high activity process, which they showed to be regulated by catabolite repression and inactivation. Genetic studies have confirmed this (Bisson, 1988) and shown that the galactose transport is also catabolite repressed (Ramos & Cirillo, 1989).

The data considered thus far lead to the following view. If we can envisage the hexose transport system in general terms, ignoring the molecular complexity, the system contains probably two components, one of which appears to be associated with (activated by?) the presence of the kinases responsible for the phosphorylation of hexose. Since the kinetic behaviour with 6-deoxy-D-glucose, which cannot be phosphorylated, seems not to be different from those hexoses that can be phosphorylated (Romano, 1982; Bisson & Fraenkel, 1983*b*), phosphorylation cannot be a part of the transport process. The inescapable conclusion is that transport of hexose across the plasma membrane of baker's yeast is by facilitated diffusion. It could be argued that biphasic kinetics are not what would be expected of such a process, but Fuhrmann *et al.* (1989) have demonstrated, from their studies on glucose transport in plasma membrane vesicles, referred to above, that biphasic kinetics are a consequence of non-steady state conditions. Indeed, using a simple kinetic model (see above) they were able to simulate the data of Bisson & Fraenkel (1983*a*) for glucose transport both in wild-type cells, with consumption by metabolism, and in the triple kinaseless mutant, in which glucose could not be metabolised. These studies again argue against the use of single flux kinetics to characterise the transport system. Fuhrmann *et al.* (1989) pointed out that, only if glucose metabolism occurs at a much faster rate than transport, such that the internal glucose concentration is kept close to zero, can single-flux kinetics be used. At high external concentrations, however, even these conditions no longer hold.

The conclusion reached here conflicts with a body of evidence that purports to show that phosphorylation is an integral part of hexose transport. Essentially the evidence for this view has been obtained by investigating the temporal order of appearance of a labelled sugar in intracellular pools. Studies of this kind have had a chequered history. Early studies (van Steveninck 1968, 1969, 1970, 1972) could be criticised for their failure to identify all the metabolic products of the labelled sugar, so that the pool sizes used to calculate specific activities could be wrong (Farcaš, Bauer & Zemek, 1969; Kotyk & Michaljaničová, 1974). Nevertheless, more recent studies avoiding these pitfalls have demonstrated that when ^{14}C-labelled 2-deoxy-D-glucose is transported into cells of baker's yeast the sugar appears first in the sugar phosphate pool (Meredith & Romano, 1977; Franzusoff & Cirillo, 1982; Beullens & Thevelein, 1990).

There is other information purporting to support this conclusion that phosphorylation of the sugar is part of the transport process.

(1) The ATP concentration of cells can be reduced by antimycin A. The uptake of 2-deoxy-D-glucose and fructose is inhibited under these conditions (Schuddemat, van den Broek & van Steveninck, 1988). The uptake velocities can be increased by raising the ATP level. The transport of 6-deoxy-D-glucose is unaffected, which might be expected; the transport of glucose and mannose is less affected, which is a little surprising.

(2) Incubation of cells with xylose and with ethanol leads to a decrease of hexokinase and glucokinase activity in the cells (Schuddematt et al., 1986). The levels of orthophosphate, ATP and polyphosphate are not inhibited by this treatment. However transport of 2-deoxy-D-glucose is inhibited; transport of 6-deoxy-D-glucose is not.

(3) When 2-deoxy-D-[1-^{14}C]glucose is transported into cells of *snf3* mutant cells, unlike in wild-type cells the first labelled compound is the free sugar and not the phosphorylated sugar.

All the above findings have been adjudged to support the view that there is transport-associated phosphorylation of sugars. But caution is needed in accepting that conclusion. First, there is uncertainty about the number of pools within the cell to which one or more of the compounds might have access. Indeed, there is good evidence, as far as hexoses are concerned, that there may be two compartments in the cell into which the sugar can enter (Kotyk & Hǎskovec, 1968; van den Broek et al., 1982). Second, one cannot assume that a sugar analogue that is either partially metabolised or indeed not metabolised does not have an effect on metabolism. Studies on the marine fungus *Dendryphiella salina* have shown that the non-metabolised sugar analogue 3-*O*-methylglucose increases the incorporation of mannitol into glycogen by preventing the turnover of this polysaccharide (McDermott & Jennings, 1976). L-Sorbose does not have this effect (Jennings & Thornton, 1984). 2-Deoxy-D-glucose has been shown to have substantial effects on the phosphorus metabolism of yeast and causes a decrease of cytoplasmic pH by 0.4 units (Loureiro-Dias & Santos, 1989). Thus, the location of metabolites in more than one compartment and the effects on metabolism could each have an effect on the pattern of incorporation of label from sugar analogues into other compounds within the cell.

Until criticisms of experiments purporting that transport involves phosphorylation have been answered, it seems wise to accept that transport of hexose into *S. cerevisiae* is by facilitated diffusion as described and argued above.

Kluyveromyces marxianus (Saccharomyces fragilis)

When *Kluyveromyces marxianus* is grown in a glucose medium, it is believed from studies using 2-deoxy-D-glucose that the hexose moves into the cell by transport-associated phosphorylation (Jaspers & van Steveninck, 1975, 1977; van den Broek & van Steveninck, 1981). While many of the problems attendant on earlier investigation of the role of phosphorylation in transport of hexoses into *S. cerevisiae* have been addressed, inevitably there must be doubt as to the correctness of the above conclusion in (a) the light of what is now known about hexose transport in *S. cerevisiae* and (b) the absence of genetic and molecular biological information of the kind used to elucidate hexose transport in that yeast. On the other hand, although alkalinisation of the medium occurs during 2-deoxy-D-glucose uptake, this has been shown not to be due to transport of the sugar by a proton symport. Indeed these particular observations are best interpreted in terms of transport involving phosphorylation of the sugar (van den Broek & van Steveninck, 1981). Confirmation of this view comes from observations of Gasnier (1987), who found that the affinity of the system for glucose and 2-deoxy-D-glucose was at least an order of magnitude greater than for 6-deoxy-D-glucose and 3-O-methylglucose, which unlike the first two sugars are either not or only slowly phosphorylated by the cells.

Under anaerobic conditions, the evidence indicates that the phosphoryl group of the 2-deoxy-D-glucose that accumulates inside the cell comes from polyphosphate; it cannot be donated by ATP (Tijssen, van den Broek & van Steveninck, 1984). It has been argued that the polyphosphate fraction involved is that located in the cell wall (Tijssen, Beekes & van Steveninck, 1981; Tijssen, Dubbelman & van Steveninck, 1983; Tijssen & van Steveninck, 1984; Tijssen, van Steveninck & de Bruijn, 1985). Under aerobic conditions on the other hand, ATP appears to be the primary phosphoryl donor (Schuddematt *et al.*, 1989*b*).

The transport system described above has been shown to be constitutive and can be considered as a glucose/fructose transporter (van den Broek *et al.*, 1986; Gasnier, 1987). The system is the only glucose transporter in cells grown on hexose until the stationary phase. During that stage, when glucose becomes exhausted from the medium, another system appears that has (unlike the glucose/fructose transporter) a high affinity for 6-deoxy-D-glucose ($K_T = 0.8$ mM, as opposed to $K_T = 90$ mM when hexose enters the cell by the glucose/fructose transporter) (van den Broek *et al.*, 1986). This high affinity system is a glucose ($K_T = 0.09$ mM)/galactose ($K_T = 0.14$ mM) transporter and is also a proton symport (van den Broek *et al.*, 1986; Gasnier, 1987). It appears to be the same as the fructose/proton symport described by van den Broek, Christianse & van Steveninck (1982). Firm evidence that the system functions as a proton symport comes from study of the system in membrane vesicles (van Leeuwen *et al.*, 1991). Under the same conditions of growth, the cells also develop a sorbose (fructose)/proton symport (de Bruijne *et al.*, 1988) and a galactoside/proton symport (van de Broek & van Steveninck, 1982; de Bruijne *et al.*, 1988).

It is believed on the basis of kinetic analysis that L-sorbose can be simultaneously transported into the cell by facilitated diffusion and by the proton symport. Kinetic equations were derived for three possible reaction sequences with respect to sugar and proton binding to the symport: random binding, and obligatory-ordered binding with either sugar or proton binding first. It was argued (van den Broek & van Steveninck, 1980) that xylose uptake by *Rhodotorula glutinis* can be explained in like manner. However, Hauer & Höfer (1982) and Severin, Langel & Höfer (1989) respectively point out that the above analysis takes no account of the possibility of an affinity change as a consequence of protonation or deprotonation of the carrier and also that the membrane potential can affect transport kinetics and is itself, as well as the transport kinetics, dependent on external pH.

The three sugar/proton transport systems described above are sensitive to catabolite inactivation, which takes place when the cells are incubated in the presence of glucose, fructose or mannose but does not require protein synthesis. Derepression of the fructose and galactoside systems occurs in the absence of these hexoses and requires protein synthesis. In the case of the glucose/galactose/proton symport there is a need for an inducer also to be present in the medium; this can be galactose, lactose, glycerol or pyruvate (de Bruijne *et al.*, 1988).

The above refers to *K. marxianus* var. *marxianus* (the strain used for the most part being IGC 3884 (CBS 397) but also IGC 2587 (Gasnier, 1987)). Other strains appear to have either facilitated transport of glucose and galactose or proton symport of these two sugars, with apparently no transport system for lactose. The disaccharide is utilised either only after prior extracellular hydrolysis or not at all (Carvalho-Silva & Spencer-Martins, 1990).

*Rhodotorula glutinis**

The lipid-forming obligate aerobic red yeast transports, by a single carrier against a concentration gradient with the involvement of metabolic energy, the following sugars and related compounds: D-xylose, L-xylose, D-glucose, 2-deoxy-D-glucose, glucosamine, D-galactose, L-rhamnose and glucuronic acid (Höfer & Kotyk, 1968; Höfer, 1970; Janda, Kotyk & Tauchová, 1976; Niemietz, Hauer & Höfer, 1981; Niemietz & Höfer, 1984; Taghikhani *et al.*, 1984). Most of the studies of the transport system have been made with D-xylose, which is not metabolised in the first 30 min of incubation. Within the limits of experimental technique, the system exhibits counter-transport for the same sugar or one sugar against another, indicating a 'mobile' carrier (Höfer, 1970). The kinetics of transport at pH 6.5 are bisphasic (Höfer & Misra, 1978), transport showing a high affinity (low K_T) system. The ability of one sugar to induce counter-transport of another and the demonstration of competitive inhibition between sugars have been the criteria by which sugars transported by the carrier have been identified.

The kinetics of the carrier are dependent upon pH (Höfer & Misra, 1978). At pH 4.5, transport of D-xylose shows only a high affinity component, whereas at pH 8.5 there is only a low affinity system and no accumulation of the sugar.

The evidence that active accumulation is by means of a proton symport is as follows:

(1) The potential across the plasma membrane determined with the lipid-soluble cations tetraphenylphosphonium (TPP$^+$) and triphenyl-

---

* Taxonomically *Rhodosporidium toruloides* but also referred to in the literature cited here as *Rhodotorula gracilis*.

phosphonium (TMP$^+$) is strongly pH dependent. Depolarisation is brought about by transportable sugars (Hauer & Höfer, 1978). Half-saturation constants for depolarisation for D-xylose and D-galactose are comparable to those for transport. Uptake of TPP$^+$ and TMP$^+$ is inhibited by anaerobic conditions, uncouplers (azide, CCCP, 2,4-dinitrophenol), and the presence of diffusible cations such as K$^+$. Transported sugars also decrease the uptake of TPP$^+$.

(2) It is possible to differentiate between the transport of sugar with protons from that without protons (Hauer & Höfer, 1982).

(a) At low pH, the stoichiometry of transport, i.e. the ratio H$^+$:sugar transported, is unity. During H$^+$ symport the influx of positive charge is compensated for by an equivalent influx of K$^+$. As the external pH is increased, there is a concomitant fall in H$^+$:sugar and K$^+$:sugar stoichiometry (Figure 5.11).

(b) At pH 7.5, when both systems (high and low affinity) are operating, it can be inferred that only the high affinity system is electrogenic. This is because the plot of the reciprocal of depolarisation ($1/\Delta V$) versus the reciprocal of sugar concentration ($1/s$) is linear, whereas the reciprocal of the rate of transport ($1/v$) versus $1/s$ gives a plot that though linear is biphasic in slope (Figure 5.12).

(3) Transport of glucosamine is energised by the membrane potential, which is partly dissipated by the transport of the amino sugar (Niemitz, Hauer & Höfer, 1981). Influx of glucosamine induces efflux of H$^+$ and to an extent K$^+$; the stoichiometry of (H$^+$ + K$^+$) efflux:glucosamine influx being unity. This seems to point to the transport of glucosamine transferring one net positive charge. The question is whether the potential is acting on the protonated carrier or the protonated substrate or both. Analysis of transport as a function of external pH that takes into account protonation of glucosamine ($pK = 7.75$) indicates that the carrier cannot distinguish between glucosamine protonated or unprotonated. A proton must therefore move in symport with the glucosamine across the membrane. In saying this, one is glossing over uncertainties about the detailed kinetics of the process in which (H$^+$ + K$^+$) efflux:glucosamine transport is unity.

(4) Glucuronate transport at the steady state, however, reflects the course of the pH gradient (Niemietz & Höfer, 1984). The transport is electroneutral, with an H$^+$:glucuronate stoichiometry of unit. The carrier – H$^+$ – glucuronate complex is electroneutral and independent of the membrane potential. Simultaneous uptake of acetic or propionic acid, which is also governed by the pH gradient across the plasma

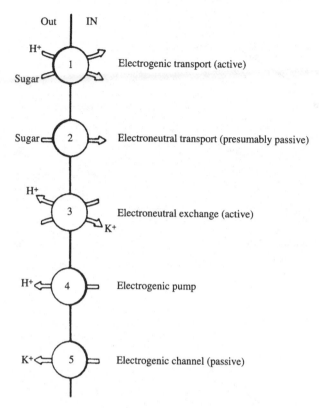

Figure 5.11. Transport systems involved in sugar transport at the plasma membrane of *Rhodotorula gracilis*. By the varying participation of the various systems it is possible to obtain the following $H^+:K^+$:sugar stoichiometries: (i) $H^+ = K^+$ = sugar if only systems 1 and 5 operate; (ii) $H^+ = K^+$ < sugar if either system 2 (at high pH) or systems 3 and/or 4 become operative. (From Hauer & Höfer, 1982.)

membrane, leads to non-competitive inhibition, i.e. the rate of transport is reduced but not the affinity of the carrier for glucuronate.

(5) Höfer, Nicolay & Robillard (1985) determined the electrochemical H^+ gradient ($\Delta\bar{\mu}_H+$) across the plasma membrane. The membrane potential difference ($\Delta\Psi$) was measured with TPP^+ at an external pH of above 4.5 (when $\Delta\Psi$ is negative) and with thiocyanate below that value (when $\Delta\Psi$ is positive). The proton gradient across the membrane was determined from internal pH measurements using ^{31}P NMR. The calculated $\Delta\bar{\mu}_H+$ was around 17–18 kJ mol^{-1}, which is sufficient to energise the 1000-fold accumulation of xylose observed by Kotyk & Höfer (1965).

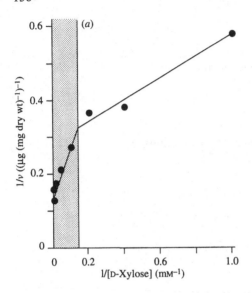

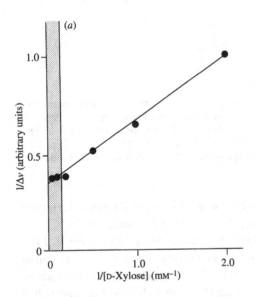

Figure 5.12. Double-reciprocal plots of (a) rate of D-xylose transport and (b) depolarisation of the plasma membrane, both as a function of D-xylose concentration in the medium for cells of *Rhodotorula gracilis*. The shaded areas in (a) and (b) indicate the region where the low affinity component was believed to be operating. K_T for the high affinity system obtained from (a) was 0.8–2 mM; K_T for depolarisation was 0.6–2 mM. (From Hauer & Höfer, 1982.)

In summary, the compounds listed above are transported into *R. glutinis* by a protonated carrier; it is only this carrier that is catalytically active. A model of the transport process has been presented by Severin *et al.* (1989). There is one observation reported above that does not seem to fit the model, namely the presence of a low affinity electroneutral component of D-xylose transport that becomes evident above pH 6 and at high substrate concentrations (Höfer & Misra, 1978; Hauer & Höfer, 1982). However, it seems likely that this represents another system transporting only D-xylose (Alcorn & Griffin, 1978; Niemietz & Höfer, 1984). Finally, it is still not certain how the transport mechanism is tightly coupled to metabolism (Höfer & Kotyk, 1968).

The alditols D-glucitol, D-mannitol, D-arabitol, L-arabitol, ribitol, xylitol and erythritol are transported by the above monosaccharide carrier, though a second carrier for the pentitols can be induced (Kloppel & Höfer, 1976*a, b*). D-Fructose is taken up by a different transport system, which is a proton symport (Höfer & Dahle, 1972; Janda *et al.*, 1976; Milbradt & Höfer, 1990; Gille, Höfer & Sigler, 1992); D-ribose seems to be transported by yet another system, which is energy dependent (Lavi, Hermiller & Griffin, 1981).

Neurospora crassa

Four transport systems for hexoses have been characterised (Scarborough, 1970*a,b*; Schneider & Wiley, 1971*a,b*; Neville, Suskind & Roseman, 1971; Klingmüller & Huh, 1972; Rand & Tatum, 1980*a,b*).

Glucose system 1 or low affinity system

The characteristics of this system have been studied with the non-metabolised 3-*O*-methylglucose (K_T in the range 8–25 mM). The sugar enters the hyphae via this system by facilitated diffusion. With glucose in the medium, the system releases previously accumulated L-sorbose.

Glucose system 2 or high affinity system

This system (with a K_T for glucose of 10–30 μM) makes its appearance in mycelium either starved or grown in low (*c.* 1 mM) glucose. The system is repressed by this hexose and to a lesser extent by fructose and xylose. Restoration of the system by transference to a low glucose medium is inhibited by cycloheximide. The system studied, using 3-*O*-methylglucose ($K_m = 70 \mu$M and $V_{max} = 4$ nmol (mg dry weight^{-1} min^{-1}), transports also

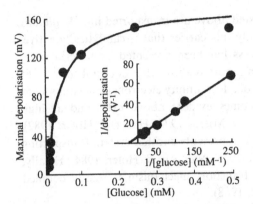

Figure 5.13. Plot of the peak (maximal) depolarisations of glucose-derepressed hyphae of *Neurospora crassa* produced by different external concentrations of glucose. Inset: double-reciprocal plot of the same data; $K_T = 42\ \mu M$. (From Slayman & Slayman, 1974.)

2-deoxy-D-glucose, 6-deoxy-D-glucose and L-sorbose. It has been argued by Rand & Tatum (1980*b*) that galactose, mannose and talose are transported by the same system.

Slayman & Slayman (1974, 1975) using intracellular microelectrodes showed that glucose entry into the hyphae via this system results in a very large depolarisation of the membrane. At saturating concentration the membrane is depolarised from -211 to -91 mV; changes in the mycelial ATP concentration could account for only 10 mV of this change in potential. A plot of peak depolarisation against extracellular glucose concentration gives a K_T of 42 μM, in good agreement with the value above (Figure 5.13). 3-*O*-methylglucose also causes depolarisation, with alkalinisation of the medium occurring at approximately the same rate as uptake of the sugar. The evidence points to a glucose/proton symport. The glucose:H^+ stoichiometry is unclear; that for 2-deoxy-D-glucose or 3-*O*-methylglucose is close to 1:1. The membrane resistance falls only slightly during glucose depolarisation, suggesting either that the system has a high internal resistance or that the current–voltage relationship in glucose-starved cells is non-linear.

Galactose system

This is a low affinity system ($K_T = 400$ mM) for galactose found in glucose-grown cells but distinct from glucose system 1 (Rand & Tatum, 1980*a*).

Fructose system

There is a high affinity system ($K_T = 0.4$ mM) for fructose that is repressed by glucose (Rand & Tatum, 1980*b*). This sugar and also 2-deoxy-D-glucose and mannose prevent derepression. However, incubation of glucose-grown mycelium with 3-*O*-methylglucose causes the appearance of five times as much fructose uptake activity as does a mineral salts medium. This might be due to the non-metabolised sugar preventing glycogen breakdown and thus reducing glucose levels as in *Dendryphiella salina* (McDermott & Jennings, 1976). The system also transports L-sorbose. Glucose, mannose and 3-*O*-methylglucose act as non-competitive inhibitors of fructose transport.

Concluding remarks

There is now a great deal of information about sugar transport in fungi. I have presented a significant amount of that information with an avowed purpose. I believe it is useful for those working with particular species to be able to locate quickly, if it is know, how sugars might be transported into that species. But a more important reason has been to demonstrate for one physiological process its complexity and diversity within the Fungal Kingdom. There is little doubt that only experimental investigation will reveal how and under what conditions a sugar is transported into a particular but as yet unstudied fungus. One cannot argue that the mechanism for the transport of one sugar is the same for another by the same species, as has been indicated in the previous section. Nor can one argue for the same mechanism for one particular sugar being present amongst species of the same genus. This is well shown by the survey by Kilian *et al.* (1991) of 21 strains representing species of the genus *Kluyveromyces* in which proton symports for one or more sugars were found in only 12 of the strains. Loureiro-Dias (1988) studied 248 strains, representing 205 yeast species, and found that only 34% of the strains, most of them belonging to the genera *Candida*, *Hansensula* and *Rhodotorula*, possessed a glucose symport. While the procedures used might not unmask a proton symport, because of inappropriate experimental conditions, the results are a useful indication of possible mechanisms.

Glucose transport in *S. cerevisiae* is the most intensively studied process. Even after nearly 40 years, there is still uncertainty about the mechanism although, as indicated, a relatively coherent picture is starting to emerge.

Molecular biological studies have indicated that more than one system may be present. One needs to stress again the need for glucose transport to be studied under a variety of growth conditions, using mutants wherever appropriate. On the other hand, there is a need to use molecular biological approaches to study those transport processes for which we seem to be obtaining a relatively clear picture of operation, e.g. glucose transport in *R. glutinis*.

Finally, there is evidence that care needs to be exercised in the use of the kinetic approach to the study of sugar transport in fungi (see next section). Anomalous results may be obtained with non-steady-state conditions (Fuhrmann *et al.*, 1989), inappropriate kinetic analysis (Fuhrmann & Völker, 1992), binding of sugar to the walls (Moore & Devadathan, 1979; Benito & Lagunas, 1992), or, possibly for filamentous fungi, when transport is being studied using non-metabolised sugars, by bursting of hyphal apices (Thornton, Galpin & Jennings, 1976). One can only stress that there is a need always for a multi-faceted approach to the study of all transport processes, not only those responsible for the uptake of sugars.

B. DISACCHARIDE UTILISATION

General

The utilisation of disaccharides and other glycosides provides a paradigm for the utilisation by fungi of many low molecular weight compounds that are not primary metabolites. The molecule may be transported intact across the plasma membrane or it may first be hydrolysed externally and the products then transported inside. One must not assume that the two possibilities are mutually exclusive.

Much of the work on the utilisation of disaccharides has been carried out on yeasts. The ability of yeasts to grow on disaccharides (and the trisaccharide raffinose) and other glycosides is given in Table 5.10. Well over half the species can utilise sucrose, maltose, trehalose and cellobiose of the naturally occurring disaccharides; a considerable number utilise raffinose. Lactose is much less frequently used. The widespread ability to utilise sucrose is probably a reflection of the natural habitat of many yeasts, namely plant nectars and exudates. Interest in disaccharide utilisation by yeasts is of biotechnological importance, so there is much information on the topic, particularly for *S. cerevisiae*. Valuable reviews on disaccharide utilisation by yeasts have been produced by Arnold (1987)

Table 5.10. *The ability of 439 species of yeast to utilise certain glycosides,*
D-glucose and D-galactose, aerobically or anaerobically

Carbohydrate	Aerobic growth[a]			Anaerobic fermentation[a]		
	+	−	?	+	−	?
D-Glucose	439	0	0	207	148	84
D-Galactose	221	148	70	49	310	80
Sucrose	258	152	29	80	295	64
Maltose	231	178	30	24	343	72
Methyl α-D-glucopyranoside	139	226	74	5	303	131
Melezitose	167	227	45	5	335	99
α,α-Trehalose	274	94	71	27	254	158
Cellobiose	243	135	61	9	290	140
Salicin	222	136	81	0	0	439
Arbutin	240	132	67	0	0	439
Melibiose	52	365	22	9	402	28
Lactose	51	342	46	3	429	7
Raffinose	144	256	39	47	343	49
Inulin	20	384	35	0	0	439
Starch	78	312	49	0	0	439

Information computed by R. W. Payne from the compilation of Barnett *et al.*
(1979). These results differ from those given by Barnett (1976) chiefly because (a)
newly described species are included, and (b) many previously described species
have been combined. The figures in the table are the numbers of species.
[a] +, Sugar utilised; −, sugar not utilised. Under the symbol '?' are those species
for which a definite + or − cannot be given, for one or more of the following
reasons: (a) some strains of the species are + and others −; (b) utilisation was
given an unequivocal qualification, such as 'weak', 'slow', or 'delayed'; or (c) there
are no results for that species in the compilation.
From Barnett (1981), From *Advances in Carbohydrate Chemistry and Biochemistry*
39, 347–404, reproduced by permission of John Wiley and Sons Limited.

and Barnett (1981). However, while much of the information that we have
is for yeasts, there is much information on the ability of filamentous fungi
to utilise disaccharides (Hawker, 1950; Cochrane, 1963). However, there
is very little information for such fungi about the steps leading to the
production of hexose molecules in the protoplasm. Where relevant,
information for filamentous fungi is indicated in the literature, but virtually
all attention is given to the information for yeasts. This imbalance needs
to be corrected by studies on filamentous fungi.

In any glucoside, there is at least one monosaccharide group – the
glycon. This is combined through an oxygenation to another residue, the
aglycon (which may be another monosaccharide). The aglycon can be

chromogenic, e.g. *p*-nitrophenol. On hydrolysis, the chromogen is released and can be measured optically. Non-metabolised analogues have been produced, however, in which the oxygen atom bridging glycon and aglycon is replaced by sulphur. Since such analogues cannot be hydrolysed they are very valuable for establishing whether uptake of a disaccharide occurs without prior hydrolysis, as is discussed below.

The evidence that external hydrolysis of the disaccharide or glycoside is required for its utilisation is as follows:

(1) The similarity of the effect of pH of the medium on *in vitro* activity of the isolated glycosidase to glycoside utilisation by the whole cell (Myrbäck & Willstaedt, 1955).

(2) Differential sensitivity to pH of glycoside loss from the medium and total glycon and aglycon absorbed by the fungus (Harley & Smith, 1956; Harley & Jennings, 1958).

(3) The lack of effect of inhibitors of transport across the plasma membrane such as UO_2^{2+} on glycoside hydrolysis (Demis *et al.*, 1954).

(4) Hexose liberated by external glycosidase activity can be trapped as hexose phosphate when hexokinase is added to the external medium (Sols & de la Fuente, 1961; de la Fuente & Sols, 1962).

(5) When protoplasts are produced, glycosidase is liberated into the external medium. The protoplasts themselves have low levels of glycosidase activity but can utilise glucose (Burger, Bacon & Bacon, 1958; Friis & Ottolenghi, 1959*a*, *b*; Sutton & Lampen, 1962; Reyes & Ruiz-Herrera, 1972).

(6) Isolated walls obtained by mechanical disruption of cells contain glycosidase (Sutton & Lampen, 1962; Reyes & Ruiz-Herrera, 1972).

One can be certain that the following glycosidases can be present in fungal cell walls (the term 'periplasm' has been used for the location of such enzymes but it is arguable that a periplasm, as applied to bacteria, exists in fungi): β-D-fructofuranosidase; α-D-galactopyranosidase; β-D-glycosidase and trehalase.

β-D-Fructofuranosidase

(EC 3.2.1.26, β-D-fructofuranoside fructohydrolase, invertase). This enzyme catalyses the reaction:

R-β-fructofuranoside + H_2O → fructose + ROH

α-Furanosides and fructopyranosides are not used as substrates, although the enzyme does act on methyl-β-D-fructofuranoside, sucrose and the oligosaccharides based on sucrose but containing successively one additional galactosyl moiety (1→6, α-linked), namely raffinose, stachyose and verbascose. Cleavage is on the fructosidic linkage on the fructose side of the oxygen bridge. The bulk of the literature concerns the enzyme from *S. cerevisiae*. The enzyme located in the wall has a molecular mass of 270 kDa and is a mannoprotein of which about 50% is carbohydrate (Neumann & Lampen, 1967; Gascón, Neumann & Lampen, 1968; Andersen & Jørgensen, 1969). The enzyme from *C. utilis* appears to be similar (Iizuka, Chiura & Yamamoto, 1978). In *S. cerevisiae* β-D-fructofuranosidase is repressed by glucose, although there is some disagreement about the effective concentration (Dodyk & Rothstein, 1964; Gascón & Ottolenghi, 1972; Meyer & Matile, 1974).

α-D-Galactopyranosidase

(EC 3.2.1.23, α-D-galactoside galactohydrolase, melibiase). It has been known for a long time that bottom (lager) yeasts, originally *Saccharomyces carlsbergensis*, now *S. cerevisiae*, are a good source of α-D-galactosidase (Fischer & Linder, 1895). Such yeasts can utilise melibiose (*O*-α-D-galactopyranosyl-(1→6)-α-D-glycopyranose) and raffinose (see above) completely (Arnold, 1981*b*). Top (ale) yeasts, which contain only β-D-furanosidase, can utilise raffinose but only the fructose moiety; melibiose accumulates in the medium (Arnold, 1981*b*). (The genetic constitution of *S. uvarum* can be such that there is a range of strains differing in the sugar or sugars produced from raffinose and in whether those sugars can be utilised (Table 5.11).) The α-D-galactosidase of *S. cerevisiae* is an inducible enzyme and is a glycoprotein containing about 60% carbohydrate, with a molecular mass of about 300 kDa (Lazo, Ochoa & Gascón, 1978). An α-D-galactosidase has been shown to be present in *Cryptococcus laurentii* (Bhanot & Brown, 1980) and *Pichia guilliermondii* (Church, Meyers & Srinivasen, 1980).

β-D-Glucosidase

(EC 3.2.1.21, β-D-glycoside glucohydrolase, cellobiase) (see also pp. 150–6). *Candida wickerhamii* has been shown to be able to ferment cellobiose to ethanol (Freer & Detroy, 1983). This is brought about by a wall β-D-glucosidase and the glucose moieties transported into the cell (Freer & Detroy, 1985). Properties of the enzyme have been described by Leclerc *et al.* (1984). The molecular mass of the enzyme determined by gel filtration is about 130 kDa and it comprises two subunits (molecular mass 98 and 38 kDa). The value for the molecular mass suggests that it might not be

Table 5.11. *Hydrolysis and utilisation of raffinose by cultures of Saccharomyces uvarum of differing genotype*[a]

	Genotype							
	MAL	MEL	SUC	MEL GAL	MEL MAL	MEL SUC	MEL MAL GEL	MEL SUC GAL
Enzymes present[b]								
α-D-Glucosidase	+	−	−	−	+	−	+	−
α-D-Galactosidase	−	+	−	+	+	+	+	+
β-D-Fructofuranosidase	−	−	+	−	−	+	−	+
?Galactokinase[c]	−	−	−	+	−	−	+	+
Sugars produced by hydrolysis of raffinose[d]								
D-Fructose	0	0	PU	0	PU	PU	PU	PU
D-Glucose	0	0	0	0	PU	PU	PU	PU
D-Galactose	0	P	0	PU	P	P	PU	PU
Sucrose	0	P	0	P	0	0	0	0
Melibiose	0	0	P	0	0	0	0	0

[a] According to Losada (1957).

[b] +, Enzyme present; −, enzyme not present.

[c] The gene *GAL*, for D-galactose utilisation, may refer to one or more genes controlling enzymes of the Leloir pathway (Barnett, 1976).

[d] 0, sugar not produced; P, sugar produced, but not utilised by yeast; PU, sugar produced, and utilised by yeast.

From Barnett (1981). From *Advances in Carbohydrate Chemistry and Biochemistry* **39**, 347–404, reproduced by permission of John Wiley and Sons Limited.

a glycoprotein; loss of the enzyme into the medium could be taken as support for this possibility. The expression of β-D-glucosidase is glucose repressed but only under aerobic conditions (Freer & Detroy, 1985). Other yeasts, namely *Brettanomyces anomolans, B. claussenii* and *Dekkera intermedia* have been shown also to ferment cellobiose to ethanol, though not as effectively as does *C. wickerhamii* (Sills & Stewart, 1982), but they have not been investigated enzymically. The value of cellobiose-fermenting yeasts lies in their ability to remove the disaccharide in any process of cellulose breakdown using a cellulase complex such as that from *Trichoderma reesei*. The glucanase activity in such cultures is inhibited by cellobiose. The presence of β-D-glucosidase activity, as provided by a yeast, in the culture removes this problem.

α,α'-Glucosidase 1-glucohydrolase

(EC 3.2.1.28, trehalase). Relatively few species assimilate trehalose without delay. An example is *Torulopsis glabrata*, which has a constitutive trehalase in the wall, as demonstrated by the use of inhibitors that block glucose uptake, but have no effect on trehalose hydrolysis by the cells (Arnold, 1979).

Disaccharide transport

In the situation in which a disaccharide can be transported unchanged into a fungus, almost all studies are on yeasts. Early studies focused on maltose and α-methylglucoside transport. It has been argued that the two glycosides are transported by the same carrier in *S. cerevisiae* (Hautera & Lövgren, 1975) but Okada & Halvorson (1964a) have argued otherwise on the basis of genetic evidence. Consider α-methylglucoside: five genes are known to be involved in its utilisation. *MGL4* in combination with *MGL1*, a maltose gene, or *MGL2* in combination with *MGL1* or *MGL3* will enable α-methylglucoside to be metabolised. *MGL2* appears to control transport (Okada & Halvorson, 1964a). *mgl2* strains can transport and ferment maltose but are impermeable to α-methylglucoside. Harris & Thompson (1961) have shown that maltose-grown cells, but not α-methylglucoside-grown cells, accumulate the disaccharide. Nevertheless α-methylglucoside inhibits maltose transport from maltose-grown cells. However, the former does not cause release of maltose into cells that are preloaded with maltose (Harris & Thompson, 1960), which is in keeping with the inability of the cell to take in α-methylglucoside. Though there are maltose transport mutants (Goldenthal, Cohen & Marmur, 1983), there has been no investigation into whether they can accumulate α-methylglucoside. However a maltose-negative strain of *K. marxianus* accumulates α-methylglucoside though impermeable to maltose (Burger, Hejmová & Kleinzeller, 1959).

Okada & Halvorson (1964*a*, *b*) studied the uptake of the non-metabolised analogue of α-methylglucoside, ethyl-1-thio-α-D-pyranoside (α-TEG). The compound moves into non-induced cells by facilitated diffusion. Accumulation is not observed and metabolic inhibitors have no effect on the process. Glucose and α-methylglucoside inhibit the uptake of α-TEG and also displace previously absorbed α-TEG from the cells. On the other hand, when cells are growing in α-methylglucoside they accumulate α-TEG, a process inhibited by azide. Loss of α-TEG previously accumulated by this active process is stimulated by exogenous α-TEG and α-methylglucose but their effect is inhibited by azide. Both facilitated diffusion and active transport obey Michaelis kinetics, with K_T values of 50 and 1.8 mM, respectively. The two transport systems have the same rate of efflux of α-TEG in the presence of α-methylglucose, indicating, along with other evidence, that the exit reaction is constitutive and energy independent, while the entry reaction is inducible and energy dependent.

van Steveninck (1970) proposed that α-methyl glucoside is transported into *S. cerevisiae* by the maltose permease involving the phosphorylation of the carbohydrate. This view has been contested by Brocklehurst, Gardner & Eddy (1977) and Kotyk & Michaljaničová (1979). The former investigators showed that, in the presence of antimycin and iodoacetamide, α-methylglucoside is absorbed but not phosphorylated. In fact the process is a proton symport with one equivalent of protons absorbed with each equivalent of α-methylglucoside molecules and charge balance maintained by the loss of one equivalent of potassium ions from the cell. Dinitrophenol and a more alkaline pH in the medium both reduce uptake of the glucoside, possibly by reducing the proton-motive force. Brocklehurst *et al.* (1977) produced evidence that α-thioethylglucoside and maltose are also substrates for proton symports. The inhibition of maltose fermentation in *S. cerevisiae* by polyene antibiotics is in keeping with the mechanism for maltose transport proposed above (Palacios & Serrano, 1978).

The evidence above indicated that maltose is transported into cells of *S. cerevisiae* by a system separate from that which transports α-methylglucoside. This has now been confirmed by the following. Fermentation of the disaccharide by the yeast requires the presence of any one of a multigene family of five unlinked loci (*MAL1*, *MAL2*, *MAL3*, *MAL4* and *MAL6*) (Barnett, 1976; Charron *et al.*, 1989). The *MAL* loci have a high degree of sequence identity exhibiting only a few restriction site polymorphisms. Each locus is complex, containing three genes required for maltose fermentation: *GENE 1, 2* and *3*, respectively those for the transport system, maltase and the *trans*-acting *MAL* activator (Needleman *et al.*, 1984; Charron *et al.*, 1989). A two-digit numbering system is used to distinguish the three *GENE*s. The gene *MAL61* is the *GENE 1* function mapping to the *MAL6* locus and it encodes the maltose transport system. The protein sequence of the system has now been established and it is very similar to the protein encoded by the *SNF3* gene of *S. cerevisiae* (see p. 127) and *Kluyveromyces lactis* lactose transport protein (see below).

It has been suggested that maltose transport in *S. cerevisiae* consists of two components with K_T values of 4 and 70 mM, respectively (Busturia & Lagunas, 1985). It is now believed that the low affinity component is an artefact due to non-specific binding of maltose to the cell wall and to the plasma membrane (Benito & Lagunas, 1992).

Maltose transport in *S. cerevisiae* is inactivated by glucose in a manner rather like glucose transport (see p. 99) (Gorts, 1969; Alonso & Kotyk, 1978; Busturia & Lagunas, 1985; Peinado & Loureiro-Dias, 1986). The inactivation in growing cells is largely a decrease in activity, specifically in the proton symport. This recovers on starvation, even in the presence of cycloheximide (Peinado & Loureiro-Dias, 1986). There is a loss of capacity but this is not due to proteolysis and therefore degradation of the carrier. The loss can be explained solely by dilution due to growth in glucose. In resting cells, there is a decrease in capacity that is enhanced non-specifically by the presence of a sugar, glucose being more effective than maltose (Busturia & Lagunas, 1985). Ethanol, however, prevents inactivation. Protein degradation clearly takes place during this inactivation, since the recovery of activity in complete medium containing maltose is inhibited by cycloheximide. As seems to be the case for glucose transport (p. 100), the amino acid sequence of the transport protein has revealed stretches that are likely to be cut by proteolytic enzymes. α-Glucosidase activity of cells, originally grown with maltose as carbon source, appears to be unaffected in cells growing on glucose (Gorts, 1969). A similar situation with respect to the maltose utilisation in *S. cerevisiae* just described appears to exist in *C. utilis* (Peinado, Barbero & van Uden, 1987).

α,α'-Trehalose also is absorbed without prior hydrolysis by *S. cerevisiae* (Kotyk & Michaljaničová, 1979). What information is available suggests that several carriers may be involved, possibly one being a proton symport. The uptake systems decay after growth on glucose but can be reactivated with trehalose, maltose, α-methyl glucoside, glucose or ethanol. This seemingly curious effect and other uncertainties demand that trehalose uptake be investigated in more detail.

In *K. marxianus* transport of lactose (and that of the non-metabolisable analogues methyl-β-D-thiogalactoside and *p*-nitrophenyl-β-D-galactoside) is via a proton symport (van den Broek & van Steveninck, 1982; van den Broek, de Bruijne & van Steveninck, 1987). The transport of lactose in *K. lactis* appears to be similar (Dickson & Barr, 1983). The gene for the transport protein has been sequenced and the protein shows similarity to other sugar transport proteins (Chang & Dickson, 1988).

C. CARBON POLYMER UTILISATION

Introduction

Green plants have the capability *par excellence* to synthesise carbon polymers using light energy from the sun. These polymers are used in striking manner for structural purposes. Cellulose, hemicelluloses and lignin are polymers used in such manner. Polysaccharides such as starch can function as stores of metabolic energy. In nature, fungi demonstrate an extensive capability to degrade such polymers – either as an immediate source of carbon or because the cell wall of the green plant has to be breached so that the nutrients in the protoplasm can be accessed, as is the case with a plant pathogen. There are, of course, carbon polymers of bacterial and animal origin that can be attacked by fungi, a prime example being chitin.

It is not the purpose of this survey to attempt to cover all aspects. For instance, there are numerous examples of the ability of fungi to attack cellulose. To attempt to cover the extensive literature would be counterproductive. More importantly, demonstrations of the ability of a fungus to degrade a particular polysaccharide do not help much in assessing how the same fungus utilises it when the sugar is incorporated into an organised structure such as the plant cell wall. Even in its primary state, the wall is a complex structure involving a number of other polysaccharides as well as cellulose. Secondary walls can be more complex, particularly if they are lignified. One might expect that with complexity there would be variation in the manner by which wood is attacked by fungi. This is in fact the case. But whatever the method of attack, the fungus faces material that is chemically complex and physically refractory, not least of the problems being the insolubility of the material.

Studies in culture using pure or relatively pure polysaccharides as carbon sources can be more helpful biotechnologically, since polysaccharide waste may be much less complex than the parent material. Nevertheless, when one is dealing with material as seemingly simple as pure cellulose, there is still sufficient complexity to make the mode of attack by a fungus difficult to analyse.

For all the above reasons, I have decided to focus on those situations in which the mechanisms used by fungi to degrade cell wall and other polymeric material have been analysed with some degree of clarity.

Breakdown of cellulose by *Trichoderma* species

Many fungi are notable for their ability to secrete cellulolytic enzymes into the culture medium. Of those fungi, *Trichoderma* has been the most studied. Virtually all work on *Trichoderma* over the past 50 years has developed from studies on a strain of *T. reesei* isolated from the cotton canvas of a US Army portable tent brought from Bougainville Island in the Solomon Islands in the Second World War and studied at the US Army Natick Laboratories (Reese & Mandels, 1984). That strain, QM6a (originally *T. viride*) (Mandels & Reese, 1964), has been used extensively and has been developed by selection and mutagenesis (El Gogary *et al.*, 1990) but *T. koningii* and *T. viride* have been investigated also. The strains developed from QM6a are now capable of yielding up to $50 \, g \, l^{-1}$ of cellobiohydrolase I, which is part of the cellulolytic system of *T. reesei*.

Other aerobic fungi known to produce cellulose-degrading enzymes in similar manner are *Aspergillus niger* (Ogawa & Toyama, 1964), *Chaetomium globosum* (Watanabe, 1968*a, b*), *Fusarium solani* (Wood, 1969), *Irpex lacteus* (Kanda, Wakabayashi & Nisizawa, 1976*a–c*), *Myrothecium verrucaria* (Mandels & Reese, 1964; Selby & Maitland, 1965), *Neurospora crassa* (Yazdi, Woodward & Radford, 1990), *Penicillium pinophilum* (Wood & McCrae, 1978), *Sclerotium rolfsii* (Lachke & Deshpande, 1988), *Sporotrichum pulverulentum* (*Chrysosporium lignosum*, Eriksson & Rzedowski, 1969*a, b*), *Talaromyces emersonii* (Folan & Coughlan, 1979).

There are many reviews concerned with the breakdown of cellulose by *Trichoderma*. The following can be recommended: Lee & Fan (1980), Lee, Fan & Fan (1980), Ladisch *et al.* (1983), Coughlan (1985), Enari & Niku-Päavola (1987), Eveleigh (1987), Wood *et al.* (1988), Wood (1989), Goyal, Ghosh & Eveleigh (1991), Walker & Wilson (1991). It should be noted that almost every review just cited contains information about the cellulolytic activities of fungi other than *Trichoderma* spp., but the bulk of

Table 5.12. Enzymes of the cellulase complex

Systematic name	Enzyme no.	Trivial names	Substrate and product
(1→4)-β-D-Glucan cellobiohydrolase	EC 3.2.1.91	Exoglucanase, cellobiohydrolase	Crystalline cellulose cellobiose
endo-(1→4)-β-D-glucan 4-glucanohydrolase	EC 3.2.1.4	Endoglucanase, β-glucanase	Amorphous cellulose cellooligosaccharides
β-D-Glucoside glucohydrolase	EC 3.2.1.21	Cellobiase, β-glucosidase	Cellobiose, triose glucose

From Jeffries (1987). Reproduced with permission from *Wood and Cellulosics: Industrial Utilisation, Biotechnology, Structure and Properties* by Kennedy et al., published in 1987 by Ellis Horwood Limited, Chichester.

the studies referred to concern this particular genus. However, there is now also a volume devoted solely to the cellulases of *T. reesei* (Kubicek *et al.,* 1990). Work on the molecular biology of cellulose degradation, with much detail for *T. reesei,* has been reviewed by Béguin (1990) and Nevalainen *et al.* (1991).

The release of cellulolytic enzymes into the medium allows the investigator more favourable opportunities to study the action of the relevant enzymes in the pure state, though even then there are still problems in enzyme purification. The consensus has been that the cellulolytic activity of *Trichoderma* is brought about by three types of enzyme (Table 5.12):

(1) $(1\rightarrow4)$-β-D-Glucan cellobiohydrolase.
(2) *endo-*$(1\rightarrow4)$-β-D-Glucan 4-glucanohydrolase.
(3) β-D-Glucoside glucohydrolase.

In actuality the situation appears to be more complex. It has been found that each type of enzyme exists in multiple forms and that the various types of enzyme act synergistically to solubilise crystalline cellulose.

Taking *T. reesei* as an example, it can be demonstrated readily that the culture medium can contain a spectrum of proteins after growth of the mycelium (Figure 5.14). To obtain a pure protein, several successive purification steps are required (Enari & Niku-Päavola, 1987). Difficulties in purification are caused by the close similarity in physicochemical properties of these extracellular proteins. The proteins can interact with each other during separation, and separation methods based on ionic properties and molecular size are not very effective. Furthermore, consideration of the possible methods is complicated by the frequent absence of mention of the growth conditions used. The composition of the culture medium affects the amount and nature of the non-cellulolytic protein produced extracellularly by the fungus. The situation is further complicated by the use of different species and strains and particularly by the use of commercial preparations, which may contain stabilisers. As well as this, the purity of a preparation is not easy to define because the substrates available to establish purity, i.e. the demonstration of clearly defined enzyme activity associated with homogeneous protein, are themselves heterogeneous and ill defined. Finally, the conditions used to assay enzyme activity lack standardisation (Table 5.13). Only the β-glucosidase assay using cellobiose can be considered acceptable in strict biochemical terms, i.e. a specific, well-defined substrate is used and the initial reaction rate is determined. In the assay for endoglucanase a change

Table 5.13. *Assay substrates for cellulases*

Enzyme	Assay substrate	Detection
Total cellulase	Crystalline cellulose[a]	RS[b,c,d]
	Practical substrates[e]	RS[b,c,d]
Cellobiohydrolase (CBH)	Avicel	RS[b,d]
	Methylumbelliferyl-β-D-cellobioside (MUC)[f]	Fluorescence
	Methylumbelliferyl-β-D-lactoside[f]	
Endoglucanase	Cellodextrins	RS[b]
	Substituted soluble celluloses (HEC, CMC)[g]	V[h], RS[b]
	Barley-β-glucan	Colour
	Trinitrophenyl-cellulose	Colour
	Ostazin Brilliant Red H3B-HEC	
Glucohydrolase	Oligodextrins	RS[b,d,i]
Cellobiase	Cellobiose	RS[b,d,i]
	p-Nitrophenyl-β-D-glucoside	Colour
	X-Glu (5-bromo-4-chloro-3-indolyl-β-D-glucoside)	Colour
	Methylumbelliferyl-β-D-glucose	Fluorescence

[a] Cotton, *Valonia* cell walls.
[b] RS, reducing sugars.
[c] Decrease in turbidity, loss of dry weight.
[d] HPLC analysis for cellobiose or glucose.
[e] Filter paper, Avicel, bacterial cellulose, acid-swollen cellulose.
[f] Methylumbelliferyl glycosides are variably susceptible to different enzymes. CBH I attacks both holo- and aglycone-bonds; CBH II binds to MUC but will not cleave it.
[g] Hydroxyethyl-cellulose and carboxymethyl-cellulose.
[h] V, Viscometric assay.
[i] Analysis of glucose production by glucose oxidase.
From Goyal *et al.* (1991).

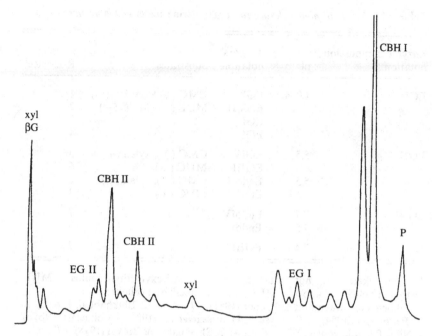

Figure 5.14. The proteins in the medium of a culture of *Trichoderma reesei*; the fractionation is according to pI values. Enzymes: xyl, xylanase; βG: β-glucosidase; EG, endoglucanase; CBH, cellobiohydrolase; P, protease. (From Enari, T.-M. & Niku-Päävola, M.-L. (1987), reprinted with permission from *CRC Critical Reviews in Biotechnology*, Copyright CRC Press, Inc., Boca Raton, FL.)

in the viscosity of the substrate is determined. An initial reaction rate can be obtained but the substrate, a substituted cellulose, lacks standardisation.

Again considering *T. reesei*, the present situation seems to indicate the following. There are certainly two cellobiohydrolases (Gritzali & Brown, 1979; Fågerstam & Pettersson, 1980; Nummi *et al.*, 1983), a number of endoglucanases (Table 5.14) and a cellobiase. A typical commercial cellulase preparation can contain three cellobiohydrolases, six endoglucanases and one cellobiase. The basis for this multiplicity of forms can be either artefactual or based on genetic factors. There can be differing degrees of glycosylation and proteolysis both intracellularly and extracellularly. It is possible that post-translational modification to give different forms might be a process by which more flexible cellulolytic activity might be generated. Clarification of this matter of multiplicity of components is likely to come from molecular and genetic studies (Knowles *et al.*, 1988, 1989; Claeyssens & Tomme, 1990; Henrissat & Mornon, 1990; Teeri *et al.*, 1990).

Table 5.14. *Range of nomenclature for* Trichoderma reesei *endoglucanases*

Enzyme commission notation	pI	Authors' notation	Substrates[a]	Reference[b]
EGI	4.0–4.7	EGI	CMC (+ +), xylan (−)	1
		EndoII	MUL (−), MUC (−)	2
		EGI		3
		EGI		4
EGII	5.5	EGIV	CMC (+), xylan (+)	5
		EGIII	MUC (+)	6
	5.5	EndoIII	CMC (+), xylan (+)	2
		EGII	MUC (+)	3
EGIII	7.7	EndoIV		7
	7.5	EndoI		8
Others	7.4	EGIII		7

[a] +, attacked; −, not attacked. CMC, carboxymethyl-cellulose; MUL, methylumbelliferyl lactoside; MUC, methylumbelliferyl cellobioside.
[b] 1, Shoemaker, Raymond & Bruner (1981) – *T. reesei*. 2, Bhikhabhai, Johansson & Pettersson (1984) – *T. reesei*. 3, Henrissat *et al.* (1985) – Celluclast (Novo). 4, Niku-Päavola *et al.* (1985) – *T. reesei*. 5, Shoemaker & Brown (1978) – *T. viride*. 6, Saloheimo *et al.* (1988) – *T. reesei*. 7, Beldman *et al.* (1985) – *Trichoderma* sp. 8, Hakansson *et al.* (1978) – *T. reesei*.
From Goyal *et al.* (1991); modified from Teeri (1987) and Kyriacou, Mackenzie & Neufeld (1987).

The first model of cellulolytic activity is that of Reese, Siu & Levinson (1950): *the* $(C_1$–$C_x)$ *concept* (Figure 5.15). The conversion of highly ordered crystalline/semi-crystalline native cellulose to soluble sugar was conceived to be a two-step process. A C_1 enzyme 'activates' or dissaggregates the cellulose chains and C_x enzymes bring about depolymerisation. This initial stage of cellulose digestion has been termed *amorphogenesis* (Coughlan, 1985); it need not be enzymic (see pp. 158–60 on cellulose degradation by brown-rot fungi).

In terms of the present position there are a number of schemes for cellulose breakdown by fungi (Goyal *et al.*, 1991). For *T. reesei*, there seem to be two possibilities.

(1) The initial attack is by endoglucanases followed by combined attack by cellobiohydrolases and endoglucanases, with the final hydrolysis of the small oligosaccharides to glucose by cellobiase (Eveleigh, 1987).
(2) The combined action of cellobiohydrolases is sufficient to yield complete hydrolysis of cellulose. In this case, the endoglucanases act

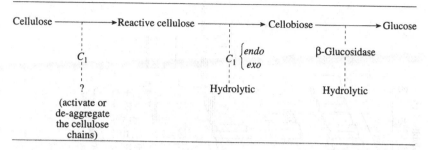

Figure 5.15. The process of degradation of cellulose according to the C_1–C_x concept. (From Lee *et al.*, 1980.)

only on the solubilised material, such as cellodextrins, to yield cellobiose. This cellobiose, together with that produced by the action of cellobiohydrolases, is converted to glucose by cellobiase (Enari & Niku-Päavola, 1987).

Descriptions of this kind of cellulose breakdown imply that action of one enzyme is independent of another; for this reason the above models are too simplistic. Indeed we have known for well over 30 years that there is synergism between the enzymes (Gilligan & Reese, 1954). Synergism is very high when highly ordered cellulose is the substrate. However, the mechanistic basis of synergism is not well understood (Wood, 1989). Figure 5.16 shows the situation to be complicated. In the example shown, both cellobiohydrolases needed to be present, as well as the four endoglucanases isolated by the investigator (Wood, 1989), for maximal activity. Interpretation of the data demands that stereochemical considerations be taken into account, clearly important when dealing with a complex structure such as cellulose.

Adsorption and desorption of the enzymes on the cellulose surface are factors which must be considered in establishing any kinetic description of the cellulolysis of cellulose (Ryu, Kim & Mandels, 1984). The situation is complicated. Traditionally, enzyme kinetic studies have been concerned primarily with homogeneous systems. However, many important enzyme reactions take place in heterogeneous systems. Such is the frequent situation with microbial enzymes involved in the solubilisation of organic material. A detailed understanding of the kinetics of the enzymic breakdown of cellulose therefore not only is important in itself and biotechnologically but also is highly relevant to other situations in which fungi produce enzymes to attack heterogeneous substrates.

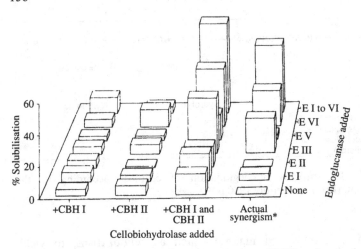

Figure 5.16. Synergism between cellobiohydrolases (CBH) and endoglucanases (E) in solubilising cellulose (cotton). *Actual synergism means solubilisation effected by CBH I + II + endoglucanases minus solubilisation effected by CBH I + II. (From Wood, 1989.)

In any kinetic analysis a number of aspects have to be addressed (Lee *et al.*, 1980):

(1) The limitation to mass transfer of the enzyme(s) to their site of action. There will be the bulk phase resistance, resistance of the boundary layer round the cellulose fibrils and resistance within the fibrils.
(2) The adsorption and desorption of the enzyme(s).
(3) The process of hydrolysis – that process closest kinetically to that occurring in a homogeneous system.
(4) Product inhibition – cellobiose appears to be the most important compound causing such inhibition (Mandels & Reese, 1963).
(5) Change of surface during hydrolysis.

As yet, there does not seem to be a completely satisfactory model for the enzymic breakdown of cellulose that fits all situations. Nevertheless, a relatively satisfactory model has been produced to describe growth and enzyme production by *T. reesei* growing on Avicel (Scheiding *et al.*, 1986).

The cellulolytic system in *Trichoderma*, as with other fungi, is adaptive. It is triggered by growth on cellulose, sophorose (see below) and to a lesser extent lactose and cellobiose but becomes repressed once a low molecular weight readily metabolisable carbon source is available (Merivuori *et al.*, 1985). What is not clear is how an insoluble molecule such as cellulose can trigger induction of the requisite enzymes. It has been proposed that physical contact between the fungus and the polysaccharide acts as the necessary stimulus (Berg & Pettersson, 1977; Binder & Ghose, 1978). An alternative possibility is that the fungus synthesises a low level of an enzyme or enzymes that, on contact with cellulose, release a soluble inducer of increased cellulolytic activity. Reports that cellulose directly induces endoglucanase activity in glucose-grown mycelium must be viewed with caution. In all likelihood such induction represents that taking place in the spores being produced by the mycelium (Kubicek, 1987).

Vaheri, Vaheri & Kauppinen (1979), Montenecourt *et al.* (1981) and Sternberg & Mandels (1982) have shown that there is a low constitutive level of β-glucosidase in mycelium of *Trichoderma*. The enzyme is tightly bound; it cannot be released by β-(1→3)-glucanase, unlike the situation in *Aspergillus phoenicus*, where 90% of the enzyme is released (Allen & Sternberg, 1980). Also the situation is different when *Trichoderma* is grown on cellulose, when 50% of the enzyme is released into the medium (Montenecourt *et al.*, 1981). If one accepts the presence of β-glucanase as a constitutive enzyme, it must be realised that hydrolases are capable of transglycosylation (Edelman, 1956). Thus, invertase under appropriate conditions can act as a fructosyl transferase producing fructosyl-sucrose and glucose. Similar transferase activity can be seen with sugars other than sucrose. In an analogous manner, the β-glucosidase can catalyse the synthesis of sophorose (2-*O*-β-D-glucopyranosyl-D-glucose) and laminaribose (3-*O*-β-D-glucopyranosyl-D-glucose) from gentiobiose (6-*O*-β-D-glucanopyranosyl-D-glucopyranoside) (Vaheri, Leisola & Kauppinen, 1978). On this basis, it has been suggested that sophorose is the inducer of cellulolytic activity (Gritzali & Brown, 1979), a suggestion supported by indirect evidence (Kubicek, 1987). Certainly, in culture this disaccharide is by far the most potent inducer of cellulolytic activity in *T. viride* (Mandels & Reese, 1959; Mandels, Parrish & Reese, 1962). However, induction by sophorose yields only two endoglucanases. On the other hand, lactose acting through relief of carbon catabolite repression yields four and cellulose five enzymes (Messner, Gruber & Kubicek, 1988). So the situation is not simple. It should be noted that *Phanerochaete chrysosporium* (see p. 178) does not respond to sophorose (Eriksson & Hamp, 1978).

Interestingly, sophorose at submicromolar concentrations represses activity of constitutive β-glucosidase. Methyl-β-glucoside has been found to be a most effective inducer of enhanced activity over that which is constitutive; cellobiose is without effect. There is also evidence from the study of a mutant of *T. reesei* without constitutive β-glucosidase activity that, when the organism is growing on cellulose, cellobiose induces the production and secretion of β-glucosidase (Strauss & Kubicek, 1990).

Other cellulolytic systems

Rumen fungi

Anaerobic zoosporic fungi belonging to the chytridiomycetes *Neocallimastix frontalis*, *Piromonas communis* and *Sphaeromonas communis* live free in the stomach contents of ruminants such as sheep (Orpin, 1975, 1976, 1977*a*) (see p. 108). *Neocallimastix frontalis* has been the most investigated and can degrade cellulose. The evidence available (Wood *et al.*, 1988; Wood, 1989) suggests that it is likely that the mode of cellulose degradation by *N. frontalis* is similar to that of *Trichoderma*, although it is not yet clear whether a cellobiohydrolase is present. Significantly cellulolytic activity of *N. frontalis* is several-fold more than that of *Trichoderma reesei* mutant strain C-30, which has one of the most active cellulases produced by an aerobic fungus (Wood *et al.*, 1986). Readers should consult Orpin (1993) for a review of present knowledge of cellulose breakdown by anaerobic fungi.

Cellulose degradation by brown-rot fungi

Brown-rot fungi preferentially attack softwoods (Liese, 1970; Wilcox, 1970). The polysaccharide of the cell wall is broken down, leaving the lignin matrix but, as shown by chemical analysis, not unchanged. The hyphae of such wood decay fungi grow mainly in the plant cell lumen (Bravery, 1971). Agents that bring about decay must diffuse into the plant cell wall to have effect. In the early stages of decay, there is considerable reduction in the mechanical strength of the wood but little commensurate weight loss (Nilsson, 1974; Doi & Saito, 1980), which is unlike the case of white rot, where loss in strength is proportional to loss of cellulose. This confirms that the cellulose is being attached by the brown-rot fungus, with little change in the amount of lignin. Further, contrasting the pore size of uninfected wood with the size of the cellulolytic enzymes and their diffusional properties (Stone & Scallan, 1965; Cowling & Brown, 1969) suggests that some small diffusible agent must be responsible for cellulose amorphogenesis.

A likely candidate is hydrogen peroxide (H_2O_2). Halliwell (1965) showed that the action of H_2O_2 and Fe(II) (Fenton's reagent) brought about catalytic degradation of cellulose. Koenigs (1972, 1974*a*, *b*) showed that wood breakdown by Fenton's reagent was very similar to that by a brown rot but unlike that by a soft or white rot or acid hydrolysis. Essentially H_2O_2 in the presence of Fe(II) produces a highly reactive radical (˙OH) that brings about oxidative breakage of bonds within the cellulose chain. Support for this hypothesis has come from studies of cellulose degradation by *Postia* (*Poria*) *placenta*.

Highley (1977) showed that the fungus produces oxidised cellulosic degradation products. Early attempts to detect H_2O_2 in the medium were nullified by inappropriate tests (Highley, 1981) and the use of excess carbohydrate and nitrogen in the growth medium (Highley, 1987). Nevertheless the evidence of H_2O_2 production in the medium is not conclusive (Veness & Evans, 1989). More compelling is the report of Kirk *et al.* (1991), who demonstrated that the end-groups of chemically pure cellulose depolymerised by *P. placenta* were sugar acids and the same as those produced by degradation with Fenton's reagent. Though the acids were the same they did not appear to be the same relative proportions in the two treatments. Hot mineral acid (but see below) and periodate/bromine/water did not have the same effect.

While it has been estimated that there is sufficient iron in wood, it would have to be oxidised to Fe(II) to become reactive. How this takes place does not appear to have been addressed, since H_2O_2 by itself acts very slowly on cellulose (Montgomery, 1982). Doubtless Fe(II) is produced metabolically, as is the case for H_2O_2. Whatever the mechanism, the metabolic production of H_2O_2/Fe(II) provides a rationale for the important observation that *P. placenta* is able to degrade cellulose only in the presence of a readily utilisable carbon source (Highley, 1977). Finally, there is also the question of how the fungus protects itself from the powerful oxidising action of H_2O_2/Fe(II). As yet, there is no answer.

Many investigators have noted that there are similarities between acid hydrolysis and brown-rot decay of wood (Birkinshaw, Findlay & Webb, 1940). The ready ability of brown-rot fungi to produce oxalic acid (Takao, 1965) (see also Chapter 11) has led to the suggestion that, in nature, this acid may be playing an important role in cellulose breakdown (Schmidt, Whitten & Nicholas, 1981). The inability of hot mineral acid to mimic the breakdown of wood by *P. placenta* in terms of the products released, referred to above, suggests that oxalate may not have a role in brown rot. However, Green *et al.* (1991) have probed the matter in some detail, again using *P. placenta*. They showed that infected wood blocks showed a fall in pH that correlated with estimates of oxalic acid in the wood. It was also shown that oxalic acid by itself could break down hemicellulose and depolymerise cellulose in the wood. On this basis, it is believed that oxalic acid is important in the initiation of decay, providing the fungus access to arabinose and galactose from hemicellulose as well as opening up the polymers. In this way, the porosity of the wood to other low molecular weight decay agents and enzymes is increased.

With respect to enzymes, there is little detailed information but sufficient to indicate that *endo*-1,4-glucanase and β-glucosidase activity is present but exoglucanase activity is absent (Highley & Wolter, 1982; Cotoras & Agosin, 1992). The relatively crude enzyme complex isolated from *P. placenta* has been found to be able to degrade xylan as well as cellulose (Highley & Wolter, 1982). It is extremely interesting that the endoglucanase activity of *Gloephyllum trabeum* (= *Lenzites trabea*) is not repressed by glucose or glycerol but is increased four to five times in the presence of cellulose or cellobiose (Cotoras & Agosin, 1992). It needs to be remembered that brown-rot fungi attack cellulose within the lignin matrix. Although these fungi do not mineralise the polymer, they are not without effect on it. There is a little ring cleavage but the major effect is demethylation of aromatic methoxyl groups (Kirk & Adler, 1970; Kirk, 1987). Whether such changes makes the cellulose more accessible remains to be seen.

Wood degradation by white-rot fungi

Introduction

It is customary to speak of the process of wood degradation by microorganisms as lignocellulose breakdown. The term lignocellulose, although well-established in the botanical literature, came into increased prominence as a consequence of the concern to make greater use of renewable energy sources, of which wood is a major one. The term, while having virtues in defining the subtrates of central interest to those concerned at present with the biochemistry of wood decay, is nevertheless restrictive. It fails to indicate that other substrates, e.g. hemicelluloses are present, nor does it reflect the complex three-dimensional nature of the substrate that is being attacked. It has been indicated above (a) that the pore size of the walls of wood cells is such as to impede the movement of enzymes into them and (b) any kinetic analysis of the rate of wood breakdown in terms of fungal catalytic capability is inevitably highly complicated because of the complex geometry of the substrate. When considering lignocellulose breakdown in wood, the complications relate not only to the access of catalytic agents but also to the access of oxygen (and possibly to the removal of carbon dioxide).

White-rot fungi are generally associated with the decay of angiosperms and have pronounced host selectivities (Eslyn & Highley, 1976). It has been mentioned above that, when white-rot fungi attack wood, the rate of weight loss and cellulose breakdown run in parallel. White-rot fungi tend to degrade cellulose and hemicelluloses in wood at the same rate and lignin at a similar rate or slightly faster (Jeffries, 1987), which suggests integration of the process of breakdown of all wood fractions (see pp. 178–81).

It will be realised from the consideration of cellulose breakdown (see above) that elucidation of the mechanism is made difficult by the complex nature of the substrate. Understanding the process of breakdown by enzymes and other biologically produced catalytic agents is not proving easy. Lignin causes even greater difficulties. This is because as a polymer it is unlike many other biological polymers in not being essentially linear and composed of repeated subunits. The subunits of lignin are linked in a non-linear and random fashion (Figure 5.17). This structure is a

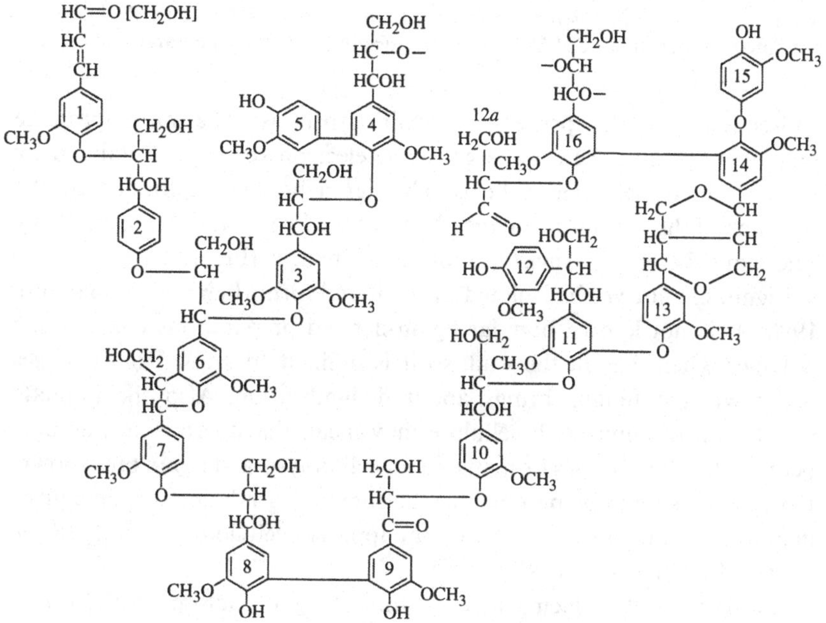

Figure 5.17. Schematic formula for a section of spruce lignin consisting of 16 units. (From Adler, 1968.)

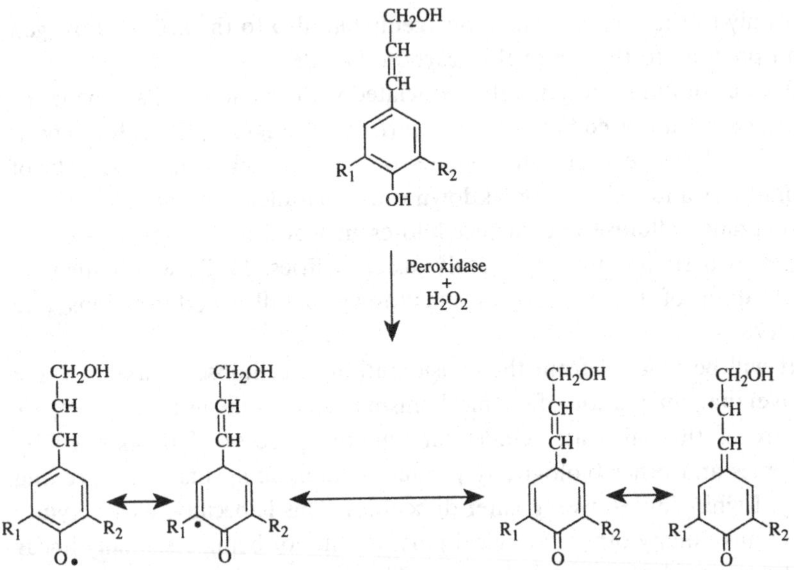

Figure 5.18. Formation of lignin precursors. The phenoxy radical precursors for chemical polymerisation of lignin are formed by the action of peroxidase on *p*-hydroxycinnamyl alcohols. $R_1 = H$, $R_2 = OCH_3$, coniferyl alcohol; $R_1 = R_2 = OCH_3$, sinapyl alcohol. (From Leisola, M. & Fiechter, A. (1985), *Advances in Biotechnological Processes*, Copyright © 1985 Alan R. Liss Inc., reprinted by permission of Wiley-Liss, a division of John Wiley and Sons, Inc.)

consequence of the process of lignin synthesis. The phenylpropane precursors (Figure 5.18) undergo a one-electron oxidation catalysed by peroxidase to yield free radicals. The lignin polymer is formed by the coupling of the various resonance forms of the free radicals (Harkin, 1976). There are therefore a variety of interunit linkages (Figure 5.19).

Lignin cannot yet be isolated in unaltered form (Lewis & Yamamoto, 1990). It is not known how far lignin-derived preparations relate to the polymer when it is in the wall, so it is difficult to know what changes occur when a fungus brings about delignification. Also the intimate association of lignin with cellulose may mean that there is an interplay between those processes bringing about cellulolysis and lignin breakdown. However, it should be pointed out that lignin degradation can take place in glucose-impregnated wood without apparent cellulolytic activity (Ruel, Barnoud & Eriksson, 1981).

Another problem facing investigators of lignocellulose breakdown is the establishment of conditions required to grow the fungus. Progress over this matter has been slow because of the initial lack of uncertainty

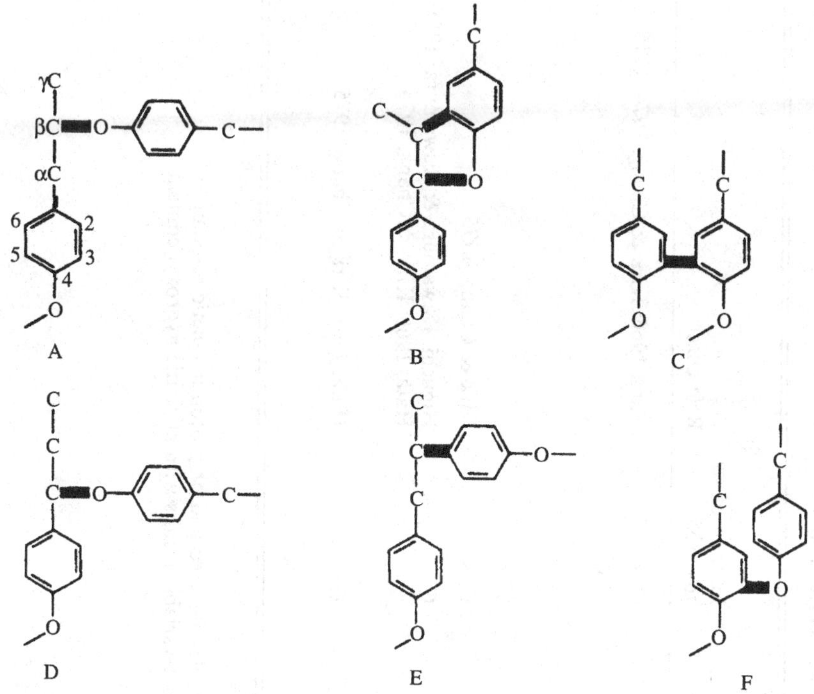

Substructure type	Proportions (% of total C₉-units)	
	Spruce	Birch
A	48	60
B	9–12	6
C	9.5–11	4.5
D	6–8	6–8
E	7	8
F	3.5–4	6.5

Figure 5.19. Frequencies of the major interunit linkages in lignin. (From Kirk, 1981.)

about both the substrate that should be used and the identity of the enzymes involved in lignin breakdown – both making for difficulties in assaying what might be happening. As is shown below, conditions have been established for growth of the relevant fungi in liquid culture, with concomitant generation of significant ligninolytic activity. Even so, it is arguable whether the growth conditions are necessarily biologically important.

Liquid culture, while having virtues in generating conditions that minimise heterogeneity, is not ideal biologically. This is because it provides

Table 5.15. *Changes in the properties of lignin caused by white-rot basidiomycete attack*

Property	Change[a]		Method of analysis[b]	References
	Increase	Decrease		
Carboxyl content	+		C, S	Hata, 1966; Kirk & Chang, 1974, 1975
Carboxyl content	+		S	
Hydroxyl content				
Aliphatic		+	C, S	
Phenolic	+[c]	+	C, S	Kirk & Chang, 1975
Aromatic content		+	C, S	Higuchi, Kawamura & Kawamura, 1955; Hata, 1966; Kirk & Chang, 1975
Yield of low mol. wt aromatic compounds on oxidative chemical degradation		+	C	
Yield of low mol. wt aromatic compounds on hydrolytic degradation		+	C	Hata, 1966; Kirk & Chang, 1975

[a] Purified sound and fungus-degraded lignins were compared.
[b] C, various chemical procedures; S, spectroscopic methods (ultraviolet, infrared and/or proton magnetic resonance).
[c] Hata (1966) reported an increase in phenolic hydroxyl content and variable results with aliphatic hydroxyl content.
From Kirk & Fenn (1982).

conditions unlike those in nature, in which the fungus is growing on a solid substrate. More significantly in nature, basidiomycete fungi produce two types of mycelium, that growing on wood and that capable of spreading away from wood through other substrates (most frequently soil) to other wood foci. The two types of mycelium are certainly anatomically different and there is evidence that they are metabolically different (Jennings, 1982, 1991b; Nuss, Jennings & Veltkamp, 1991). It is probable that, in liquid culture, both types of mycelium exist, making interpretation of results with such cultures potentially more difficult.

Finally, any mechanism of lignin degradation by a white-rot fungus must take into account the changes in the proportion of lignin occurring in wood as it rots (Table 5.15). Essentially, the attack occurs both in the aliphatic side-chains and in the aromatic nuclei still bound to the polymer. Readers should consult the references in Table 5.15 for further details.

There are numerous reviews on lignin degradation. Amongst the most recent are those by Ander & Eriksson (1978), Kirk, Higuchi & Chang (1980), Crawford (1981), Hartley, Broda & Senior (1987), Kennedy, Phillips & Williams (1987), Kirk & Farrell (1987), Zadrazil & Reininger (1988), Coughlan (1989), Blanchette (1991).

Studies with *Phanerochaete chrysosporium*

The best-studied ligninolytic fungi are *Phanerochaete chrysosporium* and *Sporotrichum pulverulentum*. The latter lacks a reliable mating system and was previously called *Chrysosporium lignorum*. On the basis of morphological comparison *S. pulverulentum* has been described as an imperfect form of *P. chrysosporium* (Burdsall & Eslyn, 1974; Eriksson & Pettersson, 1975a). DNA hybridisation studies have shown a close relationship between *P. chrysosporium* strain ME446, most used in lignin degradation studies, and *S. pulverulentum* Novobranova, again the strain most used in lignin degradation studies (Raeder & Broda, 1984). For ease of presentation, *P. chrysosporium* is used for both strains. *Phanerochaete chrysosporium* was chosen for its rapid growth in culture, its ability to degrade lignin rapidly, its prolific conidiation, its high temperature optimum for growth (40 °C) and its low phenol oxidase activity, an advantage when phenolic model compounds are being studied (Kirk, 1981). Fruiting structures can be produced in the laboratory (Gold *et al.*, 1980; Alic, Letzring & Gold, 1987).

Molecular biological studies of *P. chrysosporium* are well established (Alic & Gold, 1991; Holzbaur, Andrawis & Tien, 1991). Perhaps the most

Table 5.16. *Culture conditions that are important for the growth of and lignin degradation by Phanerochaete chrysosporium*

Parameter	Influence	References[a]
Nutritional		
Lignin	Does not influence appearance or titre of ligninolytic system: does not serve as growth substrate	1, 2
Carbon source (growth substrate)	Required for growth; not clear if lignin can support its own degradation; cellulose, glucose, xylose, glycerol, succinate, etc. are suitable carbon sources; carbon limitation can trigger ligninolytic activity	1, 3, 4, 5
Nutrient nitrogen	Cultures must be nitrogen-starved for sustained degradation of lignin; amino nitrogen, NH_4^+ are best sources for growth, although NO_3^- will serve	2, 6, 8
Other nutrients	Thiamine required for growth; balance of trace metals is important for lignin degradation; sulphur but not phosphorus limitation can trigger ligninolytic activity	5, 7
Oxygen	High concentration in culture fluid stimulatory to both development and activity of ligninolytic system	6, 8
Environmental		
pH	pH control important; optimum pH 4–4.5 for lignin degradation, probably broader for growth; certain buffers inhibit ligninolytic activity	6, 9
Temperature	Optimum for growth near 40°C; influence on ligninolytic activity not studied	10
Agitation	Growth is good in agitated or stationary cultures; agitation resulting in pellet formation strongly suppresses lignin degradation; agitation of pre-grown mats does not affect lignin degradation	6, 8

[a] 1, Kirk *et al.* (1976); 2, Keyser *et al.* (1978); 3, Hiroi & Eriksson (1976); 4, Drew & Kadam (1979); 5, Jeffries *et al.* (1981); 6, Kirk *et al.* (1978); 7, Reid (1979); 8, Yang, Effland & Kirk (1980); 9, Fenn & Kirk (1979); 10, Burdsall & Eslyn (1974). From Kirk (1981).

important observations to date from the point of view of understanding the behaviour of the fungus with respect to the development of ligninolytic activity are those of Raeder, Thompson & Broda (1989*a*, *b*). These studies have analysed the variation in ligninolytic activity by restriction fragment length polymorphism mapping and have suggested a two-stage model for the development of ligninolytic activity.

As in other fungi that bring about white rot, the hyphae of *P. chrysosporium* are surrounded by a sheath both in culture (Palmer, Murmanis & Highley, 1983) and when attacking wood (Ruel & Joseleau, 1991). Sheaths are composed of fibrillar material and in culture are most evident when hyphae are appressed against surfaces such as steel, glass, cotton or agar. Various analytical techniques have indicated the sheath is β-$(1 \rightarrow 3)$-$(1 \rightarrow 6)$-D-glucan. The use of anti-lignin peroxidase immunogold markers indicate that lignin-degrading enzymes are embedded in the matrix (Ruel & Joseleau, 1991). The sheath forms an attachment between hyphae and wood; the attachment is to an extent transitory because the sheath disappears as the attack proceeds.

At present, by far the best way to study the development of ligninolytic activity in culture is by the use of lignin labelled with ^{14}C. Classical chemical and physical methods are not suitable (Crawford, Robinson & Cheh, 1980). [^{14}C]lignins can be prepared by feeding growing plants with [^{14}C]lignin precursors (Crawford *et al.*, 1980; Haider & Trojanowski, 1980) or by synthetic methods. In the latter case [^{14}C]lignin has been prepared from ^{14}C-labelled coniferyl alcohol with peroxidase and H_2O_2 (Kirk *et al.*, 1975). Label has been inserted in the side-chain, in the aromatic nucleus and in methoxyl groups. With both types of lignin, natural and synthetic, ligninolytic activity is measured by $^{14}CO_2$ production.

Table 5.16 summarises the various characteristics of media used for growth of mycelium of *P. chrysosporium* and for lignin degradation. Some of the details given in the table are self-explanatory; others require comment.

Lignin degradation by white-rot fungi requires a co-substrate. Although the potential energy content of lignin is higher than cellulose, white-rot fungi seem unable to use lignin as either an energy or carbon source in any significant manner (Kirk, Connors & Zeikus, 1976). It has been suggested that lignin degradation rates were too low to allow substantial growth (Jeffries, Choi & Kirk, 1981). This seems less likely in view of the high rates of lignin degradation that are now achievable (Ulmer *et al.*, 1983). These increased rates are associated with increased rates of cellulose degradation (Leisola *et al.*, 1983). In all probability the energy requirement for lignin degradation is so large that a white rot fungus requires an easily accessible energy source (cellulose) for ligninolytic activity (Eriksson, Grünewald & Vallander, 1980). Although lignin does not appear to act as a carbon source, carbon from lignin can be incorporated into metabolites produced by the fungus. In particular veratryl alcohol can become labelled when *P. chrysosporium* is grown in the presence of [^{14}C]lignin (Kirk & Fenn, 1982).

When there is agitation of the culture medium leading to pellet formation (in flasks) or small fragments (in a bioreactor) there can be an absence of ligninolytic activity (Kirk *et al.*, 1978; Ulmer *et al.*, 1983). The major difference between static and agitated cultures is the much greater rate of glucose consumption in the former (Ulmer *et al.*, 1983). When an agitated culture is made stationary, a mycelial mat is produced from the pellets and then lignin breakdown commences (Leisola & Fiechter, 1985). It now seems that agitation does not prevent lignin-degrading enzymes from being produced but inactivates them when they are released into the medium (Venkatadri & Irvine, 1990). The addition of detergents such as Tween and CHAPS to the medium prevents the inactivation; the high-pressure liquid chromatography (HPLC) profile of extracellular protein in detergent-amended cultures is the same for static cultures (Jäger, Croan & Kirk, 1985). It is possible that detergents aid the release of enzymes; there is no clear evidence on this point. In terms of the protection given by detergents against activation, it has been suggested that either hydrophylic regions join to the hydrophylic head groups of the detergent molecules or hydrophobic groups join to the hydrophobic tails (Venkatadri & Irvine, 1990). Agitation also disrupts the glucan sheath found around the hyphae and described above (Bes *et al.*, 1987). If ligninolytic enzymes are within this sheath it seems probable that their stability will be maintained. Support for this view comes from the observation that a strain of *P. chrysosporium* best able to generate ligninolytic activity under agitation (Gold *et al.*, 1984) maintains its sheath under these conditions (Bes *et al.*, 1987).

There have been other studies of the effects of agitation on the development of ligninolytic activity in liquid culture (Leisola & Fiechter, 1985; Jansheker & Fiechter, 1988; Liebeskind *et al.*, 1990). These studies have shown that polypropylene glycol or polyethylene glycol also enhance production of lignin-degrading enzymes and that dimethoxybenzylamine (which is degraded to veratryl alcohol and veratryl aldehyde; see later) enhances ligninolytic activity.

Lignin is not degraded during primary growth of *P. chrysosporium*, and

Table 5.17. *The effect of various nitrogen sources on the ligninolytic and acetovanillone-oxidising activity and protein content of cultures of* Phanerochaete chrysosporium

Compound[a]	Ligninolytic activity (% of controls)	Acetovanillone-oxidising activity (% of controls)	Total protein (μg (culture)$^{-1}$ + s.d.)
None (control)	100	100	958 ± 131
Sodium nitrate	77	91	1173 ± 119
Proline	64	67	
Arginine	58	54	–
Ornithine	52	44	–
Glycine	50	64	–
Ammonium chloride	47	57	1651 ± 202
Asparagine	42	35	–
Urea	43	45	–
Serine	40	30	–
Isoleucine	30	30	–
Alanine	27	42	1969 ± 54
Histidine	24	12	969 ± 0
Glutamine	24	22	1719 ± 109
Glutamate	17	22	1479 ± 181

[a] Added in amounts sufficient to double culture nitrogen (to 5.6 mM) to 6-day-old ligninolytic cultures.
From Fenn & Kirk (1981).

does not induce ligninolytic activity. Lignin breakdown can occur only when the culture is under carbon, sulphur or nitrogen limitation, the last of these being the most important (Jeffries *et al.*, 1981). Of the various nitrogen compounds presented to the fungus and bringing about repression of ligninolytic activity, glutamic acid is by far the most effective (Table 5.17; Fenn & Kirk, 1981), though it is not clear whether glutamic acid is the compound within the mycelium that acts as the direct stimulus for repression (Fenn, Choi & Kirk, 1981). Cycloheximide, while inhibiting the development of ligninolytic activity, does not augment the ability of molecules such as ammonia or glutamate to repress the process. In essence, the production of lignin-degrading enzymes is a secondary metabolic event. Significantly, veratryl alcohol, the type of compound that might be expected to be a secondary metabolite, which is also produced by cultures of *P. chrysosporium* and is involved in lignin breakdown, has its production regulated in the same manner as the lignin-degrading enzymes.

In view of the above, it is very interesting that it has been possible to isolate a mutant (*PSBL-1*) of *P. chrysosporium* that generates ligninolytic activity under non-limiting conditions during primary metabolism. The mutant was isolated by the use of lysine covalently linked to lignin model compounds (Tien, Kersten & Kirk, 1987) from a lysine auxotroph using nutrient-rich conditions (Tien & Myer, 1990). For the mutant to grow, ligninase activity is necessary to release free lysine. Orth, Denny & Tien (1991) have provided information on the ligninolytic characteristics of *PSBL-1*.

Though nitrogen is more significant than carbon or sulphur in causing repression, it is likely that all three agents operate in an interdependent manner (Buswell, Mollet & Odier, 1984). However, *P. chrysosporium* appears able to regulate the glucose level of the medium and thus to a degree assist the lifting of carbon repression. This is achieved by the production of extracellular glucan under conditions when nitrogen is limiting (Bes *et al.*, 1987). When glucose concentrations are reduced to a level inadequate for lignin breakdown, a glucan-degrading enzyme is produced. It is significant also that the onset of conditions leading to the production of lignin-degrading enzymes is marked by a rise in cyclic AMP levels (MacDonald, Paterson & Broda, 1984) and in adenylate cyclase activity (MacDonald, Ambler & Broda, 1985). There appears to be no detectable phosphodiesterase activity in the mycelium and the level of cyclic AMP in the latter during the production of lignin-degrading enzymes appears to be regulated through excretion into the medium. These results have been added to by studies of Boominathan & Reddy (1992), indicating that cyclic AMP plays a key role in differentially regulating the production of the various lignin-degrading peroxidases by *P. chrysosporium*.

Although lignin in itself cannot induce ligninolytic activity, indeed such activity develops in the absence of lignin (Keyser, Kirk & Zeikus, 1978), the presence of lignin activates lignin degradation (Ulmer, Leisola & Fiechter, 1984; Faison & Kirk, 1985). Activation can also be brought about by low molecular weight aromatic compounds such as veratryl alcohol, veratraldehyde, *p*-coumaric acid and sinapic acid (Leisola *et al.*, 1984; Faison & Kirk, 1985). Finally lignin degradation is enhanced by increasing the ambient oxygen concentration (Reid & Seifert, 1982; Faison & Kirk, 1985), which increases the amount of enzyme activity as well as enhancing the production of H_2O_2 (see below).

There are a number of kinds of enzyme that are likely to be involved in some way with lignin breakdown. The fact that several new enzymes have been discovered in recent years suggests that the full complement of enzymes involved in ligninolytic activity of *P. chrysosporium* has yet to be revealed.

The most significant class is lignin peroxidase (ligninase) discovered in 1983 (Gold *et al.*, 1983; Kirk & Tien, 1983; Tien & Kirk, 1983, 1984), and then thought to be an H_2O_2-requiring oxygenase. Subsequent studies (Harvey *et al.*, 1985; Kuila *et al.*, 1985) showed the enzyme to be a haem-containing peroxidase. Lignin peroxidase is produced by *Bjerkandera adusta, Coriolus versicolor, Panus tigrinus, Phlebia radiata* and *Pleurotus ostreatus*, although it has not been shown to be produced by *Fomes lignosus, Lentinus edodes* and *Trametes cingulata* (Kirk & Farrell, 1987).

Lignin peroxidase of *P. chrysosporium* contains one atom of iron per enzyme molecule, is a glycoprotein and exists in a number of isoenzymic forms with molecular mass of 39–43 kDa (Kirk & Farrell, 1987). The pH optimum is around 3 (Tien & Kirk, 1984). The enzyme catalyses a very diverse range of reactions (Table 5.18). For a description of the function of the enzyme and other enzymes involved in the degradation of lignin, the reader should consult articles by Harvey, Schoemaker & Palmer (1987), Kirk (1987), Kirk & Farrell (1987), Harvey, Gilardi & Palmer (1989*a*), Leisola & Garcia (1989), Palmer, Harvey & Schoemaker (1987), Lewis & Yamamoto (1990), Schoemaker & Leisola (1990).

The apparent metabolic complexity of the action of the enzyme is a simple function of it being a peroxidase. The most likely model for the enzyme is given in Figure 5.20, which represents the redox cycle consisting of three steps. The ferric (Fe(III)) form of the enzyme is oxidised by H_2O_2 to a two-electron deficient oxyferryl state known as compound I. Compound I is then reduced by two consecutive single-electron steps to the ground state via compound II. These two steps leading to the ferric enzyme result in the production of radical cations.

The first step leads to the C_α–C_β cleavage of lignin. When this takes place, the substrate is oxygenated by molecular oxygen (Figure 5.21). The consumption of oxygen occurs because oxygen binds to a free radical to

Ligninase (Fe(III)p) + H_2O_2 $\longrightarrow$ Compound I (Fe(IV)(O)p$^{+\bullet}$) + H_2O

Compound I (Fe(IV)(O)p$^{+\bullet}$) + ED $\longrightarrow$ Compound II (Fe(IV)(O)p) + ED$^{+\bullet}$

Compound II (Fe(IV)(O)p) + ED $\longrightarrow$ Ligninase (Fe(III)p) + ED$^{+\bullet}$ + H_2O

Figure 5.20. The redox cycle in ligninase, Fe(III)p represents the haem group of the ferric form of the enzyme; ED the electron donor, e.g. lignin model dimer; ED$^{+\bullet}$, the radical cation formed by the removal of a single electron from the donor. (From Palmer *et al.*, 1987.)

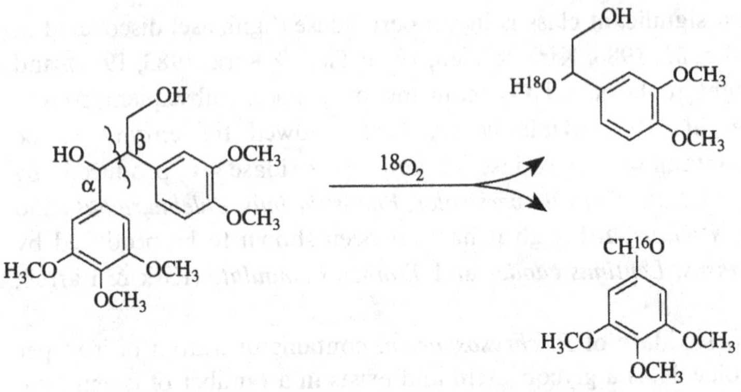

Figure 5.21. C_α–C_β cleavage in a lignin dimer brought about the ligninase of *Phanerochaete chrysosporium*. (From Leisola & Garcia, 1989.)

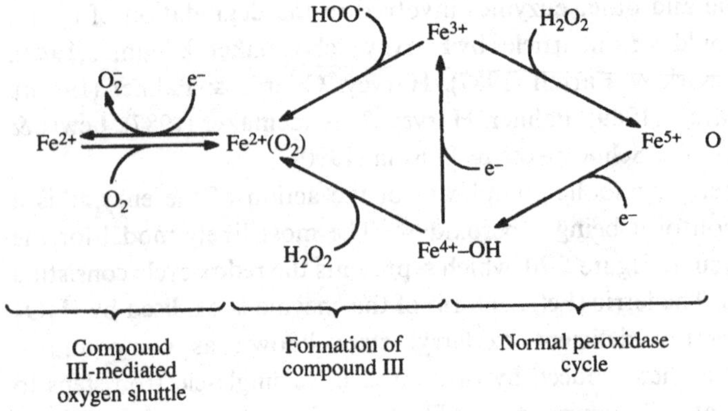

Figure 5.22. Interrelations between the redox states of horseradish peroxidase. (From Palmer, Harvey & Schoemaker, 1987.)

form an organic peroxy radical or it can participate in a shuttle mechanism in which it reacts with the ferrous (Fe(II)) form of the peroxidase to yield compound III (Figure 5.22). Compound III can then oxidise an electron donor to produce an oxidised product and superoxide. It should be noted that compound III is formed either by perhydroxy radicals or by excessive amounts of H_2O_2 from the ground state (Fe(III) forms) and compound II, respectively.

The other reactions are a consequence of the production of radical cations (Table 5.18). These undergo a variety of non-enzymic reactions, involving molecular oxygen, leading to a plethora of products (Table 5.19; Figure 5.23). Details of reaction sequences can be found elsewhere (Kirk,

Table 5.18. *Reactions characteristic of radical cations*

Reactions	Consequence
Act as one-electron oxidants	Mediators
Undergo side-chain reactions	C_α–C_β bond cleavage
	C–H bond cleavage
	Decarboxylation
Addition of solvent (water)	Hydroxylation of styrenes
	Cleavage of ether bonds
	Phenol formation
Reaction with $HO_2^\cdot$	Ring-opening

From Palmer *et al.* (1987).

Table 5.19. *Reactions catalysed by ligninase*

Cleavage of C_α–C_β bonds in lignin-model dimers
Oxidation of benzylic alcohols
Oxidation of benzylic methylene groups
Hydroxylation of benzylic methylene groups
Hydroxylation of olefinic bonds in styrenes
Decarboxylation of phenylacetic acids
Cleavage of ether bonds
Aromatic ring-opening
Polymerisation of phenols

From Palmer *et al.* (1987).

1987; Kirk & Farrell, 1987). Only two matters need to be referred to here. First, when non-phenolic radical cations undergo side-chain fragmentation reactions carbon-centred radical species are obtained and such radical species can reduce molecular oxygen to the superoxide anion. Furthermore under acidic conditions, i.e. the pH optimum of lignin peroxidase, superoxide anions will protonate to give perhydroxy radicals. These are quite powerful one-electron oxidants generating H_2O_2. Thus, oxygen increases both the rate and the extent of lignin degradation, which has been observed experimentally.

The second matter concerns veratryl alcohol, which is accumulated in the medium coincidentally with the appearance of ligninase (see above). In fact, veratryl alcohol is readily oxidised by ligninase to veratraldehyde. That oxidation appears to be a two-stage process. The first step, as with the oxidation of lignin, is the production of a radical cation that is very

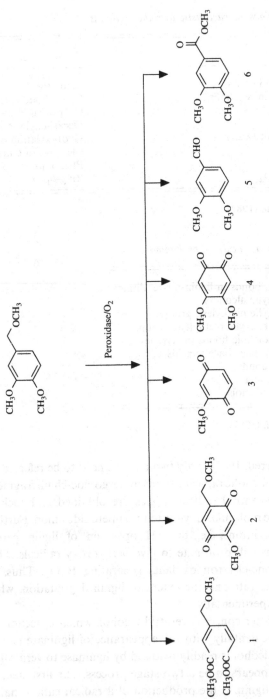

Figure 5.23. Oxidation products of 3,4-dimethoxybenzylmethylether. The action of lignin peroxidase at pH 3.0 under oxygen leads to: 1, ring cleavage compound; 2–4, quinones; 5, veratraldehyde; 6, veratryl acid methyl ether. (From Leisola & Garcia, 1989.)

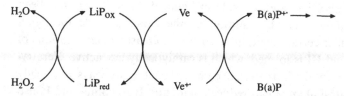

Figure 5.24. Veratryl alcohol as an electron transfer mediator: B(a) = benzo(*a*)-pyrene. (From Leisola & Garcia, 1989.)

stable in acidic media. In the presence of a second oxidisable substrate (which can be lignin) that substrate is oxidised and veratryl alcohol is regenerated (Harvey *et al.*, 1986; Figure 5.24). In less acidic conditions and in the absence of an oxidisable substrate, the deprotonated radical reacts with a molecule of veratryl alcohol and veratraldehyde is formed.

Thus it can be seen that lignin peroxidase can oxidise a wide variety of substrates not only by itself but also indirectly using veratryl alcohol as a mediator. Indeed the latter is more effective in oxidising certain aromatic compounds, e.g. monomethoxylated substrates, than is lignin peroxidase. This may explain why veratryl alcohol can increase the initial rate of lignin degradation when added to cultures of *P. chrysosporium* (Leisola *et al.*, 1984; Faison & Kirk, 1985). Perhaps of greater significance is the suggestion that veratryl alcohol, because of its size, relative to that of lignin peroxidase, may act in the wood at a site distant from the enzyme. Immunogold labelling of sections of wood attacked by *P. chrysosporium* and examined electron microscopically has shown that lignin peroxidase is located at the surface of the walls of wood cells or within areas of heavy attack (Srebotnik, Messner & Foisner, 1988; Daniel, Nilsson & Pettersson, 1989*a*; Daniel *et al.*, 1990*b*). The enzyme is not located in undecayed parts of the wall. There is no penetration of the wall when wood is infiltrated with concentrated culture filtrates. This supports the role suggested for veratryl alcohol in nature. However, the view that veratryl alcohol is oxidised in the manner described above has been discounted by Tien *et al.* (1986), who were unable with the method used (electron spin resonance) to detect radical cations of veratryl alcohol. It has been argued that this is not the best technique, NMR being more appropriate (Harvey *et al.*, 1989*a*). Kinetic studies are in keeping with the formation of radical cations of veratryl alcohol by a one-electron reduction of compound I (Harvey *et al.*, 1989*b*). Others (Leisola & Garcia, 1989) have also argued that, without the availability of a radical cation, their results on lignin degradation are hard to explain.

Veratryl alcohol has another role in lignin degradation. It has been pointed out above that, in addition to the catalytic cycle for lignin peroxidase, under certain conditions such as increased concentrations of H_2O_2 compound III is formed, which is catalytically less active. Veratryl alcohol can prevent the formation of compound III in two ways: as already indicated, veratryl alcohol can reduce compound II to compound I; also veratryl alcohol can bind to compound III, displacing $HO_2^{\bullet}/O_2^{\bullet-}$ from it, thus converting it to compound I. Importantly, veratryl alcohol is not consumed in this displacement reaction (Cai & Tien, 1989; Wariishi & Gold, 1989).

Studies with a lignin peroxidase-negative mutant (*lip*) have shown that lignin breakdown can take place in the absence of the enzyme (Boominathan *et al.*, 1990). The *lip* mutant has 16% of the activity of the wild-type. That lignin breakdown can occur in this manner is confirmed by other observations, particularly that there is no temporal correlation between lignin peroxidase appearance and lignin breakdown (Kirk *et al.*, 1986). Indeed lignin breakdown in culture can occur before the enzyme appears. So it seems very likely that other enzymes are involved.

Manganese peroxidases are also produced in the medium (Huynh & Crawford, 1985; Kuwahara *et al.*, 1984). Their properties have been established in some detail (Glenn & Gold, 1985; Paszczynska, Huynh & Crawford, 1985, 1986; Glenn, Akileswaran & Gold, 1986; Wariishi *et al.*, 1989). Though first demonstrated with *P. chrysosporium*, they have been shown to be produced by *Phanerochaete radiata*, *Phlebia tremelosa* (Bonnarme & Jeffries, 1989) and *Trametes versicolor* (Johansson & Nyman, 1987). Manganese peroxidase is a haem protein with a catalytic cycle and set of oxidation states analogous to that of lignin peroxidase. Thus, as one might expect, the native enzyme is converted to compound I with H_2O_2. However, unlike lignin peroxidase, compound I is reduced back to the native enzyme via compound II by two one-electron steps in which two atoms of Mn(II) are oxidised to Mn(III) (Figure 5.25). The enzyme, together with Mn(II) is also able to express oxidase activity against NADPH, NADH, reduced glutathione, dithiothreitol and dihydroxymaleic acid, with the generation of H_2O_2.

Although NAD, NADH, NADP and NADPH have been demonstrated in the medium of growing cultures of *P. chrysosporium* (Kuwahara *et al.*, 1984), it is believed that the production of Mn(III) is the major function of manganese peroxidases. It has been shown that, in the absence of enzyme, suitably chelated Mn(III) is a ligninolytic agent, capable of oxidising veratryl alcohol, lignin model compounds and lignin itself

$MnP + H_2O_2 \longrightarrow MnPI + H_2O$

$MnPI + Mn(II) \longrightarrow MnPII + Mn(III)$

$MnPII + Mn(II) \longrightarrow MnP + Mn(III)$

$Mn(III) + AH \longrightarrow Mn(II) + A\cdot$

Figure 5.25. The partial reactions in the production of Mn(III) with the involvement of manganese peroxidase (MnP). A, organic reducing agents. (From Wariishi *et al.*, 1989.)

(Forrester *et al.*, 1988). Significantly, as is discussed below, the chelating agent can be oxalate; pyrophosphate is also effective. The process is greatly stimulated (approximately 300 times) by reduced glutathione, but this compound has no effect on ligninolytic activity involving Mn(II) and the peroxidase. Reduction of Mn(III) in this way regenerates Mn(II). Chelated Mn(III) is a much more mobile ligninolytic agent than an enzyme and, as has been proposed for veratryl alcohol, provides a means by which lignin breakdown can take place some distance from the fungus.

Thus far, emphasis has been on agents capable of degrading lignin. But there is a conundrum. While the mycelium is in liquid culture, lignin breakdown can be demonstrated. On the other hand, isolated lignin peroxidase brings about polymerisation that is a consequence of the reactivity of the radical cations generated as a result of the one-electron oxidation described earlier. That polymerisation readily occurs in this manner has led to the suggestion that some process takes place other than that conceived of as the action of ligninase and manganese peroxidases described above (Lewis & Yamamoto, 1990). However, it is also possible that there are processes that prevent polymerisation. Harvey *et al.* (1989*a*) suggested that, if lignin peroxidase is fixed in space, processes of charge transfer through the lignin will be pre-eminent. In this situation, phenolic fragments cleaved from the lignin structure will be spatially separated from the oxidising system, no longer part of the network of charge transfer. Leisola & Garcia (1989) have suggested that the fungus removes degradation products through the hyphal membrane. These authors pointed out that Janshekar *et al.* (1982) reported tight binding of lignin to the fungal mycelia during the degradation process. The lignin seems to be inside the β-glucan layer described above. It is presumed that this binding is by covalent bonds – conceivably similar to those formed between hemicellulose and lignin in wood. The products of ligninolytic activity might not polymerise and being low molecular weight and water soluble could then be degraded further – possibly inside the fungal hyphae, though this was not stated explicitly. Much of the above is speculation and the drive must now be for identification of those processes that lead to depolymerisation rather than polymerisation after the action of lignin peroxidase and other enzymes capable of attacking the lignin molecule.

As regards the source of H_2O_2 in lignin degradation, there are a number of possibilities, not mutually incompatible. The following mechanisms have experimental evidence to support them:

(1) Oxidation of NADPH, NADH, reduced glutathione, dithiothreitol and dihydroxymaleic acid by manganese peroxidase, as described above.

(2) Glucose oxidase oxidises glucose to D-glucosone with the production of H_2O_2 (Volc, Sedmera & Musilek, 1978; Eriksson *et al.*, 1986; Kelley & Reddy, 1986*a, b*). δ-D-Gluconolactose and D-xylose are also oxidised at significant rates (60% and 37% of that of glucose). The enzyme is active over a broad pH range, including very significant activity at pH 3.5.

(3) Fatty acids when fed to mycelium of *P. chrysosporium* consume oxygen and produce extracellular H_2O_2 when incubated with fatty acyl-CoA substrates (Greene & Gould, 1984*b*). It is suggested that this is a consequence of peroxisomal fatty acyl-CoA oxidase activity.

(4) Methanol oxidase oxidises methanol to formaldehyde, with the production of H_2O_2 (Nishida & Eriksson, 1987). Ethanol is also oxidised at a significant rate. The enzyme is active over the pH range 6.0–10.5.

Of the possibilities suggested above, the last three enzyme systems are intracellular. Only glucose-2-oxidase has the ability to function in rather acid pH, a prerequisite for any extracellular enzyme involved in lignin degradation, and therefore if it were found that the process of H_2O_2 production did take place outside the mycelium, glucose-2-oxidase must be a candidate enzyme. There is some evidence from a study of mutants lacking glucose-2-oxidase that the enzyme is necessary for lignin breakdown by *P. chrysosporium* (Ramasamy, Kelley & Reddy, 1985; Kelley, Ramasamy & Reddy, 1986).

For the degradation of cellulose, *P. chrysosporium* utilises the following enzymes: five endoglucanases and one *exo*-(1→4)-β-glucanases (Almin, Eriksson & Pettersson, 1975; Eriksson & Pettersson, 1975*a, b*; Streamer, Eriksson & Pettersson, 1975; see also previous section). However, oxygen is required by *P. chrysosporium* for efficient degradation of cellulose

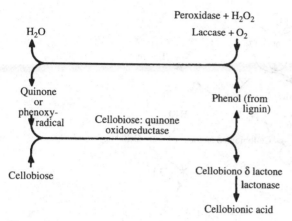

Figure 5.26. The reaction mechanism of the enzyme cellobiose:quinone oxidoreductase. (From Eriksson, 1981.)

(Eriksson, Pettersson & Westermark, 1974), and we now know that the fungus produces two extracellular enzymes that involve molecular oxygen either directly or indirectly. Cellobiose oxidase, a haem-protein of approximate molecular mass 93 kDa utilises molecular oxygen to oxidise cellodextrins to the corresponding aldonic acids. The pH optimum of the enzyme is 5.0, with 50% activity at pH 4.0 (Ayers, Ayers & Eriksson, 1978). The other enzyme is cellobiose:quinone oxidoreductase. This enzyme does not use molecular oxygen directly but reduces phenoxy radicals and quinones with the production of cellobiono-δ-lactone (Westermark & Eriksson, 1974a, b, 1975). The involvement of oxygen can be through the regeneration of the phenoxy radicals and quinones via the action of phenol oxidases. Equally the phenols might be oxidised by laccase and H_2O_2 (Figure 5.26). The enzyme has a molecular mass of around 58 kDa and a pH optimum of 4.5. Cellobionolactose is hydrolysed spontaneously or enzymically by a lactonase (Westermark & Eriksson, 1974b). It is possible that cellobionolactone is the first product of the action of cellobiose oxidase.

Cellobiose at a very low concentration is a potent inhibitor of endo- and exoglucanases. The most important mechanism for removing cellobiose is via the presence of two β-glucosidases (Eriksson, 1981). The cellobionic acid is cleaved by the two enzymes into glucose and gluconolactose, which is a powerful inhibitor of β-glucosidases. In the foregoing it is clear that there is much interaction between the products of various enzyme reactions involved in the degradation of both cellulose and lignin and other enzymes.

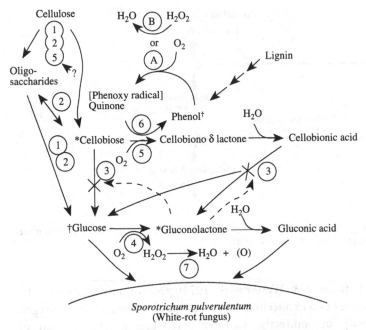

Figure 5.27. Enzyme mechanisms for cellulose degradation and their extracellular regulation in *Phanerochaete chrysosporium*. The following enzymes are involved in the reactions: 1, endo-1,4-β-glucanases; 2, exo-1,4-β-glucanase; 3, β-glucosidases; 4, glucose oxidase; 5, cellobiose oxidase; 6, cellobiose:quinone oxidoreductase; 7, catalase. Enzymes involved in lignin degradation: A, laccase; B, peroxidase. (*) Products regulating enzyme activity: gluconolactone inhibits β-glucosidases; cellobiose increases transglycosylations. (†) Products regulating enzyme synthesis: glucose, gluconic acid→catabolite repression; phenols→repression of glucanases. (From Eriksson, 1981.)

It follows that there is an interplay between the processes of cellulose and lignin degradation when a white-rot fungus such as *P. chrysosporium* attacks wood, which is what would be expected from gross studies of white-rot wood decay as indicated on p. 161. The situation has been well described diagrammatically by Eriksson (1981) (Figure 5.27).

Lignin peroxidase has its maximum activity in very acid pH – 3 or less. Although manganese peroxidase has a higher pH optimum (5.0), it is likely that for optimum breakdown of lignin, the fungus must ensure that within the wood such an acid environment is both generated and maintained. Wood itself has significant buffering capacity. The pH of the wood can be reduced by excretion of protons, in exchange for cations bound to the carboxyl groups within the wall, or by excretion of acid, the most notable possibility being oxalic acid. There is abundant evidence that this acid is produced by many wood-decay fungi, brown-rot fungi producing more in culture than white-rot fungi (Takao, 1965) (see pp. 159–60). It is known to be produced by *P. chrysosporium*, but it is not yet known whether the acid is produced at the site of ligninolytic activity. Be that as it may, oxalic acid not only may be important in maintaining the necessary low pH but also may be involved in reactions that are significant in lignin breakdown. Thus, in the presence of oxalate, veratryl alcohol and oxygen, lignin peroxidase oxidises Mn(II) to Mn(III), which, as has been shown above, may itself act as an oxidising agent for lignin (Popp, Kalyanaraman & Kirk, 1990). In the absence of Mn(II), the carbon dioxide anion radical ($CO_2^{\bullet-}$) produced along with carbon dioxide from oxalate appears to decay to carbon dioxide and is thought to be a process for removing oxalate (Akamatsu *et al.*, 1990). This process of oxalate removal will of course lead to an increase in pH in the milieu of the wood decay fungus. Such an increase in pH could be necessary for the cellulolytic enzymes to reach maximum activity.

There are two aspects to the metabolism of the phenylpropanoid and other units released by the degradation of lignin. First, there is the process by which the benzenoid nuclei are dearomatised. This can be accomplished by a wide range of microorganisms, bacteria and fungi, and has been well described elsewhere (Crawford, 1981). A key process is ring cleavage, involving oxygen and mediated by dioxygenases. While the majority of studies have been concerned with enzymes from bacteria, there are sufficient reports for the presence of dioxygenases from fungi to emphasise the ability of the latter to utilise aromatic carbon sources (Table 5.20).

The second aspect concerns the presence of enzymes colloquially known as 'phenol oxidases', which oxidise phenols to the corresponding quinones. There are three types: laccase (O_2:*p*-diphenol oxidoreductase), peroxidase (donor:H_2O_2 oxidoreductase) and tyrosinase (O_2:*o*-diphenol oxidoreductase) (Figure 5.28; Ander & Eriksson, 1978). There seems to be a correlation between lignin degradation and the ability of fungi to produce phenol oxidases. Certainly the ability of white-rot fungi to produce these enzymes extracellularly has been known for some time (Bavendamm, 1928; Davidson, Campbell & Blaisdell, 1938). That said, the exact role of these enzymes in lignin degradation is not really clear. They can or may be involved in the following:

(1) Mediation of a certain amount of direct oxidative degradation of lignin (Kirk, 1971).

Table 5.20. *Representative aromatic ring-cleaving dioxygenases of fungi*

Fungus	Dioxygenase	Reference
Tilletiopsis washingtonensis	Homoprotocatechuate-3,4	Subba Rao et al. (1971)
Penicillium patalum, Penicillium urticae	2,5-Dihydroxybenzaldehyde-4,5	Scott & Beadling (1974); Forrester & Gaucher (1972)
Aspergillus niger	Homogentisate-1,2	Sugumaran & Vaidyanathan (1978)
Candida tropicalis	4-Methylcatechol-1,6	Hashimoto (1970); Hashimoto (1973)
Tilletiopsis washingtonensis	Protocatechuate-3,4	Subba Rao et al. (1971)
Candida tropicalis	Catechol-1,2	Neujahr, Lindsjö & Varga (1974)
Neurospora crassa	Protocatechuate-3,4	Gross, Gafford & Tatum (1956)
Aspergillus niger	Protocatechuate-3,4	Cain, Bilton & Darrah (1968)
Rhodotorula mucilaginosa		
Sporobolomyces sp.		
Penicillium spinulosum		
Endomycopsis sp.		
Cylindrocephalum sp.		
Fusarium oxysporum		
Polystictus versicolor	Catechol-1,2	Cain et al. (1968)
(= *Coriolus versicolor*)		
Debaryomyces subglobosus		
Rhodotorula mucilaginosa		
Fusarium oxysporum		
Penicillium spinulosum		
Vararia granulosa		
Schizophyllum commune		
Debaryomyces hansenii		
Aureobasidium pullulans	Protocatechuate-3,4	Henderson (1961)
(= *Aureobasidium pullulans*)		

From Crawford (1981), *Lignin Biodegradation and Transformation*, Copyright © 1981, Reprinted by permission of John Wiley & Sons, Inc.

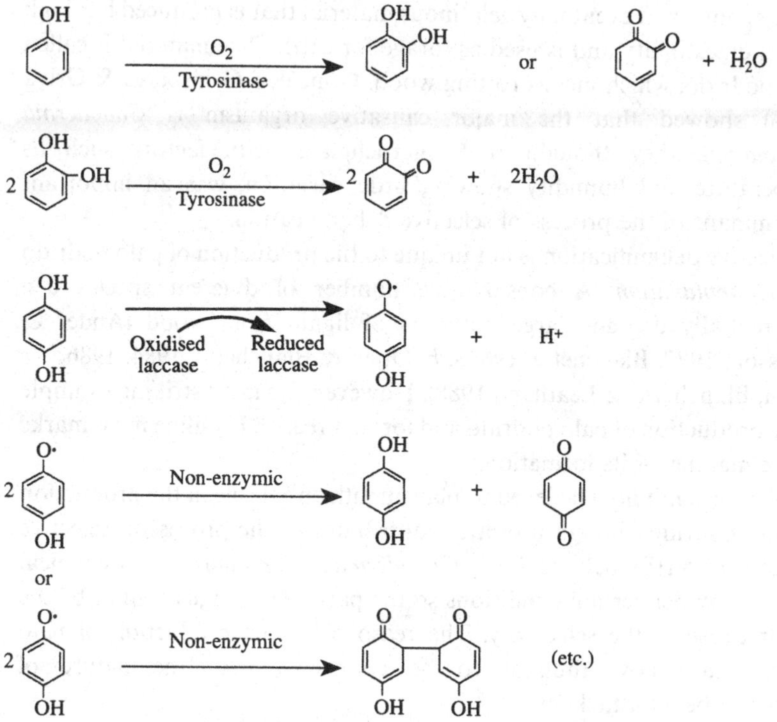

Figure 5.28. Activities of phenol oxidases. (From Ander & Eriksson, 1978.)

(2) Removal of the original asymmetric centres in the lignin molecule by catalysing the non-stereospecific oxidation of benzyl alcohol moieties to form benzyl ketones (Shimada, 1979). Further breakdown of the lignin molecule would require many fewer enzymes than would be needed to degrade the stereospecifically complex natural polymer.

(3) Generation of quinone for the oxidation of cellobiose (see above).

(4) Detoxification of low molecular weight phenols that could be injurious to the activity of the wood-decay fungus (Ander & Eriksson, 1978).

Selective delignification and resistance to delignification

By selective delignification, I mean the preferential removal of lignin from wood without concomitant removal of appreciable amounts of cellulose. There is no doubt, from the initial studies of Phillipi (1893), that selective delignification can occur under natural conditions, exemplified by its extensive occurrence in the evergreen rainforests of southern Chile. The white spongy and eventually gelatinous material that is produced has high rumen digestibility and is used as forage for cattle. The material is called palo podrido, which means rotting wood. González, Grinsbergs & Griva (1986) showed that the major causative organism is *Ganoderma applanatum*; they thought that microclimate, with factors such as temperature and humidity showing little variation, was an important determinant of the process of selective delignification.

Selective delignification is not unique to the production of palo podrido or *G. applanatum*. A considerable number of different species can preferentially degrade large amounts of lignin from wood (Ander & Eriksson, 1977; Blanchette, 1984*a*, *b*; Otjen & Blanchette, 1984, 1986*a*, *b*; Otjen, Blanchette & Leatham, 1988). However, the most striking example is the production of palo podrido and for that reason I confine my remarks to the manner of its formation.

Climate is not now believed to play an influential role in the production of palo podrido, though it clearly contributes to the process of selective delignification (Blanchette, 1991). *Ganoderma applanatum* can cause typical white rot under certain conditions so the particular fungus cannot be the major cause of the selectivity. The reasons for the production of palo podrido are now thought to be connected with the nature of the wood being attacked:

(1) Selective delignification occurs mainly in the trunks of *Eucryphia cordifolia* and *Nothofagus donbeyi* (González *et al.*, 1986; Agosin *et al.*, 1990).
(2) The wood has a very low nitrogen content (0.3037%–0.073% dry weight; Dill & Kraepelin, 1986). Addition of nitrogen to the wood leads to cellulose degradation.
(3) The wood has a very low lignin content, the largest amount of β-arylether linkages and the highest syringyl:guaiacyl ratio of all the native woods tested. The type of lignin thus present appears to be more readily attacked (Agosin *et al.*, 1990).
(4) The wood anatomy is different from those species that are not selectively delignified; the thick walls of fibres of *E. cordifolia* seem to be significant, appearing to allow maintenance of structure during decay (Barrasa, González & Martíez, 1992).

The importance of wetlands for preserving wooden artefacts for archaeology is testament to the prevention of ligninolytic activity by anaerobic conditions. But there are other circumstances in which ligninolytic activity is restricted, albeit in not such a dramatic fashion.

The most outstanding is the heartwood of tree trunks. The intriguing question is why, of several thousand wood-decay fungi, only a small number can bring about degradation of heartwood (Wagner & Davidson, 1954). Highley & Kirk (1979) have reviewed the situation. They raised the following questions:

(1) What special capabilities enable some fungi to invade and decay the hearts of living trees?
(2) Why do many heart-rot fungi cease to cause decay when the tree is cut or even when it dies?
(3) Why is the durability of wood in use not always related to its resistance to decay in living trees?

There are no easy answers to these questions. An important factor may be an unusual tolerance of heart-rot fungi to low oxygen and high carbon dioxide concentrations. Response to compounds specific to heartwood, including volatiles, may also be important. The change of durability of wood in service may not be a symptom of changed resistance but of a change in the environment in which the wood is placed.

This section emphasises that the ability of a fungus to bring about lignocellulose breakdown in the natural environment is governed by a range of factors that are only imperfectly understood. Since we have a fairly clear picture of the biochemistry of lignocellulose breakdown, it is now timely to investigate more precisely how these various factors have their effect.

D. ONE-CARBON COMPOUNDS

Until 1969, it was thought that only amongst the prokaryotes was there the capability to grow using compounds containing only one carbon atom and at a level of oxidation between that of methane and that of carbon dioxide – the so-called one-carbon compounds – as the sole source of carbon and energy. In 1969, Ogata, Nishikawa & Ohsugi were able to show that a yeast, now known as *Candida boidinii*, was able to utilise methanol as the carbon source for growth. Since then other strains of yeasts from the genera *Candida*, *Hansenula*, *Pichia* and *Torulopsis* have been shown to be able to grow on methanol (Lee & Komagata, 1980; Komagata, 1981). It is interesting that the majority of the strains are able to utilise pectin. Since pectin yields methanol on hydrolysis, the ability to utilise the alcohol may be related to this plant cell wall material. Decaying plant material or soils rich in organic matter are very good sources of yeasts capable of utilising methanol or methylotrophic yeasts. There is a report of a few species of filamentous fungi (*Penicillium* and *Trichoderma*) being able to grow on methanol (Goncharova, Babitskaya & Lobanok, 1977).

Yeasts use methanol as a one-carbon substrate. They are facultative, being able to grow on glucose, but, as far as other one-carbon compounds are concerned, the yeasts are not able to grow on methane, methylamine, formaldehyde and formate as the sole carbon and energy source. However, it should be noted that some yeasts capable of using methane have been isolated (Wolf & Hansen, 1979), though little is known about them. There are reports purporting to show the ability of yeasts to grow on methylamine as the sole source of carbon, but it has been argued that the evidence is not convincing (Harder & Veenhuis, 1989).

The topic of one-carbon metabolism in yeasts has been well reviewed by Veenhuis, van Dijken & Harder (1983), Gleeson & Sudbery (1988), Harder & Veenhuis (1989), de Koning & Harder (1992). There is also a highly relevant review on the structure and functioning of yeast microbodies by Veenhuis & Harder (1989). These reviews should be consulted for details of the processes involved. Here, only the broad outlines of the processes will be considered.

When utilisation of methanol by yeasts is considered, two separate issues need to be addressed, namely the production of energy and the process of carbon assimilation. Essentially the oxidation of methanol to carbon dioxide via formaldehyde provides the energy, while formaldehyde fixation via the xylulose monophosphate pathway leads to the production of glyceraldehyde phosphate from which cell constituents are built. The pathways involved in the two processes are shown diagrammatically in Figures 5.29 and 5.30. Figure 5.31 shows compartmentation of the whole process, indicating how energy metabolism and carbon assimilation are integrated.

Oxidation of methanol to formaldehye is brought about by alcohol oxidase, which oxidises short-chain aliphatic alcohols (C_1–C_5), with the production of H_2O_2. The enzyme has a low affinity for oxygen. The enzyme of *Hansenula polymorpha* has a K_m of between 0.24 and 0.4 mM, contrasting with the oxygen concentration in air-saturated water of 0.4 mM (Kato *et al.*, 1976; van Dijken, Otto & Harder, 1976). This high K_m accounts for the large amount of alcohol oxidase produced in cells, which can be as much as 30% of the cellular protein. The hydrogen peroxide produced is removed by catalase, though it is not clear how that occurs, i.e. by catalatic or peroxidatic reactions (see Table 5.21).

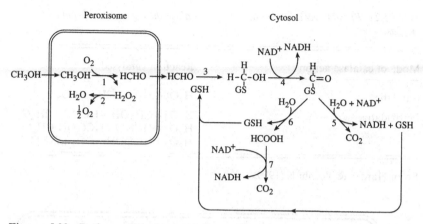

Figure 5.29. Pathway for the oxidation of methanol to carbon dioxide in methylotrophic yeasts. 1, Alcohol oxidase; 2, catalase; 3, non-enzymic formation of *S*-hydroxymethylglutathione; 4, formaldehyde dehydrogenase; 5, *S*-formylglutathione dehydrogenase; 6, *S*-formylglutathione hydrolase; 7, formate dehydrogenase; GSH, reduced glutathione. Methanol may be oxidised also in peroxisomes by the peroxidatic activity of catalase. (From Harder & Veenhuis, 1989.)

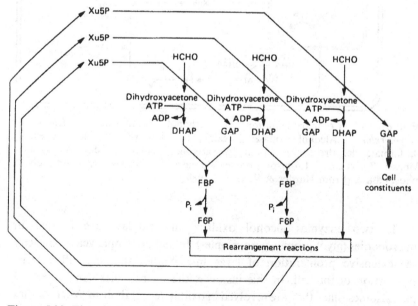

Figure 5.30. The xylulose monophosphate pathway of formaldehyde fixation in methylotrophic yeasts. Xu5P, xylulose 5-phosphate; DHAP, dihydroxyacetone phosphate; GAP, glyceraldehyde phosphate; FBP, fructose 1,6-bisphosphate; F6P, fructose 6-phosphate. (From Harder & Veenhuis, 1989.)

188

Carbon

Table 5.21. *Possible reactions catalysed by catalase during methanol metabolism in yeasts*

Mode of catalase action	Reaction catalysed
Catalatic	$H_2O_2 + H_2O_2 \rightarrow 2H_2O + O_2$
Peroxidatic	$H_2O_2 + CH_3OH \rightarrow HCHO + 2H_2O$
	$H_2O_2 + HCHO \rightarrow HCOOH + H_2O$
	$H_2O_2 + HCOOH \rightarrow CO_2 + 2H_2O$

From Harder & Veenhuis (1989).

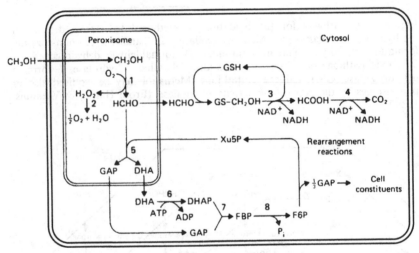

Figure 5.31. Compartmentation of methanol metabolism in *Hansenula polymorpha*. 1, Alcohol oxidase; 2, catalase; 3, formaldehyde dehydrogenase; 4, formate dehydrogenase; 5, dihydroxyacetone synthase; 6, dihydroxyacetone kinase; 7, fructose 1,6-bisphosphate aldolase; 8, fructose 1,6-bisphosphate phosphatase. (From Harder & Veenhuis, 1989.)

The two enzymes alcohol oxidase and catalase are located in microbodies (glyoxysomes/peroxisomes). In methylotropic yeasts, one can see extensive proliferation of large microbodies that can fill a large proportion of the cell. In this instance, the microbodies can be termed peroxisomes, since they are involved in oxidative metabolism. Microbodies containing the enzymes of the glyoxylate cycle are termed glyoxysomes. The development of peroxisomes after transference of cells of *Hansenula polymorpha* to media in which methanol is the sole source of carbon has been examined in some detail (Veenhuis & Harder, 1989). The peroxisomes originate from small ones already present in glucose-grown cells. Electron microscopic observations and enzyme analysis have shown that growth of the peroxisomes is associated with a high rate of synthesis of alcohol oxidase and catalase.

The complete oxidation of formaldehyde by yeasts is brought about by two steps involving the reduction of NAD^+. The enzyme for first step, formaldehyde dehydrogenase, acts not on formaldehyde per se but on S-hydroxymethylglutathione, which is formed non-enzymically by formaldehyde reacting with reduced glutathione. The products of the reaction are NADH and S-formylglutathione. In the second step, the latter can be oxidised in two ways. Either the compound is hydrolysed to formate and reduced glutathione, the former being oxidised to carbon dioxide with the reduction of NAD^+, or the compound is oxidised directly such that reduced glutathione and NADH are produced and carbon dioxide is released.

The NADH released from these oxidative reactions is oxidised by the mitochondria, with the production of ATP. The exact manner by which this occurs is uncertain (Harder & Veenhuis, 1989). It is believed that with methanol as substrate the yield of ATP from NADH is only 2 mol mol^{-1}.

Carbon fixation occurs by the condensation of formaldehyde (HCOOH) with xylulose 5-phosphate to give dihydroxyacetone and glyceraldehyde 3-phosphate (GAP). The enzyme termed dihydroxyacetone synthase (O'Connor & Quayle, 1980) was discovered by Kato et al. (1979) and by Waites & Quayle (1980). It has transketolase-like properties but is a separate enzyme from the classic transketolase. Dihydroxyacetone is phosphorylated and is subsequently condensed with glyceraldehyde 3-phosphate to give fructose 1,6-bisphosphate. There are then rearrangements to regenerate xylulose 5-phosphate. The overall reaction is

$$3HCHO + 3ATP \rightarrow GAP + 3ADP + 2P_i$$

Though one-carbon amines are unable to support growth as carbon sources they can act as the sole source of nitrogen. The ability of yeasts to use mono-, di- or triethylamine, triethylamine N-oxide and tetraethylammonium as a sole nitrogen source is much more widespread than the ability to use methanol as the sole source of carbon. Details of the metabolic pathways involved were given by Harder & Veenhuis (1989). The essential steps in the case of di- and triethylamine are their oxidation to methylamine and formaldehyde and oxidation of the former product to formaldehyde and ammonia in the peroxisome. The formaldehyde is oxidised to carbon dioxide in a manner similar to that in which methanol is oxidised to carbon dioxide.

In industrial terms, methanol can act as a carbon source for single-cell protein production. It is obtained primarily from natural gas; however, gas at oil wells or gas from crude oil or coal can also be a source. When yeasts are grown on methanol on the industrial scale, methanol concentrations must be carefully adjusted. It seems that adjustment must be finely controlled, since a small increase can lead to a decrease in yield. This decrease is seemingly a result of an inadequate rate of oxidation of formate and formaldehyde, two compounds that are very much more toxic than methanol to cell growth (Cartledge, 1987).

It is relevant here to point out that a catalase-negative mutant of *Hansenula polymorpha* is able to decompose hydrogen peroxide at a high rate (Verduyn *et al.*, 1991). The enzyme responsible appears to be cytochrome *c* peroxidase, with reducing equivalents provided by the respiratory chain. Hydrogen peroxide itself can act as an electron acceptor, as shown by oxidation of ethanol to acetate under anaerobic conditions. The presence of hydrogen peroxide is not without untoward effects on the cells, since yield is reduced.

E. HYDROCARBONS

There are two major reasons for interest in the ability of fungi to utilise hydrocarbons. First, these compounds are used as a source of carbon for the production of single-cell protein and, second, because of the ability of fungi and some microorganisms to degrade aircraft fuel and cause biodeterioration of the fuel systems (Park, 1975; Genner & Hill, 1981). Outstanding amongst the latter fungi is *Cladosporium resinae*. This fungus is also called the creosote fungus because of its ability to colonise creosote-treated timber. The dominance of *C. resinae* in aircraft fuel is probably due to the following characters in combination: (a) ability to metabolise aliphatic hydrocarbons in the C_8–C_{20} chain-length range; (b) ability to sporulate frequently when growing in fuel; (c) long-term viability when suspended in fuel; (d) ability of the fungus to withstand the successive freezing and thawing cycles in the aircraft wing (Genner & Hill, 1981).

While fungi can grow on crude and fuel oils, it is the alkanes that support the best growth. Many yeasts accept alkanes in the C_4–C_{20} range, but the most readily accepted are in the C_{10}–C_{20} range. Above C_{20} chain length, alkanes are solid at temperatures below 35 °C (Cartledge, 1987).

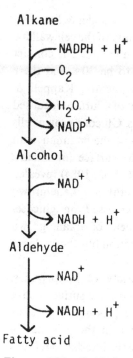

Figure 5.32. Oxidation pathway of alkanes to the corresponding fatty acids. (From Tanaka & Fukui, 1989.)

There are two steps specific to alkane utilisation: first, incorporation of alkanes into the cells and, second, oxidation therein of the alkanes to the corresponding fatty acids (Figure 5.32). Thereafter metabolism is by those steps common to fatty acid metabolism leading to the formation of acetyl-CoA: the synthesis of citric acid cycle intermediates from acetyl-CoA via the glyoxylate cycle, and the flux of carbon to carbohydrates via gluconeogenesis. The metabolism of alkanes in yeasts has been reviewed by Cartledge (1987) and Tanaka & Fukui (1989).

The most coherent picture of the process of alkane uptake is that for *Candida tropicalis*. Miura *et al.* (1977) have shown that cells absorb very little alkane when in solution form. The ability to absorb alkanes depends on the ability of yeast cells to adhere to droplets, adhesion being measured by the centrifugation force needed to remove cells from the alkane. *Saccharomyces cerevisiae*, which does not utilise alkanes, cannot adhere. Light microscopic observations have shown cells of *C. tropicalis* clinging to the oil phase (Mimura, Watanabe & Takeda, 1971; Blanch & Einsele, 1973). Electron microscopic observations have shown submicrometre droplets of various sizes adhering to the wall (Einsele, Schneider & Fiechter, 1975). Käppeli & Fiechter (1976) showed that the affinity of the cell wall for alkanes is independent of pH, temperature and chain length. The contact time for saturation of the wall surface was found to be 30 s, and there was no evidence for any enzyme reaction. Subsequently, Käppeli & Fiechter (1977) provided evidence for the involvement of a surface-localised mannan–fatty acid complex in the binding of alkanes. Glucose-grown cells contained very much less of the complex. Masking the mannan with concanavalin A reduced the binding affinity of the surface for alkane. Later, critical electron microscopic studies (Käppeli *et al.*, 1984) revealed the presence of hair-like structures (*c.* 0.1 μm in length) on alkane-grown cells. The hair-like structures were much less expressed on glucose-grown cells and were removed by protease treatment of alkane-grown cells. It is not clear how the hairs assist alkane absorption but the evidence indicates a probable important role.

Microorganisms are known to produce a variety of lipopeptides, lipoproteins, glycolipids and lipopolysaccharides that have surface-active properties (Rosenberg, 1986). Such compounds are produced by yeasts that utilise alkanes. *Candida lipolytica*, when grown in the presence of hexadecane, produces emulsifying agents that reduce the interfacial tension of the liquid (Prokop, Ludvik & Erikson, 1972; Nakahara, Erikson & Gutierrez, 1977). During the early stages, the reduction of interfacial tension is attributed mainly to release of fatty acids; at the later stages it is attributed to other compounds. Cirigliano & Carman (1984, 1985) showed that negligible emulsification activity was produced when glucose was the carbon source. On hexadecane, under the culture conditions used maximum emulsification activity was produced when cell growth had ceased. These authors isolated from the later stages of growth a water-soluble compound that could bring about emulsification. This compound consists of 83% carbohydrate and 17% protein and has a molecular mass of 27.6 kDa. It can stabilise oil-in-water emulsions produced with commercial vegetable oils.

Ito & Inoue (1982) investigated the effects of sophorolipids that are produced extracellularly by *Torulopsis bombicola* when it is grown on water-insoluble alkanes. These lipids and related model compounds, which were not themselves used for growth, were shown to stimulate markedly the growth of *T. bombicola*. What was very interesting was the lack of effect on other alkane-utilising yeasts such as *Candida* spp. and *Pichia* spp. Also, though some of the lipids used in the study had emulsifying activity, synthetic non-ionic surfactants had no effect. It seems from this study that, although a compound has the ability to emulsify a water/alkane mixture and thus increase the surface area of the substrate, its ability to enhance growth is due to other properties. Ito & Inoue (1982) alluded to the possibility that when sphorolipids are present in the wall, they make it more lipophilic allowing better contact with the alkane droplets, as indicated above.

Once inside the fungus, the alkanes are oxidised to fatty acids. There is uncertainty about the exact nature of the process. There could be three possible ways in which oxidation might occur:

(1) Hydroxylation by a monooxidase (mixed function oxidase) to form a primary alcohol.

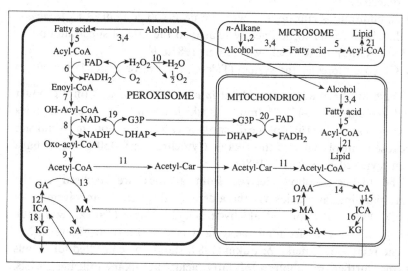

Figure 5.33. Presumptive roles of peroxisomes, mitochondria and microsomes in alkane-assimilating yeasts. 1, Cytochrome P-450; 2, NADPH-cytochrome P-450 (cytochrome *c*) reductase; 3, long-chain alcohol dehydrogenase; 4, long-chain aldehyde dehydrogenase; 5, acyl-CoA synthetase; 6, acyl-CoA oxidase; 7, enoyl-CoA hydratase; 8, 3-hydroxyacyl-CoA dehydrogenase; 9, 3-ketoacyl-CoA thiolase; 10, catalase; 11, carnitine acetyltransferase; 12, isocitrate lyase; 13, malate synthase; 14, citrate synthase; 15, aconitase; 16, NAD$^+$-linked isocitrate dehydrogenase; 17, malate dehydrogenase; 18, NADP-linked isocitrate dehydrogenase; 19, NAD$^+$-linked glycerol-3-phosphate dehydrogenase; 20, FAD-linked glycerol-3-phosphate dehydrogenase; 21, glycerol-3-phosphate acyltransferase. Abbreviations: Acetyl-Car, acetylcarnitine; CA, citrate; DHAP, dihydroxyacetone phosphate; GA, glyoxylate; G3P, glycerol 3-phosphate; iCA, isocitric acid; KG, 2-ketoglutaric acid; MA, malic acid; OAA, oxalacetic acid; OH-Acyl-CoA, 3-hydroxyacyl-CoA; Oxo-acyl-CoA, 3-ketoacyl-CoA; SA, succinic acid. (From Tanaka & Fukui, 1989.)

(2) Dehydrogenation to the corresponding alkanes.

(3) Hyperperoxidation, by which oxygen is incorporated into the alkane by a dioxygenase yielding an n-alkyl hydroperoxide that is then reduced to a primary alcohol.

In the case of yeasts capable of growing on alkanes, the first mechanism is the one that seems to be operative. The terminal oxidase involved is cytochrome P-450 utilising molecular oxygen and electrons derived either from NADPH or NADH via a flavoprotein:cytochrome P-450 reductase.

The evidence for the involvement of cytochrome P-450 in oxidation of alkanes is as follows

(1) No hydroxylation system other than cytochrome P-450 has been detected.

(2) In continuous culture, the cytochrome P-450 content of *C. tropicalis* has been found to be linearly related to the rate of substrate uptake rate (Gmünda, Käppeli & Fiechter, 1981*a*, *b*).

(3) Glucose represses the system (Takagi, Moriya & Yano, 1980*a*, *b*).

There has been a question about *in vitro* turnover rates (mol product (mol cytochrome P-450)$^{-1}$ (min)$^{-1}$), which do not match *in vivo* rates (Gmünda *et al.*, 1981*b*). But the disparity may be due to poor mass transfer because of insolubility of the alkane. Cytochromes P-450 of yeasts have been reviewed by Käppeli (1986*b*).

The higher alcohols derived from alkanes are oxidised to the corresponding fatty acids via the aldehyde, through the involvement of NAD^{+}-linked alcohol and aldehyde dehydrogenase (Tanaka & Fukui, 1989). The pathway from alkane to fatty acid is shown in Figure 5.32.

The review by Tanaka & Fukui (1989) should be consulted for details of the further metabolism of fatty acids in alkane-utilising yeasts. Figure 5.33 gives the presumptive roles of peroxisomes, mitochondria and microsomes in the essential metabolic steps in which the carbon of alkanes is fed into primary metabolism.

6
Nitrogen

Utilisation of nitrogen compounds

Fungi can utilise nitrate, nitrite, ammonia (see footnote, p. 82) and a very wide range of organic nitrogen compounds. Some idea of the diversity of the latter can be gained by reference to an older text such as that by Cochrane (1963), in which the capability of fungi to use a wide diversity of compounds is less taken for granted than is often the case now. The same volume and those of Hawker (1950) and of Lilly & Barnett (1951) give useful information about fungi which are able to utilise:

(1) Nitrate nitrogen.
(2) Ammonia or organic nitrogen but not nitrate nitrogen.
(3) Only organic nitrogen.

At the time the above authors were writing, there was debate about whether or not fungi are able to use atmospheric nitrogen. That debate was prior to the use of ^{15}N for detecting nitrogen fixation. Since the isotope became available, it has been established now that no known fungus is able to fix atmospheric nitrogen (Stewart, 1966). Of course, fungi lichenised with cyanobacteria form organisms that are able to use this source of nitrogen.

The ability to use nitrate nitrogen represents a greater capability on the part of a fungus, for all nitrate users can use ammonia but not all ammonia users can use nitrate. A list of some fungi known to be able to use nitrate is given in Table 6.1. It gives an indication only; to obtain anything like completeness would be impossible. Table 6.2 gives a list of fungi known to use ammonia but not nitrate, while Table 6.3 indicates the situation in yeasts with respect to inorganic nitrogen utilisation. The latter is interesting in that it shows that there are two genera, *Debaryomyces* and *Sporopachydermia*, that can utilise nitrite but not nitrate.

Table 6.1. *Some fungi that have been reported to utilise nitrate nitrogen*

Armillaria mellea	*Macrosporium sarcinaeforme*
Ascobolus denudata	*Marasmius fulvobulbillosus*
A. leveillei	*Neocosmopara vasinfecta*
Ascochyta pisi	*Neurospora crassa*
Aspergillus spp.	*Ophiobolus graminis*
Botryotinia convoluta	*O. miyabeanus*
Botrytis allii	*Penicillium* spp.
B. cinerea	*Phoma apiicola*
Cephalothecium roseum	*P. betae*
Cercospora apii	*Pleurage curvicolla*
C. beticola	*Pyronema confluens*
Chaetomium cochlioides	*Pythiomorpha gonapodyoides*
C. convolutum	*Pythium debaryanum*
C. globosum	*P. intermedium*
Colletotrichum lagenarium	*P. irregulare*
C. lindemuthianum	*Rhizoctonia solani*
Collybia tuberosa	*Sclerotinia minor*
C. velutipes	*S. sclerotiorum*
Cordyceps militaris	*Sclerotium bataticola*
Dendrophoma obscurans	*Scopulariopsis brevicaulis*
Dendryphiella salina	*Septoria nodorum*
Dothidella quercus	*Sordaria fimicola*
Fusarium spp.	*Sphaerobolus stellatus*
Glomerella cingulata	*Sphaeropsis malorum*
Gymnoascus setosus	*Trichoderma lignorum*
Helminthosporium spp.	*Verticillium albo-atrum*
Lambertella corni-maris	*Xylaria mali*
Lentinus tigrinus	

See also Table 6.3.
From Lilly & Barnett (1951).

Certain fungi are unable to grow on inorganic nitrogen. *Blastocladiella emersonii* (Barner & Cantino, 1952) and *Sapromyces elongatus* (Golueke, 1957) and probably *Leptomitus lacteus* (Schade, 1940) and *Thraustochytrium aureum* (Paton & Jennings, 1989) are the few authentic examples of free-living fungi unable to grow on ammonia. The list could well be longer if there were opportunities to study other lower fungi. Nevertheless, there is a need for care in such investigations. The at one time seeming inability of *Allomyces javanicus* var. *macrogymus* and *Chytridium* to grow on inorganic nitrogen has been shown to be a failure to grow at low pH consequent upon a small amount of ammonia assimilation in the early stages of the culture (Machlis, 1953*a*, *b*; Crasemann, 1954; Cochrane, 1963). Even those free-living fungi that are unable to use ammonia as the sole

Table 6.2. *Some fungi that have been reported to require ammonia or organic nitrogen and to be unable to assimilate nitrate nitrogen*

Absidia coerulea	*M. putillus*
A. cylindrospora	*M. ramealis*
A. dubia	*M. rotula*
A. glauca	*M. scorodonius*
A. orchidis	*Monilinia fructicola*
Basidiobolus ranarum	*Mortierella rhizogena*
Ceratostomella fimbriata	*Mucor flavus*
C. ulmi	*M. hiemalis*
Choanephora cucurbitarum	*M. nodosus*
Cyathus striatus	*M. pyriformis*
Endothia parasitica	*M. saturninus*
Lenzites trabea	*M. stolonifer*
Marasmius alliaceus	*M. strictus*
M. androsaceus	*Phycomyces blakesleeanus*
M. chordalis	*Pleurotus ostreatus*
M. epiphyllus	*Rhizophylyctis rosea*
M. foetidis	*Rhizopus nigricans*
M. graminum	*R. oryzae*
M. performis	*Sporodina grandis*
M. personatus	*Zygorrhynchus moelleri*

See also Table 6.3.
From Lilly & Barnett (1951).

Table 6.3. *The distribution of the ability of yeasts to assimilate nitrate*

(A) Genera containing species that assimilate nitrate

Ambrosiozyma	*Cryptococcus*	*Pachysolen*	*Sterigmatomyces*
Brettanomyces	*Dekkera*	*Rhodosporidium*	*Trichosporon*
Bullera	*Filobasidium*	*Rhodotorula*	*Wickerhamiella*
Candida	*Hansenula*	*Sporidiobolus*	
Citeromyces	*Leucosporidium*	*Sporobolomyces*	

(B) Genera containing species that assimilate nitrite (but not nitrate)
Debaryomyces   *Sporopachydermia*

(C) Some genera notable for their inability to assimilate nitrate
Kluyveromyces   *Pichia*   *Saccharomyces*   *Schizosaccharomyces*

From Hipkin (1989), by permission of Oxford University Press.

nitrogen source may still be able to assimilate the compound. While this possibility has not been tested, i.e. with studies using ^{15}N, there is evidence that *Thraustochytrium*, which requires lysine for growth (Paton & Jennings, 1988), and possibly other amino acids, is able to metabolise inorganic nitrogen (Bahnweg, 1979).

Any fungi, other than lower fungi, unable to grow when only supplied with ammonia as the sole source of nitrogen are likely to be parasites, which have evolved a specialised nitrogen nutrition due to the locale in which they live. Examples are *Phytophthora bahamensis*, *P. fragariae* and *P. infestans* (Hohl, 1983) and *Puccinia* (Maclean, 1982). But the situation is not simple. Many *Phytophthora* species while preferring organic nitrogen do not have an absolute requirement (Hohl, 1983). *Puccinia graminis* and virtually all other rust fungi tested, except *Gymnosporangium juniperi-virginianae*, require a sulphur-containing amino acid as their source of the element (Maclean, 1982) (see Chapter 8). This complicates an understanding of the nitrogen requirements of rust fungi. Certainly, neither cysteine nor methionine alone as sources of nitrogen can support growth of cultures of *P. graminis* (Howes & Scott, 1972). However, growth occurs if ammonia is present, though glutamine gives better growth (Maclean, 1982). On the other hand, *Cronartium ribicola* can use either cysteine or glutathione as the sole organic nitrogen source (Harvey & Grasham, 1974). *Puccinia graminis* cannot use nitrate and a nitrate reductase is absent from the mycelium; *G. juniperi-virginianae* can grow on nitrate (Maclean, 1982).

It is generally assumed that only bacteria are capable of denitrification. This is not so. *Fusarium oxysporium* possesses denitrifying activity and a cytochrome P-450, inducible by nitrate or nitrite, is involved in the process (Kizawa *et al.*, 1991; Shoun & Tanimoto, 1991). The fungus can bring about the anaerobic evolution of nitrous oxide (N_2O) from nitrate or nitrite. A subsequent study showed that genera closely related to *Fusarium* also have denitrifying ability (Shoun *et al.*, 1992). However, a wide variety of species such as *Mucor pursillus*, *Rhizopus javanicus*, *Aspergillus nidulans* and *Penicillium expansum* appear to be unable to bring about denitrification.

Absorption of inorganic nitrogen

Surprisingly little is known about how nitrate is transported into fungi. This is in contrast to the much larger body of information for green plant cells (Deane-Drummond, 1986; Ullrich, 1983, 1987). Undoubtedly, it has not helped that there is no usable radioisotope for nitrogen. But this should not be too great a handicap in view of the availability of ion-selective electrodes and the fact that $^{36}ClO_3^-$ can be used as a radioactive analogue (Deane-Drummond & Glass, 1983*a*, *b*).

In *Neurospora crassa*, the nitrate transport system is induced by either nitrate or nitrite (Schloemer & Garrett, 1974*a*) but is not present in mycelium grown in the presence of ammonia or Casamino acids. The appearance of the transport system is inhibited by cycloheximide. The K_T for transport is 0.25 mM; nitrite and ammonia are non-competitive inhibitors. A 50-fold accumulation has been observed and transport is inhibited by metabolic poisons. Mutants lacking a functional nitrate reductase locus can still accumulate nitrate; further, the K_T for nitrate transport is very similar to that of the wild-type.

The gene for the nitrate transporter (*crnA*) of *Aspergillus nidulans* has been characterised (Tomsett & Cove, 1979; Brownlee & Arst, 1983; Unkles *et al.*, 1990). From the nucleotide sequence, a hydropathy plot suggests that the polypeptide encoded by the gene has 10 membrane-spanning helices. The indications are that the protein is a member of a class hitherto not described. The expected molecular mass is *c.* 52 kDa. That said, there is no information about the actual process of nitrate transport and this is true also for *N. crassa*.

Candida utilis appears to possess an electrogenic nitrate/proton symport (Eddy & Hopkins, 1985). When washed cells grown on nitrate as the limiting nutrient were depleted of ATP, at pH 4–6 protons accompanied nitrate absorption in the ratio 2:1, charge balance being obtained by the expulsion of potassium from the cells.

Nitrite transport in *N. crassa* also has been studied by Schloemer & Garrett (1974*b*). The K_T is 86 μM, transport being inhibited by metabolic poisons. There is no repression or inhibition of transport with nitrate, ammonia or Casamino acids. Since no nitrate uptake occurs in the presence of the last two nitrogen compounds, it must be concluded that nitrate cannot use the nitrite transport system.

As with nitrate transport, knowledge of ammonia transport in fungi is relatively sparse. For instance, there is for no fungus the depth of knowledge about the mechanism of ammonia transport that exists for large algal coenocytes. In these organisms, ammonia and methylamine are transported by the same carrier, which is a uniport (Jennings, 1986*b*). The system has been very effectively probed using [^{14}C]methylamine. A somewhat similar system appears to be present in *N. crassa* (Slayman, 1977). The system has been investigated electrophysiologically. When ammonia is presented to the mycelium, the membrane potential is depolarised, the interior becoming more positive. The extent of the depolarisation is increased by nitrogen starvation. While the system does not appear to transport methylamine, the observed depolarisation is consistent with the transport of charge (NH_4^+) across the membrane by a uniport.

Methylamine enters the mycelium of *Penicillium chrysogenum* by a transport system that satisfies important criteria for its identification as the ammonia transport system (Hackette *et al.*, 1970). Though there are technical difficulties in carrying out the experiments, due to the low rates of uptake of methylamine in the presence of ammonia, it has been demonstrated that ammonia is a competitive inhibitor of methylamine transport, as is ethylamine. The K_T values for transport of ammonia, methylamine and ethylamine are, respectively, (approximately) 2.5×10^{-7} M, 1×10^{-5} M and 1×10^{-4} M, the increasing values being in keeping with the system for ammonia.

The transport of methylamine by *S. cerevisiae* shows biphasic Lineweaver-Burk kinetics, both phases being inhibited by ammonia, indicating that the compound enters by the methylamine systems, as suggested by Roon, Levy & Larimore (1977*a*). The two functions can be lost separately as a result of two genetically unlinked mutations (*mep-1* and *mep-2*). Double mutants grow very slowly on 10^{-3} M ammonia as the sole nitrogen source. Single mutations are little affected. *mep-1* mutations in which the cells have only the high affinity ($K_T \approx 2.5 \times 10^{-4}$ M) low capacity ($V_{max} \approx 20$ nmol (mg protein)$^{-1}$ min^{-1}) are resistant to methylamine toxicity but *mep-2* mutants, which have a low affinity ($K_T \approx 2 \times 10^{-3}$ M), high capacity system ($V_{max} \approx 50$ nmol (mg protein)$^{-1}$ min^{-1}) are not, due to accumulation of methylamine to toxic levels. Double mutants grow on 2×10^{-2} M ammonia, indicating the presence of a third transport system.

It is important that the studies on *P. chrysogenum* and *S. cerevisiae* are expanded to probe the driving forces moving ammonia into these fungi, so that the nature of transport can be established. Equally there is a need to establish the extent to which ammonia can diffuse into fungi, particularly in view of early studies by Morton & MacMillan (1954) and MacMillan (1956), which seemed to show that ammonia enters *Scopulariopsis brevicaulis* by simple diffusion.

Assimilation of nitrate

Nitrate assimilation in eukaryotic cells is a two-stage process, nitrate being reduced to nitrite, and nitrite being reduced to ammonia. The first stage is mediated by nitrate reductase and the second by nitrite reductase. Within eukaryotes, the capability for nitrate assimilation is found in nearly all algae and higher plants. The knowledge about the process in those organisms is relevant to understanding the process in those fungi possessing the capability. Thus, there is a considerable literature for the mycologist to consider. Relevant reviews are those by Dunn-Coleman, Smarrelli & Garrett (1984), Solomonson & Barber (1990), Tomsett (1989) and Wray & Kinghorn (1989).

The reduction of nitrate to ammonia by the cell represents considerable energy expenditure. Eight reducing equivalents are required for the complete reduction, viz:

$$NO_3^- \xrightarrow[\substack{\text{Nitrate} \\ \text{reductase}}]{2e^-} NO_2^- \xrightarrow[\substack{\text{Nitrite} \\ \text{reductase}}]{6e^-} NH_4^+$$

This means that cells possessing the capability of assimilating nitrate regulate the activity of the two enzymes to avoid the use of reducing power if ammonia is available. In the slow or non-growing photosynthetic tissues the enzymes are inactivated; in growing fungi, regulation is through gene expression. Indeed the genetic control of nitrate assimilation in fungi has been well studied. For a significant number of fungi a number (if not all) of the relevant genes have been identified (Table 6.4). However, by far the best studied are those of *Aspergillus nidulans* and *N. crassa* (Kinghorn, 1989; Marzluf & Fu, 1989; Scazzochio & Arst, 1989; Tomsett, 1989) and for the moment attention is focused here on what is known for these two fungi.

The synthesis of both enzymes is subject to two control mechanisms. There is induction by either nitrate or nitrite and there is nitrogen catabolite control (see pp. 230–5). When ammonia, glutamate or glutamine are absent from the medium, nitrate or nitrite switch on expression of their respective enzymes. In *A. nidulans*, the regulating gene is *nirA*, in *N. crassa*, it is *nit-4*, both encoding the regulatory protein for the process of nitrate assimilation. The genes relevant to nitrate assimilation and the growth tests used in the identification of mutant strains are listed in Table 6.5.

As far as the actual process of genetic control is concerned the following appears to be the situation in *A. nidulans* (Cove, 1970, 1979). The product of the *nirA* gene, when switched on by *areA*, is thought to switch on the structural genes for both nitrate and nitrite reductase. In the absence of nitrate, the nitrate reduction enzyme converts the *nirA* gene product to an inactive state, so that transcription is no longer stimulated. The cycle is broken by the presence of nitrate, which combines with the enzyme so

Table 6.4. *Nitrate assimilation genes in filamentous fungi*

Organism	Nitrate reductase	Nitrite reductase	Nitrate uptake	Nitrate induction control	Nitrogen metabolite control	Molybdenum cofactor (loci)	Reference
Aspergillus amstelodami	nia	nii	crn	nir	–	cnx (5 loci)	Unkles (1989)
A. nidulans	niaD	niiA	crnA	nirA	areA	cnx (6 loci)	See text
A. niger	niaD	–	–	–	–	cnx	Unkles et al. (1989)
A. oryzae	niaD	–	–	nirA	–	cnx	J. A. Macro & S. E. Unkles unpublished data
Cephalosporium acremonium	niaD	–	(crn)[a]	nir	–	cnx	Whitehead et al., unpublished data cited by Unkles (1989)
Fulvia fulva	nia	–	–	(nir)[a]	–	cnx	Talbot et al. (1988)
Fusarium oxysporum	nit-1	–	–	nit-3	–	nit (5 loci)	Correll, Klittich & Leslie (1987)
Gibberella fujikuroi	nit-1	–	–	nit-3	–	nit (4 loci)	Klittich, Leslie & Wilson (1986); Correll et al. (1987)
Neurospora crassa	nit-3	nit-6	–	nit-4	nit-2	nit (4 loci)	See text
Penicillium chrysogenum	niaA	niiA	–	nir	–	cnx (7 loci)	Birkett & Rowlands (1981)
Septoria nodorum	Nia	–	ChlT	Nir	Nit	Cnx (4 loci)	Newton & Caten (1988)
Ustilago maydis	nar-A	nir-1	–	nar-C	–	cnx (4 loci)	Lewis & Fincham (1970); Holliday (1974)

[a] Tentative designation.
From Unkles (1989), by permission of Oxford University Press.

Table 6.5. *Classification on the basis of growth tests of mutants concerned with nitrate metabolism in* Aspergillus nidulans *and* Neurospora crassa

Function	A. nidulans locus	N. crassa locus	Growth media						Plate test for nitrite production
			NH_4^+	NO_3^-	NO_2^-	Hypo-xanthine	Uric acid	Uric acid $+100\,mM$ ClO_3^-	
Wild-type			+	+	+	+	+	VS	+
NAR structural gene encoding the apoprotein	niaD	nit-3	+	−	+	+	+	Most R, few S	−
NIR structural gene encoding the apoprotein	niiA	nit-6	+	−	−	+	+	VS	++
Formation of the molybdenum-containing cofactor required by both NAR and XDH	cnxABC, E, F, G, H	nit-1, 7, 8, 9ABC	+	−	−	−	+	Most R, few S	−
Regulator gene necessary for nitrate induction of nitrate assimilation	nirA[a]	nit-4	+	−	−	+	+	SR	−
Regulator gene responsible for nitrogen metabolite control of nitrate assimilation and other pathways of nitrogen acquisition	areA[a]	nit-2	+	−	−	+	−	−	−
Nitrate permease gene	crnA	?	+	+	+	+	+	SR	+

NAR, nitrate reductase; NIR, nitrite reductase; XDH, xanthine dehydrogenase. For nitrogen source utilisation in growth media: + indicates growth on that medium, − indicates no growth (nitrogen-starved morphology). For chlorate resistance: R indicates resistance, SR an intermediate resistance characteristic of nirA⁻ and crnA, VS indicates extreme sensitivity, and − that the strain cannot grow. For the nitrite production plate test: + indicates colony stains pink, ++ indicates the colony has a pink halo from excreted nitrite, and − that the colony does not stain.

[a] The nirA and areA genes can mutate to give a variety of mutant phenotypes; these examples are representative of null alleles. Data are compiled from Cove (1976a, b), Tomsett & Garrett (1980) and see Wiame et al. (1985). From Tomsett (1989).

that it is unable to inactivate the *nirA* gene. Thus, nitrate reductase is not only important enzymically; it also plays a key (autogenous) role in regulation of the pathway. A similar situation seems to hold for *N. crassa* (Tomsett & Garrett, 1981; Fu & Marzluf, 1988).

It can be seen that in wild-type strains of *A. nidulans* and *N. crassa* both nitrate and nitrite reductase activities are not expressed in the absence of nitrate, even when reduced nitrogen is also absent. However, this is not true for all fungi; nitrate reductase activity can be expressed and sustained in ammonia-grown fungi in the absence of nitrate (Morton, 1956; Ali & Hipkin, 1985, 1986). This might be due to derepression or to the fact that the nitrate reductase in these fungi has a lower affinity than that in *A. nidulans* for the pathway-specific gene product (Hipkin, 1989). Whatever the mechanism, the extent of the nitrate reductase activity within the fungus has an effect on the temporal change in the rate of assimilation of nitrate from the medium when a fungus grown in the presence of ammonium is transferred to nitrate medium and vice versa (Hipkin, 1989). Thus, when cells of *Candida nitrophila* are grown in nitrate there is a lag in the disappearance of nitrate reductase activity when cells are transferred into a medium containing ammonium. At the initial stages after the transference, cells can assimilate nitrate as well as ammonium (Hipkin, 1989).

Nitrate reductase is a homodimer of molecular mass of 228 kDa in *N. crassa* (Garrett & Nason, 1969) and 180 kDa in *A. nidulans* (Minagawa & Yoshimoto, 1982). The enzyme is a multimeric protein containing flavin adenine dinucleotide, haem (cytochrome b_{557}) and molybdenum-pterin prosthetic groups in a 1:1:1 stoichiometry. In addition to the full nitrate reductase activity, the enzyme catalyses a variety of partial reactions involving one or more of these prosthetic groups, in the presence of artificial electron donors and acceptors (Figure 6.1).

Nitrite reductase from *N. crassa* has been purified to homogeneity (Greenbaum, Prodouz & Garrett, 1978; Prodouz & Garrett, 1981). The enzyme is a 290 kDa molecular mass homodimeric flavoprotein possessing two iron–sulphur centres organised as a tetranuclear Fe_4S_4 cluster. Also present is a sirohaem prosthetic group (Figure 6.2). As with nitrate reductase, the nitrite reductase of *N. crassa* is able to catalyse a number of partial reactions *in vitro* (Lafferty & Garrett, 1974; Vega, 1976).

As far as other fungi are concerned, there is information about nitrate reductase from *P. chrysogenum* and three species of yeast *Candida nitrophila*, *Hansenula anomola* and *Rhodotorula glutinis* (Table 6.6). As can be seen from the table, the enzyme from the first and last organisms

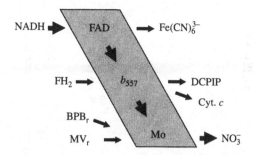

Figure 6.1. Interaction of electron donors and acceptors with nitrate reductase. The various inter- and intra-electron reactions that comprise the full and partial activities associated with nitrate reductase are shown, together with the individual prosthetic groups required for these activities. FH_2, reduced flavin; b_{557}, cytochrome b_{557}; DCPIP, dichlorophenol indophenol; BPB_r, reduced bromophenol blue; MV_r, reduced methyl viologen; FAD, flavin adenine dinucleotide. (From Solomonson & Barber, 1990.)

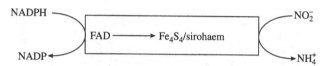

Figure 6.2. Electron-transfer reactions of nitrite reductase (From Tomsett, 1989.)

resembles that of *N. crassa*; that from *H. anomola* has a smaller subunit size. Of particular interest is the specificity of the enzyme for NADH and NADPH. The enzyme from the filamentous fungi *A. nidulans* and *N. crassa* is specific for NADPH; the same is true for the yeast *Sporobolomyces roseus* (Hipkin, 1989) and essentially the case for *Rhodotorula glutinis* and *P. chrysogenum*. However, the enzyme from many yeasts appears to be bispecific, that from *C. nitrophila* supports higher activity with NADH than with NADPH. There appears to be a similar situation with the yeasts *Brettanomyces lambicus*, *Candida utilis* and *Dekkera intermedia* (Hipkin, 1989). The extent to which there can be this difference in specificity is important in interpreting the physiology of a fungus growing in the presence of nitrate, particularly in the study of the levels of individual polyols, if such are produced in the cells (see Chapter 5).

Uptake of organic nitrogen compounds (except peptides) into *Saccharomyces cerevisiae*

Transport of organic nitrogen compounds by *S. cerevisiae* has been extensively studied. However, this has been primarily by genetic means,

Table 6.6. *The physical and biochemical properties of nitrate reductase (NR) from the yeasts* Candida nitrophila, Hansenula anomala *and* Rhodotorula glutinis, *and the filamentous fungi* Neurospora crassa *and* Penicillium chrysogenum

Organism	C. nitrophila	H. anomala	R. glutinis	N. crassa	P. chrysogenum
Molecular mass (kDa)					
Subunit	97	52	118	115	97
Dimer	200	–[a]	230	228 (290)	199
Tetramer	365	215	–	–	–
Stokes radius (nm)	8.5	6.5	7.0	7.0	6.3
$S_{20,w}$	10.25	8.0	7.9	8.0	–
Pyridine nucleotide specificity	NADH; NADPH	NADPH; NADH	NADPH; NADH	NADPH	NADPH; NADH
Specific activity (μkat mg^{-1})	7.0 (NADH–NR)	0.8 (NADPH–NR)	2.5 (NADPH–NR)	2.1	0.17–0.225 (NADPH–NR)
Turnover number (s^{-1})	1200 (dimer)	180	570	475	–
Optimum pH	7.0	–	7.5	7.5	7.2
K_m NO$_3$ (μM)					
NADH–NR	120	–	45	–	–
NADPH–NR	110	500	125	200	91
K_m NADH (μM)	17	80	160	–	130
K_m NADPH (μM)	30	25	20	62 (22)	10
Spectral absorption maxima (nm)	280; 413 (ox) 423; 525; 557 (red)	–	278; 412 (ox) 423; 527; 557 (red)	413 (ox) 423; 528; 557 (red)	–
References	Hipkin, Ali & Cannons (1986)	Zauner & Dellweg (1983)	Guerrero & Guttierez (1977)	Garrett & Amy (1978); Pan & Nason (1978); Horner (1983) (values in parentheses)	Renosto, Martin & Segel (1981)

[a] –, data not available (or not applicable).
Based on Hipkin (1989).

so that in many cases little is known about the mechanism of transport. Such compounds can be divided into two groups: those that are nitrogen sources and therefore catabolised on entry, and those that are biosynthetic precursors incorporated more or less directly into macromolecules. In both cases the mechanism of transport is subject to controls in a manner similar to the subsequent metabolic pathways used once the compound has entered the cell.

Table 6.7 summarises the catabolic uptake systems found in *S. cerevisiae*. The information in the table is for the most part explicit and does not demand further elaboration here in the text. Nevertheless a few comments are needed. The matter of nitrogen catabolite control is dealt with below (see pp. 230–5).

Urea is transported by two systems. That with a K_T of 14 μM is active, requires the presence of glucose and can lead to a 200-fold accumulation of urea within the cell. The system is induced by allophanate and the gratuitous inducer oxalurate. At concentrations in excess of 0.5 mM, urea is transported by a facilitated diffusion system, insensitive to nitrogen depression. There are some indications that the two systems are linked (Cooper, 1982). Oxalurate, which is an inducer of allantoin-degrading enzymes, is not a normal metabolite of *S. cerevisiae* and is transported by the constitutive allantoate transport system (Chisolm *et al.*, 1987).

Information on the general amino acid transport system (GAP) is based essentially on *gap* mutants. The system apparently transports all amino acids, including D-amino acids and citrulline. It is not clear whether proline, which is transported by another system (see below), is also transported by the GAP system. Control of the system appears to be complex (Cooper, 1982). There is inhibition by ammonia and nitrogen catabolite control. The system does not seem to have an inducer and is presumed to be constitutive.

The evidence available indicates that each amino acid molecule transported via the GAP system is accompanied in symport by two proton equivalents (Seaston *et al.*, 1973). Glycine transport has been studied in some detail (Ballarin-Denti *et al.*, 1984), particularly in relation to intracellular pH, as determined by ^{31}P NMR. Internal pH changes have more effect on transport than those occurring externally. Such findings urge caution in interpretations of the effects of inhibitors as being through the dissipation of the proton gradient across the membrane. Ballarin-Denti *et al.* (1984) provided a model for glycine transport that requires a finite membrane potential, which is sensitive to both external and internal pH and accommodates the observed discrepancy between the measured proton

Table 6.7. *Catabolic transport systems for nitrogenous compounds in Saccharomyces cerevisiae*

Compound	Concentration (inside/outside)	K_T	Mode of uptake	Control[a]	pH optimum	Efflux	Exchange	Inhibitors	Gene	Reference[b]
Allantoin	7700	12 μM	Active	Ind, N-rep	5.2	Very slow	No		DAL4	1
Allantoate	15 000	50 μM	Active	Const, N-rep	5.8	Very slow	No		DAL5	2
Urea	200	14 μM	Active	Ind, N-rep	3.3	Yes		Formamide, acetamide	DUR3	3
Urea		2.5 mM	Facilitated	Const					DUR4	3
Oxalurate	2–4	1.2 mM	Active	Const	3.0	Rapid	Slow			4
Methylamine (ammonia)	850	0.25 mM	Active	?	6.0–6.5				MEP2	5
Methylamine (ammonia)		2 mM	Active	?	6.0–6.5				MEP1	5
Amino acids		Variable	Active	Const, N-rep					GAP1	6
Proline		0.025 mM 13 mM		Const, N-rep	4.0				PUT4	7
Asparagine									*	8
L-Tryptophan	200	0.41 mM	Active		5.0–5.5	Very slow				9

[a] ind, inducible; N-rep, sensitive to nitrogen repression; Const, constitutive.

[b] 1, Sumadra & Cooper (1977), Sumadra et al. (1978), Cooper, Gorski & Turoscy (1979a); 2, Morlion & Domnas (1962), Turoscy & Cooper (1979); 3, Cooper & Sumadra (1975), Sumadra, Gorski & Cooper (1976); 4, Cooper, McKelvey & Sumadra (1979b); 5, Roon et al. (1975a), Roon, Larimore & Levy (1975b), Roon, Levy & Larimore (1977a), Roon, Meyer & Larimore (1977b), Dubois & Grenson (1979); 6, Grenson et al. (1966), Grenson, Hou & Crabeel (1970), Darte & Grenson (1975), Rytka (1975), Roon et al. (1977b), Larimore & Roon (1978), Larimore et al. (1980); 7, Lasko & Brandriss (1981); 8, Dunlop & Roon (1975); 9, Kotyk & Dvořáková (1990).

* Mutant exists but locus is unnamed.

Based on Cooper (1982).

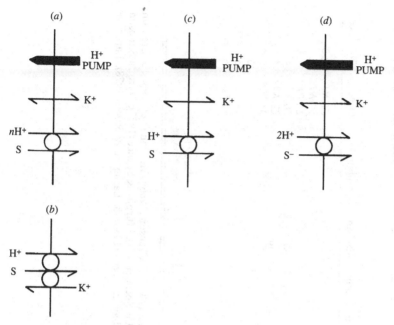

Figure 6.3. Charge neutralisation during the operation of symport mechanisms for substrates (S) (amino acids) bearing different formal charges. In (a), (c) and (d) the mechanism is inherently electrogenic. Neutralisation might initially be achieved by membrane depolarisation leading to the efflux of K^+ or by ejection of protons through the proton pump. In (b) efflux of K^+ (antiport) through the symport itself brings about the necessary neutralisation. (From Eddy, 1982.)

flux (one per glycine) and that required kinetically (two or more per glycine). This accommodation is brought about by the shunting of protons through other proton-conducting pathways in the plasma membrane.

As may be surmised from the above, the charge carried by the amino acid is an important determinant of the possible movement of other ions. Eddy (1980, 1982) provided a helpful background to the mechanism of GAP and other amino acid transport systems in *S. cerevisiae* and also *Candida* sp., in particular focusing on how charge neutralisation is established, the possibilities being indicated in Figure 6.3.

To date, virtually all studies on amino acid transport in *S. cerevisiae* have used intact cells. There has recently been a move to use plasma membrane vesicles (Naider & Becker, 1986; Opekarová, Driessen & Konings, 1987; Calahorra *et al.*, 1989). Though there have been no new findings concerning the detailed mechanism of transport, the use of such vesicles is to be recommended in that the complications of cellular metabolism are removed.

Table 6.8. *Anabolic nitrogen transport systems in Saccharomyces cerevisiae: amino acids*

Compound	K_T	Concentration (inside/outside)	Mode of uptake	Control	pH optimum	Efflux	Exchange	Inhibitors	Gene	Reference[a]
Arginine	10 μM		Active						CAN1	1, 8
Lysine	0.2 mM		Active						CAN1	2, 8
Lysine	78 μM	100	Active		c. 3				LYP1	3, 9
Histidine	20 μM					No			HIP1	4
Histidine	0.5 mM					No			CAN1?	4
Methionine	12 μM	56	Active				No		METP	5
Methionine	0.8 mM		Active				No			5
Cysteine		17	Active							5
S-adenosyl-methionine	1.6–3.3 μM	Active							SAP3	6
Glutamate	See text									
Leucine	0.5 mM, 1 mM	6	Active					S-adenosyl-methionine		7

[a] 1, Grenson et al. (1966); Chan & Cossins (1976); Whelan, Gocke & Manney (1979). 2, Grenson (1966). 3, Grenson (1966). 4, Crabeel & Grenson (1970). 5, Maw (1963); Gits & Grenson (1967). 6, Spence (1971); Murphy & Spence (1972); Petrotta-Simpson et al. (1975). 7, Bussey & Umbarger (1970); Ramos et al. (1975, 1980); Law & Ferro (1980). 8, Larimore et al. (1980). 9, Garcia & Kotyk (1988). Based on Cooper (1982).

Proline transport has been studied in some detail (Horák & Říhová, 1982; Horák & Kotyk, 1986). Two systems appear to be involved: one high affinity ($K_T = 31\,\mu M$) and the other low affinity ($K_T > 25\,mM$). The high affinity system is highly specific for proline and its analogues. Some 50%–60% of the accumulated proline is lost in 90 min and the loss is insensitive to 2,4-dinitrophenol and external pH between pH 3 and pH 7.3. This loss, probably by diffusion, is in contrast to the situation with other amino acids that are retained. Though proline is concentrated within the cells, uptake is virtually insensitive to inhibitors except antimycin and anaerobic conditions. Uptake is not accompanied by alkalinisation of the medium or by membrane depolarisation (as determined by distribution of tetraphenylphosphonium) suggesting that proton cotransport is not involved. The energisation of transport is therefore unclear.

Details of the transport systems for amino acids involved directly in biosynthesis/anabolism are summarised in Table 6.8. As with Table 6.7, only a few comments are required in the text. The arginine system is for the basic amino acids – arginine, lysine, ornithine, canavanine and probably histidine. Little is known about control of the system. As one can see from Table 6.8, lysine and probably histidine are also transported by other systems of much higher affinity. The lysine and methionine transport systems are both proton symports, with one proton equivalent accompanying each sugar molecule (Seaston *et al.*, 1973; Cockburn, Earnshaw & Eddy, 1975).

Glutamate transport is thought to occur by three different systems (Cooper, 1982). One system is likely to be the GAP because the lower K_T component of a biphasic plot based on Lineweaver-Burk kinetics disappears when proline is replaced by ammonia as the nitrogen source for growing the cells (Darte & Grenson, 1975). The high affinity system for L-leucine is also likely to be the GAP system (Ramos, de Bongioanni & Stoppani, 1980). The other two systems for glutamate, if indeed there are two other systems, are ill defined. However, it is known that glutamate transport involves the influx of three proton equivalents and the efflux of three potassium equivalents, indicating that at least one system out of the three involves proton cotransport (Cockburn *et al.*, 1975).

Table 6.9 summarises knowledge about anabolic transport systems for purines and pyrimidines. Purine transport has been shown to be a proton symport (Reichert, Schmidt & Forêt, 1975; Reichert & Forêt, 1977) and kinetically described (Forêt, Schmidt & Reichert, 1978). The amino acid sequence and folding of the polypeptide of the uracil carrier in the membrane has been predicted from the nucleotide sequence of the *FUR4*

Table 6.9. *Anabolic nitrogen transport systems in Saccharomyces cerevisiae: purines and pyrimidines*

Compound	K_T	Concentration (inside/outside)	Mode of uptake	Control[a]	pH optimum	Efflux	Exchange	Inhibitors	Gene	Reference[b]
Adenosine, guanine[c] hypoxanthine, cytosine	0.12 μM	1900	Active			Rapid	Rapid		APP1	1
Uracil	12 μM, 156 μM	40	Active		4.3–4.5	Rapid		Dihydrouracil	FUR4	2
Uridine									URD1	3
Ureido-succinate	27 μM	10 000	Active	Const, N-rep	4.2–4.7	No	No	*[d]	DAL5 UEP1	4

[a] Const, constitutive; N-rep, sensitive to nitrogen repression.
[b] 1, Grenson (1969); Jund & Lacroute (1970); Pickering & Woods (1972); Polak & Grenson (1973). 2, Jund, Chevallier & Lacroute (1977). 3, Grenson (1969). 4, Drillien & Lacroute (1972); Greth, Chevallier & Lacroute (1977).
[c] Here, the data are controversial.
[d] N-carbamyl-glutamate, -butyrate, -valine, -alanine, -glycine, and N-acetylglycine.
From Cooper (1982).

gene (Jund, Weber & Chevallier, 1988). Hopkins *et al.* (1988) have shown that both uracil and cytosine transport are by proton symports.

The tables summarising transport of nitrogenous compounds into *S. cerevisiae* indicate the extent to which there is also efflux and exchange of a particular compound, as well as the characteristics of influx. A simple carrier model suggests that amino acids would leak out of the cells if they were to be transferred to a medium free from the relevant nitrogen compounds. Eddy (1980, 1982) surveys those observations pertaining to the kinetic asymmetry of transport systems for nitrogenous compounds in yeasts such as *C. utilis* and *S. cerevisiae*. It has to be said that, although mechanisms can be envisaged, e.g. feedback inhibition through allosteric control of the transport system or transinhibition (Cuppoletti & Segel, 1974; Sanders & Hansen, 1981), the actual evidence for any one mechanism appertaining to a particular system is scanty. It has been suggested (see Eddy, 1980, 1982) that sequestration of nitrogen compounds in the vacuole can account for irreversibility of uptake. Certainly, there is clear evidence that amino acids, purines and other compounds accumulate in the vacuole of yeasts (see Chapter 4). Nevertheless the presence of two membranes through which the compound has to move in order to reach the external medium demands the same kind of mechanism to produce asymmetry as one membrane. It has to be said that there is as yet no evidence to indicate how location of a compound in the vacuole makes it less accessible to the external medium.

However, under certain conditions certain yeast strains can be made to lose a considerable proportion of their amino acids. This was first observed by Lewis & Phaff (1965) and is known as 'shock excretion'. It is not clear why it occurs and is not thought to be a particular aspect of the normal functioning of the amino acid transport systems (Eddy, 1982).

Uptake of amino acids

Achlya

The characteristics of the amino acid transport systems in this lower fungus are summarised in Table 6.10. It should be realised that the information presented there does not reveal a number of uncertainties. Of the three amino acids (methionine, threonine and lysine) transported by two systems, it is possible that methionine might be transported by a specific system and by the neutral-aromatic system ((ix) in Table 6.10) (Wolfinbarger, 1980*a*). The effect of one amino acid on the uptake of another has proved

Table 6.10. *Amino acid transport systems in* Achlya

Amino acids transported	K_T (M)	Comments	System notation[a]
Methionine	5.33×10^{-6} and 2.0×10^{-4}	Two saturable components	(i)
Cysteine	7.5×10^{-5}		(ii)
Proline	1.5×10^{-4}		(iii)
Serine	1.25×10^{-4}	Two saturable components for threonine	(iv)
Threonine	2.5×10^{-5} and 2.0×10^{-4}		
Aspartate	4.0×10^{-6}		(v)
Glutamate	2.5×10^{-5}		
Glutamine	1.5×10^{-4}		(vi)
Asparagine	7.5×10^{-5}		
Glycine	1.5×10^{-4}		(vii)
Alanine	1.67×10^{-4}		
Histidine	1.67×10^{-4}	Two saturable components for lysine	(viii)
Lysine	8.33×10^{-6} and 1.0×10^{-4}		
Arginine	3.33×10^{-5}		
Phenylalanine	5.0×10^{-5}	System (ix) transports these amino acids as two overlapping groups	(ix)
Tyrosine	3.33×10^{-5}		
Tryptophan	1.67×10^{-4}		
Leucine	1.11×10^{-4}		
Isoleucine	6.67×10^{-5}		
Valine	8.3×10^{-5}		

Ca^{2+} bound to a membrane-localised glycopeptide permits concentrative transport to take place (see text for discussion).
[a] A definitive system notation has not been proposed. The symbols employed here (Wolfinbarger, 1980*a*) correspond only to those used to delineate one system from another (Singh & Lé John, 1975).
From Singh & Lé John (1975); Wolfinbarger (1980*a*). From *Microorganisms and Nitrogen Sources*, 1980, reproduced by permission of John Wiley and Sons Limited.

difficult to analyse. The reader should consult the article by Singh & Lé John (1975) for details. The most interesting finding is that methionine inhibits in a non-competitive manner the transport of every other amino acid. None of the same amino acids inhibits the transport of methionine. There is no effect of methionine on the K_T values for transport; the amino acid must be having its effect not on the transport systems per se but indirectly.

Very interestingly, addition of calcium to calcium-deficient cultures enhances amino acid transport by about 100% (Cameron & Lé John, 1972a, b). The requirement is specific; Ba^{2+}, Mn^{2+}, Mg^{2+}, Co^{2+} and Fe^{2+} could not substitute. However, high concentrations inhibit transport, an action that could be overcome by the presence of citrate (Singh & Lé John, 1975). Earlier it had been shown that osmotic shock removes a low molecular weight glycopeptide from germlings of *Achlya*, which are then unable to concentrate amino acids (Lé John et al., 1974). Further the glycopeptide was shown to bind calcium to the extent of 20 atoms mol^{-1} (Lé John & Cameron, 1973). It has been suggested by Singh & Lé John (1975) that the glycopeptide acts as a calcium sink to regulate the concentrations of the ion made available to the fungus for its metabolic activities. These authors argued against calcium being involved in the transport process; this is supported indirectly by vibrating-probe studies, which indicated that amino acids are transported into *Achlya* by a proton symport (Kropf et al., 1984). The transport of nucleosides (Stevenson & Lé John, 1978) and sugars (Goh & Lé John, 1978) respond in a similar manner to the presence and absence of calcium and citrate in the external medium.

Aspergillus nidulans

The information available concerning amino acid transport in *A. nidulans* is summarised in Table 6.11. There are considerable similarities with what has been found for *N. crassa*, for which there is more information (see following section); a few comments are all that is necessary to amplify the material in Table 6.11.

The optimum pH for acidic amino acids is 5.0, indicating that the anionic species of aspartate, glutamate and cysteate is transported. The system is repressed by ammonia. The basic system appears to be constitutive, transporting lysine, arginine and ornithine. Histidine does not seem to be transported by this system. Characterisation of other transport systems has been helped by the availability of the *nap-3* mutant, which is unable to transport phenylalanine, methionine, serine and leucine when growing on minimal medium (Piotrowska et al., 1976). However, under nitrogen starvation, phenylalanine and methionine transport is regained but can be inhibited by a 10-fold excess of arginine. This suggests, by analogy with other moulds, the presence of a general amino acid permease. When *nap-3* is starved of sulphur, methionine transport is regained, but not that of phenylalanine, indicating the presence of a specific

Table 6.11. *Amino acid transport systems in Aspergillus nidulans*

Amino acids transported	K_T (M)	V_{max} (nmol min^{-1} mg^{-1})	Comments	Reference
Aspartate	1.0×10^{-5}	1.72	Regulated by ammonium ion, missing in mutant *aau*	Robinson, Anthony & Drabble (1973*a*, *b*), Pateman, Kinghorn & Dunn (1974)
Glutamate	1.8×10^{-5}	4.35		
Cysteate	1.9×10^{-5}	0.03		
Lysine	8×10^{-6}	—	His is not an effective inhibitor of Arg transport	Piotrowska *et al.* (1976)
Arginine	1×10^{-5}	2.8		
Ornithine	3×10^{-3} (K_i)	—		
Phenylalanine	1×10^{-4}	1	Missing in *nap-3*	Sinha (1969), Piotrowska *et al.* (1976)
Methionine	—	—		
Serine	—	—		
Leucine	—	—		
Methionine	—	—	Derepressed under sulphur starvation	Piotrowska *et al.* (1976)
General amino acids	—	—	Derepressed under nitrogen starvation	Piotrowska *et al.* (1976)

Transport systems characterised in germinated conidia and mycelia.
—, no data obtained.
From Wolfinbarger (1980*a*). From *Microorganisms and Nitrogen Sources*, 1980, reproduced by permission of John Wiley and Sons Limited.

system for the former amino acid, as is the case for *N. crassa* (see next section).

As well as the transport systems listed in Table 6.11, there are two proline transport systems characterised solely by genetic studies (Arst & MacDonald, 1975, 1978; Arst, MacDonald & Jones, 1980). One, specified by the *prnB* gene, is responsible for most of the proline uptake when the system has been induced by the amino acid; it is also subject to ammonia repression. The other is not induced by proline and is not repressed, but is inhibited, by ammonia. The amino acid sequence of the *prnB*-specified transport protein has been predicted from the nucleotide sequence of the gene (Sophianopoulon & Scazzocchio, 1989).

Neurospora crassa

This fungus possesses only five genetically and biochemically distinct transport systems (Table 6.12). Readers should consult the article by Wolfinbarger (1980*a*) for the background. The five systems are (the notation following that of Pall, 1970): (I) for neutral and aromatic amino acids; (II) for general transport of neutral, basic and acidic amino acids; (III) for basic amino acids; (IV) for acidic amino acids; and (V) for methionine. System IV exhibits low activity in growing mycelium but high activity in mycelium starved of carbon, nitrogen or sulphur (Pall, 1970). However, at low pH values, acidic amino acids are also transported by the neutral and general amino acid transport systems in both conidia and mycelia. Indeed after germination of conidia, 85%–90% of the total amino acid transport capacity resides in the general amino acid transport system; in ungerminated conidia it is 40% (Wolfinbarger, Jervis & de Busk, 1971). Deprivation in *N. crassa* of a readily utilisable source of sulphur, as well as resulting in a derepression of enzymes involved in sulphur metabolism, results also in the derepression of the methionine transport system (Pall, 1971). The system can be considered as one for scavenging sulphur. Transinhibition has been shown to take place in systems I, II and V and is system specific (Pall, 1971; Pall & Kelly, 1971); the other two systems appear not to have been studied from this point of view.

Transport of neutral and basic amino acids via the general amino acid transport system (system II) has been studied by Sanders, Slayman & Pall (1983) in a mutant of *N. crassa* (*bat mtr*) that lacks other transport systems for these solutes. All the amino acids tested, including those with no net charge, elicited rapid depolarisation of the membrane potential. This indicates coupling of transport of the amino acid to ion movement and,

Table 6.12. *Amino acid transport systems in Neurospora crassa*

System	Amino acids transported	System K_T (M)	V_{max} (nmol min^{-1} mg^{-1})	Comments	System designation
I	Phenylalanine	$2-4 \times 10^{-5}$	—	System also transports Val, Ala, Gly, Ser, Met; Asp and Glu transported optimally at acidic pH values	Neutral amino acid transport system
	Tyrosine	5×10^{-5}	—		
	Tryptophan	1×10^{-4}	—		
	Leucine	6.5×10^{-4}	12.4		
	Histidine	8×10^{-5}	1.5		
	Aspartate	—	—		
	Glutamate	—	—		
II	Tryptophan	4×10^{-5}	—	Asp and Glu optimally transported at acidic pH values	General amino acid transport system
	Methionine	3.0×10^6	—		
	Phenylalanine	4×10^{-5} (C);[a] 2×10^{-6} (M)[b]	0.85		
	Tyrosine	—	—		
	Leucine	4.9×10^{-6} (M)	—		
	Asparagine	2×10^{-5}	—		
	Aspartate	3.4×10^{-3} (C); 1×10^{-5} (M)	20 (C); 25 (M)		
	Glutamate	4×10^{-5} (M)	15 (M)		
	Citrulline	—	—		
	Arginine	3.2×10^{-6}	—		
	Lysine	—	—		
	Glycine	8×10^{-6}	—		
	Histidine	1.2×10^{-3}	3.6		

III	Arginine	2×10^{-6}	—		Basic amino acid transport system
	Lysine	5×10^{-6}	—		
	Histidine	$3.5 \times 10^{-3}, 1.6 \times 10^{-6}$ (K_i)	30.5		
	Canavanine	7×10^{-6}	—		
IV	Cysteic acid	7×10^{-6}	—	Active under nitrogen starvation conditions	Acidic amino acid transport system
	Aspartate	1.3×10^{-5}	0.2		
	Glutamate	1.6×10^{-5}	—		
V	Methionine	2.3×10^{-6}	—	Active under sulphur starvation conditions	Methionine transport system

—, no data obtained.
[a] (C) is conidial stage of development.
[b] (M) is mycelium.
From Wolfinbarger (1980a). From *Microorganisms and Nitrogen Sources*, 1980, reproduced by permission of John Wiley and Sons Limited.

since potassium and sodium have no effect on transport, it was argued that amino acids enter the hyphae in symport with protons. The depolarisation caused by basic amino acids was found to be 60%–70% greater than that for neutral amino acids. Analysis of the current v. voltage curves showed the difference to be consistent with a constant proton:amino acid stoichiometry of 2. The greater depolarisation with basic amino acids is caused by the positive charge on the amino acid itself. It should be noted that the approach used in this study of proton cotransport is to be recommended – though it will depend on the ability to insert electrodes into the cell or hypha – as the technique overcomes such problems as variations in the background proton permeability. The reader should consult the article by Sanders *et al.* (1983) for a discussion of the procedures for determining the proton:solute stoichiometry of a transport system.

When a strain of *N. crassa* devoid of any of the three constitutive amino acid transport systems (I, II and III) is provided with arginine as the sole source of nitrogen, growth can still occur (de Busk & Ogilvie-Villa, 1982). Subsequent studies showed that, under nitrogen starvation, the presence of a functional *nit-2* gene product and the presence of certain amino acids leads to the induction of an extracellular deaminase (de Busk & Ogilvie, 1984*a*, *b*, *c*). The enzyme, which is L-stereospecific, converts the amino acid to its respective keto acid plus equimolar amounts of ammonia, which then act as the nitrogen source. The relationship between the amino acid in the medium and the production of deaminase or deaminases (several may exist) is not simple. There are four groups:

(1) L-Glutamate, L-glutamine, L-asparagine, L-lysine, L-tryptophan and L-proline fail to elicit production of deaminase.
(2) L-Ornithine and L-alanine elicit production only in mycelium in which all three constitutive amino acid transport systems are non-functional.
(3) Arginine elicits enzyme production in mycelium in which the general amino acid transport system is non-functional.
(4) Several neutral amino acids elicit enzyme production irrespective of the presence or absence of transport systems.

The situation with respect to glutamine has been analysed further (de Busk & Ogilvie, 1985). This amide is transported by systems I and II for amino acids. A strain without these systems that is also unable to produce the deaminase can nevertheless utilise glutamine. de Busk & Ogilvie (1985) suggested two ways by which utilisation might take place, both involving the degradation of the amine and the release of ammonia.

Table 6.13. *Amino acid transport systems in* Penicillium chrysogenum

Transport system	K_T (M)	V_{max} (nmol mg^{-1} min^{-1})	Comments
Methionine	10^{-5}	1	Appears under sulphur deficiency
Cystine	2×10^{-5}	1.2	Appears under sulphur deficiency
Cysteine	1.4×10^{-4}	2.1	Constitutive
Glutathione (reduced)	1.7×10^{-5}	2.3	Appears under sulphur deficiency
Neutral and basic amino acids	2×10^{-5}	10	Appears under carbon or nitrogen deficiency, cysteine is a substrate of this system, cystine is not
Acidic amino acids	10^{-4}	10	Appears under carbon or nitrogen deficiency, does not transport D-stereoisomers
Basic amino acids	5×10^{-6}	1	Constitutive
Arginine	—	—	Constitutive
Lysine	—	—	Constitutive
Proline	—	—	Appears under carbon or nitrogen deficiency

Kinetic constants given were obtained using 'only fine hairy microfilamentous mycelium for transport' (Hunter & Segel, 1971).
From Hunter & Segel (1971); Wolfinbarger (1980a). From *Microorganisms and Nitrogen Sources*, 1980, reproduced by permission of John Wiley and Sons Limited.

Penicillium

Our knowledge of amino acid transport in *Penicillium* is essentially dependent on the work of I. H. Segel and his colleagues studying *P. chrysogenum* and W. Roos and colleagues studying *P. cyclopium*. The findings for the former are summarised in Table 6.13. All the transport systems listed are active. Studies on the effect of weak acids on transport of amino acids suggest that it is brought about by a proton electrochemical potential gradient across the plasma membrane (Hunter & Segel, 1973a). Transport of amino acids bearing a positive charge is increasingly inhibited by sodium chloride at concentrations greater than 10^{-5} M but particularly so at concentrations greater than 10^{-1} M (Hunter & Segel, 1971). Under the same conditions the uptake of zwitterionic methionine (Benko, Wood & Segel, 1969; Hackette *et al.*, 1970) and choline-*O*-sulphate (Bellenger

Table 6.14. *Kinetic constants of amino acid transport systems in* Penicillium cyclopium

System	Substrate(s)	K_T (M)	J_{max} (mol min^{-1} (mg protein)$^{-1}$)
(A) Specific systems, assayed in nutrient-sufficient cells			
a	L-Arginine, L-lysine (L-ornithine)[a]	Arg 1.9×10^{-6}	4.4×10^{-9}
b	L-Phenylalanine, L-tyrosine	Phe 6.6×10^{-3}	2.8×10^{-7}
c	L-Glutamic acid	Glu 4.8×10^{-4}	5.5×10^{-8}
	(L-leucine, L-alanine)[a]	Leu[b] 3.0×10^{-3}	6.0×10^{-8}
d	L-Leucine (L-alanine)[a]	Leu 1.3×10^{-2}	2.3×10^{-7}
(B) Less specific systems, assayed in nitrogen-starved cells			
I	α-Amino-α-carboxylic acids,	Arg 1.2×10^{-5}	3.2×10^{-8}
	L-isomer preferred	Glu 2.4×10^{-5}	3.1×10^{-8}
	(General amino acid permease)[a]	Phe 1.4×10^{-5}	6.3×10^{-9}
		Leu 2.4×10^{-5}	1.5×10^{-8}
II	L-Arginine, L-phenylalanine	Arg 5×10^{-3}	1.8×10^{-7}
III	L-Glutamic acid, L-phenylalanine	Glu 7×10^{-3}	2.1×10^{-7}

[a] Information in brackets indicates other possible amino acids transported.
[b] K_i (Leu) of L-glutamic acid uptake.
From Roos (1989).

et al., 1968) is unaffected. The sulphate anion shows an increasing rate of uptake as the external potassium chloride concentration is increased above 10^{-4} M, though above 10^{-1} M the rate of uptake declines (Bradfield *et al.*, 1970; Hackette *et al.*, 1970). It is likely that these changes are due to changes in the electrochemical potential gradient across the plasma membrane acting on porters involving protons.

There are some inexplicable effects of cycloheximide, phosphate and calcium on transport (Hunter, Norberg & Segel, 1973). High concentrations of phosphate are inhibitory but the effect is enhanced by cycloheximide. Calcium reverses the effect but does not prevent an inhibition of protein synthesis.

The transport systems appears to be subject to transinhibition but it is only for the general amino acid transport system that there is firm evidence that this is so (Hunter & Segel, 1973*b*).

The characteristics of the amino acid transport system reported for *P. cyclopium* are summarised in Table 6.14. The uptake of all amino acids by this fungus is accompanied by a transient acidification within the hyphae as determined by the distribution of the acid dye bromocresol blue (Roos, 1989). Short-term preaccumulation of several anions such as citrate, α-oxoglutarate and glutamate (but not glutamine) increases the initial rate of amino acid uptake at a pH above the normal optimum. Also uncouplers such as azide inhibit uptake under both anaerobic and aerobic conditions. In the former situation it appears that the ATP content is not influenced by the inhibitors. *In toto* the evidence is adduced to be in favour of the presence of proton/amino acid symports.

Penicillium cyclopium is like *P. chrysogenum* (and *N. crassa* and *S. cerevisiae* and probably *A. nidulans*) in possessing a general amino acid transport system. However, the optimum pH for the system is much lower for *P. cyclopium*, namely 3.5. This difference may explain the lower affinity of this transport system in *P. cyclopium* for acidic amino acids than in the other fungi (Roos, 1989). At the pH optimum (5–6) for the system in these other fungi, the acidic amino acids would be predominantly anionic and as such virtually unacceptable to a lower affinity general amino acid transport system. It is significant that these same fungi develop separate systems, along with the general amino acid transport system, when conditions for derepression are brought about, for transporting anionic and zwitterionic species (Hunter & Segel, 1971; Wolfinbarger & de Busk, 1971; Darte & Grenson, 1975). In *P. cyclopium* such systems could not be detected; system c (Table 6.14) may be equivalent to the acidic amino acid transport system in other fungi, but it is not under nitrogen catabolite control. The low pH optimum of the general amino acid transport system is believed to be a function of the fact that the fungus is supposed to live in a rather acid environment (Roos, 1989).

On the other hand, system c in *P. cyclopium* is like similar basic amino acid transport systems in *N. crassa* and *P. chrysogenum* (see above). A specific leucine transport system has been reported also for *S. cerevisiae* but with a 10-fold higher affinity (Ramos *et al.*, 1980) (see Table 6.8).

Peptide utilisation

Candida albicans

The knowledge that the peptide transport system in the bacterium *Escherichia coli* has very broad specificity led to the realisation that it should be possible to introduce normally impermeant molecules into the organism if they were joined to a peptide. Ames *et al.* (1973) and Fickel & Gilvarg (1973) both demonstrated that this could take place. The process has been termed 'illicit transport' (Ames *et al.*, 1983). Further the process can be used to introduce an antimicrobial agent into the cell. In view of the importance of *Candida* infections in humans, it is not surprising that there should be interest in peptide transport in *C. albicans*.

Before detailing the information for this yeast it is appropriate to amplify a little further the characteristics of effective illicit transport. To be effective the peptide conjugate must:

(1) enter the cell intact;
(2) not be degraded externally, such as by serum enzymes;
(3) after entering the cell, elicit its lethal effect directly or be enzymically cleaved to produce a toxic component;
(4) not be absorbed or, if absorbed, then only slowly so by the host cells and the natural microflora of the host;
(5) not be cleared from the host before it has time to enter and kill the pathogen.

It must be apparent from the foregoing that knowledge of the peptide transport system is a necessary adjunct to the design of successful drugs. In order to achieve this end, there is a need for specific and reproducible techniques to measure the uptake of peptides. Early studies used radiolabelled peptides. However there is now a method, based on the conjugation of peptides, to make them fluorescent with fluorescamine, which itself is not fluorescent, such that uptake can be monitored continuously (Payne & Nisbet, 1981; McCarthy *et al.*, 1985*a*). Another development of value is the use of what have been called 'detector peptides' (McCarthy *et al.*, 1985*a*). One such is L-analyl-L-thiophenylglycine, which when cleaved releases thiophenol, the rate of release of which can be determined spectrophotometrically.

Early studies on peptide transport into *C. albicans* are confusing. Logan, Becker & Naider (1979), studying the uptake of L-methionyl-L-methionyl-L-[methyl-^{14}C]methionine found that dimethionine does not compete for uptake with the tripeptide. Some tripeptides were found to be effective competitors, e.g. trileucine, but others not, e.g. trialanine. Davies (1980) reported that di- and tripeptides share a common carrier, though it has to be pointed out that only the ability of a peptide to reduce the uptake of another, labelled with ^{14}C, was taken as the criterion. Transport was found to require metabolic energy and an α-linkage between the amino acid residues.

Not only is there disagreement between the two sets of data with respect to the number of peptide transport systems but other data, concerned with whether or not the free amino or carboxyl termini are required for successful transport, also are not compatible. Thus, Logan *et al.* (1979) found that acetylated trimethionine inhibits trimethionine transport but trimethionine when methylated to form the ester or in the form of the amide is without effect. This suggests that the free carboxyl group, but not the free amino group, is required for transport. In contrast, Davies (1980) found that acetyltrialanine did not reduce the rate of transport of trialanine. Also, she found that dialanine in which the terminal carboxyl group had been modified by a tetrazole or a phosphonate group inhibited trialanine transport. In this case it seemed that the amino group was necessary for transport but not the carboxyl group. Davies (1980) argued that the difference between the two sets of observations was due to the differences between the two strains of *C. albicans* used.

More recent studies (Yadan *et al.*, 1984; McCarthy *et al.*, 1985*a*) have shown that there are two peptide transport systems in *C. albicans*. One system is able to transport dipeptides with a reduced affinity for oligopeptides, while the other transports oligopeptides with a low affinity for dipeptides. This was shown by competition studies and the use of a mutant lacking dipeptide but with unaltered oligopeptide transport. It was argued that the different results obtained by Logan *et al.* (1979) and Davies (1980) were due to differences in the concentrations of peptides used; only at low concentrations would it appear that there was only one transport system.

The polyoxins, nikkomycins and bacilysin (tetaine) are all molecules that contain a peptide moiety (Figure 6.4) and all have strong anti-*C. albicans* activity in the laboratory (Kenig & Abraham, 1976; Becker *et al.*, 1983; McCarthy, Troke & Gull, 1985*b*). The activity of all three types of compound is inhibited by various peptides (Mehta *et al.*, 1984; Yadan *et al.*, 1984; McCarthy *et al.*, 1985*a, b*). Those mutants, referred to above, that cannot transport dipeptides are nikkomycin resistant. Polyoxin-resistant mutants can also be resistant to nikkomycins and bacilysin (Mehta *et al.*, 1984; Payne & Shallow, 1985).

Neurospora crassa

Peptide utilisation has been reasonably well studied in *N. crassa* (Wolfinbarger, 1980*b*). Initial studies looked at the growth of mutants, auxotrophic for individual amino acids, on peptides that could act as a supply of the requisite amino acid. Use of another amino acid capable of blocking transport of the required amino acid allowed an identification of those peptides that could be transported into the fungus. As can be

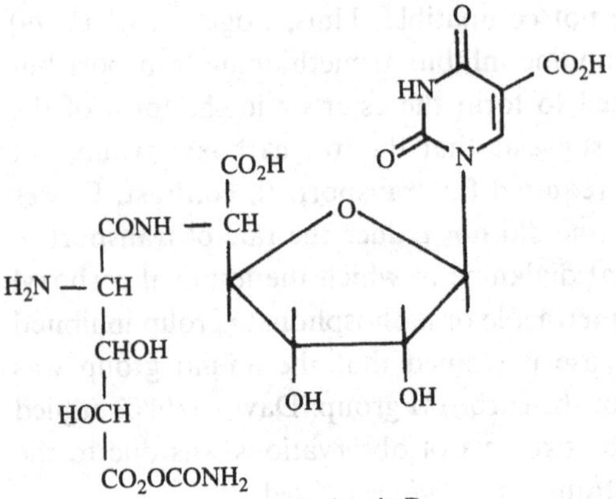

Polyoxin D

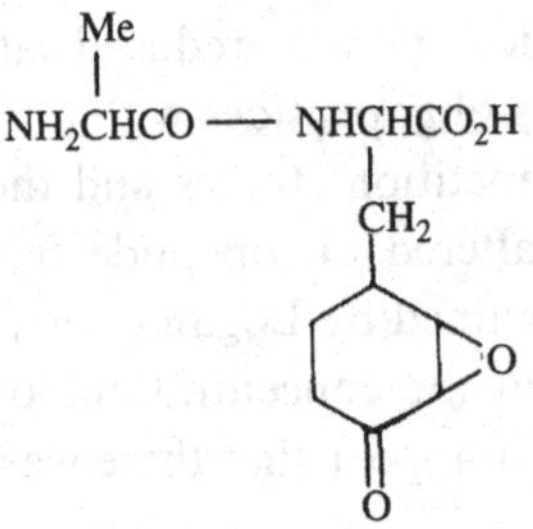

Bacilysin (tetaine, bacillin)

Figure 6.4. Formulae of polyoxin D and bacilysin.

seen in Table 6.15, where there are data for the growth of a lysine auxotroph, in the presence of arginine (which inhibits free lysine transport) there is only growth with the tripeptide trilysine. Studies with other auxotrophs are not necessarily as simple. Methionine presents difficulties because of the presence of three transport systems for the amino acid. However, the general conclusion from studies of this kind has been that dipeptides do not support growth whereas tripeptides do even in the presence of excess competing amino acids. This is similar to the situation in yeast.

Since peptides can be transported intact, it is of interest to know the upper size limit for a given peptide transport system. It would seem that the largest peptide on which an amino acid auxotroph can grow is a pentapeptide (Wolfinbarger & Marzluf, 1974, 1975*a*, *b*, 1976). This conclusion was based on studies of growth on enzymic digests of

Table 6.15. *Growth[a] of a lysine auxotroph of* Neurospora crassa (lys-1) *on free lysine and peptides containing lysine in the presence or absence of arginine*

Source of lysine[b]	Growth	
	Control	Plus Arg[c]
Lys	31	5
Lys-Gly	1	0
Gly-Lys	4	0
Lys-Lys	1	0
Lys-Lys-Lys	52	40
Lys-Leu	2	1

[a] Growth as milligrams dry weight mycelial pad following 3 days' growth at 30 °C.
[b] Source of lysine supplements at 0.33 mM.
[c] Arginine concentration at 3.3 mM.
From Wolfinbarger & Marzluf (1974); Wolfinbarger (1980b). From *Microorganisms and Nitrogen Sources*, 1980), reproduced by permission of John Wiley and Sons Limited.

Neopeptone fractionated on a Sephadex G-15 column. Use of a mutant *OS-1* with a larger porosity than the wild-type – molecules with a molecular mass of 18.5 kDa can penetrate the wall of the mutant, compared with 4.75 kDa mass limit for the wild-type (Trevithick & Metzenberg, 1966) – suggests that it is the transport system not the wall that restricts the size of the peptide used. Incidentally, though the number of amino acids has been used as a rough indication of size, it is the hydrodynamic volume which is the effective measure to use (see Chapter 3).

The only information about the kinetics of transport comes from one study of Wolfinbarger & Marzluf (1975b) using the tripeptide glycyl-D,L-leucyl-L-[³H]tyrosine. Uptake is inhibited by azide and the transport system, which is constitutive, appears to have relatively wide specificity but does not allow uptake of dipeptides or free amino acids. It is not clear whether or not the intact peptide is absorbed. Only free labelled tyrosine could be detected in the mycelium, which could be interpreted as a result of transport followed by hydrolysis by an intracellular peptidase or energy-dependent vectorial release of amino acids.

Wolfinbarger & Marzluf (1975b) questioned whether the transport system is constitutive and whether it requires either derepression or inductions for its appearance. The answer to these questions cannot be given. But it is appropriate to point out that whether or not any transport system is constitutive will depend to an extent on the nutrient conditions in the environment in which the fungus lives. Even for *N. crassa*, there is no clear view on such conditions.

Saccharomyces cerevisiae

The literature up to 1980 on peptide uptake by *S. cerevisiae* has been well reviewed by Becker & Naider (1980). It appears that, unlike *C. albicans*, *S. cerevisiae* absorbs both di- and tripeptides via a single transport system. Evidence that this is so comes from competition studies (Becker & Naider, 1977; Marder, Becker & Naider, 1977; Nisbet & Payne, 1979a) and from the generation of mutants that simultaneously have lost the ability to transport di- and tripeptides and their resistance to bacilysin (Nisbet & Payne, 1979b), L-ethionyl-L-alanine (Marder *et al.*, 1978) and nikkomycin (Monéton, Sarthou & Le Goffic, 1986). There is no evidence for the presence of extracellular peptidases (Becker, Naider & Katchalski, 1973) apart from the very specific endopeptidase integrally associated with the α-factor (Ciejek & Thorner, 1979).

The results from competition and growth studies have indicated that the transport system has a strong preference for hydrophobic peptides. Thus, peptides containing methionine compete strongly with trileucine, alanine peptides are poor competitors, and glycine or lysine peptides do not compete (Marder *et al.*, 1977). The system is able to transport certain peptides containing D-isomers but not others (Becker & Naider, 1980). As far as the importance of the amino and carboxyl termini for transport are concerned, it is difficult to generalise. Certainly, it is not obligatory that either terminus be present. Certain di- and tripeptides that have an acetyl or *t*-butyloxycarbonyl group can be transported; yet again, others, particularly those containing glycine, can not (Naider, Becker & Katchalski-Katzir, 1974). However, all the evidence indicates that the α-carboxyl group is not necessary (Naider *et al.*, 1974; Marder *et al.*, 1977). There is uncertainty about the length of the peptides that can be transported; however, the indications suggest that nothing larger than a tripeptide is transported (Becker & Naider, 1980; Marder *et al.*, 1977). The transport system can be ruled out in any consideration of the mechanism of involvement of peptides produced by haploid cells in the mating process (Naider & Becker, 1986).

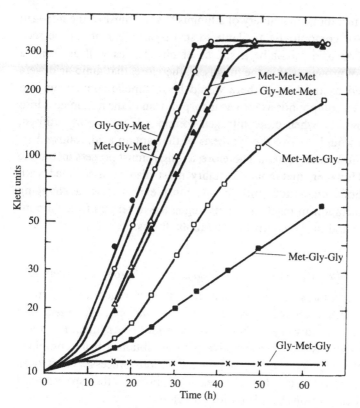

Figure 6.5. Growth of *Saccharomyces cerevisiae* on various peptides containing methionine and glycine. (From Naider *et al.*, 1974.)

It has been very interesting to find that the growth response of cells (appropriately auxotrophic) of *S. cerevisiae* to various peptides appears to be dependent upon the composition and sequence of the peptide supplied (Figure 6.5; Naider *et al.*, 1974). Cells possess the necessary enzyme activity to cleave all peptides. Also studies of the growth of cells in the presence of competing peptides have shown that competition is expressed not in reduced growth per se but in an extended lag phase. It is not clear why there should be these differences in the pattern of growth and the matter needs further investigation.

Regulation of nitrogen utilisation

Introduction

There should be little doubt from the foregoing sections that fungi are able to utilise a wide variety of nitrogen compounds. It is not difficult to see that the metabolic machinery of a fungus has to adjust to the nitrogen compound (or compounds) available to it. In particular, it can be seen that an amino acid might be a source of nitrogen as well as being a constituent of protein within the fungus. If, therefore, that amio acid were being utilised by the fungus there would be metabolic conversion into other nitrogen compounds as well as incorporation of the amino acid into protein. However, when other nitrogen compounds were being utilised, metabolism would involve the synthesis of that amino acid, followed by its incorporation into protein. Therefore, a fungus must possess metabolic flexibility. However, metabolic flexibility must also be associated with tight metabolic regulation and so one must expect both a changing pattern of metabolism inside a fungus, regulated in terms of the amount of enzyme(s) and in the control of metabolic fluxes.

Aspergillus nidulans, *Neurospora crassa* and *Saccharomyces cerevisiae*

When one considers the molecular details of nitrogen utilisation in fungi the information for three species – *A. nidulans*, *N. crassa* and *S. cerevisiae* – far outweighs what is known about other fungi. For these three fungi there is now a reasonable picture, obtained from combined biochemical and genetic studies, of how the various processes of nitrogen metabolism are regulated. However, extrapolation to other species must be undertaken cautiously as is indicated below.

The three fungi under consideration can use diverse sources of nitrogen, including ammonia, nitrite*, nitrate*, numerous amino acids, acetamide, purines and proteins (*not *S. cerevisiae*). Ammonia, glutamate and glutamine are the favoured sources, yielding maximal growth rates and the most intense depression of a whole range of activities, including membrane transport, linked to utilisation of other nitrogen compounds. This depression of activities – termed nitrogen catabolite control, regulation or repression – must be lifted if metabolism of other nitrogen compounds is to take place. Nitrogen catabolite control can be considered an adaptive mechanism that gives priority to the utilisation of nitrogen sources. Thus, an amino acid such as arginine is only utilised effectively when ammonia, glutamate and glutamine are absent from the medium.

Nitrogen catabolite control is a transcriptional control system that governs many nitrogen catabolic enzymes and all the evidence points to glutamine acting as the signal leading to metabolic control (Wiame, Grenson & Arst, 1985; Davis, 1986). In *S. cerevisiae*, control appears to be a negative system in which glutamine represses the target enzymes, though it is possible ammonia has a role (Wiame *et al.*, 1985). The effect of glutamine is presumed to be mediated by a polypeptide (GDHCR) that is the product of the *gdhCR* locus. The latter was named for the constitutive state of the NAD-linked glutamate dehydrogenase activity but is now known as *URE2* (Courchesne & Magasanik, 1988), although it is not the structural gene for *either* of the glutamate dehydrogenases. Mutants of this locus are derepressed for many glutamine-controlled enzymes. Indeed in the presence of a *gdhCR* mutation, synthesis of all the proteins (enzymes, transport proteins and probably regulatory proteins) except one has been found to be derepressed under nitrogen-repressing conditions. The exception is glutamine synthetase, synthesis of which is repressed even in the *gdhCR* mutant. The *URE2* (*gdhCR*) gene has now been shown to encode a polypeptide of 354 amino acid residues, with a molecular mass of 40 226 Da and a sequence similar to that of the glutathione S-transferases (Coschigano & Magasansik, 1991).

In the two filamentous fungi *A. nidulans* and *N. crassa*, glutamine opposes a positive regulatory element encoded by *areA* in the case of *A. nidulans* and by *nit-2* in the case of *N. crassa* (Davis, 1986; Wiame *et al.*, 1985). The two regulatory genes are very similar in sequence and *nit-2* has been shown to substitute for *areA* in *A. nidulans* and turn on the expression of the various nitrogen structural genes; the proteins encoded appear to contain a Cys-Xaa$_2$-Cys-Xaa$_{17}$-Cys-Xaa$_2$-Cys zinc finger domain (Caddick *et al.*, 1986; Davis & Hynes, 1989). Fu & Marzluf (1990*a*) have shown that the protein binds in a sequence-specific manner to three distinct sites in the 5′-promoter region of the nitrate reductase gene of *N. crassa*. As well as the positively acting *nit-2* gene there is also a negatively acting control gene *nmr* (de Busk & Ogilvie, 1984*c*; Fu, Young & Marzluf, 1988). The mutant phenotype of this latter gene exhibits loss of catabolite repression such that nitrate reductase and other nitrogen-metabolism-related enzymes are expressed even in the presence of sufficient ammonia or glutamine to prevent completely their synthesis in *nmr*[+] strains. There is no significant similarity between the sequences of *nmr* and *URE2* of *S. cerevisiae* (see above) (Coschigano & Magasansik, 1991) but there is significant sequence similarity between *nit-2* with *GLN3*, the product of which is required for the expression of NAD-dependent glutamate dehydrogenase (Mitchell & Magasanik, 1984).

Nitrogen catabolite control can be seen as one aspect of the ability of the three fungi to adapt to a changing environment with respect to the form and amount of nitrogen present in the medium. In *S. cerevisiae* there is rapid adaptation to new nitrogen sources. Thus, only 3–5 min elapse

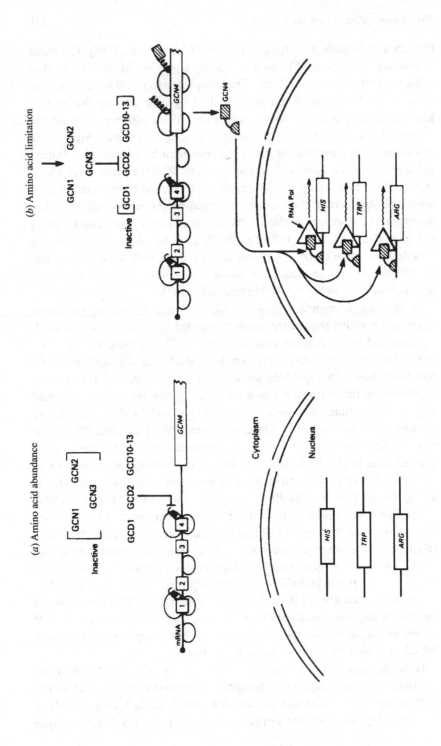

(a) Amino acid abundance

(b) Amino acid limitation

between the addition of arginine and urea to the medium and the appearance of the enzymes degrading arginine and allantoin (which is broken down to ammonia). There is induction not only of necessary catabolic enzymes within the cytosol but also of the necessary transport proteins. Indeed a prominent part of nitrogen catabolite control is via inducer exclusion, i.e. the inducer is unable to enter the protoplasm of the fungus. In particular, the general amino acid transport system possesses no activity in wild-type cells in the presence of ammonia, which inactivates the system.

Another important response to nitrogen status of the medium can be observed when a fungus responds to starvation for any one of several amino acids. In *S. cerevisiae* the expression is co-regulated with more than 30 genes encoding enzymes in ten different amino acid biosynthetic pathways. Starvation of any one of at least 10 amino acids leads to transcription of these genes, which can be stimulated from two-fold to 10-fold depending on the gene and the particular amino acid. This phenomenon has been termed 'general amino acid control' (Hinnebusch, 1986, 1988, 1990). Those enzymes that catalyse the rate-limiting step in the biosynthetic pathway are generally derepressed to the greatest extent. There are some pathways, such as that for methionine, which are not subject to general amino acid control. The model for the system is shown in Figure 6.6, in which a key feature is the presence of binding sites at each coregulated gene for the transcription activation protein GCN4.

General control of amino acid biosynthesis has been shown to be present in both *A. nidulans* and *N. crassa* (Piotrowska, 1980; Barthelmess, 1982). The derepression response of the latter fungus has been shown to result from increased expression of the mRNAs that encode the regulated enzymes (Flint, 1985; Flint & Wilkening, 1986). Significantly, it appears that as much as 20% of the total RNA species is affected by a mutation

---

Figure 6.6. Hypothetical model for the regulatory events involved in general amino acid control in *Saccharomyces cerevisiae*. Synthesis of GCN4 protein and transcription of amino acid biosynthetic genes under GCN4 control are depicted as being completely repressed under conditions of amino acid abundance (*a*). (In fact, low level GCN4 expression and transcriptional activation are expected under these conditions.) Ribosomes can traverse the upstream open reading frames in GCN4 mRNA and reach the GCN4 start codon under starvation conditions (*b*) as the result of inactivation of GCD factors by GCN1, GCN2 and GCN3. The latter are activated when the level of aminoacyl-tRNA is low. GCN4 protein is shown as a two-domain molecule, one domain being involved in DNA binding, the other making contact with a component of the transcription machinery, such as RNA polymerase (RNA Pol). (From Hinnebusch, 1990.)

of a single gene, *cpc1*, that blocks derepression of the enzymes subject to general control. This suggests that enzymes other than those involved in amino acid biosynthesis are under the same control.

As pointed out above, there is variation in the extent to which an enzyme is derepressed when control is lifted in *S. cerevisiae*. The reason for much of the variation is not clear, though it is understandable that those enzymes that catalyse the rate-limiting steps of a pathway are for the most part the ones most derepressed. One consequence of derepression is the increase in the concentration of amino acids within the cells (Hinnebusch, 1988). But this is the case for only certain amino acids. Such changes represent a greater increase in the cytoplasmic than in the vacuolar concentration of some amino acids (Messenguy, Colin & ten Have, 1980). Presumably the ensuing changes in the cytoplasmic concentration of a particular amino acid and the metabolic consequences thereof are what might have led to the evolutionary development of general amino acid control.

However, it is all too easy to assume that, under natural conditions, starvation is the consequence of a lack of one particular amino acid. It is equally possible that the mechanism has evolved because of its simplicity, since a single signal of starvation is generated under conditions in nature in which many amino acids may become limiting at one time. Thus, a single *trans*-acting regulator of transcription (the GCN4 protein) is mobilised on the appearance of the starvation signal. This regulator is able to increase the expression of all the relevant structural genes through the recognition of a *single cis*-acting regulatory sequence located upstream from each of these structural genes.

Of course there is not only control at the genetic level of nitrogen metabolism but fine control both through regulation of enzyme activity and though compartmentation of both enzymes and metabolic intermediates. Perhaps the best-studied amino acid in terms of regulation at this level is arginine. The catabolic pathway consists of the hydrolysis of arginine to ornithine and urea, the breakdown of urea to ammonia and carbon dioxide, and the conversion of ornithine to glutamate. The anabolic pathway for arginine involves the synthesis of ornithine and carbamoyl phosphate, with the conversion of these two compounds to arginine. The questions are: 'how can catabolic and anabolic processes proceed with ornithine as the common intermediate?' and 'how can both processes take place given also that ornithine and carbamoyl might be diverted into other pathways?' The metabolism of arginine has been investigated most extensively in *N. crassa* and *S. cerevisiae*; similar metabolic problems are solved in different ways in these two species (Davis, 1986).

A major difference between *N. crassa* and *S. cerevisiae* is the location of carbamoyl phosphate synthetase A (responsible for the conversion of glutamine to carbamoyl phosphate) and ornithine:carbamoyl transferase (responsible for the conversion of carbamoyl phosphate to ornithine and citrulline). In *N. crassa*, these two enzymes are in the mitochondrion; in *S. cerevisiae*, they are in the cytosol (Figure 6.7). Thus, both enzymes differ from their counterparts in kinetic characteristics as befits their different locations. Nevertheless, while there are these very distinct differences between the two fungi, it will be realised that for both organisms there is a similar role for the cell compartments in that they help to modulate metabolic activity. This modulation may occur as a result of storage in the vacuole or by separation of one pathway from another, as when enzymes are located in the mitochondrion. Thus, when arginine is added to a culture of *N. crassa* on minimal medium, the vacuole allows the rapid onset of catabolism of arginine by discharging ornithine into the cytosol, while arginine prevents the ornithine in the cytosol from reaching the ornithine carbamoyl transferase in the mitochondrion and thus minimises a futile ornithine cycle. At this point it is important to note that the vacuole acts as a store for most of the amino acids that are basic; acidic amino acids are predominantly cytoplasmic (see Chapter 4). An indication is given in Figure 6.8 of the way in which basic amino acids may distribute between the cytoplasm and the vacuole of *S. cerevisiae* with ammonia nutrition and how the fluxes of nitrogen across the tonoplast may alter as the nitrogen status of the medium is changed.

Since ammonia, glutamate and glutamine bring about nitrogen catabolite control, the enzyme glutamate dehydrogenase (GDH) must play a key role in nitrogen metabolism as well as being a branch point for the utilisation of both carbon and nitrogen. This enzyme can catalyse either the reductive amination of 2-oxoglutarate to produce glutamate or the oxidative deamination of glutamate to produce ammonia. In the three fungi considered here, there are two enzymes, one requiring NADPH and apparently serving to synthesise glutamate (NADP-GDH) and the other requiring NAD^+ and leading to the production of 2-oxoglutarate and ammonia (NAD-GDH). A great deal is known about NADP-GDH from *N. crassa*. It is a hexamer of six identical subunits, each of which is a polypeptide of 452 amino acid residues; the sequence is known (Holder *et al.*, 1975) and there is much information about its allosteric behaviour (Kinsey *et al.*, 1980). The gene for the enzyme (*am*) has been sequenced (Kinnaird & Fincham, 1983) and there is also significant information on the effects of specific amino acid replacements on enzyme properties (Fincham, Kinnaird & Burns, 1985).

(a)

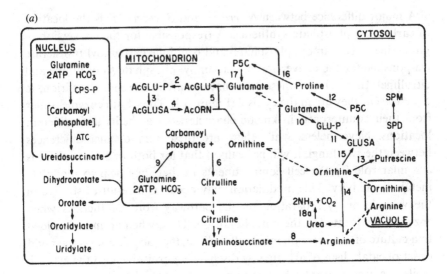

(b)

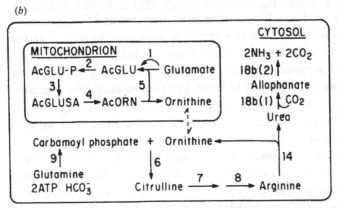

Figure 6.7. (a) Organisation of arginine, pyrimidine, proline and polyamine metabolism in *Neurospora crassa*. The locations of reactions 16 and 17 have not been proved but are shown as in *Saccharomyces cerevisiae* (see (b)). (b) Organisation of selected enzymes of arginine metabolism in *S. cerevisiae*. ATC, aspartate carbamoyltransferase; CPS-P, pyrimidine-specific carbamoyl-phosphate synthetase; Ac, acetyl; GLU, glutamate; GLUSA, glutamate semialdehyde; ORN, ornithine; P5C, pyrroline 5-carboxylate; SPD, spermidine; SPM, spermine. Identities of the enzymes: 1, acetylglutamate synthase; 2, acetylglutamate kinase; 3, acetylglutamyl-phosphate reductase; 4, acetyl-ornithine transaminase; 5, acetylornithine:glutamate acetyltransferase; 6, ornithine:carbamoyltransferase; 7, argininosuccinate synthetase; 8, argininosuccinate lyase; 9, carbamoyl-phosphate synthetase A; 10, glutamate kinase; 11, glutamylphosphate reductase; 12, pyrroline 5-carboxylate reductase; 13, ornithine decarboxylase; 14, arginase; 15, ornithine transaminase; 16, proline oxidase; 17, pyrroline 5-carboxylate dehydrogenase; 18a, urease; 18b, urea amidohydrolase. (From Davis, 1986.)

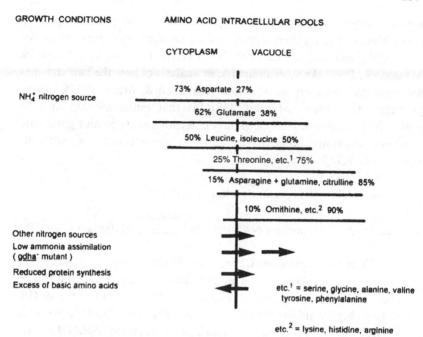

Figure 6.8. Amino acid distribution between cytoplasm and vacuole in *Saccharomycces cerevisiae* in relation to growth conditions expressed as a percentage of the total concentration. (From Messenguy *et al.*, 1980.)

The exact mechanism by which ammonia is incorporated into carbon skeletons is unclear. Although the reaction:

$$\text{glutamate} + \text{NAD(P)}^+ \longrightarrow \text{2-oxoglutarate} + \text{NH}_4^+ + \text{NAD(P)H}$$

is reversible, two lines of evidence suggest that *in vivo* NAD-GDH is a catabolic enzyme. First, under normal growth conditions, the concentration of NAD^+ in the cytoplasm is much greater than the concentration of NADH, which forces the reaction towards the degradation of glutamate. Second, cells of *S. cerevisiae* grown on glutamate as a source of nitrogen have a much higher level of NAD-GDH than when ammonia is the source of nitrogen (Grenson *et al.*, 1974). *Neurospora crassa* behaves similarly (Dantzig, Wiegmann & Nason, 1978).

However, despite considerable molecular knowledge, little is known about the metabolic regulation of NADP-GDH or of NAD-GDH. One difficulty may be that both enzymes appear to be controlled by both nitrogen and carbon circuits. In *S. cerevisiae*, there are elevated levels of NAD-GDH in cells grown with a non-fermentable carbon source of limiting amounts of glucose. In comparison with other glucose-repressed genes (Carlson, 1987) there appears to be another regulatory circuit for the *GDH2* gene which encodes NAD-GDH (Coschigano, Miller & Magasanik, 1991). Another difficulty, in analysis of how the two enzymes are regulated, certainly in *N. crassa* (Hummelt & Mora, 1980*a*, *b*) and probably *A. nidulans* (Marzluf, 1981), is that glutamate can also be synthesised via the activity of glutamine synthetase (GS) and glutamate synthase (glutamine (amide):2-oxoglutarate aminotransferase, with the acronym GOGAT). Thus:

$$\text{glutamate} + \text{ATP} + \text{NH}_3 \xrightarrow{\text{GS}} \text{glutamine} + \text{ADP} + \text{P}_i$$

$$\text{glutamine} + \text{2-oxoglutarate} + \text{NADPH} + \text{H}^+ \xrightarrow{\text{GOGAT}} \text{2 glutamate} + \text{NADP}^+$$

GOGAT is not present in *S. cerevisiae* (Brown, 1980). It is likely that glutamine synthase plays a more limited role in the regulation of nitrogen metabolism in this fungus. In *N. crassa*, in spite of the presence of GS, which has a higher affinity for ammonia (Wooton, 1983), and the presence of GOGAT, most of the glutamate is formed by the action of NADP-GDH (Lomnitz *et al.*, 1987). Nevertheless GOGAT plays an important role in the economy of *N. crassa*. Because two molecules of glutamate are formed in this reaction, one molecule is available for the continued synthesis of glutamine. The details of metabolism of glutamine in *N. crassa* have been reviewed by Mora (1990). He pointed out that, in the absence or low levels of nitrogen, glutamine turnover has an important role in protein turnover (Mora *et al.*, 1980). However, when mycelium is non-growing in the presence of nitrogen, it accumulates glutamine and arginine (Mora *et al.*, 1978). The operation of a cycle for the synthesis and catabolism of glutamine underpins these two observations (Figure 6.9). Glutamine is synthesised via the action of GS as above. The amide is converted either to glutamate via the action of GOGAT or to 2-oxoglutarate and ammonia through the concerted action of a glutamine transaminase and an ω-amidase (Calderón, Morett & Mora, 1985).

From a range of physiological and biochemical studies involving mutants, Mora and his colleagues (see e.g. Mora, 1990) have been able to show that the glutamine cycle appears to be important for growth of *N. crassa*. Nevertheless, glutamine does not appear, under normal conditions, to be a carbon and nitrogen source for growth of the fungus. This is probably because the glutamine cycle traps 2-oxoglutarate and glutamate, limiting the supply of their carbon skeletons to the citric

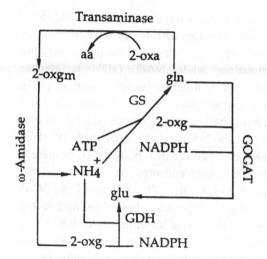

Figure 6.9. Schematic representation of the glutamine cycle. Abbreviations: glu, glutamate; gln, glutamine; 2-oxg, 2-oxoglutarate; 2-oxa, 2-oxo-acid; aa, amino acid; 2-oxgm, 2-oxoglutaramate; GDH, glutamate dehydrogenase; GOGAT, glutamate synthase. (From Mora, 1990.)

acid cycle. This seems not to be the case with a GDH-negative, GS-reduced mutant that is able to grow on glutamine as the source of carbon and nitrogen (Calderón & Mora, 1989).

The above observations pose a question as to the role of the cycle, particularly since it leads to a drainage of carbon skeletons, reducing power and ATP. Clearly nitrogen cycling will have an effect on carbon yield and energy supply. It was proposed by Mora (1990) that the glutamine cycle is futile, being involved in the dissipation of energy, though this has not been analysed in any quantitative manner (see Chapter 13). The cycle also has a determinant role in the turnover of nitrogen within the fungus. It is likely that, under conditions of low or inadequate nitrogen supply, the locking up of carbon skeletons and the dissipation of energy via the cycle leads to impairment of protein synthesis but at the same time allows protein degradation to continue and the re-entry of nitrogen into metabolism to take place.

Other fungi

Thus far, attention has been focused on the regulation of nitrogen metabolism in those fungi in which it has been studied most (by far), namely *Aspergillus*, *N. crassa* and *S. cerevisiae*. For other fungi, knowledge is more fragmentary. Nevertheless there is important information about nitrogen metabolism in fruit body development in *Coprinus cinereus*. In the mycelium, NAD-GDH is subject to some degree to nitrogen catabolite repression and urea derepression, while NADP-GDH is glucose catabolite derepressed and repressed by urea (Fawole & Casselton, 1972). In interpreting the role of these two enzymes, one should remember that the fungus is coprophilous, growing best on dung extract in the laboratory. On this basis, it is postulated that the NAD enzyme is normally involved in ammonia assimilation. However, when the fruit body develops there is amplification of activity of the citric acid and urea cycles. There is also amplification of the activities of both GDHs. The activity of the NAD enzyme in the cap (and also the stipe) increases by about three-fold during development, while that of the NADP enzyme increases some 16-fold. For the latter enzyme there is no change in the stipe. The regulation of NADP-GDH depends on a circuit involving the accumulation of acetyl-CoA (Moore, 1981) and the enzyme, together with GS, which also increases in activity (Ewaze, Moore & Stewart, 1978), probably contributes to an ammonia-scavenging system. Moore, Horner & Liu (1987) have demonstrated that the two enzymes are probably coordinately controlled. The need to scavenge ammonia is almost certainly to prevent the compound inhibiting basidiospore formation (Chiu & Moore, 1988). Glutamine is also inhibitory when presented exogenously. Presumably in the developing cap the amide is sequestered in vacuoles, although it is possible that, when entering the developing basidia, the compound is catabolised to ammonia.

The enzyme changes that occur as the cap develops lead to an increased concentration of alanine, arginine, glutamate, ornithine and urea (over that in the stipe by 1.5- to 4-fold) reaching a total concentration for these nitrogen compounds of around 16 mM (Ewaze *et al.*, 1978). The relevant metabolic reactions are identified in Figure 6.10. It should be realised that, within the developing cap itself, changes in concentration are less than changes in content, due to the continuing influx of water bringing about cap expansion. Since the concentration of the above nitrogen compounds is osmotically significant, they can be considered to play an important role in generating the basidial water potential (Ewaze *et al.*, 1978). The importance of amino acids in the water relations of the lower fungus *Thraustochytrium* is considered elsewhere (Chapter 13).

As with *S. cerevisiae* (see above), there is no evidence for the presence of GOGAT in *Coprinus cinereus* (Moore, 1984). The enzyme has been found in *Schizosaccharomyces pombe* and *S. malidevorans* but not

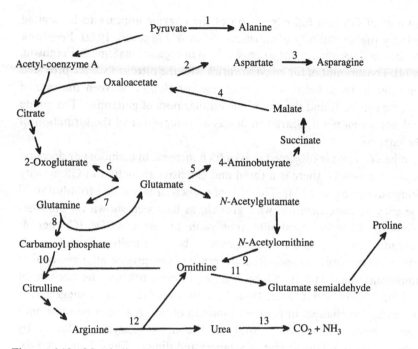

Figure 6.10. Metabolic chart summarising the reactions occurring in the fruit body of *Coprinus cinereus* to which reference is made in the text. 1, Alanine:2-oxoglutarate aminotransferase; 2, aspartate:2-oxoglutarate aminotransferase; 3, asparagine synthetase; 4, malate dehydrogenase; 5, glutamate decarboxylase; 6, glutamate dehydrogenase; 7, glutamine synthetase; 8, carbamoyl-phosphate synthetase; 9, ornithine acetyltransferase; 10, ornithine carbamoyltransferase; 11, ornithine:2-oxo-acid aminotransferase; 12, arginase; 13, urease. (From Ewaze *et al.*, 1978.)

S. versitalis (Brown, Burn & Johnson, 1973) but on the basis of physiological studies cannot be present in any significant amount in *Candida utilis* (Sims & Folkes, 1964). It would be very interesting to investigate growth of these two yeasts (*Candida nitrophila*, see below, could also be an appropriate candidate for study) under steady-state conditions with respect to nitrogen cycling and energy dissipation.

In *C. utilis*, using [^{15}N]-labelled ammonia, it was shown by Folkes & Sims (1974) that the rate of glutamate synthesis was very closely correlated with the size of the total pool of amino acids. This could be interpreted as the result of feedback inhibition, since *in vitro* all the common amino acids inhibited NADP-GDH. With respect to synthesis of glutamine from glutamate, the rate has been shown to be determined primarily by the amount of GS and full expression of the enzyme appears to be limited only by the availability of substrate (Sims & Ferguson, 1972). Feedback inhibition does not appear to be involved in regulation (Sims & Ferguson, 1974). The amount of the enzyme varies with the nitrogen source provided for growth, there being a good inverse correlation between the rate of enzyme synthesis and the size of the cellular pool of glutamine. The amide and not ammonia appeared to act as a co-repressor of the formation of the enzyme.

When *C. utilis* is subjected to a sudden increase in ammonia or decrease in glucose supply, there is a rapid and extensive reduction in GS activity (Ferguson & Sims, 1974). The rate of reduction is not due to dilution of the enzyme concentration with growth, as has been shown for enzymes following nitrogen catabolite repression in *S. cerevisiae* (Cooper & Sumadra, 1983). The enzyme appears to be specifically inactivated in a largely irreversible manner. Reappearance of the enzyme after removal of ammonia from the medium is by *de novo* protein synthesis. The mechanism of inactivation is not clear, though studies on GS *in vitro* suggest that there may be changes in the conformation of the enzyme, separation into two tetramers less relaxed than in the native enzyme, followed by dissociation into component monomers and dimers (Sims, Toone & Box, 1974).

Finally, it appears that ammonia assimilation in *Candida nitrophila* is similar to that in *N. crassa*. Under nitrogen-sufficient conditions, cells contain high activities of NADP-GDH and GS. Activities are repressed in glutamine-grown cultures but derepressed in cells starved of nitrogen or grown on nitrate. Nitrogen-deficient cultures also contain NADH-GOGAT activity, which can be inhibited by azaserine *in vitro*. This inhibitor reduces very considerably the amount of glutamate and glutamine in nitrogen-starved cells. In this yeast, it is believed that ammonia assimilation occurs via both the GDH pathway and the GS/GOGAT pathway, the latter being the major pathway in nitrogen-starved cells (Hipkin *et al.*, 1990).

Ectomycorrhiza of forest trees

It has been known for some considerable time, particularly since the experiments of Hatch (1937) on the effect of mycorrhizal infection of *Pinus strobus* seedlings, that the presence of the fungal mantle around the root is in general to increase the amount of nitrogen, phosphorus and potassium in the seedlings as well as the specific absorption of each element by the root (mg dry weight (mg root)$^{-1}$). Table 6.16 gives some relevant data. It can be seen from those of Finn (1942) that the effect on specific absorption is not always discerned; nevertheless all experiments and indeed many others show that the amount of these three nutrients per seedling weight is always greater when they are mycorrhizal (Harley & Smith, 1983).

Though the above has been known for some time, it is only in more recent years that clues have emerged as to the physiological processes involved. What is clear is that, since the fungal sheath is essentially the absorbing surface for the root, the increased content must result either from an increase in the effective absorbing area or from a greater flux of nutrients (mol cm^{-2} s^{-1}) than in the uninfected root. In the first case, the carrier molecules must be functioning in a manner similar to that of the higher plant cell. In the second case, either there are more carrier molecules per unit area of fungal plasma membrane or the carrier molecules in that membrane are more effective (lower K_T or higher V_{max}) than the equivalent molecules in the higher plant membrane.

To date, with respect to nitrogen utilisation by ectomycorrhiza it is not possible to generalise (Table 6.17) (for details concerning phosphorus utilisation, see Chapter 7). There can be stimulation of uptake of both nitrate and ammonia, particularly in the case of the presence of *Pisolithus tinctorius* on *Pinus contorta* roots (France & Reid, 1983), though it should be noted that these results are unusually high. The presence of mycorrhiza has, however, no effect on nitrate uptake by *Pinus pinaster* (Scheromm, Plassard & Salsac, 1990a) or *Picea sitchensis* (Rygiewicz, Bledsoe & Zasoski, 1984b).

In the study on nitrate and ammonia uptake by *Pinus contorta* mycorrhizae with *Pisolithus tinctorius*, it was observed that the absorption of ammonia (on a weight basis) was 3.1 and that of nitrate 7.0 times that expected from uptake calculated on the basis of what is known of the uptake capability of the fungal mycelium determined separately. But one must be cautious about these results. First, since flux values were not determined, it is not known how the increased uptake is brought about. Second, it is not known whether the mycelium grown in the laboratory is in the same physiological state as that forming the mycorrhizal sheath. So one cannot conclude that the host root is necessarily altering the transport capability of the fungal sheath.

In the few instances where the comparison is possible, ammonia is always absorbed at a greater rate than nitrate by mycorrhizal roots. Beech mycorrhizal roots (a tripartite system involving *Fagus sylvatica*, the basidiomycete *Lactarius subdulcis* and the ascomycete *Leucoscypha* sp.

Table 6.16. *Growth and specific nutrient uptake of nitrogen, phosphorus and potassium by* Pinus strobus *seedlings*

Ref.[a]	Degree of mycorrhizal infection	Dry weight mg	Root weight mg	Nitrogen[b] T	Nitrogen[b] SA	Phosphorus[b] T	Phosphorus[b] SA	Potassium[b] T	Potassium[b] SA
(1)	Mycorrhizal	448	180	5.39	0.030	0.849	0.0047	3.47	0.019
	Mycorrhizal	361	170	4.62	0.027	0.729	0.0042	2.57	0.015
	Uninfected	300	174	3.16	0.013	0.229	0.0013	1.04	0.006
	Uninfected	361	182	2.51	0.018	0.268	0.0015	1.94	0.011
		301	152	2.40	0.016	0.211	0.0014	1.17	0.008
(2)	Mycorrhizal (70.2%)	337	127	5.4	0.042	0.72	0.006	2.12	0.016
	Uninfected (9.5%)	181	75	2.2	0.029	0.13	0.002	0.81	0.011
(3)	Mycorrhizal (87%)	223	120	2.05	0.017	0.30	0.003	1.52	0.013
	Uninfected (10%)	93	56	1.17	0.021	0.24	0.004	0.98	0.018

[a] 1, Hatch (1937); 2, Mitchell, Finn & Rosendahl (1937); 3, Finn (1942).
[b] T, total absorbed mg; SA, specific absorption mg dry weight (mg root)$^{-1}$.
From Harley & Smith (1983).
From Harley & Smith (1983).

Table 6.17. *Fluxes of nitrate and ammonia into mycelia of ectomycorrhizal fungi of forest trees and into roots of such trees both non-mycorrhizal and ectomycorrhizal with the particular fungus*

Fungus	Forest tree	Units of flux	Nitrate concentration	Mycelium	Roots Non-mycorrhizal	Roots Mycorrhizal	Ammonia concentration	Mycelium	Roots Non-mycorrhizal	Roots Mycorrhizal	Reference
Lactarius subdulcis + *Leucosypha* sp.	*Fagus sylvatica*	mol h^{-1} (g d.w.)$^{-1}$					20 mM NH$_4$Cl			12.1	Carrodus (1966)
Lactarius subdulcis + *Leucosypha* sp.	*Fagus sylvatica*	mol h^{-1} (g f.w.)$^{-1}$ [a]	1 mM KNO$_3$			0.30					Smith (1972)
Pisolithus tinctorius	*Pinus contorta*	mol h^{-1} mg^{-1} [a]	10 mM KNO$_3$	34	44	280	10 mM NH$_4$Cl	46	162	320	France & Reid (1983)
Hebeloma crustuliniforme		mol h^{-1} (g d.w.)$^{-1}$	0.4 mM KNO$_3$	39			0.4 mM NH$_4$Cl	310			Littke, Bledsoe & Edmonds (1984)
Laccaria laccata		mol h^{-1} (g d.w.)$^{-1}$					0.8 mM NH$_4$Cl	360			
Pisolithus tinctorius		mol h^{-1} (g d.w.)$^{-1}$					0.8 mM NH$_4$Cl	600			
Hebeloma crustuliniforme	*Picea sitchensis*	mol h^{-1} (g d.w.)$^{-1}$	0.7 mM KNO$_3$		0.84	0.76	0.7 mM (NH$_4$)$_2$SO$_4$		5.5	7.0	Rygiewicz et al. (1984a, b)
Hebeloma crustuliniforme	*Pseudotsuga menziesii*	mol h^{-1} (g d.w.)$^{-1}$	0.7 mM KNO$_3$		0.84	1.1	0.7 mM (NH$_4$)$_2$SO$_4$		4.2	8.0	
Hebeloma crustuliniforme	*Tsuga heterophila*	mol h^{-1} (g d.w.)$^{-1}$	0.7 mM KNO$_3$		0.77	—	0.7 mM (NH$_4$)$_2$SO$_4$		2.5	5.0	
Hebeloma cylindrosporum	*Pinus pinaster*	mol h^{-1} (g f.w.)$^{-1}$	6 mM KNO$_3$	1.17	0.21	0.20					Scheromm et al. (1990a)
Laccaria bicolor		mol h^{-1} (g d.w.)$^{-1}$						22			Jongbloed, Clement & Borst-Pauwels (1991)
Lactarius hepaticus								48			
L. rufus								25.5			

[a] Not clear whether fresh (f.w.) or dry weight (d.w.).

(Lewis, 1991)) absorb very much less nitrate than ammonia than is found in other systems, the rate of uptake of nitrate being similar to the rate of uptake of chloride at the same external concentration (Carrodus, 1966; Smith, 1972; Plassard *et al.*, 1991). Also beech mycorrhiza show seasonality with respect to the rate of ammonia uptake (Carrodus, 1966). The observed rates of ammonia and nitrate uptake by beech mycorrhiza relate well to the high concentration of the former nutrient and the low concentration of the latter in the soils in which such mycorrhizae are found (Carrodus, 1966).

There are now a number of studies of the process of ammonia and nitrate assimilation by mycorrhizal roots and their constituent fungi. Interpreting what processes are taking place in the individual tissues of fungus and host root are difficult for the reasons outlined above. But, more important, in any attempt at a realistic assessment of what is happening in the two tissues, experiments need to be carried out either using separated tissues or separating the tissues after an experiment for appropriate analysis. The principles behind such approaches are exemplified in those experiments concerned with the uptake of phosphorus by beech mycorrhiza (Harley, 1959) (see Chapter 7). The fact that, at relatively low concentrations, 90% of phosphorus is retained in the sheath allows the investigator to assume that most of the observed changes in the phosphorus status of whole roots have occurred in the fungal sheath. The only information relevant to considerations of nitrogen metabolism concerns uptake of $^{14}CO_2$ by mycorrhizal roots in the presence of ammonia (Carrodus, 1967). After 4 h in the presence of 10 mM ammonium chloride and ^{14}C-labelled sodium bicarbonate, the sheath contained 62.5% of the absorbed radioactivity. This result, differing as it does from what is known about phosphorus absorption, emphasises the need for experiments, albeit difficult and time-consuming, of the kind indicated above.

Ho & Trappe (1980) have demonstrated that a number of mycorrhizal fungi grown in culture on a medium containing ammonia as a nitrogen source possess significant nitrate reductase activity, although at the most only one-sixth of that in Douglas fir (*Pseudotsuga taxifolia*). However, the data are for mycelia that had been grown for a month; there is no information about the physiological state of the mycelium after that length of time.

There is more information about one particular species studied by Ho & Trappe (1980), namely *Hebeloma crustuliniforme* (Plassard, Moussain & Salsac, 1984; Scheromm *et al.*, 1990*b*, *c*). The growth rate of this fungus on nitrate is ten times higher than that on ammonia. However, in the presence of organic acids growth in ammonia is at a level similar to that in nitrate. Organic acids help to prevent the external pH from falling, as happens when they are absent. But this effect on external pH is not the cause of the increased growth, since the pH can be maintained by phosphate without any effect on growth. The nitrate reductase activity of ammonia-grown mycelium is the same as that in mycelium grown in nitrate. When organic acids are present as well as ammonia, the level of the enzyme is much reduced. By themselves, organic acids have no effect.

It seems that nitrate reductase is regulated somewhat differently from that in *A. nidulans* and *N. crassa* (see above). There is clearly derepression when nitrogen is deficient or only slowly metabolised. Organic acids appear to be providing carbon skeletons for amino acid synthesis. In nitrate-grown mycelium this does not seem to be the case. The difference may be brought about by differences in the activity of phosphoenolpyruvate carboxykinase, that in nitrate-grown mycelium being two-fold to 15-fold that in ammonia-grown culture. Presence of an organic acid, such as pyruvate, reduces the difference very considerably.

It would be pleasing if these laboratory studies could be extended to a comparison of the ammonia and nitrate nutrition of a mycorrhizal association. At present, there is only information about the nitrate nutrition of the association of *Hebeloma cylindrosporum* with *Pinus pinaster* (Scheromm *et al.*, 1990*a*). The association modified neither the rate of nitrate absorption or reduction nor the rate of xylem transport of reduced nitrogen. Only phloem transport of reduced nitrogen (determined with ^{15}N) was different, being twice that in non-mycorrhizal plants. This increase in reduced nitrogen could be a consequence of the increased requirement of infected roots for photosynthate (combined carbon) as result of the presence of the fungal partner.

There is unambiguous evidence that in the following mycorrhizal fungi growing in culture, namely the ascomycete *Cenococeum geophilum* (Genetet, Martin & Stewart, 1984; Martin & Canet, 1986; Martin *et al.*, 1988) and the basidiomycetes *Hebeloma* sp. (Dell *et al.*, 1989) and *Laccaria bicolor* (Ahmad *et al.*, 1990), ammonia assimilation occurs via the activity of NADP-GDH and GS, the evidence for *C. geophilum* being particularly convincing. GOGAT has been shown to be present in *L. bicolor* but the activity is nearly 100-fold less than that of GS (Vézina *et al.*, 1989). However, investigations on the basidiomycete *Pisolithus tinctorius* indicate that assimilation is by the GS/GOGAT pathway (Kershaw & Stewart, 1989), though this is completely contrary to what might be inferred from studies on the enzyme complement of 7-week-old mycelium in which GOGAT was not detected (Vézina *et al.*, 1989).

Turning to ectomycorrhizal roots, both ^{15}N-labelling and enzyme studies have clearly shown that the GS/GOGAT pathway is the main route for ammonia assimilation for beech mycorrhiza collected from the forest, where the roots are said to be associated with *Lactarius* and *Russula* spp. (Martin *et al.*, 1986). The same seems to be so when seedlings of beech were made mycorrhizal with either *Hebeloma crustuliniforme* or *Paxillus involutus* (Dell *et al.*, 1989). There are indications that the path of ammonia assimilation in uninfected beech roots – and this is expected from what is known about the process in plant roots (Oaks & Hirel, 1985) – that the path is also via GS/GOGAT. On the other hand, in spruce (*Picea excelsa*)/*Hebeloma* sp. mycorrhizal roots indicate the major role of NADP-GDH and GS in ammonia assimilation.

In the light of the information available, the following seem pertinent:

(1) Ammonia might penetrate the fungal sheath round beech mycorrhizal roots more readily than that round those of spruce, such that much of the ammonia assimilation occurring in the roots of the former association is via the GS/GOGAT pathway, whereas in the latter association most of the assimilation is in the fungus via the NADH-GDH/GS pathway.

(2) Conceivably, beech roots might derepress genes for GOGAT in the fungal sheath; likewise the fungal sheath in spruce roots might repress the equivalent genes in the higher plant cells.

More information is needed on this matter of ammonia assimilation in ectomycorrhizal roots. A key candidate for investigation must be *Pisolithus tinctorius*, both in culture and in mycorrhizal association. It would assist greatly if more were known precisely about the capability of the relevant fungi to involve GOGAT. Probing for the relevant genes should provide unambiguous information.

Protein utilisation

Proteins in cells are constantly being degraded. It is this process that allows the cell to adapt its protein composition to changes in the external environment. Take cells of *S. cerevisiae* as an example. The rate of protein degradation is around 0.5%–1.0% h^{-1} (Lopez & Gancedo, 1979). The value increases to about 2%–3% h^{-1} under starvation, and proteinases are clearly involved. There is considerable information about proteinases in *S. cerevisiae* and also in *N. crassa* (Wolf, 1980, 1986; Jones, 1984; Rendules & Wolf, 1988). The proteinases of the vacuole are very non-specific and it is this non-specificity which, together with the properties of other vacuolar enzymes, has led to the vacuole being equated with a lysosome (see Chapter 4). Studies with mutants leave little doubt that the vacuolar proteinases are involved in rather non-specific protein degradation induced by nitrogen starvation and make both exogenous and endogenous peptides accessible for primary metabolism (see pp. 228–9 for details of peptide uptake). Jones (1984) pointed out that the proteinases of the vacuole are brought into play when the only source of amino acids for adaptation and restructuring is within the cell itself. The vacuole can therefore be viewed as the resting place for that machinery which needs to be mobilised in 'emergencies'.

Other proteinases play highly specific roles in metabolism and development. In *S. cerevisiae* proteinases have been implicated in a whole range of activities, including, for instance, carbon-starvation-induced degradation of NADP-GDH, enzyme secretion, production of the pheromone α-factor, and septum and spore formation (Jones, 1984).

Many fungi are capable of degrading extracellular peptides and proteins. Loll & Bollag (1983) in their review of protein transformation in soil listed 40 genera of fungi said to possess proteolytic capability, though it does not necessarily mean that they can utilise protein as a sole source of carbon. Both *Aspergillus* spp. and *N. crassa* can secrete extracellular proteinases, the production of which is regulated by carbon-, nitrogen-, sulphur- and even phosphorus-catabolite repression (Cohen, 1980, 1981; North, 1982). Starvation of any one of these elements results in the production of proteinase. In *Aspergillus*, relief of catabolite repression is sufficient for production; in *N. crassa* protein must be present.

The basidiomycetes *Agaricus bisporus*, *Coprinus cinereus* and *Volvariella volvacea* can all use protein as the sole source of carbon, nitrogen and sulphur (Kalisz, Wood & Moore, 1987, 1989). Unlike in *Aspergillus* and *N. crassa*, proteinase activity is not repressed by the presence of glucose, ammonia and sulphate in the medium. The presence of protein is required for induction of proteinases, though in the case of *C. cinereus* proteinase activity is derepressed in protein-free media in the absence of sulphate. When the three fungi were grown with protein as the sole source of carbon, very significant levels of ammonia were found in the medium. Since the amount of protein consumed contained far more nitrogen than was required for the amount of biomass produced, it must be assumed that a proportion of the protein was catabolised leading to the production of non-nitrogen-containing polymers or oxidised to provide energy. As a result of both processes, free ammonia would be produced. In the experiments referred to here, one-third to one-half of the supplied protein appears to have been assimilated and metabolised in this manner.

Frank (1894) proposed that the presence of the fungus in an ectomycorrhizal symbiosis might provide the root system of the tree with access to organic nitrogen compounds that are the major source of nitrogen in forest soils. The emphasis of green plant nutrition on inorganic nitrogen sources used to militate against this idea until the last decade. Now it is known that both ericoid mycorrhiza (Bajwa, Abuarghub & Read, 1985; Bajwa & Read, 1985, 1986; Leake & Read, 1989, 1990*a*, *b*, 1991) and ectomycorrhiza of forest trees (Abuzinadah & Read, 1986*a*, *b*, *c*, 1989*a*, *b*, *c*) are able to use proteins as a source of nitrogen. Not a great deal is known about the physiology of induction of proteinase activity. That produced by the ericoid endophyte *Hymenoscyphus ericae* is spoken of as being that of a single proteinase, though the evidence that only a single enzyme is present is not apparent (Leake & Read, 1991). Proteinase activity has a pH optimum of 2–3 and is induced by pure bovine serum albumin and also by casein hydrolysate. In both cases production of the enzyme activity was enhanced by ammonia. When protein was the sole source of carbon, ammonia was produced. Glucose repressed proteinase activity when the fungus was grown with protein but not when it was grown with the hydrosylate. The extent to which ectomycorrhizal fungi can use proteins is variable, some not being able to use them (Abuzinadah & Reid, 1986*a*).

Nitrogen storage

There is no reason to suppose that fungi should not possess means by which nitrogen might be stored in insoluble form. Indeed Blayney & Marchant (1977) showed that protein inclusions, with crystalline substructure, are formed during stipe elongation in *Coprinus cinereus* and are degraded during later stages of development. It was suggested that nitrogen in the chitin synthesised during the later stages of elongation of the stipe could be derived from degraded protein inclusions. Other candidates for stores of nitrogen are: (a) Woronin bodies, now known not to be made up of ergosterol (Markham & Collinge, 1987); (b) crystalloids, often associated with microbodies (Maxwell *et al.*, 1975); and (c) the gamma particles of *Blastocladiella emersonii* zoospores (Hohn, Lovett & Bracker, 1984).

7

Phosphorus

Introduction

Phosphorus is an essential element for all cells, being present in nucleic acids and phospholipids. As well as being an important constituent of organic molecules, very often in fungi, as in other living organisms, phosphorus can be present not only as orthophosphate but also as condensed inorganic phosphates (Kulaev, 1979; Kulaev & Vagabov, 1983). Phosphate groups in organic compounds and often those of inorganic phosphates can provide a significant contribution to the negative charge within the protoplasm.

Phosphorus is frequently not readily available in nature (Larsen, 1967; Sutton & Gunary, 1969). In soil, the pool of soluble phosphate in solution is small, much phosphate is adsorbed onto surfaces and there can be a great deal of phosphate in insoluble form as salts of calcium, aluminium and iron. There can also be a significant fraction of phosphorus in organic compounds. Fungi possess a number of mechanisms for releasing phosphorus and these are considered first.

Phosphorus solubilisation

Inorganic phosphate

The ability of microorganisms to bring mineral phosphates into solution is agriculturally important as well as having more general significance in phosphorus cycling in the natural environment. Microbial mechanisms for phosphate solubilisation include production of (a) inorganic acids, (b) organic acids, (c) chelators and (d) hydrogen sulphide, which reacts with iron phosphate to liberate the phosphate while precipitating the iron as the sulphide (Ehrlich, 1981). As far as fungi are concerned, it has been

251

assumed frequently that organic acids are the major means of solubilisation (Beever & Burns, 1980). That is because of the ready production of such acids by fungi (see Chapter 12). It should be noted that, if organic acids are involved, they may act also as chelating agents. Hydrogen sulphide can be produced by fungi (Slaughter, 1989) but it seems unlikely to be involved in solubilisation. Little is known about the production of inorganic acids by fungi. Nevertheless, according to the classic study by Conway & Brady (1950), hydrochloric acid is readily formed by fermenting cells of *Saccharomyces cerevisiae* in the presence of potassium chloride. The anion is unable to enter the yeast cell, and potassium is exchanged for hydrogen ions.

Microbial solubilisation of phosphate has been studied either by having the insoluble phosphate as dispersed particles in nutrient agar and observing the appearance of cleared zones or by determining the release of orthophosphate in liquid medium containing the insoluble phosphate (for a review of the topic, see Kucey, Janzen & Leggett, 1989). A significant investigation, as far as fungi are concerned, using the second method, has been that of Agnihotri (1970), who examined the phosphate-solubilising ability of a range of fungi. Some species, such as *Pythium ultimum* and *Rhizoctonia solani* were found to be ineffective; others such as *Aspergillus niger* could solubilise as much as 88% of added calcium phosphate, 80% of fluorapatite and 59% of hydroxyapatite. For *Sclerotium rolfsii* the values were respectively 69%, 77% and 49%. *Fusarium solani* gave 80% solubilisation of calcium phosphate but was virtually ineffective against the other two sources of phosphate, yet in both cases the pH of the medium was reduced to a value not very different from that produced by *A. niger* with the same phosphate sources.

Penicillium bilaii (orginally *P. bilaji*) has been fairly intensively studied with respect to its ability to solubilise phosphate. In experiments with wheat (*Triticum aestivum*) and bean (*Phaseolus vulgaris*) inoculation of soils containing rock phosphate with *P. bilaii* resulted in both higher plant species producing more dry matter and having increased phosphorus content (Kucey, 1987). The presence of rock phosphate without the fungus had no effect on plant growth. In other experiments it was shown that the fungus was able in pure culture to solubilise Cu(I) and Cu(II) oxides, Cu(II) carbonate and zinc metal, and to a lesser extent Fe(II) and (III) oxides and pyrite (Kucey, 1988). In soil studies, it was shown that wheat inoculated with *P. bilaii* possessed not only a greater capacity to utilise phosphorus when rock phosphate (containing significant amounts of copper, zinc and iron) was added to the soil but could contain more copper, zinc and iron. However, it should be noted that of the three metals, only zinc was at increased concentration in the plant.

More detailed pure culture studies leave a confused picture. Asea, Kucey & Stewart (1988) showed that nitrogen in the ammonium form appeared necessary for enhanced phosphate solubilisation by *P. bilaii*. A more recent study by Cunningham & Kuiack (1992) showed that in sucrose nitrate media the major acid components produced were oxalic and citric acids. The former was promoted under carbon limitation and the latter under nitrogen limitation. These acids were apparently not present when ammonia was the nitrogen source. Also, when nitrate was the nitrogen source, medium from young cultures was much more effective than that from similar cultures growing on ammonia in bringing phosphate into solution. Though it was stated by Cunningham & Kuiack (1992) that their medium was similar to those used previously, this is not so and the presence of sucrose (rather than glucose) and particularly a significant amount of manganese might be a cause of the contradictory observations (see Chapter 12).

La Peyrie, Ranger & Vairelles (1991) have shown that seven species of mycorrhizal fungi are able to solubilise insoluble mineral phosphates in culture. There is variation in their ability to use different calcium phosphates both between species and between strains of the same species. It is important that ammonia is the nitrogen source. Only a strain of *Paxillus involutus* is able to solubilise the above phosphates in the presence of nitrate.

Organic phosphates

As Beever & Burns (1980) have pointed out, phosphomonoesters, while being an important constituent of living cells, are unlikely to be free in the environment for more than a limited span of time. On cell breakdown, these esters are probably broken down by endogenous phosphatases or, if released, by similar enzymes already present in the environment, particularly the soil (Speir & Ross, 1978). In any case, these esters may undergo hydrolysis spontaneously. Because of the transitory nature of phosphomonoesters outside the cell, the need for a fungus to possess phosphatases to hydrolyse them has been questioned by Arnold (1987). He believed that acid phosphatases found in the walls of yeasts are located there for activating proteins in the wall and on the outer surface of the plasma membrane for physiological activity. Somewhat the same idea was put forward by Arst, Bailey & Penfold (1980) to explain why the mutation *pacC-5* in *Aspergillus nidulans*, which leads to loss of the activity of an acid phosphatase, results in a very considerable reduction in the level of transport of γ-amino-*n*-butyrate.

In one sense it has been a disadvantage that the phosphatases hydrolysing phosphomonoesters are easy to study. Fungal cells or mycelia often possess phosphatases that, in laboratory terms, show high activity against readily available substrates. Enzymes capable of releasing phosphate from compounds more refractory to enzymic breakdown have therefore been much less studied. Although there is a wide variety of such compounds, two groups merit comment here with respect to the ability of fungi to release phosphate from them.

The first group consists of the phosphodiesters such as phospholipids, nucleic acids, phosphorylated polysaccharides such as techoic acids, and phosphomannans. The evidence for enzymes capable of breaking down these compounds extracellularly comes for the most part from studies on the ability of fungi to grow on them as the sole source of phosphorus. However, such studies have done little to indicate whether the phosphate is released in the external medium and then absorbed or is absorbed as part of a carbon-containing fragment of the original larger molecule, the phosphate being released into the cytoplasm.

It is very easy to demonstrate the probable production of nucleic-acid-splitting enzymes by the use of solid media containing the nucleic acid but free from orthophosphate (Cazin *et al.*, 1969; Burt & Cazin, 1976; Hankin & Anagnostakis, 1975). At the time of assessment, hydrochloric acid is added to the plate. Unhydrolysed nucleic acid is precipitated, leaving a clear zone in the absence of nucleic acid. If possible, there should be confirmatory studies using liquid media in which the appropriate enzymes should be detected after a suitable period of growth (Burt & Cazin, 1976). Both DNA and RNA can be broken down by a wide variety of fungi. Not all species can break down both DNA and RNA; for example, *Trichoderma viride* (Hankin & Anagnostakis, 1975) appears to be able to break down only DNA, and a large number of ascomycetous yeasts appear unable to hydrolyse RNA (Burt & Cazin, 1976).

Properly critical studies on the ability of fungi to break down organic phosphorus molecules in the external medium necessitate the isolation of mutants. Ishikawa *et al.* (1969) isolated mutants of *Neurospora crassa* unable to grow on exogenous DNA and RNA. It would appear that these mutants (*nuc-1* and *nuc-2*) are unable to produce the necessary extracellular enzymes. In fact the mutants are regulatory mutants that fail to derepress at least seven repressible enzymes of which two nucleases, ribonuclease N_1, a 5′-nucleotidase and an alkaline phosphatase are released into the external medium (Hasunuma & Ishikawa, 1977; Metzenberg & Nelson, 1977).

A major phosphorus component of soil is phytate, a term which covers a range of inositol phosphate esters (Cosgrove, 1977). Up to 60% of the organic phosphorus in soils can be in the form of phytate (Halstead & McKercher, 1975). Shieh & Ware (1968) made a survey of the ability of saprophytic fungi to breakdown calcium phytate, which gives a white turbidity to agar that becomes clear on phytate breakdown. Some of the fungi were obtained by enrichment of culture from soil, others were from culture collections. Twenty-eight of 82 *Aspergillus* spp. but only one out of 58 *Penicillium* spp. and one out of 37 *Mucor* spp. were able to degrade phytate. None of 13 *Rhizopus* spp., 4 *Cunninghamella* spp., 4 *Neurospora* spp. and 140 yeasts from 17 genera showed any phytate-degrading capability. Jennings (1990*b*) reported that rhizomorphs of *Phallus impudicans* are able to release phosphate from phytate albeit slowly. Assay was made difficult by the concurrent release of phosphate from the tissues. Following on from the initial studies of Shieh & Ware (1968), *Aspergillus ficuum* has been shown to produce in liquid medium an extracellular non-specific phosphatase and an extracellular *myo*-inositol hexaphosphatase (phytase) that hydrolyse inositol hexaphosphates to give pentaphosphates (Shieh, Wodzinski & Ware, 1969; Irving & Cosgrove, 1972).

It should be noted that the studies just described may not be very realistic. Inositol phosphates in nature may be present as very insoluble Fe(III) and aluminium derivatives. Ferric phytate, however, is able to produce soluble complexes with ammonia and amines. Similar complexes may occur with organic nitrogen compounds and it is possible that fungi might be able partially to release orthophosphate from such complexes (Cosgrove, 1977).

Orthophosphate transport

Knowledge of the mechanism of phosphate transport into fungi comes predominantly from studies on *N. crassa* and *S. cerevisiae*. I now deal with these two species before referring to pertinent information from other fungi. Beever & Burns (1980) have provided comprehensive coverage of the literature on orthophosphate transport in their excellent review on phosphorus uptake, storage and utilisation by fungi.

Neurospora crassa

At least two phosphate transport systems are present (Lowendorf, Slayman & Slayman, 1974; Lowendorf, Bazinet & Slayman, 1975; Beever & Burns, 1977, 1978; Burns & Beever, 1979). System I has a low affinity for phosphate, is highly dependent on pH and is constitutive. System II has high affinity for phosphate, is relatively insensitive to pH and is repressed by phosphate. The latter system is encoded by the gene originally termed *van*$^+$, the name derived from the fact that the system transports vanadate (Bowman, 1983; Mann *et al.*, 1988). Now, the gene is called *pho-4*$^+$, since it is known that this gene, like the structural genes *pho-2*$^+$ and *pho-3*$^+$, is controlled by the phosphate-regulatory genes (see below). Mann *et al.* (1989) have determined the nucleotide sequence of the gene. It specifies a protein of 590 amino acid residues; the hydropathy profile of their sequence suggests 10–12 membrane-spanning helices with a large hydrophilic domain between the eighth and ninth helices.

The available evidence indicates that the two systems are genetically discrete. Though the gene for system I has not been identified, mutants have been isolated that can grow on vanadate yet still absorb phosphate (Bowman, Allen & Slayman, 1983). Vanadate is not absorbed and the derepressable high affinity phosphate transport system appears to be present in only very defective forms, with a very much lower rate of transport and K_T values different from that of the wild-type system. System I is unaffected.

System I has a K_T that increases nearly 400-fold from 0.01 to 3.62 mM as the pH is raised from 4.0 to 7.3. The flux, J_{max}, remains constant (Lowendorf *et al.*, 1974). The increase in K_T cannot be explained by (a) a single system with preference for $H_2PO_4^-$ nor (b) two systems with different pH values. It can be explained by a model in which OH^- or H^+ acts as a modifier of the transport system. The likelihood is that phosphate transport is via a proton symport for which there is unpublished evidence (Lowendorf *et al.*, 1974). Transport is stimulated by high internal concentrations of potassium. However, high extracellular concentrations of monovalent cations lead to inhibition of transport, probably due to depolarisation of the membrane.

Beever & Burns (1977) showed that at pH 6.4 the K_T of the system was constant (1.0 mM) after growth of germlings (conidia germinated over 2.5 h) in concentrations of phosphate above 1 mM. Below this concentration, the K_T value progressively decreased as the concentration of phosphate in the growth medium was reduced. For 50 μM phosphate in the growth medium, K_T was 370 μM. The V_{max} at a value of 2.33 μmol

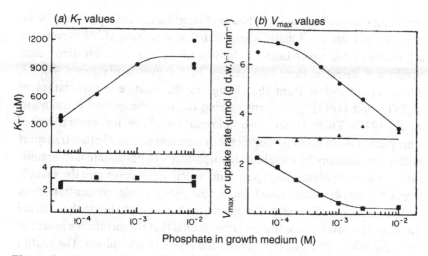

Figure 7.1. Changes in the kinetic constants of the two phosphate transport systems of *Neurospora crassa*, following growth of germlings in the presence of different phosphate concentrations. (*a*) K_T values; (*b*) V_{max} values. (●) Low affinity system; (■) high affinity system; (▲) calculated uptake rate. d.w., dry weight. (From Beever & Burns, 1980.)

(g dry weight)$^{-1}$ min^{-1} after growth in the lowest phosphate concentration rises to 0.28 μmol (g dry weight)$^{-1}$ min^{-1} after growth in 10 mM phosphate (Figure 7.1). Changes in the value of K_T and V_{max} can be brought about by transference of germlings between media of different phosphate concentration. The changes are rapid and clearly reversible. One can argue from these data, that system I is subject to feedback inhibition from phosphate within the cytoplasm. The matter needs further investigation, particularly as there are now reliable methods for estimating cytoplasmic phosphate concentrations (see pp. 269–72).

System II, the high affinity system, is derepressed during phosphate starvation, and at pH 5.8–6.4 has a K_T value of 2–3 μM (Lowendorf *et al.*, 1975; Burns & Beever, 1977). The system reaches maximal activity after about 2 h of growth. Cycloheximide presented to cultures brings about loss of activity of the system. Lowendorf *et al.* (1975) argued that this loss is due to turnover of the carrier, cycloheximide inhibiting synthesis. Beever & Burns (1978) argued persuasively, on the basis that (a) both systems I and II behave similarly to cycloheximide, and (b) the loss of uptake sensitivity can be prevented by lowering the phosphate concentration of the medium, and (c) cycloheximide by inhibiting growth reduces phosphate demand such that the internal concentration increases and brings about feedback inhibition of transport. There appears to be no effect of pH on the kinetic characteristics of system II. However, the sensitivity of the procedures at the low phosphate concentrations used would not detect a three-fold change in K_T over a pH range 4.0–7.3, which would follow from the change in the relative concentration of $H_2PO_4^-$ and HPO_4^{2-}, the former being the ion transported (Lowendorf *et al.*, 1975). There is no other information about the mechanism of transport. It is not known whether the system is active or whether transport is driven essentially by metabolic incorporation of phosphate into organic form. As a result of such incorporation, or active transport into the vacuole (see Chapter 4), there could be a low cytoplasmic concentration of orthophosphate, allowing diffusion into the hypha of the ion in the external medium. However, it needs to be remembered that the membrane potential (*c.* −200 mV) will be acting against the influx of phosphate. The matter of the cytoplasmic concentration of phosphate is considered on pp. 269–72. The alternative is that there is symport with a cation other than the proton, as can be the case with *S. cerevisiae* (see below). It is doubtful that there can be direct coupling of transport with ATP hydrolysis. The question of the mechanism of operation of system I needs resolution.

The K_T value of system II is independent of the phosphorus concentration of the growth medium. As with system I, the V_{max} is highest (2.33 μmol (g dry weight)$^{-1}$ min^{-1}) at the lowest phosphate concentration (50 μM) in the growth medium. At higher phosphate concentrations, V_{max} values are progressively lower, with a constant minimum value at 2.5 mM phosphate (0.28 μmol (g dry weight)$^{-1}$ min^{-1}) (Figure 7.1). Unlike the situation for system I, these effects are not reversible; it is not clear why this should be so.

The relationship between the two systems has been investigated by Beever & Burns (1977) and Burns & Beever (1979). As can be seen, while K_T for system II remains constant, K_T for system I and V_{max} for both systems I and II change with the phosphate concentration of the growth medium. The net effect of these changes is to alter the relative contribution that each system makes to the observed rate of phosphate uptake (Figure 7.2, Table 7.1). In this manner, the growth rate and phosphate uptake rate are able to remain constant during exponential growth over the phosphate concentration range 50 μM–100 mM (Beever & Burns, 1980). Burns & Beever (1979) provided evidence that the following occurs as a function of the phosphate status of the mycelium: system II is derepressed under conditions of phosphate starvation and irreversibly subject to feedback inhibition under conditions of oversupply. System I

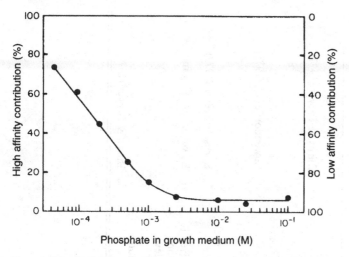

Figure 7.2. Relative contributions of the two phosphate transport systems to phosphate uptake by germlings of *Neurospora crassa* over a range of external phosphate concentrations. (From Beever & Burns, 1980.)

is also derepressed and subject to feedback inhibition under the same conditions but in this case inhibition is reversible. Though consideration of their data is essentially qualitative, the mechanisms proposed are not only consistent with what has been established from kinetic studies but are consistent with possible time-lags in transcription and translation. It should be noted that at no stage is there significant efflux, so this process can be excluded as a means of controlling phosphate levels in the mycelium.

The expression of system II and seven other proteins are regulated by at least three genes $nuc-1^+$, $nuc-2^+$ and $preg^+$. The proteins are: (a) alkaline phosphatase, (b) acid phosphatase, (c) three nucleases, (d) a 5'-nuclease and (e) phosphoethanolamine permease. This 'phosphorus family' of proteins are regulated as follows (Ishikawa *et al.*, 1969; Hasunuma & Ishikawa, 1972; Littlewood, Chia & Metzenberg, 1974; Lowendorf & Slayman, 1975; Metzenberg & Nelson, 1977):

(1) *nuc-1*, the product of which is required to turn on the transcription of the structural genes.
(2) *preg*, which is required to nullify the activity of the *nuc-1* product.
(3) *nuc-2*, the product of which is required to nullify the activity of the *preg* product or prevent its synthesis.

Phosphate or some product nullifies the activity of *nuc-2* or prevents its synthesis.

Table 7.1. *Kinetic characteristics of the two phosphate transport systems of* Neurospora crassa

	System		Reference
	Low-affinity	High-affinity	
K_T			
Germlings grown at 10 mM P_i	1029 µM	2.85 µM	Burns & Beever (1977)
Germlings grown at 50 µM P_i	370 µM	2.43 µM	
Ratio of K_T			
Determined at pH 7.3	1042	1.33	Lowendorf et al. (1975)
Determined at pH 4.0			
Ratio of V_{max}			
Germlings grown at 50 µM P_i	1.86	8.32	Burns & Beever (1977)
Germlings grown at 10 mM P_i			
Level in *nuc-1* and *nuc-2* mutants	Present	Absent irrespective of P_i concentration	Lowendorf & Slayman (1975)
Level in *pcon* and *preg* mutants	Present	Present at high levels irrespective of P_i concentration	
Feedback inhibition	Reversible	Irreversible	Burns & Beever (1979)

From Beever & Burns (1980).

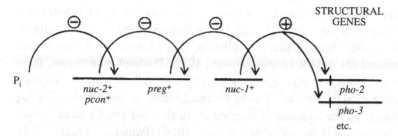

Figure 7.3. Model of the hierarchy of phosphorus regulatory genes in *Neurospora crassa*. A plus sign (+) over an arrow means 'turns on' and a minus sign (−) means 'turn off', or 'represses' or 'inhibits the action of'. Enzyme production in a strain carrying a mutation in a given control gene can be 'calculated' by multiplying the positive or negative signs of the regulatory products. Only those regulatory products that are connected to the structural genes by a sequence *unbroken by mutation* can be included in the calculation. For example, in *nuc-2* strains, only the final two regulatory steps are operative ('−' times '+') and such strains are '−', i.e. null. Regardless of the phosphate concentrations, strains carrying *pcon^c* mutations have three regulatory steps working ('−' times '−' times '+') and are '+', i.e. constitutive. Similarly, wild-type cells grown under conditions of phosphate starvation are '−' times '−' times '+', which multiplies to '+': the strains produce alkaline phosphatase and its congeners. (From Metzenberg & Nelson, 1977.)

A diagrammatic model of the somewhat complicated hierarchy of regulatory genes is given in Figure 7.3 (Metzenberg & Nelson, 1977). In addition to the genes listed above, the model takes into account *pcon*, named for *p*hosphorus *con*trol constitutive, which resides in the same cistron as *nuc-2*.

The essential features of the whole system of genes is that, under conditions of low orthophosphate supply, proteins are produced that can attack organic phosphorus compounds and also there is a transport system that can scavenge the phosphate so released.

Saccharomyces cerevisiae

There appear to be three transport systems for phosphate in *S. cerevisiae*. One system, discussed below, involves the cotransport of phosphate with sodium. Of the other two, there is one system with a high affinity ($K_T = 1$–$15\,\mu$M) for phosphate (Borst-Pauwels, 1962) and the other with low affinity ($K_T \approx 1$ mM) (Goodman & Rothstein, 1957). As with the two systems of *N. crassa*, both systems can operate simultaneously (Leggett, 1961), but otherwise there are important differences between the two fungi.

Some characteristics are common to the two transport systems not requiring sodium. Once a cellular orthophosphate concentration is achieved there is no further phosphate uptake (Holzer, 1953). Magnesium is required for uptake (Borst-Pauwels, 1967). Without magnesium, there is impairment of glycolysis such that the cellular orthophosphate concentration is increased. This impairment can be rectified, with a time-lag of 10 min, by the addition of magnesium to the medium. Calcium cannot replace magnesium in this respect (Borst-Pauwels, 1962). These observations point to feedback inhibition of phosphate on transport. In keeping with this is the finding that phosphate is rapidly incorporated (within 0.1 s) into organic form, thus helping to maintain the cellular concentration of orthophosphate at a low level (Borst-Pauwels, Loef & Havinga, 1962).

Phosphate uptake is enhanced by potassium at low medium pH and decreased at high pH (Goodman & Rothstein, 1957). Stimulation by potassium at low pH can occur by preincubation with glucose and the ion (Rothstein *et al.*, 1958). The most satisfactory interpretation of the effect of potassium is that when the ion is absorbed there is an increase of cell pH (Borst-Pauwels, 1981). It is not clear why there is the decrease of phosphate uptake with potassium at high pH.

Arsenate can also be transported by both systems (Rothstein, 1963; Jung & Rothstein, 1965). Both uptake and glycolysis are inhibited progressively by arsenate uptake. The mechanism by which inhibition occurs appears complicated. For instance, arsenate inhibits glycolysis to a maximum of only 60% in intact cells but 100% in broken cells. Inhibition of transport is irreversible but this is not so for glycolysis at lower arsenate concentrations. It is not known whether arsenate is necessarily the inhibitory agent or whether some derivative of arsenate is active. Cells of *Saccharomyces carlsbergensis* can be adapted to arsenate (Cerbón, 1970; Cerbón & Llerenas, 1976). These cells exhibit a highly reduced arsenate uptake but maintain their capacity to transport phosphate. There is also a modification of inositide metabolism.

The effect of medium pH on phosphate transport is complicated (Borst-Pauwels & Peters, 1977, 1987). It appears that cell pH is the main factor in determining the maximal rate (V_{max}) of phosphate uptake by the high affinity systems. At a cellular pH of *c.* 6.8, phosphate uptake is optimal (Figure 7.4). Reduction below or above that value leads to a reduction in maximal uptake. All this means that differences in the optima for phosphate uptake based on medium pH produced by different metabolic conditions or different buffers are due to differential effects on cell pH. The effect of

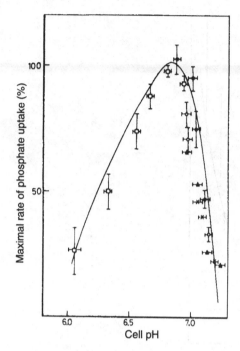

Figure 7.4. The dependence of the maximum rate of phosphate uptake by cells of *Saccharomyces cerevisiae* on the internal pH, varied by incubating the cells in butyrate or phosphate at differing concentrations at a particular external pH (represented by particular symbols). (From Borst-Pauwels & Peters, 1977.)

medium pH on K_T is more complicated (Figure 7.5). At low pH, in a medium buffered by citrate, K_T shows a dependence similar to that of V_{max}. Above pH 5.5, however, V_{max} declines with increasing medium pH, whereas K_T increases. At low pH the effect on K_T is via cell pH; above pH 5.5, the increase in K_T, or decreased affinity of the transport system for phosphate, is ascribed either to competitive inhibition from hydroxyl ions binding to the carrier sites or to the system being a proton/phosphate symport, the hydroxyl ions reducing the apparent affinity of phosphate for the carrier. Values of K_T can be corrected for the latter effect; when this is done, the relationship between K_T and cell pH is the same as that for V_{max}.

The view then is that the high affinity phosphate transport system is a proton symport. The extent to which the results described above are actually in keeping with the known functioning of such a system will depend upon a thorough kinetic analysis of the process using the approach

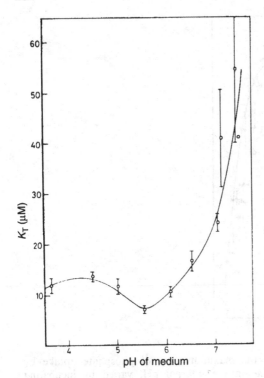

Figure 7.5. The dependence of K_T for monovalent phosphate uptake by cells of *Saccharomyces cerevisiae* on the pH of the medium. (From Borst-Pauwels & Peters, 1977.)

of Sanders (1986). However, one can be confident that the general conclusion is correct; particularly so, since Cockburn *et al.* (1975) have demonstrated that cells of *S. carlsbergensis*, grown in a chemostat in phosphate-limiting conditions, absorb phosphate accompanied by about three proton equivalents per mole of phosphate. Charge balance is maintained by the loss of two equivalents of potassium from the yeast cells. If potassium is absent from the external medium, it appears that other ions such as ammonium can be involved in maintaining charge balance (Höfeler *et al.*, 1987). Also in keeping with the high affinity system being a proton symport is the inhibitory effect on phosphate transport of uncoupling agents such as 2,4-dinitrophenol (Borst-Pauwels & Jager, 1969; Borst-Pauwels & Huygen, 1972; Huygen & Borst-Pauwels, 1972).

The low affinity phosphate transport system also depends upon cell pH, but only V_{max} is affected, a 50% change being affected by a drop from pH 6.76 to pH 6.39 (Borst-Pauwels & Peters, 1987). K_T is virtually constant under these conditions. This indicates that the system does not involve proton cotransport. There is therefore a question of how the system may be driven. Unlike the high affinity system (system I) in *N. crassa*, which operates at much lower external phosphate concentrations and for which there is a proper question mark over how it brings about facilitated diffusion of phosphate, the low affinity system in *S. cerevisiae* could well act in a passive mode. In the case of the yeast, it could be that the negative potential across the membrane acting against the influx of phosphorus is outweighed by the chemical potential gradient acting in the reverse direction. Information on the cytoplasmic orthophosphate concentration is crucial to proving or disproving this possibility (see below). As with system I of *N. crassa*, the low affinity system of *S. cerevisiae* demands further study. For both systems there is a need for information about cytoplasmic orthophosphate concentrations, the extent to which phosphate uptake is accompanied by the loss of anions such as organic acids and the role of cations. In the case of *S. cerevisiae*, potassium ions are obviously important, though not in a coupling capacity (see above).

Phosphate uptake at pH 7 and above is increased very considerably by sodium (Borst-Pauwels, Theuvenet & Peters, 1975; Roomans, Blasko & Borst-Pauwels, 1977). Kinetic analysis shows a dual mechanism consisting of two simultaneously operating transport systems. One process with a K_T of 30 μM for monovalent orthophosphate is the proton–phosphate cotransport system, the other with a K_T of 0.6 μM is a sodium–phosphate cotransport. V_{max} for the latter is increased by sodium but this is not so for the former process. Arsenate can be transported by the system and this ion together with phosphate enhances sodium uptake but inhibits rubidium uptake (Roomans & Borst-Pauwels, 1977). The value for K_i for the inhibition of rubidium transport is equal to K_T for the sodium-independent transport. Phosphate effects a transient stimulation of the efflux of the lipophilic cation dibenzyldimethylammonium from preloaded cells. The inhibition of monovalent cation uptake is best explained by a transient depolarisation of the membrane. Phosphate uptake does not appear significantly to affect intracellular pH. Finally analysis of the kinetics of sodium uptake in the presence of phosphate has shown the presence of two binding sites of the cation with K_T values of 0.04 and 29 mM, respectively (Roomans *et al.*, 1977).

Roomans *et al.* (1977) believed that in view of the internal concentrations of sodium in cells of *S. cerevisiae* it seemed unlikely that a sodium gradient across the plasma membrane is the driving force for phosphate transport by this carrier system. However, this does not disprove the probability that sodium is pumped out of cells of this yeast (see Chapter 13). A sodium/phosphate symport carrying a net positive charge is the most likely mechanism; this appears to be accepted by Borst-Pauwels & Peters (1987).

In considering the effect of cations on phosphate uptake in yeast there is a need to consider the effect on the surface potential (Borst-Pauwels & Peters, 1987; see Chapter 10).

Nieuwenhuis & Borst-Pauwels (1984) showed that the high affinity phosphate transport system is derepressed by phosphate starvation. Cycloheximide prevents derepression but is ineffective once the system is completely derepressed. The low affinity system disappears with the appearance of the high affinity system. A phosphate supplement to cells after derepression causes a reversal to some extent. There are changes to K_T and V_{max} that are explicable in terms of changes in the intracellular concentration of hydrogen ions and phosphate. The authors argue for the presence of a carrier common to both high and low affinity systems. The effect of phosphate starvation on the derepression of the high affinity system was confirmed by Tamai, Toh-e & Oshima (1985), who showed that expression of activity of this system was under the control of the genetic system that regulates the repressible acid and alkaline phosphatases. Specifically the *PHO2* gene is indispensable for the derepression of the high affinity system; it is this gene that is also essential for the derepressible acid (but not alkaline) phosphatase. The *PHO* system in *S. cerevisiae* has now been probed in some detail and the reader should consult articles by Oshima (1982, 1991) and Vogel & Hinnen (1990) for a description of the genetic hierarchy of the regulatory factors transmitting the physiological signal (cytoplasmic orthophosphate concentration).

The *PHO48* gene of the *PHO* system encodes the orthophosphate transport system. The gene has been cloned (Bun-Ya *et al.*, 1991). The sequence codes for a protein of 596 amino acid residues. The hydropathy analysis suggests that the secondary structure consists of two blocks of six transmembrane domains separated by 74 amino acid residues. The predicted structure differs significantly from the phosphate transport system of *N. crassa* encoded by *pho-4*[+]. Disruption of *PHO48* almost abolishes orthophosphate transport (Ueda & Oshima, 1975), which tends to confirm the suggestion of Nieuwenhuis & Borst-Pauwels (1984) that both high and low affinity systems share a common protein. It can be argued that this protein is encoded by *PHO48* but there is no molecular biological evidence as yet for this supposition. Even if this were so, one would still need to know how that protein was converted from acting as a low affinity system to a high affinity system and vice versa.

Up to this point, phosphate transport in *S. cerevisiae* has been treated as being essentially unidirectional. Several studies report an absence of efflux (Goodman & Rothstein, 1957; Jungnickel, 1966; Ueda & Oshima, 1975). That efflux can be undectable or very low (Swenson, 1960) is likely to be due to the low cytoplasmic concentration of orthophosphate (see below). Efflux of phosphate can be increased by glucose and is inhibited by 2,4-dinitrophenol or inhibitors of glycolysis (Borst-Pauwels, 1967; Knotová & Kotyk, 1972).

Other fungi

There is a phosphate/proton symport in *Candida utilis* with a stoichiometry of three proton equivalents per mole of phosphate (Eddy *et al.*, 1980). Studies using ^{31}P NMR on phosphate transport in *C. utilis* cells that have been phosphate-starved, are consistent with such a mechanism (Bourne, 1990, 1991). Concurrent observations on changes in the potassium and hydrogen ion concentrations in the medium have produced a helpful picture of some of the ion movements that accompany phosphate transport. The observations are essentially qualitative. However, the techniques must be commended. If it were possible to measure the membrane potential simultaneously, then the approach could allow a thorough examination of phosphate transport both in terms of the driving forces operating and the movement of other ions taking place as phosphate moves into the cell.

Candida tropicalis has two phosphate transport systems, one of which is constitutive and has a low affinity for the ion ($K_T = 1.2 \times 10^{-3}$ M) and the other which appears only under phosphate-starved conditions and has a high affinity for phosphate ($K_T = 4.5 \times 10^{-6}$ M) (Blasco, Ducet & Azoulay, 1976). Transport is sensitive to osmotic shock. Jeanjean *et al.* (1981) and Jeanjean & Fournier (1979) showed that phosphate-binding proteins were lost from cells given osmotic shock. Importantly, phosphate transport in protoplasts of *C. tropicalis* lack the high affinity transport system and phosphate-binding proteins can be found in the fluid after enzymic lysis of the walls (Jeanjean *et al.*, 1981). The hypothesis is that the high affinity transport system of *C. tropicalis* consists of a cell wall protein and a specific phosphate carrier (PiBP2), in the plasma membrane, with a molecular mass of 30 kDa (Jeanjean, Blasco & Hirn, 1984). Interestingly, it has proved possible to isolate two cell wall proteins released from *S. cerevisiae* by osmotic shock by an immunoabsorbent column coupled with antibodies against PiBP2 of *C. tropicalis* (Jeanjean *et al.*, 1986). Antibodies against the *S. cerevisiae* proteins inhibited phosphate uptake by intact cells of the yeast.

Germlings of *Phytophthora palmirora* have been shown to have three phosphate transport systems (Griffith, Atkins & Grant, 1989). One has a K_T of 1–2 μM and is repressed by phosphate, the second a K_T of 3 mM, and the third appears to be present when germlings deprived of phosphate are presented with phosphate at millimolar concentrations. The interest in phosphate transport in *Phytophthora* lies particularly in the toxic effect of phosphite (syn. phosphorus acid, phosphoric acid) to control diseases by this fungus. Phosphite enters *P. palmirora* by all three transport systems. There is competitive inhibition between phosphate and phosphite. However, while millimolar concentrations of phosphite are required to inhibit phosphate uptake, only micromolar concentrations of phosphate are required for inhibition of phosphite uptake.

It is appropriate to point out here that phosphate uptake in the obligately marine phycomycetes *Dermocystidium* spp. and *Thraustochytrium* spp. requires sodium (see Chapter 13).

Most of the other studies of phosphate uptake by fungi either free-living or in mutalistic association as lichens or mycorrhiza have been concerned with rates of uptake, from which it has been possible to extract K_T and V_{max} values (Farrar, 1976; Beever & Burns, 1980; Clipson, Cairney & Jennings, 1987; Straker & Mitchell, 1987; Cairney *et al.*, 1988; Thomson, Clarkson & Brain, 1990). In all cases but one, kinetic studies have demonstrated the presence of two transport systems for phosphate, the possible exception being the lichen *Hypogymnia physodes* (Farrar, 1976) for which the data available indicate only one system. However, the data are not sufficient to exclude the presence of a second system with a K_T higher than 33 μM, which is the K_T for the system identified. It has been implied that, since lichens are slow growing, the capacity of the system is adequate to account for phosphate uptake from rainwater (0.92 μM) by the lichen in nature (Farrar, 1976), so there is no need for a system with lower affinity. But that implication ignores the fact that the demonstration of two transport systems by kinetic means might not mean the presence of two distinct systems encoded by different genes but the kinetic consequence of random ligand binding to a transport system for phosphate with driver ions such as protons (Sanders, 1986). Equally, the implication ignores the capability of, and the necessity for, fungi to absorb phosphate at rates higher than that necessary to maintain growth, the phosphate so absorbed being stored. The matter is considered further below.

The data for K_T and V_{max} obtained from the studies referred to above are not for the most part of great value. There are two reasons for this. First, there is uncertainty in most cases about the phosphorus status of the living material used. Clipson *et al.* (1987) have shown from studies on *Phallus impudicus* and *Serpula lacrymans* that, as might be expected from the studies on *N. crassa* and *S. cerevisiae*, values of both K_T and V_{max} are dependent on the phosphate concentration of the growth medium. Second, it is unfortunate that the majority of V_{max} values are expressed in terms of dry weight and not as flux values (see Chapter 3).

Nevertheless some flux values have been obtained and these are presented in Table 7.2 along with values obtained for *C. tropicalis*, *N. crassa* and *S. cerevisiae*.

Phosphorus compartmentation in the protoplasm

It is worth reminding ourselves what the cytoplasmic concentration of the dihydrogen orthophosphate ion, $[H_2PO_4^-]$, in a fungus would be if it were to be passively distributed across the plasma membrane between external medium and cytoplasm. On the basis of the Nernst equation, for an external concentration of the ion of 1 mM, the cytoplasmic concentration will be 0.14 mM for a membrane potential difference of -50 mV (made negative) but only 4.59×10^{-4} mM for a potential difference of -200 mV. The total concentration of phosphate in the cytoplasm depends on the pH and the value of the apparent acidic dissociation constant, K_a. But assuming that the pH of the cytoplasm is the thermodynamic K_a and that value is not altered in the cytoplasmic solution, the total orthophosphate concentration in the first case above will be 0.28 mM and 9.18×10^{-4} mM in the second case. Any values for the cytoplasmic phosphate concentration greater than these mean that orthophosphate is actively transported across the plasma membrane.

Table 7.3 gives values for the cytoplasmic orthophosphate concentrations for a number of fungi, obtained by three different techniques. The values are in the range 1–34 mM but are similar to values for the cytoplasmic orthophosphate concentrations in plant tissues determined by ^{31}P NMR (Lee, Ratcliffe & Southon, 1990). Though there is uncertainty about the accuracy of the values in Table 7.3, their magnitude must cast doubt on

Table 7.2. *Phosphate fluxes into some fungi and algae*

Organism	Flux[a] nmol m^{-2} min^{-1}	Reference
Fungi		
Neurospora crassa		
Conidia germinated and grown 2.5 h in 10 mM P$_i$ medium; uptake measured at 10 mM P$_i$	738	Beever & Burns (1977)
Conidia germinated and grown 2.5 h in P-deficient medium; uptake measured in 50 μM P$_i$	2802	Beever & Burns (1980)
Candida tropicalis		
Cells grown in a 100 mM P$_i$ medium, uptake measured at 20 mM P$_i$	1680	Blasko *et al.* (1976)
Saccharomyces cerevisiae		
Maximum rate shown by P-starved cells allowed to take up P$_i$ under 'Überkompensation' conditions (29 mM P$_i$)	5532	Liss & Langen (1962)
Endogone sp.		
Hyphae of the v–a mycorrhizal fungus growing from onion roots in soil[b]	960	Beever & Burns (1980); Sanders & Tinker (1973)

Rhizomorphs of woodland basidiomycetes

Mutinus caninus
From
   100 μM phosphate   200   Clipson *et al.* (1987)
   1 mM phosphate   767

Phanerochaete sp.
From
   100 μM phosphate   190
   1 mM phosphate   1200

Stecherinum fimbriatum
From
   100 μM phosphate   288
   1 mM phosphate   1207

Algae
Nitella translucens   600–1500   Smith (1966)
Hydrodictyon africanum   108–246   Raven (1974)

[a] The surface area of 2.5 h old *N. crassa* germlings was taken as 4.6 m² (g dry weight)⁻¹ (Burns & Beever, 1977) and of *C. tropicalis* and *S. cerevisiae* as 2.2 m² (g dry weight)⁻¹ (Beever & Burns, 1980).
[b] v–a, Vesicular–arbuscular.
Based on Beever & Burns (1980).

Table 7.3. *Values for cytoplasmic orthophosphate concentrations (mM) of fungi*

Candida utilis (Nicolay *et al.*, 1983)[a]	
Prior to addition of 100 mM glucose to culture	17
After addition	4
120 min later	12
C. utilis (Bourne, 1990)[a]	
Prior to addition of 800 μM phosphate to culture	12
After addition	34
Saccharomyces carlsbergensis (Okorokov *et al.*, 1980)[b]	1
Fungal sheath of beech mycorrhizal roots (Harley & Loughman, 1963)[c]	3.12

[a] ^{31}P NMR.
[b] Differential extraction using cytochrome *c* (see p. 75).
[c] Calculated from 0.02 μg phosphorus in 100 mg dry weight of sheath tissue – the amount of phosphorus in the orthophosphate pool through which phosphorus passes in transit to host tissue. The fresh weight:dry weight ratio has been assumed to be 1:5 and cytoplasm has been assumed to be 50% of the sheath volume.

the likelihood of phosphate entering fungal cells or hyphae by facilitated diffusion except at relatively high external phosphate concentrations.

Phosphate enterning the cytoplasm may be incorporated into phosphorylated primary metabolites, structural molecules, nucleic acids and polyphosphates (Table 7.4) and may be actively transported into the vacuole (see Chapter 4). Transport into the vacuole and the formation of insoluble polyphosphate may be a major part of the mechanism by which the fungus controls the cytoplasmic orthophosphate concentration (Jennings, 1989). This matter is discussed further below.

As can be seen from Table 7.4, polyphosphates can form an appreciable proportion of the phosphate content of a fungus. Polyphosphates are found in a wide variety of fungi (see Kulaev, 1979, for the early literature) and may be present in the majority. A report (Dietrich, 1976) suggesting that polyphosphates are not present in Oomycetes can be discounted on the basis of ^{31}P NMR studies on *Phytophthora palmivora* (Niere, Griffith & Grant, 1990), though polyphosphate granules cannot be detected in members of this group of fungi but can in all the other major groups (Chilvers, Lapeyrie & Douglass, 1985).

Polyphosphates are linear polymers of orthophosphate in anhydrous linkage. The length of the polymer may range from three (tripolyphosphate) to greater than a thousand. The bond between residues has a high free energy of hydrolysis, equivalent to that for the terminal phosphate group

Table 7.4. Chemical distribution of phosphorus in actively growing fungal cells

Phosphate-containing fraction	Neurospora crassa[a]		Saccharomyces cerevisiae[b]		Mucor racemosus[c]	
	μmol P (g d.w.)$^{-1}$	% of total P	μmol P (g d.w.)$^{-1}$	% of total P	μmol P (g d.w.)$^{-1}$	% of total P
Orthophosphate	20	4	116	13	148	8
Acid-soluble poly P	55	11	213	23	603	35
Acid-insoluble poly P	60	12	97	11	517	30
Soluble organic P	80	16	39	4	86	5
Lipid P	75	15	48	5	62	3
RNA	} 220	42	379	41	311	18
DNA			25	3	9	1
Total of above	510	100	917	100	1736	100

d.w., dry weight; P, phosphate.
[a] Harold (1962a).
[b] Katchman, Fetty & Busch (1959).
[c] James & Casida (1964).
From Beever & Burns (1980).

of ATP. There are a number of reviews on polyphosphates in microorganisms or fungi by Harold (1966), Dawes & Senior (1973), Kulaev (1979), Beever & Burns (1980), Kulaev & Vagabov (1983), Wood & Clark (1988). The reader should consult these reviews for details of the metabolism of polyphosphates; only their general physiology is considered here.

In general physiological terms it has been customary to consider polyphosphates in fungi in terms of their solubility in acid. The acid-soluble polyphosphates are those with chain lengths of about 20 phosphate residues; acid-insoluble polyphosphates have chains of more than this number of residues. However, other fractions can be identified depending on their solubility in other solvents, e.g. sodium dodecyl sulphate (Harold, 1966). It should be noted that the extent of solubility depends very much on the physiological conditions. Polyphosphates are salts of acids that in solution contain two types of hydroxyl group (Kulaev, 1979). The terminal hydroxyl groups (two per polyphosphoric acid molecule) are weakly acidic, whereas the intermediate hydroxyl groups, which are equal in number to the phosphorus atoms in the molecule, are strongly acidic. All alkali metal polyphosphates are soluble in water, but those of divalent metals such as magnesium, barium and lead are either insoluble or dissolve to only a limited extent. Polyphosphates in aqueous solution of low ionic strength are capable of forming complexes with other polymers, especially proteins, polypeptides and nucleic acids. So, while Table 7.5 presents information on the polyphosphate content of fungi based on fractionation in terms of acid-solubility, the information should be viewed with a degree of caution.

Polyphosphates are found in the walls, cytoplasm and vacuoles of fungi. The presence of polyphosphate in walls has been dramatically demonstrated by comparing the phosphorus content of whole cells and that of protoplasts of *S. carlsbergensis* (Table 7.6), 25% of the polyphosphate being located in the wall. The presence of polyphosphates in isolated walls should be viewed with caution, since the polyphosphate so found might have come from the protoplasm disrupted during wall isolation (Harold & Miller, 1961; Harold, 1962a; Beever & Burns, 1980). The extent to which polyphosphates are present in fungal walls requires clarification. The most detailed study of wall-located polyphosphates has been made in *Kluyveromyces marxianus* by Tijssen, Beekes & van Steveninck (1981), Tijssen *et al.* (1983), Tijssen & van Steveninck (1984) and Tijssen, van Steveninck & de Bruijn (1985). On the basis of (a) release of polyphosphate from the cells by osmotic shock without damaging cell permeability, (b) ^{31}P NMR studies, in which the polyphosphate signal is reduced by uranyl ions, which are known not to penetrate into the protoplast, and (c) cytochemical staining, evidence is unambiguous for polyphosphate presence in the wall. The polyphosphate released by osmotic shock has a higher chain length and a lower metabolic turnover rate than does the total cellular polyphosphate. Studies of a similar kind on other fungi are needed to determine the ubiquity of polyphosphate in fungal walls. Without information of this kind, it will be difficult to assess the role of such polyphosphate – though some speculation on the matter is given below. With respect to the question of how the polyphosphate arrives in the wall, it seems likely on the basis of studies on *S. carlsbergensis* (Kulaev & Vagabov, 1983) that it occurs as part of the process of the synthesis of mannoproteins composing the wall.

The presence of polyphosphate in vacuoles has been discussed in Chapter 4. Polyphosphates of an average chain length of 14 residues have been demonstrated to be present in suspensions of purified mitochondria of *S. cerevisiae* and correspond to 10% of the total cellular polyphosphate content as determined by NMR (Beauvoit *et al.*, 1989). These observations are a possible explanation for an earlier observation by Solimene, Guerrini & Donini (1980) that in exponentially growing respiring cells of *S. cerevisiae* there was an accumulation of low molecular weight polyphosphates (three to eight residues) not seen in growing cells that were not respiring. Also polyphosphate has been detected in nuclear preparations of *Endomyces magnusii* and *N. crassa* (Kulaev & Vagabov, 1983).

Changes in polyphosphate content have been most extensively studied in *S. cerevisiae*. I do not propose to detail the studies that have been made on this yeast; the reader is advised to consult the reviews listed above. Here, I draw attention to three significant features:

(1) Polyphosphate content of cells of the yeasts is dependent on the phosphate supply (Table 7.5). The same is true for other fungi.
(2) When phosphate labelled with ^{32}P is incorporated into polyphosphate in fermenting yeast, the process takes place most rapidly into the highest molecular weight fractions, incorporation into fractions of decreasing molecular weight being progressively slower (Langen & Liss, 1958; Langen, Liss & Lohmann, 1962). Blockage of fermentation leads to the breakdown of the polyphosphate and the release of the equivalent amount of orthophosphate. If this block of fermentation is carried out after only a short period of labelling, the orthophosphate

Table 7.5. *Polyphosphate content of fungi*

	Polyphosphate content, μmol P (g dry weight)$^{-1}$			Reference[a]
	Acid-soluble	Acid-insoluble	Total	
I. Batch culture, excess P$_i$ concentration, rapid growth phase				
Aspergillus niger	42	66	108	A
Claviceps purpurea	215	77	292	B
Mucor racemosus	603	517	1120	C
Neurospora crassa	55	60	115	D
Saccharomyces cerevisiae	213	97	310	E
S. bisporus var. mellis	—	—	120–140	F
II. Non-growing cells incubated at high P$_i$ concentrations with glucose, K$^+$ and Mg^{2+} (*Überkompensation* conditions)				
Saccharomyces cerevisiae				
P-satiated cells	141	49	190	G
P-starved cells	454	1013	1467	
S. bisporus var. mellis				
P-satiated cells	—	—	212	H
P-starved cells	—	—	2940	

III. Auxotrophic mutant grown in supplemented medium and then transferred to fresh medium with or without the supplement for 24 h

Neurospora crassa histidine auxotroph

Histidine added	94	94	188	I
No histidine	258	94	352	

IV. Fungal spores

Agaricus bisporus				
Basidiospores	39	25	64	J
Aspergillus niger				
Conidiospores	0	19	19	K

[a] A, Krishnan, Damle & Bajaj (1957). B, Taber (1964). C, James & Casida (1964). D, Harold (1962a). E, Katchman et al. (1959). F, Weimberg & Orton (1965). G, Liss & Langen (1962). H, Weimberg (1975). I, Harold (1960). J, Kulaev, Kritskii & Bolozerskii (1960). K, Nishi (1961).
From Beever & Burns (1980).

Table 7.6. *Contents of polyphosphate fractions in whole cells and protoplasts of* Saccharomyces carlsbergensis

Phosphorus-containing compounds	Content (μg phosphorus (g wet cells)$^{-1}$)	
	Whole cells	Protoplasts
Acid-soluble	706	728
Salt-soluble	516	299
Alkali-soluble (pH 8–10)	208	23
Alkali-soluble (pH 12)	356	0
High molecular weight polyphosphates (total)	1786	1050

From Kulaev & Vagabov (1983).

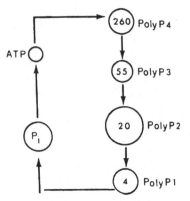

Figure 7.6. The polyphosphate cycle in *Saccharomyces cerevisiae* as proposed by Langen, Liss & Lohmann (1962). (From Harold, 1966.)

released is unlabelled, in spite of the breakdown of the first-labelled fraction (that with the highest molecular weight). The label is in fact trapped in lower molecular weight polyphosphates. These and other observations (Liss & Langen, 1962) led to the formulation of the polyphosphate cycle (Figure 7.6). Kulaev and his colleagues (see e.g. Kulaev, 1979) have made similar observations with other fungi, principally *N. crassa* and have produced a more sophisticated model (Figure 7.7). It is not clear to what degree that particular model is correct. There must be a question mark over the involvement of polyphosphate in glucose transport (see Chapter 5).

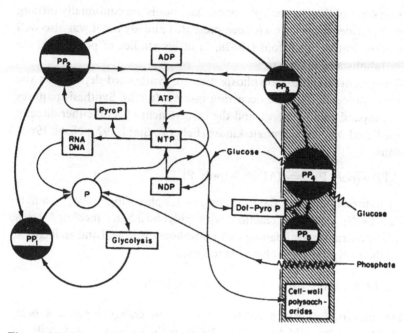

Figure 7.7. Hypothetical scheme for the reactions of high molecular weight polyphosphates (PP_3, PP_4 and PP_5) in fungi. Hatched column, the plasma membrane; Dol-PyroP, dolichol-pyrophosphate. (From Kulaev, I. S. (1979), *The Biochemistry of Inorganic Polyphosphates*, Copyright © 1979, reproduced by permission of John Wiley and Sons Limited.)

There is evidence against the polyphosphate cycle as proposed, essentially from studies showing that short-chain polyphosphates are those first formed from the phosphate entering the fungus. However, the evidence can be questioned – that by Indge (1968) on the basis that the study was carried out with protoplasts of *S. cerevisiae*, not intact cells, and that of Schuddematt *et al.* (1989*a*) on the basis that the cells of *S. cerevisiae* and *Kluyveromyces marxianus* used in the study were starved to the extent that no primer for polyphosphate synthesis was present (Bourne, 1990). In fact, the more recent evidence (Lusby & McLaughlin, 1980; Bourne, 1990) has tended to support further the proposals of Liss & Langen (1962). However, there are some indications that in cells of *S. cerevisiae* it is at low growth rates that long-chain polyphosphates are the first-formed products from orthophosphate, while at high growth rates it is short-chain (*c.* five residues) products that are formed (Núñez & Callieri, 1989). Whether or not the proposals of Liss & Langen are correct, one should not lose sight of the fact that polyphosphate pools are continually turning over. Evidence for that has been presented already but it was also well demonstrated by Harold (1962a, b) in his studies of polyphosphate metabolism in *N. crassa*.

(3) The extent to which polyphosphates are synthesised depends on the other processes that are occurring inside a fungus. Synthesis requires the expenditure of energy and the involvement of ATP either directly mediated by polyphosphate kinase (Felter & Stahl, 1973; Nishi, 1960), thus:

$$ATP + (poly\ P)_n \rightleftharpoons (ADP)_n + (poly\ P)_{n+1}$$

or indirectly through the formation of 1,3-phosphoglycerylphosphate (1,3-PGP) and then reaction with polyphosphate, mediated by a phosphotransferase, leading to 3-phosphoglyceric acid and an increase in the polyphosphate chain length, thus:

$$1,3\text{-PGP} + (poly\ P)_n \longrightarrow 3\ PGA + (poly\ P)_{n+1}$$

The importance of the requirement for an energy input has been demonstrated by Höfeler *et al.* (1987) with sodium-loaded cells of *S. cerevisiae*. Such cells, when placed in a medium containing 11.3 mM sodium bicarbonate, 3.8 mM ammonium phosphate and glucose, did not synthesise polyphosphate. It seems that the energy required for the sodium/ammonium exchange took precedence over polyphosphate synthesis. The phosphate that was metabolised was converted into sugar phosphates.

With those very general comments I now consider the role of polyphosphates in fungi. Several roles have been postulated. These are discussed here, together with some other possibilities.

(1) *Phosphate reservoir.* There is no doubt from the early literature (see Harold, 1966) that there is plenty of evidence that polyphosphate can serve as a source of phosphorus for the synthesis of nucleic acids. Harold (1966) pointed out that phosphorus as polyphosphate rather than orthophosphate is preferable osmotically. Although polyphosphate would seem to act as a phosphate reserve, it should be noted that under phosphate-deficient conditions fungi adapt in other ways to the lack of phosphate, such as reducing the phosphorus content of the wall and partially replacing phospholipid with phosphorus-free lipid (Beever & Burns, 1980). It is timely to look at

this proposed role more critically. As Wood & Clark (1988) have pointed out, the fact that when a fungus is deprived of phosphate the amount of polyphosphate decreases while nucleic acids and phospholipids increase does not mean necessarily that polyphosphates are acting as a store of phosphorus. These authors pointed out that if the rate of synthesis of polyphosphate were to decrease but the rate of utilisation remain the same and if the rate of degradation of nucleic acid were to decrease then the net result would be the same as above but the polyphosphate could not be viewed as a reserve. There is a need to determine turnover rates.

(2) *Phosphagen hypothesis.* This concept, put forward by Hoffman-Ostenhof & Weigert (1952), was that polyphosphates were comparable to animal cell phosphagens, such as creatine and arginine phosphates, in that like these compounds the phosphate residues were transferred directly only to ADP, with the synthesis of ATP. The capability of polyphosphates theoretically to act in this way follows from (a) the free energy of hydrolysis of the phosphoric anhydride bonds in polyphosphate being comparable to the terminal pyrophosphate bond in ATP and (b) the ability of polyphosphate kinase enzymes to act in a reversible manner, viz:

$$ATP + poly\ P_n \rightleftharpoons poly\ P_{n+1} + ADP$$

The general consensus (Harold, 1966; Kulaev, 1979) is that this hypothesis is not tenable in the strict sense, i.e. polyphosphate breakdown occurs without the production of ATP. Thus, polyphosphate does not break down if energy metabolism is blocked (Harold, 1966). Further the amount of energy stored in polyphosphate is very low compared with the energy liberated by the oxidation of glucose. During active growth, cells of *S. cerevisiae* exhibit rates of synthesis between 67 and 667 μmol (g dry weight)$^{-1}$ min^{-1} (Chapman & Atkinson, 1977). Beever & Burns (1980) pointed out that even the total breakdown of polyphosphate in such cells would generate only 300–400 μmol (g dry weight)$^{-1}$ min^{-1}. Thus, polyphosphate could supply the cells' energy needs for only a few minutes at most.

(3) *Control of orthophosphate concentration in the cytoplasm.* This was first suggested by Kaltwasser (1962). However, at the present juncture this must be considered non-proven. First, too little is known about how the cytoplasmic orthophosphate concentration changes. It certainly does not remain constant but the extent of variation is unclear. Nor is much known about concurrent changes in polyphosphate

concentration. In any case, it is probable that transport across the tonoplast is of more immediate importance in the control of the cytoplasmic orthophosphate concentration. Nevertheless, there is no *a priori* reason why tonoplast transport and polyphosphate metabolism should not act in concert.

(4) *Protection against alkaline stress*. Pick *et al.* (1990) have shown, using ^{31}P NMR, that cytoplasmic alkalinisation through the uptake of ammonium in cells of *Dunaliella salina* induces a breakdown of polyphosphates that is correlated kinetically with the recovery of the cytoplasmic pH. This possibility still needs to be explored with fungi, particularly since there is as yet no evidence to indicate that fungi capable of living at alkaline pH values possess transport mechanisms that appear to be required by those bacteria capable of living in similar environments (Kroll, 1990).

(5) *Energy dissipation*. In Chapter 13, I discuss the need for fungi, under certain conditions in which growth is limited by a particular environmental factor, to dissipate energy by futile cycles. While there is as yet no evidence for the polyphosphate cycle acting in this capacity, it clearly possesses the capability so to act.

Mycorrhiza

Orthophosphate uptake

In 1937, Hatch demonstrated that mycorrhizal seedlings of *Pinus strobus* absorbed more phosphorus, nitrogen and potassium than did non-mycorrhizal seedlings (see Table 6.16, p. 244). The mycorrhizal association here is of the ectomycorrhizal kind. It is now known that other kinds of mycorrhizal association are able to assist their green plant hosts in acquiring more phosphorus (Harley & Smith, 1983). Here, I explore in general terms the mechanisms by which mycorrhizal fungi are able to bring about this greater acquisition of phosphorus. In these considerations, one needs to remember that the concentration of phosphorus (for these purposes phosphate) in soil solution is low. This is because of binding of phosphate to soil and the slow release of phosphorus from other sources in the soil. Binding of phosphate to the soil and the tortuosity of the path over the soil particles means that diffusion of phosphate is much slower than in the bulk aqueous system. The low concentrations of phosphate militate against the movement of the ion by mass flow driven by the transpiration from the leaves of the green plant.

These matters are more fully discussed by Nye & Tinker (1977, Barber (1984) and Jennings (1986*b*). Finally there is no evidence that green plants can degrade organic phosphorus compounds in the soil external to the roots.

Increased influx of phosphate

This has been shown to be the case for beech mycorrhiza (Harley & McCready, 1950). Infected roots have an influx about five times greater than uninfected roots when exposed to very low concentrations of phosphate.

It was nearly 40 years before the essential features of this study were confirmed. Two investigations, one involving *Eucalyptus pilularis* (Heinrich & Patrick, 1986) and the other involving willow (*Salix viminalis*) (Jones, Durall & Tinker, 1991) on growth and phosphorus acquisition of mycorrhizal and non-mycorrhizal plants have demonstrated unambiguously that infection increases the phosphorus inflow rate (mol phosphorus absorbed per unit length of root per unit time). Jones *et al.* (1991) showed that, at the first harvest (50 days after planting of cuttings), the inflow rate for phosphorus into infected plants was $3 \times 10^{-12} \, \text{mol m}^{-1} \text{s}^{-1}$. If it is assumed that the roots have a diameter of 2 mm, the flux of phosphate into the infected roots was $1208 \, \text{nmol m}^{-2} \, \text{min}^{-1}$. This value is a minimum one (because not all the root system could have been infected) for the net flux of phosphate into *Thelophora terrestris* – the fungus producing the mycorrhiza. The value compares favourable with values for the net flux of phosphate into other fungi (see Table 7.2).

If absorption of phosphate from the soil were to occur from a source close to zero concentration, as might well be the case in some soils, it would be difficult to envisage the advantage of the mycorrhizal association. It is probable that the effectiveness of mycorrhizal beech roots in absorbing phosphate can be ascribed to the woodland soil in which they are found (Harley, 1978). The mycorrhizal roots are very closely associated with the uppermost layer of leaf litter, which is continually added to and moved around by environmental forces and animals. In this way the roots may be exposed to new sources of phosphate. In addition, because the supply of leaves to the litter is essentially seasonal and, when the leaves do fall, there are large additions to the litter, there could be seasonal flushes of nutrients, amongst those released being phosphate either directly from the leaves or indirectly as a result of microbial activity. The higher the influx into the fungal sheath, the better is the ability to scavenge the phosphate

Table 7.7. *Inflows and fluxes of phosphorus during uptake by onion plants either infected with vesicular–arbuscular mycorrhizal fungus or not infected*

Conditions	Root infected (%)	Inflows			Calculated flux in hyphae[b] ($mol\,cm^{-2}\,s^{-1} \times 10^{-3}$)
		Mean for mycorrhizal onions ($mol\,cm^{-1}\,s^{-1} \times 10^{-14}$)	Mean for non-mycorrhizal onions ($mol\,cm^{-1}\,s^{-1} \times 10^{-14}$)	Via hyphae[a] ($mol\,cm^{-1}\,s^{-1} \times 10^{-14}$)	
Expt A: 44–45 days	50	13	4.2	17.6	⎫
Expt B: 39–54 days	45	11.5	3.2	18.5	⎬ 3.8 ⎭

[a] Calculated from difference between mycorrhizal and non-mycorrhizal inflows, divided by fraction of root infected.
[b] Based on six main entry hyphae (not actual entry points) per centimetre of infected root.
From Tinker (1975).

so released. The arguments put forward here probably apply also to other forest trees.

Of the phosphorus absorbed by beech mycorrhizal roots from 1 mM phosphate and lower concentrations (though high for soil), 90% is retained by the fungal sheath (Harley & McCready, 1952a, b). The phosphorus so absorbed is incorporated into two pools of orthophosphate (one cytoplasmic, the other vacuolar?), polyphosphate and organic phosphate (Harley & Loughman, 1963; Jennings, 1964; Chilvers & Harley, 1980; Harley & McCready, 1981; Strullu *et al.*, 1993). After the phosphate has been absorbed, there is a slow movement of orthophosphate from the fungal sheath to the host tissue, which appears to be driven by the metabolic activity of the fungus (Harley & Loughman, 1963; Harley & Brierley, 1954).

Increased exploration of the soil

There is general agreement that, compared with uninfected roots, the beneficial effects of vesicular–arbuscular (v–a) mycorrhiza are due to their ability to exploit soil phosphorus when growing in those soils with a low concentration of available phosphorus (for reviews, see Mosse, 1973; Tinker, 1975; Harley & Smith, 1983; Hetrick, 1989). The main mechanism by which there is a greater uptake of phosphorus by infected roots is due to the hyphae increasing the root length (Nye & Tinker, 1969) and thus allowing the root system to exploit a greater volume of soil to acquire the necessary phosphorus. The experimental evidence for this has been provided by Sanders & Tinker (1973), who used onion (*Alium cepa*) plants and analysed their growth and ability to acquire phosphorus in phosphorus-deficient soil when the plants were both mycorrhizal and non-mycorrhizal. Data obtained by means of nutrient flux analysis (Tinker, 1969) are presented in Table 7.7. As can be seen, the inflow of phosphorus into mycorrhizal plants is greater than in those that are non-mycorrhizal and the greater uptake can be ascribed to the additional length of the root system provided by the hyphae. The flux of phosphate into a hypha is much greater than the movement of phosphorus along the hypha and into the host root. In consequence, phosphorus accumulates in the hyphae, particularly as polyphosphate (Callow *et al.*, 1978). Movement of the polyphosphate granules may be the manner in which phosphorus is translocated along the hypha (see Chapter 14).

Sources of phosphorus other than orthophosphate

There is little doubt that ectomycorrhizal fungi can utilise organic phosphorus. There is plenty of evidence that these fungi when in culture and in association with roots possess surface or extracellular phosphatases that act on organic phosphates including phytate (Theodorou, 1971; Bartlett & Lewis, 1973; Ho & Zak, 1979; Antibus, Kroeler & Linkins, 1986; Doumas *et al.*, 1986; Kroeler, Antibus & Linkins, 1988; MacFall, Slack & Iyers, 1991*a*, *b*). Activity can be increased by growth in media of low phosphorus content, though this has not been explored in much detail. Also there has yet to be a proper assessment of the extent of the contribution of organic phosphates to the phosphorus nutrition of ectotrophic mycorrhiza. The acid phosphatase activity of ericoid mycorrhizal fungi has been characterised in some detail by Straker & Mitchell (1986). There is activity against phytate. There is no evidence that v–a mycorrhizal fungi are able to release phosphorus when the latter is in organic form.

Three approaches have been used to attempt to identify whether or not v–a mycorrhizal fungi and thus the roots with which they are associated are able to utilise sources of phosphate not available to non-mycorrhizal roots (Bolan, 1991):

(1) Use of [^{32}P]phosphate.
(2) Comparison of different phosphorus fertilisers.
(3) Fractionation of soil phosphate.

The first procedure involves labelling soil with [^{32}P]phosphate and then comparing the specific activity of the phosphorus absorbed by mycorrhizal and non-mycorrhizal plants. A lower value for the former would indicate that mycorrhizal fungi are able to capture phosphate not available to non-mycorrhizal plants. The procedure assumes that with short-term labelling, only the labile phosphate fraction equilibrates with the labelled phosphate added to the soil. This assumption may not be correct and other fractions will become labelled as the duration of an experiment is prolonged. Nevertheless, in most experiments the specific activity of mycorrhizal and non-mycorrhizal plants is similar (Bolan, 1991).

Experiments concerned with phosphorus fertilisers have examined the ability of mycorrhizal and non-mycorrhizal plants to grow in the presence of those which are poorly soluble. There have been two approaches (Bolan, 1991). One involves comparing the yield and phosphorus uptake at a particular level of added phosphorus and the other involves comparing the amount of phosphorus required to produce a particular yield using the complete phosphorus response curve. Even when there is a greater growth of mycorrhizal plants, it cannot be assumed that their v–a fungi are directly bringing about the solubilisation of the insoluble fertiliser.

A very good critique of the procedures for analysing the results of this type of experiment was given by Pairunan, Robson & Abbott (1980). Their conclusions, consistent with those of a number of other investigators (Hayman & Mosse, 1972; Mosse, Hayman & Arnold, 1973; Sanders & Tinker, 1973; Barrow, Malajczuk & Shaw, 1977), are that mycorrhizal and non-mycorrhizal roots obtain their phosphorus from the same soil pool. There is no need to postulate mechanisms additional to the one described on p. 285, i.e. the greater capability of mycorrhizal fungi, compared to uninfected roots, to explore the soil, to account for the enhanced phosphorus uptake by mycorrhizal plants when insoluble phosphorus fertiliser is added to soil.

Experiments concerned with examining how soil phosphorus fractions are changed by mycorrhizal and non-mycorrhizal plants are not very satisfactory. This is because change in one fraction due to activity by the fungus or plant can alter other fractions, making it very difficult to identify the source(s) of phosphorus that the plant is using.

8
Sulphur

Introduction

The relationship between fungi and sulphur is intriguing. Though inorganic sulphur at the most oxidised state – sulphate – is the customary source of the element when fungi are grown in culture, it is well known that sulphur-containing amino acids can be used with almost equal facility. In *Neurospora crassa*, for instance, when inorganic sulphur is lacking, the methionine transport is derepressed (see Chapter 6). Fungi are capable of oxidising inorganic sulphur compounds that may be important in ways other than the provision of sulphate for assimilation. Also fungi can produce volatile sulphur compounds that may be important in the natural environment. Equally, the ability of species to tolerate sulphur dioxide may be important in environments subject to air pollution. Of course the same gas is used also to preserve beverages and food, so that there is an industrial dimension to fungal tolerance of sulphur dioxide. All of the above are considered in more detail in the sections that follow.

The emphasis of this chapter is on inorganic sulphur as the source of the element for a fungus. The evidence from culture studies is compatible with an almost universal ability to utilise inorganic sulphur. As far as I am aware, inorganic sulphur compounds cannot be released from organic sulphur compounds in the external medium of a fungus. Nevertheless fungi contain sulphatases that release sulphate from organic form. An interesting example of this is the utilisation of glucose 6-sulphate by *N. crassa* (Reinert & Marzluf, 1974). The compound is absorbed by a specific transport system and can be the sole sulphur source for the fungus. The rate of intracellular hydrolysis is slow, such that the glucose 6-sulphate cannot act as a carbon source. *Neurospora crassa* also possesses a specific transport system for choline-*O*-sulphate (Marzluf, 1972*a*) (both the above two systems are repressed by methionine, see below). Finally, while sulphate appears to be the appropriate source of sulphur for almost all fungi, there are species of rusts, e.g. *Melampsora lini* and *Puccinia graminis tritici*, that cannot use inorganic sulphate and require a reduced organic source of sulphur such as the amino acids cysteine, cystine, homocysteine or methionine or the tripeptide glutathione (Maclean, 1982).

Sulphate transport
Neurospora crassa, *Aspergillus* and *Penicillium* spp.

Most of the information about sulphate transport comes from studies using *N. crassa*. There are two distinct transport systems, one (termed system II), which predominates in the mycelium, and a second (system I), present only in conidia (Marzluf, 1970a, b). The former, encoded by the *cys-14* locus, has a K_T of 5×10^{-5} to 10×10^{-5} M, while the latter, encoded by *cys-13*, has a somewhat higher affinity, the K_T being 8×10^{-6} M (Marzluf, 1970b). This latter system can also transport chromate; mutants for the gene are very resistant to the ion (Roberts & Marzluf, 1971). Chromate competes with sulphate for system I, having a K_i of 3×10^{-5} M. Tungstate and molybdate are also inhibitory but to a lesser extent. Efflux of sulphate requires the presence of the ion (or chromate, selenate or thiosulphate) in the external medium (Marzluf, 1974). This suggests that efflux (determined by loss of ^{35}S loaded into the mycelium) is probably an exchange reaction. Efflux still occurs in the presence of azide, even though there is no longer accumulation. Both sulphate transport systems turn over with a half-life of approximately 2 h (Marzluf, 1972b).

The two transport systems and early steps of assimilation of a variety of sulphur sources are under genetic and metabolic control (Marzluf & Metzenberg, 1968). When the availability of sulphur to the fungus becomes limited, a battery of enzymes is synthesised in a concerted fashion. These include aryl sulphatase (Metzenberg & Ahlgren, 1971), a methionine permease (Pall, 1971), an extracellular protease (Hanson & Marzluf, 1973) and enzymes of the two transport systems. If sulphur sources are available, a metabolite derived from methionine, perhaps cysteine (Jacobson & Metzenberg, 1977), acts to repress the battery of enzymes. Three regulatory genes have been shown to control the sulphur circuit: *cys-3* encodes a protein that turns on the expression of the structural genes (Fu & Marzluf, 1990b); *scon-1* has a negative control function (Burton & Metzenburg, 1972); and *scon-2*, more recently identified (Paietta, 1990), is also a negative regulatory gene. The simple model of the circuit is that the *scon-1* gene product senses the cellular level of a sulphur-containing metabolite and controls the expression of *cys-3*. It is thought that *scon-2* is involved in the signal between *scon-1* and *cys-3* (Burton & Metzenburg, 1972; Jarai & Marzluf, 1991). The gene product encoded by *cys-3$^+$* has been shown to be a bZIP (basic-region leucine zipper) protein (Paietta *et al.*, 1987). This protein (CYS3) appears from further experimental evidence (Paietta, 1992) to be a transcriptional activator whose regulation is a crucial control point in the induction of sulphur-metabolising enzymes. CYS3 has substantial sequence similarity to a number of bZIP proteins, including GCN4 of yeast (see Chapter 6).

The complete nucleotide sequence of the system II for sulphate transport has been determined (Ketter *et al.*, 1991). The transport protein (CYS14) has 781 amino acid residues, a predicted molecular mass of 87 kDa and contains 12 potential hydrophobic membrane-spanning domains. The protein as isolated, however, has a determined molecular mass of 95 kDa. Use of anti-CYS14 antibody has shown the protein to be present in the plasma membrane. Using the antibody, the protein has been found to be constitutive in a *scon-1* mutant but absent from a *cys-3* mutant.

It is relevant to consider the transport of sulphate into *Aspergillus* and *Penicillium* in conjunction with what is known for transport of the anion into *N. crassa*, because the information for the former two species amplifies what is known for *N. crassa* and indicates how far one may extrapolate to sulphate transport in other genera.

Sulphate enters *Aspergillus nidulans*, *Penicillium chrysogenum* and *P. notatum* by an essentially unidirectional process (Yamamoto & Segel, 1966; Bradfield *et al.*, 1970). The K_T for transport for the three species has been found to be in the range 6×10^{-5}–1×10^{-4} M. Thiosulphate, selenate and molybdate enter the above species by a sulphate transport system (Tweedie & Segel, 1970). The evidence for this is: (a) there is reciprocal competitive inhibition between the four anions; (b) mycelia grown on sulphur sources that repress the sulphate transport system show similar reduced transport rates for the other three anions; (c) sulphur starvation results in depression of transport for all four anions; and (d) throughout the period of derepression the ratio of the rate of uptake of one anion to another remains constant. In addition to the sulphate transport system, there are distinct transport systems for sulphite and tetrathionate. It is possible that thiosulphate can also be transported by the latter system, though a separate system for thiosulphate cannot be excluded on the data available.

In *N. crassa*, there appears to be no transinhibition of sulphate transport

(Marzluff, 1974), i.e. the rate of transport is not regulated by the anion remaining bound to the transport site (where it is facing the cytoplasm) in a concentration-dependent manner and preventing the unloaded carrier returning to the outside of the plasma membrane. There was absence of transinhibition even in an ATP-sulphurylase mutant. In such a mutant, there is no incorporation of sulphur into organic form, since the first enzyme for such incorporation is absent. Cuppoletti & Segel (1974), however, have purported to demonstrate transinhibition in a similar mutant of *P. notatum*. Data on the effect of preincubating mycelium with various compounds on the activity of the sulphate and choline-*O*-sulphate (see below) transport systems are presented in Table 8.1. It can be seen that only choline-*O*-sulphate and sodium chromate (which seems to be a non-specific inhibitor of transport processes) have any marked effect on choline-*O*-sulphate transport. A whole range of compounds affect sulphate transport. At first sight it might appear that transinhibition is occurring, as sulphate is inhibited by selenate, molybdate and thiosulphate. But the extent of inhibition by an anion ought to be related to the affinity of the anion for the carrier (Pall, 1971; Sanders, 1988). Greater inhibition would be expected from internal sulphate than from internal molybdate.

While at face value transinhibition does not seem to be occurring, it needs to be realised that criticism of the interpretation of the data in Table 8.1 is based on the assumption that each anion is uniformly distributed within the protoplasm. Evidence obtained by Hunter & Segel (1985) could be taken to suggest otherwise. Efflux experiments (into 10 mM azide at pH 8.4) indicated that [^{35}S]sulphate was lost from two pools in the mycelium, which are postulated to be the cytoplasm and the vacuole. Regrettably the evidence for this putative location of the two pools is by no means convincing. On the evidence available, wall and protoplasm are equally plausible suggestions for the two pools. Nevertheless, the possibility that there is more than one pool of sulphate in the protoplasm should not be discounted.

The matter of the compartmentation of sulphate in fungi needs further investigation. The methods described in Chapter 4 may provide the way forward. Nevertheless, while there is uncertainty about whether or not sulphate is sequestered in the vacuole, if there were to be sequestration of anions there it would provide a basis for reconciling the data in Table 8.1 with transinhibition of sulphate transport. It would only need sequestration in the vacuole in the order sulphate > thiosulphate > selenate > molybdate to produce cytoplasmic concentrations of the anions in the reverse order, which, if the amounts

Table 8.1. *The effect of preincubating with various compounds (at 10^{-3} M) on the activity of the sulphate and choline-O-sulphate transport systems in* Penicillium notatum

Preincubation conditions	$^{35}SO_4^{2-}$ transport rate		Choline-O-$^{35}SO_4^{2-}$ transport rate	
	$\mu mol\ g^{-1}\ min^{-1}$	% of control	$\mu mol\ g^{-1}\ min^{-1}$	% of control
Control	1.35	100	0.69	100
+ Na$_2$SO$_4$	0.50	37	0.67	98
+ K$_2$SeO$_4$	0.24	18	0.67	98
+ Na$_2$MoO$_4$	0.11	8	0.67	98
+ Na$_2$S$_2$O$_3$	0.47	35	0.68	99
Choline-O-sulphate	0.71	53	0.18	27
+ Choline-O-phosphate	1.79	133	0.80	117
+ Choline chloride	1.25	93	0.75	109
+ L-Methionine	0.86	64	0.66	97
+ L-Cysteic acid	1.36	100	0.64	93
+ Na$_2$CrO$_4$	0.03	2	0.02	3
+ Na$_2$SO$_3$	0.67	49	0.69	100
+ L-Cysteine	0.10	7	0.48	70
+ L-Cysteine-S-sulphate	0.09	7	0.44	65

From Cuppoletti & Segel (1974).

sequestered were of the right magnitude, could account for the order of inhibition of sulphate transport in Table 8.1.

The great sensitivity of sulphate transport in *P. notatum* to cysteine (Table 8.1) suggests that the amino acid could be a feedback inhibitor. Cuppoletti & Segel (1974) suggested that sulphate transport might be inactivated non-specifically by reducing agents.

Cuppoletti & Segel (1975) provided some evidence that sulphate transport in *P. notatum* is via a proton symport with a $H^+:SO_4^{2-}$ stoichiometry of 1, but further experiments are required before one can be certain that transport is by this mechanism. Transport is stimulated by bivalent cations. Cuppoletti & Segel (1975) suggested that calcium (and other bivalent cations) bind to the carrier. It is much more likely, however, that these cations act on the surface potential (see Chapter 10).

Saccharomyces cerevisiae

Breton & Surdin-Kerjan (1977) postulated that *S. cerevisiae* may well have two sulphate transport systems. The argument in support of this

Table 8.2. *Kinetic constants of sulphate permeases I and II in* Saccharomyces cerevisiae *in a wild-type strain (WT) and the sulphate permease mutants (chr, sel, and* chr,sel: *mutants resistant to chromate, selentate and chromate and selenate, respectively)*

	Sulphate permease I		Sulphate permease II	
Strain	$K_T{}^a$	$V_{max}{}^b$	K_T	V_{max}
WT	$(4\pm1)\times10^{-6}$	7 ± 1	$(3.5\pm1.5)\times10^{-4}$	7.5 ± 1.5
chr	$(8\pm1)\times10^{-6}$	7 ± 1	$(7.5\pm2)\times10^{-4}$	6.5 ± 1
sel	$(2.0\pm0.3)\times10^{-6}$	3.00 ± 0.05	No activity	No activity
chr,sel	No activity	No activity	No activity	No activity

^{a} Molar.
^{b} nmol min^{-1} (mg dry weight)$^{-1}$.
From Breton & Surdin-Kerjan (1977).

postulate was (a) the presence of biphasic kinetics when rates of sulphate transport as a function of external sulphate concentration were subjected to a Lineweaver-Burk plot and (b) the absence of sulphate uptake in a strain bearing two independent mutations (Table 8.2). However, Breton & Surdin-Kerjan (1977) pointed out that the model is probably too simple. There may not be two independent systems but two systems that interact, or even one protein that can exist in two conformational states, the two states being encoded by two separate genes.

More detailed physiological studies by Roomans *et al.* (1979*a*) suggested only one transport system. The K_T depends on the nutritional state of the cells. In nitrogen-deficient and sulphur-deficient cells at pH 6.8 the K_T values are 200 μM and 77 μM, respectively. The sulphate transport system has properties similar to that for phosphate, i.e. the effects of 2,4-dinitrophenol (DNP), cell pH and bivalent cations on both systems are similar. The conclusion is that sulphate is transported into cells of *S. cerevisiae* accompanied by a proton influx of three H$^+$ and an efflux of one K$^+$ for each sulphate ion taken up. The effects of cations on sulphate transport are in keeping with what one would expect if the effect were on the surface potential. Sulphate transport is inhibited by related ions in the sequence $CrO_4^{2-} > SeO_4^{2-} > S_2O_3^{2-} > SO_3^{2-} > TeO_3^{2-} > MoO_4^{2-}$ (McCready & Din, 1974).

Transport (V_{max}) is stimulated about 12-fold by glucose and ethanol and the stimulation involves protein synthesis (with a half-time of 18 min) Horák, Říhová & Kotyk, 1981). Ammonia, and some amino acids (leucine,

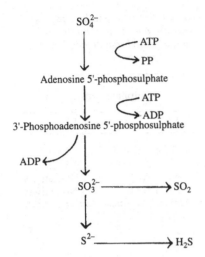

Figure 8.1. A summary of the first steps of sulphur assimilation in fungi. See text for details.

aspartate, cysteine and methionine) have been found to depress the stimulation, while cationic amino acids (typically arginine and lysine) and potassium increase it by 50%–80%. The transport system degrades with a half-life of $c.$ 10 min.

Assimilation of sulphur

The first steps of sulphur assimilation from sulphate, though only investigated for a few species, are likely to be universal amongst all fungi utilising sulphate, since the same pathway resides in all organisms investigated (Figure 8.1). Four steps are involved (Jones & Fink, 1982; Slaughter, 1989; Brunold, 1990).

(1) ATP sulphurylase (sulphate adenylyl transferase) brings about the formation of adenosine 5′-phosphosulphate (APS) from sulphate and ATP. The equilibrium is in favour of the latter two substrates but driven in the reverse direction by the hydrolysis of pyrophosphate by pyrophosphatase and the low K_m of the next enzyme in the sequence. The aryl sulphurylase from *P. chrysogenum* has been studied in detail (Renosto *et al.*, 1990). The enzyme is strongly inhibited by 3′-phosphoadenosine 5′-phosphosulphate (PAPS), which appears to be unique for fungi. Renosto *et al.* (1990) pointed out that this allosteric

inhibition by PAPS is probably not only because it is the substrate for sulphate assimilation but also because it has the role as sulphurating agent in the synthesis of choline sulphate, which is a storage compound for sulphur (see below).

(2) APS kinase catalyses the formation of PAPS from APS and ATP. The APS kinase from *P. chrysogenum* has been purified and characterised by Renosto *et al.* (1991) and Renosto, Seubert & Segel (1984), and the enzyme from *S. cerevisiae* by Schriek & Schwenn (1986).

(3) The sulphur atom is reduced by sulphotransferases using either APS or PAPS. In fungi, from what little information there is, it appears PAPS is the preferred compound. Certainly that is so for *S. cerevisiae*. The original studies of Torii & Bandurski (1967) suggested that there was a transfer of the sulphonyl group from PAPS onto an unidentified carrier

$$PAPS^{4-} + carrier\ S^- \longrightarrow carrier\text{-}SSO_3^- + PAP^{4-}$$

The most recent studies (Schwenn, Krone & Husmann, 1988) indicate that reduced thioredoxin is the first substrate. This leads to the reduction of the enzyme and oxidised thioredoxin. The second substrate is PAPS, which is reduced such that both PAP and free sulphite is produced.

(4) Sulphite reductase catalyses the reduction of sulphite to sulphide. In *S. cerevisiae* the enzyme has been purified (Yoshimoto & Sato, 1968*a*, *b*, 1970) and the source of reducing power identified as NADPH. Significantly, in view of the original suggestion that sulphite might be bound to a carrier, the enzyme acts on free sulphite.

The sulphide produced by the action of sulphite reductase is incorporated into cysteine and homocysteine, the carbon skeletons coming from *O*-acetylserine and *O*-acetylhomoserine, respectively. The genetic regulation of sulphur assimilation in *N. crassa* is mentioned above. Biochemical studies on *S. cerevisiae* have shown also that enzymes involved in sulphur assimilation are regulated by repression when cells are grown in the presence of methionine (Jones & Fink, 1982; Cherest, Kerjan & Surdin-Kerjan, 1987; Thomas, Jacquemin & Surdin-Kerjan, 1992).

It was mentioned earlier that *Puccinia* spp. require methionine as a source of sulphur, being unable to utilise inorganic sulphate. Howes & Scott (1972) showed that *P. graminis* was unable to utilise not only inorganic sulphate but also sulphite, thiosulphite and sulphide as the sole source of sulphur. Further studies (Howes & Scott, 1973) with ^{35}S-labelled sulphate and hydrogen sulphide showed that sulphate could be taken up by the fungus but no label could be incorporated into cysteine. However, sulphide was readily incorporated into cysteine and other organic compounds. It was presumed from these studies (Howes & Scott, 1973; Maclean, 1982) that the step that is blocked is the reduction of either PAPS to thiosulphonate or thiosulphonate to sulphide. There could be blockage at both steps. The reasoning behind this conclusion was the knowledge that PAPS is required by all living organisms as a sulphating agent for synthesising sulphate ester. The most interesting feature of the experiments on the assimilation of hydrogen sulphide was the loss of label into the medium in low molecular weight sulphur compounds, predominantly cysteine, glutathione and cysteinylglycine (Howes & Scott, 1973; Maclean, 1982).

It is not my intention to discuss further the metabolic pathways for sulphur in fungi. The reader should refer to the reviews cited at the start of the chapter. However, there is a need to point out that a number of volatile sulphur compounds are produced by fungi (Berry & Watson, 1987; Slaughter, 1989). In particular yeasts produce hydrogen sulphide and dimethyl sulphide. These volatile sulphur compounds at low concentrations contribute to the flavour of alcoholic beverages. At higher concentrations, the compounds are commercially undesirable because they contribute to 'off-flavour'. If there were to be a systematic investigation of volatile sulphur compounds, they might be shown to be biologically significant for the fungi that produce them (Slaughter, 1989).

Storage of sulphur

There is no doubt that inorganic sulphate can achieve high concentrations within a fungus. Wethered, Metcalf & Jennings (1985) showed that when *Dendryphiella salina* was grown for 48 h in the presence of 435 mM sodium sulphate the concentration of sulphate in the mycelium was 172 mM. Under most conditions, however, the concentration of sulphate external to a fungus is very much lower. Then the sulphur can be stored in organic form.

Choline sulphate has been found to account for 1.5% of the dry weight of spores of *Aspergillus niger* (Takebe, 1960) and about 10% of the total organic sulphur (Renosto *et al.*, 1990). For *S. cerevisiae*, there is compelling evidence that glutathione can act as an endogenous source of sulphur (Elskens, Jaspers & Penninckx, 1991). The compound can account for 1% of the dry weight (Penninckx, Jaspers & Legrain, 1983). The position of

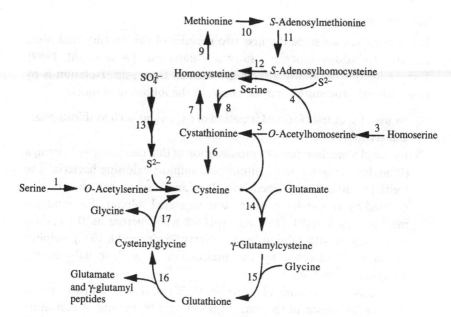

Figure 8.2. A model for the main fluxes of sulphur in *Saccharomyces cerevisiae*. 1, Serine acetyltransferase; 2, cysteine synthase; 3, homoserine acetyltransferase; 4, homocysteine synthase; 5, γ-cystathionine synthase; 6, γ-cystathionase; 7, β-cystathionase; 8, β-cystathionine synthase; 9, homocysteine methyltransferase; 10, S-adenosylmethionine synthetase; 11, S-adenosylmethionine demethylase; 12, adenosylhomocysteinase; 13, sulphate-reducing pathway; 14, γ-glutamylcysteine synthetase; 15, glutathione synthetase; 16, γ-glutamyl trans-peptidase; 17, cysteinylglycine dipeptidase. (From Elskens *et al.*, 1991.)

glutathione in relation to the probable main fluxes of sulphur in *S. cerevisiae* is shown in Figure 8.2. Elskens *et al.* (1991) showed that, in the presence of sulphate, most of the excess sulphur is incorporated into glutathione. The activity under these conditions of the enzyme degrading glutathione (γ-glutamyltranspeptidase) is low. When cells are deprived of sulphate, glutathione can act as a source of sulphur for the synthesis of sulphur-containing amino acids such as methionine and cystathionine. Under these conditions γ-glutamyltranspeptidase increases in specific activity. It seems highly likely that glutathione synthesis and breakdown could constitute a futile cycle for energy dissipation when yeast cells, under sulphur limitation, are subject to growth limitation by some other environmental factor (Jennings & Burke, 1990) (see Chapter 13).

That so little is known about storage of sulphur in fungi highlights in a pointed manner the considerable ignorance about sulphur nutrition of fungi.

Inorganic sulphur oxidation

It has been known since the first two decades of this century that fungi are able to oxidise inorganic sulphur compounds (Wainwright, 1989). Fungi with this capability are readily isolated from soils. Isolation is by conventional procedures and screening, in the following manner:

(1) By use of spot tests for the formation of oxyanions such as thiosulphate and tetrathionate.
(2) By use of a medium that is a modification of that described by Wieringa (1966) for the isolation of autotrophic sulphur-oxidising bacteria. The medium consists of a basal layer lacking a carbon source, which is covered by an overlay containing precipitated sulphur. The modified medium has Caapek-Dox agar (pH 6.8 with sucrose as the carbon source) again with sulphur in the overlay (Wainwright, 1978). Sulphur oxidation is detected by the production of a clear halo in the milky-white medium.
(3) By use of the medium of Germida (1985), which contains colloidal sulphur, oxidation of the element is detected by the use of indicators to detect acid production.

Table 8.3 gives a list of those fungi that have been reported to oxidise sulphur in culture. Fungi with this capability have been isolated from soils (Krol, 1983; Wainwright, 1984; Germida, 1985), vegegation (Wainwright, 1978) and H_2S-impregnated filter beds (van Langenhove, Wuyts & Schamp, 1986). Isolates from culture collections are able to oxidise sulphur without prior exposure to the element (Wainwright, 1989).

Oxidation of sulphur by fungi leads usually to thiosulphate and sulphate. The rate of sulphate production is low compared with thiobacilli but is similar to the rates achieved by other heterotrophs (Pepper & Miller, 1978; Wainwright, 1989). The enzymic nature of the process of sulphur oxidation in fungi has not been investigated in any detail. Studies on *Aspergillus niger* and *Trichoderma harziarum* by Grayston & Wainwright (1987) have shown that sulphur oxidation can be supported by a wide range of carbon sources including acetate and amino acids. It should be noted that, although sulphate was the major sulphur anion produced, thiosulphate was also produced in amounts that were often not much less.

The ability of fungi to oxidise sulphur may be for the following two reasons. First, it may be a means of utilising inorganic sulphur sources other than sulphate, since it is only through the latter than sulphur can be assimilated. Second, the process may yield energy that the fungus may use, mainly chemolithotrophically.

Table 8.3. *Fungi reported to oxidise sulphur in culture*

(A) Deuteromycotina and Zygomycotina

Absidia glauca	*Acremonium*
Alternaria tenuis	*Aspergillus flavus*
Aspergillus fumigatus	*Aspergillus niger*
Aureobasidium pullulans	*Epicoccum nigrum*
Fusarium episphaeria	*Fusarium tricinctum*
Monilia sp.	*Mortierella isabellina*
Mucor flavus	*Penicillium pinetorum*
Trichoderma hamatum	*Trichoderma harzianum*
Trichoderma viride	*Trichoderma* sp.
Zygorhynchus molleri	*Zygorhynchus vuilleminii*

(B) Mycorrhizal fungi

Amanita muscaria	*Hymenoscyphus ericae*
Paxillus involutus	*Pisolithus tinctorius*
Rhizopogon roseolus	*Suillus bovinus*

(C) Rhizomorph-forming Basidiomycotina

Hypholoma fasciculare	*Phanerochaete velutina*

(D) Yeasts

Rhodotorula sp.

(E) Thermophilic fungi

Geosmithia argillacea	*Geosmithia emersonii*
Myceliophthora thermophila	

From Wainright (1989), based on Germida (1985); Grayston & Wainwright (1988); Wainwright (1984).

The first reason has yet to be tested; the second is as yet unproven. However, it is relevant to report the following. Grayston & Wainwright (1987) and Wainwright & Grayston (1988) have shown that thiosulphate can significantly stimulate growth at very low carbon concentrations (including that in media with no added carbon). A more critical study on growth of *Fusarium oxysporium*, with a continuous supply of carbon at low concentration (laboratory contaminants at *c*. 13 μmol day^{-1}), by Jones *et al.* (1991) showed that the fungus could oxidise thiosulphate to sulphate. The inability to exclude organic carbon means that chemolithotrophy could not be demonstrated. While calculations of the energy gained from the oxidation of sulphur depend upon a number of assumptions, the estimate of energy gain indicated that there was sufficient to provide a biomass gain of 25% of that obtained in experiments. Certainly the energy gain from thiosulphate was more than sufficient to meet the probable amount of CO_2 fixation, based on data of Parkinson *et al.* (1991) obtained

with a somewhat different procedure. However, it needs to be emphasised
that the amount of carbon so fixed is trivial (see Chapter 12).

Tolerance to sulphur dioxide

Sulphur dioxide has been used for centuries to preserve beverages. More
recently the compound has been used to preserve food. In this respect,
the ability of fungi to tolerate sulphur dioxide is of considerable applied
interest. That interest has been enhanced by the anthropogenic release of
sulphur dioxide into the atmosphere. The topic of tolerance to sulphur
dioxide has been reviewed by Magan (1993) and Rose (1989).

There are three aspects to tolerance. The ability of a fungus to prevent
entry of sulphur dioxide, the mechanism(s) available to metabolise the
compound once it is inside the fungus and the extent to which sulphite
can be bound to other compounds. With respect to the ability of sulphur
dioxide to penetrate the plasma membrane, one needs to keep in mind
the equilibrium

$$H_2O + SO_2 \leftrightarrows SO_2.H_2O \leftrightarrows HSO_3^- + H^+ \leftrightarrows SO_3^{2-} + 2H^+$$

Ionisation of SO_2 has two pK_a values, the first leading to the bisulphite
(strictly hydrogen sulphite) ion of 1.86 and the second leading to the
sulphite ion of 7.18, both determined at $25\,°C$ and zero ionic strength
(Wedzicha, 1984). Thus studies on uptake of sulphur dioxide can be
confounded by the possible presence of SO_2 and $SO_2.H_2O$, HSO_3^- and
SO_3^{2-} in solution, the extent of their presence depending on the pH both
of the medium external to the fungus and in the wall external to the
plasma membrane. The latter could be at a much lower pH than the
external medium owing to (a) proton extrusion across the plasma
membrane and (b) the movement into the medium of the protons so
extruded being retarded by the wall and the unstirred layer external to it.

It has been proposed by Tweedie & Segel (1970) that there is a specific
carrier for sulphite in *Penicillium chrysogenum*, whereas Macris &
Markakis (1974) have proposed that there is a carrier for SO_2 in *S.
cerevisiae*. One must doubt the former proposal, since in the study of
sulphite transport by *P. chrysogenum* there was no evaluation on the effect
of pH. With respect to the information on *S. cerevisiae*, there is agreement
with the conclusion of Macris & Markakis (1974) that the cells do not
transport HSO_3^- over the pH range 3.0–5.0 used for the experiments (Hinze
& Holzer, 1985; Stratford & Rose, 1986).

Stratford & Rose (1986) argued that transport of sulphur dioxide in

S. cerevisiae is essentially not carrier mediated and that the molecule passes across the plasma membrane by simple diffusion, though there was some evidence for a carrier for HSO_3^- with a very low rate of transport. The relevant observations for this conclusion are as follows: (a) the initial velocity of uptake was unchanged after preincubation without glucose or with 2-deoxy-D-glucose or with the latter in the reaction mixture; (b) Hofstee plots for the initial velocity of uptake were near vertical; (c) insensitivity of transport to the inhibitors iodoacetamide and *p*-chloromercuribenzoate and to those inhibitors, CCCP and DNP, that will destroy the proton electrochemical gradient across the plasma membrane. There is little doubt that one cannot sustain the argument for the presence of a carrier for SO_2 in the light of these observations. Active transport is clearly excluded – the observed Q_{10} for transport of 2.5 cannot be used as an argument in favour of such a process (Jennings, 1963) – but so also is facilitated diffusion. Such a process would be susceptible to inhibition by *p*-chloromercuribenzoate and one would expect an effect of pH on transport other than that changing the concentration of the substrate for transport. The same conclusions as those above by Stratford & Rose (1986) have been reached about sulphur dioxide transport into *Saccharomycodes ludwigii* (Stratford, Morgan & Rose, 1987) and *Zygosaccharomyces bailii* (Pilkington & Rose, 1989).

The most compelling evidence for carrier-mediated transport into a fungus comes from the studies of Benítez *et al.* (1983) and García *et al.* (1983) on *Candida utilis*. It is the work of the latter group that is of most interest. García *et al.* isolated selenate-resistant mutants that did not take up sulphate, and grew poorly not only on sulphate but also on sulphite and thiosulphate. Sulphite reductase activity in the mutants was similar to that of the wild-type. This suggested that sulphite is transported on the sulphate carrier, a conclusion reached from uptake studies. These mutants need to be examined further with respect to sulphur dioxide/sulphite transport using an experimental attack similar to that of Stratford & Rose (1986). Rose (1989) pointed out that the experiments on *C. utilis* were carried out at pH 6.1, in contrast to those of Stratford & Rose (1986), which were carried out at pH 5.0 and below. At pH 6.1, molecular SO_2 concentrations are negligible. It may be that in *C. utilis* at pH 6.1 a carrier similar to the putative carrier for HSO_3^- of Stratford & Rose (1986) may be functioning but at a higher rate. This argues for studies to be made on the ability of sulphite to enter cells of *S. cerevisiae* at pH values much higher than 5.0.

Little is known of the physiological basis for the different degrees of sulphite resistance amongst yeast species (Pilkington & Rose, 1989). There seems to be little information about whether resistance resides in the plasma membrane, i.e. SO_2 or the anionic species are less able to penetrate into the cell. Indeed, in the one instance in which a realistic comparison of the rates of sulphite uptake between yeast species has been made (Stratford *et al.*, 1987), the more resistant species (*Saccharomycodes ludwigii*) exhibited a much higher uptake of sulphite than did the more tolerant species (*S. cerevisiae*). The difference in uptake was ascribed to the former yeast being richer in $C_{18:1}$ phospholipid fatty-acyl residues, which may indicate that the fluidity of the plasma membrane of *S. ludwigii* was greater than that of *S. cerevisiae*, thus allowing a faster rate of diffusion of SO_2.

At present, differences in resistance are attributed to the production of compounds, particularly acetaldehyde, that bind sulphite to form α-hydroxysulphonates (Burrows & Sparks, 1964; Stratford *et al.*, 1987), and possibly to a greater buffering capacity to counteract the decline in pH brought about by the production of HSO_3^- inside the cell from the absorbed SO_2 (Pilkingon & Rose, 1988).

Almost certainly the way forward in the study of sulphite tolerance by yeasts is isolation of mutants. The method of Brown & Oliver (1983) to isolate mutants with greater tolerance to ethanol (see Chapter 5) is likely to be the most appropriate way of isolating mutants with greater tolerance to SO_2.

9

Growth factors

General

The term 'growth factor' has been defined in two ways:

(1) A compound that is required for growth in small amounts but excludes those compounds that function as structural material and are not used for energy (Cochrane, 1963).
(2) A compound that in minute amounts is necessary for, or stimulatory to, growth and does not serve merely as an energy source (Fries, 1965).

The second definition clearly encompasses a broader range of compounds than the first. Fries (1965) in amplifying his definition pointed out that it includes certain organic compounds, e.g. amino acids, nucleotides, fatty acids. These compounds would not be covered by the first definition because they or carbon skeletons based on them are incorporated into structural material. Equally, the compounds are usually required at higher concentration ($10–1000 \, \mathrm{pg \, cm^{-3}}$) in contrast to $0.01–1 \, \mathrm{pg \, cm^{-3}}$, the concentration at which vitamins (the bulk of those compounds embraced by the first definition) are required by fungi. By vitamins I mean those compounds as understood by other biologists to come under this title, i.e. the compounds have a catalytic function in the organism as a coenzyme or constituent part of a coenzyme.

Nevertheless, it is difficult to make a clear separation between the two categories. Cochrane (1963) pointed out the arbitrary nature of the definition that he used. Thus, riboflavin is included under it but one cannot separate adenine's cofactor-like role in ATP from its part in nucleic acids. Inositol, however, is covered by the definition of Cochrane (1963), though the compound is not a vitamin *sensu stricto*.

On balance the second definition is preferable. To restrict growth factors for the most part to vitamins places the factors somewhat in a

303

straightjacket, manufactured from attempts to model fungal growth in mammalian terms. The second definition, while being somewhat ambiguous, better reflects biological reality. This is because a growth factor requirement could well have evolved from selection within the niche in which a fungus has established itself. In saying this, one must stress that, in spite of all the work on growth factors, there is minimal evidence to relate unambiguously the requirement for a growth factor by a fungus with the habitat or niche in which it is found.

Nevertheless, the tendency has been to assume that the composition of a culture medium allowing growth of a fungus does relate to the nature of the habitat in which the fungus is found. While of course this assumption is not made with respect to saprotrophic fungi of wide habitat tolerance, there is much less caution when dealing with fungi with much more fastidious nutrient requirements. Maclean (1982) favoured this assumption when considering the relevance of axenic cultures to the biology of rust fungi. He pointed out that host fluids provide a balance of nutrients to which rust fungi have adapted over a long time. The relative concentrations of essential organic nutrients in artificial media must represent an approximation to these fluids. Nevertheless he does allude to difficulties, namely the propensity of rust fungi to leak amino acids into the culture medium.

It is worth digressing a little further about extrapolating from culture experiments to the fungus in its natural habitat. There are two points to be made. First, in pure culture a fungus is often provided with carbon source in excess. In consequence, it is possible that requirement for a growth factor may be the result of a rate of synthesis too low to match the potential growth rate in the presence of other nutrients. Second, the ratio of the volume of medium to that of the mycelium could be untypical of natural conditions. Thus, leakage of a growth factor from the mycelium where it might be synthesised at a low rate could result in that rate being insufficient for growth if the leakage is into a large volume of culture medium. When the volume is small relative to the volume of mycelium, the external concentration of the growth factor could quickly equilibrate with that within the protoplasm. An example of this is the ability of uredospores of *Puccinia graminis* to grow better on nutrient media when at a high inoculum density (Maclean, 1982).

Many of the relevant publications on growth factors are from experimental work carried out two, three and four decades ago. Any search of the literature should bear this in mind. Indeed two valuable reviews of the literature are from an early date, namely those of Cochrane (1963) and Fries (1965). A more recent review of the literature is that of Garraway & Evans (1984). Suomalainen & Oura (1971) have produced a detailed review of the requirements for growth factors in yeasts, whereas Unezawa & Kishi (1989) have reviewed knowledge on vitamin metabolism in these fungi. Table 9.1 gives a list of those vitamins known to be required by fungi, the range of concentrations at which the vitamins are required and their function within the organism. Note that vitamins A, D, E and K required by higher animals do not appear to play a role in fungi. Table 9.2 lists some compounds that, for one species or another, came into the category of growth factor. The list is by no means comprehensive; it has been drawn up to indicate the variety of compounds that can be categorised within the second definition given at the start of this chapter.

With respect to the vitamins listed in Table 9.1, the following should be noted. Fungi are more often auxotrophic (see below) for thiamine. Biotin is probably second only to thiamine in the frequency with which it is required exogenously. Biotin, along with pantothenate, is essential for all strains of *Saccharomyces* (Williams, Eakin & Snell, 1940). Pyridoxine appears to be less required. For all the other vitamins listed in Table 9.1, the likelihood is that a fungus will be prototrophic and will not require the medium to be supplemented. Riboflavin, the membrane transport for which is considered below, is rarely required by microorganisms isolated from nature (Cochrane, 1963).

It is not my intention to consider at any length the extensive literature on growth factors. It is difficult to avoid a catalogue and furthermore there is no simple picture. Here I make some comments of a general kind, dwelling in detail only upon vitamin transport in fungi:

(1) A fungus that requires a growth factor is described as auxotrophic; one that can synthesise it is prototrophic.
(2) Establishing the requirement for a growth factor requires care. There is the problem of carry-over brought about by the transference of inoculum from one medium to another, such that a medium seemingly free of the factor gains it on entry of the inoculum. There can be contamination from the plugs (Sherwood & Singer, 1944). Given that this latter kind of contamination can be avoided, it is necessary to carry out serial subculture, and a series of at least three transfers is recommended (Cochrane, 1963). Thereafter, it should be shown that growth is a function of growth factor concentration. In this respect, it may be that there is growth on the basal medium lacking the added growth factor. Its concentration in the basal medium needs to be

Table 9.1. *Some vitamin requirements of fungi*

Compound	Representative concentration (M)	Function	Active Form
Thiamine (B_1)	10^{-9}–10^{-6}	Co-carboxylase	Thiamine pyrophosphate
Biotin (B_7)	10^{-10}–10^{-8}	Coenzyme for carboxylations	Covalently bound to enzyme by carboxyl
Pyridoxine (B_6)	10^{-9}–10^{-7}	Coenzyme for transaminations	Pyridoxal phosphate, pyridoxamine phosphate
Riboflavin (B_2)	10^{-7}–10^{-5}	Coenzyme for dehydrogenases	Flavin mononucleotide (FMN), flavin-adenine dinucleotide (FAD)
Nicotinic acid (B_3), nicotinamide (niacin)	10^{-8}–10^{-7}	Coenzyme for dehydrogenases	Nicotinamide-adenine dinucleotide (NAD) and its phosphate (NADP)
p-Aminobenzoic acid	10^{-8}–10^{-6}	Coenzyme in one-C transfers	Tetrahydrofolic acid
Pantothenic acid (B_5)	10^{-7}	Coenzyme in two-C transfers	Coenzyme A
Cyanocobalamin (B_{12})	10^{-12}–10^{-6}	Coenzyme for methyl transfers	Various cobalamin derivatives

From Griffin (1981), *Fungal Physiology*, Copyright © 1981, reprinted by permission of John Wiley & Sons, Inc.

Table 9.2. *A list of compounds that, for the fungi listed, may be considered to be growth factors*

Growth factor	Fungus	Reference
Adenine	*Fomes officinalis* *Poria vaillantii*	Jennison *et al.* (1955)
Coprogen	*Pilobolus* sp.	Hesseltine *et al.* (1953)
Ergosterol	*Saccharomyces cerevisiae* (anaerobic growth)	Andreasen & Stier (1953)
Histidine	*Trichophyton megninii*	Georg (1952)
Inositol	Many species	
Lysine	*Thraustochytrium aureum*	Paton & Jennings (1989)
Methionine	*Allomyces* spp.	Cantino (1955)
Myristate or palmitate	*Pityrosporium ovale*	Shifrine & Marr (1963)
L-α-phosphatidyl choline (L-α-lecithin)	*Phytophthora epistomum*	Bahnweg (1980)
Sulphur amino acids	*Blastocladiella*	Cantino (1955)

determined. At the least, presence of the factor can be demonstrated by growth of an organism known to have an absolute requirement for the factor (Robbins & Kavanagh, 1942). Finally, growth should be measured if possible as increase in biomass. Linear extension over a solid substrate as a measure of growth can be confounded by branching pattern and hyphal diameter changing with increase in growth rate.

(3) There may be only a partial requirement for a growth factor. This has been alluded to above in the discussion of the rate of synthesis of a growth factor within a fungus and its rate of growth. Partial deficiencies of vitamins are probably more common than complete deficiencies (Cochrane, 1963). Sometimes the requirement is not for the whole molecule; part of the molecule or a precursor is what is necessary. Thus, there are biotin-deficient fungi that can grow on desthiobiotin, which lacks the requisite sulphur atom, and similar fungi that require the complete biotin molecule (Lilly & Leonian, 1944). Presumably the former group have the pathway to biotin blocked prior to desthiobiotin, whereas the latter have the final step in the synthesis of biotin blocked, namely its synthesis from desthiobiotin. Finally, requirement of fungus for a growth factor can depend on environmental conditions, e.g. pH, temperature or on the growth stage.

(4) Fungi can require several growth factors. Thus *Kloeckeria brevis* has been shown to require thiamine, pyridoxine, biotin, pantothenic acid, niacin and inositol (Burkholder, McVeigh & Moyer, 1944), whereas *Poria vaillantii* requires thiamine, biotin, riboflavine and adenine (Jennison, Newcomb & Henderson, 1955).

(5) There appears to be almost complete lack of correlation between growth factor requirements and fungal taxonomy. This is not surprising, since there can be marked differences between the growth factor requirements of different strains of the same species and between species within a genus (Cochrane, 1963). Probably the only statement of a general nature that can be made is that a large proportion of members of the Agaricales require thiamine (Cochrane, 1963).

(6) No ecological generalisations can be made about growth factor requirements except that broadly the following genera of soil saprotrophs can be said to exhibit an absence of requirement *Aspergillus*, *Chaetomium*, *Fusarium*, *Mucor*, *Penicillium*, *Rhizopus*, *Zygorhynchus* (Robbins & Kavanagh, 1942).

Membrane transport

Knowledge of the transport of growth factors in fungi is not especially extensive, as will be seen below. There is a real deficiency in knowledge of the manner in which transport is driven, which needs to be rectified.

Thiamine

There is much information on transport of thiamine in *Saccharomyces cerevisiae* (Iwashima, Nishino & Nose, 1973; Iwashima & Nose, 1975; Iwashima, Nishimura & Sempunku, 1980). The vitamin is actively accumulated by a carrier-mediated process, most of the thiamine taken up remaining unphosphorylated during short-term transport experiments. Thiamine taken up by cells of *S. cerevisiae* is not lost into thiamine-free solution or exchanged with exogenous thiamine (Ruml, Šilhánková & Rauch, 1988). The K_T for transport is 1.8×10^{-7} M. Transport has a pH optimum of 4.5 and is inhibited by iodoacetate, 2,4-dinitrophenol (DNP) and DCCD. Short-chain fatty acids (C_2–C_6) strongly inhibit transport. Studies with caproate showed that the fatty acid brings about the efflux of thiamine already absorbed. All this evidence points to thiamine transport being driven by a proton electrochemical potential gradient, though direct evidence is lacking.

Transport is inhibited by the following analogues pyrithiamine, chloroethyl thiamine, dimethialium and thiamine disulphide, though only for the last two has competitive inhibition been demonstrated. Dimethialium has been found to be accumulated. Since the compound is not phosphorylated by thiamine pyrophosphokinase, there is support for transport driven by the proton gradient and not requiring phosphorylation of the transported molecule. Oxythiamine has no effect on thiamine transport even when the former is at a 25-fold greater concentration.

Uptake of both thiamine and pyrithiamine is repressed after growth of the cells in exogenous thiamine (Iwashima & Nose, 1976). Cells grown in thiamine-deficient minimal medium and preloaded with thiamine have the same rate of transport as non-preloaded cells. Nevertheless the preloaded cells can have an internal concentration of thiamine that is the same as that in cells grown in the presence of exogenous thiamine. Thus, reduction of thiamine uptake in the latter cells cannot be due to transinhibition.

The thiamine transport system can be photoinactivated by thiamine derivatives having an arylazido substituent in the thiazole moiety, e.g. 4-azido-2-nitrobenzoyl thiamine (Sempuku, Nishimura & Iwashima, 1981; Sempuku, 1988).

A thiamine-binding protein can be isolated from *S. cerevisiae* by cold osmotic shock (Iwashima & Nishimura, 1979). The protein has a molecular weight of 140 kDa, is a glycoprotein and is not present in cells grown in the presence of thiamine. Iwashima, Nishimura & Nose (1979) compared the thiamine-binding capability of the protein with the affinity of the thiamine transport system. Table 9.3 shows that, whereas several thiamine derivatives and analogues such as pyrithiamine and oxythiamine inhibit thiamine binding to the protein and thiamine transport to the same extent, other compounds such as thiamine phosphates and 2-methyl-4-amino-5-hydroxymethylpyrimidine have different effects. The apparent K_D for the binding of thiamine was found to be 29 nM, which is about six times lower than the K_T (0.18 μM) for transport. At one time the role of the soluble thiamine-binding protein coming from the periplasmic space seemed unclear. As is now known, from biochemical and genetic studies, the protein is a phosphatase encoded by the *PHO3* gene (Nosaka, Nishimura & Iwashima, 1988). Since the enzyme can hydrolyse thiamine monophosphate and thiamine pyrophosphate, it is thought that it may have a role in thiamine nutrition, given that these two compounds are available to yeast cells under natural conditions.

Tunicamycin decreases the amount of the periplasmic protein but the

Table 9.3. *Effect of thiamine derivatives and analogues on the binding of thiamine to the soluble thiamine-binding protein and thiamine transport in* Saccharomyces cerevisiae

Each thiamine derivative or analogue was added to the dialysing buffer simultaneously with [^{14}C]thiamine at molar ratios as indicated, then the thiamine-binding activity was assayed by an equilibrium dialysis.

Addition	Thiamine derivative / Thiamine	Thiamine-binding (%)	Thiamine transport (%)
Control	0	100	100
Pyrithiamine	2	26.2	26.3
	10	9.0	6.0
Chloroethylthiamine	2	34.1	35.5
	10	5.5	7.9
Dimethialium	2	51.5	44.4
	10	7.3	9.2
Oxythiamine	10	100	100
Thiamine monophosphate	10	14.0	62.2
Thiamine pyrophosphate	10	5.4	80.8
2-Methyl-4-amino-5-hydroxymethylpyrimidine	10	94.3	12.4
O-Benzoylthiamine disulphide	0.1	100	7.1

From Iwashima *et al.* (1979).

antibiotic also inhibits the activity of the thiamine-binding protein and also thiamine transport in a dose-dependent manner (Nosaka *et al.*, 1986). It is concluded from these observations that the transport system is also a glycoprotein.

Further, it is now known that not only is transport repressed when cells are grown in the presence of thiamine but several enzymes, namely the acid phosphatase referred to above, hydroxymethylpyrimidine kinase, phosphomethylpyrimidine kinase, hydroxyethylthiazole kinase and thiamine-phosphate pyrophosphorylase are also repressed (Kawasaki *et al.*, 1990). These enzymes are involved in the synthesis of thiamine from hydroxymethylpyrimidine and hydroxyethylthiazole. Further studies indicate that the utilisation and synthesis of thiamine in *S. cerevisiae* is controlled negatively by the intracellular level of thiamine pyrophosphate (Nishimura *et al.*, 1991).

In *Schizosaccharomyces pombe* it has been shown that thiamine here too represses thiamine transport and an acid phosphatase, also a glycosylated protein, that resides in the wall. As with *S. cerevisiae*, the

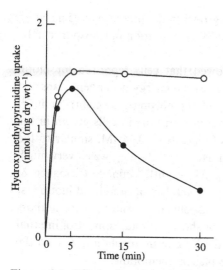

Figure 9.1. Effect of cycloheximide and anisomycin on [³H]hydroxymethyl-pyrimidine uptake. Yeast cell suspensions were preincubated for 15 min with cycloheximide (1 μg/ml (○)); without inhibitor (●). [³H]Hydroxymethyl-pyrimidine was added to the medium at 1 μM, followed by further incubation for 30 min. (From Iwashima *et al.*, 1990*b*.)

genetic basis of thiamine metabolism and its regulation are now fairly well established (Schweingruber *et al.*, 1986, 1991, 1992).

Finally, it has been shown that hydroxymethylpyrimidine (2-methyl-4-amino-5-hydroxmethylpyrimidine) is transported into cells of *S. cerevisiae* by the thiamine transport system. However, the accumulated hydroxymethylpyrimidine is lost from the cells (Iwashima, Kawasaki & Kimura, 1990*a*; Figure 9.1). The accumulation has been termed overshoot. It depends on the growth phase and is temperature sensitive (Iwashima, Kimura & Kawasaki, 1990*b*). The overshoot does not occur after preincubation of cells with cycloheximide and anisomycin. It is argued that there is an active efflux of hydroxymethylpyrimidine that is brought about by a rapidly synthesised transport system.

Biotin

In *S. cerevisiae*, biotin is transported into the cells by a system, dependent on glucose, with a K_T of 0.32 μM and a pH optimum of 4.0 (Rogers & Lichstein, 1969*a*, *b*). A number of analogues including homobiotin, desthiobiotin, oxybiotin, norbiotin and biotin sulphone are also transported. Cells grown in the presence of 0.25 ng biotin cm^{-3}

accumulated the vitamin, while cells grown in the presence of 25 ng cm^{-3} were repressed for biotin uptake. Feedback inhibition of transport can be discounted.

Cicmanec & Lichstein (1974) showed that cells repressed for biotin uptake by growth in excess biotin possess an energy-dependent transport system with a K_T of 0.66 μM and a V_{max} of 39 pmol (mg dry weight)$^{-1}$ min^{-1}. In these investigations, the K_T for biotin transport in cells grown in the presence of 0.25 ng biotin cm^{-3} was found to be 0.64 μM, similar to the value for cells grown in excess biotin. However, the V_{max} was a very much higher value, namely 530 pmol (mg dry weight)$^{-1}$ min^{-1}. Cicmanec & Lichstein (1974) showed also that cells which had accumulated biotin lost the compound when the cells were suspended in water. These authors concluded that loss was the manner of short-term adjustment of internal biotin concentration whereas, over the long term, regulation was via the number of transport proteins in the plasma membrane.

Biotin transport is specifically inhibited by biotinyl *p*-nitrophenyl ester (Becker, Wilchek & Katchalski, 1971). Though the inhibition was thought to be irreversible, conversion of cells to protoplasts or treatment with thiols restores biotin transport (Viswanatha, Bayer & Wilchek, 1975).

Vitamin B$_6$

Pyridoxine (also called pyridoxol) or vitamin B$_6$ and the related compounds pyridoxal and pyridoxamine are precursors for the coenzyme pyridoxal phosphate. Vitamin B$_6$ appears to be required by fewer fungi than is thiamine or biotin (Cochrane, 1963).

The interest in transport of this vitamin is a consequence of the observation that thiamine depresses growth, and brings about a marked decrease of cytochrome oxidase activity and cellular vitamin B$_6$ cotent in *Saccharomyces* yeasts growing in a vitamin B$_6$-free medium (Rabinowitz & Snell, 1947; Nakamura *et al.*, 1982). The inhibition is eliminated by adding vitamin B$_6$ to the growth medium. Otherwise few fungi appear to require the vitamin. The effect of thiamine just described is observed only under aerobic conditions and cells exhibit a very low respiration rate (Nakamura *et al.*, 1974). Where looked for, the effect cannot be observed with species of *Candida*, *Kluyveromyces* and *Schizosaccharomyces* (Nakamura *et al.*, 1982). Cells of *Saccharomyces* sp. accumulate thiamine to a 24-fold greater extent than cells of non-sensitive species. It is not clear how thiamine exerts its effect in the cell (Unezawa & Kishi, 1989).

Transport of pyridoxine, pyridoxal and pyridoxamine leads to their accumulation in resting cells of *Saccharomyces carlsbergensis* 4228 (now *S. uvarum* 4228; Shane & Snell, 1976). There appear to be two transport systems. Thus, uptake of pyridoxine exhibits two pH optima of 3.5 and 6.0. Pyridoxal is transported primarily at pH 3.5 and pyridoxamine at pH 6.0. Pyridoxine is transported by both systems almost equally effectively. Uptake at pH 3.5 is stimulated four-fold by potassium and to a lesser extent by lithium and sodium. Uptake at pH 6.0 is only slightly stimulated by potassium. Transport is inhibited by many unphosphorylated vitamin analogues, the most effective being 5'-deoxypyridoxine, 5'-deoxypyridoxal, toxopyrimidine, 4'-deoxypyridoxine and 3-amino-3-deoxypyridoxine. Pyridoxal 5'-phosphate is not transported.

Interestingly, there is an increased loss of pyridoxine once a certain level in the cell has been achieved (overshoot). Loss is stimulated by glucose and accelerated by exchange with external pyridoxine or its analogues. Phosphorylation is not required for activation of overshoot, which is time dependent. Influx is not inhibited as overshoot develops. There are very considerable parallels with the overshoot observed with hydroxymethylpyrimidine described earlier.

Riboflavin

Perl, Kearny & Singer (1976) showed that in riboflavin-requiring mutant cells of *S. cerevisiae* riboflavin is transported by a system with a K_T of 15 μM for the vitamin. Only 7-methylriboflavin, 8-methylriboflavin and 5-deazaflavin can also be transported. A number of flavin analogues in which the ribityl side-chain is modified are said not to be transported but are competitive inhibitors – this needs further investigation. The pH optimum for transport is 7.5 and monovalent cations, particularly potassium and rubidum, stimulate transport. As transport proceeds, glucose, mannose and fructose become increasingly inhibitory. This inhibition of uptake is due to a stimulation of efflux. This latter process has a K_T of 48 μM and a pH optimum of 5.0. This indicates that efflux is brought about by a system distinct from that involved in uptake. This is confirmed by the fact that, while significant uptake of riboflavin is seen only in mutants cultured anaerobically and at early to mid log phase, efflux is observed in wild-type cells and appears to be independent of the growth phase. Perl *et al.* (1976) suggested, on the basis of an inverse relationship between the intracellular concentration of acid-extractable flavins and the rate of riboflavin uptake, that intracellular flavin may be a repressor of carrier biosynthesis.

Table 9.4. *Microorganisms producing considerable
amounts of riboflavin and the effects of iron on
biosynthesis of the vitamin*

Microorganism	Riboflavin yield (mg l^{-1})	Optimum iron concn (mg l^{-1})
Clostridium acetobutylicum	97	1–3
Mycobacterium smegmatis	58	Not critical
Mycocandida riboflavina	200	Not critical
Candida flareri	567	0.04–0.06
Eremothecium ashbyii	2480	Not critical
Ashbya gossypii	6420	Not critical

From Perlman (1978).

Riboflavin is a unique vitamin in that it can be synthesised totally, to
a very high concentration, rather rapidly by certain microorganisms
including fungi (Perlman, 1978; Table 9.4). It is not clear how
overproduction is brought about, strikingly so in *Eremothecium ashbyii*
and *Ashbya gossypii*, but loss from the mycelium must be brought about
by an efflux system similar in kind to that just described for *S. cerevisiae*.

Inositol and choline

As is discussed below, the metabolism of inositol and that of choline in
S. cerevisiae are closely interrelated. It is not known whether this is so
for other fungi.

Transport of both growth factors has been investigated by traditional
methods – kinetic studies – and by isolation of the genes. According to
kinetic studies, cells possess a specific system for *myo*-inositol (Nikawa,
Nagumo & Yamashita, 1982). Transport requires glucose and there is no
effect of the following analogues: *scyllo*-inositol, 2-inosose, mannitol and
1,2-cyclohexanediol. The K_T for *myo*-inositol is 0.1 mM. Activity of the
system is reduced by growth in media containing *myo*-inositol. Removal
of *myo*-inositol reverses the effect, and this is completely prevented by
cycloheximide.

Nikawa, Tsukagoshi & Yamashita (1991) isolated, by complementation
of a yeast mutant defective in *myo*-inositol transport, two genes (*ITR1*
and *ITR2*) coding for proteins with molecular masses of 63 605 and
67 041 Da, respectively. The sequence similarity between the two gene

Table 9.5. *The effect of the addition of* myo-*inositol and choline to the preincubation medium on choline transport by* Saccharomyces cerevisiae

Addition[a]	Choline (nmol min^{-1} (mg protein)$^{-1}$) transport
None	7.1
Choline	0.4
myo-Inositol	5.9
Choline + *myo*-inositol	0.6

[a] 20 μg of each per ml.
From Nikawa *et al.* (1990).

products is extremely high and there is significant sequence similarity to the superfamily of sugar transport proteins including the glucose (SNF3), galactose (GAL2) and maltose (MAL61) systems of *S. cerevisiae*. The *ITR1* product appears to be the major transporter, since disruption of *ITR2* does not cause very much change in transport activity. In the converse situation with *ITR1* disrupted, cells cannot grow in the presence of 2 μg *myo*-inositol cm^{-3}. The K_T values for the two systems are very similar – hence the lack of discrimination between them in the earlier study. The *ITR1*-encoded system has a K_T of 0.1 mM and the *ITR2* system a K_T of 0.14 mM.

Choline is transported with a K_T of 0.56 μM, *N*-methylethanolamine, *N*,*N*-dimethylethanolamine and β-methylcholine being competitive inhibitors (Hosaka & Yamashita, 1980). Ethanolamine and phosphorylcholine are not transported. Glucose is required for maximal transport activity; 2,4-dinitrophenol and carbonyl *p*-trifluoromethoxyphenylhydrazone abolish transport. The ionophores valinomycin and gramicidin have no effect. The gene for the system (*CTR*) has been sequenced and it codes for a protein of 563 amino acid residues with a molecular mass of 62 055 Da (Nikawa *et al.*, 1990). The sequence has many similarities with the yeast arginine (*CAN1* product) and histidine (*HIP1* product) transport protein (Hoffman, 1985; Tanaka & Fink, 1985). The abundance of *CTR* mRNA decreases significantly in cells incubated with a combination of choline and *myo*-inositol (Table 9.5).

The abundance of *ITR1* mRNA decreases also in response to the addition of inositol and choline to the culture medium (but not *ITR2* mRNA) (Nikawa *et al.*, 1991). Similar control to that observed on *CTR* mRNA and *ITR1* mRNA has been observed for enzymes of phospholipid metabolism in *S. cerevisiae*. These include phosphatidylethanolamine methyltransferase, phospholipid methyltransferase, CDP-diacylglycerol synthase, phosphatidyl serine synthase, choline kinase and inositol-1-phosphate synthase (see Nikawa *et al.*, 1990; White, Lopes & Henry, 1991). With all of the above coregulated enzymes, choline has little or no repressive effect if present by itself in the growth medium (but note the situation for the choline transport protein in Table 9.5). However, if choline is added to the growth medium already containing inositol, the level of repression is greater for many of the enzymes than is the level of repression when only inositol is present. It has been suggested that this pattern of regulation has a role in ensuring that the total charge balance of the plasma membrane phospholipids is maintained within certain limits (White *et al.*, 1991). In this respect, it is worth remembering that phosphatidylinositol is the major negatively charged lipid in the plasma membrane of *S. cerevisiae*, whereas phosphatidylcholine is the major zwitterionic species. A similar regulation of phospholipid charge has been found in *Neurospora crassa* by Hubbard & Brody (1975).

The enzymes of phospholipid synthesis in *S. cerevisiae* are not only under the dual control of choline and inositol but also under growth phase control (Homann *et al.*, 1987). Maximum activities of the membrane-associated phospholipid biosynthetic enzymes, CDP-diacylglycerol synthase, phosphatidylserine synthase, phosphatidylinositol synthase and the phospholipid methyltransferases were found in the exponential phase of growth in complete medium. As cells entered the stationary phase the activities of the first two and last enzymes decreased by 2.5- to 5-fold.

A *myo*-inositol transport system has been reported for *Schizosaccharomyces pombe* (Cheneval, Deshusses & Posternak, 1970).

Choline is transported in *Fusarium graminearum* by a system highly specific for the compound (Robson *et al.*, 1992). The K_T is 32 μM and only *N,N*-dimethylethanolamine, betaine, aldehyde and chlorocholine can also be transported. The maximum rate is repressed about three-fold by ammonia and by glucose.

10

Potassium and other alkali metal cations

Introduction

Although this chapter deals with transport of the alkali metal cations as a group, it focuses on potassium. That is because potassium is an essential element for all fungi. Of the other alkali metal cations, sodium appears to be required at relatively high concentrations (in the millimolar range) for a small group of lower marine fungi (see Chapter 13). There is no evidence that sodium is required as a microelement for the simple reason that producing sodium-free media for use with the great majority of fungi demands an intensity of effort that is not matched by the probability of a rewarding outcome, i.e. that it will be possible to demonstrate a requirement. Interest in caesium transport and accumulation in those basidiomycetes producing large fruiting bodies, some of which might be used for human consumption, has developed as a consequence of the release of ^{137}Cs and [134]Cs isotopes from the nuclear reactor at Chernobyl, Ukraine, in 1986. In particular, comparison of the relationship between the ratios of the two isotopes found in such fungi with that of the emissions from the reactor has indicated that there might have been considerable accumulation of and maintenance within mycelium of significant levels of caesium from pre-Chernobyl weapons-testing (Byrne, 1988; Dighton & Horrill, 1988).

Evans & Sorger (1966), in their seminal review of the role of univalent cations in green plant nutrition, pointed out that potassium and other univalent cations function as cofactors for a wide variety of enzymes throughout the living world. They also pointed out that most of the enzymes requiring univalent cations from non-halophytic organisms function effectively with potassium, ammonium or rubidium in the reaction mixture, with concentrations from 0.01 to 0.1 M for maximum activity. Further, though potassium is specifically required by plants, the element

317

can be partially replaced by rubidium and sodium. Evans & Sorger (1966) hypothesised that univalent cations function in non-specific charge balance with charged groups of proteins. This accounts for the relatively high concentration of the cations required for maximal activity. Also, since potassium, ammonium and rubidium have very similar atomic radii (0.532, 0.537 and 0.509 nm, respectively) enzymes function equally well in environments of the three cations. In the light of this, it is not unexpected that sodium with a radius of 0.79 nm is not always effective and may be inhibitory and that lithium with a radius of 1.00 nm is even less effective.

Since 1966, there is nothing to dispute the general thesis put forward by Evans & Sorger. There is additional information about the effect of monovalent cations on enzyme activity increasingly indicating that the requirement of potassium for enzyme activity is often more specific than was originally thought (Wyn Jones, Brady & Spiers, 1979; Wyn Jones & Pollard, 1983). Further, it is now realised that potassium is now important for the proper functioning of ribosomes (Wyn Jones *et al.*, 1979). As with enzyme activity, other monovalent cations can substitute for potassium. Ammonium is often as effective but sodium can replace potassium only to a limited extent.

Of course, whereas monovalent cations, particularly potassium, are required for proper biochemical functioning through their effect on the stability of macromolecules, at higher concentrations of the salt containing the cation (when the electrostatic forces exerted by the macromolecule are suppressed) lyotropic effects become dominant. By that I mean that the electrolyte can bring about the precipitation of the macromolecules, the process of salting out. This was first reported in an organised manner by Hofmeister (1888), who proposed that water–solute interactions are the key to an understanding of the process. A comprehensive theory of the effect, which is dependent essentially on the nature of the anion of the salt, has been produced by Collins & Washabaugh (1985). Lyotropic effects become apparent in the range 0.1–1.0 M, a concentration range that becomes particularly significant for those fungi that have adapted to grow in media of high saline concentrations (Chapter 13). In these organisms, there must be mechanisms for keeping the salt concentrations in the cytoplasm at such a level that salting out of proteins does not occur. However, fungi growing in media of low concentrations of inorganic nutrients must accumulate monovalent cation salts to ensure that enzymes and other macromolecules have the appropriate ionic environment to function properly.

Knowledge of the relationship between fungal growth and external potassium concentration is extremely limited. Nevertheless there have been important studies on *Candida utilis*, grown in continuous culture, that provide an important insight into how a fungus responds to potassium limitation. Because of the nature of these studies (Aiking & Tempest, 1976, 1977; Aiking, Sterkenberg & Tempest, 1977*a*, *b*), the data are not confounded by carry over of potassium with the inoculum and there is no distortion due to varying growth rate, as is the case in batch culture.

It is clear that, among the alkali cations, rubidium is the only one that can substitute for potassium in *C. utilis*, giving the same yield, on a molar basis and under conditions where cation availability limits growth (Aiking & Tempest, 1977). Comparison of potassium- and rubidium-limited cultures showed them to be virtually identical in all measured characteristics, the single exceptions being the maximal growth rate. With rubidium, the value was $0.35\ h^{-1}$ as compared with $0.55\ h^{-1}$ for potassium. Two possible explanations were put forward to explain this difference. First, the maximum growth rate may be limited by the rate at which rubidium is taken up into the cell. Second, the function of potassium is fulfilled by rubidium but with decreased efficiency and here the investigators had in mind the strong correlation between intracellular potassium content and the yield value for oxygen (milligrams of cells produced per atom of oxygen consumed) (Aiking & Tempest, 1976). Hence, it can be argued that the maximal growth rate in the presence of limiting rubidium is limited by energy supply. Interestingly, it was found in the same study that, under ammonia-limited conditions with equimolar (1.5 mM) rubidium and potassium, there was a much higher rubidium:potassium ratio inside the cells than under other conditions. It was argued that rubidium could be transported into the cells via the ammonia uptake system, since the ionic radii of Rb^+ and NH_4^+ are very similar; this needs to be investigated further.

Under potassium limitation, potassium content of the cells is a linear function of growth rate. It appears that there is very little stoichiometry between potassium, RNA and phosphorus content. Thus, the requirement for potassium is not associated with the presence of certain anionic polymers. As pointed out above, the yield value for oxygen was almost linearly related to potassium content. A similar relationship was found for the yield values for glucose and potassium content under conditions in which cells were potassium limited as well as under conditions in which cells were glucose limited. This was taken to indicate an involvement of potassium with oxidative phosphorylation.

Confirmation of these ideas came from studies on isolated mitochondria

(Aiking, Sterkenberg & Tempest, 1977a). When growing under potassium limitation at dilution rates greater than $0.2\,h^{-1}$, mitochondria possessed three oxidative phosphorylation sites; at lower dilution rates there were only two sites. Mitochondria from cells grown under glucose-, ammonia-, magnesium-, phosphate- or sulphate-limited cultures always possessed three sites. Mitochondria with only two sites showed differences in the cytochrome b:cytochrome a ratios with dilution rates that were different with mitochondria from cells grown under other limiting conditions.

There seem to be no other studies in which fungi have been grown in the steady state under potassium-limiting conditions. It would be of great value to know whether the results obtained with *Candida utilis* are applicable to other fungi.

Monovalent cation transport

Saccharomyces cerevisiae

In spite of the fact that monovalent cation transport in *S. cerevisiae* has been studied intensively for nearly 50 years (Conway & O'Malley, 1946; Rothstein & Enns, 1946), there is not yet a coherent picture of the various processes involved in the governing of the distribution of monovalent cations between the medium and the cell interior. The somewhat fragmented view that has emerged is in large measure due to the various approaches used, to a degree in the isolation of one from the other, but also a consequence of the difficulty of mounting a proper biophysical attack on the problems being addressed. The lack of biophysical information is highlighted by the more coherent picture that we have for the monovalent relations of *Neurospora crassa*, which has depended heavily on biophysical studies and which is described in the next section. Here, discussion of the topic of monovalent cation relations in *S. cerevisiae* takes the form of a consideration of the approaches used.

Kinetic studies

There have been extensive kinetic studies on monovalent cation transport in *S. cerevisiae*. However, not until recently did it become clear that the kinetics are dependent upon the conditions in which the cells are grown and the experimental conditions used for studying transport (Rodríguez-Navarro & Ramos, 1984). Many studies have been made with cells separated from the growth medium and starved for a period before use (Conway & O'Malley, 1946; Armstrong & Rothstein, 1964; Theuvenet &

Borst-Pauwels, 1976a). It has been shown that the presence or absence of magnesium in the suspension medium can alter the potassium/sodium exchange characteristics of cells of *S. cerevisiae* (Rodríguez-Navarro & Sancho, 1979). The importance of calcium and other bivalent cations for the integrity of membranes of green plant cells and for uptake of potassium in the presence of sodium by these cells is well documented (Jennings, 1969, 1986b). Also, Jones & Jennings (1964, 1965) have clearly shown the importance of bivalent cations for the potassium nutrition and growth of marine fungi in saline conditions. In one species of such fungi, *Dendryphiella salina*, it has been shown that calcium reduces the efflux of potassium in the presence of 0.5 M sodium chloride (Jennings & Aynsley, 1971).

It has been suggested that growth in ammonia, which is frequently used as the nitrogen source, has an effect on potassium transport (Rodríguez-Navarro & Ramos, 1984). The evidence for this is not clear, nor is it readily apparent how ammonia might exert its effect in experimental solutions that would not normally contain the compound. Conceivably the acid medium generated by ammonia assimilation through the increased concentration of hydrogen ions displaces calcium or other bivalent cations from those sites that, when filled with the ion, allow proper functioning of the potassium transport.

It is now clear from kinetic studies using cells grown in media containing 2 mM magnesium and without added sodium or ammonia (Rodríguez-Navarro & Ramos, 1984) that *S. cerevisiae* possesses a dual system for potassium, one with a K_T of 24 μM and the other a K_T of 2 mM. The high affinity system is expressed only in cells grown in medium of low potassium concentration. The identities of these two transport systems are discussed below. At this point, it is necessary to focus on those other kinetic studies that have investigated the effects of other cations on the uptake of potassium or rubidium by *S. cerevisiae*.

The kinetics derived from these other studies appear to be rather complex. Armstrong & Rothstein (1964) were the first to show that, while monovalent cation transport by *S. cerevisiae* can be described by Michaelis–Menten kinetics, deviations occur at low pH. It was argued that protons bind to the carrier at two sites. One is the *carrier* (or transport) site for the monovalent cations, and, at this site, there is competitive inhibition of transport of alkali metal cations. This was described first by Conway & Duggan (1958); at high concentrations Mg^{2+} is capable of binding to this site and is thus transported into the cell. Hydrogen ions have a higher affinity for the carrier site than do potassium ions. Thus, at low pH (below 4.0), protons are transported into the cell rather than K^+, which is lost from the cell in a one-to-one ratio with protons.

At the second site, the inhibition of monovalent cation transport is partially non-competitive, i.e. K_T is unaffected but V_{max} is reduced. The reduction in V_{max} is not the same for each cation. This second site has been termed the 'modifier' site. At high pH (8.0) in the medium, the V_{max} for potassium transport is reduced to the same extent in the presence of relatively low concentrations of Li^+, Na^+ and Cs^+ and also of Ca^{2+} and Mg^{2+}; these ions appear to bind to the modifier site (Armstrong & Rothstein, 1967).

Later, Borst-Pauwels, Wolters & Henricks (1971) showed that there is a third site associated with monovalent cation transport. At concentrations lower than those used by Armstrong & Rothstein (1967), it was found that there is also what has been termed an *activation* site. Thus, when rubidium is being transported, the presence of rubidium or potassium or sodium on the activation site is stimulating. But when sodium is being transported, only rubidium and potassium stimulate – sodium is inhibitory. The similarity of the half-maximal concentration that gives stimulation of rubidium transport to that which gives inhibition of sodium transport indicates that sodium is binding to the same site in both cases (Borst-Pauwels, Schnetkamp & van Well, 1973).

Dissociation constants for the three sites are given in Table 10.1. It should be noted that the so-called activation site may be a transport site, as pointed out by Borst-Pauwels (1981), who summarised the theoretical basis of the kinetic procedures used. That the activation site is a transport site is given credence by the similarity of the dissociation constant for potassium (14 μM) with the K_T (24 μM) for potassium transport by the high affinity carrier of Rodríguez-Navarro & Ramos (1984) (see above).

The kinetic description given above is complicted further by the effect of cations on the surface potential. The largest effects are seen with polyvalent cations. Such ions increase the dissociation constants of the substrate and activation sites for monovalent cations such as rubidium in proportion to the square root of the divalent cation and the cube root of the trivalent cation concentrations present in the medium (Theuvenet & Borst-Pauwels, 1976c). Essentially, the polyvalent cations affect the magnitude of the membrane surface potential. The negative charges can be considered to be blocked by the polyvalent cations, so that monovalent cations are less able to reach the surface. Thus, the relationship between the concentration of a monovalent cation in the medium or bulk phase

Table 10.1. *Values of the apparent dissociation constants (mM) of monovalent and divalent cations for the potassium carrier of* Saccharomyces cerevisiae

Ion	Sites		
	Substrate	Modifier	Activation
H^+	0.2	0.02	0.006
Li^+	27	19	18
Na^+	16	14.4	0.032
K^+	0.5	1.6	0.014
Rb^+	1.0		0.020
Cs^+	7.0	1.3	2.0
Tl^+	0.025	2	0.004
NH_4^+	2.0		
$Tris^+$		130	
Mg^{2+}	500	4.0	
Ca^{2+}	600	1.5	

From Borst-Pauwels (1981).

and that at the membrane surface is changed, the relationship being that the value of the latter is a function of the surface potential of the membrane (Ψ) given by the Boltzmann distribution law. Thus:

$$C_m = C_o \, e^{-zq\Psi/kT}$$

where C_m is the concentration of the bulk phase, C_o the concentration at the membrane surface, z the valency of the cation, q the absolute value of the charge on the electron, T the temperature in kelvins, and k the Boltzmann constant. In considering the kinetics of ion transport, Michaelis–Menten-type equations thus have to be modified by substituting for C_m (s), using the relationship above. When no polyvalent cations are present with a negatively charged membrane, for a monovalent cation $C_m > C_o$; for a monovalent anion $C_m < C_o$. Polyvalent cations reduce the surface potential, making it less negative. Thus, polyvalent cations reduce monovalent cation transport (increase the dissocation constant of that ion for the carrier) and increase monovalent anion transport (reduce the dissociation constant). The data in support of this have been presented by Theuvenet & Borst-Pauwels (1976a, b); from them it was concluded that mainly phosphoryl groups determine the net charge on the membrane at pH 4.5 (see also Rothstein, 1954). In fact, the modifier site described above is probably no more than an enzyme kinetic description of the influence of polyvalent cations on the surface potential and therefore on the effective concentration of monovalent cations for transport. The case for this interpretation of the modifier site is given by Derks & Borst-Pauwels (1979, 1980).

There have been a number of other studies characterising in more detail the effect of the surface potential on monovalent cations (Theuvenet & Borst-Pauwels, 1983). In particular it has been shown that there is a direct relationship between the zeta potential (dependent upon charge density and charge potential) and the apparent K_T for rubidium transport. Equally, measured zeta potentials are to a good approximation linearly related to surface potentials calculated from the kinetic analysis of rubidium uptake. Finally, it has been shown that monovalent cations (for instance $Tris^+$) can affect the uptake of other monovalent cations through an effect on the surface potential.

Salt uptake by green plant cells is stimulated by bivalent cations – the so-called Viets effect (Jennings, 1963). This has been interpreted variously. The simplest explanation, with an experimentally sound basis, is that salt uptake is limited by anion transport across the plasma membrane and the presence of bivalent cations mask negative charges in the wall, allowing easier diffusion of anions to the transport site with increased salt uptake (Pitman, 1964). This does not seem to be the case with cells of *S. cerevisiae*, since neither removal of the wall nor plasmolysis has any effect on the kinetics of the rubidium transport or the effect of calcium on this uptake (Gage *et al.*, 1985; Gage, Theuvenet & Borst-Pauwels, 1986).

Finally, it should be noted that 9-aminoacridine appears to be useful as a probe of the surface potential (Theuvenet *et al.*, 1984).

Genetic studies

In 1985, Ramos, Contreras & Rodríguez-Navarro isolated a transport mutant of *S. cerevisiae* that required 100 times more potassium than did wild-type cells for the same half-maximal growth rate. Cells of the mutant grown in sufficient potassium showed little difference from the wild-type with respect to rubidium transport when determined in the presence of 3 mM potassium. However, in potassium-starved cells, the wild-type showed the presence of a high affinity system for rubidium, which would be expected from the observations of Rodríguez-Navarro & Ramos (1984) referred to above. This system, however, was absent from the mutant. There were no significant differences between mutant and wild-type cells with respect to sodium transport or the ability of sodium to inhibit rubidium transport. It was hypothesised from the results that there is only one carrier for potassium, rubidium and sodium transport. It has two sites only one of which has any significant affinity for sodium. Conversion to high affinity potassium or rubidium does not affect the sodium site, which is regulated only slightly by the internal potassium concentration.

The genetics of potassium transport do not appear to be simple.

More mutants of a similar kind were isolated by Gaber, Styles & Fink (1988). Only those mutations at the *TRK1* locus showed defects in the high affinity transport system. The gene encodes a polypeptide of 1235 amino acid residues that contains 23 membrane-spanning domains. Other studies (Anderson, Best & Gaber, 1991) have shown that most *Saccharomyces* sp. contain a high affinity potassium transport gene that is closely related to *TRK1*. The encoded protein of *S. uvarum* shows the highest divergence, with 78% identity but 86% similarity when conservative substitutions are included in the amino acid sequence when compared with the protein of *S. cerevisiae*. Nevertheless the membrane spanning domains show 95% identity. Also the gene from *S. uvarum* can counteract the absence of *TRK1* in *S. cerevisiae*.

Importantly, the results of the genetic studies show there is a physical and functional independence of the high affinity potassium uptake system from both the low affinity system and the plasma membrane proton pump. In particular, deletion of *TRK1* from haploid cells leaves the low affinity system intact. Further, *trk1* cells, which cannot take up potassium below 1 mM, lose the ion, possibly via the low affinity system.

Subsequent search for other mutants located the gene for the low activity potassium transport system *TRK2* (Ko, Buckley & Gaber, 1990; Vidal et al., 1990; Ko & Gaber, 1991). The gene has been cloned and encodes a polypeptide of 9889 amino acid residues that shares significant sequence identity with the polypeptide encoded by *TRK1*. The structural features of the two proteins must be very similar. The indications are that the two genes arose from a duplication event. Nevertheless the studies confirm that the two genes are functionally independent.

Cells deficient in both transport systems are able to grow in the presence of 100 mM potassium chloride but grow much less well at lower concentrations and are particularly sensitive to acid (pH 3.0) media at low potassium concentrations. This sensitivity to low pH can be counteracted by addition of potassium to the medium but not sodium. Thus, there is another transport system in *S. cerevisiae* as well as those encoded by *TRK1* and *TRK2*. *TRK2* is non-essential in *TRK1* haploid cells. Nevertheless, the *TRK2* gene possesses a large promoter conferring complex regulation, suggesting that the gene must have a significant role in the living cell.

Physiological studies

It is well established that when cells of *S. cerevisiae* absorb potassium they do so in exchange for protons (Conway & O'Malley, 1946; Rothstein & Enns, 1946). More recently, Boxman, Dobbelmann & Borst-Pauwels (1984) have provided the most critical work on potassium accumulation. Studying cells metabolising under anaerobic conditions and critically evaluating the equilibrium distribution of tetraphenylphosphonium as a measure of the membrane potential, these investigators showed that potassium accumulation can be accounted for almost quantitatively by a cotransport (symport) of potassium with protons with a 1:1 stoichiometry. Uptake of potassium is thus driven by the sum of the membrane potential and the proton-motive force. It is not clear whether the high affinity potassium transport system functions in a similar manner, for, as far as one can gather, the cells used were such that that system was not present.

As has been indicated above, the kinetics of transport are dependent upon the concentration of potassium within the cells. The relationship between transport and internal potassium concentration has been studied in more detail by Ramos & Rodríguez-Navarro (1986 and Ramos, Haro & Rodríguez-Navarro (1990). Cells with reduced potassium content were produced in four different ways: (1) cells were treated with varying concentrations of sodium azide for the same short interval of time (under these conditions there is an enhanced potassium efflux); (2) cells were grown in a medium of low potassium concentration; (3) cells grown normally were starved in potassium-free medium; and (4) the volume of cells was reduced osmotically and in consequence the concentration of internal solutes increased.

Even with these various treatments, it is not easy to disentangle what might be occurring, because reduction of the internal potassium concentration might be accompanied by either decreased cytoplasmic pH (increased hydrogen concentration) or increased internal sodium concentration. That said, it seems that the internal potassium concentration has no effect on potassium influx as determined with rubidium. The most compelling evidence comes from cell shrinkage produced osmotically (method 4), which has no effect on rubidium uptake. This is also true for cells produced by method 2 and then shrunk osmotically when the internal concentration of potassium rises to values similar to those for cells grown normally. This conclusion about the lack of an effect of internal potassium concentration is in keeping with a K^+ (Rb^+)–H^+ exchange system. However, it should be noted that at much higher molalities (greater than 0.3 molal used above) there is a reduction in the rate of potassium uptake (Gage *et al.*, 1986).

The use of method 1 leads to an increase of V_{max} for rubidium transport (Ramos & Rodríguez-Navarro, 1986). This is now thought to be due to changes in the cytoplasmic pH. Previous work by Ryan & Ryan (1972) had indicated that uptake of monovalent cations was dependent on the internal hydrogen ion concentration. Theuvenet, Roomans & Borst-Pauwels (1977), using rubidium, carried out a detailed analysis of the relationship and concluded that there could be two effects of intracellular pH on the rate of transport – a direct one on the exchange of protons for rubidium or an indirect one consequent upon the stimulation of H^+-ATPase activity leading to an increase in the membrane potential, such that the cell interior becomes more negative.

It has to be said that very few of the findings discussed above provide much further understanding of the mechanism of potassium uptake proposed by Boxman *et al.* (1984), but none of the data can be considered to be contradictory.

Potassium efflux

In potassium-free medium, cells lose potassium until a certain, albeit relatively low, external concentration (0.05 mM) is achieved (Rothstein & Bruce, 1958a). If the initial concentration is 0.05 mM, there is no change in the potassium concentration of the medium; however, with an initial concentration of 0.3 mM, uptake continues until the external concentration reaches 0.05 mM. Loss of potassium must be the difference between efflux and influx and this needs to be kept in mind when considering studies which purport to be measuring export. Unless it is clear that initial rates of loss are being determined or radiotracer is being used to discriminate between efflux and influx, some caution is needed in the interpretation of results.

Rothstein & Bruce (1958a, b) were the first to study in some detail potassium efflux from *S. cerevisiae*. Since they used ^{42}K-labelled cells for studying efflux in media containing potassium and unambiguously determined the initial rate of potassium loss into media in which potassium was absent, their findings are particularly important. The rate of potassium efflux was found to be dependent on the potassium content of the cells and is in exchange for hydrogen ions when no substrate is present – in the presence of substrate, the effect is obscured by excretion of organic acids. Above pH 4.5, sodium stimulates efflux, as does triethylamine, suggesting that it is uptake of positive charge rather than the nature of the cation that is important. Below pH 4.5, sodium reduces efflux, though this effect is dependent on a number of factors including the pH value and whether or not glucose is present. Glucose, as well as fructose and pyruvate, increases efflux, while ethanol and lactate reduce it. But the increase with glucose is reduced by sodium and still further if calcium is also present, though this particular cation is only as effective as sodium alone when that ion is absent from the medium. Thus, there is synergism between sodium and calcium. Dyes (at 10^{-4} M) with a redox potential of less than 0.01 mV produce a high rate of efflux. This suggests that part of the efflux might be driven by the plasma membrane redox system (see Chapter 1) but the efflux may also be brought about by the all-or-none effect (see below).

Subsequent studies have done little to clarify the mechanism of potassium efflux. The significant findings can be summarised:

(1) In ATP-depleted cells, potassium efflux is not stimulated by acid pH values (Ortega & Rodríguez-Navarro, 1985). It is argued that in such cells potassium efflux via a potassium/proton antiport does not function.

(2) It is clear now that a large number of compounds induce potassium efflux from cells of *S. cerevisiae* (Table 10.2). Though not established in every case there is clear evidence for all those compounds listed in Table 10.2 that they act via an all-or-none process. This means that, during the interaction of the toxic compound with a suspension of yeast cells, a gradually increasing number of cells become leaky, such that there is almost complete loss of potassium from the cells. Potassium lost from these cells can be transported into those cells that are still intact, which complicates any quantitative analysis of the process (Borst-Pauwels, 1988). It is not completely clear to what extent the membrane becomes permeable to other ions and molecules. Certainly phosphates are lost with potassium when cells are treated with methylene blue (Passow, Rothstein & Loewenstein, 1959). Also, cells under the same conditions show enhanced uptake of divalent cations (Eilam, 1983, 1984; Borst-Pauwels *et al.*, 1986). Some of the enhanced uptake may be due in some circumstances to hyperpolarisation of the membrane through a specific increase in the permeability of the plasma membrane to potassium. In other instances, the uptake of bivalent cations can be ascribed to the increased permeability of the cell.

The frequent response of yeast cells to toxic agents via an all-or-none

Table 10.2. *Compounds known to increase loss of potassium from cells of* Saccharomyces cerevisiae

Compounds	References
Basic dyes[a]	
Ethidium bromide	Passow *et al.* (1959)
Methylene blue	Peña, Mora & Carrasco (1979)
Nile blue	Peña & Ramirez (1975)
Phenosafranine	
Toluylene blue	
Calmodulin antagonists	
Calmidazolium	Borst-Pauwels & Theuvenet (1985)
Compound 48/80	Eilam (1983)
Trifluoperazine	
Detergents	
Benzalkonium chloride	Riemersma (1966)
Cetrimide	Scharff & Maupin (1960)
Cetylpyridinium chloride	
Metals	
Mercury	Passow & Rothstein (1960)
Copper	
Plasma membrane ATPase inhibitors	
N,N'-dicyclohexylcarbodiimide	Borst-Pauwels, Theuvenet & Stols (1983)
Diethylstilboestrol	
Dio-9	
Miconazole	
Suloctidil	
Triphenyltin	
Polyene antibiotics	
Nystatin	Eilam & Grossowicz (1982)
Uncouplers	
Azide	Ortega & Rodríguez-Navarro (1985)
Carbonylcyanide *m*-chlorophenylhydrazone	
Dinitrophenol	

[a] Above 0.5 mM.

phenomenon means that caution is needed in interpreting effects of such compounds on transport of solutes into and out of cells of *S. cerevisiae*. However, the toxic compounds can be having other effects on the cells. Kuypers & Roomans (1979), from an X-ray microanalysis of individual cells of *S. cerevisiae* exposed to mercuric chloride, have produced evidence that there are particular cells that lose potassium but apparently not in the distinct all-or-none manner.

(3) Potassium efflux has been studied in strains lacking the high affinity potassium carrier (Peña & Ramirez, 1991). Loss of potassium from the cells of these strains is enhanced in the presence of glucose, whereas in the wild-tupe it is reduced, suggesting that the mutant strains are unable effectively to reabsorb the potassium lost. In keeping with this, bivalent cations and organic cations such as ethidium bromide, dihydrostreptomycin, dimethylaminodextran and trivalent cations such as aluminium and lanthanides stimulate efflux both in the wild-type and in mutants. Though the percentage increase in loss over that in the controls was found not to be greater in the mutants than in wild-type, the absolute rate of loss was greater. The ions would be acting, as described earlier, on the surface potential and thus inhibiting transport.

(4) Potassium efflux is reduced in mutants with decreased ATPase activity (Peña & Ramirez, 1991). Conceivably this is due to increased pH inside the cells and increased competition of protons for a site on a carrier normally transporting potassium out of the cell, as has been suggested with respect to sodium efflux.

(5) There is a transient efflux of potassium on the addition of glucose to an anaerobic suspension of cells of *S. cerevisiae* at pH 7.0. It is accompanied by a transient hyperpolarisation of the plasma membrane (van de Mortel *et al.*, 1988; van de Mortel, Theuvenet & Borst-Pauwels, 1990). Both phenomena are suppressed by bivalent cations such as Mn^{2+} and Ca^{2+}, and also by La^{3+}. The effect of calcium can be reversed by ethylenediamine tetra-acetic acid (EDTA). It is thought that potassium is lost via potassium-selective channels but not those that are described below.

Sodium efflux

Cells of *S. cerevisiae* can be loaded with cations such as ammonium (Conway & Breen, 1945), calcium (Conway & Gaffney, 1966), magnesium (Conway & Beary, 1962) and sodium (Conway & Moore, 1954). These cations, as well as lithium, probably enter cells via the low affinity potassium transport system, though the bivalent cations also enter by other systems (see Chapter 11) (Conway & Duggan, 1958). Such loaded cells are suitable for studying, by chemical means, the efflux of these cations, though, as is indicated below, there are strong arguments for attempting studies with cells under more normal physiological conditions, perhaps labelled with radioisotopes of these cations.

Sodium efflux from *S. cerevisiae* has been investigated by Conway, Ryan & Carton (1954), Foulkes (1956), Dee & Conway (1968), Ryan & Ryan (1972), Rothstein (1974) and Rodríguez-Navarro & Ortega (1982). Efflux can either be an exponential (perhaps sigmoidal) function of internal sodium concentration (Dee & Conway, 1968; Rothstein, 1974) or a linear function (Rodríguez-Navarro & Ortega, 1982). The different results might be a consequence of using sodium-loaded cells in the first instance and non-loaded cells in the second. Equally it might be a consequence of using different strains; Rodríguez-Navarro & Ortega (1982) used a respiratory-deficient strain.

Though sodium-loaded yeasts may not be representative of cells grown in media low in sodium, nevertheless the most detailed analysis of sodium efflux has been carried out by Rothstein (1974) using such cells. In cation-free medium, sodium efflux is largely balanced by the loss of organic acids from the cells. In the presence of cations (including protons), the loss of organic acids is reduced almost completely, as observed also by Conway *et al.* (1954). It is probably the rate of loss of organic acids that limits sodium efflux into media of low cation content.

Sodium efflux is accompanied by proton efflux, though the proportion changes with time, with relatively more protons being lost initially. The total efflux of the two cations is limited by cation influx (rubidium in this study). If the efflux of protons is via the same carrier as that for the efflux of sodium, then the carrier has a very much higher (of the order of 10 000) discrimination in favour of protons. There is certainly discrimination of the carrier for sodium over potassium but the extent is hard to quantify because the latter ion when it enters the compartment between the plasma membrane and the wall will rapidly re-enter the protoplasm via the potassium transport system.

The simplest model is that there is a carrier for protons, sodium and potassium ions, in decreasing order of affinity. That sodium efflux may possess sigmoidal kinetics might mean the carrier functions in a similar manner to that responsible for potassium influx in *Neurospora crassa*, for which a two-site model has been proposed (see below). Alternatively, proton loss may be via the proton pump, which drives a proton/sodium antiport. This is the most likely possibility in view of identification of the *sod2* gene in *Schizosaccharomyces pombe*, discussed in Chapter 13. Ryan & Ryan (1972) have suggested, from a study of $K^+–H^+$, $K^+–K^+$ and $K^+–Na^+$ exchange by cells of *S. cerevisiae* as a function of internal pH, that there are two transport systems, one for K^+–H^+ and the other for K^+–Na^+ exchange. While that is a possibility, it does not need to be invoked to explain the findings to date.

Potassium channels in the plasma membrane

Such channels have been observed when the plasma membrane of *S. cerevisiae* is patch-clamped. Some are voltage dependent (Gustin *et al.*, 1986); others are ATP sensitive (Ramirez *et al.*, 1989). It is not clear what is the role of these channels. Putative potassium-selective channels have been invoked as the pathway for the transient net efflux of potassium from cells presented with glucose under anaerobic conditions (see above). If efflux is via these channels they do not seem to be the same as those observed with patch-clamping.

Neurospora crassa

Potassium transport

Studies on *N. crassa* demonstrate the power of the biophysical approach for providing an understanding of the mechanisms underlying the movement of ions between the medium and the hyphae. Nowhere is this power more evident than in understanding monovalent cation transport. Insertion of electrodes into fungi has never been easy. In *N. crassa*, as has been pointed out earlier (Chapter 1), the wider diameter hyphae have been the ones of choice for electrophysiological studies. But even these hyphae present difficulties when all the relevant measurements have to be made on the same compartment. These difficulties have been overcome by the use of large spherical cells. These are produced by inoculating media containing *c*. 15% (v/v) ethylene glycol (Bates & Wilson, 1974). The conidia grow as single cells; they do not form germ-tubes. The size of these cells is variable. They can be 20 μm in diameter and 78% can have a volume of over 1000 μm^3. Ungerminated conidia have a volume of between 1 and 200 μm^3. The problems of using KCl electrodes with such cells are discussed by Blatt & Slayman (1983).

Before I describe these electrophysiological studies, they need to be put in context. Mycelium of *N. crassa* relatively low in potassium and high in sodium absorbs potassium in exchange for protons and sodium (Slayman & Slayman, 1968). All the fluxes of the three cations are exponential with time and obey Michaelis–Menten kinetics as a function of external potassium concentration. For the most part, the fluxes of anions can be discounted. Slayman & Slayman (1970) examined the kinetics of net potassium uptake as a function of potassium concentration and pH of the medium. At low pH (4.0–6.0) the net flux is a simple exponential function of time that obeys Michaelis-Menten kinetics as a function of potassium concentration. At high pH, potassium uptake is more complex, obeying sigmoid kinetics. The data have been fitted satisfactorily by two different two-site models. In one, the transport system is thought to contain both a carrier site responsible for potassium uptake and a modifier site, for a proton at low pH and for potassium at high pH. The other model postulates a transport protein of multiple subunits, each with an active site for potassium; protons act as allosteric activators. The system described in these two studies must be considered as high velocity and low affinity. In the presence of 0.1–10 mM calcium, there is a low velocity, high affinity potassium transport system (reported by Rodríguez-Navarro, Blatt & Slayman, 1986). These results have been confirmed by a study of rubidium uptake by *N. crassa* (Ramos & Rodríguez-Navarro, 1985; Rodríguez-Navarro & Ramos, 1986). In low potassium mycelium, transport is biphasic; the kinetic characteristics are dependent on the potassium content of the mycelium. The high affinity system disappears in mycelium grown in the presence of 37 mM potassium or in low potassium mycelium allowed to absorb potassium. On the basis of Arrhenius plots of V_{max} values and the pH dependence of rubidium uptake in the two types of mycelium (low and high potassium), there are two independent transport systems.

Attention may now be turned to the electrophysiological studies directed specifically at potassium transport (Rodríguez-Navarro, Blatt & Slayman, 1986; Blatt, Rodríguez-Navarro & Slayman, 1987). Cells were, for the most part, grown in low potassium (0.3 mM), high ammonia (40 mM) media containing ethylene glycol and generally suspended for experiments in a (buffered) medium of variable potassium and 1.5 mM calcium. These electrophysiological studies have shown via current–voltage experiments that the potassium-associated current moving into the cells exceeds half that normally exiting the cells via the proton pump. The potassium influx (there is negligible efflux under the same conditions) is approximately half the inward current, both being a hyperbolic function of external potassium concentration (Figure 10.1). Inward flux of protons must account for the difference. From this it can be seen that the stoichiometry of entry is one potassium ion accompanied by one proton, but two protons are pumped out. Thus, there is a net 1:1 exchange of potassium for protons.

Further studies, involving a detailed analysis of the current–voltage characteristics of potassium–proton cotransport have clarified the

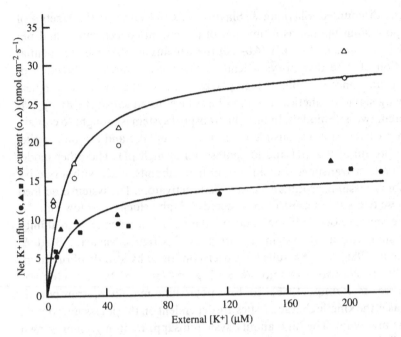

Figure 10.1. The relationship between net potassium influx into cells of *Neurospora crassa* (lower curve) or current across the plasma membrane (upper curve) and external potassium concentration. Different experiments are represented by different symbols. The stoichiometry of current flux to net potassium influx is very close to 2. (From Rodríguez-Navarro *et al.* (1986), reproduced from the *Journal of General Physiology*, 1986, **87**, 649–74, by copyright permission of the Rockefeller University Press.)

mechanism further. In particular, it has been shown that inhibition of potassium transport by inhibitors, such as vanadate, of the plasma membrane H^+-ATPase have their effect via depolarisation of the membrane. It was found that increasing the membrane potential decreases the affinity of the transport system for potassium but increases the affinity for protons. This behaviour can be explained by two models: (1) a neutral carrier binds potassium first and protons last in the forward direction of transport, or (2) there is negatively charged carrier binding of protons first and potassium last. Be that as it may, it can be seen that regulation of potassium transport by cytoplasmic pH is via the plasma membrane H^+-ATPase and not by means of the potassium transport system itself.

It should be noted that the above applies to the potassium transport system found in low potassium mycelium. It is not known how the findings apply to mycelium that has a high potassium content. Also, while the transport system in low potassium mycelium (the high affinity system) shows biphasic kinetics, these kinetics are not incompatible with the potassium/proton symport. This is readily explicable in terms of the kinetic effect of the diminishing membrane potential that occurs as the potassium concentration is raised (Gerson & Poole, 1971).

The earlier kinetic formulations for potassium transport in *N. crassa* now need interpretation in terms of the more recent biophysical findings.

Sodium transport

Neurospora crassa accumulates potassium and excludes sodium except when there is only a low concentration of the former in the external medium (Slayman & Tatum, 1964). In the absence of external potassium, the influx of sodium is independent of internal potassium concentration (Ortega & Rodríguez-Navarro, 1986). With potassium present in the external medium, there is an inhibition of sodium influx, the inhibition being greater the greater the mycelial potassium concentration. Presumably this influx is via the potassium transport systems, using the high affinity system in the absence of potassium, and the low affinity system in the presence of potassium. However, Ortega & Rodríguez-Navarro (1986) urged caution in accepting this interpretation.

Efflux of sodium into a medium of low potassium concentration is increased by both the sodium and potassium content of the mycelium (Ortega & Rodríguez-Navarro, 1986). When mycelium was grown in the absence of potassium and in the presence of 10 mM sodium, efflux of sodium was triggered by the addition of 0.5 mM potassium chloride to the medium. Cyanide had no effect. Acidification of the cell by the addition of butyric acid at an external pH of 4.5, but not at pH 5.8 even though the cytoplasmic pH was reduced by 0.6 pH units, reduced sodium efflux. Potassium uptake was not affected. It is not clear from the above whether there is one carrier system for sodium or two.

11

Multivalent metals (required or toxic)

Introduction

Magnesium, calcium, iron, copper, manganese, zinc and often molybdenum are required by fungi for growth. The order approximates to the concentration required in a normal growth medium. Magnesium is required at $c.$ 10^{-3} M, while molybdenum need only be at true trace concentrations, namely $c.$ 10^{-9} M. The latter element is required by those fungi utilising nitrate (see Chapter 6). Cobolt will be needed by those fungi that synthesise their own vitamin B_{12} (Fries, 1965), whereas nickel is required by those fungi exhibiting urease activity (Hausinger, 1987). The list might be longer but rigorous evidence for inclusion of other elements is not easy to obtain. Trace contaminants of reagents by a suspected element make it difficult to demonstrate that it is required for growth. Molecular biology and enzyme studies are likely to prove more revealing.

Copper and zinc can be toxic at concentrations higher than those required for growth. Since other multivalent metals can also be toxic, this chapter deals with the avoidance of toxicity. I have refrained, except in certain instances, from listing the role(s) of these various elements in fungi. To give a proper indication of the function of an element requires the provision of biochemical information beyond the scope of this volume.

Transport of multivalent metals (other than iron) across the plasma membrane

Specific systems

At its simplest, uptake of a multivalent metal by a fungus is biphasic; there is an initial rapid phase, which is believed to reflect binding to the wall, followed by a second, slower phase (Figure 11.1). The extent of the binding

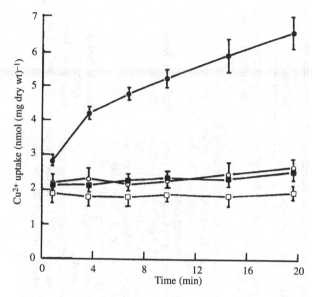

Figure 11.1. Copper uptake by *Candida utilis* at 30 °C from 10 μM Cu²⁺ in the presence (●) and in the absence (○) of glucose and in the presence of 2-deoxy-ᴅ-glucose (■), and at 4 °C in the presence of glucose (□). Note that only when there is a metabolisable substrate (glucose) and a high temperature is there continuing uptake after the first minute. The uptake during the first minute can be considered to be binding to the wall. (From Ross & Parkin, 1989.)

to the wall affects the concentration available for transport across the plasma membrane. Binding to the wall is considered below. The second phase has usually been ascribed to transport across the plasma membrane because of its disappearance in the presence of metabolic inhibitors, often when exogenous energy sources are absent and at low temperature. Also, analysis of the kinetics of the second phase show them to be enzymic in form, providing K_T values.

In recent years, it has become clear that transport systems exist specifically for the bivalent metals required for growth and considered here. Those transport systems identified to date are listed in Table 11.1, together with some of their properties.

These systems are for the most part highly specific, with very high affinity for the bivalent metal transported. Cadmium, where studied, competes with zinc for transport. Since cadmium is not normally a component of the natural environment, the ability of this metal to compete in the manner indicated does not argue against the transport systems listed in Table 11.1 being highly specific. To function, the systems require

Table 11.1. *Some of the properties of the known specific or near-specific transport systems for bivalent cations in fungi*

Fungus	Metal	K_T	Competing ions	Comments on specific system	Reference
Aspergillus niger	Mn^{2+}	3.04 μM	Zn^{2+} Cu^{2+} Cd^{2+}	Mn^{2+}-depleted mycelium used; transport inhibited by DNP	Hockertz, Schmid & Auling (1987)
Candida utilis	Zn^{2+}	1.3 μM	Cd^{2+}	Uptake inhibited by lack of glucose and by range of metabolic inhibitors. No efflux detected except with nystatin. Uptake requires prior protein synthesis	Failla et al. (1976)
	Mn^{2+}	16.4 nM		Uptake inhibited by lack of glucose and by CCCP and DNP. Inhibition of uptake by Mg^{2+}, Zn^{2+}, Co^{2+}, Ni^{2+}, Ca^{2+} and Cu^{2+} only at 1000-fold excess, not at 100-fold. Low rate of efflux inhibited by lack of glucose and by DNP.	Parkin & Ross (1986)
	Cu^{2+}	3.1 μM		No efflux observed. High external K^+ inhibits uptake	Ross & Parkin (1989)
Neocosmospora vasinfecta	Zn^{2+}	1.1 μM		Uptake is inhibited by anaerobiosis and by CCCP but not by DES; glucose has no effect. Uptake stimulated in low Mg^{2+} mycelium	Budd (1988); Paton & Budd (1972)

Organism	Metal	K value	Inhibition	Notes	Reference
Penicillium notatum	Mn^{2+}	1.29 nM (protoplasts) 4.37 nM (mycelium)	Slight inhibition with Mg^{2+} and Zn^{2+}	Uptake requires starvation of Mn^{2+}; no effect of glucose unless mycelium starved. Uptake inhibited by FCCP and by Mg^{2+}, Zn^{2+}, Ni^{2+} and Co^{2+} at 1000-fold, but not at 100-fold concentration	Starling & Ross (1990)
	Zn^{2+}	6.66 nM	Cd^{2+} (competitive) Cu^{2+} (non-competitive)	Uptake enhanced by growth in low Zn^{2+}-medium. Uptake low-affinity system stimulated in low Mg^{2+} mycelium	Starling & Ross (1991)
Saccharomyces cerevisiae	Cu^{2+}	1.13 µM	Equimolar Zn^{2+}, Mn^{2+}, Mg^{2+}, Ca^{2+} inhibit uptake	Not clear in view of inhibitory effect of other bivalent cations whether transport system truly specific. Influx of Cu^{2+} is related to K^+ efflux but not tightly coupled	De Rome & Gadd (1987)
Sclerotium rolfsii	Zn^{2+}	3.81 µM		Zn^{2+}-deficient cultures used. Mn^{2+} stimulates uptake slightly. Uptake is inhibited by DNP, DES, CCCP and azide and by anaerobiosis and low temperature	Pilz et al. (1991)

DNP, 2,4-dinitrophenol; CCCP, carbonylcyanide *m*-chlorophenylhydrazone; DES, diethylstilboestrol; FCCP, carbonylcyanide *p*-trifluoromethoxyphenylhydrazone.

metabolism. However, their mechanisms are as yet unclear. There are two possibilities put forward by Budd (1988) with respect to zinc transport in *Neocosmospora vasinfecta*, namely either a zinc/proton antiport energised by the proton extrusion pump or a zinc uniport driven by the membrane potential. The latter seems the most likely, since diethylstilboestrol (DES), which inhibits the proton pump, has little effect on zinc transport, while CCCP and anaerobiosis, which would be expected to depolarise the membrane, strongly inhibit transport. However, this argument would not hold for zinc transport by *Sclerotium rolfsii*, which is inhibited by DES (Pilz *et al.*, 1991). Any mechanism must account for the highly directional flux of the metal and the fact that the rate of uptake appears to be affected by the amount of metal already present in the fungus. As part of any further study, one needs to know the extent to which the metal once transported across the plasma membrane is then sequestered in an organelle, e.g. the vacuole.

The question of the presence of less specific systems, with special reference to *Saccharomyces cerevisiae*

Bivalent metal ions also appear to be transported into fungi by less specific systems. The exact nature of these systems is not readily elucidated from the published information. The most studied fungus with respect to transport of these ions is, as one might expect, *Saccharomyces cerevisiae*. The initial studies (Jennings, Hooper & Rothstein, 1958; Rothstein *et al.*, 1958; Fuhrmann & Rothstein, 1968) showed that:

(1) There could be a transport system (magnesium?) for the following (in order of affinity for the system): Mg^{2+}, Co^{2+}, Zn^{2+}, Mn^{2+}, Ni^{2+}, Ca^{2+}, Sr^{2+}.
(2) These ions are not transported via the potassium carrier as originally proposed for magnesium by Conway & Beary (1958).
(3) Transport requires the presence of glucose.
(4) Transport is stimulated by the prior incubation of cells in the presence of phosphate.

It is still not clear whether there are one or more transport systems. This uncertainty appears to be in part because there has not yet been a concerted study involving all the ions, using the same strain of yeast, with a coherent programme of experiments using consistent pretreatment and experimental protocols. Some idea of the unsatisfactory nature of the information is provided in Table 11.2. This shows the dissociation

Table 11.2. *Half-saturation constants* (K$_T$) *for uptake of a selection of divalent metal ions by* Saccharomyces cerevisiae

Metal ion	K_T (μM)	Reference
Zn^{2+}	3.7	White & Gadd (1987a)
	<10	Fuhrmann & Rothstein (1968)
	1150	Ponta & Broda (1970)
	5000	Mowll & Gadd (1983)
Co^{2+}	<10	Fuhrmann & Rothstein (1968)
	77	Norris & Kelly (1977)
	100	Norris & Kelly (1979)
	100	Heldwein, Tromballa & Broda (1977)
	800	Heldwein *et al.* (1977)
Mn^{2+}	<10	Fuhrmann & Rothstein (1968)
	100	Norris & Kelly (1979)
	860	Bianchi, Carbone & Lucchini (1981)

From Ross (1993). Reprinted by courtesy of Marcel Dekker Inc.

constants of divalent cations for transport into cells of *S. cerevisiae* obtained in a number of investigations. There is clearly great variation in the values found. There are several reasons for this variation (Borst-Pauwels, 1981). The manner of computing the constant, i.e. from the initial rate of uptake or the amount absorbed over a set time. Formation of complexes with the buffer can also present problems. There may also be difficulties due to binding on to the wall or membrane as well as differences brought about by differences in the surface potential (see Chapter 10). Nevertheless, in spite of the problems of reconciling the ambiguities in the published information, there are some aspects of bivalent metal cation transport about which there can be a fairly clear-cut view.

Energy dependence

The requirement for exogenous glucose to attain a high rate of uptake has been amply confirmed for cadmium (Norris & Kelly, 1977), calcium (Roomans *et al.*, 1979b), cobolt (Fuhrmann & Rothstein, 1968), manganese (Rothstein *et al.*, 1958), strontium (Roomans *et al.*, 1979b) and zinc (Ponta & Broda, 1970). There can be a small but significant uptake without glucose (Norris & Kelly, 1977), while Roomans *et al.* (1979b) observed a lag of 1–2 min before uptake of calcium and strontium commenced, and also that the uptake occurred under anaerobic conditions.

Interaction with phosphate

It has been shown for strontium that, if that bivalent cation is added to the medium along with phosphate, there is an initial inhibition of strontium uptake before the eventual stimulation of uptake ensues (Roomans *et al.*, 1979*b*). This inhibition shows saturation kinetics, the inhibition constant being equal to the affinity constant of phosphate for its transport system. The inhibition by phosphate is very similar to that for monovalent cation transport (Roomans & Borst-Pauwels, 1977), for which inhibition it was argued that phosphate affects a transient depolarisation of the plasma membrane. The cell by becoming less negative is less able to absorb cations.

Conversely, the stimulation of phosphate uptake, thought at one time to be due to the synthesis of a membrane carrier involving a phosphorylation step (Jennings *et al.*, 1958) and at another to be due to the increased synthesis of ATP (Fuhrmann, 1974), may result from a change in surface potential. The case is argued by Borst-Pauwels & Theuvenet (1984) with the theory as it relates to bivalent cation uptake presented by Borst-Pauwels & Severens (1984). Some of the theory has already been presented in Chapter 10. Essentially, the theory deals with the sequence of inhibition by a series of cationic inhibitors affecting the rate of transport of a particular bivalent cation. The sequence depends on (a) the concentration of the substrate cation, (b) the concentration of the inhibitory cation, (c) the charge density of the cell membrane, (d) the ionic strength of the medium and (e) the affinity of the substrate cation for the negatively charged groups located on the cell membrane. It is worth repeating here that an essential feature that substrate cations face as they move towards the plasma membrane comprises those same negative charges. These charges provide an attractive force for the cations. Polyvalent cations make the membrane less negative.

With respect to the stimulation of bivalent cations by phosphate, Borst-Pauwels & Theuvenet (1984) argued that loading the cells with phosphate increases the net negative charge on the cell membrane.

Bivalent cation interactions

Here it is appropriate to make some further comments about the effect of the surface potential. First, because of the valence term in the relationship between the concentration of an ion in the bulk phase and that at the membrane surface, the effect of bivalent cations on the surface potential is more pronounced than the effect of monovalent cations. Also it should be remembered that there will be self-inhibition as the concentration of

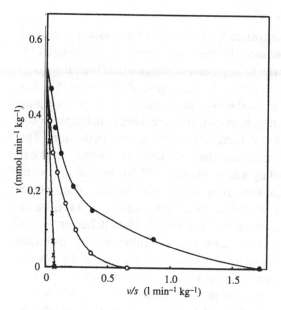

Figure 11.2. A Hofstee plot of calcium uptake by *Saccharomyces cerevisiae* showing the so-called 'concave' plot (see the text). (○) pH 6.0; (×) pH 6.0 with 4 mM strontium added to the experimental solution; (●) pH 7.2. (From Roomans *et al.*, 1979*b*.)

the substrate bivalent cation itself is increased. Thus, even though the carrier may be at a single site, the kinetics are more complex than simple Michaelis–Menten. By means of specialised kinetics Borst-Pauwels & Theuvenet (1984) were able to explain the so-called 'concave' plot obtained for calcium uptake observed by Roomans *et al.* (1979*b*) (Figure 11.2).

It one accepts the above, it is not surprising that, given the variation in experimental protocols and yeast strains used, there seems to be more than one carrier for any ion and that there is considerable variation in any affinity constants that are obtained, as demonstrated in Table 11.2. Of course, it follows also that, for those bivalent cations required for growth, there are only relatively specific transport processes. The seeming presence of one or more less specific transport systems could as well be the consequence of interactions of the various bivalent cations with negative groups on the plasma membrane as with one or more transport sites. This has been implied by Borst-Pauwels & Severens (1984) when considering differences in the sequence of inhibitory cations found for the inhibition of cobalt and cadmium uptake by *S. cerevisiae* (Norris & Kelly, 1977).

Mechanism of transport

It seems possible that, as suggested for the specific transport systems and as alluded to when I considered the transient inhibition of transport by phosphate, transport is driven by the membrane potential. The ratio Co^{2+} influx:K^+ efflux of 1:2 observed by Fuhrmann & Rothstein (1968) and Norris & Kelly (1977) supports this possible mechanism. Also in keeping with it is the fact that, when sodium-rich cells are used, cobalt uptake has the same stoichiometry to sodium efflux as that of potassium. The monovalent cation loss maintains electroneutrality. Further, a linear relationship has been found by Roomans *et al.* (1979*b*) between the extent of inhibition by monovalent cations in the external medium of strontium uptake and the extent of inhibition of dibenzylammonium uptake, which enters the cell by the thiamine carrier (Barts *et al.*, 1980) (Chapter 9). The inhibition by the monovalent cations, such as rubidium, of strontium uptake shows a K_T for the saturation kinetics that is equal to that for the transport system for rubidium. The matter of calcium transport across the plasma membrane is discussed in more detail on pp. 355–7, where more evidence is presented for inward transport across that membrane being driven by the membrane potential.

However, there is no simple relationship between zinc transport and potassium efflux (Mowll & Gadd, 1983; White & Gadd, 1987*b*). Indeed the situation is complicated. Below 25 μM zinc, the initial rate of potassium efflux is unaffected by external zinc concentration. It should be noted that, in this instance, it proved important to determine the initial rate of efflux, since, below 50 μM potassium, efflux is followed by uptake (if analysis of the medium is delayed, it can appear that there is no efflux). Above 25 μM, the initial rate of potassium uptake increases with external zinc concentration but both potassium and proton efflux are also inhibited, the former being much more sensitive to the concentration of external zinc. Viability as a result of exposure to zinc declined as the external zinc concentration was increased. With cadmium, uptake is accompanied by potassium loss but there is no stoichiometric relationship between the two processes, more potassium leaving the cell than there is cadmium entering (Norris & Kelly, 1977).

What is not clear is the extent to which the overall transport process is driven by events at the plasma membrane or at the tonoplast, where there can be active transport for at least some of the bivalent cations under consideration. The rate of strontium efflux from cells of *S. cerevisiae* loaded with radiolabel has been shown to be greater than that of manganese and this is what might be expected on the basis of the greater sequestration of manganese than strontium in the cells concerned (Nieuwenhuis, Weijers & Borst-Pauwels, 1981).

Other fungi

In general terms, the characteristics of bivalent metal cation transport in other fungi are the same as those reported for *S. cerevisiae*. Specific systems of high affinity are referred to in Table 11.1. On the basis of what has been argued for *S. cerevisiae*, there must be equal doubts about the presence of a multifunctional carrier for any one of these ions. Of the published information for other fungi, the following are significant.

Aureobasidium pullulans

This polymorphic fungus can exist as yeast-like cells, mycelium and melanin-pigmented chlamydospores, as well as various transitory forms. Cadmium uptake by the three major forms has been studied by Mowll & Gadd (1984) and copper by Gadd & Mowll (1985). The K_T uptake of cadmium (*c.* 0.10 mM) and copper (*c.* 0.21 mM) was found to be virtually the same for both yeast-like cells and mycelium, as was the V_{max}. The same held true for the K_T for potassium release. However, whereas for both cadmium and copper in yeast-like cells and copper in mycelium the ratio of uptake of each bivalent cation to potassium loss was 1:2, this was not true for cadium and mycelium when the ratio was found to be 1:6.7.

Candida utilis

For this fungus there is information about bivalent metal cation transport for cells grown in continuous culture (Lawford *et al.*, 1980*a*, *b*). It was found that cells grown under zinc-limited conditions had a zinc content of <1 nmol (mg dry weight)$^{-1}$. Such cells could accumulate zinc far in excess of 4 nmol (mg dry weight)$^{-1}$ in carbon-limited cells, 35 nmol (mg dry weight)$^{-1}$ as a maximum in growing cells and 60 nmol (mg dry weight)$^{-1}$ in cells treated with cycloheximide (cyanide or CCCP inhibit uptake almost to zero). The K_T determined from initial rates of uptake was found to be 0.36 μM, which is very close to the value (1.3 μM) for the specific transport system as determined by Failla, Benedict & Weinberg (1976). Uptake was found to be unidirectional, there being no detectable efflux or exchange of zinc. The enhanced level of accumulated zinc was not inhibitory to growth. The zinc-depleted cells that were used for the transport studies contained twice as much polyphosphate as did the carbon-limited cells. This increased level may assist maintenance of the enhanced zinc content. The location of polyphosphate was not ascertained, but a significant proportion could be in the vacuole (see Chapter 4).

Some of these findings confirm those of Failla & Weinberg (1977), who studied zinc accumulation by cells of *C. utilis* in batch culture. In zinc-sufficient cultures, zinc was accumulated in log-phase and late log-phase cells. This was not true for iron, which was accumulated at constant rate throughout the whole culture period. Growth of cells in low zinc medium appeared the same as those in zinc-sufficient media. So zinc could be accumulated beyond requirements for growth. Further, zinc accumulated could not be lost; the concentration within the cells could only decrease with cell number increase. Some very revealing facts came from studies on late exponential-phase cells resuspended in fresh growth medium. Cells exposed to 0.1–1.1 μM zinc accumulated the same amount of zinc, but at higher concentrations the amount accumulated decreased as the medium concentration was increased to 5.1 μM zinc. Cells exposed to this and higher concentrations accumulated 12 nmol (mg dry weight)$^{-1}$. Those cells exposed to the lowest zinc concentrations accumulated 96 nmol (mg dry weight)$^{-1}$. Therefore cells that were described as zinc deficient had altered physiology with respect to zinc accumulation. An allied phenomenon is the effect of medium zinc status on zinc accumulation by stationary phase cells. Such cells from glucose-limited medium containing 5.1 μM zinc did not take up zinc; cells from low zinc media did. The quantity of zinc accumulated declined the longer the cells were in the stationary phase but zinc-deficient cells always took up more than cells that had been grown in sufficient zinc. Finally, the affinity of the transport system does not appear to change during the growth cycle. There is little doubt that further research is required on the control of zinc content and transport in *C. utilis*.

Parkin & Ross (1986), also studying cells grown in continuous culture, showed that transport of manganese was repressed when cells were grown in high concentrations (100 μM) of the ion but not at much lower concentrations (0.45 μM). There was no such effect for copper over the concentration range studied (< 0.45–25 μM).

Neocosmospora vasinfecta

The level of magnesium in the mycelium of this fungus appears to be governed by the rate of influx; exogenous glucose has no effect on the rate of transport (Budd, 1979).

Schizosaccharomyces pombe

Calcium transport has all the features one would have expected from the studies on *Saccharomyces cerevisiae* (Boutry, Foury & Goffeau, 1977). Glucose stimulates uptake, uncouplers inhibit it, while Dio 9 (potassium ionophore) stimulates uptake several-fold even at 2 °C. The data are in keeping with a transport system dependent on the membrane potential.

Calcium

For many years, there was conflicting information about the requirement for calcium by fungi (Bollard & Butler, 1966). Even careful investigation produced conflicting results (Steinberg, 1948). As late as 1981, D. H. Griffin was stating 'several attempts to demonstrate the essentiality of calcium have yielded negative results but it has been shown to be required for growth and development of certain Oomycetes'. It is now accepted that, in eukaryotic cells, certainly those of plants and animals, calcium acts as a second messenger. Calcium transduces a primary stimulus at the outer membrane into intracellular events. The case for calcium functioning in this way in fungi is as follows:

(1) Some fungi, predominantly lower fungi such as *Achlya ambisexualis* (Barksdale, 1962), *Allomyces arbuscula* (Ingraham & Emerson, 1954), *Phycomyces blakesleeanus* (Ødegård, 1952), *Phytophthora fragariae* (Davies, 1959), *Saprolegnia ferax* (Jackson & Heath, 1989) and *Trichophyton interdigitale* (Mosher *et al.*, 1936) have been shown to require calcium for growth, though the requirement may not be absolute.

(2) There are a number of studies indicating that calcium is required for morphogenesis in the lower fungi such as *Lagenidium giganteum* (Kerwin & Washino, 1986), *Phytophthora cactorum* (Elliott, 1986), *P. palmivora* (Griffith, Iser & Grant, 1988), *Pythium graminicola* (Yang & Mitchell, 1965; Lenny & Klemmer, 1966) and *Saprolegnia diclina* (Fletcher, 1988).

(3) Several species of *Aspergillus* and *Penicillium* sporulate in submerged culture when nutrients are depleted. Certain species of *Penicillium* in the presence of calcium can be induced to sporulate without such nutrient depletion (Gilbert & Hickey, 1946; Pitt & Ugalde, 1984; Pitt & Kaile, 1990).

(4) Calcium is required for the differentiation of perithecia in *Chaetomium* spp. (Basu, 1951, 1952; Buston, Jabbar & Etheridge, 1953) and of

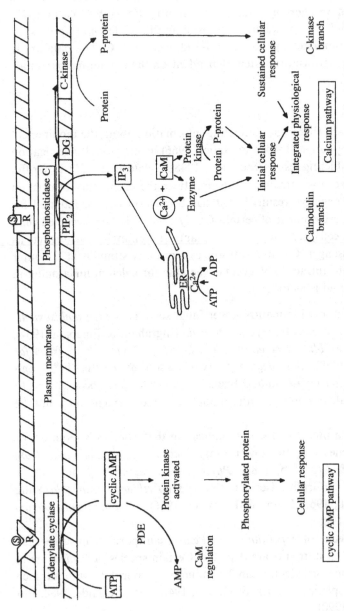

Figure 11.3. Summary of the general models of major signal transduction in eukaryotic cells. Information flow in most types of animal cell is via both the cyclic AMP (left) and Ca^{2+} (right) messenger systems, with interaction occurring between each system and the two limbs of the Ca^{2+} pathway. S, stimulus; R, receptor; PIP$_2$, phosphatidylinositol 4,5-bisphosphate; DG, diacylglycerol; IP$_3$, *myo*-inositol 1,4,5-trisphosphate; ER, endoplasmic reticulum; CaM, calmodulin; PDE, phosphodiesterase; P-protein, phosphorylated protein. (From Pitt & Kaile, 1990.)

certain strains of *Ceratocystis fimbriata* (Campbell, 1970). It is also required for the differentiation and growth of aggregated organs (aerial coremia and submerged rhizomorphs) in mycelium of surface cultures of *Sphaerostilbe repens* (Botton, 1978). Complete differentiation of fruiting bodies also requires calcium in *Cyathus stercoreus* (Lu, 1973).

(5) Mycelium of *Fusarium graminearum* growing on medium containing 1.4×10^{-8} M calcium is highly branched and has wide hyphae with irregular walls and subapical balloons, unlike mycelium growing at higher concentrations of calcium (Robson, Wiebe & Trinci, 1991).

(6) Calcium ionophores increase the branching of hyphae in *Neurospora crassa* (Reissig & Kinney, 1983) and in *Achlya bisexualis* (Harold & Harold, 1986). For the latter fungus it was shown that cytochalasins and proton ionophores have the same effect.

(7) One of the earliest physiological effects of the mating pheremones in the yeasts *Rhodosporidium toruloides* and *S. cerevisiae* is a very rapid and transient uptake of calcium, suggesting that the response appears to be involved in the mating pheromone signal transduction (Miyakawa *et al.*, 1985; Tachikawa *et al.*, 1987).

The pheromone response during the process of mating in *S. cerevisiae* is the best-studied signal transduction system in fungi (Marsh, Neiman & Herskowitz, 1991). A general model of major signal transduction in eukaryotic cells has been established (Figure 11.3). The presence of similar systems in fungi can be inferred only from scattered information. Table 11.3 lists some of the elements of the systems that have been identified as being present in a number of fungi.

In spite of the apparent strength of the case for a requirement of fungi for calcium, Youatt (1993) made the important point that many studies purporting to show such a requirement have involved the use of chelating agents such as EGTA and A23187. In the presence of these agents, growth of a range of microorganisms, including fungi, has been suppressed. Unavailability of calcium to the organism has been presumed, because of a belief that the two chelating agents were specific for calcium. But this is not so; indeed EGTA is a better chelator of iron, zinc and manganese ions than of calcium. Thus, inhibition of growth might be due to unavailability of zinc or manganese. Youatt (1993) presented much evidence for her case but did not suggest that calcium is not required. However, there is not doubt that there is a need to heed her recommendation for further experiments to clarify the issue.

There are only five determinations of the cytoplasmic calcium

Table 11.3. *A list of some of the elements of the signalling system shown in Figure 11.3 that have shown to be present in fungi. The list is by no means comprehensive, being only illustrative*

Calmodulin
Achlya ambisexualis (Suryanarayana, Thomas & Mutus, 1985)
Agaricus bisporus (Grand, Nairn & Perry, 1980)
Aspergillus niger (Muthukumar, Nickerson & Nickerson, 1987)
Blastocladiella emersonii (Gomes *et al.*, 1979)
Ceratocystis ulmi (Muthukumar, Kulkarni & Nickerson, 1985)
Coprinus lagopus (Nakamura *et al.*, 1984)
Neurospora crassa (Cox *et al.*, 1982)
Penicillium notatum (Muthukumar *et al.*, 1987)
Phycomyces blakesleeanus (Martinez-Cadena, Lucas & Gobema, 1982)
Saccharomyces cerevisiae (Ohta *et al.*, 1987)
Trichoderma lignorum (Muthukumar *et al.*, 1987) also a number of yeast species
  (Muthukumar *et al.*, 1987)

Activation of adenyl cyclase by calmodulin
Neurospora crassa (Reig *et al.*, 1984)

Activation of cAMP phosphodiesterase by calmodulin
Neurospora crassa (Téllez-Iñón *et al.*, 1985)

Ca^{2+}, calmodulin-dependent protein kinases
Aspergillus nidulans (Bartelt *et al.*, 1988)
Neurospora crassa (Techel *et al.*, 1990; van Tuinen *et al.*, 1984)
Saccharomyces cerevisiae (Londesborough & Nuutinen, 1987)

cAMP-dependent protein kinase
Coprinus cinereus (Swamy, Uno & Ishikawa, 1985)
Metarhizium anisopliae (St Leger *et al.*, 1990)
Neurospora crassa (Powers & Pall, 1986)
Saccharomyces cerevisiae (Behrens & Mazon, 1988; Cherry *et al.*, 1989;
  Kato *et al.*, 1989)

cAMP
(Pall, 1981)

Phospholipids as part of the signal transduction system in *S. cerevisiae*
(Nikoloff & Henry, 1991; White *et al.*, 1991)

concentration in fungi (Table 11.4). The values obtained by Miller & Sanders (1989) and Ohya *et al.* (1991) with wild-type strains are in the region one would expect on the basis of studies on animal and green plant cells (Williamson & Monck, 1989; Evans, Briars & Williams, 1991). The value for the *cls* mutant, which loses its viability when an attempt is made to grow the cells in medium containing 100 mM calcium chloride, is feasible. The value for *S. cerevisiae* obtained with DEAE-fractionation

Table 11.4. *Values for the cytoplasmic calcium concentration in fungi*

Fungus	Concentration (M)	Method	Reference
Neurospora crassa	9.2×10^{-8}	Ca^{2+}-selective electrode	Miller & Sanders (1989)
Saccharomyces cerevisiae	$0.58 \times 10^{-5} - 2.3 \times 10^{-5}$	DEAE-dextran fractionation	Eilam et al. (1985)
	34×10^{-8}	Fluorescent dye indo-1	Halachmi & Eilam (1989)
	15.0×10^{-8}	Fluorescent dye fura-2	Ohya et al. (1991)
S. cerevisiae cls mutant	90.0×10^{-8}	Fluorescent dye fura-2	Ohya et al. (1991)

must be suspect, for reasons outlined in Chapter 4. The value obtained with the fluorescent dye indo-1 is much higher than one would expect. This may be due to the problems of dye localisation within the cell and autofluorescence. Further determinations of cytoplasmic calcium concentrations in fungi are much needed.

The concentration of calcium in the cytoplasm of a fungus is a function of transport processes occurring at the various membranes and the affinity of molecules in the cytoplasm itself to bind calcium. The presence of a calcium/proton antiport at the tonoplast has been considered in Chapter 4. This antiport, driven by the proton-motive force acting across the tonoplast, brings about accumulation of calcium in the vacuole.

Stroobant & Scarborough (1979) showed that functionally inverted plasma membrane vesicles from *N. crassa* are able to accumulate calcium. Uptake of calcium is stimulated by the permeant anion SCN^- to a degree that varies reciprocally with the ability of the anion to dissipate the membrane potential. Conversely, uptake is inhibited by potassium in the presence of nigericin, an ionophore that facilitates the exchange of the cation for protons. The proton conductor CCCP markedly inhibits uptake. Inhibition of the H^+-ATPase also inhibits uptake of calcium. All this information was believed to point to a Ca^{2+}/H^+ antiport at the plasma membrane, pumping calcium out of the hypha.

Miller & Sanders (1989) pointed out that if a calcium concentration of around 100 nM were to be maintained in the cytoplasm, there would be a large electrochemical force tending to drive the ion into a fungus. This means that maintenance of the gradient demands considerable energy expenditure. The electrochemical driving force ($\Delta\mu_{Ca^{2+}}$) across the plasma membrane is given by:

$$\Delta\mu_{Ca^{2+}} = 2\Delta\Psi + 58.7(pCa_o - pCa_c) \tag{11.1}$$

where $\Delta\Psi$ is the membrane potential difference and pCa is strictly analogous to pH with o being the external medium and c the cytoplasm. Given an external calcium concentration of 1 mM and a potential difference of -200 mV (inside negative) across the plasma membrane (see Chapter 1) the electrochemical driving force across the membrane is -637 mV.

Miller & Sanders (1989) went on to consider the free energy of calcium efflux if it were brought about by:

(1) A plasma membrane ATPase pumping one calcium ion out of the cell per ATP hydrolysed.

(2) An $H+/Ca^{2+}$ antiport as proposed by Stroobant & Scarborough (1979).

With regard to (1) the free energy relationship is given by

$$\frac{\Delta G'}{F} = \frac{\Delta G^0_{ATP}}{F} + 58.7\left[\log\left(\frac{[ADP][P_i]}{[ATP]}\right) + pCa_c - pCa_o\right] - 2\Delta\Psi \quad (11.2)$$

where $\Delta G'/F$ and $\Delta G^0_{ATF}/F$ are, respectively, the free energy change for the Ca^{2+}-ATPase and the standard free energy for hydrolysis of ATP, F being the Faraday constant.

For (2), the free energy relationship is given by:

$$\Delta G' = (n-2)\Delta\Psi + 58.7[n(pH_o - pH_c) + pCa_c - pCa_o] \quad (11.3)$$

where n is the H^+/Ca^{2+} stoichiometry.

Table 11.5 gives the values for $\Delta G'$ (in mV) for the two systems. It is clear that a Ca^{2+}-ATPase would not be able to maintain Ca_c at the observed value, whereas the H^+/Ca^{2+} antiport must have an H^+/Ca^{2+} stoichiometry of at least 3 at pH_o 5.8 but of 4 at pH_o 8.4. The latter would represent the highest stoichiometric ratio yet observed for fungal, plant or animal transport systems.

Miller & Sanders (1989) proposed a calcium efflux brought about by a hybrid system, namely an ATP-powered H^+/Ca^{2+} exchange (Figure 11.4). Such an ATPase was proposed by Rasi-Caldogno, Pugliarello & De Michaelis (1987) from their studies of radish plasma membrane vesicles (for a review of active calcium transport by green plant cell membranes, see Evans et al., 1991). Miller & Sanders (1989) and Miller et al. (1990) provided experimental evidence for their contention as follows.

First, the free energy equation for the proposed system is

$$\frac{\Delta G'}{F} = \frac{\Delta G^0_{ATP}}{F} + 58.7\left[\log\left(\frac{[ADP][P_i]}{[ATP]}\right) + n(pH_o - pH_c) + pCa_c + pCa_o\right]$$
$$+ (n-2)\Delta\Psi \quad (11.4)$$

On the basis of this equation, Miller & Sanders (1989) predicted that even in the presence of cyanide, using known data in which the internal ATP concentration was reduced by cyanide from 2.5 to 0.5 mM, the system could still function at $pH_o = 5.8$, but not at $pH_o = 8.4$. This was shown to

Table 11.5. *Calculated $\Delta G'$ values for simple plasma membrane calcium efflux systems*

Gibbs free energy changes for reactions catalysed by a Ca^{2+}-ATPase (1 Ca^{2+}:1 ATP stoichiometry) and H^+/Ca^{2+} antiport (H^+:Ca^{2+} stoichiometry: n) are calculated according to equations (11.2) and (11.3) (see text), respectively, from the measured values of pCa_c, pCa_o for two different values of pH_o. Additional data were as follows: $pH_c = 7.2$ ($pH_o = 5.8$) or 7.4 ($pH_o = 8.4$) (Sanders & Slayman, 1982); $[ATP] = 2.5$ mM; $[ADP] = 0.5$ mM (Slayman, 1973); $[P_i] = 10$ mM (Harold, 1962b); $\Delta\Psi = -200$ mV ($pH_o = 5.8$) or -250 mV ($pH_o = 8.4$); $\Delta G^0_{ATP}/F = -300$ mV ($pH_c = 7.2$) or -308 mV ($pH_c = 7.4$) (Rosing & Slater, 1972).

	$\Delta G'$ (mV)	
	$pH_o = 5.8$	$pH_o = 8.4$
Ca^{2+}-ATPase	+173	+271
$n\ H^+/Ca^{2+}$ antiport		
$n = 1$	+355	+546
$n = 2$	+73	+355
$n = 3$	−209	+163
$n = 4$	−491	−28

From Miller & Sanders (1989).

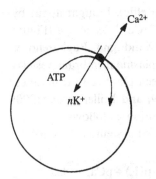

Figure 11.4. Model for an H^+–Ca^{2+}-ATPase at the plasma membrane of *Neurospora crassa*. (From Miller & Sanders, 1989.)

be true experimentally (Miller *et al.*, 1990). It was suggested that, at the latter pH, taking into account the value for influx and the binding capacity of the cytoplasmic proteins, though the cytoplasmic concentration of free calcium rises, it is not as high as it might be because of transport of calcium into the vacuole. This idea was supported by the observation that the protonophore FCCP can disrupt calcium homeostasis by acting (presumably) at the tonoplast.

As yet, the only biochemical evidence for a plasma membrane Ca^{2+}-ATPase is the isolation of such an enzyme from the two yeasts *Rhodosporidium toruloides*, a heterobasidiomycetous species, and *S. cerevisiae*. The enzyme was solubilised and purified by calmodulin affinity chromatography. In the case of *R. toruloides* the enzyme could be inhibited by physiological concentrations of rhodotorucine *A*, a lipopeptide mating pheromone (Miyakawa *et al.*, 1987) and in the case of the *S. cerevisiae* enzyme there was inhibition by physiological concentrations of the α-factor (Hiraga *et al.*, 1991). Such inhibition could explain the increase in the cytoplasmic calcium concentration that is induced by the α-factor and referred to earlier. A transient increase in calcium influx has been observed in *S. cerevisiae* in response to glucose (Eilam, Othman & Halachmi, 1990). Part of the increase appears to be due to intracellular acidification. Part, but not all, may be due to cyclic AMP, which shows a transient increase with glucose. Whether or not this relates to any switching off of a Ca^{2+}-ATPase acting as an efflux pump remains to be seen. The increased influx is not caused by an increase in the membrane potential, which is thought to drive uptake of calcium in non-energised cells (Eilam & Othman, 1990) (see below).

There are at least two other Ca^{2+}-ATPases in *S. cerevisiae* and one in *Schizosaccharomyces pombe*. In the former case, two genes *PMR1* (formerly called *SSC1*) and *PMR2* for P-type ATPases have been sequenced and the amino acid composition of the proteins encoded showed that both are very similar to Ca^{2+}-ATPases (Rudolph *et al.*, 1989). It was argued that the more intensively studied PMR1 functions as a calcium pump affecting transit of proteins through the secretory pathway. The Ca^{2+}-ATPase in *Schizosaccharomyces pombe* has also been identified from studies on its gene (*cta3*) (Ghislain *et al.*, 1990). The sequence of the encoded protein is in keeping with it being a Ca^{2+}-ATPase. Also, it contains a segment, just terminal to the phosphorylation site, that is analogous to the phospholambdan-binding domain present in the endo-(sarco-)plasmic reticulum pump. This suggests that the pump is located in some non-vacuolar organelle analogous to the endo-(sarco-)plasm in muscle or calciosomes in non-muscle cells in animals.

Thus far, I have been considering mechanisms responsible for efflux of calcium from the cytoplasm. With respect to influx of calcium into fungi, there is the evidence from studies on *S. cerevisiae* that calcium enters the cell by a uniport driven by the membrane potential, some of the data for

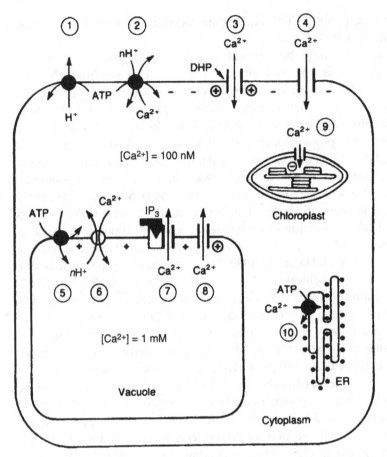

Figure 11.5. The identity and location of transport systems involved in cytosolic Ca^{2+} homeostasis in an idealised plant cell. Plasma membrane transport systems (identified as 1 to 4) are as follows: 1, primary electrogenic H^+-ATPase, which generates a highly negative membrane potential (typically $-150\,mV$); 2, primary Ca^{2+}-ATPase, which energises export of cytosolic Ca^{2+} in exchange for extracellular H^+; 3, dihydropyridine (DHP)-sensitive Ca^{2+} channel, activated by depolarising voltages ($+$); 4, other classes of Ca^{2+}-permeable channels (including a verapamil-sensitive channel) that have not been characterised in detail. Systems 3 and 4 catalyse dissipative entry of Ca^{2+} into the cytosol in accordance with the electrical and chemical potential gradients set up by transport systems 1 and 2, respectively. Tonoplast transport systems (identified as 5 to 8) are as follows: 5, primary electrogenic H^+-ATPase, which generates a vacuole-positive membrane potential between $+20$ and $+50\,mV$ and an inside-acid pH gradient of 2 units; 6, Ca^{2+}/H^+ exchanger, responsible for accumulation of Ca^{2+} within the vacuole, and driven by the chemical and electrical components of the proton-motive force; 7, inositol 1,4,5-trisphosphate (IP_3)-gated Ca^{2+} channel; 8, voltage-operated Ca^{2+} channel, opened by positive shifts in transtonoplast electrical potential ($+$). Systems 7 and 8 catalyse Ca^{2+} release into the cytosol, down the transtonoplast

which have been given already (see pp. 344–5). Other evidence comes from a study by Eilam & Chernichovsky (1987). They showed that the initial uptake after 20 s was only into the cytoplasm and therefore could be used to assess transport across the plasma membrane. With cells depleted of ATP by treatment with antimycin A and 2-deoxy-D-glucose, the effect of membrane potential could be investigated by preincubating the cells with varying concentrations of trifluoperazine. In this way cells with varying membrane potentials ($\Delta\Psi$) could be produced. Such cells had substantial calcium influx. Below a threshold value of -69.5 mV, the logarithms of the initial rates were linear with respect to $\Delta\Psi$. It was suggested that calcium influx is driven by the membrane potential via channels that open when $\Delta\Psi$ is below a threshold value. However, in cells undergoing metabolism it is possible that the influx of calcium is not solely a function of membrane potential (see above; Eilam & Othman, 1990).

Calcium-selective ion channels (as well as broadly selective anion channels) have been found to be present in *Blastocladiella emersonii* (Caldwell, van Brunt & Harold, 1986). These can be demonstrated by injection of a depolarising current into the vegetative cells, when action potentials appear that are abolished by the absence of calcium (or strontium) from the medium.

The data for the membrane processes involved in the movement of calcium into and out of the cytoplasm in fungi are fragmentary. More research is needed to give an improved picture. Though much care is needed in taking a comparative approach, it is probable that the eventual picture established with respect to calcium homeostasis in fungi will be rather similar to that being suggested for green plant cells (Figure 11.5).

One aspect of calcium metabolism that deserves note is the production of calcium oxalate. Production of oxalate in culture has been demonstrated for *Aspergillus niger* (Cleland & Johnson, 1955), *Botrytis cinerea* (Gentile, 1954), *Sclerotinia sclerotiorum* (Maxwell & Lumsden, 1970) and *Sclerotium rolfsii* (Punja & Jenkins, 1984) amongst others. Crystals of calcium oxalate have been shown to be associated with the walls of aerial hyphae of *Agaricus bisporus* (Whitney & Arnott, 1987), apothecia of *Dasyscypha*

---

electrochemical gradient for Ca^{2+}. Ca^{2+} transport systems at other endomembranes are as follows: 9, Ca^{2+} channel in the envelope membrane of chloroplasts, facilitating Ca^{2+} uptake in response to an inside-negative membrane potential developing during photosynthesis; 10, primary Ca^{2+}-ATPase at the endoplasmic reticulum (ER) operating in the direction of Ca^{2+} accumulation within the compartment. (From Johannes, Brosnan & Sanders, 1991.)

capitata (Horner, Tiffany & Cody, 1983), sporangiophores and sporangia of *Gibertella persicaria* (Whitney & Arnott, 1986, 1988), spines on sporangia of *Mucor plumbeus* and *Cunninghamella echinulata* (Jones, McHardy & Wilson, 1976), zygophores and sporangiophores of *Mucor mucedo* (Urbanus, van den Ende & Koch, 1978) and vegetative hyphae of *Sclerotium rolfsii* (Punja & Jenkins, 1984) and *Serpula lacrymans* (Bech-Andersen, 1985). Malajczuk & Cromack (1982) have demonstrated the extensive formation of crystals of calcium oxalate in the fungal mantle of ectomycorrhizal roots of *Pinus radiata* (symbiont *Rhizopogon luteolus*) and *Eucalyptus marginata* (symbiont unknown).

Oxalate production is almost certainly widespread in fungi (see Foster, 1949; Hodgkinson, 1977; Horner, Tiffany & Cody, 1983). The works cited represent the most recent studies; those on the location of oxalate associated with walls involve physical methods to attempt to characterise properly the crystals being observed as calcium oxalate. There may be a number of pathways of oxalate biosynthesis depending on the organism and cultural conditions (see Chapter 12). In *Aspergillus niger*, there appear to be two mechanisms: one involves the splitting of oxaloacetate to form oxalate (Hayaishi *et al.*, 1956), the oxaloacetate being formed by the citric acid cycle; in the other, the carboxylation of pyruvate in the cytoplasm is catalysed by a constitutive pyruvate carboxylase (Kubicek *et al.*, 1988). In *Sclerotium rolfsii*, oxalate appears to be formed by the oxidation of glycoxalate (Maxwell & Bateman, 1968; Balmforth & Thomson, 1984).

While there must be uncertainty about how oxalate is produced by almost all fungi, there is little doubt that production of oxalate is probably ecologically significant. Cromack *et al.* (1979) have shown that mycelial mats of the ectomycorrhizal basidiomycete *Hysterangium crassum* accumulate 20 times more calcium oxalate in the soil A-horizon (the uppermost developed layer) than in adjacent uncolonised soil. Production of oxalate in this way may be important in the weathering of soils. Hintikka & Naykki (1967) and Fisher (1972) found increased weathering loss of iron and phosphorus from the upper soil profile of forest soils colonised by basidiomycete fungi, and the latter author also found greater depletion of both aluminium, calcium and magnesium. All the authors believed increased organic acid production by the fungi could be responsible. Graustein, Cromack & Sollins (1977) have argued the case further.

With respect to the physiology of fungi, the production of calcium oxalate has been suggested to be important in a number of ways:

(1) Buller (1934) suggested that it provides a hydrophobic coating to aerial structures such as sporangia of *Pilobolus*.

(2) *Sclerotium cepivorum* and *S. rolfsii*, known to produce oxalate in culture (see above), which are pathogens of a number of crops, produce oxalate in association with pectin-degrading enzymes (Bateman & Beer, 1965; Stone & Armentrout, 1985). Bateman & Beer (1965) postulated that oxalic acid both lowered the pH of the infected tissue to a level optimal for polygalacturonase activity on the green plant pectic material and removed calcium that might inhibit enzyme activity. Brady *et al.* (1985) have provided evidence for the latter possibility.

(3) Sollins *et al.* (1981) have suggested that production of oxalic acid may be important to regulate cytoplasmic pH. In assessing this role for oxalate, one needs to remember that production of calcium oxalate releases hydrogen ions. But if this occurs outside the cell, movement of undissociated oxalic acid across the plasma membrane may be a way of removing hydrogen ions from the cytoplasm.

(4) Jennings (1984*b*) has suggested that prevalence of calcium oxalate in the walls of aerial hyphae or organs may be a way of removing calcium from the cytoplasm brought from mycelium in the substratum by the pressure-driven flow of solution (see Chapter 14).

(5) Whitney & Arnott (1986) have argued for a role for oxalate as a structural material in aerial hyphae.

(6) The production of oxalate may be a means of reducing the free calcium concentration of the medium to a level commensurate with that required inside the cell to trigger differentiation (Whitney & Arnott, 1988).

(7) The role of oxalate in wood decay was discussed in Chapter 5.

It can be seen from the above that the formation of calcium oxalate may be significant in calcium homeostasis in fungi. However, there is much surmise and not much experimental support for what has been suggested. Nevertheless the ideas put forward merit further investigation.

Compounds binding multivalent cations

There is a large range of compounds that can act as ligands for multivalent cations, from small organic molecules such as citric acid to polymeric material such as the components of the cell wall (see Chapter 3). Here I focus on those soluble compounds for which there seems to be a fairly specific biological role. An example of the compounds that I have in mind is calmodulin, referred to in the previous section. Most of the emphasis here is on those class B/borderline 'heavy metals'[1] (Nieboer & Richardson, 1980) required for cellular functioning but which can be toxic in excess, namely (in order of increasing atomic number) nickel, copper and zinc. This group of metals has a specific gravity greater than 4 and they form complexes primarily with ligands containing nitrogen and sulphur centres rather than with oxygen. The group also includes the elements silver, cadmium, mercury and lead, whose significance for living organisms has increased with human beings' use of these metals, such that their concentration in the environment away from where they were deposited in geological time has increased sometimes to toxic levels.

Metallothioneins

These are metal-binding polypeptides that have the following features (Rauser, 1990): low molecular mass; high metal content; high cysteine (Cys) content with an absence of aromatic amino acids and histidine; and abundance of Cys-X-Cys sequences, where X is an amino acid other than Cys; and spectroscopic features characteristic of metal thiolates and metal thiolate clusters. There are a number of reviews on metallothioneins in fungi and green plants (Robinson & Jackson, 1986; Butt & Ecker, 1987; Tomsett & Thurman, 1988; Rauser, 1990; Steffens, 1990; Mehra & Winge, 1991; Tomsett, 1993). These reviews should be consulted, particularly about the structure of the molecules and the details of their genetics.

Metallothioneins can be classified as follows (Fowler *et al.*, 1987):

Class I. Polypeptides with the locations of cysteine closely related to those in equine renal metallothionein, the first one to be discovered by Margoshes & Vallee (1957).
Class II. Polypeptides with locations of cysteine only distantly related to those in equine renal metallothionein.
Class III. Non-translationally synthesised metal thiolate polypeptides.

Classes I and II are the products of genes, synthesised by transcription and translocation via an mRNA template. Class III metallothioneins are aggregates of a heterogeneous population of polypeptides joined by γ-carboxyamide linkages. These metallothioneins have been named cadystins, γ-glutamyl peptide (Casterline & Barnett, 1977; Kondo *et al.*, 1983), phytochelatins (Grill, Winnaker & Zenk, 1985), poly-(γ-glutamylcysteinyl) glycines (Jackson *et al.*, 1987) and Cd-peptide (Reese

---

[1] Class A (oxygen seeking), class B (nitrogen/sulphur) and borderline (intermediate).

& Wagner, 1987). None of these terms is really satisfactory. Phytochelatin is the one generally preferred (Steffens, 1990). But the presence of Class III metallothioneins in fungi as well as in green plants suggests that 'phytochelatin', the term implying presence only in the latter organisms, is now outmoded.

Both Class I and Class II metallothioneins are found in fungi. There are two well-characterised Class I compounds: in *Agaricus bisporus* (Münger & Lerch, 1985) and in *Neurospora crassa* (Lerch, 1980; Beltramini & Lerch, 1981, 1982, 1983; Münger, Germann & Lerch, 1985, 1987; Münger & Lerch, 1985; Malikayil, Lerch & Armitage, 1989). Both are copper specific. In the *N. crassa* metallothionein it appears that all of the cysteines are ligated to form Cu(I) complexes, with six molecules of copper per molecule of protein. Transcription of the *MT* gene in *N. crassa* has been shown to be induced only by copper. Also copper-resistant strains of the fungus do not have an increased *MT* gene copy number.

A Class II metallothionein has been found in *S. cerevisiae* (Prinz & Weser, 1975) and extensively studied (Butt & Ecker, 1987; Fogel, Welch & Maloney, 1988; Mehra & Winge, 1991). The protein sequence was deduced from the sequence of the gene *CUP1*, which had been found to be responsible for high levels of copper resistance in the yeast (Brenes-Pomales, Lindegren & Lindegren, 1955). The basic repeat unit of the *CUP1* gene codes for two polypeptides, the metallothionein and a protein of 246 amino acid residues whose function is as yet unknown (Karin *et al.*, 1984). Regulation of *CUP1* gene expression appears to be mediated only by copper (Butt *et al.*, 1984). In resistant strains of *S. cerevisiae* there are multiple copies of the *CUP1* locus tandemly repeated with remarkable fidelity; copper-sensitive (wild-type) strains contain only a single copy (Fogel & Welch, 1982; Fogel *et al.*, 1983; Karin *et al.*, 1984). It has been proposed by Butt & Ecker (1987) that the system has many of the right qualities for cells of *S. cerevisiae* to be used for metal recovery under conditions in which other nutrients are optimal for growth.

A most interesting development has been the genetic investigation of a cadmium-resistant strain of *S. cerevisiae* (301N) (Tohoyama *et al.*, 1992). This strain produces a cadmium metallothionein with the same characteristics as the copper metallothionein encoded by *CUP1*. The sequence for the structural gene for the cadmium metallothionein resembles that of *CUP1*. Furthermore, the gene may be amplified as repeating units in both cadmium- and copper-resistant strains. Nevertheless when the gene was used to transform wild-type cells, the resulting transformants were resistant to copper but not to cadmium. It seems that *CUP1* is amplified in both cadmium- and copper-resistant cells. However, the cadmium-specific effect seems to be restricted to the cadmium-resistant strain. There is support for these observations from those of Jeyaprakash, Welch & Fogel (1991), who found that a *CUP1*-carrying plasmid enhanced, under certain conditions, the cadmium resistance induced by cadmium. Interestingly, when 301N is cultured for 48 h in the presence of 1 mM copper, its resistance to copper is enhanced but that to cadmium is depressed (Inouhe *et al.*, 1991).

Class III metallothioneins were first found in *Schizosaccharomyces pombe* by Murasugi, Wada & Hayashi (1981*a, b*) when this yeast was grown in the presence of high concentrations of cadmium. Two cadmium-binding peptides are synthesised under these conditions (Hayashi, Nakagawa & Murasugi, 1986). They have identical amino acid composition, having a basic peptide with L-glutamic acid, L-cysteine and glycine in the proportions 3:3:1 and the glutamic acid-cysteine linkage being through the carboxyl group of the γ-carbon of glutamic acid. One of the cadmium-binding peptides (Cd-BP1) contains 4 mol of the basic peptide, 6 mol of cadmium and 1 mol of labile sulphide and has a molecular mass of 4000 Da; the other peptide (Cd-PB2) contains 2 mol of the basic peptide, 2 mol of cadmium and has a molecular mass of 1800 Da (Figure 11.6). It is now known that the basic peptide can form a polymeric series $[\gamma\text{-Glu-Cys}]_n$Gly, where $n = 2$ to 8, the smaller molecules being synthesised in the earlier stages of growth in the presence of cadmium and the larger molecules at the later stages (Grill, Winnacker & Zenk, 1986; Hayashi *et al.*, 1987). Cd-BP1 becomes the major form also at the later stages. Copper triggers the synthesis of a similar polymeric series of basic peptides (Mehra & Winge, 1988). It needs to be said that interpretation of some of the studies on Class III metallothioneins in fungi is not easy because of the different assay procedures and experimental designs used (Mehra & Winge, 1991).

Candida glabrata is very interesting because it interacts differently with cadmium and copper (Mehra *et al.*, 1988). Cadmium stimulates the production of $(\gamma\text{-Glu-Cys})_n$Gly peptides (Figure 11.7), whereas copper induces the synthesis of a family of metallothioneins of the Class I/II variety that exhibits very little sequence identity with any other known metallothioneins. Within the family there are two subfamilies that themselves have very limited sequence identity (Mehra *et al.*, 1989). The Class III metallothioneins can, like those of *Schizosaccharomyces pombe*, contain labile sulphur.

The sulphide containing Cd-((γ-Glu-Cys)$_n$Gly) complexes are chemically

Cd-BP1

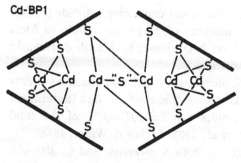

Cd-BP2

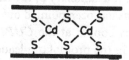

Figure 11.6. Proposed models for Cd-PB1 and Cd-BP2 from *Schizosaccharomyces pombe*. The unit peptide (γ-Glu-Cys-γ-Glu-Cys-γ-Glu-Cys-Gly) is shown as a solid bar. Sulphydryl groups of cysteine residues in cadystin bind Cd atoms in Cd-PB1 in cooperation with labile sulphide ('S'), thus increasing the molar ratio Cd:Cys. (From Hayashi *et al.*, 1986.)

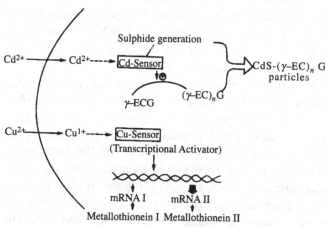

Figure 11.7. Dual metal resistance pathways in *Candida glabrata*. Cadmium salts trigger the biosynthesis of (γ-Glu-Cys)$_n$Gly((γ-EC)$_n$G) peptide complexes that incorporate sulphide ions to form CdS crystallites. The enzyme involved in (γ-Glu-Cys)$_n$Gly peptide biosynthesis appears to be the cellular Cd(II) sensor regulating Cd resistance. Copper salts induce the biosynthesis of a family of metallothionein (MT) molecules via an intracellular sensor. Only minimal amounts of the MTI molecule are synthesized relative to MTII(s). (From Mehra, R. K. & Winge, D. R. (1991), *Journal of Cellular Biochemistry*, Copyright © 1991, reprinted by permission of John Wiley & Sons, Inc.)

more stable than Cd-peptide complexes not containing sulphide (Mehra & Winge, 1991). Significantly, mutants of *S. pombe* unable to form sulphide-containing complexes have been shown by Mutoh & Hayashi (1988) to be cadmium hypersensitive.

It should be noted that both *S. pombe* and *C. glabrata* produce variants of the basic peptide of the Class III metallothioneins that lack the terminal glycine. In *S. pombe*, such peptides can be 10%–20% of the total metallothionein peptide (Mehra *et al.*, 1988; Mehra & Winge, 1988).

For metallothionein production in both *S. cerevisiae* and *C. glabrata*, there is now emerging a picture of how the processes are regulated, the most detailed being that for former fungus (for a summary of the research, see Mehra & Winge, 1991). In *S. cerevisiae*, as has been pointed out above, the process is transcriptionally regulated by copper at the *CUP1* locus. An increase in the concentration of copper in the protoplasm leads to an activation of a DNA-binding protein, which in turn brings about the accelerated transcription of the *CUP1* locus. The accelerated synthesis of metallothionein binds the copper and shuts off activation of the transcription factor and thus that of the *CUP1* locus. Thus, there is a feedback regulatory loop (Figure 11.8). How the more special case of cadmium stimulation of metallothionein synthesis in strain 301N relates to what has just been described is not clear. The possible situation in *C. glabrata* is shown diagramatically in Figure 11.7. While not proven, it seems possible that induction of both types of metallothionein is mediated by a cellular factor (Mehra & Winge, 1991).

It is clear that metallothioneins play an important role in the tolerance of fungi to metals at potentially toxic concentrations. Nevertheless, they may only be part of the machinery for dealing with toxic metals, as is indicated in the next section, which is concerned with those membrane processes that help to restrict net flux of the metal into the protoplasm. There are indications that production of metallothioneins is not absolutely essential for metal tolerance. When *C. glabrata* is grown on a medium containing yeast extract and peptone, the cells do not synthesise $(\gamma\text{-Glu-Cys})_n\text{Gly}$ or $(\gamma\text{-Glu-Cys})_n$ peptides. Nevertheless the cells are resistant to cadmium; instead the metal forms complexes with glutathione and its des-glycine derivative and with sulphide (Dameron, Smith & Winge, 1989). Tomsett (1993) has pointed out that there is a clear relationship between metal and sulphur metabolisms. Thus, some of the observed responses may have greater significance for the metabolism of the latter element than necessarily for metal tolerance.

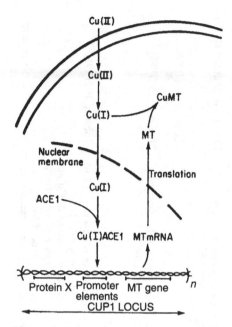

Figure 11.8. Metalloregulation of the *CUP1* locus in *Saccharomyces cerevisiae*. The CuMACE1 metalloprotein complex interacts with upstream sequences from the *CUP1* coding sequences and facilitates transcription of the metallothionein (MT) gene within the *CUP1* locus. Translation of the MT mRNA yields MT protein that buffers the cytosolic copper ion concentration. ACE1, copper-actived transcription factor 1. (From Mehra, R. K. & Winge, D. R. (1991), *Journal of Cellular Biochemistry*, Copyright © 1991, reprinted by permission of John Wiley & Sons, Inc.)

Membrane processes in relation to tolerance to metals

There have been a number of investigations showing that strains of fungi more resistant than the wild-type to the toxic effects of metals have a reduced uptake of the metals (Ross, 1993). However, many of the studies are not very illuminating. First, the status of the strains is not clear. Often there is no indication that the strains exhibit heritable resistance or have altered DNA sequences. The differences in tolerance may be due to an inducible cellular response. It has been possible therefore to 'train' *Cunninghamella blakesleeana* and *Rhizopus stolonifer* to tolerate higher levels of copper (Garcia-Toledo, Babich & Stotsky, 1985). Second, the simple measurement of an amount taken up by the fungus over a known period tells one very little. It is possible, for instance that reduced uptake might be wholly or in part due to reduced binding by the wall, with little

Table 11.6. *Transport processes involving a metal to which strains of a particular fungus are more tolerant than the wild-type strain*

Fungus	Metal	Effect on transport	Comments	Reference
Neocosmospora vasinfecta	Cd	Tolerant strain has a reduced influx but unchanged efflux		Budd (1991)
Neurospora crassa	Ni	Two strains have higher rates of uptake than wild-type; K_T values almost the same; different rates of uptake associated with differences in V_{max}	Little surface binding; uptake of nickel is inhibited by Mg, Mn, Co, Cu, Zn to different extents according to the strain	Mohan, Rudra & Sastry (1984)
	Co	Accumulation of the metal by the above three strains mirrors their accumulation of Ni	Cobalt accumulation not affected by Cu	Mohan & Sastry (1984)
	Cd	Slightly lower rate of uptake by resistant strains; also, the transport process is less sensitive to azide	Genetic basis of strains well established	Levine & Marzluf (1989)
Saccharomyces cerevisiae	Cu	Protoplasts of resistant strains take up less copper than the wild-type		Gadd et al. (1984)
	Co	Both initial binding and uptake is lower in resistant strain than in wild-type. No simple relationship between Cd content and K$^+$ efflux in resistant cells as found in wild-type	Strain may not be changed genetically	Belde et al. (1988)
	Co and Ni	Uptake of both metals less in resistant strain than in wild-type	Mg uptake is also reduced in resistant strain. The inhibitory action of Co and Ni on growth of both strains can be reduced by Mg	Joho et al. (1991)

change in the rate of plasma membrane transport. Table 11.6 provides information for those studies, which, while not necessarily free from the first criticism, do indicate what is known about membrane transport of metal ions in relation to tolerance by the fungus concerned. The information is fragmentary but it indicates that resistance does not necessarily require reduced uptake. Nevertheless given the caveats above, there is a need for detailed study of strains of known genetic provenance, focusing not only on transport across the plasma membrane but also on the ability of the fungus to produce compounds that bind the metal and the extent to which vacuolar accumulation can occur.

Penicillium ochro-chloron is a species that deserves special study because it can grow in 10 mM copper solutions, particularly at acid pH (Stokes & Lindsay, 1979; Gadd & White, 1985). Although copper is transported into the protoplasm, there is no information on how the rate compares with that of other species of *Penicillium*.

Copper tolerance of the two brown-rot wood decay fungi *Poria placenta* and *P. vaillantii* has been shown to be brought about by precipitation of copper as oxalate external to the mycelium (Rabanus, 1939; Sutter, Jones & Wälchli, 1983).

Iron

It is now well known that aerobic and facultative anaerobic microorganisms, when exposed to media of reduced iron content, produce Fe(III)-binding ligands with a very high degree of specificity for the metal. These ligands are termed siderophores, because they are 'iron bearers'. They were discovered by Francis *et al.* (1949). The first fungal siderophore to be isolated was ferrichrome produced by *Ustilago sphaerogena* (Neilands, 1952), while Hesseltine *et al.* (1952) showed that a brick-red crystalline compound produced by fungi such as *Penicillium* could replace the usual dung extract in the nutrition of *Pilobolus*. Since then, many siderophores produced by fungi have been isolated and their structures determined. The term 'siderophore' is reserved for the metal-free ligand, though in some cases the name has been given to the metal-bound ligand. When iron is present, the compound should be called a *ferric siderophore*; the term 'des-ferri-siderophore' is redundant. Information about fungal siderophores has been reviewed by Winkelmann (1991, 1992) and Winkelmann, van der Helm & Neilands (1987).

The majority of fungal siderophores are what is called the hydroxamate type. Hydroxamate groups are based on ornithine, which has been

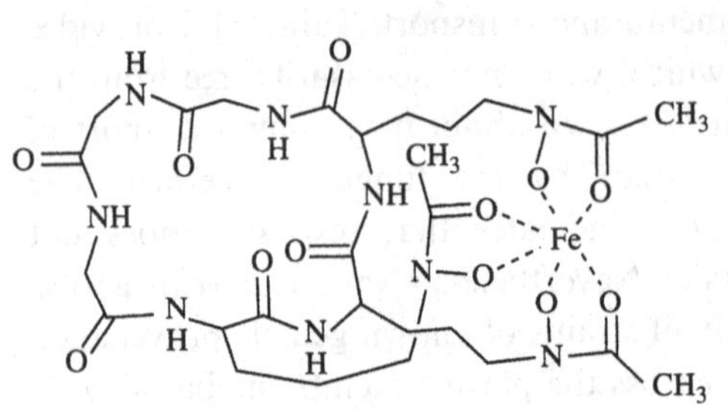

Ferrichrome

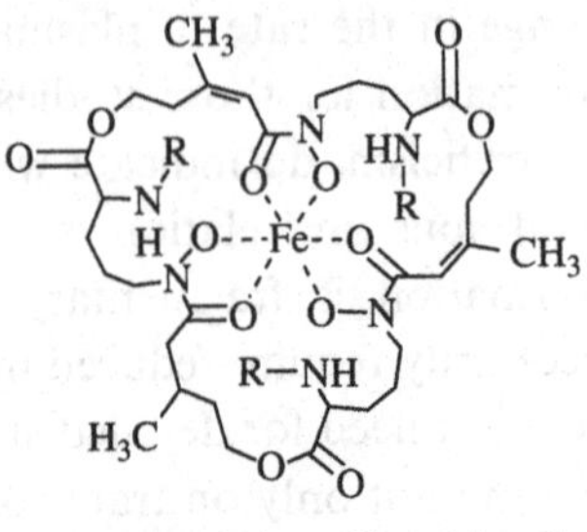

N,N',N'' -Triacetylfusarinine C
(Triacetylfusigen)

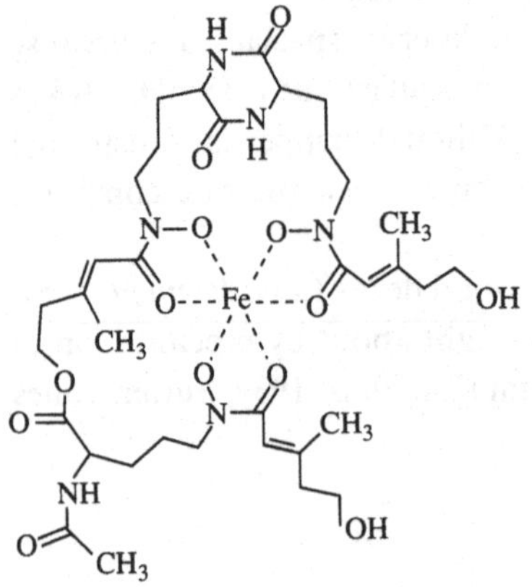

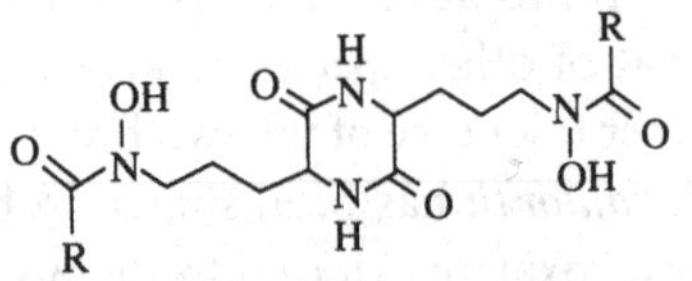

Rhodotorulic acid

Coprogen

Figure 11.9. Structures of representative members of the four hydroxamate-type siderophores found in fungi (see the text). (From Winkelmann, 1992.)

subsequently N^8-hydroxylated and N^8-acetylated. The basic reaction of the hydroxamate group with Fe(III) is given in Figure 11.9. Essentially, deprotonation leads to the formation of an anionic bidentate ligand that can form a very stable and selective complex with Fe(III). Hydroxamate siderophores may form mono-, di- or trihydroxamate iron complexes. Hexadentate trishydroxamates, in which the ratio iron:ligand is 1:1, tend to predominate in nature because of their high formation constants compared with those of monohydroxamates in which the ratio iron:ligand is 1:3. This may be so even though the formation constants of the individual ligand in the latter case may be greater than in the former. It is the location of ligands with six conveniently disposed coordinating atoms that make them more effective binding agents.

   It is routinely stated that siderophores are specific for Fe(III) (the ferrous form shows little affinity for the ligand). However, it is known that the stability of gallium–hydroxamate complexes resembles those of ferric–hydroxamate complexes (Winkelmann, 1992), although, in nature, gallium is in low abundance. Nevertheless, the ability of siderophores to complex gallium provides a means of probing their function. More important is the fact that aluminium–hydroxamate complexes are ten orders of magnitude less stable than the equivalent ferric complex.

N^8-hydroxyornithine is common to all fungal hydroxamate-type siderophores. However, the N^8-acyl residues can vary. Four major classes of fungal hydroxamate siderophore can now be recognised (Figure 11.9):

(1) Ferrichromes.
(2) Fusarinines.
(3) Coprogens.
(4) Rhodotorulic acid – this may be regarded as a subclass of (3).

A new kind of fungal siderophores has recently been discovered based on citric acid. Drechsel *et al.* (1991) showed that *Rhizopus microsporus* var. *rhizopodiformis* produces a siderophore that was called rhizoferrin and contains two citric acid residues linked via amide bonds to putrescine (butanediamine). The siderophore belongs to the complexerone type, which, though unusual in fungi, is well known in bacteria. Rhizoferrin was shown subsequently to be produced by *Mucor mucedo* (Mucoraceae), *Chaetostylum fresenii* and *Cokeromyces recurratus* (Thamnidiaceae), *Cunninghamella elegans* and *Mycotypha africana* (Thieken & Winkelmann, 1992). Thus, the Mucorales possess a class of siderophore different from those of other fungi. Interestingly, the non-storage compound ferritin has been detected in *Phycomyces blakesleeanus* (David & Easterbrook, 1971). In fungi that produce hydroxamate siderophores, it is these molecules which seem, from studies on *Aspergillus ochreaceus*, *Neurospora crassa*, *Rhodotorula minuta* and *Ustilago sphaerogena*, to act as iron storage compounds (Matzanke *et al.*, 1987b, 1988, 1990).

There is now a considerable body of information about the ability of, and the mechanisms by which, ferric siderophores enter fungi. However, not all siderophores do enter. For example, rhodotorulic acid produced by *Rhodotorula pilimanae* aids the capture of iron by the yeast cells but does not act as the vehicle for transport into them, as demonstrated by the entry of ^{56}Fe but not to any great extent that of ^{3}H when the siderophore was labelled with these isotopes (Carrano & Raymond, 1978). Gallium rhodotorulates do enter. It was proposed that in this instance the ferri-siderophore exchanges the ferric iron with a membrane-bound chelating agent that brings about transport.

As regards other siderophores, it is known that ferrichrome, coprogen and triacetylfusarinine enter the fungus with iron bound to the molecule (Emery, 1971; Winkelmann & Zähner, 1973; Adjimani & Emery, 1987). Mössbauer spectroscopic studies of ferric-coprogen uptake by *N. crassa* showed that, when absorbed, the major portion of the intracellular coprogen remains complexed with the iron, suggesting that coprogen acts not only as a siderophore but also as a storage compound (Matzanke & Winkelmann, 1981). This has been confirmed by a subsequent study using a siderophore-free mutant. Ferric-coprogen supplied to the mycelium of this mutant was absorbed, the rate of iron release being dependent upon the degree of iron starvation (Matzanke *et al.*, 1987a). Also there is evidence that *Ustilago sphaerogena* ferrichromes can be used again after the iron has been delivered (Emergy, 1971). On the other hand, extracts of a *Penicillium* sp. were found to contain an enzyme that hydrolyses the ornithine ester bond of N,N',N''-triacetylfusarinine C (Emery, 1976). The same enzyme acts also on the siderophore. A similar enzyme from *Fusarium roseum*, however, has no effect on the siderophore. The amount of the *Penicillium* enzyme extractable from the mycelium is reduced by growth in the presence of iron. It seems therefore that the enzyme is important in iron metabolism, but it is believed, in view of the inactivity of the *F. roseum* enzyme against the siderophore, that reduction of the iron has to take place before the enzyme can act. Indeed, it seems likely that the main mechanism by which iron is released from siderophores is by reduction of the metal.

The best evidence that reduction of iron is the mechanism for its release comes from work of Ernst & Winkelmann (1984). They isolated a soluble NADH-linked sideramine reductase from *N. crassa* and *Aspergillus fumigatus*. That of the former fungus was studied in most detail. The enzyme was assayed in the presence of EDTA (to trap the iron after reduction); in the cell the Fe(II) acceptor is thought to be citrate. When mycelium was grown in the presence of 10^{-5} M Fe(III), the activity of the enzyme was only 30% of that in low iron conditions. The enzyme has a pH optimum of 7.0 and is active against ferricrocin, coprogen and some other ferrichrome-type compounds. There is much less activity against ferrirubrin, coprogen B, ferrioxamine B and ferrichrome. However, the enzyme from *A. fumigatus* was active against ferrichrome and to a lesser extent against coprogen. The affinity of the enzyme from *N. crassa* does not match very well the specificity of uptake. The enzyme is more effective against ferricrocin than against coprogen but there is the reverse preference by the uptake processes. Since this appears to be the only study on an enzyme that will reduce iron in the ferric siderophore, it would be foolish at this stage to draw any firm conclusions about the significance of the substrate specificity of sideramine (siderophore) reductases.

Originally, it was thought that siderophores, with their associated iron, might diffuse across the plasma membrane in a manner similar to that of lipid-soluble ionophores. There is now plenty of evidence to the contrary. In particular, it has been shown several times that uptake is stereospecific (Winkelmann, 1979; Winkelman & Braun, 1981; Huschka *et al.*, 1985; Adjimani & Emery, 1988). From a study of this kind, it has been proposed that in *N. crassa* there are two different siderophore receptors, one recognising ferrichrome-type siderophores of a certain structure and the other coprogen (which differs from the ferrichromes both in structure and absolute configuration of the metal centre). A model of the processes involved is presented in Figure 11.10. Similar conclusions have been reached by Müller, Barclay & Raymond (1985) concerning the mechanism of transference of iron by the siderophore rhodotorulic acid from the medium into the cells of *Rhodotorula pilimanae*.

There is no doubt from such studies as those of Emery (1971) and Jalal, Love & van der Helm (1987) that siderophore uptake is dependent upon

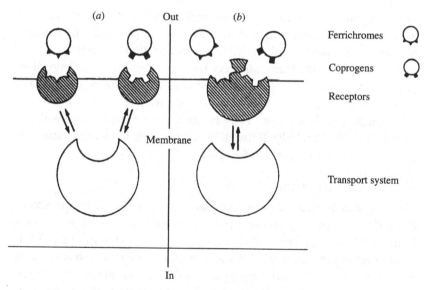

Figure 11.10. Model of siderophore transport in *Neurospora crassa* showing diferent recognition sites located (*a*) on separate receptors or (*b*) on a single receptor, both with a common transport system. (From Winkelmann & Huschka, 1987.)

metabolic energy. The exact coupling is not absolutely clear, though Huschka, Müller & Winkelmann (1983) have provided strong evidence that the membrane potential must be an important part of it, if not the sole driving force.

With respect to the regulation of transport of ferric siderophores and the production of siderophores within the fungus, very little is known, although work on bacteria may provide valuable guidance (Bagg & Neilands, 1987; Braun & Hantke, 1991). Also, there is a need for studies on iron uptake when conditions are such that siderophore production is unnecessary. With respect to iron acquisition from the soil, a fungus faces exactly the same problem as the root of a green plant, which has evolved a battery of mechanisms for enhancing the mobilisation and uptake of iron (Römheld & Marschner, 1986) (Table 11.7). As one can surmise, fungi may use similar mechanisms. At any rate, reduction of the local pH in soil by a fungus almost certainly occurs, as does the release of organic acid (Chapter 12). It is known (see Chapter 1) that fungi can possess a plasma membrane redox system but whether, as in green plants, free iron can be absorbed as Fe(II) rather than as Fe(III) remains to be seen.

Mycorrhiza

The literature on mycorrhiza and multivalent metals has increased considerably with increasing public concern about pollution, the presence of which can increase directly the concentration of metals in the environment or make them more available as a result of acid deposition. Many of the publications are concerned with the effect of a metal on the growth of the green plant partner in the presence or absence of the mycorrhizal associate. As yet, there has been little overt concern with the mechanisms that underlie those differences which have been observed. The salient facts are considered in turn.

Siderophore production

This has been demonstrated to take place in pure culture with the ericoid mycorrhizal fungi *Hymenoscyphus ericae* and *Oidiodendron griseum* (Schuler & Haselwandter, 1988; Federspiel, Schuler & Haselwandter, 1991) and the ectomycorrhizal fungi *Amanita muscaria, Cenococcum geophilum, Pisolithus tinctorius. Suillus brevipes, S. lakei, S. pinctipes, S. tomentosus* and *Boletus edulis* (which was given more intensive study (Szaniszlo *et al.*, 1981). The evidence for the role of siderophores in iron metabolism in the intact association of mycorrhizal fungus and higher plant is fragmentary,

Table 11.7. *Mechanisms for the enhanced mobilisation and uptake of iron in the rhizosphere of various plant species*

A. Nonspecific mechanisms
Root-induced pH decrease because of preferential cation uptake
Release of organic acids by the roots
Root release of photosynthates as substrate for rhizosphere microorganisms, which in turn affect pH, Fe(III) reduction, and chelator concentration (e.g. siderophores)

B. Specific mechanisms (related to the Fe nutritional status of the plants)

	Strategy I (e.g. soybean, sunflower, peanut, and most other dicots)	Strategy II (e.g., barley, wheat, oat, and probably most other grasses)
Formation of rhizodermal transfer cells	Important	Absent
Enhancement of proton release	Important	Absent
Enhancement of Fe(III) reduction at the plasma membrane	Important	Of limited importance
Enhancement of release of phenolics	Important	Of limited importance
Enhancement of release of non-proteinogenic amino acids ('phytosiderophores')	Absent as far as investigated	Of crucial importance

From Römheld & Marschner (1986).

to say the least. Certainly siderophores of fungal and also bacterial origin have been shown to provide higher plants with iron (Reid *et al.*, 1984; Bienfait, 1986, 1989; Crowley, Reid & Szaniszlo, 1988). However, there is clear evidence with respect to ericoid mycorrhiza that infected plants have significantly increased absorption of iron over uninfected plants from a range of concentrations typical of the soil solution concentration in heathland (Shaw *et al.*, 1990). The most convincing evidence of the myobiont producing siderophores is that of Cress, Johnson & Barton (1986) for the vesicular–arbuscular mycorrhizal association between the semi-arid warm season grass *Hilaria jamesii* and four *Glomus* spp. Excised mycorrhizal roots of iron-deficient plants absorbed significantly more iron than do non-infected roots. Furthermore, the mycorrhizal roots assayed positive for siderophore activity, which was not the case for non-mycorrhizal roots. There is no doubt that the relationship between siderophores and mycorrhiza is an open field for research. Essentially, a range of associations should be studied to see whether the presence of the fungus stimulates iron uptake under conditions of low iron, whether fungal siderophores are present in mycorrhizal roots, whether they are confined to the fungus or whether they move into and within the higher plant.

Heavy-metal tolerance

Bradley, Burt & Read (1982) showed unequivocally that the ericaceous plants *Calluna vulgaris*, *Rhododendron ponticum* and *Vaccinium macrocarpon*, when mycorrhizal, were able to grow with concentrations of copper as high as 1.2 mM and with zinc as high as 2.3 mM, whereas non-mycorrhizal plants failed to grow at all but the lowest concentrations. An isolated fungus, *Hymenoscyphus ericae*, grew vigorously in culture in the presence of either copper or zinc at the above concentrations. Bradley *et al.* argued that the fungus enhances tolerance by binding the metal. Only further experiments may show this to be the case.

As far as ectomycorrhiza are concerned, there are now a number of studies of the effect of infection on the uptake by, and toxicity of, metals to tree seedlings. The subject has been well reviewed by Wilkins (1991). Although care is needed in interpreting the findings, particularly those using soil, the general conclusion is that the presence of the mycorrhizal association can decrease the toxicity of nickel, copper and zinc. This has been shown in controlled experiments with seedlings of *Betula* with *Amanita*, *Paxillus* and *Scleroderma* and *Picea* and *Pinus* with *Pisolithus*, *Rhizopogon* and *Suillus*. Nevertheless there is sufficient information to indicate that the situation is not simple. Thus, when mycorrhizal associations are formed between one of *Laccaria proxima*, *Lactarius hibbardae*, *L. rufus*, and *Scleroderma flavidum* and *Betula papyrifera*, only the last association is effective in protecting the tree seedling against nickel toxicity.

One very interesting observation has been made by Denny & Wilkins (1987a–d) with the mycorrhizal association between *Betula* sp. and *Paxillus involutus*. The association enhances zinc tolerance of the higher plant. When the plants are ectomycorrhizal, the roots have a lower zinc content than those of non-mycorrhizal plants. There is no evidence of accumulation in the fungal mantle; the zinc concentration there could be lower than in the cortex of uninfected roots. By far the highest concentrations appear in the extramatrical hyphae. It was suggested that accumulation of zinc in this way reduces the concentration for transport into the higher plant root cells.

Aluminium toxicity

It has been known for many years, since the now classic studies of Rorison (1960*a*, *b*), that aluminium is an important determinant of higher plant distribution. In particular it is the soluble aluminium ion (Al^{3+}) that is the effective species and is present in soils of pH below 4.0, decreasing to virtually negligible values above pH 5.5 (Figure 11.11). Those plants capable of growing on acid soils are tolerant of aluminium. The importance of aluminium as a potentially toxic metal has been highlighted by the problem of acid rain and the probable role of aluminium in forest decline. In consequence, attention has now focused on the ability of mycorrhizal fungi to grow in the presence of aluminium and the ability of mycorrhiza to enhance the tolerance of seedlings to aluminium. Analysis of the effect of aluminium is made complicated by the fact that not only can there be a direct effect of the metal but there are important interactions with phosphorus (Foy, Chaney & White, 1978).

To date, the most satisfactory and comprehensive study of mycorrhiza and aluminium is that by Cumming (1990) and Cumming & Weinstein (1990*a–c*) on *Pinus rigida* made mycorrhizal with *Pisolithus tinctorius*. Mycorrhizal seedlings were more able to extract phosphorus from aluminium phosphate than were non-mycorrhizal seedlings. Also mycorrhizal seedlings were found to be able to grow better in the presence of aluminium than were non-mycorrhizal seedlings. However, this difference was later found to be a function of the nitrogen nutrition. The greater tolerance to aluminium occurred only if nitrate was the nitrogen source. With ammonia as the nitrogen source, there was no significant difference between the growth of non-mycorrhizal and that of mycorrhizal

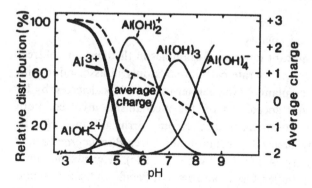

Figure 11.11. Relative distribution and average charge of soluble aluminium as a function of pH. (From Marschner, 1986.)

plants in the presence or absence of aluminium. These observations suggest further studies are needed.

The concern about the possible ability of mycorrhiza to enhance aluminium tolerance of trees has triggered investigations of the effect of aluminium on ectomycorrhizal fungi in culture (Väre, 1990; Väre & Markkola, 1990; Browning & Hutchinson, 1991; Jongbloed & Borst-Pauwels, 1992; Jongbloed, Tosserams & Borst-Pauwels, 1992). The results to date are preliminary and it is not yet possible to provide a coherent picture. There is variation in the response to aluminium. There seem to be interactions with other cations, particularly calcium, and there is also, as one would expect, an interaction with phosphorus due to complexing of phosphate by aluminium. Interaction of the latter kind can occur within the fungus as well as in the medium. Väre (1990) has shown that there can be abundant formation of aluminium polyphosphate granules in hyphae of *Suillus variegatus*. Jongbloed *et al.* (1992) have shown, after correcting for any decrease of external phosphate complexed with aluminium, that for some fungi there can be enhanced phosphate uptake due to reduction of the hyphal surface potential.

Reducing activity at the root surface

Following up a chance observation, Cairney & Ashford (1989) have shown that mycorrhizal roots of the *Eucalyptus pilularis–Pisolithus tinctorius* association are capable of reducing silver bromide, ferric EDTA and a higher oxide of manganese, while non-mycorrhizal plants can not. A range of mycorrhizal fungi in culture including *Cenococcum geophilum*, *Hebeloma crustuliniforme*, *Lacaria bicolor*, *Paxillus tinctorius*, *Suillus granulatus* and *Hymenoscyphus ericae* were screened for their ability to reduce manganese. Only *H. crustuliniforme* showed no activity. These findings may well be very relevant to iron uptake by fungi along the lines indicated earlier.

'Symbiocalcicole' woody plants

There are now a significant number of examples in the literature of trees or dwarf shrubs that can tolerate calcium soils only in association with mycorrhizal fungi. The plants have been termed 'symbiocalcicoles' by La Peyrie (1990), who has speculated on the mechanisms involved. Both vesicular–arbuscular mycorrhizal and ectomycorrhizal fungi can be involved. The ectomycorrhizal association of the tree *Pisonia grandis* may belong to this category (Ashford & Allaway, 1985). It grows on small islands in the western Indian Ocean and eastern Pacific and can be found on soil where there is a permanently high calcium level from the coral as well as a high input of organic nitrogen and phosphorus (guano).

12
Organic acids

Introduction

Organic acids are important in fungal nutrition because:

(1) They can be a source of combined carbon for growth.
(2) They can contribute to the internal osmotic potential.
(3) They can contribute to the internal charge balance.
(4) They may be involved in the control of internal pH.
(5) They can be involved in the loss of protons across the plasma membrane.
(6) It has been proposed that the fixation of carbon dioxide into organic acids may be part of the battery of processes allowing a particular species to grow oligotrophically.
(7) The production of oxalic acid appears to be important in assisting the decay of wood by fungi (see Chapter 5).
(8) Oxalic acid may also be important in reducing the inhibitory effect of too much calcium in the external environment (see Chapter 11).

As well, production of various organic acids, particularly by *Aspergillus* and *Candida* spp., is commercially important (Miall, 1978; Mattey, 1992).

Transport

Table 12.1 gives details of those carboxylic acid transport systems that are known to be present in fungi. Except for succinate transport in *Neurospora crassa*, all the information is for yeasts. The interest in the ability of yeasts to transport carboxylic acids across the plasma membrane is two-fold. First, the ability to use, for growth, carboxylic acids of the citric acid cycle is used as a taxonomic discriminator. Barnett & Kornberg (1960) examined a number of yeast species, including strains of the same

377

Table 12.1. *Properties of the known organic acid transport systems in fungi*

Acid	Organism	Glucose repression[a]	Type of transport	K_T (mM)[b]	V_{max} (nmol s^{-1} (mg dry wt)$^{-1}$)	Other acids transported[c]	Acids not transported	Diffusion of undissociated acid	Reference
L-Lactic	Saccharomyces cerevisiae	?	Electroneutral proton/anion symport			D-Lactate			Eddy & Hopkins (1985)
L-Lactic	Candida utilis	Yes (not present with ethanol or glycerol)	Proton/anion symport	0.06 (pH 5.5)		Pyruvate (0.03) Propionate (0.05) D-Lactate (0.06) Acetate (0.1)		Yes	Leão & van Uden (1986)
L-Lactic	Saccharomyces cerevisiae	Yes (not present with ethanol or glycerol)	Electroneutral proton/anion symport	0.13 (pH 5.5)	0.34	Acetate (0.05) Proprionate (0.09) D-Lactate (0.13) Pyruvate (0.34)	Fumaric	Yes	Cássio et al. (1987)
L-Lactic	Kluyveromyces marxianus	Partially (less so with glycerol, almost completely with ethanol)	Monocarboxylate uniport (loose coupling with bicarbonate?)	0.42 (irrespective of carbon source for growth) (pH 5.4)	2.7×10^4 (lactic acid-grown cells)			Not ruled out	Fonseca, Spencer-Martins & van Uden (1991)
L-Malic	Schizosaccharomyces pombe	No (constitutive)		3.0 (pH 3.5)	0.66 (on protein basis)	Maleate? Malonate? Succinate?			Osothslip & Subden (1986)
L-Malic	Saccharomyces cerevisiae (resting cells)		Active efflux					Influx?	Salmon (1987)
L-Malic	Candida sphaerica (anamorph Kluyveromyces marxianus)	Yes (not present with ethanol or glycerol)	Proton/anion symport	0.1 (pH 5.0)	0.44	Succinate Fumarate 2-Oxo-glutarate D-Malate	Maleic Malonic Oxalic L(+)-Tartaric	Yes	Côrte-Real et al. (1989)

Acid	Organism	Carbon source for growth[a]	Transport mechanism[c]	K_T (pH)[b]	K_T	Substrates (and K_T values)	Not transported	Reference
L-Malic	Hansenula anomala	Yes	Proton/anion symport	0.076 (pH 5.0)	0.2	Succinate, Fumarate, 2-Oxo-glutarate, Oxaloacetate, D-Malate	Maleic, Malonic, Oxalic, L(+)-Tartaric, Citric, Isocitric	Côrte-Real & Leão (1990)
Succinic	Neurospora crassa	Yes (sucrose as carbon source for growth)	Probably proton/anion symport	0.34 (pH 5.8)	0.22	L-Malate (0.16), Fumarate (0.34), L-Aspartate (0.46)		Wolfinbarger & Kay (1973)
Succinic	Kluyveromyces lactis	Yes (but low concentrations of glucose facilitate development of system)		0.018 (pH 5.4)	0.33	Malate (0.05), Fumarate (0.075), 2-Oxoglutarate (0.23)	Malonic, D,L-Tartaric, Glutaric, Adipic	Zmijewski & MacQuillan (1975)
Citric	Candida utilis	Yes	High affinity proton/anion symport	0.056	0.38	Isocitrate	Probably pyruvic, Acetic, Fumaric	Cássio & Leão (1991)
		Induced by Citric, Lactic, Malic	Low affinity facilitated diffusion	0.59 (pH 3.5)	1.14	Isocitric, L-Lactic, Fumaric		

[a] Not present with ethanol or glycerol: this refers to these compounds as the carbon source for growth
[b] pH value in brackets is the value at which kinetic measurements were made
[c] Where anion is named it is for consistency with nature of transport mechanism. Numbers in brackets are K_T values.

species, in terms of their ability to grow on citrate, fumarate, malate and succinate as sole sources of carbon. They were able to divide the yeasts into those that could use some or all of the acids (K^+) and those which could not (K^-). Studies with cell-free extracts as well as whole cells indicated that the differences were probably due to the presence or absence of transport systems at the plasma membrane. Second, the ability of yeasts to utilise organic acids has practical importance. L($-$)-Malic and L($+$)-tartaric acids are the predominant organic acids in grape must and wine; removal of these acids, possibly by the yeasts themselves, can be important during wine fermentation.

With respect to the information in Table 12.1 the following general comments are appropriate. All the transport systems except that for L-malic acid in *Schizosaccharomyces pombe* are repressed by glucose. However, as will be seen below in the consideration of benzoic acid transport in *Zygosaccharomyces bailii*, the interaction between sugar and organic acid transport may be more complicated. In those yeasts that have been tested, the acid transported is required for induction of the transport system. In order that induction takes place, it is highly likely that it is brought about by the undissociated organic acid diffusing into the fungus. Where glucose is present in the growth medium along with the organic acid, diauxic growth, when looked for, has been demonstrated (Leão & van Uden, 1986; Cássio, Leão & van Uden, 1987; Côrte-Real, Leão & van Uden, 1989; Côrte-Real & Leão, 1990; Cássio & Leão, 1991). Transport via a carrier system appears to involve the organic acid anion, though in not all instances has conclusive evidence been provided. Thus, a proton symport of the anion has been postulated, but, unless the critical experiments are forthcoming, facilitated diffusion of the undissociated acid is equally tenable. This latter process would lead also to an increase in the external pH not due to influx of protons, as with the symport, but to the re-establishment of the acid–base equilibrium brought about by influx across the plasma membrane of the undissociated acid (Côrte-Real *et al.*, 1989). It is not clear whether in the case of dicarboxylic and tricarboxylic acids it is a mono-, di- or triionic species that is transported. Wolfinbarger & Kay (1973) have provided evidence that for dicarboxylic acid transport in *Neurospora crassa* it is the monoionic species. The compelling evidence is from transport studies with substrate analogues. When both carboxyl groups are substituted, the analogues are poor competitors, whereas analogues with a single substituted group are good competitors. Incidentally, *N. crassa* is not permeable to citrate (Wolfinbarger & Kay, 1973).

Monocarboxylic acid transport into *Zygosaccharomyces bailii* demands separate consideration. This yeast is a common contaminant of low pH media with a high sugar content (Ingram, 1960). The yeast also grows in benzoic, sorbic and acetic acids at concentrations at which they are commonly used as preservatives. Warth (1977) showed that cells of *Z. bailii* concentrate benzoic and sorbic acids approximately as expected from the pH of the cell and the pK_a of the acid. Subsequent studies have confirmed this (Warth, 1989). On the addition of glucose, the intracellular content of acid is considerably reduced in induced cells (see below). Glucose brings about loss of the acid, which is reversed by iodoacetate. 2,4-Dinitrophenol (DNP) has the same effect as iodoacetate, though at a relatively high (1.0 mM) concentration. It was argued (Warth, 1977) that the process of uptake of the acids is due to diffusion of the undissociated acid but that in the presence of glucose an energy-requiring efflux system for the acids is induced, removing them from the cell. However, this conclusion about the active extrusion of benzoic acid has been criticised by Cole & Keenan (1987). They pointed out that, no matter how rapidly benzoic acid is pumped out of the cell, it should be counteracted by extremely rapid passive penetration in the reverse direction. These authors also question the method of determination of the cell pH. Be that as it may, the matter of rapid equilibration across the membrane must not be ignored. One has the feeling that the effect of glucose is in some way through the membrane potential, which will become more negative in the presence of the hexose (see Chapter 1) as has been implied by Warth (1991). Under such circumstances there could be less benzoate ion in the cells in the presence of hexose compared with when it is absent. There would need to be some transport system to allow benzoate to traverse the plasma membrane.

The plasma membrane of cells of *Z. bailii* has a much lower permeability to benzoic and propanoic acids than have other equivalent membranes, e.g. erythrocyte and toad bladder (Warth, 1989). Growth in benzoic acid decreases the permeability of the cells to benzoic acid still further, with no effect on that for propanoic acid. The cells grown in this way are the induced cells used in the above experiments. Warth (1989) has shown that the greater the tolerance of yeast species to benzoic acid the lower the permeability of their cells to the preservative. Nevertheless, it is unlikely that reduced permeability can completely account for tolerance to benzoic acid. This is similar to what has been concluded for the mechanism of tolerance to sulphur dioxide (see Chapter 8).

Malic acid is transported into cells of *Z. bailii* by a system that is induced by glucose (Baranowski & Radler, 1984). The system is repressed by fructose, which both inhibits transport and inactivates the transport system. Its presence in cells growing in glucose allows the yeast to degrade malic acid during fermentation, malate being decarboxylated to pyruvate by malic enzyme. This capability does not exist in *Saccharomyces cerevisiae* because malate transport is repressed by glucose (Table 12.1). Further studies (Herzberger & Radler, 1988) have shown that, whereas glucose induces the malic acid carrier, the sugar inhibits its function. Thus, in this latter respect glucose behaves like fructose; mannose and 2-deoxy-D-glucose behave similarly. Non-metabolisable sugars have no effect. Thus, sugar metabolism seems in some way to inactivate transport. In keeping with this view, it was demonstrated that pretreatment of cells with 5 mM DNP counteracted to some extent the inhibitory action of sugars and in their absence could stimulate malic acid uptake. The underlying mechanism is completely obscure. Clearly, *Z. bailii* is a yeast that merits further study.

Metabolism of organic acids

Many fungi can grow on organic acids and also on ethanol, both of which types of substance are either intermediates of the citric acid cycle or are converted to such an intermediate as one of the steps in metabolism, an example other than ethanol being propionate. It is not a function of this text to discuss the biochemical details by which such metabolism occurs. The reader should consult articles by Casselton (1976), McCullough *et al.* (1986) and Gancedo & Serrano (1989) for further information, though it should be noted that there is still much speculation as to the details. Nevertheless it is appropriate here to make some general points.

First, with respect to acetate and ethanol, if growth is to proceed it is necessary for the organism to provide:

(1) NADPH, normally provided by the pentose phosphate pathway when the organism is growing in glucose, for reductive biosynthesis.
(2) Malate and its derivatives aspartate, phosphoenolpyruvate (PEP) and pyruvate for the synthesis of hexoses, amino acids and pyrimidines.

Figure 12.1 gives a proposed scheme for filamentous fungi. In that scheme, as in other eukaryotic organisms, the glyoxylate bypass plays a key role. Nevertheless, although structures resembling the glyoxosomes in higher plants have been seen in filamentous fungi (McCullough *et al.*, 1986) the enzyme composition of these structures is uncertain.

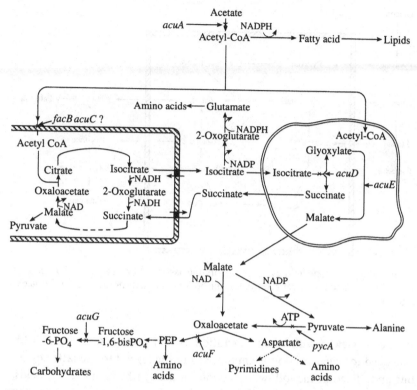

Figure 12.1. Proposed scheme for the subcellular organisation of acetate metabolism in *Aspergillus nidulans* and *Neurospora crassa*. Enzymes located in the mitochondria are shown in the box on the left; those thought to be located in glyoxosomes are enclosed on the right. A considerable degree of uncertainty is inherent in this figure because a reasonable yield of intact glyoxosomes has not been isolated and thus quantitative subcellular fractionation data are not available. Ethanol is oxidised to acetate by alcohol dehydrogenase and acetaldehyde dehydrogenase (From McCullough *et al.*, 1986.)

For growth on substrates utilised by the citric acid cycle, it is necessary for the organism to provide:

(1) NADPH for reductive biosynthesis.
(2) Acetyl CoA for the continued operation of the citric acid cycle and for fatty acid and steroid synthesis.
(3) Pyruvate for the synthesis of amino acids such as alanine and valine.
(4) PEP for the synthesis of both hexose and amino acids.

Figure 12.2 gives for filamentous fungi a proposed scheme of metabolic processes for substances utilised via the citric acid cycle. The scheme

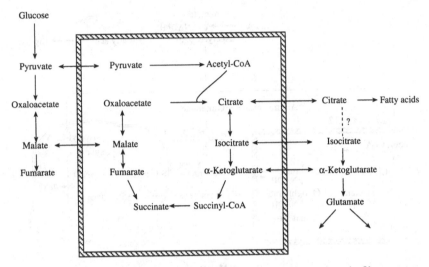

Figure 12.2. The citric acid cycle and probable associated reactions in filamentous fungi. The hatched area represents the mitochondrial membrane. (From Kubicek, 1988a.)

should be viewed with as much caution as that presented in Figure 12.1. The need for such caution is emphasised in the legend to the latter figure but underlined in the following section on the synthesis of organic acids, where it becomes clear that there is a great deal of ignorance about the operation of the citric acid cycle and mitochondrial functioning in fungi, except for a few species.

Synthesis of organic acids

It was Foster (1949) who asked the question: Why, when cultivated under laboratory or industrial conditions, do moulds produce large amounts of metabolic products, other than cell material and carbon dioxide from sugar, such as organic acids and carbohydrates? He pointed out that the common situation is one where carbohydrate substrate in the medium is in excess of requirements for growth. He suggested that, as a result of this excess, the enzyme mechanisms normally involved in the complete oxidation of the carbon substrate become saturated, leaving incompletely oxidised compounds. Either these compounds are lost to the medium or they are *shunted* to secondary metabolism. Foster termed the production of incompletely oxidised compounds as a consequence of the presence of excess carbohydrate substrate 'overflow metabolism'. This term is now common currency (McCullough *et al.*, 1986). Such metabolism will be expected to occur under conditions of nitrogen or phosphate limitation. Foster (1949) pointed out that trace metal limitation might also play an important role in overflow and shunt metabolism.

Mattey (1992) has pointed out that there are two groups of organic acids produced by fungi when overflow metabolism occurs. One group contains those that have a 'long' biosynthetic path from glucose, while the other has a 'short' path, essentially the direct transformation of glucose. In the former group are citric, fumaric, itaconic and malic acids, while in the latter group are gluconic and 2-oxogluconic acid. All these acids are produced or could potentially be produced commercially by fungi.

Most of the interest in the ability of fungi to produce organic acids has arisen from commercial considerations. Though it is probable that all fungi will produce organic acids under appropriate conditions, the commercial interest in organic acid production could mean that the biochemical events leading to the high yields of an acid in an industrial process may not necessarily be those responsible for the production of the same acid in the same or similar species of fungus growing under more natural conditions. High yields will have been brought about by strain improvement by selection and mutation and by manipulation of the growth medium.

These aspects need to be borne in mind when one is extrapolating from knowledge of the production of organic acids by fungi in industrial processes to what might be occurring in other species not selected for industrial use. Nevertheless it is appropriate to start this consideration of the mechanisms by which organic acids might be synthesised by focusing on citric acid production in the species that is used industrially, *Aspergillus niger*. The considerable amount of information on citric acid production by the fungus indicates the complexities of the metabolic processes involved.

Before summarising that information about citric acid synthesis, it needs to be stressed that much of the previous uncertainty about the pathways leading to the synthesis of organic acids in filamentous fungi has been due to the difficulties that have been encountered in the isolation of fully functional (coupled) mitochondria from those fungi (Kubicek, 1988*a, b*). It is probably correct to say that anything approaching comprehensive knowledge of mitochondrial functioning in filamentous fungi is only available for *A. nidulans/niger*, *Neurospora crassa* and *Rhizopus arrhizus*. This situation is in marked contrast to that for yeasts, particularly *Saccharomyces cerevisiae*, for which organism a very great deal is known about the biochemistry of its mitochondria (Gancedo & Serrano, 1989; Guérin, 1991).

It is estimated that some 400 000 tons citric acid year^{-1} are produced by fungi, for the most part by *A. niger*, though an increasing amount is produced by *Yarrowia lypolytica* (= *Candida lypolytica*). The biochemistry and physiology of the process in the two fungi has been reviewed in detail by Kubicek & Röhr (1986).

In considering the biochemical mechanisms for the synthesis of citrate by *A. niger*, account must be taken of the fact that production of citric acid can be accompanied by that of oxalic acid. The latter will accumulate when the medium is not sufficiently acid for the induction of oxalate decarboxylase (Hayaishi *et al.*, 1956). With the fact that oxalic acid is produced in mind, there is a need to discriminate between two possible pathways:

Pathway 1. Citrate arises from the citric acid cycle within the mitochondria from acetyl CoA and oxaloacetate. Since the cycle cannot supply all the oxaloacetate because of the drain of carbon out of the cycle by the accumulation of citrate, there must be another source of supply for oxaloacetate. This arises from pyruvate in the cytosol through carboxylation mediated by pyruvate carboxylase:

$$\text{Pyruvate} + \text{Mg-ATP} + \text{HCO}_3^- \longrightarrow \text{Oxaloacetate} + \text{Mg-ADP} + \text{P}_i$$

This enzyme from *Aspergillus nidulans* has been characterised and shown to be located in the cytosol (Osmani *et al.*, 1991; Osmani & Scrutton, 1983). It is proposed that oxaloacetate is then reduced to malate, which is transported into the mitochondria, wherein the malate is oxidised to oxaloacetate as part of the citric acid cycle. Oxalate is formed (along with acetate) from oxaloacetate through the action of oxaloacetate hydrolase in the cytosol (Kubicek *et al.*, 1988). Production of oxalate can be seen as a way for removing excess oxaloacetate.

Pathway 2. Citrate arises entirely in the cytosol from oxaloacetate, which is cleaved to oxalate and acetate by oxaloacetase, an enzyme that is induced under conditions when citrate is produced (Lenz, Wunderwald & Eggerer, 1976). Acetate is converted to acetyl-CoA, which together with oxaloacetate is converted to citrate by the action of a cytoplasmic citrate lyase (Verhoff & Spradlin, 1976).

The first pathway is now believed to be the one responsible for citric

Table 12.2. *The general requirements for citric acid production by* Aspergillus niger

Parameter	Component	Level	Range[a]
Carbon source	Sucrose, molasses, etc.	High	60–240
Nitrogen source	Ammonium salts	Low	2–3.0
	Urea		
Phosphate source	Potassium phosphate	Low	1–3.0
Trace metals	Manganese	Deficiency	<0.000001
	Zinc	Low	0.00025
	Iron	Low	0.0013
Oxygen	Air	High	1 p.p.m.
pH		Low	<2

[a] $g l^{-1}$ except for oxygen and pH.
From Mattey (1992). Reprinted with permission from *CRC Critical Reviews in Biotechnology* **12**, 87–132. Copyright CRC Press, Inc., Boco Raton, FL.

acid production. The evidence is as follows:

(1) Citrate yields can approach 90% of the carbohydrate supplied. This is consistent with pathway 1 but not with that proposed for pathway 2 to occur in the cytosol, which has a theoretical maximum yield of 66%.
(2) The pattern of label in citrate after feeding $^{14}CO_2$ can only be explained by pathway 1 (Cleland & Johnson, 1954).
(3) Although the level of oxaloacetase rises prior to active citrate synthesis it declines to an undetectable level during the phase of rapid citric acid accumulation (Joshi & Ramakrishnan, 1959).

The general requirements for citric acid production are given in Table 12.2. The high levels of the necessary carbon source and low levels of nitrogen and phosphorus are what would be expected of a process that is part of the overflow metabolism. However, the need for deficiency levels of manganese is very striking (Shu & Johnson, 1948; Tomlinson, Campbell & Trussell, 1950). The effects of deficiency levels of manganese on *A. niger* can be catalogued (Table 12.3). It would seem that the effects are not ascribable to the effect of a particular enzyme. It has been argued by Kubicek & Röhr (1986) that increased protein breakdown in the absence of sufficient manganese leads to increased levels of ammonia in the mycelium. The ammonia activates phosphofructokinase (Habison, Kubicek & Röhr, 1979) and thus diminishes the negative feedback from citrate. This allows the necessary flux of glucose to maintain the production

Table 12.3. *Metabolic and morphological alterations induced by manganese deficiency in* Aspergillus niger *during citric acid fermentation*

Alteration	Reference
Accumulation of amino acids and NH_4^+	Kubicek, Hampel & Röhr (1979)
Elevated proteinase activity	
Elevated rate of protein degradation	Ma, Kubicek & Röhr (1985)
Increased monosome portion of ribosomes	
Decreased cellular phospholipid content	Orthofer, Kubicek & Röhr (1979)
Plasma membrane phospholipid and fatty acid alterations	Meixner et al. (1985)
Cell wall chitin elevation and decrease in β-1,3-glucan	Kisser, Kubicek & Röhr (1980)
Crippled, branched mycelial development, single giant cells	
Increased glycolytic flux at the expense of the pentose phosphate pathway	Kubicek & Röhr (1977)

From Kubicek & Röhr (1986). Reprinted with permission from *CRC Critical Reviews in Biotechnology* 3, 331–73. Copyright CRC Press, Inc., Boco Raton, FL.

of citric acid (Figure 12.3). The regulation of carbohydrate metabolism in relation to citric acid accumulation has been discussed by Steinböck et al. (1991).

For other aspects of the regulation of citric acid production one should consult articles by Kubicek & Röhr (1986), McCullough et al. (1986), Kubicek (1988a, b), Kubicek & Scrutton (1991a) and Mattey (1992). However, there is no information about how citric acid leaves the mycelium.

The conditions required for high yields of itaconic acid (methylene succinic acid) by *Aspergillus terreus* are very similar to those for high yields of citric acid by *A. niger* (Winskill, 1983; Mattey, 1992). Also high external concentrations of calcium and magnesium are required but relatively low (360 mM) concentrations of potassium. However, there is no information about the effect of the presence or absence of manganese. Until fairly recently, much of the biochemical information about the synthesis of itaconic acid was equivocal.

Kinoshita (1932) was the first to isolate itaconic acid as a metabolite from *A. terreus*. He postulated that the acid was formed by the decarboxylation of cis-aconitate, which suggested that itaconic acid is formed via the citric acid cycle. This route was supported by later studies (Bentley & Thiessen, 1957a, b, c). However, other mechanisms, not

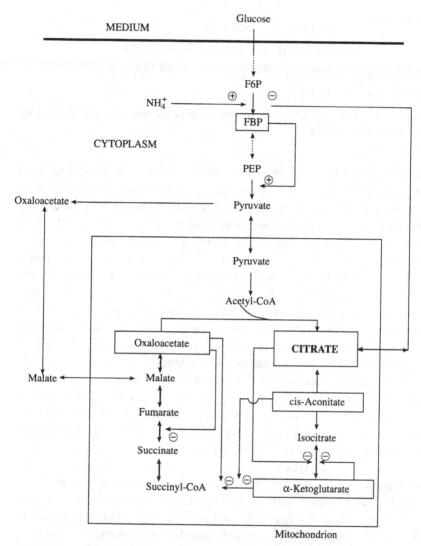

Figure 12.3. Metabolic scheme of interrelationships leading to citrate accumulation: ⊕ indicates activation by the respective metabolite; ⊖ indicates inhibition. F6P, Fructose 6-phosphate; FBP, fructose 1,6-bisphosphate; PEP, phosphoenolpyruvate. (From Kubicek & Röhr, 1986.)

involving the citric acid cycle, have been proposed (Shimi & Nour El Dein, 1962; Nowakowska-Waszezuck, 1973; Jakubowska, Metodiewa & Zakowska, 1974). A major problem in the biochemical investigations has been the isolation of active fully functional mitochondria. More recent

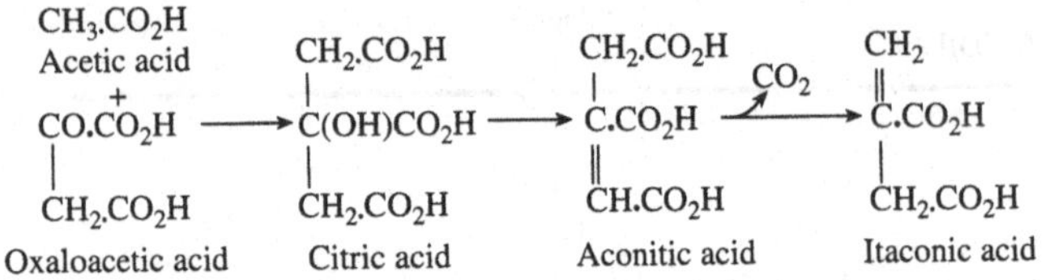

Figure 12.4. The metabolic path for formation of itaconic acid from acetic acid. (From Winskill, 1983.)

studies in which active intact mitochondria have been isolated indicate that itaconic acid is formed via the citric acid cycle (Winskill, 1983). Jaklitsch, Kubicek & Scrutton (1991b) have found two striking differences between mycelium of *A. terreus* separated from the medium used for its growth and mycelium exposed to a medium to maximise production of itaconic acid. The first difference was the absence of cis-aconitate decarboxylase in the growing-medium mycelium but present in the production-medium mycelium exclusively in the cytosol. The other difference was the at least two-fold increase in aconitase activity in production-medium mycelium. There appeared to be no other differences in other enzymes, particularly citric acid cycle enzymes. Figure 12.4 shows the pathway proposed for itaconic acid synthesis.

Takahashi & Sakaguchi (1927) were the first to show that a number of *Rhizopus* spp. produce fumaric acid and this mould has since been used for commercial production of the acid. Studies by Romano and his colleagues (Romano, Bright & Scott, 1967; Overman & Romano, 1969) showed by the use of radiolabelled glucose that carbon dioxide fixation is involved an acetyl-CoA-dependent pyruvate carboxylase. Such an enzyme is found in the mitochondria of animal cells (Böttger *et al.*, 1969). As indicated above with respect to citric acid synthesis, pyruvate carboxylase in *A. nidulans* is located in the cytosol. Also it is known to be activated by oleoyl-CoA but not by acetyl-CoA (Osmani *et al.*, 1981).

Pyruvate carboxylase has been shown by Osmani & Scrutton (1985) also to be located in the cytosol of *Rhizopus arrhizus*. Those authors also showed that, while it would be possible for fumaric acid to be produced via the citric acid cycle, it was much more probable that the metabolic pathways involved are located in the cytosol (Figure 12.5). Location in this compartment allows synthesis to occur in a self-contained manner with respect to both carbon flow and ATP and NADH production and utilisation.

Peleg, Stieglitz & Goldberg (1988b) showed that, when *Aspergillus flavus*

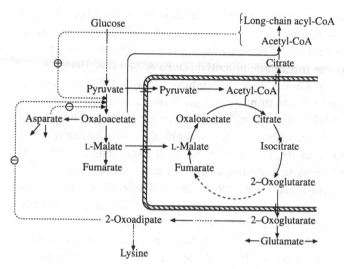

Figure 12.5. A scheme describing the possible organisation of pyruvate metabolism in *Rhizopus arrhizus* and incorporating the possible regulatory mechanisms for pyruvate carboxylase. +, Positive feedback; −, negative feedback. (From Osmani & Scrutton, 1985.)

is grown in glucose, inorganic salts and high concentrations of calcium carbonate, but with limiting amounts of nitrogen, large amounts of malic acid are formed in the medium. Indeed crystals, essentially of calcium malate, aggregate on the mycelium in the early stages of the production phase (Peleg *et al.*, 1988*a*). The evidence from enzyme (Peleg *et al.*, 1988*b*) and ^{13}C NMR studies (Peleg *et al.*, 1989) indicates that malic acid is synthesised by the same steps by which it is formed in *R. arrhizus* (see above). However, for *A. flavus* the total picture is not completely clear but there is no doubt that a major element of the ability to synthesise malic acid is the increase of NAD-dependent malate dehydrogenase relative to fumarase (Peleg *et al.*, 1989). Both are present as isoenzymes but their location (cytosolic or mitochondrial) is not clear.

Succinic acid together with glycerol are the major by-products of ethanol fermentation by *S. cerevisiae*. The evidence supports the thesis that succinic acid is produced by the citric acid cycle acting oxidatively. The evidence for this is as follows (Gancedo & Serrano, 1989): (a) the redox relationship between succinic acid and glycerol synthesis can be maintained only by succinic acid produced oxidatively (Oura, 1977); (b) production of succinic acid is stimulated by glutamic acid, which must act through deamination to 2-oxoglutaric acid then oxidation to succinic acid (Heerde & Radler,

1978); (c) succinic acid production is not connected with high NADH levels, since it is not formed in association with glycerol fermentation brought about by sulphite. Succinate transport out of the yeast cell appears to be driven by the membrane potential (Duro & Serrano, 1981).

Also, it has been known since the studies of Dakin (1924) that cells of *S. cerevisiae* can be made to produce significant concentrations of malic acid during ethanol fermentation. The fermentation of the acid is favoured by a medium with low nitrogen content, a pH of 5.5 and a concentration of biotin of less than $10 \, g \, l^{-1}$ (Radler & Lang, 1982). There is no effect of trace metals (Schwartz & Radler, 1988). The biochemical events leading to malate production are unclear. Schwartz & Radler (1988) found that feeding $^{14}CO_2$ to cells producing malic acid led to the formation of equal quantities of labelled malate and succinate. Carbon dioxide fixation is clearly involved and, on the basis of inhibition of malic acid formation by avidin, the fixation is believed to be by a biotin-dependent carboxylation, presumably involving pyruvate carboxylase. Experiments where various organic acids and inhibitors of malic acid formation were used as substrates were hard to interpret.

I have considered here the pathways of synthesis of those acids that fungi are known to be capable of producing in amounts significant enough to make their production by fungal fermentation a commercial proposition. Indeed it is actual or possible industrial application that has led to research on the metabolic pathways involved. I have focused on those acids that have a long biosynthetic path from glucose. For the metabolic pathways to the 'short' path group of organic acids, the reader should consult articles by Mattey (1992) and Miall (1978). This focus on the long biosynthetic path acids has been because of the much greater uncertainty about how they are synthesised. Finally, there is a need to keep present knowledge of the biosynthesis of organic acids in perspective. Knowledge is limited to only a few acids; fungi can synthesise around 75 different organic acids (Wright & Vining, 1973), about which little is known except their identification as a fungal product. Also, as the next section shows, similarly little is known about the physiological role of organic acids in fungi.

The role of organic acids in charge balance and control of cytoplasmic pH in fungi

The above details of those metabolic pathways leading to the synthesis of carboxylic acids are for fungi exhibiting overflow metabolism and concern the synthesis of individual acids, with little consideration for the

Table 12.4. *The organic acids and their amounts produced by various species and strains of* Aspergillus

Fungal strains	Amount of organic acid produced (g l^{-1})[a]			
	Gluconic	L-Malic	Citric	Succinic
Gluconic acid producers				
A. wentii ATCC 1023	68	11		
A. niger ATCC 9029	67	14	4	
A. wentii NRRL 377	64	20		
A. niger ATCC 12846	62	4	5	
A. niger ATCC 10577	48	4	3	
A. niger ATCC 9142	23	3	5	
Malic acid producers				
A. flavus ATCC 13697		39		13
A. oryzae ATCC 56747	8	35		6
A. sojae ATCC 46250		30		3
A. ochraceus ATCC 18500		17		5
A. foetidus ATCC 10254		11	8	
A. nidulans ATCC 16848		8		3
Citric acid producer				
A. aculeatus NRRL 358		3	17	

[a] Organic acids were determined in the fermentation broth after 135 h.
From Bercovitz *et al.* (1990).

other acids that might be formed at the same time. Table 12.4 presents data for organic acid production in fermentation by a range of *Aspergillus* spp. Though the fungi have been selected for their ability to produce either gluconic or malic acid, it can be seen that they produce other acids, sometimes in significant amounts.

It is not known to what extent the concentrations of acids in the medium represent the concentrations inside the fungus concerned. Nor is much known about organic acid synthesis in a fungus exhibiting balanced growth. Production of organic acids within the fungus could be important in maintaining charge balance when there is excess uptake of cations over anions, a situation which is both well recognised and studied in green plants (Jennings, 1986b). The evidence for the same role in fungi is much more slender. Shere & Jacobson (1970) showed that, when potassium is absorbed by *Fusarium oxysporium*, ion balance is maintained not only by loss of hyphal magnesium and hydrogen ions but by an increase in the total organic acid concentration in the mycelium. Acetic, citric and succinic acids increase in concentration, malic and pyrolidine carboxylic acids decline. Hobot & Jennings (1981) with *Debaryomyces hansenii* and *S. cerevisiae* and Peña (1975) and Ryan, Ryan & O'Connor (1971) with *S. cerevisiae* have shown that the presence of acids, including citric acid cycle acids, in the external medium can enhance potassium uptake by these two yeasts. Wethered *et al.* (1985) showed that the marine hyphomycete *Dendryphiella salina*, when growth in various media of low osmotic potential generated by different salts or with inositol, often had significant concentrations of organic acid within the mycelium. There was a high concentration of organic acid when magnesium chloride was the external osmolyte at 0.8 osmol (kg water)$^{-1}$, whereas, when sodium chloride was external to the mycelium at the same osmolar concentration, organic acid was not detectable in the mycelium.

Organic acids can therefore be important in maintaining charge balance in fungi. Further information is needed to find out what acids are involved. Also, in certain instances organic acids, as is the case in *D. salina*, can make a significant contribution to the protoplasmic osmotic potential.

Davies (1973) has pointed out that carboxylation and decarboxylation can be a means by which a cell can control its internal pH. He proposed a biochemical pH-stat that can be represented thus:

Decarboxylating   Carboxylating

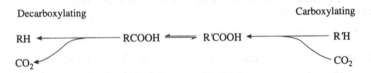

If the decarboxylating enzyme has an acid pH optimum and the carboxylating enzyme an alkaline pH optimum, then the system will behave as a pH-stat. Davies (1973) produced circumstantial evidence for such a pH-stat in green plants, but the idea has never been comprehensively tested in one plant to demonstrate the biochemical pH-stat unambiguously (Kurkdjian & Guern, 1989).

Knowledge about the role of organic acids in the control of the internal pH of fungi is even more sketchy than for green plants. The presence of the biochemical pH-stat must be considered speculative. It would seem from studies on the proton pump discussed in Chapter 1 that proton extrusion by the pump cannot exclusively control cytoplasmic pH; oxidative metabolism also appears to be involved. It is not clear what such metabolism might be; organic acid synthesis and breakdown might be involved. Studies by Sigler & Höfer (1991) on *S. cerevisiae* suggest that transfer of succinate across the plasma membrane after the addition of glucose to a cell suspension is an important part of the cellular mechanism for controlling cytoplasmic pH of that fungus. As with *N. crassa* (see Chapter 1) proton extrusion is a dominant part of the process and Sigler & Höger (1991) pointed out that in *S. cerevisiae*, as with the filamentous fungus, the intracellular buffering capacity (carbonate, phosphate, organic acids, polyelectrolytes) will also be important. Also possible is transfer of protons to the vacuole as well as conversion of acids and bases to neutral compounds – of which the biochemical pH-stat would be a special case, though it is not mentioned specifically.

In all this, the assumption is made that the fungus is only responding to changing hydrogen ion concentration and that hydrogen ions themselves do not have a direct regulatory effect. There is now evidence to suggest otherwise. Caddick, Brownlee & Arst (1986) have provided evidence that in *A. nidulans* the levels of a number of enzymes whose location is at least in part extracellular (e.g. acid phosphatase, alkaline phosphatase, phosphodiesterase) and of certain transport proteins (e.g. that for γ-amino-*n*-butyrate) are controlled by the pH of the growth medium. Furthermore, from the same fungus mutations have been isolated, of which some confer resistance to acid media of both high and low buffering capacity and others confer resistance only in media of low buffering capacity (Rossi & Arst, 1990). When the mutants were grown in media of more moderate pH mycelia of both classes led to reduced acidification of the medium compared with the wild-type.

Carbon dioxide fixation and oligotrophy

Carbon dioxide fixation is a necessary adjunct to anaplerotic metabolism and can be important in the synthesis of organic acids (see above). Fixation is brought about by pyruvate carboxylase. The report of the involvement of phosphoenolpyruvate carboxylase by Bushell & Bull (1981) has not been substantiated (McCullough *et al.*, 1986). Pyruvate carboxylase from *A. nidulans* has apparent K_m values for pyruvate, Mg-ATP^{2-} and HCO$_3^-$ of 0.1, 0.5 and 1.5 mM, respectively (Osmani *et al.*, 1981). The effect of key metabolites on the enzyme is complicated. Thus, L-aspartate is a competitive inhibitor to HCO$_3^-$ but non-competitive to pyruvate and MgATP^{2-}. 2-Oxoglutarate induces complex kinetic behaviour, whereas acetyl-CoA causes activation in the presence of these two acids but not in their absence. One presumes that the concentration of bicarbonate at the locus of the enzyme is that in equilibrium with carbonate and carbon

Table 12.5. *Fungi capable of growing on silica gel*

A. Growth on gel containing N but no added C	
Aspergillus flavus	*Mucor flavus*
A. repens	*M. rouxii*
A. niger	*Penicillium chrysogenum*
Cladosporium herbarum	*P. notatum*
Fusarium solani	*P. rubrum*
Gliocladium roseum	
B. Growth on gel lacking both added C and N	
Aspergillus flavus	*Fusarium* (5 soil isolates)
A. niger	*Mucor rouxii*
Fusarium solani	*Penicillium chrysogenum*

All fungi tested showed growth. Not all species which grew on gel A were tested
for growth on gel B.
From Parkinson, Wainwright & Killham (1989).

dioxide, both in the fungus and outside. However, Hobot & Jennings
(1981) have speculated about whether there might be a bicarbonate pump
at the plasma membrane in fungi growing under the alkaline conditions
of seawater.

  Oligotrophs are microorganisms that can grow when nutrients are
present at very low concentrations. Wainwright (1993) has pointed out
that the term oligotrophy is usually used to describe the strategy used by
prokaryotes to grow under low nutrient conditions (Poindexter, 1981;
Fry, 1990). There are now a number of reports of fungi being able to grow
under oligotrophic conditions (Wainwright, 1993). The phylloplane, i.e.
the surface of leaves, and that part of the soil away from the rhizosphere
and decomposing wood (Wainwright, 1988) can both be considered
oligotrophic. In the case of soil, not all nutrients are at low concentration
but this is likely to be so for carbon, such that strictly one is dealing with
what has been termed 'oligocarbonotrophy' (Wainwright, 1993). Table
12.5 gives examples of oligocarbonotrophs isolated from soil and also of
those that can grow under even more stringent conditions, namely that
the solid substrate on which ability to grow was determined, silica gel,
contained no added nitrogen as well as no added carbon.

  It can be argued that growth is possible because the fungi are able to
scavenge combined carbon compounds and nitrogen compounds from the
air. Nitrogen fixation is extremely unlikely (see Chapter 6). Mirocha &
De Vay (1971) investigated the ability of *Cephalosporium* sp. and
Fusarium sp. to grow on a totally inorganic medium. As far as can be

seen, though the details are not completely clear, all necessary precautions were taken to exclude combined carbon from the growth vessels. At the initial stages of growth, there was a linear relation between the rate of growth and CO_2 gas or bicarbonate concentrations supplied. Direct evidence for assimilation of carbon dioxide was obtained by using $^{14}CO_2$ followed by combustion analysis of the harvested mycelium. There is very serious doubt that growth can be by autotrophic carbon fixation involving ribulose bisphosphate carboxylase/oxygenase; certainly Parkinson, Killham & Wainwright (1990) could detect no ribulose bisphosphate carboxylase activity in *Fusarium oxysporium*. This study needs to be repeated, which may be difficult (Jones *et al.*, 1991). That said, if as I am inclined to believe not all organic carbon was excluded from the cultures, nevertheless the evidence highlights the importance of carbon dioxide fixation and anaplerotic metabolism under conditions of carbon limitation. This has been well shown by Bushell & Bull (1981), who demonstrated that the effect of increasing the bicarbonate concentration in carbon-limited cultures of *A. nidulans* was substantial. Thus, an increase in concentration from 0.7 to 2.8 mM increased the biomass synthesis by 22%. However, Parkinson *et al.* (1991) have shown that, when 7 day-old cultures of *F. oxysporium* are grown in the absence of any exogenous carbon source in the presence of $^{14}CO_2$ over 21 days, only 0.78% of the carbon in the cell is derived from carbon dioxide.

13
Water relations and salinity

Introduction

Water is required by fungi for growth and for movement of solutes within the mycelium (translocation). Many aspects of reproduction are governed by water availability; for example, spore release and dispersal are dependent on it, as discussed in the classic text by Ingold (1971). In the past 20 years, there has been little to modify what is described there. However, now there is a better idea of spore release in some of the lower fungi (see, for example, Duniway, 1979; Gisi, Hemmes & Zentmyer, 1979; Gisi & Zentmyer, 1980; Money & Webster, 1988, 1989; Money, Webster & Ennos, 1988; Horn, 1989, 1990). Nevertheless, the water relations of basidiospore discharge are still a matter of controversy (Webster *et al.*, 1988; Ingold, 1990; Corner, 1991). In this volume, I consider only the water relations of growth (this chapter) and translocation (Chapter 14). For terminology and background to this chapter the reader is referred to reviews by D. M. Griffin (1981), Eamus & Jennings (1986*a*), Papendick & Mulla (1986) and Jennings (1990*c*).

Growth

The general picture

When a fungal cell grows or a hypha extends, it increases its volume. Part of that increase in volume is brought about by the inflow of water. The rate of flow must be such that the pressure generated does not cause the wall to burst (Robertson, 1959, 1965). However, there must be sufficient pressure to cause the wall to expand, though it has to be admitted that there is considerable debate as to how that expansion takes place: either (a) via the action of lytic enzymes, a balance being maintained between synthesis of polymers and their lysis (Bartnicki-Garcia, 1973; Gooday, 1977; Rast *et al.*, 1991) or (b), as a consequence of the bringing together of an assemblage of polymers being inherently plastic, rigidity developing due to an increase of coupling interactions with time (Wessels, 1986, 1988). It is not proposed to enter into that debate here.

Essentially, the fungal cell or hypha generates a water potential that is less than that of the surrounding medium. It is this difference in water potential that is the driving force for water inflow. Focusing for the most part on hyphae, I consider, below, how that driving force is generated. Before doing that, I should draw attention to some observations made on the bursting tendency of hyphal tips. These observations by Bartnicki-Garcia & Lipmann (1972) and Dow & Rubery (1975) need to be kept in mind when one tries to understand the relationship between water inflow and growth.

Bartnicki-Garcia & Lipmann (1972) showed that, when hyphae of *Mucor rouxii* are grown on what was termed 'full-strength agar', the hyphae burst readily at their apices on being flooded with distilled water. However, when the hyphae were grown on diluted medium, bursting did not occur. That occurred only in dilute solutions (particularly of acids, but also of neutral salts, EDTA, alcohols, acetone and detergents). Bursting was generally inhibited by alkaline solutions, while divalent cations such as Ca^{2+}, Mg^{2+} and Mn^{2+} caused no bursting but produced apical swelling. Hyphae in both very young and older colonies did not burst readily. The hyphae of some fungi did not burst when colonies were flooded with distilled water – these included non-spreading mutants of *N. crassa* (but not the wild-type), four *Phytophthora* sp. studied and *Schizophyllum commune*. Bursting of hyphal apices in both *M. rouxii* and *N. crassa* exhibited a temperature coefficient significantly well above unity (1.37–2.03).

The results are extremely difficult to interpret in a coherent manner but are referred to below. Dow & Rubery (1975) believed that the effects of dilute acids and bivalent cations were not on the balance between the activity of enzymes bringing about wall synthesis and those causing lysis of polymers, as proposed by Bartnicki-Garcia & Lipmann (1972), but were a modification of the physical properties of the matrix wall polysaccharides. First, polyethyleneglycol (PEG, molecular weight 400) prevented the bursting in dilute acid, and this could be reversed by exposure to PEG-free acid solution. But similar hyphae when transferred to distilled water did not burst, indicating that the acid treatment did not appear to weaken the wall. Calcium ions were antagonistic to acid, there being an increase in the time to bursting with an increase in the external calcium concentration.

Turgor

The foregoing makes clear some of the complexities that need to be taken into account when one considers the water relations of growing hyphae. Some of the facts are referred to subsequently below. Here, I address the issue of the relationship between the rate of growth of a hypha and its turgor potential. The relationship is not easily investigated, there being methodological difficulties in the determination of turgor potential in fungal hyphae. So the experimental procedures used need to be kept in mind when one considers the data.

Though it is not the first investigation into the relationship between turgor and growth (two earlier ones are considered below), I focus first on work by Kaminskyj, Garrill & Heath (1992). They studied hyphal tip extension in *Saprolegnia ferax* as affected by the presence of varying concentrations of polyethylene glycol or sorbitol in the external medium. They determined turgor potential by two means. The first depended on finding a solution of known osmotic potential that would induce plasmolysis in 50% of the hyphae. The second was based on determining the osmotic potential of the mycelium by thermocouple psychrometry. The turgor potential was calculated by assuming that the water potential of the hyphae is equal to (in equilibrium with) that of the medium. It was found that, irrespective of the method of measuring turgor and the osmoticum used to reduce the growth rate, over most of the range of rates there was an almost constant turgor potential (Figure 13.1). Only at the highest growth rate was there an increase in turgor potential, twice that at the lower growth rates. Thus, the relationship found between turgor potential and growth rate was not simple.

The two other investigations of the relationship between growth and turgor potential are that by Adebayo, Harris & Gardner (1971) using *Mucor hiemalis* and *Aspergillus wentii* (Figure 13.2) and that by Luard & Griffin (1981) using a range of fungi (Table 13.1). In both cases, turgor potential was determined psychrometrically, in a manner essentially similar to that of Kaminskyj *et al.* (1992). Both sets of workers believed that they demonstrated no relationship between growth rate and turgor potential when the former was reduced by adding osmotica to the growth medium. But this is not correct. The majority of the data indicate that reduced growth rate as the result of an unfavourable osmotic environment is

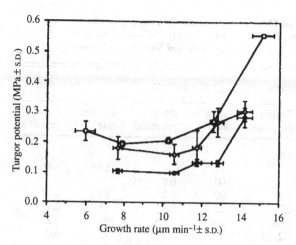

Figure 13.1. The relation between turgor potential and hyphal growth rate in *Saprolegnia ferax*. Growth rate altered by decreasing the osmotic potential of the medium with either sorbitol (turgor estimated by (○) plasmolysis or (●) osmometry) or polyethyleneglycol (turgor estimated by plasmolysis (□)). (From Kaminskyi *et al.*, 1992.)

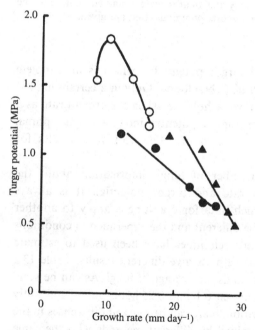

Figure 13.2. The relation between turgor potential and colony radial growth rate in *Aspergillus wentii* (○), *Mucor hiemalis* ((●) potassium-chloride-adjusted osmotic potential of medium; (▲) sucrose-adjusted). (From Adebayo *et al.*, 1971.)

Table 13.1. *Turgor potential (MPa) as a function of colony growth rate (% maximum), altered by changing the osmotic potential of the medium, of* Phytophthora cinnamomi, Fusarium equisesti *and* Xeromyces bisporus

Osmotic potential of medium (−MPa)	P. cinnamomi		F. equisesti		X. bisporus	
	Growth rate	Turgor potential	Growth rate	Turgor potential	Growth rate	Turgor potential
0.33	99	0.86				
6.0					17.5	1.4
1.2			100	0.6		
	100	0.93	90	0.95	87.5	2.25
	90	0.71	55	0.75	75	2.9
	60	1.05	35	1.25	100	3.25
	40	1.2	27	1.7	42	4.0
	25	1.1			73	
3.75	24	1.35				
16.3			10	2.1		
47.5					32	3.4

Data obtained from Figure 2 of Luard & Griffin (1981). Thus, the figures are not precise and it is because of the way the figures were obtained that only the maximum and minimum medium osmotic potentials used are given.

accompanied by an increased turgor potential – which is in complete contrast to what Kaminskyj et al. (1992) found. Only in a certain limited number of instances does the reverse hold, i.e. when the growth rate as a function of decreasing medium osmotic potential increases to an optima. The most striking example is shown in Table 13.1 with results for *Xeromyces bisporus*.

I have presented, for a number of fungi, information about the relationship between growth rate and turgor potential. It is always problematic the extent to which data for one fungus apply to another when the growth media may be different and the experimental conditions not the same. Further different techniques have been used to estimate turgor potential, and different methods give different results. Table 13.2 gives turgor and osmotic potentials for a range of fungi. As can be seen, turgor potentials can vary between 0.1 and 3.25 MPa but, significantly for the matter under consideration, there can be marked differences in the value for turgor potential obtained by different methods with the same fungus, as demonstrated by the data for *Achlya bisexualis* (Money, 1990b) and *Saprolegnia ferax* (Kaminskyj et al., 1992). As is discussed below, there can also be doubts about turgor potential values determined by thermocouple psychrometry, when it is assumed that the water potential of the mycelium is equal to that of the medium. Nevertheless, while there must be a question mark about the quantitative picture, as I indicate below, I believe that there is a relatively coherent qualitative picture.

Some theoretical considerations

The problem with interpreting the relationship between turgor and growth as just described is that so little is known about the water relations of fungi. Even for green plants there is not a completely clear picture of the water relations of growing cells, in spite of investigations over several decades (Cosgrove, 1986). Nevertheless, it is worth referring to some of the key equations that underpin the water relations of growing green plant cells. Water uptake by such a cell can be defined in terms of the relative rate of volume increase and is given by:

$$(1/V)\frac{dV}{dt} = AL_p(\sigma\Delta\pi - P)/V \tag{13.1}$$

The definitions of the symbols used in this and the following two equations are as follows:

L_p membrane hydraulic conductivity (cm s^{-1} MPa^{-1})
σ solute reflection coefficient (dimensionless)
$\Delta\pi$ osmotic pressure difference between the cell and its surroundings (MPa)
P hydrostatic pressure (MPa)
V cell volume (cm^3)
A cell surface area (cm^2)
L volumetric hydraulic conductance (s^{-1} MPa^{-1}); for single cells this is equal to AL_p/V
Ψ water potential (MPa)
ϕ cell wall yielding coefficient (extensibility) (s^{-1} MPa^{-1})
Y minimum turgor necessary for growth (yield threshold) (MPa)

Growth will occur only when the wall yields to internal pressure, which will arise from cell turgor. This means that at the simplest:

$$(1/V)\frac{dV}{dt} = \phi(P - Y) \tag{13.2}$$

Table 13.2. *Values for the water potential (Ψ), osmotic potential (Ψ_π) and turgor potential (Ψ_P) of a range of fungi, together with the technique used for determining the potentials. Where known, the growth rate is given, as are values for a different growth rate produced by changing the osmotic potential of the medium*

Fungus	Growth rate (μm min^{-1})	Osmoticum to change the growth rate	Ψ ($-$MPa)	Ψ_π (MPa)	Ψ_P (MPa)	Technique for determining Ψ_P	Reference
Neurospora crassa	4.1				1.24	Incipient plasmolysis[a] Change in hyphal diameter[b]	Robertson & Rizvi (1968)
Mucor hiemalis	1.9	KCl	0.62[c]	1.29	0.65	Psychrometry	Adebayo et al. (1971)
	1.3		3.12[c]	4.25	1.13		
Aspergillus wentii	2.1	Sucrose	0.62[c]	2.15	1.58		
	1.7[d]	Sugar[e]	3.12[c]	4.43	1.31		
Phellinus noxius			1.1[c]	1.7	0.6[g]	Psychrometry	Luard & Griffin (1981)
Fusarium equisesti	1.5[d]	[f]	0.8[c]	1.4	0.72[g]		
Eurotium amstelodani	0.83[d]	Sugar	11.2[c]	12.75	1.55[g]		
Chrysosporium fastidium	0.27[d]	Sugar	12.4[c]	14.9	2.5[g]		
Xeromyces bisporus	0.27	Sugar	26.5[g]	29.75	3.25[g]		
Phallus impudicus	18.0	Sugar		2.92	0.41	Psychrometry	Eamus & Jennings (1986b)

Species		Solute	Ψ_p[a]	Ψ[b]		Method	Reference
Serpula lacrymans	6.5		1.68	2.13	0.425		
	3.8	Sucrose	5.77	5.96	0.19		
Saccharomyces cerevisiae		—	0.72[h]	1.53	0.81[h,i]		Meikle et al. (1988)
			0.72[j]	1.33[j]	0.61[i,j]		Money (1990b)
Achlya bisexualis	3.0				0.74	Psychrometry	
					0.92	Pressure probe	
			1.02–1.27	1.0–1.2	Incipient plasmolysis		Kaminskyj et al. (1992)
Saprolegnia ferax	15		0.19[c]	0.75	0.56	Incipient plasmolysis	
	14.2	Sorbitol	0.19[c]	0.48	0.29	Psychrometry	
	7.8		1.3[c]	1.44	0.14	Incipient plasmolysis	
	7.8		1.3[c]	1.4	0.1	Psychrometry	

[a] Ψ_p.
[b] Ψ.
[c] Osmotic potential of medium (assumed to be equal to mycelial water potential).
[d] Maximum growth rate exhibited in relation to changed water potential of medium.
[e] Glucose or sucrose used – publication unclear as to which sugar used.
[f] Maximum growth observed in control treatment.
[g] Determined from published graphs.
[h] Stationary phase cells.
[i] Determined as difference between Ψ_π and Ψ_{medium}.
[j] Mid-exponential phase cells.

It needs to be clear that ϕ, the wall yielding coefficient, is not the same as some other terms, such as elastic extensibility, that have been obtained with dead cell walls (Taiz, 1984).

As Cosgrove (1986) has pointed out, equation (13.2) is a model for cell growth, turgor potential acting as a link between water uptake and the ability of the cell wall to yield to internal pressure. The extent to which the wall does yield governs the extent to which turgor and water potentials are displaced from their equilibrium values. The disequilibrium of water potential, i.e. the extent to which $\Delta\Psi = \sigma\Delta\pi - P > 0$, under conditions of steady-state growth is given by:

$$\Delta\Psi = \frac{\phi}{\phi + L}(\sigma\Delta\pi - Y) \tag{13.3}$$

This equation is derived from equations (13.1) and (13.2). Cosgrove (1986) has pointed out the ratio $\phi/(\phi + L)$ describes the limitation of growth by water uptake. Thus, when L is much larger than ϕ, the ratio approaches zero, $\Delta\Psi$ approximates to zero and growth is limited by ϕ. When L is much smaller than ϕ, the ratio moves towards unity and P drops to Y and growth is limited by L.

In the green plant cell growing at a rapid rate, the driving force for water absorption is only 0.3 kPa (Cosgrove, 1986). In other words, turgor is reduced by as little as a thousandth of the turgor typical of such cells and thus water transport does not limit growth. The cells are nearly at perfect equilibrium and growth is controlled entirely by the ability of the wall to yield under pressure.

The only values for L_p of fungal hyphae come from a study by Cosgrove, Ortega & Shropshire (1987), using the pressure probe, of the water relations properties of the sporangiophores of *Phycomyces blakesleeanus*. While L_p can increase by a factor of 3 between stages I and IV (Castle, 1942), the average value of 4.6×10^{-5} cm s^{-1} MPa^{-1} is very similar to that for green plant cells (Jennings, 1991c). Some calculations with respect to water flow along hyphae of *Serpula lacrymans* to form droplets at the apex (see Chapter 14) suggest that the hyphae of this fungus could have a similar value for L_p (Jennings, 1991c). The conclusion is that, as with higher plant cells, growth in fungi is controlled by wall yielding processes, i.e. equation (13.2) describes what is occurring.

In the light of these considerations the following comments on the results that have been presented above are in order. Consider first the experiments of Kaminskyj *et al.* (1992). In the hyphae of *Saprolegnia ferax*, reduction of hyphal extension by decreased osmotic potential of the

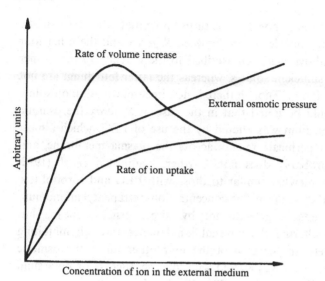

Figure 13.3. The relation between the osmotic potential within a fungus, solute uptake, and growth, as measured by increase in volume when the fungus is growing in a range of external water (= osmotic) potentials. (From Jennings, 1990c.)

medium is accompanied by a reduction of hyphal turgor potential until a certain value is obtained; at lower growth rates the potential is constant. It is presumed that the drop in turgor is because solutes cannot be absorbed in sufficient amount to maintain the osmotic potential required (Figure 13.3). If yield turgor (Y) remains the same, then water uptake and growth decrease. One would expect turgor and growth rate to decline together. Thus, it can be argued that the constant turgor that is observed below a certain medium water potential is due to a decrease in ϕ (and is not a consequence of turgor regulation as proposed by Kaminskyj *et al.* (1992).

If one focuses on the data for *Mucor hiemalis* and *Aspergillus wentii* (Figure 13.2) and for *Phytophthora cinnamomi* and *Fusarium equisesti* (Table 13.1), in virtually every instance (the exception of *A. wentii* is dealt with below) it can be seen that reduction of the osmotic potential in the medium leads to a reduction in the growth rate. However, this reduction is accompanied by an increase in turgor potential. One must presume that, though sufficient solutes can be absorbed to decrease the hyphal osmotic potential, the hyphal wall becomes less able to extend – there is a decrease in the value of ϕ. As with *S. ferax*, reduced growth due to osmotic stress is brought about by a change in the properties of the wall.

The difference in the response of the turgor potential, when the osmotic potential of the medium decreases, between *S. ferax* and the other four fungi referred to above has been ascribed to the inability of the former fungus to absorb sufficient solutes, whereas the latter four fungi are not subject to this limitation. Though there is no direct information on solute absorption it should be noted that in the case of *S. ferax* the osmotic potential of the medium was altered by the use of PEG, which almost certainly does not permeate readily across the plasma membrane, and sorbitol, which probably does not. Further, Kaminskyj *et al.* (1992) obtained results somewhat similar to those with PEG and sorbitol (see Figure 13.1) with the use of varying concentrations tetrapentyl ammonium chloride, which reduces growth not by direct osmotic means but in some way by altering the internal ion balance through inhibiting potassium channels. In the case of the four other fungi, the osmotic potential of the medium was altered with glucose, sucrose or potassium chloride, all three of which are readily absorbed by fungi.

Turning to *Xeromyces bisporus* (Table 13.1) it can be seen that, when the osmotic potential of the medium is decreased, growth increases, as does the turgor potential. Clearly the absorption of solutes is increased as the osmotic potential of the medium is decreased thus increasing the turgor potential, but the extensibility of the wall is such that hyphal extension is increased with the increased turgor potential. Undoubtedly, the rate of absorption of osmotically active solute is able to match the increased growth rate (Figure 13.3). If the osmotic potential of the medium is reduced sufficiently, it is possible that the rate of hyphal extension is reduced by a change in wall properties (but results are too few for one to be certain on this point). The stimulation of growth of *Aspergillus wentii* as the osmotic potential of the medium is reduced by a small amount can be explained in the same way as for *X. bisporus*.

The consideration earlier of the observations of Bartnicki-Garcia & Lipmann (1972) suggested that changing wall properties are a very important element in the osmotic behaviour of fungi. The discussion above indicates similar conclusions. Satisfactorily, there is now other experimental evidence that changes in wall properties can underpin either how a fungus behaves osmotically or changes in growth rate brought about by other environmental factors. Money & Harold (1992) have provided evidence that hyphae of *Achlya bisexualis* adapt to changes in external osmotic potential by changing the properties of the walls of their apical cells. When hyphae were exposed to external osmotic stress, the turgor potential (measured by a pressure probe) fell in proportion to the drop in osmotic potential of the medium (produced either by sucrose or PEG-300). Yet growth rate was unaffected until turgor was a third of the normal level. Hyphae responsed by softening their apical cell walls, determined by measurements of the pressure at which hyphae burst. When osmolytes were excluded from the wall, e.g. PEG-6000, there was reduced turgor. However, growth was also inhibited. Under these circumstances, wall softening was not induced. While these results are somewhat in conflict with those for *Saprolegnia ferax* obtained by Kaminskyj *et al.* (1992), it needs to be remembered that not only are the two fungi different but the growth media used were very different also. Nevertheless the thesis that wall properties are important in the osmotic properties of a fungal hypha is supported. Another study, indicating that changes in growth rate are associated with changed mechanical properties, is that of Ortega, Manica & Keanini (1988*b*). They showed that the light and avoidance response (movement away from a surface) of sporangiophores of *Phycomyces blakesleeanus* can be mimicked by increasing the turgor potential, yet neither response itself increases turgor. Incidentally, it has been proposed that the avoidance response is mediated by a rise in water vapour pressure between the sporangiophore and the surface. This increased water vapour pressure maintains a more hydrated wall, assisting with the movement of the sporangiophore (Gamow & Bottger, 1982). The inhibitory effects of PEG-6000 on the hyphal extension of *A. bisexualis* referred to above may be due to dehydration of the wall.

More information is much needed about the mechanical properties of hyphal cell walls. There is a significant body of information for the wall of the sporangiophore of *Phycomyces blakesleeanus* (Cerda-Olmedo & Lipson, 1987; Ortega, Manica & Keanini, 1988*b*) but that wall is not like that of the extending hyphal apex, e.g. the presence of a cuticle and a high solute content (Cosgrove *et al.*, 1987). The proper evaluation of the wall mechanical properties must be made *in vivo* and the correct approach must be by stress relaxation (Cosgrove, 1985, 1986). Essentially pressure is introduced into the cell under conditions in which there is minimum water uptake and then the rate of stress relaxation is examined.

The theory behind stress relaxation is as follows (Cosgrove, 1985). When turgor pressure changes in a non-growing cell, the change is related to the change in volume via the volumetric elastic modulus ε, which is given by:

$$\varepsilon = \frac{dP}{dV} V \tag{13.4}$$

It is important to realise that ε is a property of the ability of a cell to take up or lose water, leading to the volume change dV. Because of this change, the wall is deformed elastically, its stress is altered and, as a consequence, turgor pressure is altered. ε should not be confused with bulk modulus K, which is formally similar to ε, but depends on the direct mechanical analysis of the wall (Cosgrove, 1988). ε should be considered as a capacitance factor with respect to water movement between the cell and its exterior.

Equation (13.4) can be rearranged, and with both sides divided by dt, we obtain:

$$\frac{1}{V}\frac{dV}{dt}=-\frac{1}{\varepsilon}\frac{dP}{dt} \tag{13.5}$$

It should be apparent that equation (13.2) (p. 403) for a growing cell needs modification to include the elastic term. Thus:

$$\frac{1}{V}\frac{dV}{dt}=\phi(P-Y)+\frac{1}{\varepsilon}\frac{dP}{dt} \tag{13.6}$$

If there is no volume change, $(1/V)\,dV/dt=0$; then:

$$-\frac{1}{\varepsilon}\frac{dP}{dt}=\phi(P-Y) \tag{13.7}$$

This is the equation central to a stress relaxation experiment. If the turgor falls and the overall dimensions do not change, the elastic elements in the wall shrink as the wall yields. Equation (13.7) can be rearranged and, using $dP=d(P-Y)$ for the constant Y, one obtains:

$$\frac{d(P-Y)}{dt}=\phi\varepsilon(P-Y), \tag{13.8}$$

or, by differentiation:

$$P(t)=Y+(P_0-Y)\times e^{-\varepsilon\phi t} \tag{13.9}$$

where $\varepsilon\phi$ is the rate constant for the decay of $P(t)$, the turgor pressure at time t after the start of stress relaxation and P_0 is turgor pressure at zero time, i.e. at the beginning of the experiment. $\varepsilon\phi$ defines in an unambiguous manner the properties of the wall in terms of expansion.

A study making use of the above theory has been used in an examination of water relations of sporangiophores of *P. blakesleeanus* (Ortega, Keanini

Table 13.3. *Water* (Ψ), *osmotic* (Ψ_π) *and turgor* (Ψ_P) *potentials* (*MPa*) *of basal and apical regions of hyphae and rhizomorphs of four species of basidiomycetes grown under different conditions*

Species/tissue	Growing conditions	Basal			Tip		
		Ψ	Ψ_π	Ψ_P	Ψ	Ψ_π	Ψ_P
Armillaria mellea							
Rhizomorphs	Field-grown	−2.1	−3.1	1.0	−3.1	−3.6	0.5
Phallus impudicus							
Hyphae	Over agar	−1.1	−1.3	0.2	−3.3	−3.4	0.1
Rhizomorphs	Over agar	−1.5	−1.9	0.4	−1.6	−1.9	0.3
Rhizomorphs	Field-grown	−0.85	−1.2	0.35	−1.7	−1.9	0.2
Phanerochaete velutina							
Rhizomorphs	Over agar	−0.7	−0.95	0.25	−2.1	−2.3	0.2
Rhizomorphs	Field-grown	−0.9	−1.3	0.4	−1.6	−1.7	0.1
Serpula lacrymans							
Hyphae	Over agar	−0.5	−1.3	0.8	−2.1	−2.2	0.1
Rhizomorphs	Over Perspex	−1.0	−1.5	0.5	−1.8	−2.0	0.2

From Eamus & Jennings (1984).

& Manica, 1988*a*). Whether stress relaxation studies are possible with hyphae remains questionable at the moment.

Conditions far from equilibrium

The discussion to this point has been based on the premise that there is near equilibrium between a growing hypha and the surrounding medium. However, many fungi, particularly wood decay basidiomycetes, which can grow from a nutrient source over a solid non-nutrient substratum incapable of supplying moisture to the growing hyphae. As shown below, such mycelia maintain turgid extending hyphal apices by hydrostatic pressure-driven flow of solution along the hyphae. Even on agar, the necessary gradient of pressure for this flow can be demonstrated (Eamus & Jennings, 1984; Table 13.3).

Eamus & Jennings (1986*b*) examined the relationship between turgor and growth in the mycelium on agar of two wood decay basidiomycetes, *Phallus impudicus* and *Serpula lacrymans*. Colony radial growth rate, altered either by changing the osmotic potential of the medium with sucrose or by changing the concentration of the nutrient (malt) agar, was linearly related to turgor potential of the mycelial front (Figure 13.4). The water potential of the mycelium as determined by thermocouple

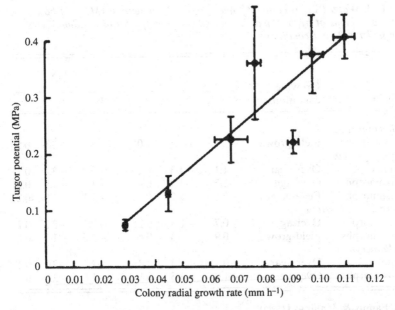

Figure 13.4. The relation between turgor potential at the colony margin and colony radial growth rate of *Phallus impudicus* growing on sucrose-adjusted media of different osmotic potentials. (From Eamus & Jennings, 1986*b*.)

psychrometry, was shown to be less than that of the medium. Turgor potential was taken as the difference between the water potential so determined and the osmotic potential determined in the same way but after freezing and thawing of the mycelium.

Eamus & Jennings (1986*b*) conceded that their data must be questioned, since the lines of best fit between colony radial growth rate and turgor potential, when extrapolated to zero growth rate, gave negative values for turgor potential. Theoretically, as discussed above, one would expect that, even at zero growth rate, there must be positive turgor equal to the yield turgor.

There are three possibilities for the seemingly anomalous values for the turgor potential at low growth rates. One possibility, and this applies to all determinations of turgor potential based on thermocouple psychrometry, is that as the water potential of the medium decreases the solute potential values become an increasing overestimate owing to a significant volume of solution external to the plasma membrane, i.e. in the wall. If there were to be such a volume of solution, the potential of its solutes would be higher than that of the osmotically active material within the plasma membrane. On freezing and thawing, the protoplasmic solute potential would be increased owing to dilution by the solution external to the plasma membrane. It is also possible that the values for turgor potential, while representing the bulk of the sample used, nevertheless underestimate that potential in the apical compartment of the extending hyphae; there is some evidence for this in data presented in Table 13.2. The third possibility is that the turgor potential values are an underestimate because turgor is maintained at the hyphal front at least in part as a result of hydrostatic pressure generated within the mycelium. When hyphae are severed when the sample is prepared for thermocouple psychrometry, the hydrostatic pressure gradient is dissipated.

In addition to the above possible sources of error or causes of misinterpretation, it needs to be said that there can be errors arising from thermocouple psychrometry, particularly when the technique is used for determining water potential, since there can be uncertainty as to whether or not there is equilibrium between the tissue and the vapour phase above it (Brown & van Haveren, 1972). However, the study of Thompson, Eamus & Jennings (1985), which used the same methods as Eamus & Jennings (1986b) (see Chapter 14), showed changes in water and turgor potential that related well to growth rates and to measured changes of direction of water movement within the mycelium.

Eamus & Jennings (1986b) favoured the view that turgor at the apex of *P. impudicus* and *S. lacrymans* growing over agar was due in part to the hydrostatic pressure generated in the mycelium further back. Results consistent with this view were obtained in an experiment concerned with growth from a medium of high water potential to one of lower water potential and vice versa. Thus, the increase in turgor with increasing growth rate on a medium of more favourable osmotic potential is due in part to the hydrostatic pressure within the mycelium. The observed relationship between growth rate and turgor is therefore not simple in the above two fungi. It is possible that a similar complexity underlies the relationship between turgor and growth in other fungi. While Money (1990b) has shown by pressure probe analysis that there do not appear to be turgor gradients in hyphae of *Achlya bisexualis*, this observation alone should not be taken to discount the possibility in fungi other than the wood decay basidiomycetes. Some early studies on translocation in hyphae by Schütte (1956) suggested that the oomycete *Rhizopus* is able to translocate solutes by pressure-driven flow of solution; there is very good evidence that this can be so in hyphae of *Neurospora crassa* (Girvin & Thain, 1987; Thain & Girvin, 1987; Chapter 14).

Table 13.4. Concentrations (mM) of what are believed to be the major solutes in four fungi growing on media used for normal growth

Solute(s)	Saccharomyces cerevisiae	Neurospora crassa	Aspergillus nidulans	Neocosmospora vasinfecta[a]
K^+	261	37	161	197
Na^+	7	152	21	8
Mg^{2+}	14	—	15	37
Trehalose	—	—	—	16
Organic acids and P_i	—	—	—	69
Organic acids	44	—	—	—
P_i	26	12	—	—
P	14	300	11	302
Amino acids	224	—	313	180
Polyols	—	—	—	—
Reference	Conway & Armstrong (1961)	Slayman & Slayman (1968)	Beever & Laracy (1986)	Kelly & Budd (1990)

—, no data obtained.
[a] Values in mmolal.

Generation of the osmotic (solute) potential

The osmotic (or sometimes solute) potential of fungi is ultimately dependent on the absorption of solutes from the external medium. Though matric forces must make a contribution to the cell or hyphal water potential, that contribution is likely to be rather small. Therefore the generation by a fungus of water potential appropriate to its environment is heavily dependent on the availability of appropriate solutes. In order to determine what solutes contribute to the osmotic potential of a fungus one must analyse a suitable amount of its cells or mycelium and, on the basis of a knowledge of the water content of the protoplasm, determine the concentration of individual solutes within it. The osmotic potential obtained by totalling the potentials due to individual solutes should be equal to the potential measured independently. In a wall-less cell the osmotic potential should be equal to the water potential of the medium. For walled cells or hyphae, however, there must be sufficient turgor potential, i.e. the osmotic potential must be significantly higher than the water potential of the medium.

Eamus & Jennings (1986a), Jennings (1990c) and Blomberg & Adler (1992) have considered at length the problems of determining those solutes contributing to the osmotic potential. Essentially, there are two major problems. Determining the amount of water in the protoplasm and establishing what proportion of the solutes present in the cell or hypha are located in the wall external to the protoplasm.

Surprisingly, there is little information about those solutes contributing to the osmotic potential in fungi growing on normal media (osmotic potential close to zero). Table 13.4 presents those data that are available. Only for one fungus is it known to what extent all the osmotically active species have been accounted for (some 63% in *Neocosmospora vasinfecta*). The message from the data is that ionic species probably make the major contribution to the osmotic potential, the particular species involved almost certainly depending on the medium used. Thus, sodium appears to be more important than potassium in *Neurospora crassa*, but that is due to the fungus being grown on a medium low in potassium and high in sodium. The capability of amino acids to make a more significant contribution than carbohydrates should be noted, as also should the low concentration of organic acids in the fungi examined. However, it is known that organic acids can play a more significant role in fungi (see Chapter 12).

With respect to growth at low osmotic potential, Table 13.5 presents information about the percentage contribution of the major classes of

Table 13.5. Percentage contribution, where known, to the osmotic potential (Ψ_π) of major classes of solute within the protoplasm of various fungi together with the value for the osmotic potential either calculated from the concentrations of the known solutes or determined directly on the mycelium

Organism osmoticum added to basal medium (water potential)	Contribution as					Identity of major organic solute	Ψ_π		Reference
	Carbohydrates	α-Amino nitrogen	Inorganic ions	Organic acid	Major organic		Calculated (−MPa)	Measured (−MPa)	
Aspergillus nidulans[a]									
NaCl (−3.86 MPa)	67	—	32	—	26	Mannitol	4.63	—	Beever & Laracy (1986)
Chrysosporium fastidium[b]									
Glucose (−10 MPa)	53	—	11	—	24.5	Glycerol	7.6	11.85	Luard (1982a)
Dendryphiella salina[a]									
NaCl									Wethered et al. (1985)
(−0.96 MPa)	28	31	41	0	14	Glycerol	1.63	—	
(−1.92 MPa)	27	14	59	0	14	Glycerol	1.63	—	
MgCl₂									
(−0.96 MPa)	29	20	51	0	19	Mannitol	1.88	—	
(−1.92 MPa)	19	11	66	4	8	Mannitol	4.32	—	
Inositol									
(−0.96 MPa)	26	34	32	8	14	Inositol	1.58	—	
(−1.92 MPa)	33	26	33	8	23	Inositol	2.30	—	

Neocosmospora vasinfecta[b]								
KCl (−2.5 MPa)	60	18	5	28[c]	Arabitol	2.69	—	Kelly & Budd (1990)
Penicillium chrysogenum[b]								
Glucose (−10 MPa)	44	15	—	24	Glycerol	7.8	13.4	Luard (1982a)
KCl (−10 MPa)	20	74	—	14	Glycerol	11.9	12.7	
Phytophthora cinnamomi[b]								
Sucrose (−2 MPa)	33	58	—	33	Sucrose	3.9	3.15	Luard (1982b)
KCl (−2 MPa)	0	175	—	35	Proline	7.7	3.43	
Thraustochytrium aureum[a]								
Seawater								
50% (−1.24 MPa)	1	9	0	0.5	Proline	1.36	—	Wethered & Jennings (1985)
75% (−1.85 MPa)	1	11	0	0.5	Proline	1.78	—	

—, Not determined
[a] Percentage of calculated.
[b] Percentage of measured.
[c] Estimated.
From Jennings (1990c).

solutes to the protoplasmic solute potential, either calculated from the concentrations of the major solutes or determined directly. As with Table 13.4, the full complement of solutes has not been determined. Only for *Aspergillus nidulans* and *Dendryphiella salina* has an attempt been made to correct for ions associated with the wall. There must be some doubt about the data for *Chrysosporium fastidium* and *Penicillium chrysogenum* grown at −10 MPa glucose, since the calculated protoplasmic solute potential is greater than both the measured potential and the water potential of the medium. On the other hand, it seems likely in the case of *Phytophthora cinnamomi* that there are considerable errors in the determination of some of the solutes, since the calculated solute potential is so much less than the water potential of the medium. The most likely possibility is that a considerable amount of potassium was bound to the wall. In this position the ion would not contribute to the solute potential of the protoplasm and would not have an osmotically active counter-ion, the concentration of which would have been assumed in calculating the solute potential.

With respect to the data in Table 13.5, the following general points can be made. Under conditions of water stress, both ions and organic solutes can make a significant contribution to the osmotic potential. The extent to which either class of solutes plays a major role depends on the organism. Thus, in *Dendryphiella salina* and *Thraustochytrium aureum*, ions make the bigger contribution to the osmotic potential; in *Aspergillus nidulans* it is organic solutes, particularly carbohydrates, that are more relevant. When the osmotic potential is lowered by ions, the latter can be absorbed in order to help to generate the requisite solute potential. The same is true for organic solutes. There is a considerable contribution of sorbitol to the osmotic potential of *Walemia sebi* and *Xeromyces bisporus* grown in a medium with an osmotic potential of −30.7 MPa. After 60 days' growth there was in the two fungi five and six times, respectively, more sorbitol than glycerol, the next highest solute in the protoplasm. Likewise there was as much glucose as glycerol in *Chrysosporium fastidium* grown in a medium with an osmotic potential of −10 MPa generated with glucose. Glucose was at a high concentration in *Penicillium chrysogenum* similarly grown but absent when the fungus was grown in a medium with an osmotic potential of −10 MPa generated with potassium chloride. Likewise *Phytophthora cinnamomi* accumulates sucrose when grown in high concentrations of the sugar.

Finally in this section, there is a need to comment on the polyols that accumulate within higher fungi when the latter are exposed to low osmotic

Table 13.6. *Concentrations (M) of individual polyols and inositol in mycelium of* Dendryphiella salina *after 48 h growth in media containing various osmotica at two osmolal concentrations*

External osmoticum (osmol (kg water)$^{-1}$)	NaCl		MgCl$_2$		Na$_2$SO$_4$		Inositol	
	0.4	0.8	0.4	0.8	0.4	0.8	0.4	0.8
Glycerol	0.098	0.197	0.038	0.121	0.039	0.062	—	—
Erythritol	—	0.006	0.006	—	0.002	0.002	—	—
Arabitol	0.007	0.048	0.034	0.065	0.033	0.051	—	0.013
Mannitol	0.086	0.102	0.142	0.128	0.126	0.129	0.082	0.082
Inositol	—	—	—	—	—	—	0.091	0.214
Total	0.191	0.353	0.220	0.314	0.200	0.244	0.173	0.309

—, None detected.
Wethered *et al.* (1985).

potentials. As can be seen from Table 13.6, glycerol is often an important and sometimes (see pp. 426–33) by far the most important polyol contributing to the protoplasmic solute potential in fungi under water stress. Nevertheless, there are conditions when glycerol is less important and also when the mix of polyols can change. Wethered *et al.* (1985) showed that, when *Dendryphiella salina* is grown at low osmotic potentials of the same magnitude, the amounts of individual polyols can vary depending on the external osmoticum but the total soluble carbohydrate concentration remains relatively constant (Table 13.6). Glycerol may not be present. Equally, Nobre & da Costa (1985) have shown that *Debaryomyces hansenii*, which normally accumulates glycerol under saline conditions, does not produce it when erythritol replaces glucose as the carbon source, the tetritol being accumulated itself.

The constancy of the total soluble carbohydrates (probably almost entirely polyols under most conditions of water stress) in *Dendryphiella salina* may reflect a more general situation in fungi. There is evidence for a similar constancy in *N. crassa* when glucose and fructose have been compared as carbon sources for growth in a medium where the osmotic potential was reduced by 0.4 M NaCl (Ellis, Grindle & Lewis, 1991). With both carbon sources, glycerol and mannitol increase with decreased osmotic potential of the medium to give the same total concentration, but while with glucose to the ratio glycerol:mannitol is *c.* 9:1 that with fructose is *c.* 2:1. There are indications of a similar constancy also in

Table 13.7. Approximate values for the lowest water potentials (Ψ) allowing growth of algae, fungi, bacteria in culture

Group	Ψ (MPa)	Algae	Fungi	Bacteria
1	−2	Freshwater species Chlorella emersonii Scenedesmus obliquus	Many wood decay species Serpula lacrymans	Mycoplasma gallisepticum
2	−5	Brackish and marine species	Many phycomycetous and coprophilous species Phytophthora cinnamomi Pythium ultimum	Many Gram-negative species Pseudomonas spp. Enterobacteriaceae
3	−5 to −20		Soil fungi such as Alternaria alternata Cladosporium cladosporoides Fusarium culmorium Rhizoctonia solani Saccharomyces cerevisiae	Many Micrococcaceae Staphylococcus aureus Most Bacillus spp. Many Lactobacillaceae Clostridium botulinum Moderate halophiles Paracoccus (Micrococcus) halodenitrificans Vibrio parahaemolyticus
4	−20 to −40		Many Aspergillus, Eurotium and Penicillium spp. Osmophilic yeasts on salt Debaryomyces hansenii Zygosaccharomyces rouxii	
5	Below −40	Dunaliella parva and D. viridis	Chrysosporium fastidium Monascus bisporus Some isolates of Aspergillus restrictus Zygosaccharomyces rouxii on sugars	Actinospora halophila Halophilic bacteria Halobacterium halobium

From Jennings (1990c).

Aspergillus nidulans, though there is an exception when glycerol is the carbon source (Beever & Laracy, 1986).

The role of polyols in the fungal response to water stress is discussed on pp. 443–6.

Tolerance to low water potential

Given that fungi are a significant component of the marine biota, it can be seen that they have been remarkably successful in adapting to environments of low water potential. The usual term for those fungi which have so adapted is 'osmophile' (von Richter, 1912; Jennings, 1990c). There are synonyms such as osmotolerant (Anand & Brown, 1968), osmotophilic (van der Walt, 1970), osmoduric (van der Walt, 1970), osmotrophic (Sand, 1973), xerophilic (Pitt, 1975) and xerotolerant (Brown, 1976). I leave for the moment the question of whether or not the various terms can be considered as true synonyms and, if so, whether one or more terms should be used in preference to the others. However, there is no doubt that the Fungal Kingdom contains species that can live at some of the lowest water potentials able to allow the growth of microbes (Table 13.7).

Consideration of the physiology of fungi able to grow at low water potentials is complicated by those that are able to grow in saline environments. Fungi are able to absorb sodium and chloride. If these two ions were to be used almost exclusively to generate the required osmotic potential in those fungi growing in saline environments, then the concentration of sodium chloride in the protoplasm would be toxic (the evidence for which is presented on pp. 435–8). For that reason, fungi attempt to exclude sodium and chloride from their protoplasm. The mechanisms by which exclusion of sodium occurs, and some of the consequences, are also considered in Chapter 10. However, virtually nothing is known about chloride transport in fungi, except for some interesting studies on the process in *Neocosmospora vasinfecta* (Miller & Budd, 1971, 1975a,b, 1976a,b). In that fungus, there is an inducible transport system inhibited by exogenous glucose, and the transport of the ion is against the electrochemical potential gradient. However, apart from the matter of dealing with sodium chloride, those fungi living in saline environments have many features similar to those of fungi living in other environments of low water potential. The physiology of the whole group of fungi has been reviewed by Brown (1976, 1978, 1990), D. M. Griffin (1977, 1981), Jennings (1990c), Blomberg & Adler (1992, 1993) and Hocking (1993).

There is no doubt that fungi differ very considerably in their ability to withstand low water potentials. It ought to be possible to discriminate between those fungi that grow well in solutions of low osmotic potential and those that are better able to withstand low matric potentials, namely the measure of those forces holding water within a solid matrix. D. M. Griffin (1981) gave a very good brief account of the physical chemistry of systems developing strong negative matric potentials. I do not propose to say much about how fungi grow in media in which matric forces predominate. This is because I believe that as yet there are no satisfactory techniques for studying how a fungus grows in media of low matric potential in which the problem for the fungus may not be the seemingly unfavourable conditions for water absorption but rather the acquisition of nutrients. For these reasons it is preferable to classify fungi with respect to water stress by their growth response to osmotic potential. D. M. Griffin (1981) has produced such a classification. He pointed out that the groups are in no way discrete; each merges into the next. However, the groups do provide a useful guide to the likely response of fungi belonging to broad ecological groupings.

Before listing these groups it is necessary to define some terms. This is best done with reference to Figure 13.5. Here, the growth rate of a fungus is given as a function of the water (for our purposes here osmotic) potential of the medium. There are three so-called cardinal osmotic potentials (Blomberg & Adler, 1992): maximum, optimum and minimum. The maximum is usually (but not always, see below) close to zero, growth being limited then by availability of carbon or some other major element for growth. The optimum is that potential which gives optimal growth and may be that osmotic potential at which the flux of solute(s) contributing to the osmotic potential of the fungus limits growth (see Figure 13.3). As can be seen from Figure 13.5, it is not difficult to conceive that there may be no or only a very indistinct optimum. The minimum osmotic potential is that at which growth is barely detectable.

The groups proposed by D. M. Griffin (1981) are as follows:

Group 1. Extremely sensitive species with an optimum at *c.* −0.1 MPa and a minimum at *c.* −0.2 MPa. Wood decay fungi – but not those growing on dead, attached branches (Griffith & Boddy, 1991) – and some soil basidiomycetes belong to this group.

Group 2. Sensitive species with an optimum of *c.* −1 MPa and a minimum of *c.* −5 MPa. Most phycomycetous and coprophilous fungi, as well as many other filamentous fungi, belong to this group.

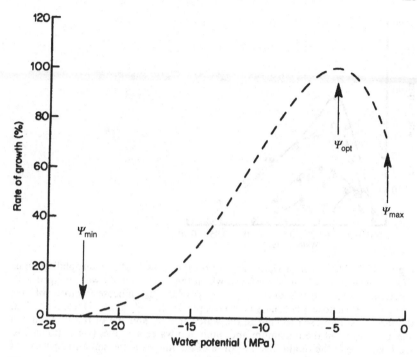

Figure 13.5. A generalised graph of the relative rate of growth in relation to the water potential of the growth medium. Indicated are the cardinal water potentials of growth, namely Ψ_{max}, Ψ_{opt} and Ψ_{min} (respectively maximum, optimum and minimum). (From Blomberg & Adler, 1992.)

Group 3. Species of moderate sensitivity with an optimum of c. -1 MPa and a minimum of c. -10 to -15 MPa. Species of *Fusarium*, a number of other soil fungi and *Saccharomyces cerevisiae* belong to this group.

Group 4. So-called 'xerotolerant' species, with an optimum of -5 MPa or lower and a minimum of -20 to -40 MPa or even -50 MPa. Many yeasts such as *Debaryomyces hansenii* and *Zygosaccharomyces rouxii* and a number of filamentous fungi, of which those within the genera *Aspergillus*, *Eremascus*, *Eurotium* and *Penicillium* and the species *Wallemia sebi* are characteristic.

Group 5. The so-called 'xerophilic' species, which do not grow at high osmotic potentials, namely c. -4 MPa and have a minimum at -40 MPa or less (Hocking, 1993; Figure 13.6). All isolates of the species *Chrysosporium fastidium* and *Monascus bisporus* and some isolates of *Aspergillus restrictus* and *Zygosaccharomyces rouxii* fall into this group, as does *Xeromyces bisporus* (Table 13.1). The reason why these fungi do not

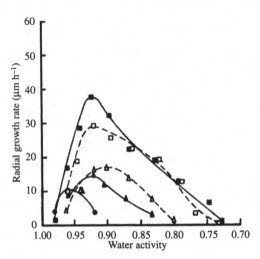

Figure 13.6. The effect of water activity on the growth of the xerophilic fungus *Chrysosporium fastidium* on media in which the water activity was changed with different solutes. (●) Sodium chloride, pH 4.0; (■) glucose/fructose, pH 4.0; (□) glucose/fructose, pH 6.5; (▲) glycerol, pH 4.0; (△) glycerol, pH 6.5. Note that, though the solute used to change the water activity and pH of the medium can affect the value of the maximum and minimum water activities (water potentials, see Figure 13.5) the qualitative response of the fungus is the same irrespective of the medium. (From Pitt & Hocking, 1977.)

grow at high osmotic potentials is not clear, though it has been argued earlier with respect to *X. bisporus* that, at relatively high osmotic potentials of the medium, there may be problems of generating the necessary net influx of osmotically active solute to realise the required turgor potential for growth. In this respect it is very interesting to find that *X. bisporus* has been shown to leak large quantities of glycerol, particularly at high osmotic potentials (Hocking, 1993). It has been suggested that this fungus may have a metabolism that is energetically inefficient, with a requirement for a high concentration of sugars. The fungi in this group are clearly candidates for further study.

Internal osmotic potential as a function of the external water potential

Some theoretical considerations

So far, I have been concerned with growth of fungi in media of constant water (osmotic) potential. While such conditions are not strictly steady state, the time-scale of change is such that they can be accepted as such.

Under natural conditions there might be more rapid changes in the water potential. If the water potential of the medium decreases, water will be drawn out of the organism; conversely if the water potential of the medium rises, water will enter the fungus. In either case, for normal functioning there needs to be an adjustment of the fungal water potential. Of the two situations, entry of water can result in the starker consequences, namely bursting of the cell. Reference has been made earlier to the bursting of hyphal apices when the mycelium of a filamentous fungus is exposed to a higher external water potential. However in hyphae, the consequences need not be irremediable, since in septate hyphae the subapical or older compartments can be sealed off by blockage of the septal pore by a Woronin body, a proteinaceous inclusion within the cytoplasm (Markham *et al.*, 1987). Under such circumstances, renewed growth is achieved by older compartments producing new branch hyphae (Luard, 1982*b*; Trinci & Collinge, 1974).

When a walled cell adjusts to a new osmotic potential in the medium, there are two points of reference for that adjustment. The cell can regulate to a set volume or it can regulate to a set turgor. The matter has been well discussed with particular reference to green plant cells (Cram, 1976; Zimmermann, 1978; Reed, 1984). When walled cells are investigated, it is found frequently that, when the water potential of the medium is changed, the osmotic potential alters such that the turgor potential is restored close to a value achieved in the original medium. However, in wall-less cells, there is evidence that it is the volume that is restored to the initial value. This has led to the use of the two terms 'turgor regulation' and 'volume regulation' (Cram, 1976). But these two terms assume that there is no relation between turgor and cell volume. That is not so; change in both turgor and volume depend on the elastic properties of the wall (Dainty, 1976; Zimmermann, 1978; Zimmermann & Steudle, 1978). Thus:

$$\Delta P = \varepsilon \Delta V / V_r \qquad (13.10)$$

where V_r is the reference volume, the other terms being defined as above. It should be noted that the more elastic the wall the smaller the value of ε. Further, ε is not constant and varies with the pressure. Since this is so, the more appropriate formulation is:

$$\varepsilon = (dP/dV)V_r \qquad (13.11)$$

As one can see from equations (13.10) and (13.11), for a cell with relatively elastic walls, a small change in the external osmotic potential leads to a considerable change in osmotic volume. Therefore there is a great deal to be said for having a rigid wall if the fungus is to face only minor fluctuations in the external osmotic potential. Conversely, an elastic wall will be beneficial if the cell is exposed to a sudden fall in the external osmotic potential. But that is assuming that there is no corrective action with respect to the protoplasmic osmotic potential. If the osmotic potential is decreased, turgor potential will be increased, thus helping to minimise cell volume decrease.

It can be seen that it is only with wall-less cells that one can speak with confidence about volume regulation. In walled cells it is not clear whether turgor or volume is the output of any control process. This means, as Reed (1984) has pointed out, that it is not easy to produce terms that are necessarily appropriate to all situations. He has suggested that qualifying adjectives can be used. Thus, one can speak of 'full' or 'complete' and 'incomplete turgor'. Equally, 'turgor maintenance' might be used if positive turgor is still generated subsequent to a change in the osmotic potential of the medium. However, it is clear that 'osmoregulation' (or 'osmotic regulation'), a term frequently used by plant physiologists and microbiologists alike, is not appropriate unless it can be unequivocally demonstrated that the osmotic potential of the protoplasm is the reference point. Reed favours the more general use of the term 'osmo-acclimation' – a term which I commend – using other terms such as 'turgor regulation' ('full' or 'partial') and 'volume regulation' ('complete' or 'incomplete') only when the data are sufficient to indicate that such regulation is taking place.

Two other terms also need to be defined. They are 'upshock' and 'downshock'. Both refer to changes in the *concentration* of osmotica in the external medium. 'Upshock' (or 'hyperosmotic shock') indicates an *decrease* in osmotic potential; 'downshock' (or 'hypoosmotic' shock) is the converse.

Experimental observations

First a word about the extent to which cell or hyphal volume is maintained when the osmotic potential of the medium is reduced. The few facts on this matter have been well considered by Blomberg & Adler (1992). Protoplasm must be considered osmotically in terms of the active volume, which changes according to whether there is inflow or outflow of water, and the inactive or non-osmotic volume. The osmotically inactive volume is the value of the intercept on the y-axis of a plot of cell volume as a function of the reciprocal of the osmotic potential of the medium, assuming that the cells behave as perfect osmometers at lower osmotic potentials,

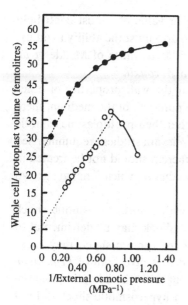

Figure 13.7. Boyle–van't Hoff plots for (●) whole cells and (○) protoplasts of *Saccharomyces cerevisiae*, showing the relationsip between volume and osmotic pressure (potential) of the medium. From (Meikle *et al.*, 1988.)

i.e. the cells obey the Boyle–van't Hoff relationship. This holds for cells of *S. cerevisiae* below an external osmotic potential of −2.48 MPa (Meikle, Reed & Gadd, 1988; Figure 13.7). That potential is the point at which the turgor potential (pressure) is dissipated, though why the cells should behave as prefect osmometers when seemingly plasmolysed is not clear – it could be presumed that the plasma membrane is anchored to the wall such that there is coordinated shrinkage (but see pp. 433–5). The plot can also be used to determine, for any external osmotic potential, the turgor and internal osmotic potentials and the volumetric elastic modulus (ε) via equation (13.10). From these studies Meikle *et al.* (1988) found ε for stationary phase cells to be 1.95 MPa, whereas ε from mid-exponential phase was found to be much higher, namely 3.25 MPa.

The above values mean that the walls of cells of *S. cerevisiae* are relatively elastic and therefore weakly buffered against water loss with transference to media of low water potential. Rose (1975) showed that, within 30 min of exposure of cells grown in normal nutrient medium to a medium of osmotic potential −17 MPa, their volume was reduced to 33% of the original – probably the inactive osmotic volume. Conversely, the much more osmotolerant species *Zygosaccharomyces rouxii* retain 80% of their volume under the same conditions. A higher value of ε must contribute to the greater resistance to water loss but so also must the ability to adjust the internal osmotic potential. Studies similar to those of Meikle *et al.* (1988) on *S. cerevisiae* need to be made on *Z. rouxii*.

Though almost nothing is known about the wall properties of other fungi subjected to a change in the osmotic potential of the medium, there is a body of information about how fungi alter their protoplasmic osmotic potential in response to such a change. The relevant studies are summarised in Table 13.8. The information presented therein should not be treated in isolation; there is a need to refer to other studies in which fungi are grown *ab initio* in media of low water potential.

The first matter to draw attention to with respect to osmotic shock studies is that either an upshock or a downshock may be detrimental to growth. It has already been noted that hypoosmotic shock may cause hyphal tip bursting. Luard (1982*b*) showed that mycelium of *Penicillium chrysogenum* exposed to a new medium but with the osmotic potential reduced to $-10\,\text{MPa}$ and therefore given a hyperosmotic shock did not resume growth. The same occurred with *Chrysosporium fastidium* when the mycelium was transferred from a medium with osmotic potential lowered to $-10\,\text{MPa}$ to one with a potential lowered to $-20\,\text{MPa}$.

With hyperosmotic shock, there are transient changes, the nature of which appear to depend upon the fungus, the medium and the solute used to change the osmotic potential. Thus, *Debaryomyces hansenii*, growing in continuous culture, with the concentration of sodium changed in a stepwise manner in the medium, contains much more sodium in these semi-transient states than in the final steady state (Burke & Jennings, 1990) (Table 13.9). Likewise, as shown in Table 13.8, when *P. chrysogenum* is exposed to an upshock with glucose that solute is the highest in concentration in the initial stages of adjustment to the lowered osmotic potential of the medium. Other examples are given in the same table. In general terms, the major solute within a fungus exposed to a low osmotic potential is likely to be the solute used to generate that potential, a possibility that was mentioned earlier when I considered the concentrations of solutes in fungi growing in media of low osmotic potential. Beever & Laracy (1986), in confirmation of this, with a variety of solutes in growing mycelium of *Aspergillus nidulans*, showed that the solute used to reduce the osmotic potential of the medium achieved the highest concentration in the mycelium.

As well as the development of high concentrations of the major external solute in the fungus during the first stages of adjustment to a reduced osmotic potential in the medium, there can be transient changes in the internal organic solutes. Thus, in *Neocosmospora vasinfecta* mannitol concentration is doubled after 1 h in mycelium exposed to −6.7 MPa potassium chloride, and then declines to the initial level after 5 h (Kelly & Budd, 1991).

Without exception, in the higher fungi the internal polyol concentrations increase as the external osmotic potential is decreased, the extent of the increase and when it occurs depending on the time-scale of study after the start of the upshock. In many instances, glycerol shows the greatest increase, and there is evidence that it is the polyol likeliest to achieve the highest internal concentration the more the external osmotic potential is reduced (Table 13.8). But there are conditions in which glycerol is not produced in response to low external water potential. In non-growing mycelium of *Dendryphiella salina*, only mannitol and arabitol increase in concentration in response to increasing external sodium chloride concentration (Jennings, 1973), while stationary phase cells of *Debaryomyces hansenii* synthesise only arabitol (Adler & Gustafsson, 1980). Nevertheless, it appears that, when a range of species of yeasts capable of growing in sodium chloride with glucose as the carbon source have been examined, glycerol is the sole polyol whose concentration responds to increases in sodium chloride in the external medium (Meikle *et al.*, 1991). Arabitol appears to be present in most osmotolerant species. In a careful study of cells of *D. hansenii*, *S. cerevisiae* and *Zygosaccharomyces rouxii* growing exponentially with glucose as the carbon source, Reed *et al.* (1987) showed that net glycerol synthesis within the cell can counterbalance osmotically between 63% and 95% of the concentration of sodium chloride in the external medium.

When there is downshock, there is a need for the fungus to increase its osmotic potential by reducing the internal concentrations of solutes. In the case of ions, they must be lost to the medium and there is good evidence for expulsion of potassium. However, with organic solutes, the destination of the carbon in a solute that declines in concentration is not clear. Some must be channelled into insoluble material, but other solutes are lost to the medium. This is often the case with glycerol. In *D. hansenii*, the loss of glycerol with downshock from a medium of 2 M sodium chloride is proportional to the decrease in concentration of sodium chloride; there is no conversion of glycerol to other compounds (André, Nilsson & Adler, 1988). But glycerol is not always the organic solute that is lost. In *N. vasinfecta* with downshock from a medium with a high concentration of potassium chloride to one which has a low concentration, there is very

Table 13.8. *Summary of experiments in which fungi have been subjected to osmotic upshock or downshock and, respectively, the major solutes increased in concentration in the protoplasm or reduced in concentration and were lost to the medium*

Fungus	Method	Upshock or downshock	Treatment[a]	Upshock: Major solute(s) in fungus		Downshock: solute(s) in fungus showing a decline	Solutes lost to the medium	Reference
				Initial phase	After a period			
Penicillium chrysogenum	Mycelium on Cellophane	Upshock	−10 MPa with glucose	Glucose	Glycerol			Luard (1982a)
	Mycelium on Cellophane	Downshock	−10 MPa with glucose			Glycerol, glucose		
Chrysosporium fastidium	Mycelium on Cellophane	Downshock	−10 MPa with glucose			Glycerol, glucose		Luard (1982c)
Phytophthora cinnamomi	Mycelium on Cellophane	Upshock	−2 MPa with KCl	K^+, Cl^-	Proline			
	Mycelium on Cellophane	Downshock	−2 MPa with KCl			K^+, Cl^-, proline		
Saccharomyces cerevisiae	Separated cells	Upshock	−8.5 MPa with NaCl	Glycerol (much less than *Z. rouxii*)			Glycerol	Edgely & Brown (1983)
Zygosaccharomyces rouxii	Separated cells	Upshock	−8.5 MPa with NaCl	Glycerol, arabitol	Glycerol, arabitol		Glycerol, arabitol (both less than *S. cerevisiae*)	
	Separated cells	Upshock	5% glucose to 2% glucose + 3.1 M NaCl	Arabitol	Glycerol		K^+	Onishi & Shiromaru (1984)
Aspergillus nidulans	Mycelial mass	Upshock	−3.3 MPa with NaCl	Na^+, glycerol	Na^+, glycerol, erythritol			Beever & Laracy (1986)
	Mycelial mass	Downshock	−3.3 MPa with NaCl			Glycerol, erythritol, arabitol, mannitol		

Organism	Addition of medium		Osmotic change	Solutes			Reference
Debaryomyces hansenii	Addition of medium	Upshock	NaCl, KCl, Na$_2$SO$_3$, sucrose, range of equimolar concentrations	Glycerol		Glycerol	André et al. (1988)
	Separated cells	Downshock	Range of NaCl concentrations	Glycerol			
Saccharomyces cerevisiae	Loop inoculum	Upshock	−1.76 MPa with sorbitol	Glycerol	Trehalose		Meikle et al. (1988)
Debaryomyces hansenii	Continuous culture	Upshock	Stepwise increase of −5.17 MPa NaCl	Na$^+$, glycerol	Glycerol, reduced Na$^+$		Burke & Jennings (1990)
Neocosmospora vasinfecta	Mycelial transfer	Upshock	−6.7 MPa with KCl		Polyols		Kelly & Budd (1990)
	Mycelial transfer	Downshock	−6.7 MPa with KCl		K$^+$, polyols		
	Mycelial transfer	Upshock	−3.0 MPa with KCl	Arabitol, mannitol	Arabitol		
	Mycelial transfer	Downshock	−3.0 MPa with KCl into growth medium			Glucose, fructose mannitol	Kelly & Budd (1991)
	Mycelial transfer	Downshock	−3.0 MPa with KCl into CaCl$_2$ equimolar with growth medium			Mannitol, arabitol erythritol, glycerol	
Saccharomyces cerevisiae	Separated cells	Upshock	−6.46 MPa with NaCl	K$^+$, Na$^+$, glycerol, trehalose	Na$^+$, trehalose then Na$^+$, glycerol	K$^+$	Singh & Norton (1991)
Zygosaccharomyces rouxii	NaCl added	Upshock	2 M NaCl	Na$^+$, Cl$^-$ (glycerol only in adapted cells)			Yagi (1991)
	Separated cells	Downshock	1 M NaCl	Na$^+$, Cl$^-$, glycerol			

[a] Where given as the osmotic potential change together with the solute, upshock is the consequence of lowering the osmotic potential with the solute, whereas downshock is the replacement of the medium with one of higher osmotic potential due to the absence of the solute.

Table 13.9. *Percentage contribution of potassium, sodium, magnesium, glycerol and arabitol to the total osmotic potential of these solutes within cells of* Debaryomyces hansenii *under different growth conditions*

Growth condition	K^+	Na^+	Mg^{2+}	Glycerol	Arabitol
Chemostat (7.6 mM NaCl)	46 ± 9.6	1.2 ± 0.36	13.3 ± 4.6	15 ± 8.5	24.6 ± 11.2
Step-up[a] (25 mM NaCl)	65	15.7	11.9	—	7.4
Step-up[a] (1.28 M NaCl)	19.7	66.3	2.4	11.5	—
Chemostat (1.5 M NaCl)	13.9 ± 5.2	32.5 ± 8	4.4 ± 1.63	48.1 ± 5.6	1.73 ± 0.17

—, None detected.
[a] No replicate samples taken.
From Burke & Jennings (1990).

little loss of glycerol, the major solutes lost being glucose, fructose and mannitol. However, to date, it has to be admitted that knowledge of what solutes are lost from a fungus given a downshock has been governed by the analytical techniques used. Most analyses have been limited in scope. When cells of *S. cerevisiae* are given a cold osmotic shock, similar to that used on bacterial cells to investigate the periplasmic proteins (Heppel, 1969), i.e. growing cells are suspended in 0.8 M mannitol and then transferred to ice-cold 0.5 mM magnesium chloride, the cells can lose 6%–8% of their cell protein (Patching & Rose, 1971). While the cells are viable, they grow more slowly and solutes are absorbed at slower rates. However, when cells of *Z. rouxii* growing in medium containing 2.56 M sodium chloride are transferred to 100 mM bicarbonate–carbonate buffer at pH 9.0, protein and ultraviolet-absorbing material is released to about one-third of the amount liberated by mechanical disruption or alkaline digestion. Such cells are no longer viable (Hosono, 1991). How far such studies can be related to the more customary downshock treatments being considered here cannot be established but it seems likely that loss of high molecular weight compounds ought to be expected.

Very little is known about the mechanisms controlling loss of solutes during downshock. Some clues have come from a study on uptake of glycerol by cells of *Z. rouxii* during upshock (van Zyl, Kilian & Prior, 1990). The rate of uptake is increased as a result of upshock. For the same external osmotic potential more glycerol is accumulated when sodium chloride is the external osmoticum than when PEG-400 is used. 2,4-Dinitrophenol brings about virtually complete loss of the glycerol so accumulated. However, no proton flux was detected during glycerol accumulation. It is suggested that there is a sodium–glycerol cotransport. Presumably the system can reverse during downshock.

In *D. hansenii*, which can maintain a glycerol gradient of 5000–10 000-fold during steady-state growth in the presence of 0.68–1.35 M sodium chloride (Larsson *et al.*, 1990), there is clear evidence for a sodium–glycerol cotransporter (Lucas, da Costa & van Uden, 1990). Indeed, the ability to retain glycerol inside the cell is an important feature of yeasts tolerant to sodium chloride (Brown, 1978). For instance, when cells of *Z. rouxii* are compared with those of *S. cerevisiae*, those of the latter yeast are much less able to retain glycerol in sodium chloride media.

Cell wall and cell volume changes – a comment

There is no doubt that fungal cells and hyphae change volume as a result of changes in the osmotic potential of the medium (see e.g. Robertson & Rizvi, 1968; Niedermeyer, Parish & Moor, 1977; Meikle *et al.*, 1988). Figure 13.8 shows the volume changes that occur in cells of *S. cerevisiae* when they are subjected to upshock with either glycerol or sorbitol. With sorbitol there appears to be no recovery, probably due to the inability of this polyol to enter the cells (Niedermeyer *et al.*, 1977). Recovery with glycerol appears to be a two-stage process and, even though this solute is able to enter the cell, the rate of recovery of the original volume is slow. It is not clear why there is this slow rate. However, the changes in volume, which can be of considerable magnitude (see Figure 13.8), result in changes of concentration of solutes within the protoplast. The actual magnitude of the changes is dependent upon the non-osmotic volume and whether the solutes are uniformly distributed or located in a particular compartment. There are clear indications that the vacuole of cells of *S. cerevisiae* decreases in volume when they are exposed to a hypertonic solution (Niedermeyer *et al.*, 1977; Morris *et al.*, 1986). It is not clear whether the reduction in volume is greater than that of the osmotic volume of the cytoplasm. But if there is a differential change, then the volume that changes most will have the greater effect on changes in the concentrations of solutes contained therein. Such changes in concentration of a particular solute might act as a signal for the fungus to initiate corrective action so that there is a return to the original volume.

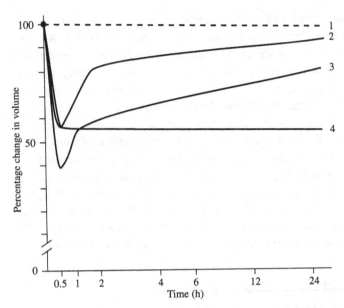

Figure 13.8. Volume changes of cells of *Saccharomyces cerevisiae* during incubation in hypertonic media. 1, Control (nutrient medium minus carbon and nitrogen sources); 2, 8.7% (v/v) glycerol; 3, 17.4% (v/v) glycerol; 4, 1.2 M sorbitol. (From Niedermeyer *et al.*, 1977.)

One of the most interesting features of the shrinkage of cells of *S. cerevisiae* is that the wall shrinks as well as the protoplast, such that, as the osmotic potential decreases below a certain value, the cell obeys the Boyle–van't Hoft law (see p. 427). Thus plasmolysis is not observed, which is unlike what happens in many green plant cells and those filamentous fungi that have been studied, namely *Botrytis cinerea*, *Sclerotinia sclerotorium* and haustoria of rusts (Thatcher, 1939), *Aspergillus niger* (Park & Robinson, 1966), *Neurospora crassa* (Robertson & Rizvi, 1968) and *Achlya bisexualis* (Money, 1990b). However, in saying this about filamentous fungi, the indications are that in some instances there does not seem to be a clear-cut picture of what is happening. The wall of *S. cerevisiae* must shrink as a result of dehydration. The plasma membrane appears from both light and electron microscopical evidence to be closely appressed to the wall at all stages of volume decrease, though Arnold & Lacy (1977) from a study of changes in the fractional space accessible to various solutes argue to the contrary. Be that as it may, as will be shown, the lack of plasmolysis does not necessarily mean that, when the protoplasm decreases in volume, it draws with it the wall such that the wall decreases in volume in concert with the protoplast.

The plasma membrane will be kept in contact with the cell wall, if that structure allows solutes only very slowly through itself. Oertli (1984, 1986) pointed out that, if a walled cell is subjected to a water stress by exposure to a solute that cannot penetrate the wall, a negative pressure is generated in the wall. The theory is straightforward. Thus if w represents the wall and i the protoplast and Ψ_P and Ψ_π, respectively turgor and osmotic potentials, then at equilibrium:

$$\Psi_{Pw} - \Psi_{\pi w} = \Psi_{Pi} - \Psi_{\pi i} \qquad (13.12)$$

Since $\Psi_{\pi w} > \Psi_{\pi i}$, because the solute cannot penetrate the wall, $\Psi_{Pw} < \Psi_{Pi}$, i.e. the pressure in the wall is less than the pressure inside the protoplast. This pressure difference maintains the plasma membrane against the wall, but there is also negative pressure in the wall. This negative pressure might be responsible for the shrinkage of the wall of the yeast cell.

The argument against this happening in yeast cells is that studies on wall porosity (de Nobel & Barnett, 1991) indicated that the solutes used by Niedermeyer *et al.* (1977) to generate hypertonic conditions should penetrate the wall readily, though there is evidence that the wall can change porosity depending on the physiological and growth conditions (de Nobel *et al.*, 1991; see Chapter 3). Certainly, it is conceivable that the porosity of the wall could decrease markedly as it shrinks. Within the context of this discussion, it is interesting to note for two marine fungi that the walls are *c.* 30% of the volume of the cells for *Debaryomyces hansenii*, calculated from electron micrographs (Gezelius & Norkrans, 1970) and 40% of that of hyphae for *Dendryphiella salina* (Clipson *et al.*, 1989). Is the size of the wall such as to prevent its collapse under conditions where there could be negative pressure? There are, however, other reasons why there may be walls of a volume large relative to that of the protoplasm and this is considered in the next section.

The special case of sodium chloride

The need to exclude sodium

There are microorganisms, the bacteria *Synechococcus* and *Halobacterium* and the unicellular alga *Dunaliella* being examples, that require sodium chloride to remain viable and grow (Ginzburg, 1987; Gilmour, 1990). These microorganisms are termed halophiles. The only known group of halophiles in the Fungal Kingdom appear to be those lower marine fungi of uncertain affinities such as the thraustochytrids (Moss, 1985; Chamberlain & Moss, 1988). These fungi require sodium for growth and this requirement can be met only partly by potassium (Jennings, 1983; Garrill *et al.*, 1992). There is evidence that sodium is required in the cotransport of solutes into these organisms (Siegenthaler, Belsky & Goldstein, 1967; Siegenthaler *et al.*, 1967; Belsky, Goldstein & Menna, 1970). Nevertheless, tolerance to sodium chloride per se is not high; studies with *Thraustochytrium aureum* have shown that in 1 M sodium chloride growth and cell division are severely inhibited (Garrill *et al.*, 1992). As well as this group of fungi, there are particular genera that would repay further study. In particular, there is a member of the Chytridiales, *Phlyctochytrium*, isolated from the seaweed *Bryopsis plumosa*, that requires sodium to complete its life cycle. Sodium also appears to act in cotransport, in this particular instance, of amino acids, the sodium gradient being maintained possibly by a K^+/Na^+ ATPase (Amon, 1976; Amon & Arthur, 1980, 1981; Kazama, 1972), though the presence of such an enzyme in a fungus must be viewed with considerable caution (see Chapter 1 and pp. 3–19).

Though the evidence is limited, there is nothing to suggest that the bulk of other fungi, particularly ascomycetes and basidiomycetes, require sodium for growth, even at low concentrations, though one must keep in mind the difficulty of producing media, possibly of a complex nature, which are sodium-free (Brownell, 1979). However, there are many fungi that grow well in saline conditions, for example in the sea (Kohlmeyer & Kohlmeyer, 1979; Moss, 1986) and as frequent contaminants of brine used for preserving food (Onishi, 1963). Such environments also possess low water potentials. Thus, the fungi living in saline media must be as capable of withstanding these low water potentials as they are of withstanding the potentially toxic concentrations of sodium and chloride. Not surprisingly, these fungi that grow in saline environments have features typical of those fungi shown to be capable of withstanding water stress. In saying that, one needs to remember that the evidence that a particular physiological response is brought about by water stress per se is not always straightforward, since it is possible that sodium chloride has been used to produce the water stress. Since sodium chloride has been used many times to generate water stress, it is clear that many fungi, particularly ascomycetes and fungi imperfecti capable of growing in saline conditions. Thus, while there is a fungal flora in the sea with its own characteristic taxonomy (Kohlmeyer & Kohlmeyer, 1979), it is not entirely surprising to isolate from marine habitats fungi that one would classify as being terrestrial (Tresner & Hayes, 1971). It is very probable that such fungi are growing vegetatively; there is good evidence that reproductive processes are much more sensitive to salinity (Jennings, 1983).

Though the evidence is limited, it would seem that the enzymes of halotolerant fungi do not have special features that differentiate them from the enzymes of green plant halophytes (Flowers, Hajibagheri & Clipson, 1986) and allow them to carry out their catalytic activity more effectively in concentrated saline media. The most detailed study is that of Paton & Jennings (1988). It showed that the activities *in vitro* of malate dehydrogenase and glucose-6-phosphate dehydrogenase from *Dendryphiella salina* at pH 7.4, a value close to that of the cytoplasm (Davies, Brownlee & Jennings, 1990b), were increasingly inhibited by sodium and potassium chlorides as the concentrations of these salts in the assay medium were raised. Malate dehydrogenase was particularly sensitive, activity being reduced to less than 30% by 400 mM sodium chloride. Neither glycerol nor mannitol, at their probable concentration inside the mycelium, could protect the enzymes from the inhibitory effect of the two cation chlorides. Similar results were obtained by Brown & Simpson (1972) for isocitrate dehydrogenase of *Zygosaccharomyces rouxii*. Somewhat similarly, Onishi (1963) found that glucose consumption and glycerol production by cell-free extracts of the salt-tolerant *Pichia miso* were completely inhibited by sodium chloride although the latter did not inhibit growth. Higher-plant physiologists have continuously urged caution in extrapolating the results from *in vitro* to what might be occurring *in vivo* (Flowers *et al.*, 1986), as indeed have Blomberg & Adler (1993) when considering the enzymes of halotolerant fungi. The latter authors pointed to the study by Nilsson & Adler (1990) on the NAD^+-dependent glycerol-3-phosphate dehydrogenase of *Debaryomyces hansenii*. The enzyme is strongly inhibited by monovalent cations accompanied by inorganic anions at 0.5 M. However, when the counter-ion was glutamate, the enzyme showed no diminution of activity when the salt concentration was 0.9 M.

These data by themselves suggest that it may be unnecessary to exclude sodium from fungal cells unless it rises to very high concentrations, i.e. above 1.0–1.5 M, to avoid unnecessary metabolic complications. It could be argued, as has been done by Paton & Jennings (1988), that differential effects on enzymes as seen in *in vitro* may be important to the fungus in allowing it to respond to different types of salt because of unequal absorption of cation and anion (see pp. 392–4) and conceivably leading to the different mix of polyols referred to above. But this comment is based on what may be an unrepresentative set of data;

if one were to argue about the ionic milieu of the cytoplasm from enzyme studies, many more such studies would be required. In any case, too high a concentration of sodium in the cytoplasm tends both to reduce the concentration of potassium and possibly compete with that present, such that potassium-requiring processes such as protein synthesis are inhibited. It seems likely that a similar kind of situation holds for chloride, though there is no clear idea where the anion has its inhibitory effect.

Exclusion of sodium

It would therefore seem on *a priori* grounds that there will be mechanisms to counter the possible deleterious effect of too high a concentration of sodium within the cytoplasm of fungi. The ion might be pumped out of the cytoplasm into the medium or sequestered at a higher concentration in the vacuole. On balance, the evidence put forward in Chapter 4 would suggest the latter does not occur. However, there is good evidence that sodium is actively transported out of halotolerant fungi. Three examples will suffice. Clipson & Jennings (1990) showed this to be so for *Dendryphiella salina*, where sodium and also potassium distribution across the plasma membrane resulted in much lower concentrations in the protoplasm than one would expect on the basis of the Nernst equation. Both Wethered & Jennings (1985) and Larsson *et al.* (1990) with *Thraustochytrium aureum* and *Debaryomyces hansenii*, respectively, have shown that there is much less sodium inside the cells than outside and that the selectivity of the cell for potassium over sodium is very much more than the ratio of the two ions in the external medium.

In Chapter 1 it was pointed out that the evidence available indicates that in terms of transport across the plasma membrane by far the majority of fungi function in a proton economy. That evidence included information from the following three halotolerant and one (*T. aureum*) halophilic species, *Debaryomyces hansenii, Dendryphiella salina, Thraustrochytrium aureum* and *Zygosaccharomyces rouxii*, though it has to be admitted that only for *D. salina* is the evidence more than superficial. However, there is possible evidence for a primary sodium pump. Haro, Garciadeblas & Rodríguez-Navarro (1991) have shown that the product of the *ENA1* locus in *S. cerevisiae* is required for the efflux of lithium, sodium and probably potassium from the cells. The protein encoded by *ENA1* (and that by *ENA2*) appears from nucleotide sequence analysis of DNA to be a plasma membrane ATPase. Haro *et al.* (1991) argued that because cells altered with a single-step gene disruption (*ena1* and *ena2*) do not lose these cations but do so when transformed with *ENA1*, the ATPase is acting as a pump for the cations, because 'it [is] unlikely that the product of *ENA1* pumps another cation whose concentration regulates the activity of the transport system [for the cations]'. The results could, however, be explained along the lines that the cells containing the gene disruption possessed ATPases whose activity became more susceptible to the inhibitory effects of the above-mentioned monovalent cations. Primary active transport would be less effective and in consequence so would secondary transport such as proton/cation symports and antiports. Poor growth of the gene-disrupted strain in alkaline media is in keeping with this possibility.

In any case, some very simple but conceptually clear experiments with *Z. rouxii* argue strongly for a key role for the proton-extrusion pump (H^+-ATPase) intolerance to sodium chloride. Watanabe & Tamai (1992) grew the yeast in media containing varying concentrations of sodium chloride in the presence of either 2.5 μM CCCP or varying concentrations of vanadate. In the presence of either inhibitor, growth in sodium chloride was severely inhibited. When sorbitol replaced sodium chloride at similar concentrations, there was no effect with these inhibitors. Further, growth in sodium chloride at pH 6.5 was less than at pH 4.5 or 5.5; there was not this difference with sorbitol instead of sodium chloride. The conclusion an appropriate proton electrochemical potential gradient generated by an H^+-ATPase across the plasma membrane is required to remove sodium from the cell.

Thus, for now, the presence of a sodium pump in fungi must be considered non-proven. In all likelihood, sodium is moved out of halotolerant fungi by a sodium efflux system of the kind described for *S. cerevisiae* in Chapter 10. Some confirmation, indeed enhancement, of this view comes from the recent identification by Jia *et al.* (1992) of a new locus (*sod2*) in *Schizosaccharomyces pombe*, based on increased lithium tolerance. The predicted product of the gene can be placed in the broad class of transport proteins that possess 12 hydrophobic transmembrane domains. Further, the product shows some sequence similarity to the human and bacterial Na^+/H^+ antiport. Overexpression of *sod2* increased sodium export capacity and strains with increased sodium export capacity and strains with increased lithium tolerance showed enhanced pH-dependent efflux. Importantly, sodium efflux was accompanied by a rise in the extracellular pH. The evidence for an Na^+/H^+ antiport in this instance is very strong and furthermore this antiport is involved in halotolerance. Lucas *et al.* (1990) have provided some indirect evidence for a sodium/proton antiport in *Debaryomyces hansenii*.

If one believes that halotolerant fungi function predominantly on a proton economy in terms of their transport processes, then under alkaline conditions – as indicated above with respect to growth experiments with *Z. rouxii* – it will be increasingly difficult to generate the necessary pH gradient to drive a sodium/proton antiport. Also, there will not be the necessary proton-motive force to drive organic solute transport via proton symports. With respect to the latter force, the generation of a large negative potential between the medium and the cell interior, as is the case in *Dendryphiella salina* in which the potential is $-300\,\mathrm{mV}$ (Brownlee, 1984; Davies *et al.*, 1990*a*, *b*; Clipson & Jennings, 1992) will help to mitigate the fall in the pH gradient. But the above transport processes will also be assisted by a large volume of wall – as is the case in *D. salina* (see p. 435 and Chapter 3). The larger the volume of the wall, the more likely that the environment outside the plasma membrane, particularly the presence of an acid pH, will be buffered from the effect of the bulk external medium. Mucilage external to the wall will assist even further (see Chapter 3).

Adaptation for growth in the presence of high concentrations of sodium chloride

The time taken for cells of *Debaryomyces hansenii* to commence growth when transferred to media containing sodium chloride increases as the concentration of the salt is increased (Hobot & Jennings, 1981). At 3 M sodium chloride, the lag phase can be around 50 h. Some of the changes that take place during the period of adaptation have now been identified.

In *S. cerevisiae*, protein synthesis is required (Blomberg & Adler, 1989). The specific activity of NAD^+-dependent glycerol-3-phosphate dehydrogenase is enhanced. In keeping with this enzyme having a major impact on the control of glycerol synthesis, the amount of glycerol in the cells increases, though part of the increase is brought about by increased retention in the cell. The increase in glycerol synthesis may be aided by enhanced activities of phosphofructokinase and key enzymes of the pentose phosphate pathway (Brown & Edgely, 1980). Acetate production is also increased (Blomberg & Adler, 1989); in keeping with this there is also an increase in the specific activity of acetaldehyde dehydrogenase and a decrease in that of alcohol dehydrogenase.

However, what I have just described is a somewhat simplified view of what is occurring. One of the earliest events in cells transferred to media containing sodium chloride is the inhibition of glycolysis (Singh & Norton,

1991). This appears to be due to an increased cell content of potassium and sodium. The amounts of glycerol and ethanol produced are decreased and that of trehalose increased. In the experiments of Singh & Norton (1991), it was only after 8 h that the intracellular glycerol content increased in the manner expected of cells able to grow in high concentrations of sodium chloride. These results suggest that a net synthesis of trehalose, as the necessary compatible solute, is required to produce the requisite environment for the processes of adaptation to take place, i.e. the production of the enhanced internal glycerol concentration. The role of trehalose in stress tolerance is alluded to towards the end of this chapter. It is probable that the increase in trehalose content is not due to the effect of salt per se but to the entry of the cells into stationary phase brought about by water stress (Thevelein, 1984; Bellinger & Larher, 1987). It should be noted that the data of Singh & Norton (1991) for trehalose, and indeed also for the potassium and sodium content, did not properly reflect the actual concentrations in the cells because osmotic shrinkage was not accounted for.

The process of adaptation of *Z. rouxii* to increased sodium chloride, though not studied in so much detail, appears to be similar to what occurs in *S. cerevisiae* (van Zyl, Prior & Kilian, 1989; van Zyl & Prior, 1990; Yagi, 1988). We know also that, with growth in sodium chloride, there is a significant change in the membrane fatty acid composition of cells of *Z. rouxii* (see Chapter 3).

When *D. hansenii* is grown in 1.5 M, the cells have been found to contain an alkaline ATPase activity that has a different sensitivity to inhibitors and to sodium from that of the plasma membrane H^+-ATPase (see Chapter 1) (Comerford et al., 1985). There was no evidence for such an ATPase in *S. cerevisiae*. It is almost certain that the sodium chloride-induced ATPase in cells of *D. hansenii* is part of the armoury of transport processes required for growth under saline conditions. However, that said, the role of the ATPase is completely unclear, nor is there any proper evidence as to its location.

It should be noted that only a very small proportion (1 in 10^4) of growing cells of *S. cerevisiae* survive when moved from salt-free media on to agar containing c. 2 M sodium chloride (Mackenzie, Blomberg & Brown, 1986). This has been called 'water stress plating hypersensitivity' (Brown, Mackenzie & Singh, 1986). The phenomenon is not restricted to plated cells: exponentially grown cells take 100 h to adapt to 1.5 M sodium chloride, whereas cells in transition phase take only 15 h (Blomberg & Adler, 1992). The generation of resistant cells develops in the second half of the exponential phase and is closely related to the accumulation of trehalose inside the cells but unrelated to the glycerol content. A mutant that has diminished ability to produce trehalose and glycogen shows much less resistance (Mackenzie, Singh & Brown, 1988). The results confirm the view expressed above about the role of trehalose in the process of adaptation.

Genetic studies

One can have little doubt that many genes underlie the ability of a fungus to withstand water stress. Isolation of mutants can do no more at present than provide pieces for the genetic jigsaw whose overall design cannot yet be discerned, but some important features have been forthcoming from studies on mutants.

(1) Strains of *Z. rouxii* that are tolerant of water stress may show sensitivity to high concentrations of sodium chloride (Rodríguez-Navarro, 1971; Ushio & Nakata, 1989).

(2) From a study of 20 salt-sensitive (*ss*) mutants of *Z. rouxii*, Yagi & Tada (1988) concluded that the following factors, amongst other(s), are necessary for growth in the presence of high concentrations of sodium chloride: (a) the ability to produce glycerol, (b) the ability to maintain a defined concentration of glycerol in the cells, and (c) the ability to take up glycerol that has leaked into the medium.

(3) Defects in vacuole biogenesis in *S. cerevisiae* lead to osmotic sensitivity, cells being unable to survive 1.0 or 1.5 M sodium chloride, 1.0 M potassium chloride or 2.5 M glycerol (Banta *et al.*, 1988). This should not be taken to mean that vacuoles are necessary for sequestering solutes absorbed from the medium during growth in media of low osmotic potential. The more likely explanation is that effective responses to water stress involve calcium (as a signal?), which is known to be accumulated in the vacuole (see Chapters 4 and 11).

(4) A respiratory-deficient mutant (lacking cytochrome aa_3 and possessing a reduced level of cytochrome *b*) of *Z. rouxii* has been found to be almost as tolerant as wild-type cells in media containing up to 2.5 M sodium chloride (Yagi, Nogami & Nishi, 1992). Only at concentrations containing more than 3 M sodium chloride was there a difference, the mutant retaining less glycerol. This is in keeping with the view that fungi can adapt their growth rate to the internal energy supply (Gradmann & Slayman, 1975) and also with the view (see next section)

that, when the fungi are growing at low osmotic potentials, energy dissipation may be as important a process as energy production. So it should be expected that respiratory-deficient cells of the kind discussed here would be as tolerant to osmotic stress as wild-type cells.

(5) Matsutani *et al.* (1992) have derived two salt-tolerant mutants of *S. cerevisiae*, one from a haploid strain and another from a diploid. Both mutants can grow in high concentrations of sodium chloride, the one from the diploid strain particularly so, being able to grow in 3 M sodium chloride. The cell size of this latter mutant (1.0 μM in diameter) is comparable with that of bacteria. The physiological basis for tolerance is unknown.

(6) Morales, André & Adler (1990) have devised an elegant method for the enrichment and isolation of mutants (in their case those of *Debaryomyces hansenii*) having increased glycerol production. The procedure is based on the increased buoyant density of cells with increasing content of glycerol. As yet no physiological studies appear to have been made on glycerol-accumulating mutants obtained in this way.

Compatible solutes

Those organic solutes that accumulate in fungi under conditions of water stress are known to be compatible solutes (Brown & Simpson, 1972). These are compounds that not only make an important contribution to the protoplasmic osmotic potential but at the same time have minimal effect on the confirmation of proteins. Readers should consult the review by Collins & Washabaugh (1985) for a comprehensive consideration of the way in which solutes may effect the potential difference at a water interface, the interaction between compatible solutes and proteins being through the water/protein interface (see also Chapter 10). *In vitro* studies on enzymes have shown that sometimes they are protected to a degree, by compatible solutes, from the deleterious effects of salts, particularly sodium chloride. This latter property may be termed 'haloprotection' (Trüper & Galinski, 1986). Table 13.10 lists those compounds produced within fungi and other microorganisms that are thought to act as compatible solutes.

Although the concept of a compatible solute is valuable, it is almost certain that it is simplistic to consider these solutes as having solely an osmotic role. Thus polyols, whose physiology and metabolism in fungi and green plants had been studied in much detail prior to any consideration of them having a role as compatible solutes, were thought to serve the

Table 13.10. *Details of compounds, thought to act as compatible solutes, produced by microorganisms under stress*

Compound	Species or class of microorganisms
Glycerol	*Debaryomyces hansenii* (osmophilic yeast)
	Dendryphiella salina (marine mould)
	Zygosaccharomyces rouxii (osmophilic yeast)
	Xerophilic fungi
Arabitol	*D. hansenii*
	D. salina
Erythritol	} *D. salina*
Mannitol	
Glycosyl glycerol	Marine cyanobacteria
Sucrose	Freshwater cyanobacteria
Trehalose	Freshwater cyanobacteria
	Escherichia coli
Proline	Lower fungi, including the marine fungus *Thraustochytrium*
Glutamate	Bacteria
Glycine betaine	Hypersaline cyanobacteria
Proline betaine	*E. coli* in the presence of choline
Glutamate betaine	Hypersaline cyanobacteria
Ectoine	*Ectothiorhodospira halochloris*

From Jennings (1990*c*).

following roles: (a) as carbohydrate reserves, (b) as translocating compounds, (c) in coenzyme regulation, as well as (d) helping to maintain turgor (Lewis & Smith, 1967).

In fungi, any role for polyols in translocation is uncertain (see Chapter 14). They can certainly act as reserve metabolites and sources of reducing power (Jennings, 1984*a*), as also can proline (Bellinger & Larher, 1987) – although there is no evidence for the amino acid playing this role in the fungi, particularly those in which it can make a significant contribution to the osmotic potential. But some of the compatible solutes may be involved in energy spillage (Neijssel & Tempest, 1975, 1976; Tempest & Neijssel, 1980, 1984) through being involved in futile cycles. Energy spillage is important when the growth of cells moves out of being subject to nutrient limitation into energy limitation. This occurs for instance when growth is no longer limited by carbon supply but by some kind of stress such as an unfavourable concentration of sodium chloride in the external medium (Burke & Jennings, 1990; Jennings & Burke, 1990). Under such conditions carbon is channelled away from anabolic to

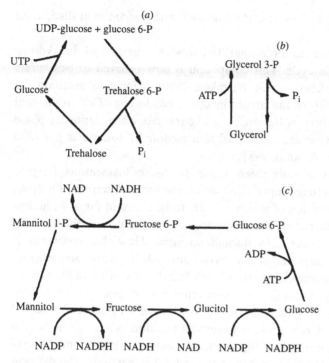

Figure 13.9. Futile cycles that have been proposed to be important for energy dissipation in fungi (see text for details). (a) Trehalose cycle; (b) mannitol cycle; (c) glycerol cycle. (From Jennings, 1993.)

catabolic conditions. Thus, though energy generation still occurs it has to be dissipated because cell synthesis is curtailed.

Futile cycles occur when two compounds can be interconverted by two different irreversible reactions. When the enzymes responsible for both reactions are functional there is futile cycling in which there is dissipation of energy. To date, four metabolic cycles have been suggested to have the role of energy dissipation in fungi (Figure 13.9), namely: (a) a trehalose cycle in heat-shocked cells of *S. cerevisiae* (Hottiger, Schmutz & Wiemken, 1987), (b) the mannitol cycle in many fungi (Jennings & Burke, 1990), (c) a glycerol cycle in *D. hansenii* growing in saline conditions (Adler, Blomberg & Nilsson, 1985; Jennings & Burke, 1990), and (d) the glutamine cycle in *N. crassa* under carbon limitation (Mora, 1990; see Chapter 6). It is arguable whether these cycles are operating in a strictly futile mode, since, as in the case of glycerol, there can be considerable accumulation under saline conditions. Nevertheless, though the cycles may not be strictly futile, there is considerable cyclic turnover with concomitant dissipation of energy.

It is interesting to note that trehalose synthesis and breakdown constitute a futile cycle. This compound is now believed to be a stress metabolite (van Laere, 1989; Wiemken, 1990), helping to maintain the structural integrity of the organism under conditions of environmental stress, particularly dehydration, as can be the case when stationary phase cells of *S. cerevisiae* are transferred to a medium of low water potential (Gadd, Chalmers & Reed, 1987).

Reference has already been made to those filamentous fungi – ascomycetes and fungi imperfecti – which produce a spectrum of polyols, the total concentration of which appears to be constant for any one low external osmotic potential, irrespective of the nature of the solute(s) generating that lower than normal potential. How that constancy is achieved is not clear and it may occur only when glucose and fructose are the carbon sources. Nevertheless, the fact that the ratios of the various polyols can vary within the total may represent a considerable metabolic flexibility for the fungus. Thus, overall synthesis from glucose of a hexitol such as mannitol is a proton-consuming reaction, while synthesis of a pentitol is a proton-producing reaction (Jennings, 1984a). Thus, polyol synthesis can drive both oxidative and reductive reactions. The different protoplasmic concentrations of individual polyols may be associated with different concentrations of other organic solutes (particularly organic acids), depending on whether or not they too are synthesised by oxidative or reductive reactions. If one considers organic acids, the extent to which they accumulate within the fungus is governed by the extent to which cations are absorbed from the external medium in excess of anions (see Chapter 12). For a given salt as the external osmoticum, the amount of organic acid produced can vary and, as a result, so can the amount of individual polyols, consequent upon the oxidative or reductive nature of the metabolic linkages with the synthesis of individual organic acids. So, irrespective of the details of the metabolic processes involved, it is possible to envisage a flexible metabolic response (in which the various individual polyols have an important role) by the fungus to different types of external osmotica, which allows the generation of the appropriate protoplasmic solute potential (Wethered *et al.*, 1985).

14

Nutrient movement within the colony

Introduction

Some relatively simple observations suggest that there must be movement of nutrients within the hyphae of mould mycelium from the growing margin to the centre of the colony. When such mycelium is growing on agar, the concentration of glucose under the colony decreases markedly as one moves from the margin towards the centre and increasingly moves towards zero with time (Robson et al., 1987). If the central portion of the colony is to function normally, there must be movement to it (translocation) of metabolic substrates from those regions where glucose (and other nutrients) are being absorbed.

In making the above statement, one is making the assumption that the central portion of the colony is functioning normally. It is a reasonable assumption in the experiments referred to but it should not be taken as axiomatic, particularly in cultures that have been growing for more than a few days. Then it is possible that autolysis may take place in the centre of the colony. Much of what is known about changes taking place in a colony on agar comes from a variety of sources and a range of fungi. There is little doubt that metabolic activity declines as the colony ages – all the various observations support this (Trinci & Thurston, 1976). There are also morphological changes, particularly increased vacuolation (Buller, 1933; Park & Robinson, 1967) and septal plugging (Trinci & Collinge, 1973). However, there is no integrated picture of what is happening over a realistic distance from the colony margin. The ideal experimental situation would be akin to the growth tube used almost exclusively for the study of the mycelial extension of Neurospora crassa (Ryan, Beadle & Tatum, 1943; Gillie, 1968). Under such conditions growth can take place over many centimetres and one can sample mycelium of very distinct ages for chemical and ultrastructural study.

447

For fungi growing in agitated or stirred liquid culture, there is a much more coherent body of information on temporal changes in the mycelium, though not of spatial changes. In such culture, the formation of pellets is the norm. As pellets grow, hyphae in the interior are subjected increasingly to anaerobic conditions (Pirt, 1966; Trinci, 1970). The extent to which these conditions occur depends on the density of the hyphae. The looser the hyphae are within the pellet, the more convective flow can take place in the interior. At any rate, it is known that the specific respiration rate of pellets of *Aspergillus niger* decreases with pellet diameter (Yano, Kodama & Yamada, 1961; Kobayashi, Vandedem & Moo-Young, 1973). The inner part of a pellet undergoes autolysis such that it becomes hollow (Camici, Sermonti & Chain, 1952; Trinci & Thurston, 1976).

Autolysis also occurs throughout a pellet when the latter is starved. This has been well demonstrated electron microscopically in *Dendryphiella salina* by Holligan & Jennings (1972a) and in *Penicillium chrysogenum* by Trinci & Righelato (1970). In both cases autolysis leads to complete breakdown of the hyphal contents such that only walls are left, though membranes are also rather resistant to breakdown. But within the hyphal mass there can be found cytologically normal hyphae. These are capable of regrowth in fresh media (Holligan & Jennings, 1972a). One presumes that in aged culture the cytologically normal hyphae are maintained on the breakdown products of the other hyphae. Eventually the walls within an autolysing culture are degraded. The medium of such cultures has been shown to contain a wide range of wall-degrading enzymes (Reyes & Byrde, 1973; Lahoz, Reyes & Perez Leblic, 1976; Lahoz *et al.*, 1983; Reyes, Calatayud & Martínez, 1988). Indeed the medium, free from mycelium, can be used to lyse the walls of other, actively growing mycelium to yield protoplasts (Reyes & Lahoz, 1976; Isaac & Gokhale, 1982).

The process of autolysis raises the question of the extent to which the products of autolysis can be translocated from the region where autolysis is occurring to the growing hyphal front. Increased blockage of septa in the older parts of the colony suggests that such movement could be restricted. Nevertheless, a study by Hanson (1980) has indicated that, while autolysis products *sensu stricto* may not be translocated forward into the younger parts of the mycelium from the older parts, this direction of translocation should not therefore be discounted. In the study, it was found that, when tritium-labelled inositol was presented to older parts of mycelium of the *inl* (inositol-requiring mutant) of *N. crassa* growing across a barrier onto non-radioactive medium, label was translocated to the extending margin of the colony. It would seem that inositol is translocated unchanged, though the data are not absolutely clear on this point. Interestingly, very little label moves in the opposite direction when radioactive inositol is presented to the younger part of the mycelium. This is so even when the older part is on a medium with limiting concentrations of inositol.

There are a number of investigations using Petri-dish cultures in which experiments of a similar kind have been carried out to study the movement of a radioactive tracer within the mycelium (Lucas, 1960; Monson & Sudia, 1963; Littlefield, Wilcoxson & Sudia, 1965a; Littlefield, 1967; Wilcoxson & Subbarayadu, 1968; Lyon & Lucas, 1969a, b; Milne & Cooke, 1969). These investigations leave one in no doubt that the radiotracers used, ^{32}P or ^{14}C, can be readily translocated through mycelia of a variety of fungi and translocation can take place in both directions. Often the experiment is so designed that the mycelium has crossed a gap or a solid barrier (often glass) between the two portions of agar on which the mycelium is growing. This means that diffusion of label through the agar can be discounted. Nevertheless, there is evidence that label may cross a solid barrier in solution external to the hyphae by capillary forces (Read & Stribley, 1975).

The above is actually not very informative. It tells us little else other than that translocation, i.e. movement of nutrients from one part of the mycelium to another, can take place. But it tells little about the quantitative aspects of such movement, whether it can occur in both directions in one hypha or whether there are specific routes within the mycelium which translocation may follow. At the very least, experiments are needed on development of mycelium under conditions in which the nutrients available to it are heterogeneously distributed, e.g. one nutrient such as carbon located in a space separate from other nutrients.

Mechanism of translocation

This section is concerned with the forces bringing about the movement of nutrients within various hyphal systems. I have avoided the use of the term 'mycelium' in this context because some of the information relates to movement in single hyphae. Strictly speaking, there are three possible mechanisms by which nutrients may be translocated in hyphae, namely:

(1) Diffusion, the energy for movement residing in the concentration gradient.

(2) Expenditure of energy in contractile systems that move nutrients in 'packages' along hyphae.

(3) Flow of solution within the hyphae, the energy for movement coming from either (a) the generation of an osmotic gradient to drive an inflow of water and thus generate a hydrostatic pressure within a hypha or hyphae at the source of translocation or (b) evaporation of water drawing water from the hyphae. In both cases solution is caused to move along hyphae.

The following reviews are concerned with translocation in fungi, focusing particularly on mechanisms (Jennings *et al.*, 1974; Jennings, 1987*a*, 1994).

Mould mycelium

Most studies of translocation in mould mycelium have provided little information of value to our understanding of the mechanism of translocation (Jennings, 1987*a*). Nearly all studies have been made with Petri-dish cultures. The major reasons for the lack of information are as follows:

(1) In nearly all instances, the mycelium has been allowed to colonise the entire dish before movement of isotope has been studied. The implied reason is the need to have translocation occurring over as great a distance as possible. However, the colony is likely to be suffering from nutrient depletion, with consequences for the viability of the hyphae, particularly in the oldest parts of the mycelium. Moreover, the geometry of the system is both complicated and uncertain. The colony cannot be considered simply as hyphae radiating out and branching in a simple manner. There are anastomoses that increase as the colony ages and further there is often a tendency for hyphae to spiral so that they cross other hyphae (Buller, 1931; Jennings, 1986*a*, 1991*a*). The processes could lead to pathways that allow translocation to take place in a direction other than that being studied. Even if the branching pattern were simple, there could still be this problem when investigating translocation from the front to older parts of the mycelium. This would be a consequence of translocation along a hypha from the apical region being directed in the reverse direction at a branch.

(2) Radiotracers have been used extensively, with little attempt to relate movement of isotope to changes in chemical concentration of the compound or compounds translocated. The failure to do this is particularly significant when the isotope is incorporated into a

compound that may be converted into other compounds (Lyon & Lucas, 1969*b*).

(3) Little attention has been given to ensuring that the water relations of the system is either understood or controlled. When the translocation of ^{14}C has been studied, it has been customary to have a well containing alkali to absorb $^{14}CO_2$. The presence of alkali has consequences for the vapour pressure over different parts of the mycelium. There is clear evidence that evaporation from a localised area of the mycelium can direct translocation to that area (Lucas, 1977; Jennings, 1987*b*).

Fortunately there are two investigations that are relatively free from these criticisms. One is by Girvin & Thain (1987) and Thain & Girvin (1987) and the other is by Olsson & Jennings (1991*a*, *b*). In the former investigation, *N. crassa* was grown from a nutrient source across a large gap on to nutrient-free agar covered with cellophane (Figure 14.1). There was no evidence of movement of the isotope from the nutrient source by capillarity. Throughout the experiments, the mycelium was still extending

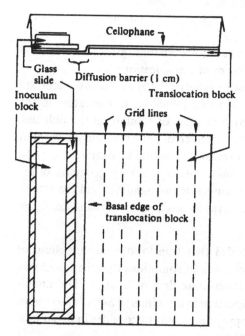

Figure 14.1. The apparatus used by Girvin & Thain (1987) to study translocation in the mycelium of *Neurospora crassa*. The mycelium was allowed to grow from the inoculum block over the cellophane lying on top of nutrient-depleted agar.

from the nutrient source, though there was an exponential decrease in growth rate with time as extension proceeded. However, the fresh weight:dry weight ratio stayed constant over the experimental period used, indicating that the mycelium was probably functioning relatively normally throughout its whole extent.

The results obtained are consistent with translocation in mycelium of *N. crassa* under these circumstances being by bulk flow of solution driven by an osmotically produced hydrostatic pressure gradient in the hyphae:

(1) Protoplasmic streaming (presumably towards the growing margin) occurred strongly in all hyphae behind the margin but confined to particular hyphae in the older portion (those which are not undergoing vacuolation).

(2) Water droplets were formed at the hyphal apices; each droplet ceased to expand when the apex grew away from it. This suggests that the hydraulic conductivity of the apical region is greater than further back along the hyphal system.

(3) Translocation appeared to be unselective, the pattern for dry matter, rubidium and 3-*O*-methylglucose being similar.

(4) The velocity of translocation was found to be much greater than the extension of the mycelial margin.

(5) The diffusion coefficient would have to be ten times greater than normal for diffusion to be the mechanism of translocation.

(6) Analysis of the profiles of radioactivity (^{86}Rb) in the mycelium are consistent with the model of Horwitz (1958) for translocation in the higher plant (known to be by bulk flow of solution in the phloem), though that analysis must be considered only preliminary without more detail about the path of translocation and its characteristics. In particular, there is a need for information about the geometry of the translocation pathway in relation to its surroundings and the rate of transfer of translocate from the translocation stream into the surroundings.

Girvin & Thain (1987) concluded that translocation in mycelium of *N. crassa* growing under the experimental conditions that they used was brought about by osmotically driven bulk flow of solution. The similar and other observations on translocation in mycelium of *Serpula lacrymans* described below are the most appropriate point to consider in detail how such bulk flow might be brought about.

The other study by Olsson & Jennings (1991*a*, *b*) focused on translocation in mycelium of various fungi growing within glass fibre on

(a)

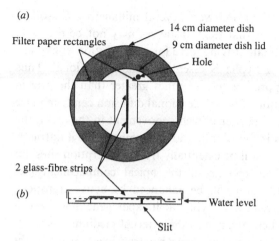

Figure 14.2. System used by Olsson & Jennings (1991*a*) for growing fungal mycelium in strips of glass fibre filters (of equal length joined end to end) as seen (a) from above and (b) from the side. The hole and slit in the inner Petri-dish lid are to give continuity between the gas and liquid phases inside and outside the inner lid and thus prevent it floating. Radioactivity might be present in a filter strip at the very start of an experiment or radioactivity might be applied at a discrete position along a strip after a defined period of mycelial growth.

opposing carbon and mineral nutrient gradients (Figure 14.2). [^{14}C]Glucose and [^{33}P]orthophosphate were used as tracers. By a variety of procedures, it was possible to determine radioactivity in the medium and within the mycelium after a known period of time. Essentially the experiments were concerned with the translocation of a nutrient (or its metabolic products) in the hyphae from a place of ample supply to those parts of the mycelium growing in the absence of the nutrient. It was found that, at the end of the experimental period, there was a similar distribution of both isotopes in strips with and without mycelium when presented in medium containing an ample supply of the nutrient containing the radiolabel. Further the distribution of radioactivity determined experimentally agreed well with that calculated on the basis of one-dimensional diffusion with two boundaries (the edges of the filter strips). Translocation could occur in both directions simultaneously. The conclusion was that translocation within the hyphae of the three moulds used *Rhizopus nigricans, Stemphylium* sp. and *Trichoderma viride* under the experimental conditions was by diffusion.

This conclusion needs comment. It is generally accepted that relatively speaking diffusion is rapid over short (micrometre) distances. In the above experiments, the distances involved were several millimetres. Olsson & Jennings (1991*b*) pointed out that these distances may not be excessive for diffusion to be an adequate driving force for translocation in hyphae. Importantly, diffusion in a hypha is essentially one-dimensional. Thus the rate of diffusion is approximately 12 times greater than the rate in three dimensions. In addition, if one-dimensional diffusion can take place only in one direction, and a barrier prevents movement in the other, the rate will be doubled. This is the situation at the hyphal apex if nutrients are adsorbed there and will still be essentially true if absorption sites are located in somewhat older regions of the apical compartment (see Chapter 2). Of course diffusion will be enhanced by active transport, i.e. accumulation of a nutrient in a hypha at its apex generates a much higher concentration gradient relative to the external medium.

   Ecologically, it is possible to see that, even when translocation is brought about only by diffusion, absorption at the apex into a narrow tube will be a very effective way of capturing nutrients for the whole colony. The rate of their movement in the hyphae to the older parts of the colony will be very much greater than their movement by diffusion in the external medium. Not only is the older part of the colony more effectively supplied with nutrients but their uptake from the medium removes them from being used by competing microorganisms. Finally, diffusion allows simultaneous translocation of nutrients in both directions.

Vesicular–arbuscular mycorrhiza

The ability of vesicular–arbuscular (v–a) mycorrhiza to enhance the phosphorus status of higher plants growing on soils of low phosphorus content has been considered in Chapter 7. Because the hyphae of v–a mycorrhizal fungi not only absorb nutrients but also transmit them to their host translocation of those nutrients within hyphae has naturally merited considerable attention. Equally important but very much less considered is the fact that the fungal hyphae – which can be up to around 16 cm in length as calculated from the data of Sanders & Tinker (1973) on the total hyphal length in the soil and number of hyphal entry points, both per unit infected root of onion (*Allium cepa*) – are believed to gain all their carbon from the host plant. This supposition still needs to be properly tested. Nevertheless it is not without good basis. Thus, it is likely that there is translocation in both direction along single hyphae. The matter of translocation in v–a fungal hyphae has been reviewed by Harley & Smith (1983).

In terms of translocation of inorganic nutrients along the hyphae, by far the most effort has gone, for obvious reasons, into the translocation of phosphorus. From a series of studies on v–a mycorrhiza of onion, it has been concluded that phosphorus moves as polyphosphate, the granules presumably within vesicles being drawn along by some contractile process (Cooper & Tinker, 1978; Callow *et al.*, 1978; Cox *et al.*, 1980). A key observation is believed to be the inhibition of the process by cytochalasin. There has been no investigation to discover whether, as is quite possible, this inhibitor has other effects on the physiology of the hyphae. That apart, the evidence for the active movement of polyphosphate is indirect (Jennings, 1987*b*). There is as much evidence for the movement of phosphorus (and other nutrients) to be mediated by the flow of solution along the hyphae brought about by the transpiration flux in the host plant:

(1) Increasing the rate of transpiration in clover increases the rate of phosphorus translocation to the root in the v–a fungal hyphae; the maximal rate at full transpiration was 2.3 times that at low transpiration (Cooper & Tinker, 1981).

(2) The presence of v–a fungal hyphae in roots can increase the amount of water absorbed by the higher plant (Allen, 1982; Hardie, 1985). The velocity of water movement and the volume moving would be sufficient to maintain the velocity and flux ($mol\,cm^{-2}\,s^{-1}$) of phosphate, the former presumed to be the observed rate of cytoplasmic streaming and the latter calculated on the basis that all the polyphosphate moves (Jennings, 1987*b*).

(3) There are also values for the translocation flux for sulphur and zinc into onion (Cooper & Tinker, 1978) and nitrogen into celery (*Apium graveolens*) (Ames *et al.*, 1983). The values for sulphur and zinc are at least an order of magnitude lower than those for phosphorus, possibly due – and this is speculation – to the lower concentration of mobile forms of these two elements within the hyphal cytoplasm. However, the flux values for nitrogen, supplied either in inorganic (ammonia) or organic (ground plant tissue) form, were very similar to each other and similar to the highest values for phosphorus, namely those based on the idea that the element is translocated as polyphosphate granules and that movement is by a contractile mechanism. If nitrogen is translocated by the same mechanism as phosphorus, then the nitrogen moves either as some positively charged compound such as arginine, which is associated with the polyphosphate (see Chapter 4), or in the form of protein granules. Conversely, similarity of the flux values for

phosphorus and nitrogen could be a consequence of translocation occurring by bulk flow of solution. There is a need to keep these considerations in perspective. The values being discussed are for different plants and different experimental procedures. Also the values being discussed depend upon a number of assumptions.

It has been pointed out above that it is likely that carbon is translocated in the opposite direction to phosphorus in the same hypha. This needs to be proved. There is no doubt that movement of material can be observed to occur in both directions within the same hypha (Buller, 1933). Further, there is no *a priori* reason why solution flow in both directions cannot take place within the same hypha. Given that there is turgor-driven bulk flow in one direction, this flow could set up a streaming potential such that there is osmotic flow in the reverse direction dependent on the fixed charges (negative at the probable pH of the cytoplasm) in the peripheral cytoplasm. Of course, it is also equally possible that any stream of material flowing in a direction opposite to bulk flow could be brought about by contractile elements in the cytoplasm. However, the nature of the carbon compounds moving from the host along the hyphae needs to be elucidated. There is a possibility from what is known about the carbon metabolism of v–a mycorrhizal fungi (Cooper, 1984) that lipids, not carbohydrates, may be the carbon compounds translocated. One can conceive of lipid being moved by contractile processes if it is in vesicles.

Serpula lacrymans

Serpula lacrymans is capable of growing for many metres over non-nutrient surfaces, as exemplified by brick, stone, cement and other similar non-timber building material. It is this capability that has made the fungus the major timber decay organism in buildings in northern Europe (Jennings & Bravery, 1991).

There is now a wealth of information that leaves little doubt that the translocation of nutrients through mycelium of *S. lacrymans*, spreading over a non-nutrient surface away from an infected piece of timber, is brought about by a pressure-driven or bulk flow of solution. The evidence is as follows (Jennings, 1987b, 1991b).

(1) The velocity of translocation determined by Brownlee & Jennings (1982a) as the time taken for radioactivity to move between two high efficiency Geiger–Müller detectors set at a fixed distance apart over the mycelium (Figure 14.3) was found to be the same irrespective of

Table 14.1. *Ranges for the velocities of translocation of radioactivity within mycelium of* Serpula lacrymans *supplied with labelled compounds at the nutrient source*

Compound	Velocity ($cm\,h^{-1}$)
$[^{14}C]$glucose	20–25
$[^{32}P]KH_2PO_4$	25–30
$[^{14}C]$3-O-methylglucose	20–25
$[^{14}C]$aspartic acid	25–30
$[^{42}K]K_2CO_3$	20–30

From Brownlee & Jennings (1982a).

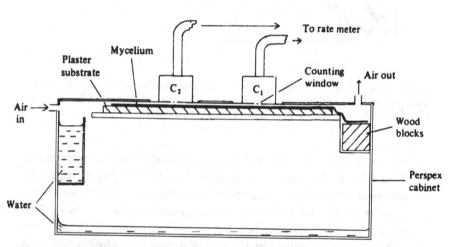

Figure 14.3. Apparatus used by Brownlee & Jennings (1982a) for the direct continuous monitoring of radioactivity in the mycelium of *Serpula lacrymans*. The radioactivity was supplied initially to the mycelium on the wood blocks.

the isotope used (Table 14.1). It seems unlikely that the various ions and molecules might all be packaged together in an organelle or indeed that the movement of all might be by movement of different vesicles.

(2) The velocity of translocation was reduced to zero by 100 mM sodium azide, 2 M glucose or 2 M potassium chloride, when presented to the mycelium where the loading of the isotopic compound took place. With the radioactivity under each detector being continuously recorded, it was found that the rise at both detectors, taking place

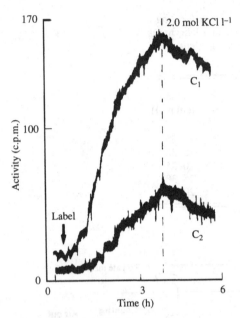

Figure 14.4. Effect on translocation in mycelium of *Serpula lacrymans* of the addition of 2 M potassium chloride to the nutrient source labelled with [^{14}C]aspartic acid. (From Brownlee & Jennings, 1982*a*.)

before the addition of the inhibitory compound, ceased instantaneously (Brownlee & Jennings, 1982*a*). Indeed with 2 M glucose and 2 M potassium chloride the amount of radioactivity decreased, indicating a reversal in the direction of translocation (Figure 14.4). The simultaneous inhibition of the rise of radioactivity at both detectors is most simply explained by translocation being by solution flow. The effect of azide must be presumed to be via the inhibition of loading of solute into the hyphae, while the effect of a high concentration of glucose or potassium chloride must be presumed to be osmotic, reversing the water flow into the mycelium.

(3) Droplets are produced at the hyphal apices at the mycelial front and their development is readily observed when hyphae are on a transparent non-absorptive surface (Coggins, Jennings & Clarke, 1980; Brownlee & Jennings, 1981*a*). The location of the droplets has many similarities to that of the droplets formed on mycelium of *Neurospora crassa* (Girvin & Thain, 1987) described above. The rate of increase in volume of the droplets in *S. lacrymans* can be altered in a manner analogous to the alteration of the velocity of translocation of

radiotracers, i.e. metabolic inhibitors and high concentrations of solutes added to the nutrient source for the mycelium prevent the droplets increasing in size. Hyphal continuity is required for the droplets to increase in size. The osmotic pressure of droplets was found to be less than that of the nutrient medium, indicating that they are formed by the ultrafiltration of solution, such that the droplet has an osmotic pressure lower than that of the hyphal protoplasm and of the nutrient medium. Secretion of solute, with water following, would not lead to osmotic values for droplets as just described.

(4) When mycelium is growing over plaster (calcium carbonate or sulphate) in a chamber similar to that shown in Figure 14.3, the water content of the substratum increases very considerably (Clarke, Jennings & Coggins, 1986). Thus, water can be transferred from the mycelium to change the water activity/water potential of the substratum. Indeed the mycelium can grow over substrate with water activity/water potential at values which, if they appertained to nutrient agar, would inhibit growth. Water can be transferred from a substratum osmotically favourable to growth (wood) to a substratum that, in osmotic terms, would not appear to be favourable to growth (plaster). The water lost from the mycelium can only come from the solution flowing along the mycelium. In keeping with this, the growth rate (dependent on the rate of translocation) over a non-absorptive surface is very much greater than over a surface capable of drawing water osmotically from the mycelium.

The above observations can be interpreted in terms of the following mechanism (Figure 14.5). Mycelium attacks the cellulose in the wood, producing glucose, which is taken into the hyphae by active transport (see Chapter 5). Inside the hypha, glucose is converted to trehalose, which is the major carbohydrate translocated (Brownlee & Jennings, 1981b). The accumulation of trehalose leads to the hyphae having a water potential lower than that outside. There is a flux of water into the hyphae and the hystostatic pressure so generated drives the solution through the mycelium. The sink for translocated material is the new protoplasm and wall material produced at the extending mycelial front and the material deposited in walls to increase their thickness in already extant hyphae. The mechanism of translocation in *S. lacrymans* is thus the same as that now accepted for translocation in the phloem of higher plants, namely osmotically driven mass flow. It is therefore satisfactory to find that when a band of mycelium between Geiger–Müller detectors is exposed to low temperature, the

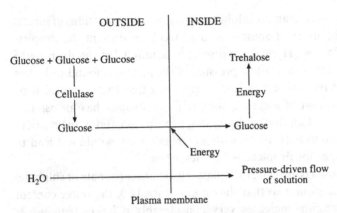

Figure 14.5. The processes underlying the translocation of solutes in mycelium of *Serpula lacrymans*. For further details see text. Glucose + glucose + glucose = cellulose. (From Jennings, 1991*b*.)

changes in radioactivity under each detector, observed when the temperature is reduced and when it returns to ambient, are very similar to the changes observed when a localised region of the phloem of a higher plant is exposed to similar changes of temperature (Thompson *et al.*, 1987).

A firm test of the hypothesis just presented was provided by Thompson, Eamus & Jennings (1985), who related the uptake (and loss) of water by the mycelium on the wood food-source to the water potential of the mycelial front and the most mature part of the mycelium. When mycelium was growing normally across a non-nutrient surface there was a turgor gradient across the mycelium such that water would move to the growing front (Table 14.2). Addition of 2 M glucose to the mycelium on the wood reversed both the water flow (water was lost from mycelium on the wood) and also the turgor gradient (Table 14.2). This is in keeping with the reversal of translocation that has been observed under similar conditions in studies on the movement of radiotracers (see above). Growth was not inhibited under the experimental conditions used. Positive turgor was maintained throughout in the growing region (Table 14.2) so that the driving force for growth could be maintained. It has been argued that the water for such extension comes from the droplets previously formed when translocation was occurring normally (Jennings, 1991*c*).

Basidiomycetes forming linear hyphal aggregates

There are a considerable number of basidiomycetes particularly those living in woodlands or forests that form linear hyphal aggregates or

Table 14.2. Water, osmotic and turgor potentials (in MPa) of mycelium of Serpula lacrymans growing from wood blocks over Perspex prior to, and after, exposure of the mycelium at the blocks to 2 M glucose

| | Prior to the addition of glucose | | Time after addition of glucose (h) | | | | | | | |
| | | | 4.5 | | 24 | | 48 | | 72 | |
	Base	Front	Base	Front	Base	Front	Base	Front	Base	Front
Water potential	-2.0	-2.7	-2.5	-4.46	-2.98	-5.85	-3.85	-7.30	-4.21	-6.80
Osmotic potential	-2.37	-2.94	-2.70	-4.78	-3.17	-6.85	-4.15	-7.40	-4.82	-7.21
Turgor potential	0.37	0.24	0.19	0.32	0.19	1.0	0.30	0.60	0.61	0.41
Turgor gradient	+0.13		-0.13		-0.81		-0.3		+0.2	

Samples were taken from mycelium close to the blocks (base) and at the growing front. From Thompson et al. (1985).

rhizomorphs (Cairney, Jennings & Agerer, 1991). *Serpula lacrymans* can be so classified, since as the mycelium ages it also forms rhizomorphs, though the experiments described above were carried out with relatively undifferentiated mycelium. The mature rhizomorphs can vary from simple aggregations of relatively unspecialised hyphae to aggregations of many hyphae that are differentiated into several types located preferentially in particular regions (Townsend, 1954; Cairney, 1991a, b; Cairney et al., 1991). Rhizomorphs growing under natural conditions spread from one piece of fallen timber to another or from one ectomycorrhizal root to another.

There is no doubt that rhizomorphs are capable of translocation. A simple demonstration of the fact is the ability of rhizomorphs of *Armillaria mellea sensu lato* to grow out for considerable distances from decayed wood into humid air, the translocation flux of dry matter being $1.1 \times 10^{-5}\,\mathrm{g\,cm^{-2}\,s^{-1}}$ (Eamus et al., 1985; Jennings, 1991d). There have been many experiments with radiotracers demonstrating that rhizomorphs are capable of translocation. Skinner & Bowen (1974) showed that ^{32}P could be translocated in rhizomorphs of *Rhizopogon luteolus* growing from mycorrhizal roots of *Pinus radiata*. Translocation in this case was towards the root. Finlay & Read (1986b) showed that translocation in mycelium of *Suilus bovinus* mycorrhizal with *Pinus* spp. can occur in both directions both towards and away from roots. In another study, Finlay & Read (1986a) demonstrated that ^{14}C could move along mycorrhizal rhizomorphs from one root system to another, such that one plane became a donor of carbon and another the receiver. The direction could be influenced by shading a plant; those plants forming part of the mycelial network in the soil that were shaded received more ^{14}C than those that were unshaded when ^{14}CO$_2$ was supplied to one particular donor plant. Equally, there are now several studies reporting translocation of radiotracers along rhizomorphs of saprophytic fungi (Anderson & Ullrich, 1982; Granlund, Jennings & Thompson, 1985; Wells & Boddy, 1990; Wells, Hughes & Boddy, 1990).

As yet the mechanism of translocation in rhizomorphs is not clear. Nevertheless, the following features must be taken into account in any mechanism that might be proposed:

(1) Translocation of carbon has been demonstrated to occur simultaneously in both directions along rhizomorphs of *Armillaria mellea* (Granlund et al., 1985). Cairney (1992) has argued for the same simultaneous bidirectional translocation occurring in mycorrhizal rhizomorphs. Certainly, there is every indication that carbon from the host plant

and phosphorus absorbed from the soil could be moving simultaneously in opposite directions in mycorrhizal rhizomorphs.

(2) Duddridge, Malibari & Read (1980) have demonstrated the ability of mycorrhizal rhizomorphs of *Suilus bovinus* to translocate water, the path of movement being along wide-diameter or vessel hyphae.

(3) Eamus *et al.* (1985) measured the longitudinal water flux through segments of rhizomorphs of *Armillaria mellea* and *Phallus impudicus* under a known hydrostatic pressure gradient. The values obtained were compared with those calculated on the supposition that the vessel hyphae that run for considerable distances along the rhizomorph are the sole channels for water movement along the organs. The calculated values were only slightly lower than the experimental values.

(4) Finlay & Read (1986*b*) showed that when ^{32}P was fed to rhizomorphs (or mycelium differentiating into them) of *Suillus bovinus* mycorrhizal with *Pinus*, there was translocation to the roots, but not exclusively so, there being translocation to other parts of the mycelial system. On the basis of this observation and the fact that the rate of translocation seemed to be independent of the direction of movement and also transposition of the host plant, it was argued that translocation was brought about by cyclosis (= contractile processes) within hyphae with fully functional cytoplasm.

Cairney (1992) has proposed how translocation might be occurring in rhizomorphs of ectomycorrhizal and saprotrophic basidiomycetes. The system can be represented diagrammatically (Figure 14.6). Essentially, bidirectional flow occurs in spatially separated hyphal elements. It is suggested that translocation of carbon from the resource supplying it is by mass flow of solution along vessel hyphae. Nutrients obtained from the soil are believed to be translocated along those hyphae that contain fully functional cytoplasm by contractile processes. Water movement in the reverse direction to that carrying carbon is thought to occur in the walls of the functional hyphae.

Though it is certain from studies of translocation in *Serpula lacrymans* that mass flow of solution can occur in fully functional hyphae as evidenced by autoradiographic evidence (Brownlee & Jennings, 1982*b*) and droplet studies (see previous section), Cairney (1992) believes that translocation by mass flow along fully functional hyphae does not occur in rhizomorphs because flow will be impeded by dolipore septa. There are two points to make here. First, there is a lack of information about septal pores in rhizomorphs and it is not at all certain that the dolipore type is necessarily

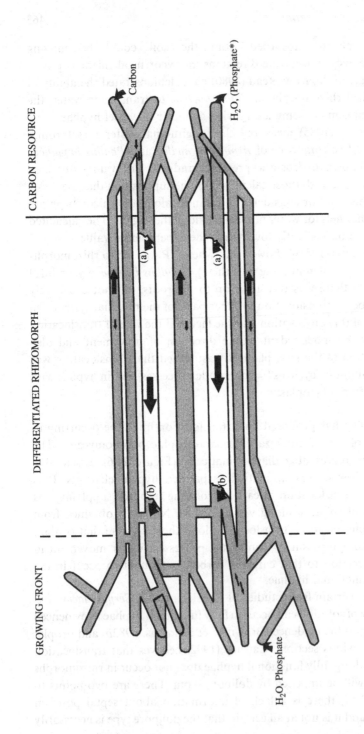

frequent. Second, if contractile processes are responsible for translocation, it is unlikely that they will be taking material across the septum if a dolipore is present. This means there must be unloading from the translocation stream in one hyphal compartment and loading into the translocation stream in the next compartment. If there is no bulk flow, movement from one compartment to another will be by diffusion. The situation is akin to solute flow along large algal coenocytes, which has been studied in some detail (Jennings, 1986*b*). Whether what I have just described for hyphae in rhizomorphs will allow the necessary rate of translocation can be decided only after further study.

One needs to be cautious in accepting the interpretation of Finlay & Read (1986*b*) of their data referred to above. First, the rate of movement of phosphorus in the mycelium may not be determined by the rate of translocation per se but could be heavily influenced by the rate of loading and unloading of the translocation stream, which two processes could severely limit the amount of phosphorus moved. It is known that, when

---

Figure 14.6. Diagram of a rhizomorph growing from a unit of carbon resource (ectomycorrhizal root or woody substrate), showing putative routes of translocation as hypothesised by Cairney (1992). Shaded hyphae are septate and retain their cytoplasmic contents. Unshaded lengths of vessel hyphae ('defined apoplast of vessel hyphae') are without cytoplasm and have undergone septal dissolution. Carbon absorbed by loading hyphae at the carbon resource is envisaged as being translocated acropetally by cytoplasmic streaming in the hyphal symplast to the sink at the apoplastic junction (a). Turgor pressure created by carbon efflux would generate a mass flow of solution in the apoplastic phase of 'vessel' hyphae, maintained by absorption of carbon by hyphae at the symplastic junction of the growing front (b). Carbon compounds could then move by cytoplasmic streaming to sinks at the apices of extending hyphae.

Nutrients absorbed from soil by hyphae at the growing front are suggested to be translocated basipetally by streaming in the cytoplasm along the entire rhizomorph length. Water may flow along the rhizomorph apoplast (which excludes the 'defined apoplast' of vessel hyphae) by mass flow along a gradient of water potential generated at the base (see the text). Removal of water at the carbon resource (see the text) may increase rates of cytoplasmic streaming, permitting rapid basipetal movement in the symplast. Some acropetal movement of carbon may also occur in the symplast of these hyphae.

The apoplastic phase of vessel hyphae may be isolated from the remainder of the apoplast, and also the cytoplasmic translocatory stream, by the presence of hydrophobic vessel hyphal walls and deposition of poorly permeable extracellular material. Hyphal branches (not shown) connecting cytoplasmic hyphae with apoplastic regions of vessel hyphae may occur at points within the differentiated rhizomorph.

The asterisk denotes efflux of phosphate to the interface with the host plant (ectomycorrhizal fungi only).

phosphorus is absorbed by rhizomorphs of *Armillaria mellea*, a considerable proportion of the phosphorus goes into the vacuole and into polyphosphate (Cairney *et al.*, 1988). Thus, the extent to which the translocation stream can become labelled will depend on the rates of transport across the plasma membranes of both absorbing and translocating hyphae, across the tonoplast and the rate of formation and breakdown of polyphosphate. It should be noted that a lack of effect of transpiration on ion movement into a plant does not argue against movement of the ion in the transpiration stream (Tanner & Beevers, 1990). So one cannot argue necessarily that lack of an effect of transpiration on translocation in the mycorrhizal rhizomorph means that the mechanism is not the mass flow of solution. Finally, one needs to remember that circulatory movement of solutes in the higher plant occurs by mass flow in both directions in the plant in tissues (phloem and xylem) that may be not much further apart than the hyphae responsible for translocation in opposite directions in a rhizomorph. All the above argues for an open mind on the mechanism of translocation in rhizomorphs; more work is needed before what might be happening is clarified. Most particularly, there is a need to attempt to relate what is observed physiologically with hyphal organisation within the rhizomorphs.

Basidiomycete sporophores

Water can be lost very readily for basidiomycete sporophores (Jennings, 1987*b*). The rate of loss can be similar to that from a free water surface. Control over loss is in many cases related to the depth of the boundary layer over the sporophore. The depth of this layer is course dependent on the turbulence of the air.

There is clear evidence that evaporation (transpiration) from a sporophore can increase the rate of translocation into the organ. This was well demonstrated for dry matter increment, potassium and ascent of dye within the sporophore of *Polyporus brumalis* (Plunkett, 1958). The movement of ^{32}P in sporophores of *Lentinus tigrinus* has been shown to be increased by high transpiration (Littlefield, Wilcoxson & Sudia, 1965*b*). On the other hand, Plunkett (1956) showed that growth of the sporophore of *Collybia velitupes* does not respond to increased water loss, whereas Badham (1985) showed that, for *Psilocybe cubensis*, the fastest growth rate and lowest transpiration occurred at the highest humidities.

Thus, water can be lost readily from basidiomycete sporophores and under certain circumstances the water flow through the structures can enhance growth, Under these circumstances one can presume that the increased water flow draws nutrients up the sporophore. However, sporophore growth can still occur in water-saturated air. Under such conditions, droplets can be seen, as described by Buller (1931), who also pointed out that upward growth of the sporophore generates significant internal pressure. This indicates turgor-driven movement of water into the sporophore. There is no reason why this flow of water should not carry nutrients. The location of the pathway of water movement up a sporophore is not clear.

All the available information to date has been obtained with dyes but their use can lead to ambiguities (Jennings, 1987*b*). Nevertheless the indications that there is significant water movement upwards via hyphal walls should not be discounted. This matter needs to be pursued. The elegant study by Young & Ashford (1992) of changes that occur in the permeability of sclerotia of *Scerotinia minor* during their development using the fluorochrome sulphorhodamine G has demonstrated how the extent to which walls and protoplasm of a fungal organ are accessible to the external medium can be properly assessed. Studies on water flow through higher plants offer guidance as to other possible tracers for water movement (Canny, 1990).

Concluding remarks

It is only for *Serpula lacrymans* that there is a fair degree of certainty about the mechanism of translocation, namely that it is brought about by pressure-driven mass flow of solution. It seems likely that, under conditions in which mycelium of *Neurospora crassa* is made to grow over a non-nutrient surface, translocation is by the same mechanism. But diffusion is likely to be an important means of moving solutes in mould colonies a few centimetres in diameter. For other translocating systems, the situation is less clear, though for rhizomorphs with a well-defined system of vessel hyphae it is probable that here too pressure-driven mass flow is important. The extent to which water does flow within many fungal colonies could well be important for integrated functioning. Evidence suggests that integration within a colony is not hormonal and it could well be that it is water relations that determine the functioning of the organised colony (Jennings, 1991*a*). It is clear that more knowledge is needed about the water relations of fungal colonies.

Literature cited

Abuzinadah, R. A. & Read, D. J. (1986a). The role of proteins in the nitrogen
nutrition of ectomycorrhizal plants. I. Utilization of peptides and proteins
by ectomycorrhizal fungi. *New Phytologist* **103**, 481–94.

Abuzinadah, R. A. & Read, D. J. (1986b). The role of proteins in the nitrogen
nutrition of ectomycorrhizal plants. II. Utilization of protein by
mycorrhizal plants of *Pinus contorta*. *New Phytologist* **103**, 495–506.

Abuzinadah, R. A. & Read, D. J. (1986c). The role of proteins in the nitrogen
nutrition of ectomycorrhizal plants. III. Protein utilization by *Betula, Picea*
and *Pinus* in mycorrhizal association with *Hebeloma crustuliniforme*. *New
Phytologist* **103**, 507–14.

Abuzinadah, R. A. & Read, D. J. (1989a). The role of proteins in the nitrogen
nutrition of ectomycorrhizal plants. IV. The utilization of peptides by birch
(*Betula pendula* L.) infected with different mycorrhizal fungi. *New
Phytologist* **112**, 55–60.

Abuzinadah, R. A. & Read, D. J. (1989b). The role of proteins in the nitrogen
nutrition of ectomycorrhizal plants. V. Nitrogen transfer in birch (*Betula
pendula*) grown in association with mycorrhizal and non-mycorrhizal fungi.
New Phytologist **112**, 61–8.

Abuzinadah, R. A. & Read, D. J. (1989c). Carbon transfer associated with
assimilation of organic nitrogen sources by silver birch (*Betula pendula*
Roth.). *Trees* **3**, 17–23.

Achstetter, T. & Wolf, D. H. (1985). Proteinases, proteolysis and biological
control in the yeast *Saccharomyces cerevisiae*. *Yeast* **1**, 139–57.

Addison, R. (1986). Primary structure of the *Neurospora crassa* plasma
membrane H^+-ATPase deduced from the gene sequence. Homology to
Na^+/K^+-, Ca^{2+}-, and K^+-ATPase. *Journal of Biological Chemistry* **261**,
14896–901.

Addison, R. & Scarborough, G. A. (1982). Conformational changes of the
Neurospora plasma membrane H^+ ATPase during its catalytic cycle.
Journal of Biological Chemistry **257**, 10421–6.

Addison, R. & Scarborough, G. A. (1986). Interactions of *Neurospora crassa*
plasma membrane H^+-ATPase with *N*-(ethoxycarbonyl)-2-ethoxy-1,2-
dihydroquinoline. *Biochemistry* **25**, 4071–6.

Adebayo, A. A., Harris, R. F. & Gardner, W. R. (1971). Turgor pressure of
fungal mycelia. *Transactions of the British Mycological Society* **57**, 145–51.

Adjimani, J. P. & Emery, T. (1987). Iron uptake in *Mycelia sterilia* EP-76.
Journal of Bacteriology **169**, 3664–8.

Adjimani, J. P. & Emery, T. (1988). Stereochemical aspects of iron transport in *Mycelia sterilia* EP-76. *Journal of Bacteriology* **170**, 1377–9.

Adler, E. (1968). [The chemical structure of lignin.] (In Swedish.) *Svensk Kemiska Tidskrifft* **80**, 279–90.

Adler, L., Blomberg, A. & Nilsson, A. (1985). Glycerol metabolism and osmoregulation in the salt-tolerant yeast *Debaryomyces hansenii*. *Journal of Bacteriology* **162**, 300–6.

Adler, L. & Gustafsson, L. (1980). Polyhydric alcohol production and intracellular amino acid pool in relation to halotolerance in the yeast *Debaryomyces hansenii*. *Archives of Microbiology* **126**, 123–30.

Agnihotri, V. P. (1970). Solubilization of insoluble phosphates by some soil fungi isolated from nursery seedbeds. *Canadian Journal of Microbiology* **16**, 877–80.

Agosin, E., Blanchette, R. A., Silva, H., Lapierre, C., Cease, K. R., Ibach, R. E., Abad, A. R. & Muga, P. (1990). Characterization of palo podrido, a natural process of delignification of wood. *Applied and Environmental Microbiology* **56**, 65–74.

Aguilera, A. & Benitez, T. (1986). Ethanol-sensitive mutants of *Saccharomyces cerevisiae*. *Archives of Microbiology* **143**, 337–44.

Ahmad, I., Carleton, T. J., Malloch, D. W. & Hellebust, J. A. (1990). Nitrogen metabolism in the ectomycorrhizal fungus *Laccaria bicolor* (R. Mre) Orton. *New Phytologist* **116**, 431–42.

Aiking, H., Sterkenberg, A. & Tempest, D. W. (1977*a*). Influence of simple growth limitation and dilution rate on the phosphorylation efficiency and cytochrome content of mitochondria of *Candida utilis* NCYC 321. *Archives of Microbiology* **113**, 65–72.

Aiking, H., Sterkenberg, A. & Tempest, D. W. (1977*b*). The occurrence of polyphosphates in *Candida utilis* NCYC 321, grown in chemostat cultures under conditions of potassium- and glucose-limitation. *FEMS Microbiology Letters* **1**, 251–4.

Aiking, H. & Tempest, D. W. (1976). Growth and physiology of *Candida utilis* NCYC 321 in potassium-limited chemostat culture. *Archives of Microbiology* **108**, 117–24.

Aiking, H. & Tempest, D. W. (1977). Rubidium as probe for function and transport of potassium in the yeast *Candida utilis* NCYC 321, grown in chemostat culture. *Archives of Microbiology* **115**, 215–21.

Akamatsu, Y., Ma, D. N., Higuchi, T. & Shimada, M. (1990). A novel enzymatic decarboxylation of oxalic acid by the lignin peroxidase system of white-rot fungus *Phanerochaete chrysosporium*. *FEBS Letters* **269**, 261–3.

Alcorn, M. E. & Griffin, C. C. (1978). A kinetic analysis of D-xylose transport in *Rhodotorula glutinis*. *Biochimica et Biophysica Acta* **510**, 361–71.

Aldermann, B. & Höfer, M. (1981). The active transport of monosaccharides by the yeast *Metschnikowia reukaufii*: evidence for an electrochemical gradient of H^+ across the cell membrane. *Experimental Mycology* **5**, 120–32.

Alderman, D. J. & Jones, E. B. G. (1971). Physiological requirements of two marine phycomycetes *Althornia crouchii* and *Ostracoblabe implexa*. *Transactions of the British Mycological Society* **57**, 213–25.

Ali, A. H. & Hipkin, C. R. (1985). Nitrate assimilation in the basidiomycete yeast *Sporobolomyces roseus*. *Journal of General Microbiology* **131**, 1867–74.

Ali, A. H. & Hipkin, C. R. (1986). Nitrate assimilation in *Candida nitrophila* and other yeasts. *Archives of Microbiology* **144**, 263–7.

Alic, M. & Gold, M. H. (1991). Genetics and molecular biology of the lignin-degrading basidiomycete *Phanerochaete chrysosporium*. In *More Gene Manipulation in Fungi*, ed. J. W. Bennett & L. L. Lasure, pp. 319–41. Academic Press: San Diego.

Alic, M., Letzring, C. & Gold, M. H. (1987). Mating system and basidiospore formation in the lignin-degrading basidiomycete *Phanerochaete chrysosporium*. *Applied and Environmental Microbiology* **53**, 1464–9.

Allen, A. & Sternberg, D. (1980). β-Glucosidase production by *Aspergillus phoenicus* in stirred tank fermentors. *Biotechnology and Bioengineering Symposium* **10**, 189–87.

Allen, M. F. (1982). Influence of vesicular–arbuscular mycorrhizae on water movement through *Bouteloua gracilis* (H.B.K.) Lag ex Steud. *New Phytologist* **91**, 191–6.

Almin, K. E., Eriksson, K.-E. & Pettersson, B. (1975). Extracellular enzyme system utilized by the fungus *Sporotrichum pulverulentum* (*Chrysosporium lignorum*) for the breakdown of cellulose. 2. Activities of the five endo-1,4-β-glucanases towards carboxymethyl cellulose. *European Journal of Biochemistry* **51**, 207–11.

Alonso, A. & Kotyk, A. (1978). Apparent half-lives of sugar transport proteins in *Saccharomyces cerevisiae*. *Folia Microbiologica* **23**, 118–25.

Ames, B. N., Ames, G. F., Young, J. D., Tsuchiya, D. & Lecocq, J. (1973). Illicit transport: the oligopeptide permease. *Proceedings of the National Academy of Sciences, USA* **70**, 456–8.

Ames, R. N., Reid, C. P. P., Porter, L. K. & Cambardella, C. (1983). Hyphal uptake and transport of nitrogen from two ^{15}N-labelled sources by *Glomus mosseae* a vesicular–arbuscular mycorrhizal fungus. *New Phytologist* **95**, 381–96.

Ammann, D., Lanter, F., Steiner, R. A., Schulthess, P., Shijo, Y. & Simon, W. (1981). Neutral carrier based hydrogen ion selective microelectrodes for extra- and intracellular studies. *Analytical Chemistry* **53**, 2267–9.

Amon, J. P. (1976). An estuarine species of *Phlyctochytrium* (Chytridiales) having a transient requirement for sodium. *Mycologia* **68**, 470–80.

Amon, J. P. & Arthur, R. D. (1980). The requirement for sodium in marine fungi: uptake and incorporation of amino acids. *Botanica Marina* **23**, 639–44.

Amon, J. P. & Arthur, R. D. (1981). Nutritional studies of a marine *Phlyctochytrium* sp. *Mycologia* **73**, 1049–55.

Anand, J. C. & Brown, A. D. (1968). Growth rate patterns of the so-called osmophilic and non-osmophilic yeasts in solutions of polyethylene glycol. *Journal of General Microbiology* **52**, 205–12.

Ander, P. & Eriksson, K.-E. (1977). Selective delignification of wood components by white rot fungi. *Physiologia Plantarum* **41**, 239–48.

Ander, P. & Eriksson, K.-E. (1978). Lignin degradation and utilization by microorganisms. *Progress in Industrial Microbiology* **14**, 1–58.

Andersen, B. & Jørgensen, O. S. (1969). Purification of external invertase from brewer's yeast. *Acta Chemica Scandinavica* **23**, 2270–6.

Anderson, J. A., Best, L. A. & Gaber, R. F. (1991). Structural and functional conservation between high-affinity K[+] transport in *Saccharomyces uvarum* and *Saccharomyces cerevisiae*. *Gene* **99**, 39–46.

Anderson, J. B. & Ullrich, R. C. (1982). Translocation in rhizomorphs of *Armillaria mellea*. *Experimental Mycology* **6**, 31–40.

André, L., Nilsson, A. & Adler, L. (1988). The role of glycerol in the osmotolerance of the yeast *Debaryomyces hansenii*. *Journal of General Microbiology* **134**, 669–77.

Andreasen, A. A. & Stier, T. J. B. (1953). Anaerobic nutrition of *Saccharomyces cerevisiae*. I. Ergosterol requirement for growth in a defined medium. *Journal of Cellular and Comparative Physiology* **41**, 22–36.

Andreasen, A. A. & Stier, T. J. B. (1954). Anaerobic nutrition of *Saccharomyces cerevisiae*. II. Unsaturated fatty acid requirement for growth in a defined medium. *Journal of Cellular and Comparative Physiology* **43**, 271–82.

Anraku, Y., Umemoto, N., Hirata, R. & Wada, Y. (1989). Structure and function of the yeast vacuolar membrane proton ATPase. *Journal of Bioenergetics and Biomembranes* **21**, 589–604.

Antibus, R. K., Kroehler, C. J. & Linkins, A. E. (1986). Effects of external pH, temperature and substrate concentration on acid phosphatase activity of ectomycorrhizal fungi. *Canadian Journal of Botany* **64**, 2383–7.

ap Rees, T. (1974). Pathways of carbohydrate breakdown in plants. In *MTP International Review of Science, Biochemistry Series 1*, vol. XI *Plant Biochemistry*, ed. D. H. Northcote, pp. 89–127. Butterworth: London.

ap Rees, T. (1980*a*). Assessment of the contributions of metabolic pathways to plant respiration. In *Biochemistry of Plants: A Comprehensive Treatise*, vol. 2 *Metabolism and Respiration*, ed. D. D. Davies, pp. 1–29. Academic Press: New York.

ap Rees, T. (1980*b*). Integration of pathways of synthesis and degradation of hexose phosphates. In *Biochemistry of Plants: A Comprehensive Treatise*, vol. 3 *Carbohydrates: Structure and Function*, ed. J. Preiss, pp. 1–42. Academic Press: New York.

Armbruster, B. L. & Weisenseel, M. H. (1983). Ionic currents traverse growing hyphae and sporangia of the water mold *Achlya debaryana*. *Protoplasma* **115**, 65–9.

Armstrong, W. McD. & Rothstein, A. (1964). Discrimination between alkali metal cations by yeast. I. Effect of pH on uptake. *Journal of General Physiology* **48**, 61–71.

Armstrong, W. McD. & Rothstein, A. (1967). Discrimination between alkali metal cations by yeast. II. Cation interactions in transport. *Journal of General Physiology* **50**, 967–88.

Arnold, W. N. (1979). Trehalose assimilation and turnover in *Torulopsis glabrata*. *Current Microbiology* **2**, 109–12.

Arnold, W. N. (ed.) (1981*a*). *Yeast Cell Envelopes: Biochemistry, Biophysics and Ultrastructure*, vols. I and II. CRC Press: Boca Raton, FL.

Arnold, W. N. (1981*b*). Enzymes. In *Yeast Cell Envelopes: Biochemistry, Biophysics and Ultrastructure*, vol. II, ed. W. N. Arnold, pp. 1–46. CRC Press: Boca Raton, FL.

Arnold, W. N. (1987). Hydrolytic enzymes. In *Yeast Biotechnology*, ed. D. R. Berry, I. Russell & G. G. Stewart, pp. 369–400. Allen & Unwin: London.

Arnold, W. N. & Lacy, J. S. (1977). Permeability of cell envelope and osmotic behaviour in *Saccharomyces cerevisiae*. *Journal of Bacteriology* **131**, 564–71.

Arst, H. N. Jr, Bailey, C. R. & Penfold, H. A. (1980). A possible role for acid phosphatase in γ-amino-*n*-butyrate uptake in *Aspergillus nidulans*. *Archives of Microbiology* **125**, 153–8.

Arst, H. N. Jr & MacDonald, D. W. (1975). A gene cluster in *Aspergillus nidulans* with an internally located *cis*-acting regulatory region. *Nature, London* **254**, 26–31.

Arst, H. N. Jr & MacDonald, D. W. (1978). Reduced expression of a distal gene of the *prn* gene cluster in deletion mutants of *Aspergillus nidulans*: genetic evidence for a dicistronic messenger in a eukaryote. *Molecular and General Genetics* **163**, 17–22.

Arst, H. N. Jr, MacDonald, D. W. & Jones, S. A. (1980). Regulation of proline transport in *Aspergillus nidulans*. *Journal of General Microbiology* **116**, 285–94.

Asea, P. E. A., Kucey, R. M. N. & Stewart, J. W. B. (1988). Inorganic phosphate solubilization by two *Penicillium* species in solution culture and soil. *Soil Biology and Biochemistry* **20**, 459–64.

Ashford, A. E. & Allaway, W. G. (1985). Transfer cells and Hartig net in the root epidermis of the sheathing mycorrhiza of *Pisonia grandis* R. Br. from Seychelles. *New Phytologist* **100**, 595–612.

Ayers, A. R., Ayers, S. B. & Eriksson, K.-E. (1978). Cellobiose oxidase, purification and partial characterization of a hemoprotein from *Sporotrichum pulverulentum*. *European Journal of Biochemistry* **90**, 171–81.

Azam, F. & Kotyk, A. (1969). Glucose-6-phosphate as a regulator of monosaccharide transport in baker's yeast. *FEBS Letters* **2**, 333–5.

Badham, E. R. (1985). The influence of humidity upon transpiration and translocation in the mushroom *Psilocybe cubensis*. *Bulletin of the Torrey Botanical Club* **111**, 159–64.

Bagg, A. & Neilands, N. B. (1987). Molecular mechanism of regulations of siderophore-mediated iron assimilation. *Microbiological Reviews* **51**, 509–18.

Bahnweg, G. (1979). Studies on the physiology of Thraustochytriales. I. Growth requirements and nitrogen nutrition of *Thraustochytrium* spp., *Schizochytrium* sp., *Japonochytrium* sp., *Ulkenia* spp. and *Labryinthuloides* spp. *Veroffentlichungen der Instituts für Meeresforschung, Bremerhaven* **17**, 245–68.

Bahnweg, G. (1980). Phospholipid and steroid requirements of *Haliphthorus milfordensis*, a parasite of marine crustaceans and *Phytophthora epistomium*, a facultative parasite of marine fungi. *Botanica Marina* **23**, 209–18.

Bajwa, R., Abuarghub, S. & Read, D. J. (1985). The biology of mycorrhiza in the Ericaceae. X. The utilization of proteins and the proteolytic enzymes by the mycorrhizal endophyte and by mycorrhizal plants. *New Phytologist* **101**, 469–86.

Bajwa, R. & Read, D. J. (1985). The biology of mycorrhiza of the Ericaceae. IX. Peptides as nitrogen sources for the ericoid endophyte and for mycorrhizal and non-mycorrhizal plants. *New Phytologist* **101**, 459–68.

Bajwa, R. & Read, D. J. (1986). Utilization of mineral and amino N sources by the ericoid mycorrhizal endophyte *Hymenoscyphus ericae* and by mycorrhizal and non-mycorrhizal seedlings of *Vaccinium*. *Transactions of the British Mycological Society* **87**, 269–77.

Ballarin-Denti, A., Den Hollander, J. A., Sanders, D., Slayman, C. W. & Slayman, C. L. (1984). Kinetics and pH-dependence of glycine-protein symport in *Saccharomyces cerevisiae*. *Biochimica et Biophysica Acta* **778**, 1–16.

Ballou, C. E. (1988). Organisation of the *Saccharomyces cerevisiae* cell wall. In *Self-Assembling Architecture*, ed. J. E. Varner, pp. 105–17. Alan R. Liss: New York.

Balmforth, A. J. & Thomson, A. (1984). Isolation and characterization of glyoxylate dehydrogenase from the fungus *Sclerotium rolfsii*. *Biochemical Journal* **218**, 113–18.

Banta, L. M., Robinson, J. S., Klionsky, D. J. & Emr, S. D. (1988). Organelle assembly in yeast: characterization of yeast mutants defective in vacuolar biogenesis and protein sorting. *Journal of Cell Biology* **107**, 1369–84.

Baranowski, K. & Radler, F. (1984). The glucose-dependent transport of L-malate in *Zygosaccharomyces bailii*. *Antonie van Leeuwenhoek* **50**, 329–40.

Barber, S. A. (1984). *Soil Nutrient Bioavailability – A Mechanistic Approach.* Wiley Interscience: New York.

Barksdale, A. W. (1962). Effect of nutritional deficiencies on growth and sexual reproduction of *Achlya ambisexualis*. *American Journal of Botany* **49**, 633–8.

Barner, H. D. & Cantino, E. C. (1952). Nutritional relationships in a new species of *Blastocladiella*. *American Journal of Botany* **39**, 746–51.

Barnett, J. A. (1976). The utilization of sugars by yeasts. *Advances in Carbohydrate Chemistry and Biochemistry* **32**, 135–234.

Barnett, J. A. (1981). The utilization of disaccharides and some other sugars by yeasts. *Advances in Carbohydrate Chemistry and Biochemistry* **39**, 347–404.

Barnett, J. A. & Kornberg, H. L. (1960). Utilization by yeasts of acids of the tricarboxylic acid cycle. *Journal of General Microbiology* **23**, 65–82.

Barnett, J. A., Payne, R. W. & Yarrow, D. (1979). *A Guide to Identifying and Classifying Yeasts.* Cambridge University Press: Cambridge.

Barnett, J. A., Payne, R. W. & Yarrow, D. (1983). *Yeasts. Characterization and Identification.* Cambridge University Press: Cambridge.

Barnett, J. A. & Sims, A. P. (1982). The requirement of oxygen for the active transport of sugars into yeasts. *Journal of General Microbiology* **128**, 2303–12.

Barrasa, J. M., González, A. E. & Martínez, A. T. (1992). Ultrastructural aspects of fungal delignification of Chilean woods by *Ganoderma australe* and *Phlebia chrysocrea* – a study of natural and *in vivo* degradation. *Holzforschung* **46**, 1–8.

Barrow, N. J., Malajczuk, N. & Shaw, T. C. (1977). A direct test of the ability of vesicular–arbuscular mycorrhiza to help plants to take up fixed soil phosphate. *New Phytologist* **78**, 269–76.

Bartelt, D. C., Fidel, S., Farber, L. H., Wolff, D. J. & Hammell, R. L. (1988). Calmodulin-dependent multifunctional protein kinase in *Aspergillus nidulans*. *Proceedings of the National Academy of Sciences, USA* **85**, 3279–83.

Barthelmess, I. B. (1982). Mutants affecting amino acid cross-pathway control in *Neurospora crassa*. *Genetical Research* **39**, 169–85.

Bartlett, E. M. & Lewis, D. H. (1973). Surface phosphatase activity of mycorrhizal roots of beech. *Soil Biology and Biochemistry* **5**, 249–57.

Bartnicki-Garcia, S. (1973). Fundamental aspects of hyphal morphogenesis. *Symposium of the Society for General Microbiology* **23**, 245–67.

Bartnicki-Garcia, S. (1987). The cell wall: a crucial structure in fungal evolution. In *Evolutionary Biology of the Fungi*, ed. A. D. M. Rayner, C. M. Brasier & D. M. Moore, pp. 389–403. Cambridge University Press: Cambridge.

Bartnicki-Garcia, S. & Lipmann, E. (1972). The bursting tendency of hyphal tips: presumptive evidence for a delicate balance between wall synthesis and wall lysis in apical growth. *Journal of General Microbiology* **73**, 487–500.

Bartnicki-Garcia, S. & Lipmann, E. (1982).Fungal cell wall composition. In *CRC Handbook of Microbiology*, vol. IV *Microbial Composition:*

Carbohydrates, Lipids and Minerals, ed. A. I. Laskin & H. A. Lechevalier, pp. 229–52. CRC Press: Boca Raton, FL.

Barts, P. W. J. A., Hoeberichts, J. A., Klaassen, A. & Borst-Pauwels, G. W. F. H. (1980). Uptake of the lipophilic cation dibenzyldimethylammonium into *Saccharomyces cerevisiae*. Interaction with the thiamine transport system. *Biochimica et Biophysica Acta* **597**, 125–36.

Basu, S. N. (1951). Significance of calcium in the fruiting of *Chaetomium* sp., particularly *Chaetomium globosum*. *Journal of General Microbiology* **5**, 231–8.

Basu, S. N. (1952). *Chaetomium basiliensis* Batista & Pontual: nutritional requirements for growth and fruiting. *Journal of General Microbiology* **6**, 199–204.

Bateman, D. F. & Beer, S. V. (1965). Simultaneous production and synergistic action of oxalic acid and polygalacturonase during pathogenesis by *Sclerotium rolfsii*. *Phytopathology* **55**, 204–11.

Bates, W. & Wilson, J. (1974). Ethylene glycol-induced alteration of conidial germination in *Neurspora crassa*. *Journal of Bacteriology* **117**, 560–7.

Bauchop, T. (1989). Biology of gut anaerobic fungi. *BioSystems* **23**, 53–64.

Bavendamm, W. (1928). Über das Vorkommen und der Nachweis von Oxygen bei holzzerstorenden Pilzen. I. Mitteilung. *Zeitschrift für Pflanzenkrankheiten* **38**, 257–76.

Beauvoit, B., Rigoulet, M., Guerin, B. & Canioni, P. (1989). Polyphosphates as a source of high energy phosphates in yeast mitochondria: a ^{31}P NMR study. *FEBS Letters* **252**, 17–21.

Beavan, M. J., Belk, D. M., Stewart, G. G. & Rose, A. H. (1979). Changes in electrophoretic mobility and lytic enzyme activity associated with development of flocculating ability in *Saccharomyces cerevisiae*. *Canadian Journal of Microbiology* **25**, 888–95.

Bech-Andersen, J. (1985). [Why is *Serpula lacrymans* only found in houses?] (In Danish.) *Svampe* **12**, 60–4.

Becker, J. M., Covert, N. L., Shenbagamurthi, P., Steinfield, A. S. & Naider, F. (1983). Polyoxin D inhibits growth of zoopathogenic fungi. *Antimicrobial Agents and Chemotherapy* **23**, 926–9.

Becker, J. M. & Naider, F. (1977). Peptide transport in yeast: uptake of radioactive trimethionine in *Saccharomyces cerevisiae*. *Archives of Biochemistry and Biophysics* **178**, 245–55.

Becker, J. M. & Naider, F. (1980). Transport and utilization of peptides by yeast. In *Microorganisms and Nitrogen Sources*, ed. J. W. Payne, pp. 257–79. John Wiley: Chichester.

Becker, J. M., Naider, F. & Katchalski, E. (1973). Peptide utilization in yeast. Studies on methionine and lysine auxotrophs of *Saccharomyces cerevisiae*. *Biochimica et Biophysica Acta* **291**, 388–97.

Becker, J. M., Wilchek, M. & Katchalski, E. (1971). Irreversible inhibition of biotin transport in yeast by biotinyl-*p*-nitrophenyl ester. *Proceedings of the National Academy of Sciences, USA* **68**, 2604–7.

Beever, R. E. & Burns, D. J. W. (1977). Adaptive changes in phosphate uptake by the fungus *Neurospora crassa* in response to phosphate supply. *Journal of Bacteriology* **132**, 520–5.

Beever, R. E. & Burns, D. J. W. (1978). Does cycloheximide-induced loss of phosphate uptake activity in *Neurospora crassa* reflect rapid turnover? *Journal of Bacteriology* **134**, 1176–8.

Beever, R. E. & Burns, D. J. W. (1980). Phosphorus uptake, storage and utilization by fungi. *Advances in Botanical Research* **8**, 128–219.

Beever, R. E. & Laracy, E. P. (1986). Osmotic adjustment in the filamentous fungus *Aspergillus nidulans*. *Journal of Bacteriology* **168**, 1358–65.

Béguin, P. (1990). Molecular biology of cellose degradation. *Annual Review of Microbiology* **44**, 219–48.

Behrens, M. M. & Mazon, M. J. (1988). Yeast cAMP-dependent protein kinase can be associated with the plasma membrane. *Biochemical and Biophysical Research Communications* **151**, 561–7.

Belde, P. J. M., Kessels, B. G. F., Moelans, I. M. & Borst-Pauwels, G. W. F. H. (1988). Cd^{2+} uptake, Cd^{2+} binding and loss of cell K^+ by a Cd-sensitive and a Cd-resistant strain of *Saccharomyces cerevisiae*. *FEMS Microbiology Letters* **49**, 493–8.

Beldman, G., Searle-van Leeuwen, M. F., Rombouts, F. M. & Voragen, F. G. J. (1985). The cellulase of *Trichoderma viride*: purification, characterization and comparison of all detectable endoglucanases, exoglucanases and β-glucosidases. *European Journal of Biochemistry* **146**, 301–8.

Bellenger, N., Nissen, P. W., Wood, T. C. & Segel, I. H. (1968). Specificity and control of choline-*O*-sulfate transport in filamentous fungi. *Journal of Bacteriology* **96**, 1574–85.

Bellinger, Y. & Larher, F. (1987). Proline accumulation in higher plants: a redox buffer? *Plant Physiology (Life Sciences Advances)* **6**, 23–7.

Bellinger, Y. & Larher, F. (1988). A ^{13}C comparative nuclear magnetic resonance study of organic solute production and excretion by the yeasts *Hansenula anomala* and *Saccharomyces cerevisiae* in saline media. *Canadian Journal of Microbiology* **34**, 605–12.

Belsky, M. M., Goldstein, S. & Menna, M. (1970). Factors affecting phosphate uptake in the marine fungus *Dermocystidium* sp. *Journal of General Microbiology* **62**, 399–402.

Beltramini, M. & Lerch, K. (1981). Luminescence properties of *Neurospora* copper metallothionein. *FEBS Letters* **127**, 201–3.

Beltramini, M. & Lerch, K. (1982). Copper transfer between *Neurospora* copper metallothionein and type 3 copper apoproteins. *FEBS Letters* **142**, 219–22.

Beltramini, M. & Lerch, K. (1983). Spectroscopic studies on *Neurospora* copper metallothionein. *Biochemistry* **22**, 2043–8.

Benítez, J. A., Alonso, A., Delgado, J. & Kotyk, A. (1983). Sulphate transport in *Candida utilis*. *Folia Microbiologica* **28**, 6–11.

Benito, B. & Lagunas, R. (1992). The low-affinity component of *Saccharomyces cerevisiae* maltose transport is an artefact. *Journal of Bacteriology* **174**, 3065–9.

Benko, P. V., Wood, T. C. & Segel, I. H. (1969). Multiplicity and regulation of amino acid transport in *Penicillium chrysogenum*. *Archives of Biochemistry and Biophysics* **129**, 498–508.

Bentley, R. & Thiessen, C. P. (1957*a*). Biosynthesis of itaconic acid in *Aspergillus terreus*. I. Tracer studies with ^{14}C-labelled substrates. *Journal of Biological Chemistry* **226**, 673–87.

Bentley, R. & Thiessen, C. P. (1957*b*). Biosynthesis of itaconic acid in *Aspergillus terreus*. II. Early stages in glucose dissimilation and the role of citrate. *Journal of Biological Chemistry* **226**, 689–701.

Bentley, R. & Thiessen, C. P. (1957*c*). Biosynthesis of itaconic acid in *Aspergillus terreus*. III. The properties and reaction of mechanism of *cis*-aconitate decarboxylase. *Journal of Biological Chemistry* **226**, 703–20.

Bercovitz, A., Peleg, Y., Battat, E., Rokem, J. S. & Goldberg, I. (1990). Localization of pyruvate carboxylase in organic acid-producing *Aspergillus* strains. *Applied and Environmental Microbiology* **56**, 1594–7.

Berdicevsky, I. & Grossowicz, N. (1977). Effect of polyene antibiotics on growth and phosphate uptake by *Candida albicans*. *Journal of General Microbiology* **102**, 299–304.

Berg, B. & Pettersson, L. G. (1977). Location and formation of cellulases in *Trichoderma viride*. *Journal of Applied Bacteriology* **42**, 65–75.

Berry, D. R. & Watson, D. C. (1987). Production of organoleptic compounds. In *Yeast Biotechnology*, ed. D. R. Berry, I. Russell & G. G. Stewart, pp. 345–68. Allen & Unwin: London.

Bes, B., Pettersson, B., Lennholm, H., Iversen, T. & Eriksson, K.-E. (1987). Synthesis structure and enzymatic degradation of an extracellular glucan produced in nitrogen-starved cultures of the white rot fungus *Phanerochaete chrysosporium*. *Biotechnology and Applied Biochemistry* **9**, 310–18.

Beullens, M. & Thevelein, J. M. (1990). Investigation of transport-associated phosphorylation of sugar in yeast mutants (*snf3*) lacking high-affinity glucose transport and in a mutant (*fdp1*) showing deficient regulation of initial sugar metabolism. *Current Microbiology* **21**, 39–46.

Bhanot, P. & Brown, R. G. (1980). Effect of 3-*O*-methyl-D-glucose on the production of glycosidases by *Cryptotoccus laurentii* and *Saccharomyces cerevisiae*. *Canadian Journal of Microbiology* **26**, 1289–95.

Bhargava, K. S. (1945). Physiological studies on some members of the family Saprolegniaceae. IV. Carbohydrate requirements. *Lloydia* **8**, 60–8.

Bhikhabhai, R., Johansson, G. & Pettersson, G. (1984). Isolation of cellulolytic enzymes from *Trichoderma reesei* QM 9414. *Journal of Applied Biochemistry* **6**, 336–45.

Bianchi, M. E., Carbone, M. L. & Lucchini, G. (1981). Mn^{2+} and Mg^{2+} uptake in Mn-sensitive and Mn-resistant yeast strains. *Plant Science Letters* **22**, 345–52.

Bicho, P. A., Runnals, P. L., Cunningham, J. D. & Lee, H. (1988). Induction of xylose reducase and xylitol dehydrogenase activities in *Pachysolen tannophilus* and *Pichia stipitis* on mixed sugars. *Applied and Environmental Microbiology* **54**, 50–4.

Bienfait, H. F. (1986). Iron-efficiency of monocotyledonous and dicotyledonous plants. In *Iron Siderophores and Plant Diseases*, ed. T. R. Swinburne, pp. 21–7. Plenum Press: New York.

Bienfait, H. F. (1989). Prevention of stress in iron metabolism of plants. *Acta Botanica Neerlandica* **38**, 105–29.

Bilinski, C. A. & Miller, J. J. (1983). Translocation of zinc from vacuole to nucleus during yeast meiosis. *Canadian Journal of Genetics and Cytology* **25**, 415–19.

Binder, A. & Ghose, T. K. (1978). Adsorption of cellulose by *Trichoderma viride*. *Biotechnology and Bioengineering* **20**, 1187–99.

Birkett, J. A. & Rowlands, R. T. (1981). Chlorate resistance and nitrate assimilation in industrial strains of *Penicillium chrysogenum*. *Journal of General Microbiology* **123**, 281–5.

Birkinshaw, J. H., Findlay, W. P. K. & Webb, R. A. (1940). Biochemistry of wood-rotting fungi. A study of acids produced by *Coniophora cerebella* Pers. *Biochemical Journal* **34**, 906–16.

Bisson, L. F. (1988). High-affinity glucose transport in *Saccharomyces cerevisiae* is under general glucose repression control. *Journal of Bacteriology* **170**, 4838–45.

Bisson, L. F., Coons, D. M., Kruckeburg, A. L. & Lewis, D. A. (1993). Yeast sugar transporters. *Critical Reviews in Biochemistry and Molecular Biology* **28**, 259–308.

Bisson, L. F. & Fraenkel, D. G. (1983*a*). Transport of 6-deoxyglucose in *Saccharomyces cerevisiae. Journal of Bacteriology* **155**, 995–1000.

Bisson, L. F. & Fraenkel, D. G. (1983*b*). Involvement of kinases in glucose and fructose uptake by *Saccharomyces cerevisiae. Proceedings of the National Academy of Sciences, USA* **80**, 1730–4.

Bisson, L. F. & Fraenkel, D. G. (1984). Expression of kinase-dependent glucose uptake in *Saccharomyces cerevisiae. Journal of Bacteriology* **159**, 1013–17.

Bisson, L. F., Neigeborn, L., Carlson, M. & Fraenkel, D. G. (1987). The *SNF3* gene is required for high-affinity glucose transport in *Saccharomyces cerevisiae. Journal of Bacteriology* **169**, 1656–62.

Blanch, H. W. & Einsele, A. (1973). The kinetics of yeast growth on pure hydrocarbons. *Biotechnology and Bioengineering* **15**, 861–77.

Blanchette, R. A. (1984*a*). Screening wood decayed by white rot fungi for preferential lignin degradation. *Applied and Environmental Microbiology* **48**, 647–53.

Blanchette, R. A. (1984*b*). Selective delignification of eastern hemlock by *Ganoderma tsugae. Phytopathology* **74**, 153–60.

Blanchette, R. A. (1991). Delignification by wood-decay fungi. *Annual Review of Phytopathology* **29**, 381–98.

Blasco, F., Chapuis, J.-P. & Giordani, R. (1981). Characterization of the plasma membrane ATPase of *Candida tropicalis. Biochimie* **63**, 507–14.

Blasco, F., Ducet, G. & Azoulay, E. (1976). Mise en évidence de deux systèmes de transport du phosphate chez *Candida tropicalis. Biochimie* **58**, 351–7.

Blasco, F. & Gidrol, X. (1982). The proton translocating ATPase of the *Candida tropicalis* plasma membrane. *Biochimie* **64**, 531–6.

Blatt, M. R., Rodríguez-Navarro, A. & Slayman, C. L. (1987). Potassium-proton symport in *Neurospora*: kinetic control by pH and membrane potential. *Journal of Membrane Biology* **98**, 169–90.

Blatt, M. R. & Slayman, C. L. (1983). KCl leakage from microelectrodes and its impact on the membrane parameters of a nonexcitable cell. *Journal of Membrane Biology* **72**, 223–34.

Blayney, G. P. & Marchant, R. (1977). Glycogen and protein inclusions in elongating stipes of *Coprinus cinereus. Journal of General Microbiology* **98**, 467–76.

Blomberg, A. & Adler, L. (1989). Roles of glycerol and glycerol-3-phosphate dehydrogenase (NAD$^+$) in acquired osmotolerance of *Saccharomyces cerevisiae. Journal of Bacteriology* **171**, 1087–92.

Blomberg, A. & Adler, L. (1992). Physiology of osmotolerance in fungi. *Advances in Microbial Physiology* **33**, 145–212.

Blomberg, A. & Adler, L. (1993). Tolerance of fungi to NaCl. In *Stress Tolerance of Fungi*, ed. D. H. Jennings, pp. 209–32. Marcel Dekker: New York.

Blumenthal, H. J. (1965). Carbohydrate metabolism. I. Glycolysis. In *The Fungi, an Advanced Treatise*, vol. 1 *The Fungal Cell*, ed. G. C. Ainsworth & A. S. Sussman, pp. 229–68. Academic Press: New York.

Blumenthal, H. J. (1975). Glycolysis. In *The Filamentous Fungi*, vol. 2 *Biosynthesis and Metabolism*, eds. J. E. Smith & D. R. Berry, pp. 65–91.

Edward Arnold: London.

Bolan, N. S. (1991). A critical review on the role of mycorrhizal fungi in the uptake of phosphorus by plants. *Plant and Soil* **134**, 189–208.

Bollard, E. G. & Butler, G. W. (1966). Mineral nutrition of plants. *Annual Review of Plant Physiology* **17**, 77–112.

Boller, T., Dürr, M. & Wiemken, A. (1976). Asymmetric distribution of concanavalin A binding sites on yeast plasmalemma and vacuolar membrane. *Archives of Microbiology* **109**, 115–18.

Bonnarme, P. & Jeffries, T. W. (1989). Mn (II) regulation of lignin peroxidase and manganese-dependent peroxidases from lignin-degrading white rot fungi. *Applied and Environmental Microbiology* **56**, 210–17.

Boominathan, K., Balachandradass, S., Randall, T. A. & Reddy, C. A. (1990). Nitrogen deregulated mutants of *Phanerochaete chrysosporium* – a lignin-degrading basidiomycete. *Archives of Microbiology* **153**, 521–7.

Boominathan, K. & Reddy, C. A. (1992). cAMP-mediated differential regulation of lignin peroxidase and manganese-dependent peroxidase production in the white rot basidiomycete *Phanerochaete chrysosporium. Proceedings of the National Academy of Sciences, USA* **89**, 5586–90.

Borst-Pauwels, G. W. F. H. (1962). The uptake of radioactive phosphate by yeast. I. The uptake of phosphate by yeast compared with that by higher plants. *Biochimica et Biophysica Acta* **65**, 403–6.

Borst-Pauwels, G. W. F. H. (1967). A study of the release of phosphate and arsenate from yeast. *Journal of Cell Physiology* **69**, 241–6.

Borst-Pauwels, G. W. F. H. (1981). Ion transport in yeast. *Biochimica et Biophysica Acta* **650**, 88–127.

Borst-Pauwels, G. W. F. H. (1988). Simulation of all-or-none K^+ efflux from yeast provoked by xenobiotics. *Biochimica et Biophysica Acta* **937**, 88–92.

Borst-Pauwels, G. W. F. H., Boxman, A. W., Theuvenet, A. P. R., Peters, P. H. J. & Dobbelmann, J. (1986). A study of the mechanism by which inhibitors of the plasma membrane ATPase enhance uptake of divalent cations in yeast. *Biochimica et Biophysica Acta* **861**, 413–19.

Borst-Pauwels, G. W. F. H. & Huygen, D. L. M. (1972). Comparison of the effect of acidic inhibitors upon anaerobic phosphate uptake and dinitrophenol extrusion by metabolizing yeast cells. *Biochimica et Biophysica Acta* **288**, 166–71.

Borst-Pauwels, G. W. F. H. & Jager, S. (1969). Inhibition of phosphate and arsenate uptake in yeast by monoiodoacetate, fluoride, 2,4-dinitrophenol and acetate. *Biochimica et Biophysica Acta* **172**, 399–406.

Borst-Pauwels, G. W. F. H., Loef, H. W. & Havinga, A. E. (1962). The uptake of radioactive phosphate by yeast. II. The primary phosphorylation products. *Biochimica et Biophysica Acta* **65**, 407–11.

Borst-Pauwels, G. W. F. H. & Peters, P. H. J. (1977). Effect of the medium pH and the cell pH upon the kinetical parameters of phosphate uptake by yeast. *Biochimica et Biophysica Acta* **466**, 488–95.

Borst-Pauwels, G. W. F. H. & Peters, P. H. J. (1987). Phosphate uptake in *Saccharomyces cerevisiae*. In *Phosphate Metabolism and Cellular Regulation in Microorganisms*, ed. A. Torriani-Gorini, F. G. Rothman, S. Silver, A. Wright & E. Yagil, pp. 205–9. American Society for Microbiology: Washington, DC.

Borst-Pauwels, G. W. F. H., Schnetkamp, P. & van Well, P. (1973). Activation of Rb^+ and Na^+ uptake into yeast by monovalent cations. *Biochimica et Biophysica Acta* **291**, 274–9.

Borst-Pauwels, G. W. F. H. & Severens, P. P. J. (1984). Effect of the surface potential upon ion selectivity found in competitive inhibitors of divalent cation uptake – a theoretical approach. *Physiologia Plantarum* **60**, 86–91.

Borst-Pauwels, G. W. F. H. & Theuvenet, A. P. R. (1984). Apparent saturation kinetics of divalent cation uptake in yeast caused by a reduction in the surface potential. *Biochimica et Biophysica Acta* **768**, 171–86.

Borst-Pauwels, G. W. F. H. & Theuvenet, A. P. R. (1985). All-or-none K^+ efflux from yeast cells induced by calmodulin antagonists. *FEMS Microbiology Letters* **29**, 221–4.

Borst-Pauwels, G. W. F. H., Theuvenet, A. P. R. & Peters, P. H. J. (1975). Uptake by yeast: interaction of Rb^+, Na^+ and phosphate. *Physiologia Plantarum* **33**, 8–12.

Borst-Pauwels, G. W. F. H., Theuvenet, A. P. R. & Stols, A. L. H. (1983). All-or-none interactions of inhibitors of the plasma membrane ATPase with *Saccharomyces cerevisiae*. *Biochimica et Biophysica Acta* **732**, 186–92.

Borst-Pauwels, G. W. F. H., Wolters, G. H. J. & Henricks, J. J. G. (1971). The interaction of 2,4-dinitrophenol with anaerobic Rb^+ transport across the yeast cell membrane. *Biochimica et Biophysica Acta* **225**, 269–72.

Böttger, I., Wieland, O., Brdiczka, D. & Pette, D. (1969). Intracellular localization of pyruvate carboxylase and phosphoenol pyruvate carboxykinase in rat liver. *European Journal of Biochemistry* **8**, 113–19.

Botton, B. (1978). Influence of calcium on the differentiation and growth of aggregated organs in *Sphaerostilbe repens*. *Canadian Journal of Botany* **24**, 1039–47.

Bourne, R. M. (1990). A ^{31}P-NMR study of phosphate transport and compartmentation in *Candida utilis*. *Biochimica et Biophysica Acta* **1055**, 1–9.

Bourne, R. M. (1991). Net phosphate transport in phosphate-starved *Candida utilis*: relationships with pH and K^+. *Biochimica et Biophysica Acta* **1067**, 81–8.

Bourret, J. A. (1985). Glucose transport in germinating *Pilobolus longupes* spores. *Experimental Mycology* **9**, 48–55.

Boutry, M., Foury, F. & Goffeau, A. (1977). Energy-dependent uptake of calcium by the yeast *Saccharomyces pombe*. *Biochimica et Biophysica Acta* **464**, 602–12.

Bowman, B. J. (1983). Vanadate uptake in *Neurospora crassa* occurs via phosphate transport system II. *Journal of Bacteriology* **153**, 292–6.

Bowman, B. J., Allen, K. E. & Slayman, C. W. (1983). Vanadate-resistant mutants of *Neurspora crassa* are deficient in a high affinity phosphate transport system. *Journal of Bacteriology* **153**, 292–6.

Bowman, B. J., Allen, R., Wechser, M. A. & Bowman, E. J. (1988). Isolation of genes encoding the *Neurospora* vacuolar ATPase. Analysis of *vma-2* encoding the 57-kDa polypeptide and comparison to *vma-1*. *Journal of Biological Chemistry* **263**, 14002–7.

Bowman, B. J., Beranski, C. J. & Jung, C. Y. (1985). Size of the plasma membrane H^+-ATPase from *Neurospora crassa* determined by radiation activation and comparison with the sarcoplasmic reticulum Ca^{2+}-ATPase from skeletal muscle. *Journal of Biological Chemistry* **260**, 8726–30.

Bowman, B. J., Blasco, F. & Slayman, C. W. (1981). Purification and characterization of the plasma membrane ATPase of *Neurospora crassa*. *Journal of Biological Chemistry* **256**, 12343–9.

Bowman, B. J. & Bowman, E. J. (1986). H^+-ATPases from mitochondria, plasma membrane and vacuoles of fungal cells. *Journal of Membrane Biology* **94**, 83–97.

Bowman, B. J., Dschida, W. J., Harris, T. & Bowman, E. J. (1989). The vacuolar ATPase of *Neurospora crassa* contains an F_1-like structure. *Journal of Biological Chemistry* **264**, 15606–12.

Bowman, B. J., Mainzer, S. E., Allen, K. E. & Slayman, C. W. (1978). Effect of inhibitors on the plasma membrane and mitochondrial adenosine triphosphatases of *Neurospora crassa*. *Biochemica et Biochimica Acta* **512**, 13–28.

Bowman, B. J. & Slayman, C. W. (1977). Characterization of plasma membrane adenosine triphosphatase. *Journal of Biological Chemistry* **252**, 3357–63.

Bowman, E. J. (1983). Comparison of the vacuolar membrane ATPase of *Neurospora crassa* with the mitochondrial and plasma membrane ATPases. *Journal of Biological Chemistry* **258**, 15238–44.

Bowman, E. J. & Bowman, B. J. (1982). Identification and properties of an ATPase in vacuolar membranes of *Neurospora crassa*. *Journal of Bacteriology* **151**, 1326–37.

Bowman, E. J., Bowman, B. J. & Slayman, C. W. (1981). Isolation and characterisation of plasma membrane from wild type *Neurospora crassa*. *Journal of Biological Chemistry* **256**, 12336–42.

Bowman, E. J., Mandala, S., Taiz, L. & Bowman, B. J. (1986). Structural studies of the vacuolar membrane ATPase from *Neurospora crassa* and comparison with the tonoplast membrane ATPase from *Zea mays*. *Proceedings of the National Academy of Sciences, USA* **83**, 48–52.

Bowman, E. J., Tenney, K. & Bowman, B. J. (1988). Isolation of genes encoding *Neurospora* vacuolar ATPase. Analysis of *vma-1* encoding the 67-kDa subunit reveals homology to other ATPases. *Journal of Biological Chemistry* **263**, 13994–14001.

Boxman, A. W., Dobbelmann, J. & Borst-Pauwels, G. W. F. H. (1984). Possible energization of K^+ accumulation into metabolising yeast by the proton motive force. Binding correction to be applied in the calculation of the yeast membrane potential from the tetraphenylphosphonium distribution. *Biochimica et Biophysica Acta* **792**, 51–7.

Bradfield, G., Somerfield, P., Meyn, T., Holby, M., Babcock, D., Bradley, D. & Segel, I. H. (1970). Regulation of sulfate transport in filamentous fungi. *Plant Physiology* **46**, 720–7.

Bradley, R., Burt, A. J. & Read, D. J. (1982). The biology of mycorrhiza in the Ericaceae. VIII. The role of mycorrhizal infection in heavy metal resistance. *New Phytologist* **91**, 197–209.

Brady, C. J., McGlasson, W. B., Pearson, J. A., Meldrum, S. K. & Kupeliovitch, E. (1985). Interactions between the amount and molecular forms of polygalacturonase, calcium and firmness in tomato fruit. *Journal of the American Society for Horticultural Science* **110**, 254–8.

Brady, R. J. & Chambliss, G. H. (1967). The lack of phosphofructokinase in several species of *Rhodotorula*. *Biochemical and Biophysical Research Communications* **29**, 898–903.

Braun, V. & Hantke, K. (1991). Genetics of bacterial iron transport. In *Handbook of Microbial Metal Chelates*, ed. G. Winkelmann, pp. 107–38. CRC Press: Boca Raton, FL.

Bravery, A. F. (1971). The application of scanning electron microscopy in the study of timber decay. *Journal of the Institute of Wood Science* **5**, 13–19.

Brenes-Pomales, A., Lindegren, G. & Lindegren, C. C. (1955). Gene control of copper sensitivity in *Saccharomyces. Nature, London* **136**, 841–2.

Breton, A. & Surdin-Kerjan, Y. (1977). Sulfate uptake in *Saccharomyces cerevisiae*: biochemical and genetic study. *Journal of Bacteriology* **132**, 224–32.

Brocklehurst, R., Gardner, D. & Eddy, A. A. (1977). The absorption of protons with α-methyl glucoside and α-thioethyl glucoside by the yeast NCYC 240. Evidence against the phosphorylation hypothesis. *Biochemical Journal* **162**, 591–9.

Brooker, R. J. & Slayman, C. W. (1982). Inhibition of plasma membrane [H⁺]-ATPase of *Neurospora crassa* by N-ethylmaleimide. Protection by nucleotides. *Journal of Biological Chemistry* **257**, 12051–5.

Brooker, R. J. & Slayman, C. W. (1983). Effects of Mg^{2+} ions on the plasma membrane [H⁺]-ATPase of *Neurospora crassa*. II. Kinetic studies. *Journal of Biological Chemistry* **258**, 8833–8.

Brown, A. D. (1976). Microbial water stress. *Bacteriological Reviews* **40**, 803–46.

Brown, A. D. (1978). Compatible solutes and extreme water stress in eukaryotic micro-organisms. *Advances in Microbial Physiology* **17**, 181–242.

Brown, A. D. (1990). *Microbial Water Stress Physiology*. John Wiley: Chichester.

Brown, A. D. & Edgely, M. (1980). Osmoregulation in yeast. In *Genetic Engineering of Osmoregulation*, ed. D. W. Rains, R. C. Valentine & A. Hollaender, pp. 75–90. Plenum Press: New York.

Brown, A. D., Mackenzie, K. F. & Singh, K. K. (1986). Selected aspects of microbial osmoregulation. *FEMS Microbiology Reviews* **39**, 31–6.

Brown, A. D. & Simpson, J. R. (1972). Water relations of sugar-tolerant yeasts: the role of intracellular polyols. *Journal of General Microbiology* **72**, 589–91.

Brown, C. M. (1980). Ammonia assimilation and utilization in bacteria and fungi. In *Microorganisms and Nitrogen Sources*, ed. J. W. Payne, pp. 511–35. John Wiley: Chichester.

Brown, C. M., Burn, V. J. & Johnson, B. (1973). Glutamate synthase in fission yeasts and its role in ammonia assimilation. *Nature New Biology* **246**, 115–16.

Brown, R. W. & van Haveren, B. P. (1972). *Psychrometry in Water Relations Research*. Utah Agricultural Experimental Station: Logan.

Brown, S. W. & Oliver, S. G. (1983). Isolation of ethanol tolerant mutants of yeast by continuous culture. *European Journal of Applied Microbial Biotechnology* **16**, 116–22.

Brownell, P. F. (1979). Sodium as an essential micronutrient element for plants and its possible role in metabolism. *Advances in Botanical Research* **7**, 117–224.

Browning, M. H. R. & Hutchinson, T. C. (1991). The effects of aluminium and calcium on the growth and nutrition of selected mycorrhizal fungi of Jack Pine. *Canadian Journal of Botany* **69**, 1691–706.

Brownlee, A. G. & Arst, H. N. Jr (1983). Nitrate uptake in *Aspergillus nidulans* and involvement of the third gene of the nitrate assimilation gene cluster. *Journal of Bacteriology* **115**, 1138–46.

Brownlee, C. (1984). Membrane potential components of the marine fungus *Dendryphiella salina* (Suth.) Pugh et Nicot. Possible involvement of calmodulin in electrophysiology and growth. *New Phytologist* **97**, 15–23.

Brownlee, C. & Jennings, D. H. (1981a). Further observations on tear or drop formation by mycelium of *Serpula lacrimans. Transactions of the British*

Mycological Society **77**, 33–40.
Brownlee, C. & Jennings, D. H. (1981*b*). The content of soluble carbohydrates and their translocation in mycelium of *Serpula lacrimans*. *Transactions of the British Mycological Society* **77**, 516–19.
Brownlee, C. & Jennings, D. H. (1982*a*). Long distance translocation in *Serpula lacrimans*: velocity estimates and the continuous monitoring of induced perturbations. *Transactions of the British Mycological Society* **79**, 143–8.
Brownlee, C. & Jennings, D. H. (1982*b*). Pathway of translocation in *Serpula lacrymans*. *Transactions of the British Mycological Society* **79**, 401–7.
Bruinenberg, P. M., de Bot, P. H. M., van Dijken, J. P. & Scheffers, W. A. (1983). The role of redox balances in the anaerobic fermentation of xylose by yeasts. *European Journal of Applied Microbiology and Biotechnology* **18**, 287–92.
Bruinenberg, P. M., de Bot, P. H. M., van Dijken, J. P. & Scheffers, W. A. (1984). NADH-linked aldose reductase: the key to anaerobic alcoholic fermentation of xylose by yeasts. *Applied Microbiology and Biotechnology* **19**, 256–60.
Bruinenberg, P. M., van Dijken, J. P. & Scheffers, W. A. (1983*a*). A theoretical analysis of NADPH production and consumption in yeasts. *Journal of General Microbiology* **129**, 953–64.
Bruinenberg, P. M., van Dijken, J. P. & Scheffers, W. A. (1983*b*). An enzymic analysis of NADPH production and consumption in *Candida utilis*. *Journal of General Microbiology* **129**, 965–75.
Brunold, C. H. (1990). Reduction of sulfate to sulfide. In *Sulfur Nutrition and Sulfur Assimilation in Higher Plants*, ed. H. Rennenberg, C. H. Brunold, L. J. DeKok & I. Stulen, pp. 13–31. SPB Academic Publishing: The Hague.
Budd, K. (1976). Uptake and metabolism of D-glucose by *Neocosmospora vasinfecta* E. F. Smith. *Plant Physiology* **58**, 193–8.
Budd, K. (1979). Magnesium uptake in *Neocosmospora vasinfecta*. *Canadian Journal of Botany* **57**, 491–6.
Budd, K. (1988). A high-affinity system for the transport of zinc in *Neocosmospora vasinfecta*. *Experimental Mycology* **12**, 195–202.
Budd, K. (1991). A cadmium-tolerant strain of *Neocosmospora vasinfecta* shows reduced cadmium influx. *Canadian Journal of Botany* **69**, 1296–301.
Buller, A. H. R. (1931). *Researches on Fungi*, vol. IV. Longmans Green: London.
Buller, A. H. R. (1933). *Researches on Fungi*, vol. V. Longmans Green: London.
Buller, A. H. R. (1934). *Researches on Fungi*, vol. VI. Longmans Green: London.
Bulman, R. A. & Chittenden, G. J. F. (1976). The mycelial cell wall of *Penicillium charlesii* G. Smith NRRL 1887. *Biochimica et Biophysica Acta* **444**, 202–11.
Bun-Ya, M., Nishimura, N., Harashima, S. & Oshima, Y. (1991). The *PHO84* gene of *Saccharomyces cerevisiae* encodes an inorganic phosphate transporter. *Molecular and Cellular Biology* **11**, 3229–38.
Burdsall, H. H. Jr & Eslyn, W. E. (1974). A new *Phanerochaete* with a *chrysosporium* imperfect state. *Mycotaxon* **1**, 123–33.
Burger, M., Bacon, E. E. & Bacon, J. S. D. (1958). Liberation of invertase from disintegrated yeast cells. *Nature, London* **182**, 1508.
Burger, M., Hejmová, L. & Kleinzeller, A. (1959). Transport of some mono- and disaccharides into yeast cells. *Biochemical Journal* **71**, 233–42.
Burke, R. M. & Jennings, D. H. (1990). Effect of sodium chloride on growth characteristics of the marine yeast *Debaryomyces hansenii* in batch and continuous culture under carbon and potassium limitation. *Mycological Research* **94**, 378–88.

Burkholder, P. T., McVeigh, I. & Moyer, D. (1944). Studies on some growth factors of yeasts. *Journal of Bacteriology* **48**, 385–91.

Burns, D. J. W. & Beever, R. E. (1977). Kinetic characterization of the two phosphate uptake systems in the fungus *Neurospora crassa*. *Journal of Bacteriology* **132**, 511–19.

Burns, D. J. W. & Beever, R. E. (1979). Mechanisms controlling the two phosphate uptake systems in *Neurospora crassa*. *Journal of Bacteriology* **139**, 195–204.

Burrows, L. F. & Sparks, A. H. (1964). The identification of sulfur dioxide-blending compounds in apple juices and ciders. *Journal of the Science of Food and Agriculture* **15**, 176–85.

Burt, W. R. & Cazin, J. (1976). Production of extracellular ribonuclease by yeasts and yeastlike fungi and its repression by orthophosphate in species of *Cryptococcus* and *Tremella*. *Journal of Bacteriology* **125**, 955–60.

Burton, E. G. & Metzenberg, R. L. (1972). Novel mutation causing derepression of several enzymes of sulfur metabolism in *Neurospora*. *Journal of Bacteriology* **109**, 140–51.

Bushell, M. E. & Bull, A. T. (1981). Anaplerotic metabolism of *Aspergillus nidulans* and its effect of biomass synthesis in carbon limited chemostats. *Archives of Microbiology* **128**, 282–7.

Bussey, H. & Umbarger, H. E. (1970). Biosynthesis of branched-chain amino acids in yeast – a trifluroleucine resistant mutant with altered regulation of leucine uptake. *Journal of Bacteriology* **103**, 256–94.

Buston, W. H., Jabbar, A. & Etheridge, D. E. (1953). The influence of hexose phosphates, calcium and jute extract on the formation of perithecia by *Chaetomium globosum*. *Journal of General Microbiology* **8**, 302–6.

Busturia, A. & Lagunas, R. (1985). Identification of two forms of the maltose transport system in *Saccharomyces cerevisiae* and their regulation by catabolic inactivation. *Biochimica et Biophysica Acta* **820**, 324–6.

Busturia, A. & Lagunas, R. (1986). Catabolic inactivation of the glucose transport system in *Saccharomyces cerevisiae*. *Journal of General Microbiology* **132**, 379–85.

Buswell, J. A., Mollet, B. & Odier, E. (1984). Ligninolytic enzyme production by *Phanerochaete chrysosporium* under conditions of nitrogen sufficiency. *FEMS Microbiology Letters* **25**, 295–9.

Butt, T. R. & Ecker, D. J. (1987). Yeast metallothionein and its applications in biotechnology. *Microbiological Reviews* **15**, 351–64.

Butt, T. R., Sternberg, E. J., Gorman, J. A., Clark, P., Hamer, D., Rosenberg, M. & Crooke, S. T. (1984). Copper metallothionein of yeast, structure of the gene and regulation of expression. *Proceedings of the National Academy of Sciences, USA* **81**, 3332–6.

Byrne, A. R. (1988). Radioactivity in fungi in Slovenia, Yugoslavia, following the Chernobyl accident. *Journal of Environmental Radioactivity* **6**, 177–83.

Caddick, M. X., Arst, H. N. Jr, Taylor, L. H., Johnson, R. & Brownlee, A. G. (1986*a*). Cloning of the regulatory gene *areA* mediating nitrogen metabolite repression in *Aspergillus nidulans*. *EMBO Journal* **5**, 1087–90.

Caddick, M., Brownlee, A. G. & Arst, H. N. Jr (1986*b*). Regulation of gene expression by pH of the growth medium in *Aspergillus nidulans*. *Molecular and General Genetics* **203**, 346–53.

Cai, D. & Tien, M. (1989). On the regulation of peroxidase compound III (isoenzyme K8). *Biochemical and Biophysical Research Communications* **162**, 464–9.

Cain, R. B., Bilton, R. F. & Darrah, J. A. (1968). Metabolism of aromatic acids in microorganisms. Metabolic pathways in fungi. *Biochemical Journal* **108**, 797–828.

Cairney, J. W. G. (1991*a*). Rhizomorphs: organs of exploration or exploitation. *Mycologist* **5**, 5–10.

Cairney, J. W. G. (1991*b*). Structural and ontogenic study of ectomycorrhizal rhizomorphs. In *Methods in Microbiology*, vol. 23 *Techniques in the Study of Mycorrhiza*, ed. J. R. Norris, D. J. Read & A. K. Varma, pp. 331–40. Academic Press: London.

Cairney, J. W. G. (1992). Translocation of solutes in ectomycorrhizal and saprotrophic rhizomorphs. *Mycological Research* **96**, 135–41.

Cairney, J. W. G. & Ashford, A. E. (1989). Reducing activity at the root surface in *Eucalyptus pilularis–Pisolithus tinctorius* ectomycorrhizas. *Australian Journal of Plant Physiology* **16**, 99–105.

Cairney, J. W. G., Jennings, D. H. & Agerer, R. (1991). The nomenclature of fungal multi-hyphal linear aggregates. *Cryptogamic Botany* **2/3**, 246–51.

Cairney, J. W. G., Jennings, D. H., Ratcliffe, R. G. & Southon, T. E. (1988). The physiology of basidiomycete linear organs. II. Phosphate uptake by rhizomorphs of *Armillaria mellea*. *New Phytologist* **109**, 327–33.

Calahorra, M., Opekarová, M., Ramirez, J. & Peña, A. (1989). Leucine transport in plasma membrane vesicles of *Saccharomyces cerevisiae*. *Biochimica et Biophysica Acta* **247**, 235–8.

Calderbank, J., Keenan, M. H. J. & Rose, A. H. (1984*a*). Plasma membrane phospholipid unsaturation affects expression of the general amino acid permease in *Saccharomyces cerevisiae* Y185. *Journal of General Microbiology* **131**, 57–65.

Calderbank, J., Keenan, M. H. J., Rose, A. H. & Holman, G. D. (1984*b*). Accumulation of amino acids by *Saccharomyces cerevisiae* Y185 enriched in different fatty-acyl residues: a statistical analysis of data. *Journal of General Microbiology* **130**, 2817–24.

Calderón, J. & Mora, J. (1989). Glutamine assimilation pathways in *Neurospora crassa* growing on glutamine as sole nitrogen and carbon source. *Journal of General Microbiology* **135**, 2699–707.

Calderón, J., Morett, E. & Mora, J. (1985). ω-Amidase pathway in degradation of glutamine in *Neurospora crassa*. *Journal of Bacteriology* **161**, 807–9.

Caldwell, J. H., van Brunt, J. & Harold, F. M. (1986). Calcium-dependent anion channel in the water mold *Blastocladiella emersonii*. *Journal of Membrane Biology* **89**, 85–98.

Calleja, G. B. (1984). *Microbial Aggregation*. CRC Press: Boca Raton, FL.

Calleja, G. B. (1987). Cell aggregation. In *The Yeasts*, vol. 2 *Yeasts and the Environment*, ed. A. H. Rose & J. S. Harrison, pp. 165–238. Academic Press: London.

Callow, J. A., Capaccio, L. C. M., Parish, G. & Tinker, P. B. (1978). Detection and estimation of polyphosphate in vesicular–arbuscular mycorrhizas. *New Phytologist* **80**, 125–34.

Cameron, L. E. & Lé John, H. B. (1972*a*). Ca^{2+} is a specific regulator of amino acid transport and protein synthesis in the water-mould *Achlya*. *Biochemical and Biophysical Research Communications* **48**, 181–9.

Cameron, L. E. & Lé John, H. B. (1972*b*). On the involvement of calcium in amino acid transport and growth of the fungus *Achlya*. *Journal of Biological Chemistry* **247**, 4729–39.

Camici, L., Sermonti, G. & Chain, E. B. (1952). Observations on *Penicillium chrysogenum* in submerged culture. I. Mycelial growth and autolysis. *Bulletin of the World Health Organization* **6**, 265–75.

Campbell, R. N. (1970). Perithecium formation by *Ceratocystis fimbriata* in response to nitrogen and calcium. *Mycologia* **62**, 192–5.

Canny, M. J. (1990). What becomes of the transpiration stream? *New Phytologist* **114**, 341–68.

Cantino, E. C. (1949). The physiology of the aquatic phycomycete *Blastocladiella pringheimii* with emphasis on its nutrition and metabolism. *American Journal of Botany* **36**, 95–112.

Cantino, E. C. (1955). Physiology and phylogeny in the water molds – a re-evaluation. *Quarterly Review of Biology* **30**, 138–49.

Carlson, M. (1987). Regulation of sugar utilization in *Saccharomyces* species. *Journal of Bacteriology* **169**, 4873–7.

Carrano, C. J. & Raymond, K. N. (1978). Coordination chemistry of microbial iron transport compounds: rhodotorulic acid and iron uptake in *Rhodotorula pilimanae*. *Journal of Bacteriology* **136**, 69–74.

Carrodus, B. B. (1966). Absorption of nitrogen by mycorrhizal roots of beech. I. Factors affecting the assimilation of nitrogen. *New Phytologist* **65**, 358–71.

Carrodus, B. B. (1967). Absorption of nitrogen by mycorrhizal roots of beech. II. Ammonium and nitrate as sources of nitrogen. *New Phytologist* **66**, 1–4.

Cartledge, T. G. (1987). Substrate utilization, non-carbohydrate substrates. In *Yeast Biotechnology*, ed. D. L. Berry, I. Russell & G. G. Stewart, pp. 311–42. Allen & Unwin: London.

Cartwright, C. P., Juroszek, J.-R., Beavan, M. J., Ruby, F. M. S., de Morais, S. M. F. & Rose, A. H. (1986). Ethanol dissipates the proton-motive force across the plasma membrane of *Saccharomyces cerevisiae*. *Journal of General Microbiology* **132**, 369–77.

Cartwright, C. P., Veazey, F. J. & Rose, A. H. (1987). Effect of ethanol on activity of the plasma-membrane ATPase in and accumulation of glucose by *Saccharomyces cerevisiae*. *Journal of General Microbiology* **133**, 857–65.

Carvalho-Silva, M. & Spencer-Martins, I. (1990). Modes of lactose uptake in the yeast species *Kluyveromyces marxianus*. *Antonie van Leeuwenhoek* **57**, 77–81.

Casey, G. P. & Ingledew, W. M. (1986). Ethanol tolerance in yeasts. *CRC Critical Reviews in Microbiology* **13**, 219–80.

Casselton, P. J. (1976). Anaplerotic pathways. In *The Filamentous Fungi*, vol. 2 *Biosynthesis and Metabolism*, ed. J. E. Smith & D. R. Berry, pp. 121–58. Edward Arnold: London.

Cássio, F. & Leão, C. (1991). Low-affinity and high-affinity transport systems for citric acid in the yeast *Candida utilis*. *Applied and Environmental Microbiology* **57**, 3623–8.

Cássio, F., Leão, C. & van Uden, N. (1987). Transport of lactate and other short-chain monocarboxylates in the yeast *Saccharomyces cerevisiae*. *Applied and Environmental Microbiology* **53**, 509–13.

Casterline, J. L. Sr & Barnett, N. M. (1977). Isolation and characterization of cadmium-binding components in soy bean plants. *Plant Physiology* **59** (Suppl.), 124.

Castle, E. S. (1942). Spiral growth and reversal of spiralling in *Phycomyces* and their bearing on primary cell wall structure. *American Journal of Botany* **29**, 664–72.

Cazin, J., Kozel, T. R., Lupan, D. M. & Burt, W. R. (1969). Extracellular deoxyribonucleic acid production by yeasts. *Journal of Bacteriology* **100**, 760–2.

Celenza, J. L., Marshall-Carlson, L. & Carlson, M. (1988). The yeast *SNF3* gene encodes a glucose transporter homologous to the mammalian protein. *Proceedings of the National Academy of Sciences, USA* **85**, 2130–4.

Cerbón, J. (1970). Relationship between the phospholipid and the efficiency of the arsenate transport system in yeasts. *Journal of Bacteriology* **102**, 97–105.

Cerbón, J. & Llerenas, E. (1976). Evidence for a functional change in the plasma membrane of yeasts associated with the mechanism of arsenate resistance. *Archives of Biochemistry and Biophysics* **174**, 373–80.

Cerda-Olmedo, E. & Lipson, E. D. (eds.) (1987). *Phycomyces*. Cold Spring Harbor Laboratory Press: Cold Springer Harbor, NY.

Chamberlain, A. H. L. & Moss, S. T. (1988). The thraustochytrids: a protist group with mixed affinities. *BioSystems* **21**, 341–50.

Chan, P. Y. & Cossins, F. A. (1976). General properties and regulation of arginine transporting systems in *Saccharomyces cerevisiae*. *Plant and Cell Physiology* **17**, 341–53.

Chang, A. & Slayman, C. W. (1990). A structural change in the *Neurospora* plasma membrane [H$^+$]ATPase involves phosphorylation during intracellular transport. *Journal of Cell Biology* **115**, 289–96.

Chang, A. & Slayman, C. W. (1991). Maturation of the yeast plasma membrane [H$^+$]ATPase involves phosphorylation during intracellular transport. *Journal of Cell Biology* **115**, 289–96.

Chang, P. L. Y. & Trevithick, J. R. (1974). How important is secretion of exoenzymes through apical cell walls of fungi? *Archives of Microbiology* **101**, 281–93.

Chang, Y.-D. & Dickson, R. C. (1988). Primary structure of the lactose permease gene from the yeast *Kluyveromyces lactis*. Presence of an unusual transcript structure. *Journal of Biological Chemistry* **263**, 16696–703.

Chanson, A., Fickmann, J., Spear, D. & Taiz, L. (1985). Pyrophosphate-driven proton transport by microsomal vesicles from corn coleoptiles. *Plant Physiology* **79**, 159–64.

Chapman, A. G. & Atkinson, D. E. (1977). Adenine concentrations and turnover rates. Their correlation with biological activity in bacteria and yeast. *Advances in Microbial Physiology* **1**, 253–306.

Charron, M. J., Read, E., Haut, S. R. & Michels, C. A. (1989). Molecular evolution of the telomere-associated *MAL* loci of *Saccharomyces*. *Genetics* **122**, 307–16.

Chen, X. J., Wésolowski-Louvel, M. & Fukuhara, H. (1992). Glucose transport in *Kluyveromyces lactis* 2. Transcriptional regulation of the glucose transporter gene *RAG1*. *Molecular and General Genetics* **233**, 97–105.

Cheneval, J. P., Deshusses, J. & Posternak, T. (1970). Sur le transport de l'inositol chez *Schizosaccharomyces pombe*. *Biochimica et Biophysica Acta* **203**, 348–50.

Cheng, Q. & Michels, C. A. (1989). The maltose permease encoded by the *MAL61* gene of *Saccharomyces cerevisiae* exhibits both sequence and structural homology to other sugar transporters. *Genetics* **123**, 477–84.

Cherest, H., Kerjan, P. and Surdin-Kerjan, Y. (1987). The *Saccharomyces cerevisiae MET3* gene: nucleotide sequence and relationship of the 5' non-coding region to that of *MET25*. *Molecular and General Genetics* **210**, 307–13.

Cherry, J. R., Johnson, T. R., Dollard, C., Shuster, J. R. & Denis, C. L. (1989). Cyclic AMP-dependent protein kinase phosphorylates and inactivates the yeast transcriptional activator ADR1. *Cell* **56**, 409–19.

Chilvers, G. A. & Harley, J. L. (1980). Visualization of phosphate accumulation in beech mycorrhizas. *New Phytologist* **84**, 319–26.

Chilvers, G. A., Lapeyrie, F. F. & Douglass, P. A. (1985). A contrast between Oomycetes and other taxa of mycelial fungi in regard to metachromatic granule formation. *New Phytologist* **99**, 203–10.

Chisholm, V. T., Lea, H. Z., Rai, R. & Cooper, T. G. (1987). Regulation of allantoate transport in wild-type and mutant strains of *Saccharomyces cerevisiae*. *Journal of Bacteriology* **169**, 1684–90.

Chiu, S. W. & Moore, D. (1988). Ammonium ions and glutamine inhibit sporulation of *Coprinus cinereus* basidia assayed *in vitro*. *Cell Biology International Reports* **12**, 519–26.

Christensen, M. S. & Cirillo, V. P. (1972). Yeast membrane vesicles: isolation and general characteristics. *Journal of Bacteriology* **110**, 1190–205.

Church, F. C., Meyers, S. P. & Srinivasen, V. R. (1980). Isolation and characterization of α-galactosidase from *Pichia guilliermondii*. *Developments in Industrial Microbiology* **21**, 339–48.

Cicmanec, J. F. & Lichstein, H. C. (1974). Biotin uptake by cold-shocked cells, spheroplasts and repressed cells of *Saccharomyces cerevisiae*: lack of feedback control. *Journal of Bacteriology* **119**, 718–25.

Ciejek, E. & Thorner, J. (1979). Recovery of *S. cerevisiae a* cells from G1 arrest by α-factor pheromone requires endopeptidase action. *Cell* **18**, 625–35.

Cirigliano, M. C. & Carman, G. M. (1984). Isolation of a bioemulsifier from *Candida lipolytica*. *Applied and Environmental Microbiology* **48**, 747–50.

Cirigliano, M. C. & Carman, G. M. (1985). Purification and characterization of liposan, a bioemulsifier from *Candida lipolytica*. *Applied and Environmental Microbiology* **50**, 846–50.

Cirillo, V. P. (1962a). Mechanism of glucose transport across the yeast cell membrane. *Journal of Bacteriology* **84**, 485–91.

Cirillo, V. P. (1962b). Sugar transport by *Saccharomyces cerevisiae* protoplasts. *Journal of Bacteriology* **84**, 1251–3.

Cirillo, V. P. (1968a). Relationship between sugar structure and competition for the sugar transport system in baker's yeast. *Journal of Bacteriology* **95**, 603–11.

Cirillo, V. P. (1968b). Galactose transport in *Saccharomyces cerevisiae*. 1. Nonmetabolised sugars as substrates and inducers of the galactose transport system. *Journal of Bacteriology* **95**, 1727–31.

Claeyssens, M. & Tomme, P. (1990). Structure-function relationships of cellulolytic proteins from *Trichoderma reesei*. In Trichoderma reesei *Cellulases*, ed. C. P. Kubicek, D. E. Eveleigh, H. Esterbauer, W. Steiner & E. M. Kubicek-Pranz, pp. 1–11. Royal Society of Chemistry: Cambridge.

Clarke, R. W., Jennings, D. H. & Coggins, C. R. (1986). Growth of *Serpula lacrimans* in relation to water potential of substrate. *Transactions of the British Mycological Society* **75**, 271–80.

Cleland, W. W. & Johnson, M. J. (1954). Tracer experiments on the mechanism of citric acid formation in *Aspergillus niger*. *Journal of Biological Chemistry* **208**, 679–89.

Cleland, W. W. & Johnson, M. J. (1955). Studies on the formation of oxalic acid by *Aspergillus niger*. *Journal of Biological Chemistry* **201**, 595–606.

Clipson, N. J. W., Cairney, J. W. G. & Jennings, D. H. (1987). The physiology of basidiomycete linear organs. I. Phosphate uptake by cords and mycelium in the laboratory and in the field. *New Phytologist* **105**, 449–57.

Clipson, N. J. W., Hajibagheri, M. A. & Jennings, D. H. (1990). Ion compartmentation in the marine fungus *Dendryphiella salina* in response to salinity: X-ray microanalysis. *Journal of Experimental Botany* **41**, 199–202.

Clipson, N. J. W. & Jennings, D. H. (1990). Role of potassium and sodium in generation of osmotic potential in the marine fungus *Dendryphiella salina*. *Mycological Research* **94**, 1017–22.

Clipson, N. J. W. & Jennings, D. H. (1992). *Dendryphiella salina* and *Debaryomyces hansenii*: models for ecophysiological adaptation to salinity by fungi which grow in the sea. *Canadian Journal of Botany* **70**, 2097–105.

Clipson, N. J. W., Jennings, D. H. & Smith, J. L. (1989). The response of salinity at the microscopic level of the marine fungus *Dendryphiella salina* Nicot and Pugh as investigated stereologically. *New Phytologist* **113**, 21–7.

Cochrane, V. W. (1956). The anaerobic dissimilation of glucose by *Fusarium*. *Mycologia* **48**, 1–12.

Cochrane, V. W. (1963). *Physiology of Fungi*. John Wiley: New York.

Cockburn, M., Earnshaw, P. & Eddy, A. A. (1975). The stoichiometry of the absorption of protons with phosphate and L-glutamate by yeasts of the genus *Saccharomyces*. *Biochemical Journal* **146**, 705–12.

Coggins, C. R., Jennings, D. H. & Clarke, R. W. (1980). Tear or drop formation by mycelium of *Serpula lacrimans*. *Transactions of the British Mycological Society* **75**, 63–7.

Cohen, B. L. (1980). Transport and utilisation of proteins by fungi. In *Microorganisms and Nitrogen Sources*, ed. J. W. Payne, pp. 411–30. John Wiley: Chichester.

Cohen, B. L. (1981). Regulation of protease production in *Aspergillus*. *Transactions of the British Mycological Society* **76**, 447–50.

Cole, M. B. & Keenan, M. H. J. (1987). Effects of weak acids and external pH on the intracellular pH of *Zygosaccharomyces bailii* and its implications in weak-acid resistance. *Yeast* **3**, 23–32.

Collins, K. D. & Washabaugh, M. W. (1985). The Hofmeister effect and the behaviour of water at interfaces. *Quarterly Review of Biophysics* **18**, 323–422.

Comerford, J. G., Spencer-Phillips, P. T. H. & Jennings, D. H. (1985). Membrane-bound ATPase activity, properties of which are altered by growth in saline conditions, isolated from the marine yeast *Debaryomyces hansenii*. *Transactions of the British Mycological Society* **85**, 431–8.

Conway, E. J. & Armstrong, W. McD. (1961). The total intracellular concentration of solutes in yeast and other plant cells and the distensibility of the plant cell wall. *Biochemical Journal* **81**, 631–9.

Conway, E. J. & Beary, M. E. (1958). Active transport of magnesium across the yeast cell membrane. *Biochemical Journal* **69**, 275–80.

Conway, E. J. & Beary, M. (1962). A magnesium yeast and its properties. *Biochemical Journal* **84**, 328–33.

Conway, E. J. & Brady, T. G. (1950). Biological production of acid and alkali. I. Quantitative relations of succinic and carbonic acids to the potassium hydrogen exchange in fermenting yeast. *Biochemical Journal* **47**, 360–9.

Conway, E. J. & Breen, E. J. (1945). An 'ammonia'-yeast and some of its properties. *Biochemical Journal* **39**, 368–71.

Conway, E. J. & Duggan, P. F. (1958). A cation carrier in the yeast cell wall. *Biochemical Journal* **69**, 265–74.

Conway, E. J. & Gaffney, H. M. (1966). The further preparation of inorganic cationic yeasts and some of their chief properties. *Biochemical Journal* **101**, 385–91.

Conway, E. J. & Kernan, R. P. (1955). The effect of redox dyes on the active transport of hydrogen, potassium and sodium ions across the yeast cell wall. *Biochemical Journal* **61**, 32–6.

Conway, E. J. & Moore, P. T. (1954). A sodium-yeast and some of its properties. *Biochemical Journal* **57**, 523–8.

Conway, E. J. & O'Malley, E. (1946). The nature of the cation exchanges during yeast fermentation with the formation of 0.02 N H ion. *Biochemical Journal* **40**, 69–67.

Conway, E. J., Ryan, H. & Carton, E. (1954). Active transport of sodium ions from the yeast cell. *Biochemical Journal* **58**, 158–67.

Cooke, D. T. & Burden, R. S. (1990). Lipid modulation of plasma membrane-bound ATPases. *Physiologia Plantarum* **78**, 153–9.

Cooper, K. M. (1984). Physiology of va mycorrhizal associations. In *VA Mycorrhiza*, ed. C. L. I. Powell & D. J. Bagyaraj, pp. 155–86. CRC Press: Boca Raton, FL.

Cooper, K. M. & Tinker, P. B. (1978). Translocation and transfer of nutrients in vesicular–arbuscular mycorrhizas. II. Uptake and translocation of phosphorus, zinc and sulphur. *New Phytologist* **81**, 43–52.

Cooper, K. M. & Tinker, P. B. (1981). Translocation and transfer of nutrients in vesicular–arbuscular mycorrhizas. IV. Effect of environmental variables on movement of phosphorus. *New Phytologist* **88**, 327–39.

Cooper, T. G. (1982). Transport in *Saccharomyces cerevisiae*. In *The Molecular Biology of the Yeast* Saccharomyces: *Metabolism and Gene Expression*, ed. J. N. Strathern, E. W. Jones & J. R. Broach, pp. 399–461. Cold Spring Harbor Laboratory Press: Cold Spring Harbor, NY.

Cooper, T. G., Gorski, M. & Turoscy, V. (1979a). A cluster of three genes responsible for allantoin degradation in *Saccharomyces cerevisiae*. *Genetics* **92**, 383–96.

Cooper, T. G., McKelvey, J. & Sumadra, R. (1979b). Oxalurate transport in *Saccharomyces cerevisiae*. *Journal of Bacteriology* **139**, 917–23.

Cooper, T. G. & Sumadra, R. (1975). Urea uptake in *Saccharomyces cerevisiae*. *Journal of Bacteriology* **121**, 571–6.

Cooper, T. G. & Sumadra, R. A. (1983). What is the function of nitrogen catabolite repression in *Saccharomyces cerevisiae*? *Journal of Bacteriology* **155**, 623–7.

Cornelius, G. & Nakashima, H. (1987). Vacuoles play a decisive role in calcium homeostasis in *Neurospora crassa*. *Journal of General Microbiology* **133**, 231–7.

Corner, E. J. H. (1991). The active basidium. *Mycologist* **5**, 69.

Correll, J. C., Klittich, C. J. R. & Leslie, J. F. (1987). Nitrate nonutilising mutants of *Fusarium oxysporium* and their use in vegetative compatibility tests. *Phytopathology* **77**, 1640–6.

Côrte-Real, M. & Leão, C. (1990). Transport of malic acid and other dicarboxylic acids in the yeast *Hansenula anomala*. *Applied and Environmental Microbiology* **56**, 1109–13.

Côrte-Real, M., Leão, C. & van Udén, N. (1989). Transport of L-(−)malic acid and other dicarboxylic acids in the yeast *Candida sphaerica*. *Applied Microbiology and Biotechnology* **31**, 551–5.

Coschigano, P. W. & Magasansik, B. (1991). The *URE2* gene product of *Saccharomyces cerevisiae* plays an important role in the cellular response to the nitrogen source and has homology to glutathione S-transferases. *Molecular and Cellular Biology* **11**, 822–32.

490   *Literature cited*

Coschigano, P. W., Miller, S. M. & Magasansik, B. (1991). Physiological and genetic analysis of the carbon regulation of the NAD-dependent glutamate dehydrogenase of *Saccharomyces cerevisiae*. *Molecular and Cellular Biology* **11**, 4455–65.

Cosgrove, D. J. (1977). Microbial transformations in the phosphorus cycle. *Advances in Microbial Ecology* **1**, 95–134.

Cosgrove, D. J. (1985). Cell wall yield properties of growing tissue. Evaluation by *in vivo* stress relaxation. *Plant Physiology* **78**, 347–56.

Cosgrove, D. J. (1986). Biophysical control of plant growth. *Annual Review of Plant Physiology* **37**, 377–405.

Cosgrove, D. J. (1988). In defence of the cell volumetric modulus. *Plant, Cell and Environment* **11**, 67–70.

Cosgrove, D. J., Ortega, J. K. E. & Shropshire, W. Jr (1987). Pressure probe study of the water relations of *Phycomyces blakesleeanus* sporangiophores. *Biophysical Journal* **51**, 413–24.

Cotoras, M. & Agosin, E. (1992). Regulatory aspects of endoglucanase production by the brown-rot fungus *Gloeophyllum trabeum*. *Experimental Mycology* **16**, 253–60.

Coughlan, M. P. (1985). The properties of fungal and bacterial cellulases with comment on their production and application. *Biotechnology and Genetic Engineering Reviews* **3**, 39–109.

Coughlan, M. P. (ed.) (1989). *Enzyme Systems for Lignocellulose Degradation*. Elsevier Applied Science: London.

Courchesne, W. E. & Magansik, B. (1988). Regulation of nitrogen assimilation in *Saccharomyces cerevisiae*: roles of the *URE2* and *GLN3* genes. *Journal of Bacteriology* **170**, 708–13.

Cove, D. J. (1970). Control of gene action in *Aspergillus nidulans*. *Proceedings of the Royal Society* B **54**, 291–37.

Cove, D. J. (1976a). Chlorate toxicity in *Aspergillus nidulans*: the selection and characterization of chlorate resistant mutants. *Heredity* **36**, 191–203.

Cove, D. J. (1976b). Chlorate toxicity in *Aspergillus nidulans*: studies of mutants altered in nitrate assimilation. *Molecular and General Genetics* **146**, 147–59.

Cove, D. J. (1979). Genetic studies of nitrate assimilation in *Aspergillus niger*. *Biological Reviews* **54**, 291–327.

Cowling, E. B. & Brown, W. (1969). Structural features in relation to enzymic hydrolysis. *Advances in Chemistry Series* **95**, 157–87.

Cox, G., Moran, K. J., Sanders, F., Nockolds, C. & Tinker, P. B. (1980). Translocation and transfer of nutrients in vesicular–arbuscular mycorrhizas. III. Polyphosphate granules and phosphorus translocations. *New Phytologist* **84**, 649–59.

Cox, J. A., Ferraz, C., Demaille, G., Perez, R. O., van Tuinen, D. & Marmé, D. (1982). Calmodulin from *Neurospora crassa*. General properties and conformational changes. *Journal of Biological Chemistry* **257**, 10694–700.

Crabeel, M. & Grenson, M. (1970). Regulation of histidine uptake by specific feedback inhibition of two histidine permeases in *Saccharomyces cerevisiae*. *European Journal of Biochemistry* **14**, 197–204.

Cram, W. J. (1976). Negative feedback regulation of transport in cells. The maintenance of turgor, volume and nutrient supply. In *Encyclopedia of Plant Physiology*, vol. II *Transport in Plants*, Part A *Cells*, ed. U. Lüttge & M. G. Pitman, pp. 284–316. Springer-Verlag: Berlin, Heidelberg, New York.

Cramer, C. L. & Davis, R. H. (1984). Polyphosphate-cation interaction in the amino acid containing vacuole of *Neurospora crassa. Journal of Biological Chemistry* **259**, 5152–7.

Cramer, C. L., Vaughn, L. E. & Davis, R. H. (1980). Basic amino acids and inorganic polyphosphates in *Neurospora crassa*: independent regulation of vacuolar pools. *Journal of Bacteriology* **142**, 945–52.

Crane, F. L. (1989). Plasma membrane redox reactions in signal transduction. In *Second Messengers in Plant Growth and Development*, ed. W. F. Boss & D. F. Morré, pp. 115–44. Alan R. Liss: New York.

Crane, F. L. & Barr, R. (1989). Plasma membrane oxidoreductases. *Critical Reviews in Plant Sciences* **8**, 273–308.

Crane, F. L., Roberts, H., Linnane, A. W. & Löw, H. (1982). Transmembrane ferricyanide reduction by the cells of the yeast *Saccharomyces cerevisiae. Journal of Bioenergetics and Biomembranes* **14**, 191–205.

Crasemann, J. M. (1954). The nutrition of *Chytridium* and *Macrochytrium. American Journal of Botany* **41**, 302–10.

Crawford, R. L. (1981). *Lignin Biodegradation and Transformation*. John Wiley: New York.

Crawford, R. L., Robinson, L. E. & Cheh, A. M. (1980). ^{14}C-labelled lignins as substrates for the study of lignin biodegradation and transformation. In *Lignin Biochemistry: Microbiology, Chemistry and Potential Applications*, vol. 1, ed. T. K. Kirk, T. Higuchi & H.-M. Chang, pp. 215–30. CRC Press: Boca Raton, FL.

Cress, W. A., Johnson, G. V. & Barton, L. L. (1986). The role of endomycorrhizal fungi in iron uptake by *Hilaria jamesii. Journal of Plant Nutrition* **9**, 547–56.

Cromack, K. Jr, Sollins, P., Graustein, W. C., Speidel, K., Todd, W. A., Spycher, G., Li, C. Y. & Todd, R. L. (1979). Calcium oxalate accumulation and soil weathering in mats of the hypogeous fungus *Hysterangium crassum. Soil Biology and Biochemistry* **11**, 463–8.

Crooke, W. M. (1964). The measurement of cation-exchange capacity of plant roots. *Plant and Soil* **21**, 43–9.

Crowley, D. E., Reid, C. P. P. & Szaniszlo, P. J. (1988). Utilization of microbial siderophores in ion acquisition by oat. *Plant Physiology* **87**, 680–5.

Cumming, J. R. (1990). Nitrogen source effects on Al-toxicity in nonmycorrhizal and mycorrhizal pitch pine (*Pinus rigida*) seedlings. 2. Nitrate reduction and NO_3-uptake. *Canadian Journal of Botany* **68**, 2653–9.

Cumming, J. R. & Weinstein, L. H. (1990a). Nitrogen source effects on Al-toxicity in nonmycorrhizal and mycorrhizal pitch pine (*Pinus rigida*) seedlings. 1. Growth and nutrition. *Canadian Journal of Botany* **68**, 2644–52.

Cumming, J. R. & Weinstein, L. H. (1990b). Utilization of $AlPO_4$ as a phosphorus source by ectomycorrhizal *Pinus rigida* Mill seedlings. *New Phytologist* **116**, 99–106.

Cumming, J. R. & Weinstein, L. H. (1990c).Aluminium–mycorrhizal interactions in the physiology of pitch pine seedlings. *Plant and Soil* **125**, 7–18.

Cunningham, J. E. & Kuiack, C. (1992). Production of citric and oxalic acids and solubilization of calcium phosphate by *Penicillium bilaii. Applied and Environmental Microbiology* **58**, 1451–8.

Cuppoletti, J. & Segel, I. H. (1974). Transinhibition kinetics of the sulfate transport system of *Penicillium notatum*: analysis based on an iso uni uni velocity equation. *Journal of Membrane Biology* **17**, 239–52.

Cuppoletti, J. & Segel, I. H. (1975). Kinetics of sulfate transport by *Penicillium notatum.* Interactions of sulfate, protons and calcium. *Biochemistry* **14**, 4712–18.

Curtis, P. J. (1969). Anaerobic growth of fungi. *Transactions of the British Mycological Society* **53**, 299–302.

Dainty, J. (1976). Water relations of plant cells. In *Encyclopedia of Plant Physiology*, vol. II *Transport in Plants*, Part A *Cells*, ed. U. Lüttge & M. G. Pitman, pp. 284–316. Springer-Verlag: Berlin, Heidelberg, New York.

Dakin, H. D. (1924). Formation of L-malic acid as a product of alcoholic fermentation by yeast. *Journal of Biological Chemistry* **61**, 139–45.

Dame, J. B. & Scarborough, G. A. (1980). Identification of the hydrolytic moiety of the *Neurospora* plasma membrane H^+-ATPase and demonstration of a phosphoryl-enzyme intermediate in its catalytic mechanism. *Biochemistry* **19**, 2931–7.

Dameron, C. T., Smith, B. R. & Winge, D. R. (1989). Glutathione-coated cadmium-sulfide crystallites in *Candida glabrata*. *Journal of Biological Chemistry* **264**, 17355–60.

D'Amore, T., Panchal, C. J., Russell, I. & Stewart, G. G. (1990). A study of ethanol tolerance in yeast. *Critical Reviews in Biotechnology* **9**, 287–304.

Daniel, G., Nilsson, T. & Pettersson, B. (1989a). Intra- and extracellular localization of lignin peroxidase during the degradation of solid wood and wood fragments by *Phanerochaete chrysosporium* by using transmission electron microscopy and immuno-gold labelling. *Applied and Environmental Microbiology* **55**, 871–81.

Daniel, G., Pettersson, B., Nilsson, T. & Volc, J. (1989b). Use of immunogold cytochemistry to detect Mn (II)-dependent and lignin peroxidases in wood degraded by the white rot fungi *Phanerochaete chrysosporium* and *Lentinula edodes*. *Canadian Journal of Botany* **68**, 920–33.

Dantzig, A. H., Wiegmann, F. L. Jr & Nason, A. (1978). Regulation of glutamate dehydrogenase in *nit-2* and *am* mutants of *Neurospora crassa*. *Journal of Bacteriology* **137**, 1333–9.

Darte, C. & Grenson, M. (1975). Evidence for three glutamic acid transporting systems with specialized physiological functions in *Saccharomyces cerevisiae*. *Biochemical and Biophysical Research Communications* **67**, 1028–33.

Davenport, J. W. & Slayman, C. W. (1988). The plasma membrane H^+-ATPase of *Neurospora crassa*. Properties of two reactive sulfhydryl groups. *Journal of Biological Chemistry* **263**, 16007–13.

David, C. N. & Easterbrook, K. (1971). Ferritin in the fungus *Phycomyces*. *Journal of Cell Biology* **48**, 15–28.

Davidson, R. W., Campbell, W. A. & Blaisdell, D. J. (1938). Differentiation of wood-decaying fungi by their reactions on gallic or tannic acid medium. *Journal of Agricultural Research* **57**, 683–95.

Davies, D. D. (1973). Control of and by pH. *Symposium of the Society for Experimental Biology* **27**, 513–29.

Davies, J. M., Brownlee, C. & Jennings, D. H. (1990a). Electrophysiological evidence for an electrogenic proton pump and the proton symport of glucose in the marine fungus *Dendryphiella salina*. *Journal of Experimental Botany* **41**, 449–56.

Davies, J. M., Brownlee, C. & Jennings, D. H. (1990b). Measurement of intracellular pH in fungal hyphae using BCECF and digital imaging spectroscopy. Evidence for a primary proton pump in the plasmalemma of a marine fungus. *Journal of Cell Science* **96**, 731–6.

Davies, M. B. (1980). Peptide uptake in *Candida albicans. Journal of General Microbiology* **121**, 181–6.

Davies, M. E. (1959). The nutrition of *Phytophthora fragariae. Transactions of the British Mycological Society* **42**, 193–200.

Davis, M. A. & Hynes, M. J. (1989). Regulatory genes in *Aspergillus nidulans. Trends in Genetics* **5**, 14–19.

Davis, R. H. (1986). Compartmentation and regulatory mechanisms in the arginine pathways of *Neurospora crassa* and *Saccharomyces cerevisiae. Microbiological Reviews* **50**, 280–313.

Davis, R. H. & Ristow, J. W. L. (1991). Polyamine toxicity in *Neurospora crassa* – protective role of the vacuole. *Archives of Biochemistry and Biophysics* **285**, 306–11.

Dawes, E. A. & Senior, P. Y. (1973). The role and regulation of energy reserve polymers in micro-organisms. *Advances in Microbial Physiology* **10**, 135–266.

de Bruijne, A. W., Schuddemat, J., van den Broek, P. J. A. & van Steveninck, J. (1988). Regulation of sugar transport systems of *Kluyveromyces marxianus*: the role of carbohydrates and their catabolism. *Biochimica et Biophysica Acta* **939**, 569–76.

de Busk, R. M. & Ogilvie, S. (1984a). Participation of an extracellular deaminase in amino acid utilization by *Neurospora crassa. Journal of Bacteriology* **159**, 583–9.

de Busk, R. M. & Ogilvie, S. (1984b). Nitrogen regulation of amino acid utilization by *Neurspora crassa. Journal of Bacteriology* **160**, 493–8.

de Busk, R. M. & Ogilvie, S. (1984c). Regulation of amino acid utilization in *Neurspora crassa*; effect of *nmr-1* and *ms-5* mutations. *Journal of Bacteriology* **160**, 656–61.

de Busk, R. M. & Ogilvie, S. (1985). Transport of L-glutamine by *Neurspora crassa. Journal of General Microbiology* **131**, 905–12.

de Busk, R. M. & Ogilvie-Villa, S. (1982). Physiological adaptation to the loss of amino acid transport ability. *Journal of Bacteriology* **152**, 545–8.

de Deken, R. H. (1966). The Crabtree effect: a regulatory system in yeast. *Journal of General Microbiology* **44**, 159–56.

de Koning, W. & Harder, W. (1992). Methanol-utilizing yeasts. In *Methane and Methanol Utilizers*, ed. J. C. Murrell & H. Dalton, pp. 207–44. Plenum Press: New York.

de la Fuente, G. & Sols, A. (1962). Transport of sugars in yeasts. II. Mechanism of utilization of disaccharides and related glycosides. *Biochimica et Biophysica Acta* **56**, 49–62.

de Nobel, J. G. & Barnett, J. A. (1991). Passage of molecules through yeast cell walls – a brief essay review. *Yeast* **7**, 313–24.

de Nobel, J. G., Dijkers, C., Hooijberg, E. & Klis, F. M. (1989). Increased cell wall porosity in *Saccharomyces cerevisiae* after treatment with dithiothreitol or EDTA. *Journal of General Microbiology* **135**, 2077–84.

de Nobel, J. G., Klis, F. M., Munnik, T., Priem, J. & van den Ende, H. (1990a). An assay of relative wall porosity in *Saccharomyces cerevisiae, Kluyveromyces lactis* and *Schizosaccharomyces pombe. Yeast* **6**, 483–90.

de Nobel, J. G., Klis, F. M., Priem, J., Munnik, T. & van den Ende, H. (1990b). The glucanase-soluble mannoproteins limit cell wall porosity in *Saccharomyces cerevisiae. Yeast* **6**, 491–9.

de Nobel, J. G., Klis, F. M., Ram, A., van Unen, H., Priem, J., Munnik, T. & van den Ende, H. (1991). Cyclic variations in the permeability of the cell wall of *Saccharomyces cerevisiae*. *Yeast* **7**, 589–98.

de Rome, L. & Gadd, G. M. (1987). Measurement of copper uptake in *Saccharomyces cerevisiae* using a Cu^{2+}-selective electrode. *FEMS Microbiology Letters* **43**, 283–7.

Deane-Drummond, C. E. (1986). Nitrate uptake into *Pisum sativum* L. cv. Feltham First seedlings: commonality with nitrate uptake into *Chara corallina* and *Hordeum vulgare* through a substrate cycling. *Plant, Cell and Environment* **9**, 44–8.

Deane-Drummond, C. E. & Glass, A. D. M. (1983a). Short term studies of nitrate uptake into barley plants using ion-specific electrodes and $^{36}ClO_3^-$. I. Control of net uptake by NO_3^- efflux. *Plant Physiology* **73**, 100–4.

Deane-Drummond, C. E. & Glass, A. D. M. (1983b). Short term studies of nitrate uptake into barley plants using ion-specific electrodes and $^{36}ClO_3^-$. II. Regulation of NO_3^- efflux by NH_4^+. *Plant Physiology* **73**, 105–10.

Dee, E. & Conway, E. J. (1968). The relation between sodium ion content and efflux of labelled sodium ions from yeast. *Biochemical Journal* **107**, 265–71.

Del Castillo Agudo, L. (1985). Genetic aspects of ethanol tolerance and production by *Saccharomyces cerevisiae*. *Current Microbiology* **12**, 41–4.

Delgenes, J. P., Moletta, R. & Navarro, J. M. (1986). The effect of aeration on D-xylose fermentation by *Pachysolen tannophilus*, *Pichia stipitis*, *Kluyveromyces marxianus* and *Candida shehatae*. *Biotechnology Letters* **8**, 897–900.

Delhez, J., Dufour, J. P., Thines, D. & Goffeau, A. (1977). Comparison of the properties of plasma membrane-bound and mitochondria-bound ATPases in the yeast *Schizosaccharomyces pombe*. *European Journal of Biochemistry* **79**, 319–28.

Dell, B., Botton, B., Martin, F. & Le Tacon, F. (1989). Glutamate dehydrogenases in ectomycorrhizas of spruce (*Picea excelsa* L.) and beech (*Fagus sylvatica* L.). *New Phytologist* **111**, 683–92.

Demis, D. J., Rothstein, A. & Meier, R. (1954). The relationship of the cell surface to metabolism. X. The location and function of invertase in the yeast cell. *Archives of Biochemistry and Biophysics* **48**, 55–62.

Denny, H. J. & Wilkins, D. A. (1987a). Zinc tolerance in *Betula* spp. I. Effect of external concentration of zinc on growth and uptake. *New Phytologist* **106**, 517–24.

Denny, H. J. & Wilkins, D. A. (1987b). Zinc tolerance in *Betula* spp. II. Microanalytical studies of zinc uptake into root tissue. *New Phytologist* **106**, 525–34.

Denny, H. J. & Wilkins, D. A. (1987c). Zinc tolerance in *Betula* spp. III. Variation in response to zinc among ectomycorrhizal associates. *New Phytologist* **106**, 535–44.

Denny, H. J. & Wilkins, D. A. (1987d). Zinc tolerance in *Betula* spp. IV. The mechanism of ectomycorrhizal amelioration of zinc toxicity. *New Phytologist* **106**, 545–54.

Derks, W. J. G. & Borst-Pauwels, G. W. F. H. (1979). Apparent three-site kinetics of Cs^+-uptake by yeast. *Physiologia Plantarum* **46**, 241–6.

Derks, W. J. G. & Borst-Pauwels, G. W. F. H. (1980). Interactions of monovalent cations with Rb^+ and Na^+ uptake in yeast. *Biochimica et Biophysica Acta* **596**, 381–92.

Di Pietro, A. & Goffeau, A. (1985). Essential arginyl residues in the H^+-translocating ATPase of plasma membrane from the yeast *Schizosaccharomyces pombe*. *European Journal of Biochemistry* **148**, 35–9.

Dickson, R. C. & Barr, K. (1983). Characterization of lactose transport in *Kluyveromyces lactis*. *Journal of Bacteriology* **154**, 1245–51.

Dietrich, S. M. C. (1976). Presence of polyphosphate of low molecular weight in zygomycetes. *Journal of Bacteriology* **127**, 1408–13.

Dighton, J. & Horrill, A. D. (1988). Radiocaesium accumulation in the mycorrhizal fungi *Lactarius rufus* and *Inocybe longicystis* in upland Britain following the Chernobyl incident. *Transactions of the British Mycological Society* **91**, 335–7.

Dijkema, C., Kester, H. C. M. & Visser, J. (1985). ^{13}C NMR studies of carbon metabolism in the hyphal fungus *Aspergillus nidulans*. *Proceedings of the National Academy of Sciences, USA* **82**, 14–18.

Dijkema, C., Rijcken, R. P., Kester, H. C. M. & Visser, J. (1986). ^{13}C-NMR studies on the influence of pH and nitrogen source on polyol pool formation in *Aspergillus nidulans*. *FEMS Microbiology Letters* **33**, 125–32.

Dijkema, C. & Visser, J. (1987). ^{13}C-NMR analysis of *Aspergillus* mutants disturbed in pyruvate metabolism. *Biochimica et Biophysica Acta* **931**, 311–19.

Dill, I. & Kraepelin, G. (1986). Palo podrido: model for extensive delignification of wood by *Ganoderma applanatum*. *Applied and Environmental Microbiology* **52**, 1305–12.

Dodyk, F. & Rothstein, A. (1964). Factors influencing the appearance of invertase in *Saccharomyces cerevisiae*. *Archives of Biochemistry and Biophysics* **104**, 478–86.

Does, A. L. & Bisson, L. F. (1990). Isolation and characterization of *Pichia heedii* mutants defective in xylose uptake. *Applied and Environmental Microbiology* **56**, 3321–8.

Doi, S. & Saito, M. (1980). The bending strength loss of wood decayed by *S. lacrimans*. *Journal of the Hokkaido Forest Products Research Institute* no. 344, 18.

Dombek, K. M. & Ingram, L. O. (1986a). Determination of the intracellular concentration of ethanol in *Saccharomyces cerevisiae* during fermentation. *Applied and Environmental Microbiology* **51**, 197–200.

Dombek, K. M. & Ingram, L. O. (1986b). Magnesium limitation and its role in apparent toxicity of ethanol during yeast fermentation. *Applied and Environmental Microbiology* **52**, 975–81,

Dons, J. J. M., Mulder, H.G. H., Rouwendal, G. J. A., Springer, J., Bremer, W. & Wessels, J. G. H. (1984a). Sequence analysis of a split gene involved in fruiting from the fungus *Schizophyllum commune*. *EMBO Journal* **3**, 2101–6.

Dons, J. J. M., Springer, J., de Vries, S. C. & Wessels, J. G. H. (1984b). Molecular cloning of a gene abundantly expressed during fruiting body initiation in *Schizophyllum commune*. *Journal of Bacteriology* **157**, 802–8.

Dorn, A. & Weisenseel, M. H. (1982). Advances in vibrating probe techniques. *Protoplasma* **113**, 89–96.

Doumas, P., Berjaud, C., Calléja, M., Coupé, M., Espiau, C. & D'Auzac, J. (1986). Phosphatase extracelulaires et nutrition phosphatée chez le champignons ectomycorrhiziens et les plantes hôtes. *Physiologie Végétale* **24**, 173–84.

Dow, J. M. & Rubery, P. H. (1975). Hyphal tip bursting in *Mucor rouxii*: antagonistic effects of calcium ions and acid. *Journal of General Microbiology* **91**, 425–8.

Doyle, R. J. & Rosenberg, M. (1990). *Microbial Cell Surface Hydrophobicity*.

American Society for Microbiology: Washington, DC.

Drechsel, H., Metzger, J., Freund, S., Jung, G., Boelaert, J. & Winkelmann, G. (1991). Rhizoferrin – a novel siderophore from the fungus *Rhizopus microsporus* var. *rhizopodiformis*. *Biology of Metals* **4**, 238–43.

Drew, S. W. & Kadam, K. L. (1979). Lignin metabolism in *Aspergillus fumigatus* and white-rot fungi. *Developments in Industrial Microbiology* **20**, 153–61.

Drillien, R. & Lacroute, F. (1972). Ureidosuccinic acid uptake in yeast and some aspects of its regulation. *Journal of Bacteriology* **109**, 203–8.

du Preez, J. C., Bosch, M. & Prior, B. A. (1987). Temperature profiles of growth and ethanol tolerance of the xylose-fermenting yeasts *Candida shehatae* and *Pichia stipitis*. *Applied Microbiology and Biotechnology* **25**, 521–5.

du Preez, J. C., Prior, B. A. & Monteiro, A. M. (1984). The effect of aeration on xylose fermentation by *Candida shehatae* and *Pachysolen tannophilus*. *Applied Microbiology and Biotechnology* **19**, 261–6.

du Preez, J. C., van Driessel, B. & Prior, B. A. (1989). Effects of aerobiosis on fermentation and key enzyme levels during growth of *Pichia stipitis*, *Candida shehatae* and *Candida tenuis* on D-xylose. *Archives of Microbiology* **152**, 143–7.

Dubois, E. & Grenson, M. (1979). Methylamine/ammonia uptake systems in *Saccharomyces cerevisiae*: multiplicity and regulation. *Molecular and General Genetics* **175**, 67–76.

Duddridge, J. A., Malibari, J. & Read, D. J. (1980). Structure and function of mycorrhizal rhizomorphs with special reference to their role in water transport. *Nature, London* **287**, 834–6.

Dufour, J.-P., Amory, A. & Goffeau, A. (1988). Plasma membrane ATPase from the yeast *Schizosaccharomyces pombe*. *Methods in Enzymology* **157**, 513–28.

Dufour, J.-P. & Goffeau, A. (1978). Solubilization by lysolecithin and purification of the plasma membrane ATPase of the yeast *Schizosaccharomyces pombe*. *Journal of Biological Chemistry* **253**, 7026–32.

Duniway, J. M. (1979). Water relations of water molds. *Annual Review of Phytopathology* **17**, 431–60.

Dunlop, P. C. & Roon, R. J. (1975). L-Aspargine of *Saccharomyces cerevisiae*: an extracellular enzyme. *Journal of Bacteriology* **122**, 1017–24.

Dunn-Coleman, N. S., Smarrelli, J. Jr & Garrett, R. H. (1984). Nitrate assimilation in eukaryotic cells. *International Review of Cytology* **92**, 1–50.

Duro, A. F. & Serrano, R. (1981). Inhibition of succinate production during yeast fermentation by deenergization of the plasma membrane. *Current Microbiology* **6**, 111–13.

Dürr, M., Boller, T. & Wiemken, A. (1975). Polybase induced lysis of yeast sphaeroplasts. A new gentle method for preparation of vacuoles. *Archives of Microbiology* **105**, 319–27.

Eamus, D. & Jennings, D. H. (1984). Determination of water, solute and turgor potentials of mycelium of various basidiomycete fungi causing wood decay. *Journal of Experimental Botany* **35**, 1782–6.

Eamus, D. & Jennings, D. H. (1986a). Water, turgor and osmotic potentials of fungi. In *Water, Fungi and Plants*, ed. P. G. Ayres & L. Boddy, pp. 27–48. Cambridge University Press: Cambridge.

Eamus, D. & Jennings, D. H. (1986b). Turgor and fungal growth: studies on water relations of mycelia of *Serpula lacrimans* and *Phallus impudicus*. *Transactions of the British Mycological Society* **86**, 527–35.

Eamus, D., Thompson, W., Cairney, J. W. G. & Jennings, D. H. (1985). Internal structure and hydraulic conductivity of basidiomycete translocating organs. *Journal of Experimental Botany* **36**, 1110–16.

Eddy, A. A. (1980). Some aspects of amino acid transport in yeast. In *Microorganisms and Nitrogen Sources*, ed. J. W. Payne, pp. 35–62. John Wiley: Chichester.

Eddy, A. A. (1982). Mechanisms of solute transport in selected eukaryotic microorganisms. *Advances in Microbial Physiology* **23**, 1–78.

Eddy, A. A. & Hopkins, P. G. (1985). The putative electrogenic nitrate–proton symport of the yeast *Candida utilis*. Comparison with the systems absorbing glucose or lactate. *Biochemical Journal* **231**, 291–7.

Eddy, A. A. & Rudin, A. D. (1958*a*). The structure of the yeast cell wall. I. Identification of charged groups at the surface. *Proceedings of the Royal Society* B **148**, 419–32.

Eddy, A. A. & Rudin, A. D. (1958*b*). Comparisons of the respective electrophoretic and flocculation characteristics of different strains of *Saccharomyces*. *Journal of the Institute of Brewing* **64**, 139–42.

Eddy, A. A., Seaston, A., Gardner, D. & Hacking, C. (1980). The thermodynamic efficiency of cotransport mechanisms with special reference to proton and anion transport in yeast. *Annals of the New York Academy of Sciences* **341**, 494–509.

Edelman, J. (1956). The formation of oligosaccharides by enzymic transglycosylation. *Advances in Enzymology* **17**, 189–232.

Edgely, M. & Brown, A. D. (1983). Yeast water relations: physiological changes induced by solute stress in *Saccharomyces cerevisiae* and *Saccharomyces rouxii*. *Journal of General Microbiology* **129**, 3453–63.

Ehrlich, H. L. (1981). *Geomicrobiology*. Marcel Dekker: New York.

Ehwald, R., Mavrina, L. & Wilken, B. (1981). Active transport and mediated diffusion of glucose and other monosaccharides in *Endomyces magnusii*. *Folio Microbiologica* **26**, 388–93.

Eilam, Y. (1983). Membrane effects of phenothiazines in yeasts. I. Stimulation of calcium and potassium fluxes. *Biochimica et Biophysica Acta* **733**, 242–8.

Eilam, Y. (1984). Effects of phenothiazines on inhibition of plasma membrane ATPase and hyperpolarization of cell membranes in the yeast *Saccharomyces cerevisiae*. *Biochimica et Biophysica Acta* **769**, 601–10.

Eilam, Y. & Chernichovsky, D. (1987). Uptake of Ca^{2+} driven by the membrane potential in energy-depleted yeast cells. *Journal of General Microbiology* **133**, 161–9.

Eilam, Y. & Grossowicz, N. (1982). Nystatin effects of cellular calcium in *Saccharomyces cerevisiae*. *Biochimica et Biophysica Acta* **692**, 238–43.

Eilam, Y., Lavi, H. & Grossowicz, N. (1985). Cytoplasmic Ca^{2+} homeostasis maintained by a vacuolar Ca^{2+} transport system in the yeast *Saccharomyces cerevisiae*. *Journal of General Microbiology* **131**, 623–9.

Eilam, Y. & Othman, M. (1990). Activation of Ca^{2+} influx by metabolic substrates in *Saccharomyces cerevisiae*: role of membrane potential and cellular ATP levels. *Journal of General Microbiology* **136**, 861–6.

Eilam, Y., Othman, M. & Halachmi, D. (1990). Transient increase in Ca^{2+} influx in *Saccharomyces cerevisiae* in response to glucose: effects of intracellular acidification and cAMP levels. *Journal of General Microbiology* **136**, 2537–43.

Einsele, A., Schneider, H. & Fiechter, A. (1975). Characterization of microemulsions in a hydrocarbon fermentation by electron microscopy. *Journal of Fermentation Technology* **53**, 241–3.

El Gogary, S., Leite, A., Crivellaro, O., El Dorry, H. & Eveleigh, D. E. (1990). *Trichoderma reesei* cellulase – from mutations to induction. In *Trichoderma reesei: Biochemistry, Genetics, Physiology and Applications*, ed. C. P. Kubicek, D. E. Eveleigh, H. Esterbauer, W. Steiner & E. M. Kubicek-Pranz, pp. 200–11. Royal Society of Chemistry: Cambridge.

Elliott, C. G. (1977). Sterols in fungi: their functions in growth and reproduction. *Advances in Microbial Physiology* **15**, 121–73.

Elliott, C. G. (1979). Influence of the structure of the sterol molecule on sterol-induced reproduction in *Phytophthora cactorum. Journal of General Microbiology* **115**, 117–26.

Elliott, C. G. (1986). Inhibition of reproduction of *Phytophthora* by the calmodulin-interacting compounds trifluoperazine and ophiobolin A. *Journal of General Microbiology* **132**, 2781–5.

Elliott, C. G., Hendrie, M. R. & Knights, B. A. (1966). The sterol requirement of *Phytophthora cactorum. Journal of General Microbiology* **42**, 425–35.

Elliott, C. G., Hendrie, M. R., Knights, B. A. & Parker, W. (1964). A steroid growth factor requirement in a fungus. *Nature, London* **203**, 427.

Ellis, S. W., Grindle, M. & Lewis, D. H. (1991). Effect of osmotic stress on yield and polyol content of dicarboximide-sensitive and -resistant strains of *Neurospora crassa. Mycological Research* **95**, 457–64.

Elskens, M. T., Jaspers, C. J. & Penninckx, M. J. (1991). Glutathione as an endogenous sulphur source in the yeast *Saccharomyces cerevisiae. Journal of General Microbiology* **137**, 637–44.

Emerson, R. & Held, A. A. (1969). *Aqualinderella fermentans* gen. et sp. n. adapted to stagnant waters. II. Isolation, cultural characteristics, and gas relations. *American Journal of Botany* **56**, 1103–20.

Emerson, R. & Natvig, D. O. (1981). Adaptation of fungi to stagnant waters. In *The Fungal Community: Its Organisation and Role in the Ecosystem*, ed. D. T. Wicklow & G. C. Carroll, pp. 109–28. Marcel Dekker: New York.

Emery, T. (1971). Role of ferrichrome as a ferric ionophore in *Ustilago sphaerogena. Biochemistry* **10**, 1483–8.

Emery, T. (1976). Fungal ornithine esterases: relationship to transport. *Biochemistry* **15**, 2723–8.

Enari, T.-M. & Niku-Päavola, M.-L. (1987). Enzymatic hydrolysis of cellulose: is the current theory of the mechanism of hydrolysis valid? *CRC Critical Reviews in Biotechnology* **5**, 67–87.

Entian, K.-D. (1986). Glucose repression: a complex regulatory system. *Microbiological Sciences* **3**, 366–71.

Eraso, P. & Gancedo, C. (1987). Activation of yeast plasma membrane ATPase by acid pH during growth. *FEBS Letters* **224**, 187–92.

Eriksson, K.-E. (1981). Cellulases in fungi. In *Trends in the Biology of Fermentations for Fuels and Chemicals*, ed. A. Hollaender, pp. 19–32. Plenum Press: New York.

Eriksson, K.-E., Grünewald, A. & Vallander, L. (1980). Studies of growth conditions in wood for three white-rot fungi and their celulaseless mutants. *Biotechnology and Bioengineering* **22**, 363–76.

Eriksson, K.-E. & Hamp, S. G. (1978). Regulation of *endo*-1,4-β-glucanase production in *Sporotrichum pulverulentum. European Journal of Biochemistry* **90**, 183–90.

Eriksson, K.-E. & Pettersson, B. (1975a). Extra-cellular enzyme system utilized by the fungus *Sporotrichum pulverulentum* (*Chrysosporium lignorum*) for the breakdown of cellulose. 1. Separation, purification and physico-chemical characterization of five *endo*-1,4-β-glucanases. *European Journal of Biochemistry* **51**, 193–206.

Eriksson, K.-E. & Pettersson, B. (1975*b*). Extracellular enzyme system utilized by the fungus *Sporotrichum pulverulentum* (*Chrysosporium lignorum*) for the breakdown of cellulose. 3. Purification and physico-chemical characterization of an *exo*-1,4-β-glucanase. *European Journal of Biochemistry* **51**, 213–18.

Eriksson, K.-E., Pettersson, B., Volc, J. & Musilek, V. (1986). Formation and partial characterization of glucose-2-oxidase, a H_2O_2 producing enzyme in *Phanerochaete chrysosporium*. *Applied Microbiology and Biotechnology* **23**, 257–62.

Eriksson, K.-E., Pettersson, B. & Westermark, U. (1974). Oxidation: an important enzyme reaction in fungal degeneration of cellulose. *FEBS Letters* **49**, 282–5.

Eriksson, K.-E. & Rzedowski, W. (1969*a*). Extracellular enzyme system utilized by the fungus *Chrysosporium lignorum* for the breakdown of cellulose. I. Studies on the enzyme production. *Archives of Biochemistry and Biophysics* **129**, 683–8.

Eriksson, K.-E. & Rzedowski, W. (1969*b*). Extracellular enzyme system utilized by the fungus *Chrysosporium lignorum* for the breakdown of cellulose. II. Separation and characterization of three cellulase peaks. *Archives of Biochemistry and Biophysics* **129**, 689–95.

Ernst, J. F. & Winkelmann, G. (1984). Enzymatic release of iron from sideramines in fungi. NADH: sideramine oxidoreductase in *Neurospora crassa*. *Biochimica et Biophysica Acta* **500**, 27–41.

Esfahani, M. & Devlin, T. M. (1982). Effects of lipid fluidity on quenching characteristics of tryptophan fluorescence in yeast plasma membranes. *Journal of Biological Chemistry* **257**, 9919–21.

Esfahani, M., Solomon, D. J., Mele, L. & Teter, M. N. (1979). Lipid-protein interactions in membranes: effect of lipid composition on mobility of spin-labelled cysteine residues in yeast plasma membranes. *Journal of Supramolecular Structure* **10**, 277–86.

Eslyn, W. E. & Highley, T. L. (1976). Decay resistance and susceptibility of sapwood of fifteen tree species. *Phytopathology* **66**, 1010–17.

Evans, D. E., Briars, S.-A. & Williams, L. E. (1991). Active calcium transport by plant cell membranes. *Journal of Experimental Botany* **42**, 285–303.

Evans, J. J. & Sorger, G. J. (1966). Role of mineral elements with emphasis on univalent cations. *Annual Review of Plant Physiology* **17**, 47–76.

Eveleigh, D. H. (1987). Cellulase: a perspective. *Philosophical Transactions of the Royal Society of London* A **321**, 435–47.

Ewaze, J. O., Moore, D. & Stewart, G. R. (1978). Co-ordinate regulation of enzymes involved in ornithine metabolism and its relation to sporophore morphogenesis in *Coprinus cinereus*. *Journal of General Microbiology* **107**, 343–57.

Eze, J. N. O. (1975). Translocation of phosphate in mould mycelia. *New Phytologist* **75**, 579–82.

Fågerstam, L. G. & Pettersson, L. G. (1980). The 1,4-β-glucan cellobiohydrolases of *Trichoderma reesei* QM9414. *FEBS Letters* **119**, 97–100.

Failla, M. L., Benedict, C. D. & Weinberg, E. D. (1976). Accumulation and storage of Zn^{2+} by *Candida utilis*. *Journal of General Microbiology* **94**, 23–36.

Failla, M. L. D. & Weinberg, E. D. (1977). Cyclic accumulation of zinc by

Literature cited

Candida utilis during growth in batch culture. *Journal of General Microbiology* **99**, 85–97.

Faison, B. D. & Kirk, T. K. (1985). Factors involved in the regulation of a ligninase activity in *Phanerochaete chrysosporium*. *Applied and Environmental Microbiology* **49**, 299–304.

Farcaš, V. (1979). Biosynthesis of cells walls of fungi. *Microbiological Reviews* **43**, 117–44.

Farcaš, V. (1990). Fungal cell walls: their structure, biosynthesis and biotechnological aspects. *Acta Biotechnologica* **10**, 225–38.

Farcaš, V., Bauer, S. & Zemek, J. (1969). Mechanism of 2-deoxy-D-glucose in baker's yeast. III. Formation of 2,2'-dideoxy-α,α'-trehalose. *Biochimica et Biophysica Acta* **184**, 77–82.

Farrar, J. F. (1976). The uptake and metabolism of phosphate by the lichen *Hypogymnia physoides*. *New Phytologist* **77**, 127–34.

Fawole, M. O. & Casselton, P. J. (1972). Observations on the regulation of glutamate dehydrogenase activity in *Coprinus lagopus*. *Journal of Experimental Botany* **23**, 530–51.

Federspiel, A., Schuler, R. & Haselwandter, K. (1991). Effect of pH, L-ornithine and L-proline on the hydroxamate siderophore production by *Hymenoscyphus ericae*, a typical ericoid mycorrhizal fungus. *Plant and Soil* **130**, 259–62.

Felter, S. & Stahl, A. J. C. (1973). Enzymes du métabolism des polyphosphates dans la levure. III. Purification et propriétés de la polyphosphate-ADP-phosphotransférase. *Biochimie* **55**, 245–51.

Fenn, P., Choi, S. & Kirk, T. K. (1981). Ligninolytic activity of *Phanerochaete chrysosporium*: physiology and suppression by NH_4^+ and L-glutamate. *Archives of Microbiology* **130**, 66–71.

Fenn, P. & Kirk, T. K. (1979). Ligninolytic system of *Phanerochaete chrysosporium*: inhibition by *o*-phthalate. *Archives of Microbiology* **123**, 307–9.

Fenn, P. & Kirk, T. K. (1981). Relationship of nitrogen to the onset and suppression of ligninolytic activity and secondary metabolism in *Phanerochaete chrysosporium*. *Archives of Microbiology* **130**, 59–65.

Fergus, C. L. (1952). The nutrition of *Penicillium digitatum* Sacc. *Mycologia* **44**, 183–9.

Ferguson, A. R. & Sims, A. P. (1974). The regulation of glutamine metabolism in *Candida utilis*: the role of glutamine in the control of glutamine synthetase. *Journal of General Microbiology* **80**, 159–61.

Fickel, T. E. & Gilvarg, C. (1973). Transport of impermeant substances in *E. coli* by way of oligopeptide permease. *Nature New Biology* **241**, 161–3.

Fiechter, A., Fuhrmann, G. F. & Käppeli, O. (1981). Regulation of glucose metabolism in growing yeast cells. *Advances in Microbial Physiology* **22**, 123–83.

Fincham, J. R. S., Kinnaird, J. H. & Burns, P. A. (1985). The *am* (NADP-specific glutamate dehydrogenase) gene of *Neurospora crassa*. In *Molecular Genetics of Filamentous Fungi*, ed. W. Timberlake, pp. 117–25. Alan R. Liss: New York.

Finlay, R. D. & Read, D. J. (1986a). The structure and function of the vegetative mycelium of ectomycorrhizal plants. I. Translocation of ^{14}C-labelled carbon between plants interconnected by a common mycelium. *New Phytologist* **103**, 143–56.

Finlay, R. D. & Read, D. J. (1986b). The structure and function of the vegetative mycelium of ectomycorrhizal plants. II. Tree uptake and distribution of phosphorus by mycelial strands interconnecting host plants. *New Phytologist* **103**, 157–66.

Finn, R. F. (1942). Mycorrhizal inoculation in soil of low fertility. *Black Rock Forest Papers* **1**, 116–17.

Fischer, E. & Linder, P. (1895). Ueber die Enzyme eininger Hefen. *Berichte der Deutschen Chemischen Gesellschaft* **28**, 3034–9.

Fisher, R. F. (1972). Spodosol development and nutrient distribution under Hydnaceae fungal mats. *Proceedings of the Soil Science Society of America* **36**, 492–5.

Fleet, G. H. (1991). Cell walls. In *The Yeasts*, vol. 4 *Yeast Organelles*, ed. A. H. Rose & J. S. Harrison, pp. 199–277. Academic Press: London.

Fletcher, J. (1988). Effects of external Ca^{2+} concentrations and of the ionophore A23187 on development of oogonia, oospores and gemmae of *Saprolegnia diclina*. *Annals of Botany* **62**, 445–48.

Flint, H. J. (1985). Changes in gene expression elicited by amino acid limitation in *Neurospora crassa* strains having normal or mutant cross-pathway amino acid control. *Molecular and General Genetics* **200**, 283–90.

Flint, H. J. & Wilkening, J. (1986). Cloning of the *arg-12* gene of *Neurospora crassa* and regulation of its transport via cross-pathway amino acid control. *Molecular and General Genetics* **203**, 110–16.

Flowers, T. J., Hajibagheri, M. A. & Clipson, N. J. W. (1986). Halophytes. *Quarterly Review of Biology* **61**, 313–37.

Fogel, S. & Welch, J. H. (1982). Tandem gene amplification mediates copper resistance in yeast. *Proceedings of the National Academy of Sciences, USA* **79**, 5342–6.

Fogel, S., Welch, J. W., Cathala, G. & Karin, M. (1983). Gene amplification in yeast: *CUP1* copy number regulates copper resistance. *Current Genetics* **7**, 347–55.

Fogel, S., Welch, J. W. & Maloney, D. H. (1988). The molecular genetics of copper resistance in *Saccharomyces cerevisiae* – a paradigm for non-conventional yeasts. *Journal of Basic Microbiology* **28**, 147–60.

Folan, M. A. & Coughlan, M. P. (1979). The cellulase complex in the culture filtrate of the thermophilic fungus *Talaromyces emersonii*. *International Journal of Biochemistry* **10**, 505–10.

Folkes, B. F. & Sims, A. P. (1974). The significance of amino acid inhibition of NADP-linked glutamate dehydrogenase in the physiological control of glutamate synthesis in *Candida utilis*. *Journal of General Microbiology* **82**, 77–95.

Fonseca, A., Spencer-Martins, S. & van Uden, N. (1991). Transport of lactic acid in *Kluyveromyces marxianus* – evidence for a monocarboxylate uniport. *Yeast* **7**, 775–80.

Forêt, M., Schmidt, R. & Reichert, U. (1978). On the mechanism of substrate binding to the purine-transport system of *Saccharomyces cerevisiae*. *European Journal of Biochemistry* **82**, 33–43.

Forrester, I. T., Grabski, A. C., Burgess, R. R. & Leatham, G. F. (1988). Manganese, Mn-dependent peroxidases and the biodegradation of lignin. *Biochemical and Biophysical Research Communications* **157**, 992–9.

Forrester, P. I. & Gaucher, G. M. (1972). Conversion of 6-methylsalicyclic acid into patulin by *Penicillium urticae*. *Biochemistry* **11**, 1102–7.

Foster, J. W. (1949). *Chemical Activities of Fungi*. Academic Press: New York.

Foulkes, E. C. (1956). Cationic transport in yeast. *Journal of General*

Physiology **39**, 687–704.

Foury, F. (1990). The 31-kDa polypeptide is an essential feature of the vacuolar ATPase in *Saccharomyces cerevisiae*. *Journal of Biological Chemistry* **265**, 18554–60.

Fowler, B. A., Hildebrand, C. E., Kojima, Y. & Webb, M. (1987). Nomenclature of metallothioneins. In *Metallothionein II*, ed. J. H. R. Kägi & Kojima, pp. 19–22. Birkhäuser Verlag: Basel.

Foy, C. D., Chaney, R. L. & White, M. C. (1978). The physiology of metal toxicity in plants. *Annual Review of Plant Physiology* **29**, 511–66.

Fraenkel, D. G. (1982). Carbohydrate metabolism. In *The Molecular Biology of the Yeast* Saccharomyces: *Metabolism and Gene Expression*, ed. J. N. Strathern, E. W. Jones & J. R. Broach, pp. 1–37. Cold Spring Harbor Laboratory Press: Cold Spring Harbor, NY.

France, R. C. & Reid, C. P. P. (1983). Interactions of nitrogen and carbon in the physiology of ectomycorrhizae. *Canadian Journal of Botany* **61**, 964–84.

Francis, J., Madinaveita, J., MacTurk, H. M. & Snow, G. A. (1949). Isolation from acid-fast bacteria of a growth factor for *Mycobacterium johnei* and a precursor of phthiocol. *Nature, London* **163**, 365–6.

Frank, A. B. (1894). Die Bedentung der Mykorrhiza-pilze für die gemeine Kiefer. *Forstwissenschaftliches Zentralblatt* **16**, 1852–90.

Franzusoff, A. & Cirillo, V. P. (1982). Uptake and phosphorylation of 2-deoxy-D-glucose by wild-type and single-kinase strains of *Saccharomyces cerevisiae*. *Biochimica et Biophysica Acta* **688**, 295–304.

Franzusoff, A. & Cirillo, V. P. (1983a). Glucose transport activity in isolated plasma membrane vesicles from *Saccharomyces cerevisiae*. *Journal of Biological Chemistry* **285**, 3608–14.

Franzusoff, A. & Cirillo, V. P. (1983b). Solubilization and reconstitution of the glucose transport system in *Saccharomyces cerevisiae*. *Biochimica et Biophysica Acta* **734**, 153–9.

Freeman, J. A., Manis, P. B., Snipes, G. J., Mayes, B. N., Samson, P. C., Wikswo, J. P. & Freeman, D. B. (1985). Steady growth cone currents revealed by a novel circularly vibrating probe: a possible mechanism underlying neurite growth. *Journal of Neuroscience Research* **13**, 257–84.

Freer, S. N. & Detroy, R. W. (1983). Characterization of cellobiose fermentations to ethanol by yeasts. *Biotechnology and Bioengineering* **25**, 541–57.

Freer, S. N. & Detroy, R. W. (1985). Regulation of β-1,4-glucosidase expression by *Candida wickerhamii*. *Applied and Environmental Microbiology* **50**, 152–9.

Friedman, K. J. (1977). Role of lipids in the *Neurospora crassa* membrane. II. Membrane potential and resistance studies: the effect of altered fatty acid composition. *Journal of Membrane Biology* **36**, 175–90.

Friedman, K. J. & Glick, D. (1980). Role of lipids in the *Neurospora crassa* membrane. III. Lipid composition and phase transition properties of the plasma membrane and its components. *Journal of Membrane Biology* **54**, 183–90.

Fries, N. (1965). The chemical environment for fungal growth. 3. Vitamins and growth factors. In *The Fungi, An Advanced Treatise*, vol. 1 *The Fungal Cell*, ed. G. C. Ainsworth & A. S. Sussman, pp. 491–523. Academic Pres: London.

Friis, J. & Ottolenghi, P. (1959a). Localisation of invertase in a strain of yeast. *Comptes rendus des travaux Carlsberg* **31**, 259–71.

Friis, J. & Ottolenghi, P. (1959*b*). Localisation of melibiase in a strain of yeast. *Comptes rendus des travaux Carlsberg* **31**, 272–81.

Fry, J. C. (1990). Oligotrophs. In *Microbiology of Extreme Environments*, ed. C. Edwards, pp. 93–116. Open University Press: Milton Keynes.

Fu, Y.-H. & Marzluf, G. A. (1988). Metabolic control and autogenous regulation of *nit-3*, the nitrate structural gene of *Neurospora crassa*. *Journal of Bacteriology* **170**, 657–61.

Fu, Y.-H. & Marzluf, G. A. (1990*a*). *nit-2*, the major positive-acting nitrogen regulatory gene of *Neurospora* encodes a sequence-specific DNA-binding protein. *Proceeings of the National Academy of Sciences, USA* **87**, 5331–5.

Fu, Y.-H. & Marzluf, G. A. (1990*b*). *cys-3*, the positive-acting sulfur regulatory gene of *Neurospora crassa* encodes a sequence-specific DNA-binding protein. *Journal of Biological Chemistry* **265**, 11942–7.

Fu, Y.-H., Young, J. L. & Marzluf, G. A. (1988). Molecular cloning and characterization of a negative-acting nitrogen regulatory gene of *Neurspora crassa*. *Molecular and General Genetics* **214**, 74–9.

Fuhrmann, G. F. (1974). Transport of divalent cations in *Saccharomyces cerevisiae*. *Proceedings of the Fourth International Symposium on Yeasts* **1**, H1.

Fuhrmann, G. F., Boehm, C. & Theuvenet, A. P. R. (1976). Sugar transport and potassium permeability in yeast plasma membrane vesicles. *Biochimica et Biophysica Acta* **433**, 583–96.

Fuhrmann, G. & Rothstein, A. (1968). The transport of Zn^{2+}, Co^{2+} and Ni^{2+} into yeast cells. *Biochimica et Biophysica Acta* **163**, 325–30.

Fuhrmann, G. & Völker, B. 1992). Regulation of glucose transport in *Saccharomyces cerevisiae*. *Journal of Biotechnology* **27**, 1–15.

Fuhrmann, G., Völker, B., Sander, S. & Potthast, M. (1989). Kinetic analysis and simulation of glucose transport in plasma membrane vesicles of glucose-repressed and derepressed *Saccharomyces cerevisiae*. *Experientia* **45**, 1018–23.

Gaber, R. F., Styles, C. A. & Fink, G. R. (1988). *TRK1* encodes a plasma membrane protein required for high-affinity potassium transport in *Saccharomyces cerevisiae*. *Molecular and Cellular Biology* **8**, 2848–59.

Gadd, G. M., Chalmers, K. & Reed, R. H. (1987). The role of trehalose in dehydration resistance of *Saccharomyces cerevisiae*. *FEMS Microbiology Letters* **48**, 249–54.

Gadd, G. M. & Mowll, J. L. (1985). Copper uptake by yeast-like cells, hyphae and chlamydospores of *Aureobasidium pullulans*. *Experimental Mycology* **9**, 230–40.

Gadd, G. M., Stewart, A., White, C. & Mowll, J. L. (1984). Copper uptake by whole cells and protoplasts of a wild-type copper-resistant strain of *Saccharomyces cerevisiae*. *FEMS Microbiology Letters* **24**, 231–4.

Gadd, G. M. & White, C. (1985). Copper uptake by *Penicillium ochro-chloron*: influence of pH on toxicity and demonstration of energy-dependent copper influx using protoplasts. *Journal of General Microbiology* **131**, 1875–9.

Gage, R. A., Theuvenet, A. R. P. & Borst-Pauwels, G. W. F. H. (1986). Effect of plasmolysis upon monovalent cation uptake, 9-aminoacridine binding and the zeta potential of yeast cells. *Biochimica et Biophysica Acta* **845**, 77–83.

Gage, R. A., van Wijngaarden, W., Theuvenet, A. P. R., Borst-Pauwels, G. W. F. H. & Verkleij, A. J. (1985). Influence of Rb^+ uptake in yeast by Ca^{2+} is caused by a reduction in the surface potential and not in the Donnan potential of the cell wall. *Biochimica et Biophysica Acta* **812**, 1–8.

Galpin, M. F. J. & Jennings, D. H. (1975). Histochemical study of the hypha and distribution of adenosine triphosphatase in *Dendryphiella salina*. *Transactions of the British Mycological Society* **65**, 477–83.

Galpin, M. F. J. & Jennings, D. H., Oates, K. & Hobot, J. A. (1978). Localization by X-ray microanalysis of soluble ions, particularly potassium and sodium in fungal hyphae. *Experimental Mycology* **2**, 258–69.

Gamow, R. I. & Bottger, B. (1982). Avoidance and theotropic responses in *Phycomyces*. Evidence for an 'avoidance gas' mechanism. *Journal of General Physiology* **79**, 835–48.

Gancedo, C. & Serrano, R. (1989). Energy-yielding metabolism. In *The Yeasts*, vol. 3 *Metabolism and Physiology of Yeasts*, ed. A. H. Rose & J. S. Harrison, pp. 205–59. Academic Press: London.

Gancedo, J. M. (1986). Carbohydrate metabolism in yeast. In *Carbohydrate Metabolism in Cultured Cells*, ed. M. J. Morgan, pp. 245–86. Plenum Publishing Co.: New York.

Gancedo, J. M. (1992). Carbon catabolite repression in yeast. *European Journal of Biochemistry* **206**, 297–314.

Garcia, J. C. & Kotyk, A. (1988). Uptake of L-lysine by a double mutant of *Saccharomyces cerevisiae*. *Folia Microbiologica* **33**, 285–91.

García, M., Benitez, J., Delgardo, J. & Kotyk, A. (1983). Isolation of sulphate transport defective mutants of *Candida utilis* – further evidence for a common transport system for sulphate, sulphite and thiosulphate. *Folia Microbiologica* **28**, 1–25.

Garcia-Toledo, A., Babich, H. & Stotsky, G. (1985). Training of *Rhizopus stolonifer* and *Cunninghamella blakesleeanus* to copper: cotolerance to cadmium, cobalt, nickel and lead. *Canadian Journal for Microbiology* **31**, 485–92.

Garraway, M. O. & Evans, R. C. (1984). *Fungal Nutrition and Physiology*, John Wiley: New York.

Garrett, R. H. & Amy, N. K. (1978). Nitrate assimilation in fungi. *Advances in Microbial Physiology* **18**, 1–65.

Garrett, R. H. & Nason, A. (1969). Further purifications and properties of *Neurospora* nitrate reductase. *Journal of Biological Chemistry* **244**, 2870–82.

Garrill, A., Clipson, N. J. W. & Jennings, D. H. (1992). Preliminary observations on the monovalent cation relations of *Thraustochytrium aureum*, a fungus requiring sodium for growth. *Mycological Research* **96**, 295–304.

Garrill, A. & Jennings, D. H. (1991). Isolation of a plasma membrane ATPase with H^+-ATPase-like properties from the marine fungus *Dendryphiella salina*. *Experimental Mycology* **15**, 351–5.

Garrill, A., Lew, R. R. & Heath, I. B. (1992). Stretch-activated Ca^{2+} and Ca^{2+}-activated K^+ channels in the hyphal tip plasma membrane of the oomycete *Saprolegnia ferax*. *Journal of Cell Science* **101**, 720–30.

Gascón, S., Neumann, N. P. & Lampen, J. O. (1968). Comparative study of the properties of the purified internal and external invertases from yeast. *Journal of Biological Chemistry* **243**, 1573–7.

Gascón, S. & Ottolenghi, P. (1972). Influence of glucose concentration of the medium on the invertase content of a strain of *Saccharomyces cerevisiae* bearing the *SUC2* gene. *Comptes rendus des travaux Carlsberg* **39**, 15–24.

Gasnier, B. (1987). Characterization of low- and high-affinity glucose transports in the yeast *Kluyveromyces marxianus*. *Biochimica et Biophysica Acta* **903**, 425–33.

Genetet, I., Martin, F. & Stewart, G. R. (1984). Nitrogen assimilation in mycorrhizas. Ammonium assimilation in the N-starved ectomycorrhizal fungus *Cenoccocum graniforme*. *Plant Physiology* **76**, 395–9.

Genner, G. & Hill, E. C. (1981). Fuels and oils. In *Economic Microbiology*, vol. 6 *Microbial Biodeterioration*, ed. A. H. Rose, pp. 295–306. Academic Press: London.

Gentile, A. C. (1954). Carbohydrate metabolism and oxalic acid synthesis by *Botrytis cinerea*. *Plant Physiology* **29**, 257–61.

Georg, L. K. (1952). Cultural and nutritional studies of *Trichophyton gallinae* and *Trichophyton megninii*. *Mycologia* **44**, 470–92.

Gerhardt, P. & Judge, J. A. (1964). Porosity of isolated cell walls of *Saccharomyces cerevisiae* and *Bacillus megaterium*. *Journal of Bacteriology* **87**, 945–51.

Germida, J. J. (1985). Modified sulphur-containing media for studying sulphur oxidising micro-organisms. In *Planetary Ecology*, ed. D. E. Caldwell, J. A. Brierley & C. L. Brierley, pp. 333–44. Nostrand Reinhold: New York.

Gerson, D. F. & Poole, R. J. (1971). Anion absorption by plants. A unary interpretation of 'dual mechanisms'. *Plant Physiology* **48**, 509–11.

Gezelius, K. & Norkrans, B. (1970). Ultrastructure of *Debaryomyces hansenii*. *Archiv für Mikrobiologie* **70**, 14–25.

Ghareib, M., Youssef, K. A. & Khalil, A. A. (1988). Ethanol tolerance of *Saccharomyces cerevisiae* and its relationship to lipid content and composition. *Folia Microbiologica* **33**, 447–52.

Ghislain, M., Goffeau, A., Halachmi, D. & Eilam, Y. (1990). Calcium homeostasis and transport are affected by disruption of *cta3*, a novel gene encoding Ca^{2+}-ATPase in *Schizosaccharomyces pombe*. *Journal of Biological Chemistry* **265**, 18400–7.

Ghislain, M., Schlesser, A. & Goffeau, A. (1987). Mutation of a conserved glycine residue modifies the vanadate sensitivity of the plasma membrane H^+-ATPase from *Schizosaccharomyces pombe*. *Journal of Biological Chemistry* **262**, 17549–55.

Giannini, J. L., Holt, J. S. & Briskin, D. P. (1988). Isolation of sealed plasma membrane vesicles from *Phytophthora megasperma* f. sp. *glycinea*. I. Characterization of proton pumping and ATP activity. *Archives of Biochemistry and Biophysics* **265**, 337–45.

Gibbs, M. & Gastel, R. (1953). Glucose dissimilation by *Rhizopus*. *Archives of Biochemistry and Biophysics* **43**, 33–8.

Gilbert, W. J. & Hickey, R. J. (1946). Production of conidia in submerged cultures of *Penicillium notatum*. *Journal of Bacteriology* **51**, 731–3.

Gille, G., Höfer, M. & Sigler, K. (1992). Evidence for a specific fructose carrier in *Rhodotorula glutinis* based on kinetic studies with a mutant defective in glucose transport. *Folia Microbiologica* **37**, 31–8.

Gillie, O. J. (1968). Observations on the tube method of measuring growth rate in *Neurospora crassa*. *Journal of General Microbiology* **51**, 185–94.

Gillies, R. J., Ugurbil, K., den Hollander, J. A. & Shulman, R. G. (1981). P-31 NMR studies of intracellular pH and phosphate metabolism during cell division cycle of *Saccharomyces cerevisiae*. *Proceedings of the National Academy of Sciences, USA* **78**, 2125–9.

Gilligan, W. & Reese, E. T. (1954). Evidence for multiple components in microbial celulases. *Canadian Journal of Microbiology* **1**, 90–107.

Gilmour, D. (1990). Halotolerant and halophilic microorganisms. In

506   *Literature cited*

Microbiology of Extreme Environments, ed. C. Edwards, pp. 147–77. Open University Press: Milton Keynes.

Ginzburg, M. (1987). *Dunaliella*: a green alga adapted to salt. *Advances in Botanical Research* **14**, 93–183.

Girvin, D. & Thain, J. F. (1987). Growth of and translocation in mycelium of *Neurospora crassa* on a nutrient deficient medium. *Transactions of the British Mycological Society* **88**, 237–46.

Gisi, U., Hemmes, D. E. & Zentmoyer, G. A. (1979). Origin and significance of the discharge vesicle in *Phytophthora*. *Experimental Mycology* **3**, 321–39.

Gisi, U. & Zentmyer, G. A. (1980). Mechanism of zoospore release in *Phytophthora* and *Pythium*. *Experimental Mycology* **4**, 362–77.

Gits, J. & Grenson, M. (1967). Multiplicity of the amino acid permeases in *Saccharomyces cerevisiae*. III. Evidence for a specific methionine-transporting system. *Biochimica et Biophysica Acta* **135**, 507–16.

Glaser, H.-U. & Höfer, M. (1986). Effect of cations on the plasma-membrane-bound ATPase from the yeast *Metschnikowia reukaufii*. *Journal of General Microbiology* **132**, 2615–20.

Gläser, H.-U. & Höfer, M. (1987). Ion-dependent generation of the electrochemical proton gradient $\Delta\mu_{H^+}$ in reconstituted plasma membrane vesicles from the yeast *Metschnikowia reukaufii*. *Biochimica et Biophysica Acta* **905**, 287–94.

Gleason, F. H. (1968). Nutritional comparison in the Leptomitales. *American Journal of Botany* **55**, 1003–10.

Gleason, F. H. & Price, J. S. (1969). Lactic acid fermentation in lower fungi. *Mycologia* **61**, 945–56.

Gleeson, M. A. & Sudbery, P. E. (1988). The methylotropic yeasts. *Yeast* **4**, 1–16.

Glenn, J. K., Akileswaran, L. & Gold, M. H. (1986). Mn(II) oxidation is the principal function of the extracellular Mn-peroxidase from *Phanerochaete chrysosoporium*. *Archives of Biochemistry and Biophysics* **251**, 688–96.

Glenn, J. K. & Gold, M. H. (1985). Purification and characterization of an extracellular Mn(II)-dependent peroxidase from the lignin-degrading basidiomycete *Phanerochaete chrysosporium*. *Archives of Biochemistry and Biophysics* **242**, 329–41.

Gmünda, F. K., Käppeli, O. & Fiechter, A. (1981a). Chemostat studies on the assimilation of hexadecane by the yeast *Candida tropicalis*. I. Influence of the dilution rate and specific growth limitations on the basic culture parameters. *European Journal of Applied Microbiology and Biotechnology* **12**, 129–34.

Gmünda, F. K., Käppeli, O. & Fiechter, Z. (1981b). Chemostat studies on the hexadecane assimilation by the yeast *Candida tropicalis*. II. Regulation of cytochrome and enzyme. *European Journal of Applied Microbiology and Biotechnology* **12**, 135–42.

Goffeau, A., Coddington, A. & Schlesser, A. (1989). Plasma membrane H^+-ATPase: ion and metabolite transport in the yeast *Schizosaccharomyces pombe*. In *Molecular Biology of the Fission Yeast*, ed. A. Nassim, P. Young & B. F. Johnson, pp. 397–429. Academic Press: San Diego, CA.

Goffeau, A. & Dufour, J.-P. (1988). Plasma membrane ATPase from the yeast *Saccharomyces cerevisiae*. *Methods in Enzymology* **157**, 528–33.

Goffeau, A. & Green, M. N. (1990). The H^+-ATPases from fungal plasma

membranes. In *Monovalent Cations in Biological Systems*, ed.
C. A. Pasternak, pp. 155–69. CRC Press: Boca Raton, FL.
Goffeau, A. & Slayman, C. W. (1981). The proton-translocating ATPase of the
fungal plasma membrane. *Biochimica et Biophysica Acta* **639**, 197–223.
Goh, S. H. & Lé John, H. B. (1978). Glucose transport in *Achlya*:
characterization and possible regulatory aspects. *Canadian Journal of
Biochemistry* **56**, 246–56.
Gold, M., Chang, T., Krishnagkura, K., Mayfield, M. & Smith, L. (1980).
Genetics and biochemical studies of *P. chrysosporium* and their relation to
lignin biodegradation. In *Lignin Biodegradation: Microbial, Chemical and
Potential Applications*, vol. II, ed. T. K. Kirk, T. Higuchi & H.-M. Chang,
pp. 65–72. CRC Press: Boca Raton, FL.
Gold, M. H., Glenn, J. K., Mayfield, B. B., Morgan, M. A. & Kutsuki, H.
(1983). Biochemical and genetic studies on lignin degradation by
Phanerochaete chrysosporium. In *Recent Advances in Lignin Biodegradation*,
ed. T. Higuchi, H.-M. Chang & T. K. Kirk, pp. 219–33. Tokyo University
Press: Tokyo.
Gold, M. H., Kuwahara, M., Chiu, A. A. & Glenn, J. K. (1984). Purification
and characterization of an extracellular H_2O_2-requiring diarylpropane
oxygenase from the white-rot basidiomycete *Phanerochaete chrysosporium*.
Archives of Biochemistry and Biophysics **234**, 353–62.
Goldenthal, M. J., Cohen, J. D. & Marmur, J. (1983). Isolation and
characterization of a maltose transport mutant in the yeast *Saccharomyces
cerevisiae*. *Current Genetics* **7**, 195–9.
Golueke, C. G. (1957). Comparative studies of the physiology of *Sapromyces*
and related genera. *Journal of Bacteriology* **74**, 337–44.
Gomes, S. L., Mennucci, L. & Maia, J. C. C. (1979). A calcium dependent
protein activator of mammalian cyclic nucleotide phosphodiesterase from
Blastocladiella emersonii. *FEBS Letters* **99**, 39–42.
Goncharova, I. A., Babitskaya, V. G. & Lobanok, A. G. (1977). Growth and
formation of protein biomass by fungi *Trichoderma* and *Penicillium* on
methanol. In *Microbial Growth on C_1 Compounds*, ed. G. K. Skryabin,
M. V. Ivanov, E. N. Kondratjeva, G. A. Zavarin, Y. A. Trotsenko &
A. I. Nasterov, pp. 187–8. USSR Academy of Sciences: Moscow.
González, A., Grinsbergs, J. & Griva, E. (1986). Biologische Umwandlung von
Holz in Rinderfutter – 'Palo podrido'. *Zentralblatt für Mikrobiologie* **141**,
181–6.
Gooday, G. W. (1977). The enzymology of hyphal growth. In *The Filamentous
Fungi*, vol. 3 *Developmental Mycology*, ed. J. E. Smith & D. R. Berry,
pp. 51–77. Edward Arnold: London.
Gooday, G. W. & Trinci, A. P. J. (1980). Wall structure and biosynthesis in
fungi. *Symposium of the Society for General Microbiology* **30**, 207–51.
Goodman, J. & Rothstein, A. (1957). The active transport of phosphate into the
yeast cell. *Journal of General Physiology* **40**, 915–23.
Goormaghtigh, E., Chadwick, C. & Scarborough, G. A. (1986). Monomers of
Neurospora plasma membrane H^+-ATPase catalyze efficient proton
translocation. *Journal of Biological Chemistry* **261**, 7466–71.
Gorts, C. P. M. (1969). Effect of glucose on the activity and the kinetics of the
maltose-uptake system and of α-glucosidase in *Saccharomyces cerevisiae*.
Biochimica et Biophysica Acta **184**, 299–305.
Gottlieb, D. (1963). Carbohydrate catabolism by fungi. In *The Chemistry and
Biochemistry of Fungi and Yeasts*, Symposium of the International Union

of Pure and Applied Chemistry, pp. 603–9. Butterworths: London.

Gow, N. A. R. (1984). Transhyphal electrical currents in fungi. *Journal of General Microbiology* **130**, 3313–18.

Gow, N. A. R. (1987). Polarity and branching in fungi induced by electrical fields. In *Spatial Organisation in Eukaryotic Microbes*, ed. R. K. Poole & A. P. J. Trinci, pp. 25–41. IRL Press: Oxford.

Gow, N. A. R. (1989a). Circulating ionic currents in micro-organisms. *Advances in Microbial Physiology* **30**, 89–123.

Gow, N. A. R. (1989b). Relationship between growth and the electrical current of fungal hyphae. *Biological Bulletin* **176** (suppl.), 31–5.

Gow, N. A. R., Kropf, D. L. & Harold, F. M. (1984). Growing hyphae of *Achlya bisexualis* generate a longitudinal pH gradient in the surrounding medium. *Journal of General Microbiology* **130**, 2967–74.

Goyal, A., Ghosh, B. & Eveleigh, D. (1991). Characteristics of fungal cellulases. *Bioresource Technology* **36**, 37–50.

Gradmann, D., Hansen, U.-P., Long, W. S., Slayman, C. L. & Warncke, J. (1978). Current–voltage relationships for the plasma membrane and its principal electrogenic pump in *Neurospora crassa*. 1. Steady-state conditions. *Journal of Membrane Biology* **39**, 333–67.

Gradmann, D., Hansen, U.-P. & Slayman, C. L. (1982). Reaction-kinetic analysis of current–voltage relationships for electrogenic pumps in *Neurospora* and *Acetabularia*. *Current Topics in Membranes and Transport* **16**, 257–76.

Gradmann, D. & Slayman, C. L. (1975). Oscillations of an electrogenic pump in the plasma membrane of *Neurospora*. *Journal of Membrane Biology* **23**, 181–212.

Grand, R. J. A., Nairn, A. C. & Perry, S. V. (1980). The preparation of calmodulins from barley (*Hordeum* sp.) and basidiomycete fungi. *Biochemical Journal* **185**, 755–60.

Granlund, H. I., Jennings, D. H. & Thompson, W. (1985). Translocation of solutes along rhizomorphs of *Armillaria mellea*. *Transactions of the British Mycological Society* **84**, 111–19.

Graustein, W. C., Cromack, K. Jr & Sollins, P. (1977). Calcium oxalate: occurrence in soils and effect on nutrient and geochemical cycles. *Science* **198**, 1252–4.

Grayston, J. & Wainwright, M. (1987). Fungal sulphur-oxidation: effect of carbon source and growth stimulation by thiosulphate. *Transactions of the British Mycological Society* **88**, 213–19.

Grayston, S. J. & Wainwright, M. (1988). Sulphur oxidation by soil fungi including some species of mycorrhizal and wood-rotting basidiomycetes. *FEMS Microbiology Letters* **53**, 1–8.

Green, F. III, Larsen, M. J., Winandy, J. E. & Highley, T. L. (1991). Role of oxalic acid in incipient brown-rot decay. *Material und Organismen* **26**, 181–216.

Greenbaum, P., Prodouz, K. N. & Garrett, R. H. (1978). Preparation and some properties of homogeneous *Neurospora crassa* assimilatory NADPH-nitrite reductase. *Biochimica et Biophysica Acta* **526**, 52–64.

Greene, R. V. & Gould, M. J. (1984a). Electrogenic symport of glucose and protons in membrane vesicles of *Phanerochaete chrysosporium*. *Archives of Biochemistry and Biophysics* **228**, 97–104.

Greene, R. V. & Gould, M. J. (1984b). Fatty acyl-coenzyme A oxidase activity and H_2O_2 production in *Phanerochaete chrysosporium* mycelia.

Biochemical and Biophysical Research Communications **118**, 437–43.

Gregory, J. (1989). Fundamentals of flocculation. *CRC Critical Reviews in Environmental Control* **19**, 185–230.

Grenson, M. (1966). Multiplicity of the amino acid permeases in *Saccharomyces cerevisiae*. II. Evidence for a specific lysine-transporting system. *Biochimica et Biophysica Acta* **127**, 339–46.

Grenson, M. (1969). The utilization of exogenous pyrimidines and the recycling of uridine-5'-phosphate derivatives in *Saccharomyces cerevisiae*, as studied by mutants affected by pyrimidine metabolism. *European Journal of Biochemistry* **11**, 249–60.

Grenson, M., Dubois, E., Piotrowska, M., Drillien, R. & Aigle, A. (1974). Ammonium assimilation in *Saccharomyces cerevisiae* as mediated by two glutamate dehydrogenases. Evidence for the *gdhA* locus being a structural gene for the NADP-dependent glutamate dehydrogenase. *Molecular and General Genetics* **128**, 73–85.

Grenson, M., Hou, C. & Crabeel, M. (1970). Multiplicity of the amino acid permeases in *Saccharomyces cerevisiae*. IV. Evidence for a general amino acid permease. *Journal of Bacteriology* **103**, 770–7.

Grenson, M., Mousset, M., Wiame, J. M. & Bechet, J. (1966). Multiplicity of the amino acid permeases in *Saccharomyces cerevisiae*. I. Evidence for a specific arginine-transporting system. *Biochimica et Biophysica Acta* **127**, 325–38.

Greth, M. L., Chevallier, M. R. & Lacroute, F. (1977). Ureidosuccinic acid permeation in *Saccharomyces cerevisiae*. *Biochimica et Biophysica Acta* **465**, 138–51.

Griffin, D. H. (1981). *Fungal Physiology*. John Wiley : New York.

Griffin, D. M. (1977). Water potential and wood-decay fungi. *Annual Review of Phytopathology* **15**, 319–29.

Griffin, D. M. (1981). Water and microbial stress. *Advances in Microbial Ecology* **5**, 91–136.

Griffith, G. S. & Boddy, L. (1991). Fungal decomposition of attached angiosperm twigs. III. Effect of water potential and temperature on fungal growth, survival and decay of wood. *New Phytologist* **117**, 259–69.

Griffith, J. M., Atkins, L. A. & Grant, B. R. (1989). Properties of the phosphate and phosphite transport systems of *Phytophthora palmivora*. *Archives of Microbiology* **152**, 430–6.

Griffith, J. M., Iser, J. R. & Grant, B. R. (1988). Calcium control of differentiation in *Phytophthora palmivora*. *Archives of Microbiology* **149**, 565–71.

Grill, E., Winnacker, E.-L. & Zenk, M. H. (1985). Phytochelatins: the principal heavy-metal complexing peptides of higher plants. *Science* **230**, 674–6.

Grill, E., Winnaker, E.-L. & Zenk, M. H. (1986). Synthesis of seven different homologous phytochelatins in metal-exposed *Schizosaccharomyces pombe* cells. *FEBS Letters* **197**, 115–20.

Gritzali, M. & Brown, R. D. Jr (1979). The cellulase system of *Trichoderma*. Relationship between purified extracellular enzymes from induced or cellulose grown cells. *Advances in Chemistry Series* **18**, 237–60.

Grootjen, D. R. J., Jansen, M. L., van der Lans, R. G. J. M. & Luyben, K. C. A. M. (1991a). Reactors in series for the complete conversion of glucose xylose mixtures by *Pichia stipitis* and *Saccharomyces cerevisiae*. *Enzyme and Microbial Technology* **13**, 828–33.

Grootjen, D. R. J., Meijlink, L. H. H. M., Vleesenbeck, R., van der Lans,

R. G. J. M. & Luybn, K. C. A. M. (1991*b*). Cofermentation of glucose and xylose with immobilised *Pichia stipitis* in combination with *Saccharomyces cerevisiae*. *Enzyme and Microbial Technology* **13**, 530–6.

Gross, S. R., Gafford, R. D. & Tatum, E. L. (1956). The metabolism of protocatechuic acid by *Neurospora*. *Journal of Biological Chemistry* **219**, 781–96.

Guérin, B. (1991). Mitochondria. In *The Yeasts*, vol. 4 *Yeast Organelles*, ed. A. H. Rose & J. S. Harrison, pp. 541–600. Academic Press: London.

Guern, J., Felle, H., Mathieu, Y. & Kurkdjian, A. (1991). Regulations of intracellular pH in plant cells. *International Review of Cytology* **127**, 111–73.

Guerrero, M. G. & Gutierez, M. (1977). Purification and properties of the NAD(P)H:nitrate reductase of the yeast *Rhodotorula glutinis*. *Biochimica et Biophysica Acta* **482**, 272–85.

Guijarro, J. M. & Lagunas, R. (1984). *Saccharomyces cerevisiae* does not accumulate ethanol against a concentration gradient. *Journal of Bacteriology* **160**, 874–8.

Gustin, M. C., Martinac, B., Saimi, Y., Culbertson, M. R. & Kung, C. (1986). Ion channels in yeast. *Science* **233**, 1195–6.

Habison, A., Kubicek, C. P. & Röhr, M. (1979). Phosphofructokinase as a regulatory enzyme in citric acid producing *Aspergillus niger*. *FEMS Microbiology Letters* **5**, 39–42.

Hackette, S. L., Skye, G. E., Burton, C. & Segel, I. H. (1970). Characterization of an ammonium transport system in filamentous fungi with methylammonium-^{14}C as the substrate. *Journal of Biological Chemistry* **245**, 4241–50.

Hager, K. M., Mandala, S. M., Davenport, J. W., Speicher, D. W., Benz, E. J. Jr & Slayman, C. W. (1986). Amino acid sequence of the plasma membrane ATPase of *Neurospora crassa*: deduction from genomic and cDNA sequences. *Proceedings of the National Academy of Sciences, USA* **87**, 7693–7.

Halder, K. & Trojanowski, J. (1980). A comparison of the degradation of ^{14}C-labelled DHP and corn stalk lignins by micro- and macrofungi and bacteria. In *Lignin Biodegradation: Microbiology, Chemistry and Potential Applications*, vol. I, ed. T. K. Kirk, T. Higuchi & H.-M. Chang, pp. 111–34. CRC Press: Boca Raton, FL.

Hakansson, U., Fågerstam, L., Pettersson, G. & Andersson, L. (1978). Purification and characterization of a low molecular weight 1,4-β-glucanohydrolase from cellulolytic fungus *Trichoderma viride* QM9414. *Biochimica et Biophysica Acta* **524**, 385–92.

Halachmi, D. & Eilam, Y. (1989). Cytosolic and vacuolar Ca^{2+} concentrations in yeast cells measured with the Ca^{2+}-sensitive fluorescent dye indo-1. *FEBS Letters* **256**, 55–61.

Halliwell, G. (1965). Catalytic decomposition of cellulose under biological conditions. *Biochemical Journal* **95**, 35–40.

Halstead, R. L. & McKercher, R. B. (1975). Biochemistry and cycling of phosphorus. In *Soil Biochemistry*, vol. 4, ed. E. A. Paul & A. D. McLaren, pp. 31–63. Marcel Dekker: New York.

Hankin, L. & Anagnostakis, S. L. (1975). The use of solid media for detection of enzyme production by fungi. *Mycologia* **67**, 597–607.

Hankinson, O. & Cove, D. J. (1974). Regulation of the pentose phosphate pathway in the fungus *Aspergillus nidulans*. The effect of growth with nitrate. *Journal of Biological Chemistry* **249**, 2344–53.

Hankinson, O. & Cove, D. J. (1975). Regulation of mannitol-1-phosphate dehydrogenase in *Aspergillus nidulans*. *Canadian Journal of Microbiology* **21**, 99–101.

Hanson, B. A. (1980). Inositol-limited growth, repair and translocation in an inositol-requiring mutant of *Neurospora crassa*. *Journal of Bacteriology* **143**, 18–26.

Hanson, M. A. & Marzluf, G. A. (1973). Regulation of a sulfur-controlled protease in *Neurospora crassa*. *Journal of Bacteriology* **116**, 785–9.

Harder, W. & Veenhuis, M. (1989). Metabolism of one-carbon compounds. In *The Yeasts*, vol. 3 *Metabolism and Physiology of Yeasts*, ed. A. H. Rose & J. S. Harrison, pp. 289–316. Academic Press: London.

Hardie, K. (1985). The effect of removal of extraradical hyphae on water uptake by vesicular–arbuscular mycorrhizal plants. *New Phytologist* **101**, 677–84.

Harkin, J. N. (1976). Lignin – a natural polymeric product of phenol oxidation. In *Oxidative Coupling of Phenols*, ed. A. R. Battersby & W. I. Taylor, pp. 243–321. Marcel Dekker: New York.

Harley, J. L. (1959). *The Biology of Mycorrhiza*. Leonard Hill: London.

Harley, J. L. (1978). Ectomycorrhizas as nutrient absorbing organs. *Proceedings of the Royal Society* B **203**, 1–21.

Harley, J. L. & Brierley, J. K. (1954). The uptake of phosphate by excised mycorrhizal roots of the beech. VI. Active transport of phosphorus from the fungal sheath into the host tissue. *New Phytologist* **53**, 240–52.

Harley, J. L. & Jennings, D. H. (1958). The effects of sugars on the respiratory response of beech mycorrhizas to salts. *Proceedings of the Royal Society* B **148**, 403–18.

Harley, J. L. & Loughman, B. C. (1963). The uptake of phosphate by excised mycorrhizal roots of the beech. IX. The nature of the phosphate compounds passing into the host. *New Phytologist* **62**, 350–9.

Harley, J. L. & McCready, C. C. (1950). The uptake of phosphate by excised mycorrhizal roots of the beech. *New Phytologist* **49**, 388–97.

Harley, J. L. & McCready, C. C. (1952a). The uptake of phosphate by excised mycorrhizal roots of the beech. II. Distribution of phosphorus between host and fungus. *New Phytologist* **51**, 56–64.

Harley, J. L. & McCready, C. C. (1952b). The uptake of phosphate by excised mycorrhizal roots of the beech. III. The effect of the fungal sheath on the availability of phosphate to the core. *New Phytologist* **51**, 342–8.

Harley, J. L. & McCready, C. C. (1981). Phosphate accumulation in *Fagus* mycorrhizas. *New Phytologist* **89**, 75–80.

Harley, J. L. & Smith, D. C. (1956). Sugar absorption and surface carbohydrase activity of *Peltigera polydactyla* (Neck.) Hoffm. *Annals of Botany* N.S. **20**, 513–43.

Harley, J. L. & Smith, S. E. (1983). *Mycorrhizal Symbiosis*. Academic Press: London.

Haro, R., Garciadeblas, B. & Rodríguez-Navarro, A. (1991). A novel P-type ATPase from yeast involved in sodium transport. *FEBS Letters* **291**, 189–91.

Harold, F. M. (1960). Accumulation of inorganic polyphosphate in mutants of *Neurospora crassa*. *Biochimica et Biophysica Acta* **45**, 172–88.

Harold, F. M. (1962a). Binding of inorganic polyphosphate to the cell wall of *Neurospora crassa*. *Biochimica et Biophysica Acta* **57**, 59–66.

Harold, F. M. (1962b). Depletion and replenishment of the inorganic polyphosphate pool in *Neurospora crassa. Journal of Bacteriology* **83**, 1047–57.

Harold, F. M. (1966). Inorganic polyphosphates in biology: structure, metabolism and function. *Bacteriological Reviews* **30**, 772–94.

Harold, F. M., Caldwell, J. H. & Schreurs, W. J. A. (1987). Endogenous electric currents and polarized growth of fungal hyphae. In *Spatial Organisation in Eukaryotic Microbes*, ed. R. K. Poole, & A. P. J. Trinci, pp. 11–23. IRL Press: Oxford.

Harold, F. M., Kropf, D. L. & Caldwell, J. H. (1985a). Why do fungi drive electric currents through themselves? *Experimental Mycology* **9**, 183–6.

Harold, F. M. & Miller, A. (1961). Intracellular localization of inorganic polyphosphate in *Neurospora crassa Biochimica et Biophysica Acta* **50**, 261–70.

Harold, F. M., Schreurs, W. J., Harold, R. L. & Caldwell, J. H. (1985b). Electrophysiology of fungal hyphae. *Microbiological Sciences* **2**, 363–8.

Harold, R. L. & Harold, F. M. (1980). Oriented growth of *Blastocladiella emersonii* in gradients of ionophores and inhibitors. *Journal of Bacteriology* **144**, 1159–67.

Harold, R. L. & Harold, F. M. (1986). Ionophores and cytochalasins modulate branching in *Achlya bisexualis. Journal of General Microbiology* **132**, 213–19.

Harris, G. & Thompson, C. C. (1960). Uptake of nutrients by yeasts. II. Maltotriose permease and the utilization of maltotriose by yeasts. *Journal of the Institute of Brewing* **66**, 293–7.

Harris, G. & Thompson, C. C. (1961). The uptake of nutrients by yeasts. III. The maltose permease system of a brewing yeast. *Biochimica et Biophysica Acta* **52**, 176–83.

Harris, J. O. (1959). Possible mechanism of yeast flocculation. *Journal of the Institute of Brewing* **65**, 5–6.

Hartley, B. S., Broda, P. M. A. & Senior, P. J. (1987). *Technology in the 1990s: Utilization of Lignocellulose Wastes.* The Royal Society: London.

Harvey, A. E. & Grasham, J. L. (1974). Axenic culture of the mononucleate stage of *Cronatium ribicola. Phytopathology* **64**, 1028–35.

Harvey, D. M. R., Hall, J. L., Flowers, T. J. & Kent, B. (1981). Quantitative ion localization within *Suaeda maritima* leaf mesophyll cells. *Planta* **151**, 550–60.

Harvey, P. J., Gilardi, G. F. & Palmer, J. M. (1989a). Importance of charge transfer reactions in lignin degradation. In *Enzyme Systems for Lignocellulose Degradation*, ed. M. P. Coughlan, pp. 111–20. Elsevier Applied Science: London.

Harvey, P. J., Palmer, J. M., Schoemaker, H. E., Dekker, H. L. & Wever, R. (1989b). Pre-steady-state kinetic study on the formations of compounds I and II of ligninase. *Biochimica et Biophysica Acta* **994**, 59–63.

Harvey, P. J., Schoemaker, H. E., Bowen, R. M. & Palmer, J. M. (1985). Single-electron transfer processes and the reaction mechanism of enzymic degradation of lignin. *FEBS Letters* **183**, 13–16.

Harvey, P. J., Schoemaker, H. E. & Palmer, J. M. (1986). Veratryl alcohol as a mediator and the role of radical cations in lignin biodegradation by *Phanerochaete chrysosporium. FEBS Letters* **195**, 242–6.

Harvey, P. J., Schoemaker, H. E. & Palmer, J. M. (1987). Lignin degradation by white rot fungi. *Plant, Cell and Environment* **10**, 709–14.

Hashimoto, K. (1970). Oxidation of phenols by yeast. I. A new oxidation product from *p*-cresol by an isolated strain of yeast. *Journal of General and Applied Microbiology* **16**, 1–13.

Hashimoto, K. (1973). Oxidation of phenols by yeast. II. Oxidation of cresols by *Candida utilis*. *Journal of General and Applied Microbiology* **19**, 171–87.

Håskovec, C. & Kotyk, A. (1973). Transport systems for acyclic polyols and monosaccharides in *Torulopsis candida*. *Folia Microbiologica* **18**, 118–24.

Hasunuma, K. & Ishikawa, T. (1972). Properties of two nuclease genes in *Neurospora crassa*. *Genetics* **70**, 371–84.

Hasunuma, K. & Ishikawa, T. (1977). Control of the production and partial characterization of repressible extracellular 5'-nucleotidase and alkaline phosphatase in *Neurospora crassa*. *Biochimica et Biophysica Acta* **480**, 178–93.

Hata, K. (1966). Investigations on lignin and lignification. XXXIII. Studies on lignins isolated from spruce decayed by *Poria subacida* B11. *Holzforschung* **20**, 142–7.

Hatch, A. B. (1937). The physical basis of mycotrophy in the genus *Pinus*. *Black Rock Forest Bulletin* **6**, 1–168.

Hauer, R. & Höfer, M. (1978). Evidence for interactions between the energy-dependent transport of sugars and the membrane potential in the yeast *Rhodotorula gracilis* (*Rhodosporidium toruloides*). *Journal of Membrane Biology* **43**, 335–49.

Hauer, R. & Höfer, M. (1982). Variable H^+/substrate stoichiometries in *Rhodotorula gracilis* are caused by a pH-dependent protonation of the carrier(s). *Biochemical Journal* **208**, 469–64.

Hausinger, R. P. (1987). Nickel utilization by microorganisms. *Microbiological Reviews* **51**, 22–42.

Hautera, P. & Lövgren, T. (1975). α-Glucosidase, α-glucosidase permease, maltose fermentation and the leavening ability of baker's yeast. *Journal of the Institute of Brewing* **81**, 309–13.

Hawker, L. E. (1950). *Physiology of Fungi*. University of London Press: London.

Hayaishi, O., Shimazono, H., Katagiri, M. & Saito, Y. (1956). Enzymatic formation of oxalate and acetate from oxaloacetate. *Journal of the American Chemical Society* **78**, 5126–7.

Hayashi, Y., Nakagawa, C. W. & Murasugi, A. (1986). Unique properties of Cd-binding peptides induced in fission yeast *Schizosaccharomyces pombe*. *Environmental Health Perspectives* **65**, 13–20.

Hayashi, Y., Nakagawa, C., Uyakul, D., Imai, K., Isobe, M. & Goto, T. (1987). The change of cadystin components in Cd-binding peptides from the fission yeast during their induction by cadmium. *Biochemistry and Cell Biology* **66**, 288–95.

Hayashida, S., Feng, D. D. & Hongo, M. (1974). Function of the high concentration alcohol-producing factor. *Agricultural and Biological Chemistry* **38**, 2001–6.

Hayashida, S., Feng, D. D., Ohta, K., Chaitiumvong, S. & Hongo, M. (1976). Composition and role of *Aspergillus oryzae*-proteolipid as a high concentration alcohol-producing factor. *Agricultural and Biological Chemistry* **40**, 73–8.

Hayashida, S. & Ohta, K. (1978). Cell structure of yeasts grown anaerobically in *Aspergillus*-proteolipid supplemented media. *Agricultural and Biological Chemistry* **42**, 1139–45.

Hayashida, S. & Ohta, K. (1980). Effects of phosphatidylcholine or ergosterol oleate on physiological properties of *Saccharomyces cerevisiae*. *Agricultural and Biological Chemistry* **44**, 2561–7.

Hayman, D. S. & Mosse, B. (1972). Plant growth responses to vesicular–arbuscular mycorrhizas. III. Increased uptake of labile P from the soil. *New Phytologist* **71**, 41–7.

Hazen, K. C. (1990). Cell surface hydrophobicity of medically important fungi including *Candida* species. In *Microbial Cell Surface Hydrophobicity*, ed. R. J. Doyle & M. Rosenberg, pp. 249–95. American Society for Microbiology: Washington, DC.

Heath, E. C., Nasser, O. & Koffler, H. (1956). Biochemistry of filamentous fungi. III. Alternative routes for the breakdown of glucose by *Fusarium lini*. *Archives of Biochemistry and Biophysics* **64**, 80–7.

Heath, I. B. (ed.) (1990). *Tip Growth on Plant and Fungal Cells*. Academic Press: Orlando, FL.

Heerde, E. & Radler, F. (1978). Metabolism of the anerobic formatives of succinic acid by *Saccharomyces cerevisiae*. *Archives of Microbiology* **117**, 269–76.

Heinrich, P. A. & Patrick, J. W. (1986). Phosphorus acquisition in the soil–root system of *Eucalyptus pilularis* Smith seedlings. II. The effect of ectomycorrhizas in seedling phosphorus and dry weight acquisition. *Australian Journal of Botany* **34**, 445–54.

Held, A. A., Emerson, R., Fuller, M. S. & Gleason, F. H. (1969). *Blastocladiella* and *Aqualinderella*: fermentative water molds with high carbon dioxide optima. *Science* **165**, 706–9.

Heldwein, R., Tromballa, H. W. & Broda, E. (1977). Aufnahme von Cobolt, Blei und Cadmium durch Bäckerhefer. *Zeitschrift für Algemeine Mikrobiologie* **17**, 299–309.

Henderson, M. E. K. (1961). The metabolism of aromatic compounds by soil fungi. *Journal of General Microbiology* **26**, 155–65.

Hennessey, J. P. Jr & Scarborough, G. A. (1990). Direct evidence for the cytoplasmic location of the NH_2- and COOH-terminal ends of the *Neurospora crassa* plasma membrane H^+-ATPase. *Journal of Biological Chemistry* **265**, 532–7.

Henrissat, B., Driguez, H., Veit, C. & Schulein, M. (1985). Syngergism of cellulase formation in *Trichoderma reesei* in degradation of cellulose. *Bio/Technology* **3**, 722–26.

Henrissat, B. & Mornon, J.-P. (1990). Comparison of *Trichoderma* cellulases with other glucanases. In Trichoderma reesei *Cellulases: Biochemistry, Genetics, Physiology and Applications*, ed. C. P. Kubicek, D. E. Eveleigh, H. Esterbauer, W. Steiner & E. M. Kubicek-Pranz, pp. 12–19. Royal Society of Chemistry: Cambridge.

Henry, S. A. (1982). Membrane lipids of yeast: biochemical and genetic studies. In *The Molecular Biology of the Yeast* Saccharomyces: *Metabolism and Gene Expression*, ed. J. N. Strathern, E. W. Jones & J. R. Broach, pp. 101–58. Cold Spring Harbor Laboratory Press: Cold Spring Harbor, NY.

Henschke, P. A. & Rose, A. H. (1991). Plasma membranes. In *The Yeasts*, vol. 4 *Yeast Organelles*, ed. A. H. Rose & J. S. Harrison, pp. 297–435. Academic Press: London.

Heppel, L. A. (1969). The effect of osmotic shock on the release of bacterial proteins and active transport. *Journal of General Physiology* **54**, 953–1093.

Herbert, D., Elsworth, R. & Telling, R. C. (1956). The continuous culture of bacteria: a theoretical and experimental study. *Journal of General Microbiology* **14**, 601–2.

Heredia, C. F., Sols, A. & de la Fuente, G. (1968). Specificity of the constitutive hexose transport system in yeast. *European Journal of Biochemistry* **5**, 321–9.

Herrick, J. A. (1940). The growth of *Stereum gausapatum* Fries in relation to temperature and acidity. *Ohio Journal of Science* **40**, 123–9.

Herzberger, E. & Radler, F. (1988). How hexoses and inhibitors influence the malate transport system in *Zygosaccharomyces bailii*. *Archives of Microbiology* **150**, 37–41.

Hesseltine, C. W., Pidacks, C., Whitehill, A. R., Bohonos, N., Hutchings, B. L. & Williams, J. H. (1952). Coprogen, a new growth factor for coprophilous fungi. *Journal of the American Chemical Society* **74**, 1362.

Hesseltine, C. W., Whitehill, A. R., Pidacks, C., ten Hagen, M., Bohonos, N., Hutchings, B. L. & Williams, J. H. (1953). Coprogen: a new growth factor present in dung, required by *Pilobolus* species. *Mycologia* **45**, 7–19.

Hetrick, B. A. D. (1989). Acquisition of phosphorus by VA mycorrhizal fungi and the growth responses of their hosts. In *Nitrogen, Phosphorus and Sulphur Utilization by Fungi*, ed. L. Boddy, R. Marchant & D. J. Read, pp. 205–66. Cambridge University Press: Cambridge.

Highley, T. L. (1977). Requirements for cellulose degradation by a brown rot fungus. *Material und Organismen* **12**, 25–36.

Highley, T. L. (1981). Catalase-aminotriazole assay, an invalid method for measurement of hydrogen peroxide production by wood decay fungi. *Applied and Environmental Microbiology* **42**, 925–7.

Highley, T. L. (1987). Effect of carbohydrate and nitrogen on hydrogen peroxide formation by wood decay fungi in solid medium. *FEMS Microbiology Letters* **48**, 373–7.

Highley, T. L. & Kirk, T. K. (1979). Mechanism of wood decay and the unique features of heartrots. *Phytopathology* **69**, 1151–7.

Highley, T. L. & Wolter, K. E. (1982). Propeties of a carbohydrate degrading enzyme complex from the brown rot fungus *Poria placenta*. *Material und Organismen* **17**, 127–34.

Higuchi, T., Kawamura, I. & Kawamura, H. (1955). Properties of the lignin in decayed wood. *Journal of the Japan Forestry Society* **37**, 298–302.

Hinnebusch, A. G. (1986). The general control of amino acid biosynthesis genes in the yeast *Saccharomyces cerevisiae*. *Critical Reviews in Biochemistry* **21**, 277–317.

Hinnebusch, A. G. (1988). Mechanism of gene regulation in the general control of amino acid biosynthesis in *Saccharomyces cerevisiae*. *Microbiological Reviews* **52**, 248–73.

Hinnebusch, A. G. (1990). Transcriptional and translational regulation of gene expression in the general control of amino acid biosynthesis in *Saccharomyces cerevisiae*. *Progress in Nucleic Acid Research and Molecular Biology* **38**, 195–240.

Hintikka, V. & Naykki, O. (1967). Notes on the effects of the fungus *Hydrellum ferrugineum* (Fr.) Karst on forest soil and vegetation. *Communicationes, Institute Forestalis Fenniae* **62**, 1–12.

Hinze, H. & Holzer, H. (1985). Effect of sulfite or nitrate on the ATP content and the carbohydrate metabolism of yeast. *Zeitschrift für Lebensmittel-Unterschung und -Forschung* **181**, 87–91.

Hipkin, C. R. (1989). Nitrate assimilation in yeast. In *Molecular and Genetic*

Aspects of Nitrate Assimilation, ed. J. L. Wray & J. R. Kinghorn, pp. 51–68. Oxford Science Publications: Oxford.

Hipkin, C. R., Ali, A. H. & Cannons, A. (1986). Structure and properties of assimilatory nitrate reductase from the yeast *Candida nitrophila*. *Journal of General Microbiology* **132**, 1997–2003.

Hipkin, C. R., Flynn, K. J., Marjot, E., Hamoudi, Z. S. & Cannons, A. C. (1990). Ammonium assimilation by the nitrate-utilizing yeast *Candida nitrophila*. *New Phytologist* **114**, 429–34.

Hiraga, K., Tahara, H., Taguchi, N., Tsuchiya, E., Fukui, S. & Miyakawa, T. (1991). Inhibition of membrane Ca^{2+}-ATPase of *Saccharomyces cerevisiae* by mating pheromone α-factor *in vitro*. *Journal of General Microbiology* **137**, 1–4.

Hirata, R., Ohsumi, Y. & Anraku, Y. (1989). Functional molecular masses of vacuolar membrane H^+-ATPase from *Saccharomyces cerevisiae* as studied by radiation inactivation analysis. *FEBS Letters* **244**, 397–401.

Hirata, R., Ohsumi, Y., Nakano, A., Kawasaki, H., Suzuki, K. & Anraku, Y. (1990). Molecular structure of a gene, *VMA1*, encoding the catalytic subunit of H^+-translocating adenosine triphosphatase from vacuolar membranes of *Saccharomyces cerevisiae*. *Journal of Biological Chemistry* **265**, 6726–33.

Hiroi, T. & Eriksson, K.-E. (1976). Microbiological degradation of lignin. Part I. Influence of cellulose on the degradation of lignins by the white-rot fungus *Pleurotus ostreatus*. *Svensk Papperstidining* **79**, 157–61.

Ho, I. & Tappe, J. M. (1980). Nitrate reductase activity of non-mycorrhizal Douglas-fir rootlets and some associated mycorrhizal fungi. *Plant and Soil* **54**, 395–8.

Ho, I. & Zak, B. (1979). Acid phosphatase activity of six mycorrhizal fungi. *Canadian Journal of Botany* **57**, 1203–5.

Hobot, J. A. & Jennings, D. H. (1981). Growth of *Debaryomyces hansenii* and *Saccharomyces cerevisiae* in relation to pH and salinity. *Experimental Mycology* **5**, 217–28.

Hockertz, S., Schmid, J. & Auling, G. (1987). A specific transport system for manganese in the filamentous fungus *Aspergillus niger*. *Journal of General Microbiology* **133**, 3513–19.

Hocking, A. D. (1993). Response of xerophilic fungi to changes in water activity. In *Stress Tolerance of Fungi*, ed. D. H. Jennings, pp. 233–56. Marcel Dekker: New York.

Hodgkinson, A. (1977). *Oxalic Acid in Biology and Medicine*. Academic Press: New York.

Höfeler, H., Jensen, D., Pike, M. M., Delayre, J. L., Cirillo, V. P., Springer, C. S. Jr, Fossel, E. T. & Balaschi, J. A. (1987). Sodium transport and phosphorus metabolism in sodium loaded yeast: simultaneous observations with sodium-23 and phosphorus-31 NMR spectroscopy *in vivo*. *Biochemistry* **26**, 4953–62.

Höfer, M. (1970). Mobile membrane carrier for monosaccharide transport in *Rhodotorula gracilis*. *Journal of Membrane Biology* **3**, 73–82.

Höfer, M., Betz, A. & Becker, J.-U. (1970). Metabolism of the obligately aerobic yeast *Rhodotorula gracilis*. I. Changes in metabolite concentrations following D-glucose and D-xylose addition to the cell suspension. *Archiv für Mikrobiologie* **71**, 99–110.

Höfer, M. & Dahle, P. (1972). Glucose repression of inducible enzyme synthesis in the yeast *Rhodotorula gracilis*. Effect of cell-membrane transport.

European Journal of Biochemistry **29**, 326–32.

Höfer, M., Huh, H. & Künemund, A. (1983). Membrane potential and cation permeability. A study with a nystatin-resistant mutant of *Rodotorula gracilis*. *Biochimica et Biophysica Acta* **735**, 211–14.

Höfer, M. & Kotyk, A. (1968). Tight coupling of monosaccharide transport and metabolism in *Rhodotorula gracilis*. *Folia Microbiologica* **13**, 197–204.

Höfer, M. & Misra, P. C. (1978). Evidence for a proton/sugar symport in the yeast *Rhodotorula gracilis (glutinis)*. *Biochemical Journal* **172**, 15–22.

Höfer, M. & Nassar, F. R. (1987). Aerobic and anaerobic uptake of sugars in *Schizosaccharomyces pombe*. *Journal of General Microbiology* **133**, 2163–73.

Höfer, M., Nicolay, K. & Robillard, G. (1985). The electrochemical H⁺ gradient in the yeast *Rhodotorula gracilis*. *Journal of Bioenergetics and Biomembranes* **17**, 175–82.

Höfer, M., Thiele, O. W., Huh, H., Hunneman, D. H. & Mracek, M. (1982). A nystatin-resistant mutant of *Rhodotorula gracilis*. Transport properties and sterol content. *Archives of Microbiology* **132**, 313–16.

Hoffman, W. (1985). Molecular characterization of the *CAN1* locus in *Saccharomyces cerevisiae*. A transmembrane protein without N-terminal hydrophobic signal sequence. *Journal of Biological Chemistry* **260**, 11831–7.

Hoffman-Ostenhof, O. & Weigert, W. (1952). Uber die mögliche Funktion der polymeren Metaphosphate als Speicher energiereichen Phosphats der Hefe. *Naturwissenschaften* **39**, 303–4.

Hofmeister, F. (1888). Zur Lehre von der Wirkung der Salze. II. Mittheilung. *Archiv für Experimentale Pathologie und Pharmakologie* **24**, 247–60.

Hofstee, B. H. J. (1952). On the evaluation of the constants V_m and K_m in the enzyme reactions. *Science* **116**, 329–31.

Hohl, H. R. (1983). Nutrition of *Phytophthora*. In Phytophthora: *Its Biology, Taxonomy, Ecology and Pathology*, ed. D. C. Erwin, S. Bartnicki-Garcia & P. H. Tsao, pp. 41–54. American Phytopathological Society: St Paul, MN.

Hohn, T. M., Lovett, J. S. & Bracker, C. E. (1984). Characterization of the major proteins in gamma particles, cytoplasmic organelles in *Blastocladiella emersonii* zoospores. *Journal of Bacteriology* **158**, 253–63.

Holder, A. A., Wootton, J. C., Baron, A. J., Chambers, G. J. K. & Fincham, J. R. S. (1975). The amino acid sequence of *Neurospora* NADP-specific glutamate dehydrogenase. *Biochemical Journal* **149**, 757–73.

Holliday, R. (1974). *Ustilago maydis*. In *Handbook of Genetics*, ed. R. C. King, pp. 575–95. Plenum: London.

Holligan, P. M. (1970). The analysis of soluble carbohydrates in plant tissues by gas chromatography with particular reference to the metabolism of sugar alcohols in the fungus *Dendryphiella salina*. Ph.D. thesis, University of Leeds.

Holligan, P. M. & Jennings, D. H. (1972a). Carbohydrate metabolism in the fungus *Dendryphiella salina*. I. Changes in the levels of soluble carbohydrates during growth. *New Phytologist* **71**, 569–82.

Holligan, P. M. & Jennings, D. H. (1972b). Carbohydrate metabolism in the fungus *Dendryphiella salina*. II. The influence of different carbon and nitrogen sources on the accumulation of mannitol and arabitol. *New Phytologist* **71**, 583–94.

Holligan, P. M. & Jennings, D. H. (1972c). Carbohydrate metabolism in the fungus *Dendryphiella salina*. III. The effect of the nitrogen source on the metabolism of [1-¹⁴C]- and [6-¹⁴C]-glucose. *New Phytologist* **71**, 1119–33.

Holligan, P. M. & Jennings, D. H. (1972d). Unexpected labelling patterns from radioactive sugars fed to plants containing mannitol. *Phytochemistry* **11**, 3347–51.

Holzbaur, E. L. F., Andrawis, A. & Tien, M. (1991). Molecular biology of lignin peroxidases from *Phanerochaete chrysosporium*. In *Molecular Industrial Mycology: Systems and Applications for Filamentous Fungi*, ed. S. Leong & R. M. Berka, pp. 197–223. Marcel Dekker: New York.

Holzer, H. (1953). Zur Penetration von Orthophosphat in lebende Hefezellen. *Biochemische Zeitschrift* **324**, 144–55.

Holzer, H. (1961). Regulation of carbohydrate metabolism by enzyme competition. *Cold Spring Harbor Symposium on Quantitative Biology* **26**, 277–88.

Holzer, H. (1976). Catabolite inactivation in yeast. *Trends in Biochemical Sciences* **1**, 178–81.

Homann, M. J., Poole, M. A., Gaynor, P. M. Ho, C.-T. & Carman, G. M. (1987). Effect of growth phase on phospholipid biosynthesis in *Saccharomyces cerevisiae*. *Journal of Bacteriology* **169**, 533–9.

Hope, A. B. & Walker, N. A. (1975). *The Physiology of Giant Algal Cells*. Cambridge University Press, Cambridge.

Hopkins, P., Chevallier, N. R., Jund, R. & Eddy, A. A. (1988). Use of plasmid vectors to show that the uracil and cytosine permeases of the yeast *Saccharomyces cerevisiae* are electrogenic proton symports. *FEMS Microbiology Letters* **49**, 173–8.

Horák, J. & Kotyk, A. (1986). Energetics of L-proline uptake by *Saccharomyces cerevisiae*. *Biochimica et Biophysica Acta* **857**, 173–9.

Horák, J. & Říhová, L. (1982). L-Proline transport in *Saccharomyces cerevisiae*. *Biochimica et Biophysica Acta* **691**, 144–50.

Horák, J., Říhová, L. & Kotyk, A. (1981). Energization of sulfate transport in yeast. *Biochimica et Biophysica Acta* **649**, 436–40.

Horan, D. P. & Chilvers, G. A. (1990). Chemotropism – the key to ectomycorrhizal formation? *New Phytologist* **116**, 297–301.

Horn, B. W. (1989). Requirement for potassium and pH shift in host-mediated sporangiospore extrusion from trichospores of *Smittium culisetae* and other *Smittium* species. *Mycological Research* **93**, 303–13.

Horn, B. W. (1990). Physiological changes associated with sporangiospore extrusion from trichospores of *Smittium culisetae*. *Experimental Mycology* **14**, 113–23.

Horner, H. T., Tiffany, L. H. & Cody, A. M. (1983). Formation of calcium oxalate crystals associated with apothecia of the discomycete *Dasyscypha capitata*. *Mycologia* **75**, 423–35.

Horner, R. D. (1983). Purification and comparison of *nit-1* and wild-type NADPH-nitrate reductases of *Neurospora crassa*. *Biochimica et Biophysica Acta* **744**, 7–15.

Horwitz, B. A., Weisenseel, M. H., Dorn, A. & Gressel, J. (1984). Electric currents around growing *Trichoderma* hyphae before and after induction of conidiation. *Plant Physiology* **74**, 912–16.

Horwitz, L. (1958). Some simplified mathematical treatments of translocation in plants. *Plant Physiolgy* **33**, 81–93.

Hosaka, K. & Yamashita, S. (1980). Choline transport in *Saccharomyces cerevisiae*. *Journal of Bacteriology* **143**, 176–81.

Hosono, K. (1991). The release of proteins and UV-absorbing materials from salt-tolerant yeast *Zygosaccharomyces rouxii* by osmotic shock. *Journal of Fermentation and Bioengineering* **72**, 445–9.

Hosono, K. (1992). Effect of salt stress on lipid symposition and membrane fluidity of the salt-tolerant yeast *Zygosaccharomyces rouxii. Journal of General Microbiology* **138**, 91–6.

Hossack, J. & Rose, A. H. (1976). Fragility of plasma membranes in *Saccharomyces cerevisiae* enriched with different sterols. *Journal of Bacteriology* **127**, 67–75.

Hossack, J. A., Sharpe, V. J. & Rose, A. H. (1977). Stability of the plasma membrane of *Saccharomyces cerevisiae* enriched with phosphatidylcholine or phosphatidylethanolamine. *Journal of Bacteriology* **129**, 1144–7.

Hottiger, T., Schmutz, P. & Wiemken, A. (1987). Heat-induced accumulation and futile cycling of trehalose in *Saccharomyces cerevisiae. Journal of Bacteriology* **169**, 5518–22.

Howes, N. K. & Scott, K. J. (1972). Sulfur nutrition of *Puccinia graminis* f. sp. *tritici* in axenic culture. *Canadian Journal of Botany* **50**, 1165–70.

Howes, N. K. & Scott, K. J. (1973). Sulphur metabolism of *Puccinia graminis* f. sp. *tritici* in axenic culture. *Journal of General Microbiology* **76**, 345–54.

Hubbard, M. J., Surarit, R., Sullivan, D. A. & Shepherd, M. G. (1986). The isolation of plasma membrane and characterization of the plasma membrane ATPase from the yeast *Candida albicans. European Journal of Biochemistry* **154**, 375–81.

Hubbard, S. C. & Brody, S. (1975). Glycerophospholipid variation in choline and inositol auxotrophs of *Neurospora*. Internal compensation among zwitterionic and anionic species. *Journal of Biological Chemistry* **250**, 7173–81.

Huber-Walchli, V. & Wiemken, A. (1979). Differential extraction of soluble pools for the cytosol and the vacuoles of yeast (*Candida utilis*) using DEAE dextran. *Archives of Microbiology* **120**, 141–9.

Hult, K. & Gatenbeck, S. (1978). Production of NADPH in the mannitol cycle and its relation to polyketide formation in *Alternaria alternata. European Journal of Biochemistry* **88**, 607–17.

Hummelt, G. & Mora, J. (1980a). NADH-dependent glutamate synthase and nitrogen metabolism in *Neurospora crassa. Biochemical and Biophysical Research Communications* **92**, 127–33.

Hummelt, G. & Mora, J. (1980b). Regulation and function of glutamate synthase in *Neurospora crassa. Biochemical and Biophysical Research Communications* **96**, 1688–94.

Hunter, D. R., Norberg, C. L. & Segel, I. H. (1973). Effect of cycloheximide on L-leucine transport by *Penicillium chrysogenum*: involvement of calcium. *Journal of Bacteriology* **114**, 956–60.

Hunter, D. R. & Segel, I. H. (1971). Acidic and basic amino acid transport systems of *Penicillium chrysogenum. Archives of Biochemistry and Biophysics* **144**, 168–83.

Hunter, D. R. & Segel, I. H. (1973a). Control of the general amino acid permease of *Penicillium chrysogenum* by transinhibition and turnover. *Archives of Biochemistry and Biophysics* **154**, 387–99.

Hunter, D. R. & Segel, I. H. (1973b). Effect of weak acids on amino acid transport by *Penicillium chrysogenum*: evidence for a proton or charge gradient as the driving force. *Journal of Bacteriology* **113**, 1184–92.

Hunter, D. R. & Segel, I. H. (1985). Evidence for two distinct intracellular pools of inorganic sulfate in *Penicillium notatum. Journal of Bacteriology* **162**, 881–7.

Hunter, K. & Rose, A. H. (1970). Yeast lipids and membranes. In *The Yeasts*, vol. 2 *Yeasts and the Environment*, ed. A. H. Rose & J. S. Harrison, pp. 211–70. Academic Press: London.

Huschka, H. G., Müller, G. & Winkelmann, G. (1983). The membrane potential is the driving force for siderophore iron transport in fungi. *FEMS Microbiology Letters* **20**, 125–9.

Huschka, H. G., Naegeli, H. U., Leuenberger-Ryf, H., Keller-Schierlein, W. & Winkelmann, G. (1985). Evidence for a common siderophore transport system but different siderophore receptors in *Neurospora crassa*. *Journal of Bacteriology* **162**, 715–21.

Huygen, P. L. M. & Borst-Pauwels, G. W. F. H. (1972). The effect of *N,N'*-dicyclohexyl-carbodiimide on anaerobic and aerobic phosphate by baker's yeast. *Biochimica et Biophysica Acta* **283**, 234–8.

Huynh, V.-B. & Crawford, R. L. (1985). Novel extracellular enzymes (ligninases) of *Phanerochaete chrysosporium*. *FEMS Microbiology Letters* **28**, 119–23.

Iida, H., Yagawa, Y. & Anraku, Y. (1990). Essential role for induced Ca^{2+} influx followed by $[Ca^{2+}]_i$ rise in maintaining viability of yeast cells late in the mating response pathway. A study of $[Ca^{2+}]_i$ in single *Saccharomyces cerevisiae* cells with imaging of fura-2. *Journal of Biological Chemistry* **265**, 13391–9.

Iizuka, M., Chiura, H. & Yamamoto, T. (1978). Amino acid and sugar composition of invertase of *Candida utilis*. *Agricultural and Biological Chemistry* **42**, 1207–11.

Indge, K. J. (1968). Polyphosphates of the yeast vacuole. *Journal of General Microbiology* **51**, 447–55.

Ingold, C. T. (1971). *Fungal Spores*. Oxford University Press: Oxford.

Ingold, C. T. (1990). How did the active basidium arise? *Mycologist* **4**, 149.

Ingraham, J. L. & Emerson, R. (1954). Studies on the nutrition and metabolism of the aquatic phycomycete *Allomyces*. *American Journal of Botany* **41**, 146–52.

Ingram, L. O. (1986). Microbial tolerance to alcohols: role of the cell membrane. *Trends in Biotechnology* **4**, 40–4.

Ingram, L. O. & Buttke, T. M. (1984). Effects of alcohols on microorganisms. *Advances in Microbial Physiology* **25**, 253–300.

Ingram, M. (1960). Studies on benzoate resistant yeasts. *Acta Microbiologica* **7**, 95–105.

Inouhe, M., Inagawa, A., Morita, M., Tuhoyama, H., Joho, M. & Murayama, T. (1991). Native cadmium-metallothionein from the yeast *Saccharomyces cerevisiae* – its primary structure and function in heavy-metal resistance. *Plant and Cell Physiology* **32**, 475–82.

Irving, G. C. J. & Cosgrove, D. J. (1972). Inositol phosphate phosphatases of microbiological origin: the inositol pentaphosphate products of *Aspergillus ficuum* phytases. *Journal of Bacteriology* **112**, 434–8.

Irving, G. C. J. & Cosgrove, D. J. (1974). Inositol phosphate phosphatases of microbiological origin. Properties of the partially purified phosphatases of *Aspergillus ficuum* NRRL 3135. *Australian Journal of Biology* **27**, 361–8.

Isaac, S., Briarty, L. G. & Peberdy, J. F. (1979). The stereology of protoplasts from *Aspergillus nidulans*. *Proceedings of the 5th International Protoplast Symposium, Szeged, Hungary*, pp. 213–19. Publishing House of the Hungarian Academy of Sciences: Budapest.

Isaac, S. & Gokhale, A. V. (1982). Autolysis: a tool for protoplast production from *Aspergillus nidulans*. *Transactions of the British Mycological Society* **78**, 389–94.

Isaac, S., Gokhale, A. V. & Wyatt, A. M. (1986). Selectivity to K$^+$ and Na$^+$ of protoplast fractions isolated from different regions of *Aspergillus nidulans* hyphae. *Journal of General Microbiology* 132, 1173–9.

Isaac, S., Ryder, N. S. & Peberdy, J. F. (1978). Distribution and activation of chitin synthase in protoplast fractions released during the lytic digestion of *Aspergillus nidulans* hyphae. *Journal of General Microbiology* 105, 45–70.

Ishikawa, T., Toh-e, A., Uno, I. & Hasunuma, K. (1969). Isolation and characterization of nuclease mutants in *Neurospora crassa*. *Genetics* 63, 75–92.

Ito, S. & Inoue, S. (1982). Sophorolipids from *Torulopsis bombicola*: possible relation to alkane uptake. *Applied and Environmental Microbiology* 43, 1278–83.

Iwashima, A., Kawasaki, Y. & Kimura, V. (1990a). Transport of 2-methyl-4-amino-5-hydroxymethylpyrimidine in *Saccharomyces cerevisiae*. *Biochimica et Biophysica Acta* 1022, 211–14.

Iwashima, A., Kimura, Y. & Kawasaki, Y. (1990b). Transport overshoot during 2-methyl-4-amino-5-hydroxymethylpyrimidine uptake by *Saccharomyces cerevisiae*. *Biochimica et Biophysica Acta* 1028, 161–4.

Iwashima, A. & Nishimura, H. (1979). Isolation of a thiamine-binding protein from *Saccharomyces cerevisiae*. *Biochimica et Biophysica Acta* 577, 217–20.

Iwashima, A., Nishimura, H. & Nose, Y. (1979). Soluble and membrane-bound thiamine-binding proteins from *Saccharomyces cerevisiae Biochimica et Biophysica Acta* 557, 460–8.

Iwashima, A., Nishimura, H. & Sempuku, K. (1980). Active transport of dimethialium in *Saccharomyces cerevisiae*. *Experientia* 36, 385–6.

Iwashima, A., Nishino, H. & Nose, Y. (1973). Carrier-mediated transport of thiamine in baker's yeast. *Biochimica et Biophysica Acta* 330, 222–34.

Iwashima, A. & Nose, Y. (1975). Inhibition of thiamine transport in anaerobic baker's yeast by iodoacetate, 2-4-dinitrophenol, *N,N'*-dicyclohexylcarbodiimide and fatty acids. *Biochimica et Biophysica Acta* 399, 375–83.

Iwashima, A. & Nose, Y. (1976). Regulation of thiamine transport in *Saccharomyces cerevisiae*. *Journal of Bacteriology* 128, 855–7.

Jackl, G. & Sebald, W. (1975). Identification of two products of mitochondrial protein synthesis associated with mitochondrial adenosine triphosphatase from *Neurospora crassa*. *European Journal of Biochemistry* 54, 97–106.

Jackson, D. J., Unkefer, C. J., Doolen, J. A., Watt, K. & Robinson, N. J. (1987). Poly (γ-glutamylcysteinyl) glycine: its role in cadmium resistance in plant cells. *Proceedings of the National Academy of Sciences, USA* 84, 6619–23.

Jackson, S. L. & Heath, I. B. (1989). Effects of exogenous calcium ions on tip growth, intracellular Ca^{2+} concentration and actin arrays in hyphae of the fungus *Saprolegnia ferax*. *Experimental Mycology* 13, 1–12.

Jacobson, E. S. & Metzenberg, R. L. (1977). Control of arylsulfatase in a serine auxotroph of *Neurospora*. *Journal of Bacteriology* 130, 1397–8.

Jaffe, L. F. & Nuccitelli, R. (1974). An ultrasensitve vibrating probe for measuring steady extracellular currents. *Journal of Cell Biology* 63, 614–28.

Jäger, A., Croan, S. & Kirk, T. K. (1985). Production of ligninases and degradation of lignin in agitated submerged cultures of *Phanerochaete chrysosporium*. *Applied and Environmental Microbiology* 50, 1274–8.

Jaklitsch, W. M., Kubicek, C. P. & Scrutton, M. C. (1991a). Intracellular location of enzymes involved in citrate production by *Aspergillus niger*. *Canadian Journal of Microbiology* 37, 823–7.

Literature cited

Jaklitsch, W. M., Kubicek, C. P. & Scrutton, M. C. (1991*b*). The subcellular organisation of itaconate biosynthesis in *Aspergillus terreus. Journal of General Microbiology* **137**, 533–9.

Jakubowska, J. D., Metodiewa, D. & Zakowska, Z. (1974). Studies on the metabolic pathway for itatartaric acid formation by *Aspergillus terreus.* 1. Metabolism of glucose and some C_5 and C_6-carboxylic acids. *Acta Microbiologica Polonica* Ser. B **6**, 43–50.

Jalal, M. A. F., Love, S. K. & van der Helm, D. (1987). Siderophore mediated iron (III) uptake in *Gliocladium virens.* 2. Role of ferric mono- and dehydroxamates as iron transport agents. *Journal of Bioinorganic Chemistry* **29**, 259–68.

James, A. W. & Casida, L. E. (1964). Accumulation of phosphorus compounds by *Mucor racemosus. Journal of Bacteriology* **87**, 150–5.

Janda, S., Kotyk, A. & Thauchová, R. (1976). Monosaccharide transport systems in the yeast *Rhodotorula gracilis. Archives of Microbiology* **111**, 151–4.

Janshekar, H., Brown, C., Haltmeier, T., Leisola, M. & Fiechter, A. (1982). Bioalteration of Kraft pine lignin by *Phanerochaete chrysosporium.* Carbohydrate metabolism in the fungus. *Archives of Microbiology* **132**, 14–21.

Janshekar, H. & Fiechter, A. (1988). Cultivation of *Phanerochaete chrysosporium* and production of lignin peroxidase in submerged stirred tank reactors. *Journal of Biotechnology* **8**, 97–112.

Janssens, J. H., Burris, N., Woodward, A. & Baily, R. B. (1983). Lipid-enhanced ethanol production of *Kluyveromyces fragilis. Applied and Environmental Microbiology* **45**, 598–602.

Jarai, G. & Marzluf, G. A. (1991). Sulfate transport in *Neurospora crassa* – regulation, turnover and cellular localization of the CYS-14 protein. *Biochemistry* **30**, 4768–73.

Jaspers, H. T. A. & van Steveninck, J. (1975). Transport-associated phosphorylation of 2-deoxy-D-glucose in *Saccharomyces fragilis. Biochimica et Biophysica Acta* **406**, 370–85.

Jaspers, H. T. A. & van Steveninck, J. (1977). Pulse-labelling experiments in transmembrane transport studies. *Journal of Theoretical Biology* **68**, 599–605.

Jayatissa, P. M. & Rose, A. H. (1976). Role of wall phosphomannan in flocculation of *Saccharomyces cerevisiae. Journal of General Microbiology* **96**, 165–74.

Jeanjean, R., Attia, A., Jarry, T. & Colle, A. (1981). On the involvement of P_i-binding proteins in P_i-uptake in the yeast *Candida tropicalis. FEBS Letters* **125**, 69–71.

Jeanjean, R., Bédu, S., Attia, S. & Rocca-Serra, J. (1982). Inorganic phosphate uptake by protoplasts and whole cells of *Candida tropicalis.* Absence of high affinity transport system in protoplasts. *Biochimica* **64**, 75–8.

Jeanjean, R., Bédu, S., Nieuwenuis, B. J. W. M. & Hirn, M. (1986). Immunological evidence for the involvement of cell wall proteins in phosphate uptake in the yeast *Saccharomyces cerevisiae. Archives of Microbiology* **144**, 207–12.

Jeanjean, R., Blasko, F. & Hirn, M. (1984). Identification of a plasma membrane protein involved in P_i transport in the yeast *Candida tropicalis. FEBS Letters* **165**, 83–7.

Jeanjean, R. & Fournier, N. (1979). Characterization and partial purification of phosphate binding proteins in *Candida tropicalis. FEBS Letters* **105**, 163–6.

Jeffery, J. & Jornvall, H. (1983). Enzyme relationships in a sorbital pathway that bypasses glycolysis and pentose phosphates in glucose metabolism. *Proceedings of the National Academy of Sciences, USA* **80**, 901–5.

Jeffries, T. W. (1987). Physical, chemical and biochemical considerations in the biological degradation of wood. In *Wood and Cellulosics: Industrial Utilisation, Biotechnology, Structure and Properties*, ed. J. F. Kennedy, G. O. Phillips & P. A. Williams, pp. 213–30. Ellis Horwood: Chichester.

Jeffries, T. W., Choi, S. & Kirk, T. K. (1981). Nutritional regulation of lignin degradation by *Phanerochaete chrysosporium. Applied and Environmental Microbiology* **42**, 290–6.

Jennings, D. H. (1963). *The Absorption of Solutes by Plant Cells.* Oliver & Boyd: Edinburgh.

Jennings, D. H. (1964). Changes in the size of orthophosphate pools in mycorrhizal roots of beech with reference to absorption of the ion from the external medium. *New Phytologist* **63**, 181–93.

Jennings, D. H. (1969). The physiology of the uptake by the growing plant cell. In *Ecological Aspects of the Mineral Nutrition of Plants*, ed. I. H. Rorison, pp. 261–79. Blackwell Scientific Publications: Oxford.

Jennings, D. H. (1973). Cations and filamentous fungi: invasion of the sea and hyphal functioning. In *Ion Transport in Plants*, ed. W. P. Anderson, pp. 323–35. Academic Press: London.

Jennings, D. H. (1974). Sugar transport into fungi: an essay. *Transactions of the British Mycological Society* **62**, 1–24.

Jennings, D. H. (1976). Transport in fungal cells. In *Encyclopaedia of Plant Physiology*, vol. 2 *Transport in Plants*. Part A *Cells*, ed. U. Lüttge & M. G. Pitman, pp. 189–228. Springer-Verlag: Berlin, Heidelberg, New York.

Jennings, D. H. (1979). Membrane transport and hyphal growth. In *Fungal Walls and Hyphal Growth*, ed. J. H. Burnett & A. P. J. Trinci, pp. 279–94. Cambridge University Press: Cambridge.

Jennings, D. H. (1982). The movement of *Serpula lacrimans* from substrate to substrate over nutritionally inert surfaces. In *Decomposer Basidiomycetes: their Biology and Ecology*, ed. J. C. Frankland, J. H. Hedger & M. J. Swift, pp. 91–108. Cambridge University Press: Cambridge.

Jennings, D. H. (1983). Some aspects of the physiology and biochemistry of marine fungi. *Biological Reviews* **58**, 423–59.

Jennings, D. H. (1984a). Polyol metabolism in fungi. *Advances in Microbial Physiology* **25**, 149–93.

Jennings, D. H. (1984b). Water flow through mycelia. In *The Ecology and Physiology of the Fungal Mycelium*, ed. D. H. Jennings & A. D. M. Rayner, pp. 143–64. Cambridge University Press: Cambridge.

Jennings, D. H. (1986a). Morphological plasticity in fungi. *Symposium of the Society for Experimental Biology* **40**, 329–46.

Jennings, D. H. (1986b). Salt relations in cells, tissues and roots. In *Plant Physiology: A Treatise*, vol. IX, *Water and Solutes in Plants*, ed. F. C. Steward, J. F. Sutcliffe & J. E. Dale, pp. 225–379. Academic Press: Orlando, FL.

Jennings, D. H. (1987a). The medium is the message. *Transactions of the British Mycological Society* **89**, 1–11.

Jennings, D. H. (1987b). Translocation of solutes in fungi. *Biological Reviews* **62**, 215–43.

Jennings, D. H. (1989). Some perspectives on nitrogen and phosphorus metabolism in fungi. In *Nitrogen, Phosphorus and Sulphur Utilization by Fungi*, ed. L. Boddy, R. Marchart & D. J. Read, pp. 1–31. Cambridge University Press: Cambridge.

Jennings, D. H. (1990a). Nutrient acquisition by fungi – the relation between physiological understanding and ecological reality. *Proceedings of the Indian Academy of Sciences* **100**, 153–63.

Jennings, D. H. (1990b). The ability of basidiomycete mycelium to move nutrients through the soil ecosystem. In *Nutrient Cycling in Terrestrial Ecosystems*, ed. A. F. Harrison, P. Ineson & D. W. Heal, pp. 233–45. Elsevier Applied Science: London.

Jennings, D. H. (1990c). Osmophiles. In *Microbiology of Extreme Environments*, ed. C. Edwards, pp. 117–46. Open University Press: Milton Keynes.

Jennings, D. H. (1991a). The spatial aspects of fungal growth. *Science Progress, Oxford* **75**, 141–56.

Jennings, D. H. (1991b). The physiology and biochemistry of the vegetative mycelium. In Serpula lacrymans: *Fundamental Biology and Control Strategies*, ed. D. H. Jennings & A. F. Bravery, pp. 55–79. John Wiley: Chichester.

Jennings, D. H. (1991c). The role of droplets in helping to maintain a constant growth rate of aerial hyphae. *Mycological Research* **95**, 883–4.

Jennings, D. H. (1991d). Techniques for studying the functional aspects of rhizomorphs of wood-rotting fungi: some possible applications to ectomycorrhiza. In *Methods in Microbiology*, vol. 23 *Techniques in the Study of Mycorrhiza*, ed. J. R. Norris, D. J. Read & A. K. Varma, pp. 309–29. Academic Press: London.

Jennings, D. H. (1993). Understanding tolerance to stress: laboratory culture versus environmental actuality. In *Stress Tolerance of Fungi*, ed. D. H. Jennings, pp. 1–12. Marcel Dekker: New York.

Jennings, D. H. (1994). Translocation in fungal mycelia. In *The Mycota*, vol. II *Growth, Differentiation and Sexuality*, ed. J. G. H. Wessels & F. Meinhardt, pp. 163–73. Springer-Verlag: Berlin, Heidelberg, New York.

Jennings, D. H. & Aynsley, J. S. (1971). Compartmentation and low temperature fluxes of potassium in mycelium of *Dendryphiella salina*. *New Phytologist* **70**, 713–23.

Jennings, D. H. & Bravery, A. F. (eds.) (1991). Serpula lacrymans: *Fundamental Biology and Control Strategies*. John Wiley: Chichester.

Jennings, D. H. & Burke, R. M. (1990). Compatible solutes – the mycological dimension and their role as physiological buffering agents. *New Phytologist* **116**, 277–83.

Jennings, D. H. & Garrill, A. (1994). Techniques used in the physiological study of marine fungi. In *The Isolation and Study of Marine Fungi*, ed. E. B. Gareth Jones. John Wiley: Chichester (in press).

Jennings, D. H., Hooper, D. C. & Rothstein, A. (1958). The participation of phosphate in the formation of a 'carrier' for the transport of Mg^{++} and Mn^{++} ions into yeast cells. *Journal of General Physiology* **41**, 1019–26.

Jennings, D. H. & Thornton, J. D. (1984). Carbohydrate metabolism in the fungus *Dendryphiella salina*. VII. The effect of L-sorbose on ethanol-soluble carbohydrates. *New Phytologist* **98**, 399–403.

Jennings, D. H., Thornton, J. D., Galpin, M. F. J. & Coggins, C. R. (1974). Translocation in fungi. *Symposium of the Society for Experimental Biology* **28**, 139–56.

Jennison, M. W., Newcomb, M. D. & Henderson, R. (1955). Physiology of the wood-rotting basidiomycetes. 1. Growth and nutrition in submerged culture in synthetic media. *Mycologia* 47, 275–304.

Jeyaprakash, A., Welch, J. W. & Fogel, S. (1991). Multicopy *CPU1* plasmids enhance cadmium and copper resistance levels in yeast. *Molecular and General Genetics* 225, 363–8.

Jia, Z.-P., McCullough, N., Martel, R., Hemmingsen, S. & Young, P. G. (1992). Gene amplification of a locus encoding a putative Na$^+$/H$^+$ antiporter confers sodium and lithium tolerance in fission yeast. *EMBO Journal* 11, 1631–40.

Jimenez, J. & van Uden, N. (1985). Use of extracellular acidification for rapid testing of ethanol tolerance in yeasts. *Biotechnology and Bioengineering* 27, 1596–8.

Jin, C. K., Chiang, H. L. & Wang, S. S. (1981). Steady state analysis of the enhancement in ethanol productivity of a continuous fermentation process employing protein–phospholipid complex as protective agent. *Enzyme and Microbial Technology* 3, 249–57.

Johannes, E., Brosnan, J. M. & Sanders, D. (1991). Calcium channels and signal transduction in plant cells. *BioEssays* 13, 331–6.

Johansson, T. & Nyman, P. O. (1987). A manganese (II)-dependent extracellular peroxidase from the white-rot fungus *Trametes versicolor*. *Acta Chimica Scandinavica* 41B, 762–5.

Johnson, G. T. & Jones, A. C. (1941). Data on the cultural characteristics of a species of *Coprinus*. *Mycologia* 33, 423–33.

Johnston, J. H. & Barford, J. P. (1991). Continuous growth of *Saccharomyces cerevisiae* on a mixture of glucose and fructose. *Journal of General and Applied Microbiology* 37, 133–40.

Joho, M., Tarumi, K., Inouhe, M., Tohoyama, H. & Murayama, T. (1991). Co^{2+} and Ni^{2+} resistance in *Saccharomyces cerevisiae* associated with a reduction in the accumulation of Mg^{2+}. *Microbios* 67, 177–86.

Jones, D., McHardy, W. J. & Wilson, M. J. (1976). Ultrastructure and chemical composition of spines in Mucorales. *Transactions of the British Mycological Society* 66, 153–7.

Jones, E. B. G. (1994). Presidential address, 1992. Fungal adhesion. *Mycological Research* 98, 961–81.

Jones, E. B. G. & Jennings, D. H. (1964). The effect of salinity on the growth of marine fungi in comparison with non-marine species. *Transactions of the British Mycological Society* 47, 619–25.

Jones, E. B. G. & Jennings, D. H. (1965). The effect of cations on the growth of fungi. *New Phytologist* 64, 86–100.

Jones, E. W. (1984). The synthesis and functions of proteases in *Saccharomyces*: genetic approaches. *Annual Review of Genetics* 18, 233–70.

Jones, E. W. & Fink, G. R. (1982). Regulation of amino acid and nucleotide biosynthesis in yeast. In *The Molecular Biology of the Yeast Saccharomyces: Metabolism and Gene Expression*, ed. J. N. Strathern, E. W. Jones & J. R. Broach, pp. 181–297. Cold Spring Harbor Laboratory Press: Cold Spring Harbor, NY.

Jones, M. D., Durall, D. M. & Tinker, P. B. (1991). Fluxes of carbon and phosphorus between symbionts in willow ectomycorrhizas and their changes with time. *New Phytologist* 119, 99–106.

Jones, R., Parkinson, S. W., Wainwright, M. & Killham, K. (1991). Oxidation of thiosulphate by *Fusarium oxysporium* grown under oligotrophic conditions. *Mycological Research* 95, 1169–74.

Jones, R. P. (1988). Intracellular ethanol-accumulation and exit from yeast and other fungi. *FEMS Microbiology Reviews* **54**, 239–58.

Jones, R. P. (1990). Roles for replicative deactivation in yeast-ethanol fermentations. *Critical Reviews in Biotechnology* **10**, 205–22.

Jones, R. P. & Greenfield, P. F. (1987). Ethanol and the fluidity of the plasma membrane. *Yeast* **3**, 223–32.

Jongbloed, R. H. & Borst-Pauwels, G. W. F. H. (1992). Effects of aluminium and pH on growth and potassium uptake by 3 ectomycorrhizal fungi in liquid culture. *Plant and Soil* **140**, 157–66.

Jongbloed, R. H., Clement, J. M. A. M. & Borst-Pauwels, G. W. F. H. (1991). Kinetics of NH_4^+ and K^+ uptake by ectomycorrhizal fungi effect of NH_4^+ on K^+ uptake. *Physiologia Plantarum* **83**, 427–32.

Jongbloed, R. H., Tosserams, M. W. A. & Borst-Pauwels, G. W. F. H. (1992). The effect of aluminium on phosphate uptake by 3 isolated ectomycorrhizal fungi. *Plant and Soil* **140**, 167–74.

Joshi, A. P. & Ramakrishnan, C. V. (1959). Mechanism of formation and accumulation of citric acid in *Aspergillus niger*. I. Citric acid formation and oxaloacetic hydrolase in the citric acid producing strain of *Aspergillus niger*. *Enzymologia* **7**, 43–51.

Jund, R., Chevallier, M. R. & Lacroute, F. (1977). Uracil uptake in *Saccharomyces cerevisiae*. *Journal of Membrane Biology* **36**, 233–514.

Jund, R. & Lacroute, F. (1970). Genetic and physiological aspects of resistance to 5-fluoropyrimidines in *Saccharomyces cerevisiae*. *Journal of Bacteriology* **102**, 607–15.

Jund, R., Weber, E. & Chevallier, M.-R. (1988). Primary structure of the uracil transport protein of *Saccharomyces cerevisiae*. *European Journal of Biochemistry* **171**, 417–24.

Jung, C. & Rothstein, A. (1965). Arsenate uptake and release in relation to the inhibition of transport and glycolysis in yeast. *Biochemical Pharmacology* **14**, 1093–112.

Jungnickel, F. (1966). Phosphatinduziete Aufnahme und Exkretion von Kalium bei Phosphatmangelzellen von *Candida utilis*. *Naturwissenschaften* **53**, 231–2.

Kakinuma, Y., Ohsumi, Y. & Anraku, Y. (1981). Properties of H^+-translocating adenosine triphosphatase in vacuolar membranes of *Saccharomyces cerevisiae*. *Journal of Biological Chemistry* **256**, 10859–63.

Kalisz, H. M., Wood, D. A. & Moore, D. (1987). Production, regulation and release of extracellular proteinase activity in basidiomycete fungi. *Transactions of the British Mycological Society* **88**, 221–7.

Kalisz, H. M., Wood, D. A. & Moore, D. (1989). Some characteristics of extracellular proteinases from *Coprinus cinereus*. *Mycological Research* **92**, 278–85.

Kaltwasser, H. (1962). Die Rolle der Polyphosphate im Phosphatstoff wechel eines Knallgasbakteriums (*Hydrogenomonas* Stamm 20). *Archiv für Mikrobiologie* **41**, 283–306.

Kaminskyj, S. G. W., Garrill, A. & Heath, I. B. (1992). The relationship between turgor and tip growth in *Saprolegnia ferax*: turgor is necessary but not sufficient to explain apical extension rates. *Experimental Mycology* **16**, 64–75.

Kanda, T., Wakabayashi, K. & Nisizawa, K. (1976a). Purification and properties of an endo-cellulase of avicelase type from *Irpex lacteus* (*Polyporus tulipiferae*). *Journal of Bacteriology* **79**, 989–95.

Kanda, T., Wakabayashi, K. & Nisizawa, K. (1976b). Xylanase activity of an endo-cellulase of the carboxymethyl-cellulose type from *Irpex lacteus* (*Polyporus tulipiferae*). *Journal of Bacteriology* **79**, 989–95.

Kanda, T., Wakabayashi, K. & Nisizawa, K. (1976c). Synergistic action of two different types of endo-cellulose components from *Irpex lacteus* (*Polyporus tulipiferae*) in the hydrolysis of some insoluble celluloses. *Journal of Bacteriology* **79**, 997–1006.

Kane, P. M., Yamashiro, C. T. & Stevens, T. H. (1989). Biochemical characterization of the yeast vacuolar H^+-ATPase. *Journal of Biological Chemistry* **264**, 19236–44.

Käppeli, O. (1986a). Regulation of carbon metabolism in *Saccharomyces cerevisiae* and related yeasts. *Advances in Microbial Physiology* **27**, 181–209.

Käppeli, O. (1986b). Cytochromes P-450 in yeasts. *Microbiological Reviews* **50**, 244–58.

Käppeli, O. & Fiechter, A. (1976). The mode of interaction between the substrate and cell surface of the hydrocarbon-utilizing yeast *Candida tropicalis*. *Biotechnology and Bioengineering* **18**, 967–74.

Käppeli, O. & Fiechter, A. (1977). Component from the cell surface of the hydrocarbon-utilizing yeast *Candida tropicalis* with possible relation to hydrocarbon transport. *Journal of Bacteriology* **131**, 917–21.

Käppeli, O. & Sonnleitner, B. (1986). Regulation of sugar metabolism in *Saccharomyces*-type yeast: experimental and conceptual considerations. *CRC Critical Reviews in Biotechnology* **4**, 299–326.

Käppeli, O., Walther, P., Müller, M. & Fiechter, A. (1984). Structure of the cell surface of the yeast *Candida tropicalis* and its relation to hydrocarbon transport. *Archives of Microbiology* **138**, 279–82.

Karin, M., Najarian, R., Haslinger, A., Valenzuela, P., Welch, J. & Fogel, S. (1984). Primary structure and transcription of an amplified genetic locus: the *CUP1* locus of yeast. *Proceedings of the National Academy of Sciences, USA* **81**, 337–41.

Kasher, J. S., Allen, K. E., Kasamo, K. & Slayman, C. W. (1986). Characterization of an essential arginine residue in the plasma membrane H^+-ATPase of *Neurospora crassa*. *Journal of Biological Chemistry* **261**, 10808–13.

Kasho, V. N. & Avaeva, S. M. (1978). Isolation of the inorganic pyrophosphatase from brewer's yeast and studies on the localization of the enzyme in brewer's and baker's yeast. *International Journal of Biochemistry* **9**, 51–6.

Katchman, B. J., Fetty, W. O. & Busch, K. A. (1959). Effect of cell division inhibition on the phosphorus metabolism of growing cultures of *Saccharomyces cerevisiae Journal of Bacteriology* **77**, 331–8.

Kato, N., Omori, Y., Tai, Y. & Ogata, K. (1976). Alcohol oxidases of *Kloechera* sp. and *Hansenula polymorpha*. Catalytic properties and subunit structures. *European Journal of Biochemistry* **64**, 341–50.

Kato, H., Uno, I., Ishikawa, T. & Takenawa, T. (1989). Activation of phosphatidylinositol kinase and phosphatidyl inositol-4-phosphate kinase by cAMP in *Saccharomyces cerevisiae*. *Journal of Biological Chemistry* **264**, 3116–21.

Kato, N., Nishizawa, T., Sakazawa, C., Tani, Y. & Yamada, H. (1979). Xylulose 5-phosphate dependent fixation of formaldehyde in a methanol-utilizing yeast *Kloeckera* sp. no. 2201. *Agricultural and Biological Chemistry* **43**, 2013–15.

Kawasaki, K., Nosaka, K., Kaneko, Y., Nishimura, H. & Iwashima, A. (1990).

Regulation of thiamine biosynthesis in *Saccharomyces cerevisiae*. *Journal of Bacteriology* **172**, 6145–7.

Kazama, F. (1972). Development and morphology of a chytrid isolated from *Bryopsis plumosa*. *Canadian Journal of Botany* **50**, 499–505.

Keenan, M. H. J. & Rose, A. H. (1979). Plasma membrane lipid unsaturation can affect the kinetics of solute accumulation by *Saccharomyces cerevisiae*. *FEMS Microbiology Letters* **6**, 133–7.

Keenan, M. H. J., Rose, A. H. & Silverman, B. W. (1982). Effect of plasma membrane phospholipid unsaturation on solute transport into *Saccharomyces cerevisiae* NCYC 366. *Journal of General Microbiology* **128**, 2547–56.

Kelley, R. L., Ramasamy, K. & Reddy, C. A. (1986). Characterization of glucose oxidase-negative mutants of a lignin degrading basidiomycete *Phanerochaete chrysosporium*. *Archives of Microbiology* **144**, 254–7.

Kelley, R. L. & Reddy, C. A. (1986a). Identification of glucose oxidase activity as the primary source of hydrogen peroxide production in ligninolytic cultures of *Phanerochaete chrysosporium*. *Archives of Microbiology* **144**, 248–53.

Kelley, R. L. & Reddy, C. A. (1986b). Purification and characterization of glucose oxidase from ligninolytic cultures of *Phanerochaete chrysosporium*. *Journal of Bacteriology* **166**, 269–74.

Kelly, D. J. A. & Budd, K. (1990). Osmotic adjustment in the mycelial ascomycete *Neocosmospora vasinfecta*. *Experimental Mycology* **14**, 136–44.

Kelly, D. J. A. & Budd, K. (1991). Polyol metabolism and osmotic adjustment in the mycelial ascomycete *Neocosmospora vasinfecta*. *Experimental Mycology* **15**, 55–64.

Kenig, M. & Abraham, E. P. (1976). Antimicrobial activities and antagonists of bacilysin and anticapsin. *Journal of General Microbiology* **94**, 37–45.

Kennedy, J. F., Phillips, G. O. & Williams, P. A. (eds.) (1987). *Wood and Cellulosics: Industrial Utilisation*. Ellis Horwood: Chichester.

Kershaw, J. L. & Stewart, G. R. (1989). The role of glutamine synthetase, glutamate, synthase and glutamate dehydrogenase in ammonia assimilation in the mycorrhizal fungus *Pisolithus tinctorius*. *Annales des Science Forestières* **46**, 706–10.

Kerwin, J. L. & Washino, R. K. (1986). Oosporogenesis by *Lagenidium giganteum*: induction and maturation are regulated by calcium and calmodulin. *Canadian Journal of Microbiology* **32**, 663–72.

Ketter, J. S., Jarai, G., Fu, Y.-H. & Marzluf, G. A. (1991). Nucleotide sequence, messenger RNA stability, and DNA recognition elements of *cys-14*, the structural gene for sulfate permease II in *Neurospora crassa*. *Biochemistry* **30**, 1780–7.

Keyser, P., Kirk, T. K. & Zeikus, J. G. (1978). Ligninolytic enzyme system of *Phanerochaete chrysosporium*: synthesised in the absence of lignin in response to nitrogen starvation. *Journal of Bacteriology* **135**, 790–7.

Kilian, S. G., du Preez, J. C. & Gericke, M. (1989). The effects of ethanol on growth rate and passive proton diffusion in yeasts. *Applied Microbiology and Biotechnology* **32**, 90–4.

Kilian, S. G., van Deemter, A., Kock, J. L. F. & du Preez, J. C. (1991). Occurrence and taxonomic aspects of proton movements coupled to sugar transport in the yeast genus *Kluyveromyces*. *Antonie van Leeuwenhoek* **59**, 199–206.

Kilian, S. G. & van Uden, N. (1988). Transport of xylose and glucose in the xylose-fermenting yeast *Pichia stipitis. Applied Microbiology and Biotechnology* **27**, 545–8.

Kinghorn, J. R. (1989). Genetic, biochemical and structural organisation of the *Aspergillus nidulans crnA–nilA–niaD* gene cluster. In *Molecular and Genetic Aspects of Nitrate Assimilation*, ed. J. L. Wray & J. R. Kinghorn, pp. 69–87. Oxford Science Publications: Oxford.

Kinnaird, J. H. & Fincham, J. R. S. (1983). The complete nucleotide sequence of the *Neurospora crassa am* (NADP-specific glutamate dehydrogenase) gene. *Gene* **26**, 253–60.

Kinoshita, K. (1932). Uber die produktion von itaconsaüre und Mannit durch einen neuen Schimmelpiltz *Aspergillus itaconicus. Acta Phytochimica* **5**, 271–87.

Kinsey, J. A., Fincham, J. R. S., Siddig, M. A. M. & Keighren, M. (1980). New mutational variants of *Neurospora* NADP-specific glutamate dehydrogenase. *Genetics* **95**, 305–16.

Kirk, T. K. (1971). Effects of microorganisms on lignin. *Annual Review of Phytopathology* **9**, 185–210.

Kirk, T. K. (1981). Towards elucidating the mechanism of action of the ligninolytic systems in basidiomycetes. In *Trends in the Biology of Fermentations for Fuels and Chemicals*, ed. A. Hollaender, pp. 131–49. Plenum Press: New York.

Kirk, T. K. (1987). Lignin-degrading enzymes. *Philosophical Transactions of the Royal Society* A **321**, 461–74.

Kirk, T. K. & Adler, E. (1970). Methoxyl-deficient structural elements in lignin of sweetgum decayed by a brown-rot fungus. *Acta Chimica Scandinavica* **24**, 3379–90.

Kirk, T. K. & Chang, H.-M. (1974). Decomposition of lignin by white-rot fungi. I. Isolation of heavily degraded lignins from decayed spruce. *Holzforschung* **28**, 217–22.

Kirk, T. K. & Chang, H.-M. (1975). Decomposition of lignin by white-rot fungi. II. Characterization of heavily degraded lignins from decayed spruce. *Holzforschung* **29**, 56–64.

Kirk, T. K., Connors, W. J., Bleam, R. D., Hackett, W. F. & Zeikus, J. G. (1975). Preparations and microbial decomposition of synthetic [^{14}C]lignins. *Proceedings of the National Academy of Sciences, USA* **72**, 2515–19.

Kirk, T. K., Connors, W. J. & Zeikus, J. G. (1976). Requirement for a growth substrate during lignin decomposition by two wood-rooting fungi. *Applied and Environmental Microbiology* **32**, 192–4.

Kirk, T. K. & Farrell, R. L. (1987). Enzymatic 'combustion': the microbial degradation of lignin. *Annual Review of Microbiology* **41**, 465–506.

Kirk, T. K. & Fenn, P. (1982). Formation and action of the ligninolytic system in basidiomycetes. In *Decomposer Basidiomycetes: Their Biology and Ecology*, ed. J. C. Frankland, J. N. Hedger & M. J. Swift, pp. 67–90. Cambridge University Press: Cambridge

Kirk, T. K., Higuchi, T. & Chang, H.-M. (eds.) (1980). *Lignin Biodegradation: Microbiology, Chemistry and Potential Applications*, vols. I and II. CRC Press: Boca Raton, FL.

Kirk, T. K., Ibach, R., Mozuch, M. D., Conner, A. H. & Highley, T. L. (1991). Characteristics of cotton cellulose depolymerised by a brown-rot fungus by acid or by chemical oxidants. *Holzforschung* **45**, 239–44.

Kirk, T. K., Schultz, E., Connors, W. J., Lorenz, L. F. & Zeikus, J. G. (1978). Influence of culture parameters on lignin metabolism by *Phanerochaete chrysosporium. Archives of Microbiology* **117**, 277–85.

Kirk, T. K. & Tien, M. (1983). Biochemistry of lignin degradation by *Phanerochaete chrysosporium*: investigation with non-phenolic model compounds. In *Recent Advances in Lignin Biodegradation Research*, ed. T. Higuchi, H.-M. Chang & T. K. Kirk, pp. 233–45. Tokyo University Press: Tokyo.

Kirk, T. K., Tien, M., Johnsrud, S. C. & Eriksson, K.-E. (1986). Lignin degrading activity of *Phanerochaete chrysosporium* Burds.: comparison of cellulase-negative and other strains. *Enzyme and Microbial Technology* **8**, 75–80.

Kisser, M., Kubicek, C. P. & Röhr, M. (1980). Influence of manganese on morphology and cell wall composition on *Aspergillus niger* during citric acid fermentation. *Archives of Microbiology* **128**, 26–33.

Kitamoto, K., Yoshizawa, K., Ohsumi, Y. & Anraku, Y. (1988). Dynamic aspects of vacuolar and cytosolic amino acid pools of *Saccharomyces cerevisiae. Journal of Bacteriology* **170**, 2683–6.

Kizawa, H., Tomura, D., Oda, M., Fukamizu,U., Hoshino, T., Gotoh, O., Yasui, T. & Shoun, T. (1991). Nucleotide sequence of the unique nitrate/nitrite-inducible cytochrome P-450 cDNA from *Fusarium oxysporum. Journal of Biological Chemistry* **266**, 10632–7.

Klingmüller, W. & Huh, H. (1972). Sugar transport in *Neurospora crassa. European Journal of Biochemistry* **25**, 141–6.

Klionsky, D. J., Herman, P. K. & Emr, S. D. (1990). The fungal vacuole: composition, function and biogenesis. *Microbiological Reviews* **54**, 266–92.

Klittich, C. J. R., Leslie, J. F. & Wilson, J. D. (1986). Spontaneous chlorate resistant mutants of *Gibberella fujikuroi* (*Fusarium moniliforme*). *Phytopathology* (Abstracts) **76**, 1142.

Klöppel, R. & Höfer, M. (1976a). Transport um Umsatz von Polyalkoholen, I. Konstitutiver Polyalkoholtransport. *Archives of Microbiology* **107**, 329–34.

Klöppel, R. & Höfer, M. (1976b). Transport um Umsatz von Polyalkoholen, II. Induzierbarar Transport und Abban von Pentitolen. *Archives of Microbiology* **107**, 335–42.

Kluyver, A. J. & Custers, M. T. J. (1940). The suitability of disaccharides as respiration and assimilation substrates for yeasts which do not ferment these sugars. *Antonie van Leeuwenhoek* **6**, 121–62.

Knotová, A. & Kotyk, A. (1972). Role of sugars in phosphate transport in baker's yeast. *Folia Microbiologica* **17**, 251–60.

Knowles, J., Teeri, T. T., Lehtovaara, P., Penttilä, M. & Saloheino, M. (1988). The use of gene technology to investigate fungal cellulolytic enzymes. In *Biochemistry and Genetics of Cellulose Degradation*, ed. J.-P. Aubert, P. Béguin & J. Millet, pp. 153–69. Academic Press: New York.

Knowles, J. K. C., Teeri, T. T., Penttilä, M., Saloheino, M., Harkki, A. & Lehtovaara, P. (1989). Molecular genetics of fungal celluloses. In *Enzyme Systems for Lignocellulose Degradation*, ed. M. P. Coughlan, pp. 51–6. Elsevier Applied Science: London.

Ko, C. H., Buckley, A. M. & Gaber, R. F. (1990). *TRK2* is required for low affinity K^+ transport in *Saccharomyces cerevisiae. Genetics* **125**, 305–12.

Ko, C. H. & Gaber, R. F. (1991). *TRK1* and *TRK2* encode structurally related K^+ transporters in *Saccharomyces cerevisiae. Molecular and Cellular Biology* **11**, 4266–73.

Kobayashi, T., Vandedem, G. & Moo-Young, M. (1973). Oxygen transfer in mycelial pellets. *Biotechnology and Bioengineering* **15**, 27–45.

Koenigs, J. W. (1972). Production of extracellular hydrogen peroxide and peroxidase by wood-rotting fungi. *Phytopathology* **62**, 100–10.

Koenigs, J. W. (1974*a*). Hydrogen peroxide and iron: a proposed system for decomposition of wood by brown-rot fungi. *Wood and Fibre* **6**, 66–79.

Koenigs, J. W. (1974*b*). Production of hydrogen peroxide by wood-rooting fungi in wood and its correlation with weight loss, depolymerization and pH changes. *Archives of Microbiology* **99**, 129–45.

Kohlmeyer, J. & Kohlmeyer, E. (1979). *Marine Mycology: The Higher Fungi*. Academic Press: New York.

Kolarov, J., Kulpa, J. & Baijot, M. & Goffeau, A. (1988). Characterization of a protein serine kinase from yeast plasma membrane. *Journal of Biological Chemistry* **263**, 10613–19.

Komagata, K. (1981). Taxonomic studies of methanol-utilizing yeasts. In *Microbial Growth on C_1 Compounds*, ed. H. Dalton, pp. 301–11. Heyden & Sons: London.

Kondo, M., Isobe, M., Imai, K., Goto, T., Murasugi, A. & Hayashi, Y. (1983). Structure of cadystin, the unit-peptide of cadmium-binding peptides induced in a fisson yeast, *Schizosaccharomyces pombe*. *Tetrahedron Letters* **24**, 925–8.

Kotyk, A. (1967). Mobility of the free and of the loaded monosaccharide carrier in *Saccharomyces cerevisiae*. *Biochimica et Biophysica Acta* **135**, 112–19.

Kotyk, A. & Dvořáková, M. (1990). Transport of L-tryptophan in *Saccharomyces cerevisiae*. *Folia Microbiologica* **35**, 209–17.

Kotyk, A. & Hàskovec, C. (1968). Properties of the sugar carrier in baker's yeast. III. Induction of the galactose carrier. *Folia Microbiologica* **13**, 12–19.

Kotyk, A. & Höfer, M. (1965). Uphill transport of sugars in the yeast *Rhodotorula gracilis*. *Biochimica et Biophysica Acta* **102**, 410–22.

Kotyk, A. & Michaljaničová, D. (1979). Uptake of trehalose by *Saccharomyces cerevisiae*. *Journal of General Microbiology* **110**, 323–32.

Kramer, R., Kopp, F., Niedermeyer, W. & Fuhrmann, G. F. (1978). Comparative studies on the structure and composition of the plasmalemma and the tonoplast in *Saccharomyces cerevisiae*. *Biochimica et Biophysica Acta* **507**, 369–80.

Kreger-van Rij, N. J. W. (1984). *The Yeasts, a Taxonomic Study*. Elsevier Biomedical Press: Amsterdam.

Krishnan, P. S., Damle, S. P. & Bajaj, V. (1957). Studies on the role of 'metaphosphate' in molds. II. The formation of 'soluble' and 'insoluble metaphosphate' in *Aspergillus niger*. *Archives of Biochemistry and Biophysics* **67**, 35–52.

Kroeler, C. J., Antibus, R. K. & Linkins, R. K. (1988). The effects of organic and inorganic phosphorus concentrations on the acid phosphatase activity of ectomycorrhizal fungi. *Canadian Journal of Botany* **66**, 750–6.

Krol, M. (1983). Occurrence in soils and activity of sulphur oxides. *Pamietnik Pulawski Prace Iung Zeszyt* **79**, 45–61.

Kroll, R. G. (1990). Alkalophiles. In *Microbiology of Extreme Environments*, ed. C. Edwards, pp. 55–92. Open University Press: Milton Keynes.

Kropf, D. L. (1986). Electrophysiological properties of *Achlya* hyphae: ionic currents studied by intracellular potential recording. *Journal of Cell Biology* **102**, 1209–16.

Kropf, D. L., Caldwell, J. H., Gow, N. A. R. & Harold, F. M. (1984).

Transcellular ion currents in the water mold *Achlya*: amino acid/proton symport as a mechanism for current entry. *Journal of Cell Biology* **99**, 486–96.

Kropf, D. L. & Harold, F. M. (1982). Selective transport of nutrients via the rhizoids of the water mold *Blastocladiella emersonii*. *Journal of Bacteriology* **151**, 429–37.

Kropf, D. L., Lupa, M. D. A., Caldwell, J. H. & Harold, F. M. (1983). Cell polarity: endogenous ion currents precede and predict branching in the water mold *Achlya*. *Science* **220**, 1385–7.

Kruckeberg, A. L. & Bisson, L. F. (1990). The *HXT2* gene of *Saccharomyces cerevisiae* is required for high-affinity glucose transport. *Molecular and Cellular Biology* **10**, 5903–13.

Kubicek, C. P. (1987). Involvement of a conidial endoglucanase and a plasma-membrane-bound β-glucosidase in the induction of endoglucanase synthesis by cellulose in *Trichoderma reesei*. *Journal of General Microbiology* **133**, 1481–7.

Kubicek, C. P. (1988*a*). Regulatory aspects of the tricarboxylic acid cycle in filamentous fungi – a review. *Transactions of the British Mycological Society* **90**, 339–49.

Kubicek, C. P. (1988*b*). The role of the citric acid cycle in fungal organic acid fermentations. *Biochemical Society Symposia* **54**, 113–26.

Kubicek, C. P., Eveleigh, D. E., Esterbauer, H., Steiner, W. & Kubicek-Pranz, E. M. (1990). Trichoderma reesei *Cellulases: Biochemistry, Genetics, Physiology and Applications*. Royal Society of Chemistry: Cambridge.

Kubicek, C. P., Hampel, W. & Röhr, M. (1979). Manganese deficiency leads to elevated amino acid pool in citric acid producing *Aspergillus niger*. *Archives of Microbiology* **123**, 73–9.

Kubicek, C. P. & Röhr, M. (1977). Influence of manganese on enzyme synthesis and citric acid accumulation in *Aspergillus niger*. *European Journal of Applied Microbiology* **4**, 167–75.

Kubicek, C. P. & Röhr, M. (1986). Citric acid fermentation. *CRC Critical Reviews in Biotechnology* **3**, 331–74.

Kubicek, C. P., Schreferl-Kunar, G., Wöhrer, W. & Röhr, M. (1988). Evidence for a cytoplasmic pathway of oxalate biosynthesis in *Aspergillus niger*. *Applied and Environmental Microbiology* **54**, 633–7.

Kucey, R. M. N. (1987). Increased phosphorus uptake by wheat and field beans inoculated with a phosphorus-solubilizing *Penicillium bilaji* strain and with vesicular–arbuscular mycorrhizal fungi. *Applied and Environmental Microbiology* **53**, 2699–703.

Kucey, R. M. N. (1988). Effect of *Penicillium bilaji* on the solubility and uptake of P and micronutrients from soil by wheat. *Canadian Journal of Soil Science* **68**, 261–70.

Kucey, R. M. N., Janzen, H. H. & Leggett, M. E. (1989). Microbially mediated increases in plant-available phosphorus. *Advances in Agronomy* **42**, 199–228.

Kuila, D., Tien, M., Fee, J. A. & Ondrias, M. R. (1985). Resonance Raman spectra of extracellular ligninase: evidence for a heme active site similar to those of peroxidases. *Biochemistry* **24**, 3394–9.

Kulaev, I. S. (1979). *The Biochemistry of Inorganic Polyphosphates*. John Wiley: Chichester.

Kulaev, I. S., Kritskii, M. S. & Bolozerskii, A. N. (1960). The metabolism of polyphosphates and other phosphorus compounds during the development of fruiting bodies of the mushroom. *Biochemistry* (USSR) **25**, 563–72.

Kulaev, I. S. & Vagabov, V. M. (1983). Polyphosphate metabolism in microorganisms. *Advances in Microbial Physiology* **24**, 83–171.

Kulakovskaya, T. V., Matys, S. V. & Okorokov, L. A. (1991). Transport of organic acid anions and guanosine into vacuoles of *Saccharomyces cerevisiae. Yeast* **7**, 495–502.

Künemund, A. & Höfer, M. (1983). Passive fluxes of K and H in wild strain and nystatin-resistant mutant of *Rhodotorula gracilis. Biochimica et Biophysica Acta* **735**, 203–10.

Kuo, S.-G., Christensen, M. S. & Cirillo, V. P. (1970). Galactose transport in *Saccharomyces cerevisiae*. II. Characteristics of galactose uptake and exchange in galactokinaseless cells. *Journal of Bacteriology* **103**, 671–8.

Kuo, S.-G. & Cirillo, V. P. (1970). Galactose transport in *Saccharomyces cerevisiae*. III. Characteristics of galactose uptake in transferase-less cells: evidence against transport-associated phosphorylation. *Journal of Bacteriology* **103**, 679–85.

Kurkdjian, A. & Guern, J. (1989). Intracellular pH: measurement and importance in cell activity. *Annual Review of Plant Physiology and Plant Molecular Biology* **40**, 271–303.

Kurtz, M. B. (1980). Regulation of fructose transport during growth of *Aspergillus nidulans. Journal of General Microbiology* **118**, 389–96.

Kuwahara, M., Glenn, J. K., Morgan, M. A. & Gold, M. H. (1984). Separation and characterization of two extracellular H_2O_2-dependent oxidases from ligninolytic cultures of *Phanerochaete chrysosporium. FEBS Letters* **169**, 247–50.

Kuwahara, M., Ishida, Y., Miyagawa, Y. & Kawakami, C. (1984). Production of extracellular NAD and NADP by a lignin-degrading fungus *Phanerochaete chrysosporium. Journal of Fermentation Technology* **62**, 237–42.

Kuypers, G. A. J. & Roomans, G. M. (1979). Mercury-induced loss of K^+ from yeast cells investigated by electron probe X-ray microanalysis. *Journal of General Microbiology* **115**, 13–18.

Kyriacou, A., Mackenzie, C. R. & Neufeld, R. J. (1987). Detection and the characterization of the specific and non-specific endoglucanases of *Trichoderma reesei*: evidence demonstrating endoglucanase activity by cellobiohydrolase. II. *Enzyme and Microbial Technology* **9**, 25–31.

La Peyrie, F. (1990). The role of ectomycorrhizal fungi in calcareous soil tolerance by symbiocalcicole woody plants. *Annales des Sciences Forrestières* **47**, 579–90.

La Peyrie, F., Ranger, J. & Vairelles, D. (1991). Phosphate-solubilizing activity of ectomycorrhizal fungi *in vitro. Canadian Journal of Botany* **69**, 342–5.

La Polla, R. J. & Lambowitz, A. M. (1977). Mitochondrial ribosome assembly in *Neurospora crassa*. Chloramphenicol inhibits the maturation of small ribosomal subunits. *Journal of Molecular Biology* **116**, 189–205.

Lachke, A. H. & Deshpande, M. V. (1988). *Sclerotium rolfsii*: status in cellulase research. *FEMS Microbiology Reviews* **54**, 177–94.

Ladisch, M. R., Lin, K. W., Voloch, M. & Tsao, G. T. (1983). Process considerations in the enzymatic hydrolysis of biomass. *Enzyme and Microbial Technology* **5**, 82–102.

Lafferty, M. A. & Garrett, R. H. (1974). Purification and properties of *Neurospora crassa* assimilatory nitrite reductase. *Journal of Biological Chemistry* **249**, 7555–67.

Lafon-Lafourcade, S., Geneix, C. & Ribéreau-Gayon, P. (1984). Inhibition of alcoholic fermentation of grape juice must by fatty acids produced by yeast and their elimination by yeast ghosts. *Applied and Environmental Microbiology* **47**, 1246–9.

Lagunas, R. (1986). Misconceptions about energy metabolism of *Saccharomyces cerevisiae*. *Yeast* **2**, 221–8.

Lagunas, R., Dominguez, C., Busturia, A. & Saez, M. J. (1982). Mechanisms of appearance of the Pasteur effect in *Saccharomyces cerevisiae*. Inactivation of sugar transport systems. *Journal of Bacteriology* **152**, 19–25.

Lagunas, R. & Gancedo, C. (1983). Role of phosphate in the regulation of the Pasteur effect in *Saccharomyces cerevisiae*. *European Journal of Biochemistry* **137**, 479–83.

Lagunas, R. & Gancedo, J. M. (1973). Reduced pyridine-nucleotides balance in glucose-growing *Saccharomyces cerevisiae*. *European Journal of Biochemistry* **37**, 90–4.

Lahoz, R., Reyes, F., Gómez, P. & Martinez, M. J. (1983). Lytic enzyme activity in autolysing mycelium of *Aspergillus niger*. *Zeitschrift für Allgemeine Mikrobiologie* **23**, 17–25.

Lahoz, R., Reyes, F. & Perez Leblic, M. I. (1976). Lytic enzymes in the autolysis of filamentous fungi. *Mycopathologia* **60**, 45–9.

Lambowitz, A. M., Slayman, C. W., Slayman, C. L. & Bonner, W. D. (1972). The electron transport components of wild type and *poky* strains of *Neurospora crassa*. *Journal of Biological Chemistry* **247**, 1536–45.

Lang, J. M. & Cirillo, V. P. (1987). Glucose transport in a kinaseless *Saccharomyces cerevisiae* mutant. *Journal of Bacteriology* **169**, 2932–7.

Langen, P. & Liss, E. (1958). Über bindung und Umsatz der Polyphosphate der Hefe. *Biochemische Zeitschrift* **330**, 455–66.

Langen, P., Liss, E. & Lohmann, K. (1962). Art Bildung Umsatz der Polyphosphate der Hefe. In *Acides Ribonucleiques et Polyphosphates. Structure, Synthase et Fonctions*, Colloque International du CNRS, Strasburg, ed. P. Ebel & M. Grunberg-Manago, pp. 603–12. Centre National de la Recherche: Paris.

Larimore, F. S., Kuist, I., Korokowski, P. M. & Roon, R. J. (1980). Transport and metabolism of N-δ-chloracetyl-L-ornithine by *Saccharomyces cerevisiae*. *Archives of Biochemistry and Biophysics* **204**, 234–40.

Larimore, F. S. & Roon, R. J. (1978). Possible site-specific reagent for the general amino acid transport system of *Saccharomyces cerevisiae*. *Biochemistry* **17**, 431–6.

Larsen, S. (1967). Soil phosphorus. *Advances in Agronomy* **19**, 151–210.

Larsson, C., Morales, C., Gustafsson, L. & Adler, L. (1990). Osmoregulation of the salt-tolerant yeast *Debaryomyces hansenii* grown in a chemostat at different salinities. *Journal of Bacteriology* **172**, 1769–74.

Larue, F., Lafon-Lafourcade, S. & Ribereau-Gayon, P. (1980). Relationship between the sterol content of yeast cells and their fermentation activity in grape must. *Applied and Environmental Microbiology* **39**, 808–11.

Lasko, P. F. & Brandriss, M. C. (1981). Proline transport in *Saccharomyces cerevisiae*. *Journal of Bacteriology* **148**, 241–7.

Lavi, L. E., Hermiller, J. B. & Griffin, C. C. (1981). Carrier-mediated transport of D-ribose by *Rhodotorula glutinis*. *Biochimica et Biophysica Acta* **648**, 1–5.

Law, R. E. & Ferro, A. J. (1980). Inhibition of leucine transport in *Saccharomyces* by S-adenosylmethionine. *Journal of Bacteriology* **143**, 427–31.

Lawford, H. G., Pik, J. R., Lawford, G. R., Williams, T. & Kilgerman, A. (1980*a*). Physiology of *Candida utilis* yeast in zinc-limited chemostat culture. *Canadian Journal of Microbiology* **26**, 64–70.

Lawford, H. G., Pik, J. R., Lawford, G. R., Williams, T. & Kilgerman, A. (1980*b*). Hyperaccumulation of zinc by zinc-depleted *Candida utilis* grown in chemostat culture. *Canadian Journal of Microbiology* **26**, 71–6.

Lazo, P. S., Ochoa, A. G. & Gascón, S. (1978). α-Galactosidase (melibiase) from *Saccharomyces carlsbergensis*. Structural and kinetic properties. *Archives of Biochemistry and Biophysics* **191**, 316–24.

Lé John, H. B. & Cameron, L. E. (1973). Cytokinins regulate calcium binding to a glycoprotein from fungal cells. *Biochemical and Biophysical Research Communications* **54**, 1053–60.

Lé John, H. B., Cameron, L. E., Stevenson, R. M. & Menser, R. U. (1974). Influence of cytokinins and sulfhydryl group-reacting agents on calcium transport in fungi. *Journal of Biological Chemistry* **249**, 4016–22.

Leake, J. R. & Read, J. D. (1989). The biology of mycorrhiza in the Ericaceae. XIII. Some characteristics of the extracellular proteinase activity of the ericoid endophyte *Hymenoscyphus ericae*. *New Phytologist* **112**, 69–76.

Leake, J. R. & Read, D. J. (1990*a*). Proteinase activity in mycorrhizal fungi. I. The effect of extracellular pH on the production of proteinase by ericoid endophytes from soils of contrasted pH. *New Phytologist* **115**, 243–50.

Leake, J. R. & Read, D. J. (1990b). Proteinase activity in mycorrhizal fungi. II. The effects of mineral and nitrogen source on induction of extracellular proteinase in *Hymenoscyphus ericaceae*. *New Phytologist* **116**, 123–8.

Leake, J. R. & Read, D. J. (1991). Proteinase activity in mycorrhizal fungi. III. Effects of protein, protein hydrolysate, glucose and ammonium on production of extracellular proteinase by *Hymenoscyphus ericae* (Read) Korf and Kernan. *New Phytologist* **117**, 309–18.

Leão, C. & van Uden, N. (1982). Effects of ethanol and other alkanols on the glucose transport system of *Saccharomyces cerevisiae*. *Biotechnology and Bioengineering* **24**, 2601–4.

Leão, C. & van Uden, N. (1983). Effect of ethanol and other alkanols on the ammonium transport system of *Saccharomyces cerevisiae*. *Biotechnology and Bioengineering* **25**, 2085–9.

Leão, C. & van Uden, N. (1984*a*). Effect of ethanol and other alkanols on the general amino acid permease of *Saccharomyces cerevisiae*. *Biotechnology and Bioengineering* **26**, 403–5.

Leão, C. & van Uden, N. (1984*b*). Effects of ethanol and other alkanols on passive proton influx in the yeast *Saccharomyces cerevisiae*. *Biochimica et Biophysica Acta* **774**, 43–8.

Leão, C. & van Uden, N. (1986). Transport of lactate and other short-chain monocarboxylates in the yeast *Candida utilis*. *Applied Microbiology and Biotechnology* **23**, 389–93.

Leclerc, M., Gondé, P., Arnaud, A., Ratomahenina, R., Galzy, P. & Nicolas, M. (1984). The enzymes systems in a strain of *Candida wickerhamii* Meyer and Yarrow participating in the hydrolysis of cellodextrins. *Journal of General and Applied Microbiology* **30**, 509–21.

Lee, J.-D. & Komagata, K. (1980). Taxonomic study of methanol-assimilating yeasts. *Journal of General and Applied Microbiology* **26**, 133–58.

Lee, R. B., Ratcliffe, R. G. & Southon, T. E. (1990). ^{31}P NMR measurements of the cytoplasmic and vacuolar P_i content of mature maize roots: relationships with phosphorus status and phosphate fluxes. *Journal of*

Experimental Botany **41**, 1063–78.
Lee, Y.-H. & Fan, L. T. (1980). Properties of mode of action of cellulase. *Advances in Biochemical Engineering* **17**, 101–29.
Lee, Y.-H., Fan, L. T. & Fan, L.-S. (1980). Kinetics of hydrolysis of insoluble cellulose by cellulase. *Advances in Biochemical Engineering* **17**, 131–68.
Legerton, T. L., Kanamori, K., Weiss, R. L. & Roberts, J. D. (1983). Measurement of cytoplasmic and vacuolar pH in *Neurospora* using nitrogen-15 nuclear magnetic resonance spectroscopy. *Biochemistry* **22**, 899–903.
Leggett, J. E. (1961). Entry of phosphate into yeast cells. *Plant Physiology* **36**, 277–84.
Lehninger, A. L. (1970). *Biochemistry*. Worth Publishers: New York.
Leisola, M. S. A. & Fiechter, A. (1985). New trends in lignin biodegradation. *Advances in Biotechnological Processes* **5**, 59–90.
Leisola, M. S. A. & Garcia, S. (1989). The mechanism of lignin degradation. In *Enzyme Systems for Lignocellulose Degradation*, ed. M. P. Coughlan, pp. 89–99. Elsevier Applied Science: London.
Leisola, M. S. A., Ulmer, D. C., Haltmeier, T. & Fiechter, A. (1983). Rapid solubilization and depolymerization of purified kraft lignin by thin layers of *Phanerochaete chrysosporium. European Journal of Applied Microbiology and Biotechnology* **17**, 117–20.
Leisola, M., Ulmer, D. C., Waldner, R. & Fiechter, A. (1984). Role of veratryl alcohol in lignin degradation by *Phanerochaete chrysosporium. Journal of Biotechnology* **1**, 331–40.
Lenny, J. F. & Klemmer, H. W. (1966). Factors controlling sexual reproduction and growth in *Pythium graminicola. Nature, London* **209**, 1365–6.
Lenz, H., Wunderwald, P. & Eggerer, H. (1976). Partial purification and some properties of oxalacetase from *Aspergillus niger. European Journal of Biochemistry* **65**, 225–36.
Lerch, K. (1980). Copper metallothionein, a copper-binding protein from *Neurospora crassa. Nature, London* **284**, 368–70.
Levine, W. B. & Marzluf, G. A. (1989). Isolation and characterization of cadmium resistant mutants of *Neurospora crassa. Canadian Journal of Microbiology* **35**, 359–65.
Lewis, C. M. & Fincham, J. R. S. (1970). Genetics of nitrate reductase in *Ustilago maydis. Genetical Research, Cambridge* **16**, 151–63.
Lewis, D. A. & Bisson, L. F. (1991). The *HXT1* gene product of *Saccharomyces cerevisiae* is a new member of the family of hexose transporters. *Molecular and Cellular Biology* **11**, 3804–13.
Lewis, D. H. (1991). Fungi and sugars – a suite of interactions. *Mycological Research* **95**, 897–904.
Lewis, D. H. & Smith, D. C. (1967). Sugar alcohols (polyols) in fungi and green plants. I. Distribution, physiology and metabolism. *New Phytologist* **66**, 143–84.
Lewis, G. & Yamamoto, E. (1990). Lignin – occurrence, biogenesis and degradation. *Annual Review of Plant Physiology and Plant Molecular Biology* **41**, 455–96.
Lewis, M. J. & Phaff, H. J. (1965). Release of nitrogenous substances by brewer's yeast. IV. Energetics in shock excretion of amino acids. *Journal of Bacteriology* **89**, 960–6.
Lichko, L. P. & Okorokov, L. A. (1985). What family of ATPases does the vacuolar H^+-ATPase belong to? *FEBS Letters* **187**, 349–53.

Lichko, L. P. & Okorokov, L. A. (1991). Purification and some properties of membrane-bound and soluble pyrophosphatases of yeast vacuoles. *Yeast* 7, 805–12.

Liebeskind, M., Höcker, M., Wandrey, C. & Jäger, A. G. (1990). Strategies for improved lignin peroxidase production in agitated pellet cultures of *Phanerochaete chrysosporium* and the use of a novel inducer. *FEMS Microbiology Letters* 71, 325–30.

Liese, W. (1970). Ultrastructural aspects of woody tissue disintegration. *Annual Review of Phytopathology* 8, 231–57.

Lievense, J. C. & Lim, H. C. (1982). The growth and dynamics of *Saccharomyces cerevisiae*. *Annual Reports on Fermentation Processes* 5, 211–62.

Ligthelm, M. E., Prior, B. A. & du Preez, J. C. (1988a). The oxygen requirement of yeasts for the fermentation of D-xylose and D-glucose to ethanol. *Applied and Microbial Biotechnology* 28, 63–8.

Ligthelm, M. E., Prior, B. A. & du Preez, J. C. (1988b). The induction of D-xylose catabolising enzymes in *Pachysolen tannophilus* and the relationship to anaerobic D-fermentation. *Biotechnology Letters* 10, 207–12.

Lilly, V. G. & Barnett, H. L. (1951). *Physiology of the Fungi*. McGraw-Hill Book Company: New York.

Lilly, V. G. & Leonian, L. H. (1944). The anti-biotin of dithiobiotin. *Science* 99, 205–6.

Lipke, P. N. & Hull-Pillsbury, C. (1984). Flocculation of *Saccharomyces cerevisiae tup1* mutants. *Journal of Bacteriology* 159, 797–9.

Liss, E. & Langen, P. (1962). Versuche zur Polyphosphate – Überkompensation in Hefezellen nach Phosphatverarmung. *Archiv für Mikrobiologie* 41, 383–92.

Littke, W. R., Bledsoe, C. S. & Edmonds, R. L. (1984). Nitrogen uptake and growth *in vitro* by *Hebeloma crustuliniforme* and other Pacific Northwest mycorrhizal fungi. *Canadian Journal of Botany* 62, 647–52.

Littlefield, L. J. (1967). Phosphorus-32 accumulation in *Rhizoctonia solani* sclerotia. *Phytopathology* 57, 1053–5.

Littlefield, L. J., Wilcoxson, R. D. & Sudia, T. W. (1965a). Translocation of phosphorus-32 in *Rhizoctonia solani*. *Phytopathology* 55, 536–42.

Littlefield, L. J., Wilcoxson, R. D. & Sudia, T. W. (1965b). Translocation in sporophores of *Lentinus tigrinus*. *American Journal of Botany* 52, 599–605.

Littlewood, B. S., Chia, W. & Metzenberg, R. L. (1974). Genetic control of phosphate-metabolizing enzymes in *Neurospora crassa*: relationships among mutations. *Genetics* 79, 419–34.

Lloyd, D. & Cartledge, T. G. (1991). Separation of yeast organelles. In *The Yeasts*, vol. 4 *Yeast Organelles*, ed. A. H. Rose & J. S. Harrison, pp. 121–74. Academic Press: London.

Logan, D. A., Becker, J. M. & Naider F. (1979). Peptide transport in *Candida albicans*. *Journal of General Microbiology* 114, 179–86.

Loll, M. J. & Bollag, J.-M. (1983). Protein transformation in soil. *Advances in Agronomy* 36, 351–82.

Lomnitz, Q. J., Calderón, J., Hernández, G. & Mora, J. (1987). Functional analysis of ammonium assimilation enzymes in *Neurospora crassa*. *Journal of General Microbiology* 133, 2333–40.

Londesborough, J. & Nuutinen, M. (1987). Ca^{2+}/calmodulin-dependent protein kinase in *Saccharomyces cerevisiae*. *FEBS Letters* 219, 249–53.

Lopez, S. & Gancedo, J. M. (1979). Effect of metabolic conditions on protein turnover in yeast. *Biochemical Journal* **178**, 769–76.

Losada, M. (1957). The hydrolysis of raffinose by yeast melibiase and the fermentation of raffinose by complementary gene action. *Comptes rendus des travaux Carlsberg* **25**, 460–82.

Losel, D. M. (1988). Fungal lipids. In *Microbial Lipids*, ed. C. Ratlege & S. G. Wilkinson, pp. 699–806. Academic Press: London.

Loughman, B. C. & Ratcliffe, R. G. (1984). Nuclear magnetic resonance and the study of plants. *Advances in Plant Nutrition* **1**, 241–83.

Loureiro-Dias, M. C. (1987). Glucose and polyol transport systems in *Candida intermedia* and their regulation. *Journal of General Microbiology* **133**, 2737–42.

Loureiro-Dias, M. C. (1988). Movement of protons coupled with glucose transport in yeasts. A comparative study among 248 yeast strains. *Antonie van Leeuwenhock* **54**, 331–43.

Loureiro-Dias, M. C. & Santos, H. (1989). Effects of 2-deoxyglucose on *Saccharomyces cerevisiae* as observed by in vito ^{31}P-NMR. *FEMS Microbiology Letters* **57**, 25–8.

Lowendorf, H. S., Bazinet, G. F. Jr & Slayman, C. W. (1975). Phosphate transport in *Neurospora*. Derepression of a high-affinity transport system during phosphorus starvation. *Biochimica et Biophysica Acta* **389**, 541–9.

Lowendorf, H. S., Slayman, C. L. & Slayman, C. W. (1974). Phosphate transport in *Neurospora*. Kinetic characterization of constitutive low-affinity transport system. *Biochimica et Biophysica Acta* **373**, 369–82.

Lowendorf, H. S. & Slayman, C. W. (1975). Genetic regulation of phosphate transport system II in *Neurospora*. *Biochimica et Biophysica Acta* **413**, 95–103.

Lu, S. H. (1973). The effect of calcium on fruiting of *Cyanthus stercoreus*. *Mycologia* **65**, 329–34.

Luard, E. J. (1982a). Accumulation of intracellular solutes by two filamentous fungi in response to growth at low steady state osmotic potential. *Journal of General Microbiology* **128**, 2563–74.

Luard, E. J. (1982b). Effect of osmotic shock on some intracellular solutes in two filamentous fungi. *Journal of General Microbiology* **128**, 2575–81.

Luard, E. J. (1982c). Growth and accumulation of solutes by *Phytophthora cinnamoni* and other fungi in response to changes in external osmotic potential. *Journal of General Microbiology* **128**, 2583–90.

Luard, E. J. & Griffin, D. M. (1981). Effect of water potential on fungal growth and turgor. *Transactions of the British Mycological Society* **76**, 33–40.

Lucas, C., da Costa, M. & van Uden, N. (1990). Osmoregulatory active sodium–glycerol co-transport in the halotolerant yeast *Debaryomyces hansenii*. *Yeast* **6**, 187–92.

Lucas, C. & van Uden, N. (1986). Transport of hemicellulose monomers in the xylose-fermenting yeast *Candida shehatae*. *Applied Microbiology and Biotechnology* **23**, 491–5.

Lucas, R. L. (1960). Transport of phosphorus in fungal mycelium. *Nature, London* **188**, 763–4.

Lucas, R. L. (1977). The movement of nutrients through fungal mycelium. *Transactions of the British Mycological Society* **69**, 1–9.

Lusby, E. W. Jr & McLaughlin, C. S. (1980). The metabolic properties of acid soluble polyphosphates in *Saccharomyces cerevisiae*. *Molecular and General Genetics* **178**, 69–76.

Lyon, A. J. E. & Lucas, R. L. (1969a). The effect of temperature on the translocation of phosphorus by *Rhizopus stolonifer*. *New Phytologist* **68**, 963–70.

Lyon, A. J. E. & Lucas, R. L. (1969b). The phosphorus metabolism of *Rhizopus stolonifer* and *Chaetomium* sp. with respect to phosphorus translocation. *New Phytologist* **68**, 971–6.

Ma, H., Kubicek, C. P. & Röhr, M. (1985). Metabolic effects of manganese deficiency in *Aspergillus niger*: evidence for increased degradation. *Archives of Microbiology* **141**, 266–8.

MacDonald, M. J., Ambler, R. & Broda, P. (1985). Regulation of intracellular cyclic AMP levels in the white-rot fungus *Phanerochaete chrysosporium* during the onset of idiophasic metabolism. *Archives of Microbiology* **142**, 152–6.

MacDonald, M. J., Paterson, A. & Broda, P. (1984). Possible relationship between cyclic AMP and idiophasic metabolism in the white-rot fungus *Phanerochaete chrysosporium*. *Journal of Bacteriology* **160**, 470–2.

MacFall, J., Slack, S. A. & Iyers, J. (1991a). Effects of *Hebeloma arenosa* and phosphorus fertility on growth of red pine (*Pinus resinosa*) seedlings. *Canadian Journal of Botany* **69**, 372–9.

MacFall, J., Slack, S. A. & Iyers, J. (1991b). Effects of *Hebeloma arenosa* and phosphorus fertility on root acid phosphatase of red pine (*Pinus resinosa*) seedlings. *Canadian Journal of Botany* **69**, 372–9.

Machlis, L. (1953a). Growth and nutrition of water molds in the subgenus *Euallomyces*. I. Growth factor requirements. *American Journal of Botany* **40**, 189–95.

Machlis, L. (1953b). Growth and nutrition of water molds in the subgenus *Euallomyces*. II. Optimum composition of the minimal medium. *American Journal of Botany* **40**, 450–9.

Mackenzie, K. F., Blomberg, A. & Brown, A. D. (1986). Water stress plating hypersensitivity. *Journal of General Microbiology* **132**, 2053–6.

Mackenzie, K. F., Singh, K. K. & Brown, A. D. (1988). Water stress plating hypersensitivity of yeasts: protective role of trehalose in *Saccharomyces cerevisiae*. *Journal of General Microbiology* **134**, 1661–6.

Maclean, D. J. (1982). Axenic culture and metabolism in rust fungi. In *The Rust Fungi*, ed. K. J. Scott & A. K. Chakravorty, pp. 37–84. Academic Press: London.

MacMillan, A. (1956). The entry of ammonia into fungal cells. *Journal of Experimental Botany* **7**, 113–26.

Macris, B. J. & Markakis, P. (1974). Transport and toxicity of sulfur dioxide in *Saccharomyces cerevisiae* var. *ellipsoideus*. *Journal of the Science of Food and Agriculture* **25**, 21–9.

Magan, N. (1993). Tolerance of fungi to sulfur dioxide. In *Stress Tolerance of Fungi*, ed. D. H. Jennings, pp. 173–87. Marcel Dekker: New York.

Mainzer, S. E. & Slayman, C. W. (1978). Mitochondrial adenosine triphosphatase of wild-type and *poky Neurospora crassa*. *Journal of Bacteriology* **133**, 584–92.

Maiorella, B., Blanch, H. W. & Wilke, C. R. (1983). By-product inhibition effects on ethanolic fermentation by *Saccharomyces cerevisiae*. *Biotechnology and Bioengineering* **25**, 103–21.

Mair, T. & Höfer, M. (1988). ATP-induced generation of pH gradient and/or membrane potential in reconstituted membrane vesicles from *Schizosaccharomyces pombe*. *Biochemistry International* **17**, 593–604.

Malajckuk, N. & Cromack, K. Jr (1982). Accumulation of calcium oxalate in the mantle of ectomycorrhizal roots of *Pinus radiata* and *Eucalyptus marginata*. *New Phytologist* **92**, 527–31.

Malikayil, J. A., Lerch, K. & Armitage, I. M. (1989). Proton NMR studies of a metallothionein from *Neurospora crassa*: sequence-specific assignments by NOE measurements in a rotating frame. *Biochemistry* **28**, 2991–4.

Malpartida, F. & Serrano, R. (1980). Purification of the yeast plasma membrane ATPase solubilized with a novel zwitterionic detergent. *FEBS Letters* **111**, 69–72.

Malpartida, F. & Serrano, R. (1981a). Proton translocation catalyzed by the purified yeast plasma membrane ATPase reconstituted in liposomes. *FEBS Letters* **131**, 351–4.

Malpartida, F. & Serrano, R. (1981b). Reconstitution of the proton-translocating adenosine triphosphatase of yeast plasma membranes. *Journal of Biological Chemistry* **256**, 4175–7.

Manavathu, E. K. & Thomas, D. de D. (1985). Chemotropism of *Achlya ambisexualis* to methionine and methionyl compounds. *Journal of General Microbiology* **131**, 751–6.

Mandels, M., Parrish, F. W. & Reese, E. T. (1962). Sophorose as an inducer of cellulase in *Trichoderma viride*. *Journal of Bacteriology* **83**, 400–8.

Mandels, M. & Reese, E. T. (1959). Biologically active impurities in reagent glucose. *Biochemical and Biophysical Research Communications* **1**, 338–40.

Mandels, M. & Reese, E. T. (1963). Inhibition of enzymatic hydrolysis of cellulose and related materials. In *Advances in Enzymatic Hydrolysis of Cellulose and Related Materials*, ed. E. T. Reese, pp. 115–57. Pergamon Press: Oxford.

Mandels, M. & Reese, E. T. (1964). Fungal cellulases and the microbial decomposition of cellulosic fibres. *Developments in Industrial Microbiology* **5**, 5–12.

Mann, B. J., Akins, R. A., Lambowitz, A. M. & Metzenberg, R. L. (1988). The structural gene for a phosphorus-repressible phosphate permease in *Neurospora crassa* can complement a mutation in positive regulatory gene *mic-1*. *Molecular and Cellular Biology* **8**, 1376–9.

Mann, B. J., Bowman, B. J., Grotelueschen, J. & Metzenberg, R. L. (1989). Nucleotide sequence of *pho-4*[+], encoding a phosphate-repressible phosphate permease of *Neurospora crassa*. *Gene* **83**, 281–9.

Marder, R., Becker, J. M. & Naider, F. (1977). Peptide transport in yeast: utilization of leucine- and lysine-containing peptides by *Saccharomyces cerevisiae*. *Journal of Bacteriology* **131**, 906–16.

Marder, R., Rose, B., Becker, J. M. & Naider, F. (1978). Isolation of a peptide transport-deficient mutant of yeast. *Journal of Bacteriology* **136**, 1174–7.

Margoshes, M. & Vallee, B. L. (1957). A cadmium protein from equine kidney cortex. *Journal of the American Chemical Society* **78**, 4813–14.

Margulies, M. & Vishniac, W. (1961). Dissimilation of glucose by the Mx strain of *Rhizopus*. *Journal of Bacteriology* **81**, 1–9.

Marin, B. (ed.) (1986). *Plant Vacuoles: Their Importance in Solute Compartmentation in Cells and their Application in Plant Biotechnology*. Plenum Press: New York.

Mark, C. G. & Romano, A. H. (1971). Properties of the hexose transport systems of *Aspergillus nidulans*. *Biochimica et Biophysica Acta* **240**, 216–26.

Markham, P. & Collinge, A. J. (1987). Woronin bodies of filamentous fungi. *FEMS Microbiology Letters* **46**, 1–12.

Markham, P., Collinge, A. J., Head, J. B. & Poole, R. K. (1987). Is the spatial organisation of fungal hyphae maintained and regulated by Woronin bodies? In *Spatial Organisation in Eukaryotic Microbes*, ed. R. K. Poole & A. P. J. Trinci, pp. 79–99. IRL Press: Oxford.

Marschner, H. (1986). *Mineral Nutrition of Higher Plants*. Academic Press: London.

Marsh, L., Neiman, A. M. & Herskowitz, I. (1991). Signal transduction during pheromone response in yeast. *Annual Review of Cell Biology* **7**, 699–728.

Martin, F. & Canet, D. (1986). Biosynthesis of amino acids during [^{13}C]glucose utilization by the ectomycorrhizal ascomycete *Cenococcum geophylum* monitored by ^{13}C nuclear magnetic resonance. *Physiologie Végétale* **24**, 209–18.

Martin, F., Stewart, G. R., Genetet, I. & Le Tacon, F. (1986). Assimilation of $^{15}NH_4^+$ by beech (*Fagus sylvatica* L.) ectomycorrhizas. *New Phytologist* **102**, 85–94.

Martin, F., Stewart, G. R., Genetet, I. & Mourot, B. (1988). The involvement of glutamate dehydrogenase and glutamine synthetase in ammonium assimilation by the rapidly growing ectomycorrhizal ascomycete *Cenococcum geophylum* Fr. *New Phytologist* **110**, 541–50.

Martinez-Cadena, G., Lucas, M. & Gobema, R. (1982). Biological properties of a calcium-modulator protein from *Phycomyces blakesleeanus*. *Comparative Biochemistry and Physiology* **71B**, 515–18.

Marzluf, G. A. (1970a). Genetical and biochemical studies of distinct sulfate permease species in different development stages of *Neurospora crassa*. *Archives of Biochemistry and Biophysics* **138**, 254–63.

Marzluf, G. A. (1970b). Genetic and metabolic controls for sulfate metabolism in *Neurospora crassa*: isolation and study of chromate-resistant and sulfate transport-negative mutants. *Journal of Bacteriology* **102**, 716–21.

Marzluf, G. A. (1972a). Genetic and metabolic controls for sulfate metabolism in *Neurospora crassa*: a specific permease for choline-O-sulfate. *Biochemical Genetics* **7**, 219–33.

Marzluf, G. A. (1972b). Control of the synthesis activity and turnover of enzymes of sulfur metabolism in *Neurospora crassa*. *Archives of Biochemistry and Biophysics* **150**, 714–24.

Marzluf, G. A. (1974). Uptake and efflux of sulfate in *Neurospora crassa*. *Biochimica et Biophysica Acta* **339**, 374–81.

Marzluf, G. A. (1981). Regulation of nitrogen metabolism and gene expression in fungi. *Microbiological Reviews* **45**, 437–61.

Marzluf, G. A. & Fu, Y.-H. (1989). Genetics, regulation and molecular studies of nitrate assimilation in *Neurospora crassa*. In *Molecular and Genetic Aspects of Nitrate Assimilation*, ed. J. L. Wray & J. R. Kinghorn, pp. 314–27. Oxford Science Publications: Oxford.

Marzluf, G. A. & Metzenberg, R. L. (1968). Positive control by the *CYS-3* locus in regulation of sulfur metabolism in *Neurospora*. *Journal of Molecular Biology* **33**, 423–37.

Matile, P. & Wiemken, A. (1967). The vacuole as the lysosome of the yeast cell. *Archiv für Mikrobiologie* **56**, 148–55.

Matsutani, K., Fukuda, Y., Murata, K., Kimura, A. & Yajima, N. (1992). Adaptation mechanism of yeast to extreme environments – constructions of salt-tolerance mutants of the yeast *Saccharomyces cerevisiae*. *Journal of Fermentation and Bioengineering* **73**, 228–9.

Mattey, M. (1992). Production of organic acids. *Critical Reviews in Biotechnology* **12**, 87–132.

Matzanke, B. & Winkelmann, G. (1981). Siderophore iron transport followed by Mössbauer spectroscopy. *FEBS Letters* **130**, 50–3.

Matzanke, B. F., Bill, E., Müller, G. I., Trautwein, A. X. & Winkelmann, G. (1987a). Metabolic utilization of ^{57}Fe-labelled coprogen in *Neurospora crassa* – an *in vivo* Mössbauer study. *European Journal of Biochemistry* **162**, 643–50.

Matzanke, B. F., Bill, E., Trautwein, A. X. & Winkelmann, G. (1987b). Role of siderophores in iron storage in spores of *Neurospora crassa* and *Aspergillus ochraceous*. *Journal of Bacteriology* **169**, 5873–6.

Matzanke, B. F., Bill, E., Trautwein, A. X. & Winkelmann, G. (1988). Ferricrocin functions as the main intracellular iron-storage compound in mycelium of *Neurospora crassa*. *Biology of Metals* **1**, 18–25.

Matzanke, B. F., Bill, E., Trautwein, A. X. & Winkelmann, G. (1990). Siderophores as iron storage compounds in the yeast *Rhodotorula minuta* and *Ustilago sphaerogena* detected by *in vivo* Mössbauer spectroscopy. *Hyperfine Interactions* **58**, 2359–64.

Maw, G. A. (1963). The uptake of some sulphur-containing amino acids by brewer's yeast. *Journal of General Microbiology* **31**, 247–59.

Maxwell, D. P. & Bateman, D. F. (1968). Oxalic acid biosynthesis is by *Sclerotium rolfsii*. *Phytopathology* **58**, 1635–42.

Maxwell, D. P. & Lumsden, R. D. (1970). Oxalic acid production by *Sclerotinia sclerotium* in infected bean and in culture. *Phytopathology* **60**, 1395–8.

Maxwell, D. P., Maxwell, M. D., Hänssler, G., Armentrout, V. N., Myrray, G. M. & Hoch, H. C. (1975). Microbodies and glyoxylate-cycle enzyme activities in filamentous fungi. *Planta* **124**, 109–23.

Mazon, M. J., Gancedo, J. M. & Gancedo, C. (1982). Inactivation of yeast fructose-1,6-bisphosphatase. *In vivo* phosphorylation of the enzyme. *Journal of Biological Chemistry* **257**, 1128–30.

McCarthy, P. J., Nisbet, L. J., Boehm, J. C. & Kingsbury, W. D. (1985a). Multiplicity of peptide permeases in *Candida albicans*: evidence from novel chromophoric peptides. *Journal of Bacteriology* **162**, 1024–9.

McCarthy, P. J., Troke, P. F. & Gull, K. (1985b). Mechanisms of action of nikkomycin and the peptide transport system of *Candida albicans*. *Journal of General Microbiology* **131**, 775–80.

McCready, R. G. L. & Din, G. A. (1974). Active sulfate transport in *Saccharomyces cerevisiae*. *FEBS Letters* **38**, 361–3.

McCullough, W., Payton, M. A. & Roberts, C. F. (1977). Carbon metabolism in *Aspergillus nidulans*. In *Genetics and Physiology of Aspergillus*, ed. J. E. Smith & J. A. Pateman, pp. 97–129. Academic Press: London.

McCullough, W., Roberts, C. F., Osmani, S. A. & Scrutton, M. C. (1986). Regulation of carbon metabolism in filamentous fungi. In *Carbohydrate Metabolism in Cultured Cells*, ed. M. J. Morgan, pp. 287–356. Plenum Publishing Co.: New York.

McDermott, J. C. B. & Jennings, D. H. (1976). The relationship between the uptake of glucose and 3-O-methylglucose and soluble carbohydrate and polysaccharide in the fungus *Dendryphiella salina*. *Journal of General Microbiology* **97**, 193–209.

McDonough, J. P. & Mahler, H. P. (1982). Covalent phosphorylation of the Mg-dependent ATPase of yeast plasma membranes. *Journal of Biological Chemistry* **257**, 14579–81.

McGillivray, A. M. & Gow, N. A. R. (1987). The transhyphal electrical current of *Neurospora crassa* is carried principally by protons. *Journal of General*

Microbiology **133**, 2875–81.

McKnight, K. B., McKnight, K. H. & Harper, K. T. (1990). Cation exchange capacities and mineral element concentrations of macrofungal stipe tissue. *Mycologia* **82**, 91–8.

Mehotra, B. S. (1951). Physiological studies of some *Phytophthoras*. III. Carbon requirements. *Lloydia* **14**, 122–8.

Mehra, R. K., Garey, J. R., Butt, T. R., Gray, W. R. & Winge, D. R. (1989). *Candida glabrata* metallothioneins. Cloning and sequence of the genes and characterization of the proteins. *Journal of Biological Chemistry* **264**, 19747–53.

Mehra, R. K., Tarbet, E. B., Gray, W. R. & Winge, D. R. (1988). Metal-specific synthesis of two metallothioneins and γ-glutamyl peptides in *Candida glabrata*. *Proceedings of the National Academy of Sciences, USA* **85**, 8815–19.

Mehra, R. K. & Winge, D. R. (1988). Cu(I) binding to the *Schizosaccharomyces pombe* γ-glutamyl peptides varying in chain length. *Archives of Biochemistry and Biophysics* **265**, 381–9.

Mehra, R. K. & Winge, D. R. (1991). Metal ion resistance in fungi-molecular mechanisms and their regulated expression. *Journal of Cellular Biochemistry* **45**, 30–40.

Mehta, R. J., Kingsbury, W. D., Valenta, J. & Actor, P. (1984). Anti-*Candida* activity of polyoxin: example of peptide transport in yeasts. *Antimicrobial Agents and Chemotherapy* **25**, 373–4.

Meikle, A. J., Chudek, J. A., Reed, R. J. & Gadd, G. M. (1991). Natural abundance C-13-nuclear magnetic resonance spectroscopic analysis of acyclic polyol and trehalose accumulation by several yeast species in response to salt stress *FEMS Microbiology Letters* **82**, 163–8.

Meikle, A. J., Reed, R. J. & Gadd, G. M. (1988). Osmotic adjustment and the accumulation of organic solutes in whole cells and protoplasts of *Saccharomyces cerevisiae*. *Journal of General Microbiology* **134**, 3049–60.

Meixner, O., Mischak, H., Kubicek, C. P. & Röhr, M. (1985). Effect of manganese deficiency on plasma-membrane lipid composition and glucose uptake in *Aspergillus niger*. *FEMS Microbiology Letters* **26**, 271–4.

Mellman, I., Fuchs, R. & Helenius, A. (1986). Acidification of the endocytic and exocytic pathways. *Annual Review of Biochemistry* **55**, 663–700.

Meredith, S. A. & Romano, A. H. (1977). Uptake and phosphorylation and 2-doexy-D-glucose by wild-type and single-kinase strains of *Saccharomyces cerevisiae*. *Biochimica et Biophysica Acta* **497**, 745–9.

Merivuori, H., Siegler, K. M., Sands, J. A. & Montenecourt, B. S. (1985). Regulation of cellulase biosynthesis and secretion in fungi. *Biochemical Society Transactions* **13**, 411–14.

Merkel, G. J., Naider, F. & Becker, J. M. (1980). Amino acid uptake by *Saccharomyces cerevisiae* plasma membrane. *Biochimica et Biophysica Acta* **595**, 109–20.

Messenguy, F., Colin, D. & ten Have, J. (1980). Regulation of compartmentation of amino pools in *Saccharomyces cerevisiae* and its effects on metabolic control. *European Journal of Biochemistry* **108**, 439–47.

Messner, R., Gruber, F. & Kubicek, C. P. (1988). Differential regulation of endoglucanase synthesis in *Trichoderma reesei* QM 9414. *Journal of Bacteriology* **170**, 3689–93.

Metzenberg, R. L. & Ahlgren, S. K. (1971). Structural and regulatory control of aryl sulfatase in *Neurospora*: the use of interspecific differences in structural genes. *Genetics* **68**, 369–81.

Metzenberg, R. L. & Nelson, R. E. (1977). Genetic control of phosphorus metabolism in *Neurospora*. In *Molecular Approaches to Eucaryotic Genetic Systems*, ed. G. Wilcox, J. Abelson & C. F. Fox, pp. 253–68. Academic Press: New York.

Meyer, J. & Matile, P. (1974). Regulation of isoenzymes and secretion of invertase in baker's yeast. *Biochemie und Physiologie der Pflanzen* **166**, 377–85.

Miall, L. M. (1978). Organic acids. In *Economic Microbiology*, vol. 2 *Primary Products of Metabolism*, ed. A. H. Rose, pp. 47–119. Academic Press: London.

Miki, B. L. A., Poon, N. H., James, A. P. & Seligy, V. L. (1982a). Possible mechanism for flocculation interactions governed by gene *FLO1* in *Saccharomyces cerevisiae*. *Journal of Bacteriology* **150**, 878–89.

Miki, B. L. A., Poon, N. H. & Seligy, V. L. (1982b). Repression and induction of flocculation interactions in *Saccharomyces cerevisiae*. *Journal of Bacteriology* **150**, 890–9.

Milbradt, B. & Höfer, M. (1990). Isolation and characterization of a *Rhodotorula glutinis* mutant defective in glucose transport. *Journal of Microbiology* **136**, 1961–5.

Mill, P. J. (1964). The nature of the interactions between flocculent cells in the flocculation of *Saccharomyces cerevisiae*. *Journal of General Microbiology* **35**, 61–8.

Millar, D. G., Griffiths-Smith, K., Algar, E. & Scopes, R. K. (1982). Activity and stability of glycolytic enzymes in the presence of ethanol. *Biotechnology Letters* **4**, 601–6.

Miller, A. G. & Budd, K. (1971). Chloride uptake by mycelium of *Neocosmospora vasinfecta* and its inhibition by glucose. *Journal of General Microbiology* **66**, 243–5.

Miller, A. G. & Budd, K. (1975a). Halide uptake by the filamentous ascomycete *Neocosmospora vasinfecta*. *Journal of Bacteriology* **121**, 91–8.

Miller, A. G. & Budd, K. (1975b). The development of an increased rate of Cl⁻ uptake in the ascomycete *Neocosmospora vasinfecta*. *Canadian Journal of Microbiology* **21**, 1211–16.

Miller, A. G. & Budd, K. (1976a). Influence of exogenous sugars and polyols on Cl⁻ influx and efflux by the ascomycete *Neocosmospora vasinfecta*. *Journal of Bacteriology* **126**, 690–8.

Miller, A. G. & Budd, K. (1976b). Evidence for a negative membrane potential and for movement of Cl⁻ against its electrochemical gradient in the ascomycete *Neocosmospora vasinfecta*. *Journal of Bacteriology* **132**, 741–8.

Miller, A. J. & Sanders, D. (1989). The energetics of cytosolic calcium homeostasis in fungal cells. *Plant Physiology and Biochemistry* **27**, 551–6.

Miller, A. J., Vogg, G. & Sanders, D. (1990). Cytosolic calcium homeostasis in fungi – roles of plasma membrane transport and intracellular sequestration of calcium. *Proceedings of the National Academy of Sciences, USA* **87**, 9348–52.

Miller, D. M. & Harun, S. H. (1978). The kinetics of the active and de-energized transport of O-methyl glucose in *Ustilago maydis*. *Biochimica et Biophysica Acta* **514**, 320–31.

Milne, L. & Cooke, R. C. (1969). Translocation of [^{14}C]glucose by *Rhizoctonia solani*. *Transactions of the British Mycological Society* **53**, 279–89.

Mimura, A., Watanabe, S. & Takeda, I. (1971). Biochemical engineering analysis of hydrocarbon fermentation. III. Analysis of emulsification phenomena. *Journal of Fermentation Technology* **49**, 255–71.

Minagawa, N. & Yoshimito, A. (1982). Purification and characterization of the assimilatory NADPH-reductase of *Aspergillus nidulans*. *Journal of Biochemistry* **91**, 761–9.

Mirocha, C. J. & de Vay, J. E. (1971). Growth of fungi on an inorganic medium. *Canadian Journal of Microbiology* **17**, 1373–8.

Mishra, P. (1993). Tolerance of fungi to ethanol. In *Stress Tolerance of Fungi*, ed. D. H. Jennings, pp. 189–208. Marcel Dekker: New York.

Mishra, P. & Kaur, S. (1991). Lipids as modulators of ethanol tolerance in yeast. *Applied Microbiology and Biotechnology* **34**, 697–702.

Mishra, P. & Prasad, R. (1988). Role of phospholipid head groups in ethanol tolerance of *Saccharomyces cerevisiae*. *Journal of General Microbiology* **134**, 3205–11.

Mishra, P. & Prasad, R. (1989). Relationship between ethanol tolerance and fatty acid composition of *Saccharomyces cerevisiae*. *Applied Microbiology and Biotechnology* **30**, 294–8.

Mitchell, A. P. & Magasanik, B. (1984). Regulation of glutamate-repressible gene products by the GLN3 function in *Saccharomyces cerevisiae*. *Molecular and Cellular Biology* **4**, 2758–66.

Mitchell, H. L., Finn, R. F. & Rosendahl, R. O. (1937). The growth and nutrition of white pine (*Pinus strobus* L.) seedlings in culture. *Black Rock Forest Paper* **1**, 58–73.

Mitchell, P. (1967). Translocation through natural membranes. *Advances in Enzymology* **29**, 33–87.

Mitchell, P. (1979). Compartmentation and communication in living systems. Ligand conduction: a general catalytic principle in chemical, osmotic and chemosmotic reaction systems. *European Journal of Biochemistry* **95**, 1–20.

Miura, Y., Okazaki, M., Hamada, S.-I., Murakawa, S.-I. & Yugen, R. (1977). Assimilation of liquid hydrocarbon by micoorganisms. I. Mechanism of hydrocarbon uptake. *Biotechnology and Bioengineering* **19**, 701–4.

Miyakawa, T., Tachikawa, T., Jeong, Y. K., Tsuchiya, E. & Fukui, S. (1985). Transient increase of Ca^{2+} uptake as a signal for mating pheromone-induced differentiation in the heterobasidiomycete yeast *Rhodosporidium toruloides*. *Journal of Bacteriology* **162**, 1304–6.

Miyakawa, T., Tachikawa, T., Jeong, Y. K., Tsuchiya, E. & Fukui, S. (1987). Inhibition of membrane Ca^{2+}-ATPase *in vitro* by mating pheromone in *Rhodosporidium toruloides*, a heterobasidiomycetous yeast. *Biochemical and Biophysical Research Communications* **143**, 893–900.

Mohan, P. M., Rudra, M. P. P. & Sastry, K. S. (1984). Nickel transport in nickel-resistant strains of *Neurospora crassa*. *Current Microbiology* **10**, 125–8.

Mohan, P. M. & Sastry, K. S. (1984). Cobolt transport in nickel-resistant strains of *Neurospora crassa*. *Current Microbiology* **11**, 137–42.

Monéton, P., Sarthou, P. & Le Goffic, F. (1986). Role of the nitrogen source in peptide transport in *Saccharomyces cerevisiae*. *FEMS Microbiology Letters* **36**, 95–8.

Money, N. P. (1990a). Measurement of pore size in the hyphal cell wall of *Achlya bisexualis*. *Experimental Mycology* **14**, 234–42.

Money, N. P. (1990b). Measurement of hyphal turgor. *Experimental Mycology* **14**, 416–25.

Money, N. P. & Harold, F. M. (1992). Extension growth in the water mold *Achlya* – interplay of turgor and wall strength. *Proceedings of the National*

Academy of Sciences, USA **89**, 4245–9.

Money, N. P. & Webster, J. (1988). Cell wall permeability and its relationship to spore release in *Achlya intricata*. *Experimental Mycology* **12**, 169–79.

Money, N. P. & Webster, J. (1989). Mechanism of sporangial emptying in *Saprolegnia*. *Mycological Research* **92**, 45–9.

Money, N. P., Webster, J. & Ennos, R. (1988). Dynamics of sporangial emptying in *Achlya intricata*. *Experimental Mycology* **12**, 13–27.

Monk, B. C., Kurtz, M. B., Marrinan, J. A. & Perlin, D. S. (1991). Cloning and characterization of the plasma membrane H^+-ATPase from *Candida albicans*. *Journal of Bacteriology* **173**, 6826–36.

Monson, A. M. & Sudia, T. W. (1963). Translocation in *Rhizoctonia solani*. *Botanical Gazette* **124**, 440–3.

Montenecourt, B. S., Nhlapo, S. D., Trimiño-Vaquez, H., Cuskey, S., Schamhart, D. H. J. & Eveleigh, D. E. (1981). Regulatory controls in relation to overproduction of fungal cellulases. In *Trends in the Biology of Fermentation for Fuels and Chemicals*, ed. A. Hollaender, pp. 33–55. Plenum Press: New York.

Montgomery, R. A. P. (1982). The role of polysaccharidase enzymes in the decay of wood by basidiomycetes. In *Decomposer Basidiomycetes: Their Biology and Ecology*, ed. J. C. Frankland, J. N. Hedger & M. J. Swift, pp. 51–65. Cambridge University Press: Cambridge.

Moore, D. (1981). Evidence that the NADP-linked glutamate dehydrogenase of *Coprinus cinereus* is regulated by acetyl-CoA and ammonium levels. *Biochimica et Biophysica Acta* **661**, 247–54.

Moore, D. (1984). Developmental biology of the *Coprinus cinereus* carpophore: metabolic regulation in relation to cap morphogenesis. *Experimental Mycology* **8**, 283–97.

Moore, D. & Devadathan, M. S. (1979). Sugar transport in *Coprinus cinereus*. *Biochimica et Biophysica Acta* **550**, 515–26.

Moore, D., Horner, J. & Liu, M. (1987). Co-ordinate control of ammonium scavenging enzymes in the fruit body cap of *Coprinus cinereus* avoids inhibition of sporulation by ammonium. *FEMS Microbiology Letters* **44**, 239–42.

Mora, J. (1990). Glutamine metabolism and cycling in *Neurospora crassa*. *Microbiological Reviews* **54**, 293–304.

Mora, J., Dávila, G., Espín, G., González, A., Guzman, J., Hernández, G., Hummelt, G., Lara, M., Martínez, E., Mora, Y. & Romero, D. (1980). Glutamine metabolism in *Neurospora crassa*. In *Glutamine: Metabolism, Enzymology and Regulation*, ed. J. Mora & R. Palacios, pp. 185–211. Academic Press: New York.

Mora, Y., Espín, G., Willms, K. & Mora, J. (1978). Nitrogen accumulation in mycelium of *Neurospora crassa*. *Journal of General Microbiology* **104**, 241–50.

Morales, C., André, L. & Adler, L. (1990). A procedure for enrichment and isolation of mutants of the salt-tolerant yeast. *FEMS Microbiology Letters* **69**, 73–8.

Morales, M. & Ruiz-Herrera, J. (1989). Subcellular localization of calcium in sporangiophores of *Phycomyces blakesleeanus*. *Archives of Microbiology* **152**, 468–72.

Morlion, M. & Domnas, A. (1962). Uptake and use of allantoin and allantoic acid by yeasts. *Natuurwetenschappelijk Tijdschrift* **44**, 100–21.

Morris, G. J., Winters, L., Coulson, G. E. & Clarke, K. J. (1986). Effect of osmotic stress on the ultrastructure and viability of the yeast *Saccharomyces cerevisiae. Journal of General Microbiology* **132**, 2023–34.

Morton, A. G. (1956). A study of nitrate reduction in mold fungi. *Journal of Experimental Botany* **7**, 97–112.

Morton, A. G. & MacMillan, A. (1954). Assimilation of nitrogen from ammonium salts and nitrate by fungi. *Journal of Experimental Botany* **5**, 232–52.

Mosher, W. A., Saunders, D. H., Kingery, L. B. & Williams, R. J. (1936). Nutritional requirements of the pathogenic mold *Trichophyton interdigitale. Plant Physiology* **11**, 795–806.

Moss, S. T. (1985). The ultrastructural study of taxonomically significant characters of the Thraustochytriales and the Labyrinthulales. *Botanical Journal of the Linnean Society* **91**, 329–57.

Moss, S. T. (ed.) (1986). *The Biology of Marine Fungi.* Cambridge University Press: Cambridge.

Mosse, B. (1973). Advances in the study of vesicular–arbuscular mycorrhiza. *Annual Review of Phytopathology* **11**, 171–96.

Mosse, B., Hayman, D. S. & Arnold, D. J. (1973). Plant growth responses to vesicular–arbuscular mycorrhiza. V. Phosphate uptake by three plant species from P-deficient soils labelled with ^{32}P. *New Phytologist* **72**, 809–15.

Mowll, J. L. & Gadd, G. M. (1983). Zinc uptake and toxicity in the yeasts *Sporobolomyces roseus* and *Saccharomyces cerevisiae. Journal of General Microbiology* **129**, 3421–5.

Mowll, J. L. & Gadd, G. M. (1984). Cadmium uptake by *Aureobasidium pullulans. Journal of General Microbiology* **130**, 279–84.

Mozes, N. & Rouxhet, P. G. (1990). Microbial hydrophobicity and fermentation technology. In *Microbial Cell Surface Hydrophobicity*, ed. R. J. Doyle & M. Rosenberg, pp. 75–105. American Society for Microbiology: Washington, DC.

Müller, D. & Holzer, H. (1981). Regulation of fructose-1,6-bisphosphatase in yeast by phosphorylation/dephosphorylation. *Biochemical and Biophysical Research Communications* **103**, 926–33.

Müller, G., Barclay, S. J. & Raymond, K. N. (1985). The mechanism and specificity of iron transport in *Rhodotorula piliminae* probed by synthetic analogs of rhodotorulic acid. *Journal of Biological Chemistry* **260**, 13916–20.

Münger, K., Germann, U. A. & Lerch, K. (1985). Isolation and structural organization of the *Neurospora crassa* copper metallothionein gene. *EMBO Journal* **4**, 2665–8.

Münger, K., Germann, U. A. & Lerch, K. (1987). The *Neurospora crassa* metallothionein gene. Regulation of expression and chromosomal location. *Journal of Biological Chemistry* **262**, 7363–7.

Münger, K. & Lerch, K. (1985). Copper metallothionein from the fungus *Agaricus bisporus*: chemical and spectroscopic properties. *Biochemistry* **24**, 6751–5.

Munoz, E. & Ingledew, M. W. (1989). Effect of yeast hulls on stuck and sluggish wine fermentation: importance of lipid component. *Applied and Environmental Microbiology* **55**, 1560–4.

Murasugi, A., Wada, C. & Hayashi, Y. (1981a). Cadmium-binding peptide induced in the fission yeast *Schizosaccharomyces pombe. Journal of Bacteriology* **90**, 1561–4.

Murasugi, A., Wada, C. & Hayashi, Y. (1981b). Purification and properties in UV and CD spectra of Cd-binding peptide 1 from *Schizosaccharomyces pombe*. *Biochemical and Biophysical Research Communications* **103**, 1021-8.

Murphy, J. T. & Spence, K. D. (1972). Transport of S-adenosylmethionine in *Saccharomyces cerevisiae*. *Journal of Bacteriology* **109**, 499-504.

Musgrave, A., Ero, L., Scheffer, R. & Oehlers, E. (1977). Chemotropism of *Achlya bisexualis* germ tube hyphae to casein hydrolysate and amino acids. *Journal of General Microbiology* **101**, 65-70.

Muthukumar, G., Kulkarni, R. J. & Nickerson, K. W. (1985). Calmodulin levels in the yeast and mycelial phases of *Ceratocystis ulmi*. *Journal of Bacteriology* **162**, 47-9.

Muthukumar, G., Nickerson, A. W. & Nickerson, K. W. (1987). Calmodulin levels in yeast and filamentous fungi. *FEMS Microbiology Letters* **41**, 253-6.

Mutoh, N. & Hayashi, Y. (1988). Isolation of mutants of *Schizosaccharomyces pombe* unable to synthesise cadystin, small cadmium-binding peptides. *Biochemical and Biophysical Research Communications* **151**, 32-9.

Myrbäck, K. & Willstaedt, E. (1955). Studies on yeast invertase (saccharase). Localisation of the enzyme in the cell wall and its liberation. *Archiv für Kemi* **8**, 367-74.

Nabais, R. C., Sá-Correia, I., Viegas, C. A. & Novais, J. M. (1988). Influence of calcium ion on ethanol tolerance of *Saccharomyces bayanus* and alcoholic fermentation by yeasts. *Applied and Environmental Microbiology* **53**, 2439-46.

Nagodawithana, T. W. & Steinkrauss, K. H. (1976). Influence of the rate of ethanol production and accumulation on the viability of *Saccharomyces cerevisiae* in 'rapid fermentations'. *Applied and Environmental Microbiology* **31**, 158-62.

Naider, F. & Becker, J. M. (1986). Structure-activity relationsips of the yeast α-factor. *CRC Critical Reviews in Biochemistry* **21**, 225-48.

Naider, F., Becker, J. M. & Katchalski-Katzir, E. (1974). Utilization of methionine-containing peptides and their derivatives by a methionine-requiring auxotroph of *Saccharomyces cerevisiae*. *Journal of Biological Chemistry* **249**, 9-20.

Nakahara, T., Erikson, L. E. & Gutierrez, N. R. (1977). Characteristics of hydrocarbon uptake in cultures with two liquid phases. *Biotechnology and Bioengineering* **19**, 9-25.

Nakamoto, R. K., Rao, R. & Slayman, C. W. (1991). Expression of the yeast plasma membrane [H$^+$]ATPase in secretory vesicles. A new strategy for directed mutagenesis. *Journal of Biological Chemistry* **266**, 7940-7.

Nakamoto, R. K. & Slayman, C. W. (1989). Molecular properties of the fungal plasma-membrane [H$^+$]-ATPase. *Journal of Bioenergetics and Biomembranes* **21**, 621-32.

Nakamura, I., Nishikawa, Y., Kamihara, T. & Fukui, S. (1974). Respiratory deficiency in *Saccharomyces carlsbergensis* 4228 caused by thiamine and its prevention by pyridoxine. *Biochemical and Biophysical Research Communications* **59**, 771-6.

Nakamura, I., Ohmura, Y., Nagami, Y., Kamihara, T. & Fukui, S. (1982). Thiamin accumulation and growth inhibition in yeasts. *Journal of General Microbiology* **128**, 2601-9.

Nakamura, T., Fujitak, K., Eguchi, Y. & Yazawa, M. (1984). Properties of calcium-dependent regulatory proteins from fungi and yeast. *Journal of*

Bacteriology **95**, 1551–7.

Natvig, D. O. (1982). Comparative biochemistry of oxygen toxicity in lactic acid-forming aquatic fungi. *Archives of Microbiology* **132**, 107–14.

Natvig, D. O. & Gleason, F. H. (1983). Oxygen uptake by obligately-fermentative aquatic fungi: absence of a cyanide-sensitive component. *Archives of Microbiology* **134**, 5–8.

Nawata, T. (1984). A simple method for making a vibrating probe system. *Plant and Cell Physiology* **25**, 1089–94.

Needleman, R. B., Kaback, D. B., Dubin, R. A., Perkins, E. L., Rosenberg, N. L., Sutherland, K. A., Forrest, D. B. & Michels, C. A. (1984). *MAL6* of *Saccharomyces*: a complex locus containing three genes for maltose fermentation. *Proceedings of the National Academy of Sciences, USA* **81**, 2811–15.

Neijssel, O. M. & Tempest, D. W. (1975). The regulation of carbohydrate metabolism in *Klebsiella aerogenes* NCTC 418 organism growing in chemostat culture. *Archives of Microbiology* **106**, 251–8.

Neijssel, O. M. & Tempest, D. W. (1976). Bioenergetic aspects of aerobic growth of *Klebsiella aerogenes* NCTC 418 in carbon-limited and carbon-sufficient chemostat culture. *Archives of Microbiology* **107**, 215–21.

Neilands, J. B. (1952). A crystalline organo-iron pigment from the smut fungus *Ustilago sphaerogena*. *Journal of the American Chemical Society* **74**, 4846–7.

Neirinck, L., Maleska, R. & Schneider, H. (1982). Alcohol production from sugar mixtures by *Pachysolen tannophilus*. *Biotechnology and Bioengineering Symposium* **12**, 161–9.

Nelson, H., Mandiyan, S. & Nelson, N. (1989). A conserved gene encoding the 57-kDa subunit of the yeast vacuolar H^+-ATPase. *Journal of Biological Chemistry* **264**, 1775–8.

Nelson, H. & Nelson, N. (1989). The progenitor of ATP synthases was closely related to the current vacuolar H^+-ATPases. *FEBS Letters* **247**, 147–53.

Nelson, N. (1989). Structure, molecular genetics and evolution of vacuolar H^+-ATPases. *Journal of Bioenergetics and Biomembranes* **21**, 553–72.

Neujahr, H. V., Lindsjö, S. & Varga, J. M. (1974). Oxidation of phenols by cells and cell-free enzymes from *Candida tropicalis*. *Atonie van Leeuwenhoek* **40**, 209–16.

Neumann, N. P. & Lampen, J. O. (1967). Purification and properties of yeast invertase. *Biochemistry* **6**, 468–75.

Nevalainen, K. M. H., Penttilä, M. E., Harkki, A., Teeri, T. T. & Knowles, J. (1991). The molecular biology of *Trichoderma* and its application to the expression of both homologous and heterologous genes. In *Molecular Industrial Mycology: Systems and Applications for Filamentous Fungi*, ed. S. A. Leong & R. M. Berka, pp. 129–48. Marcel Dekker: New York.

Neville, M. M., Suskind, S. R. & Roseman, S. (1971). A derepressible active transport system for glucose in *Neurospora crassa*. *Journal of Biological Chemistry* **246**, 1294–301.

Newton, A. C. & Caten, C. E. (1988). Auxotrophic mutants of *Septoria nodorum* isolated by direct screening and by selection for resistance to chlorate. *Transactions of the British Mycological Society* **90**, 199–207.

Nicolay, K., Scheffers, W. A., Bruinenberg, P. M. & Kaptein, R. (1983). *In vivo* ^{31}P NMR studies on the role of the vacuole in phosphate metabolism in yeast. *Archives of Microbiology* **134**, 270–5.

Nieboer, E. & Richardson, D. H. S. (1980). The replacement of the nondescript term 'heavy metals' by a biologically and chemically significant classification of metal ions. *Environmental Pollution* series B **1**, 3–26.

Niedermeyer, W. (1976). The elasticity of the yeast cell tonoplast related to its ultrastructure and chemical composition. I. Induced swelling and shrinking; a freeze-etch membrane study. *Cytobiologie* **13**, 380–93.

Niedermeyer, W., Parish, G. R. & Moor, H. (1977). Reactions of yeast cells to glycerol treatment. Alterations to membrane structure and glycerol uptake. *Protoplasma* **92**, 177–93.

Niemietz, C., Hauer, R. & Höfer, M. (1981). Active transport of charged substrates by a proton-sugar co-transport system. *Biochemical Journal* **194**, 433–41.

Niemietz, C. & Höfer, M. (1984). Transport of an anionic substrate by the H^+/monosaccharide symport in *Rhodotorula gracilis*: only the protonated form of the carrier is catalytically active. *Journal of Membrane Biology* **80**, 235–42.

Niere, J. O., Griffith, J. M. & Grant, B. R. (1990). ^{31}P NMR studies on the effect of phosphite on *Phytophthora palmivora*. *Journal of General Microbiology* **136**, 147–56.

Nieuwenhuis, B. J. W. M. & Borst-Pauwels, G. W. F. H. (1984). Derepression of the high-affinity phosphate uptake in the yeast *Saccharomyces cerevisiae*. *Biochimica et Biophysica Acta* **770**, 40–6.

Nieuwenhuis, B. J. W. M., Weijers, C. A. G. M. & Borst-Pauwels, G. W. F. H. (1981). Uptake and accumulation of Mn^{2+} and Sr^{2+} in *Saccharomyces cerevisiae*. *Biochimica et Biophysica Acta* **649**, 83–8.

Nikawa, J.-I., Hosaka, K., Tsukagoshi, Y. & Yamashita, S. (1990). Primary structure of the yeast choline transport gene and regulation of its expression. *Journal of Biological Chemistry* **265**, 15996–6003.

Nikawa, J.-I., Nagumo, T. & Yamashita, S. (1982). *myo*-Inositol transport in *Saccharomyces cerevisiae*. *Journal of Bacteriology* **150**, 441–6.

Nikawa, J.-I., Tsukagoshi, Y. & Yamashita, S. (1991). Isolation and characterization of two distinct *myo*-inositol transporter genes in *Saccharomyces cerevisiae*. *Journal of Biological Chemistry* **266**, 11184–91.

Nikoloff, D. M. & Henry, S. A. (1991). Genetic analysis of yeast phospholipid biosynthesis. *Annual Reviews of Genetics* **25**, 559–83.

Niku-Päavola, M.-L., Lappalainein, A., Enari, T.-M. & Ronnio, V. (1985). A new appraisal of endoglucanases of the fungus *Trichoderma reesei*. *Biochemical Journal* **231**, 75–81.

Nilsson, A. & Adler, L. (1990). Purification and characterization of glycerol-3-phosphate dehydrogenase (NAD^+) in the salt-tolerant yeast *Debaryomyces hansenii*. *Biochimica et Biophysica Acta* **1034**, 180–5.

Nilsson, T. (1974). Comparative study on the cellulolytic activity of white-rot and brown-rot fungi. *Material und Organismen* **9**, 173–8.

Nisbet, T. M. & Payne, J. W. (1979a). Peptide uptake in *Saccharomyces cerevisiae*: characteristics of transport system shared by di- and tripeptide. *Journal of General Microbiology* **115**, 127–33.

Nisbet, T. M. & Payne, J. W. (1979b). Specificity of peptide uptake in *Saccharomyces cerevisiae* and isolation of a bacilysin-resistant peptide-transport-deficient mutant. *FEMS Letters* **6**, 193–6.

Nishi, A. (1960). Enzymatic studies on the phosphorus metabolism in germinating spores of *Aspergillus niger*. *Journal of Biochemistry, Tokyo* **48**, 758–67.

Nishi, A. (1961). Role of polyphosphate and phospholipid in germinating spores of *Aspergillus niger*. *Journal of Bacteriology* **81**, 10–19.

Literature cited 551

Nishida, A. & Eriksson, K.-E. (1987). Formation, purification and partial characterization of methanol oxidase, a H_2O_2-producing enzyme in *Phanerochaete chrysosporium. Biotechnology and Applied Biochemistry* 9, 325–38.

Nishihara, H. & Toraya, T. (1987). Essential roles of cell surface protein and carbohydrate components in flocculation of a brewer's yeast. *Agricultural and Biological Chemistry* 51, 2721–6.

Nishihara, H., Toraya, T. & Fukui, S. (1977). Effect of chemical modification on cell surface components of a brewer's yeast on the floc-forming ability. *Archives of Microbiology* 115, 19–23.

Nishihara, H., Toraya, T. & Fukui, S. (1982). Flocculation cell walls of brewer's yeast and effects of metal ions, protein denaturants and enzyme treatments. *Archives of Microbiology* 131, 112–15.

Nishimura, H., Kawasaki, Y., Nosaka, K., Yoshinobu, K. & Iwashima, A. (1991). A constitutive thiamine mutation *thio80*, causing reduced thiamine pyrophosphokinase activity in *Saccharomyces cerevisiae. Journal of Bacteriology* 173, 2716–19.

Nobre, M. F. & da Costa, M. S. (1985). The accumulation of polyols by the yeast *Debaryomyces hansenii* in response to water stress. *Canadian Journal of Microbiology* 31, 1061–4.

Norris, P. R. & Kelly, D. P. (1977). Accumulation of cadmium and cobalt by *Saccharomyces cerevisiae. Journal of General Microbiology* 99, 317–24.

Norris, P. R. & Kelly, D. P. (1979). Accumulation of metals by bacteria and yeasts. *Developments in Industrial Microbiology* 20, 299–308.

North, M. J. (1982). Comparative biochemistry of the proteinases of eukaryotic microorganisms. *Microbiological Reviews* 46, 308–40.

Nosaka, K., Nishimura, H. & Iwashima, A. (1986). Effect of tunicamycin on thiamine transport in *Saccharomyces cerevisiae. Biochimica et Biophysica Acta* 858, 309–11.

Nosaka, K., Nishimura, H. & Iwashima, A. (1988). Identity of soluble thiamine-binding protein with thamin-repressible acid phosphatase in *Saccharomyces cerevisiae Biochimica et Biophysica Acta* 967, 49–55.

Nowakowska-Waszezuck, A. (1973). Utilization of some tricarboxylic-acid-cycle intermediates by mitochondria and growing mycelium of *Aspergillus terreus. Journal of General Microbiology* 79, 19–29.

Nuccitelli, R. (1986). A two-dimensional vibrating probe with a computerized graphics display. In *Ionic Currents in Development*, ed. R. Nuccitelli, pp. 13–20. Alan R. Liss: New York.

Nummi, M., Niku-Päavala, M.-L., Lappalainen, A., Enari, T.-M. & Ronnio, V. (1983). Cellobiohydrolase from *Trichoderma reesei. Biochemical Journal* 215, 677–83.

Núñez, C. G. & Callieri, D. A. S. (1989). Studies on the polyphosphate cycle in *Candida utilis*. Effect of dilution rate and nitrogen source in continuous culture. *Applied Microbiology and Biotechnology* 31, 562–6.

Nuss, I., Jennings, D. H. & Veltkamp, C. J. (1991). Morphology of *Serpula lacrymans*. In Serpula lacrymans: *Fundamental Biology and Control Strategies*, eds. D. H Jennings & A. F. Bravery, pp. 9–38. John Wiley: Chichester.

Nye, P. H. & Tinker, P. B. (1969). The concept of a root demand coefficient. *Journal of Applied Ecology* 6, 293–300.

Nye, P. H. & Tinker, P. B. (1977). *Solute Movement in the Soil–Root System.* Blackwell Scientific Publications: Oxford.

Oaks, A. & Hirel, B. (1985). Nitrogen metabolism in roots. *Annual Review of Plant Physiology* **36**, 345–65.

Obayashi, A., Yorifuji, H., Jamagata, T., Ijichi, T. & Kanie, M. (1966). Respiration in organic acid-forming molds. I. Purification of cytochrome *c*, coenzyme Q_a and L-lactate dehydrogenase from lactate-forming *Rhizopus oryzae. Agricultural and Biological Chemistry* **30**, 717–24.

O'Connor, M. L. & Quayle, J. R. (1980). Pentose phosphate-dependent fixation of formaldehyde by methanol-grown *Hansenula polymorpha* and *Candida boidinii. Journal of General Microbiology* **120**, 219–25.

Ødegård, K. (1952). The physiology of *Phycomyces blakesleeanus* Burgeff. I. Mineral requirements on a glucose-asparagine medium. *Physiologia Plantarum* **5**, 583–609.

Oertli, J. J. (1984). Water relations in cell walls and cells in the intact plant. *Zeitschrift für Pflanzenernährung und Bodenkunde* **147**, 60–7.

Oertli, J. J. (1986). Negative turgor pressures in plant cells. *Zeitschrift für Pflanzenernährung und Bodenkunde* **147**, 60–7.

O'Fallon, J. V., Wright, R. H. & Calza, R. E. (1991). Glucose metabolic pathways in the anaerobic rumen fungus *Neocallismastix fontalis. Biochemical Journal* **274**, 595–600.

Ogata, K., Nishikawa, H. & Ohsugi, M. (1969). A yeast capable of utilizing methanol. *Agricultural and Biological Chemistry* **33**, 1519–20.

Ogawa, K. & Toyama, N. (1964). Resolution of *Trichoderma viride* cellulolytic complex. I. Isolation of a celluloytic component capable of degrading filter paper. *Journal of Fermentation Technology* **42**, 199–206.

Ohsumi, Y. & Anraku, Y. (1981). Active transport of basic amino acids driven by a proton motive force in vacuolar membrane vesicles of *Saccharomyces cerevisiae. Journal of Biological Chemistry,* **256**, 2079–82.

Ohsumi, Y. & Anraku, Y. (1983). Calcium transport driven by a proton motive force in vacuolar membrane vesicles of *Saccharomyces cerevisiae. Journal of Biological Chemistry* **258**, 5614–17.

Ohsumi, Y., Kitamoto, K. & Anraku, Y. (1988). Changes induced in the permeability barrier of the yeast plasma membrane by cupric ion. *Journal of Bacteriology* **170**, 2672–82.

Ohta, K. & Hayashida, S. (1983). Role of Tween 80 and monolein in a lipid-sterol-protein complex which enhances ethanol tolerance of sake yeasts. *Applied and Environmental Microbiological* **46**, 821–5.

Ohya, Y., Ohsumi, Y. & Anraku, Y. (1986). Isolation and characterization of Ca^{2+}-sensitive mutants of *Saccharomyces cerevisiae. Journal of General Microbiology* **132**, 979–88.

Ohya, Y., Umemoto, N., Tanida, I., Ohta, A., Iida, H. & Anraku, Y. (1991). Calcium-sensitive *cls* mutants of *Saccharomyces cerevisiae* showing a Pet⁻ phenotype are ascribable to defects of vacuolar H^+-ATPase activity. *Journal of Biological Chemistry* **266**, 13971–7.

Ohya, Y., Uno, I., Ishikawa, T. & Anraku, Y. (1987). Purification and biochemical properties of calmodulin from *Saccharomyces cerevisiae. Journal of Biochemistry* **168**, 13–19.

Okada, H. & Halvorson, H. O. (1964a). Uptake of α-thioethyl D-glucopyranoside by *Saccharomyces cerevisiae. Biochimica et Biophysica Acta* **82**, 538–46.

Okada, H. & Halvorson, H. O. (1964b). Uptake of α-thioethyl D-glucopyranoside by *Saccharomyces cerevisiae*. II. General characteristics of an active transport system. *Biochimica et Biophysica Acta* **82**, 547–55.

Okorokov, L. A., Kulakovskaya, T. V., Lichko, L. P. & Polorotova, E. V. (1985). H$^+$/ion antiport as the principal mechanism of transport systems in the vacuolar membrane of the yeast *Saccharomyces cerevisiae*. *FEBS Letters* **192**, 303–6.

Okorokov, L. A., Lichko, L. P., Kadomtseva, V. M., Kholodenko, V. P., Titovsky, V. T. & Kulaev, I. S. (1977). Energy-dependent transport of manganese into yeast cells and distribution of accumulated ions. *European Journal of Biochemistry* **75**, 373–7.

Okorokov, L. A., Lichko, L. P. & Kulaev, I. S. (1980). Vacuoles: main compartments of potassium, magnesium and phosphate ions in *Saccharomyces carlsbergensis*. *Journal of Bacteriology* **144**, 661–5.

Olsson, S. & Jennings, D. H. (1991a). A glass fibre filter technique for studying nutrient uptake by fungi: the technique used on colonies grown in nutrient gradients. *Experimental Mycology* **15**, 292–301.

Olsson, S. & Jennings, D. H. (1991b). Evidence for diffusion being the mechanism of translocation in the hyphae of three molds. *Experimental Mycology* **15**, 302–9.

Ongjoco, R., Szkutnicka, K. & Cirillo, V. P. (1987). Glucose transport in vesicles reconstituted from *Saccharomyces cerevisiae* membranes and liposomes. *Journal of Bacteriology* **169**, 2926–31.

Onishi, H. (1963). Osmophilic yeasts. *Advances in Food Research* **12**, 53–96.

Onishi, H. & Shiromaru, Y. (1984). Physiological changes induced by salt stress in a salt-tolerant soy-yeast *Saccharomyces rouxii*. *FEMS Microbiology Letters* **25**, 175–8.

Opekarová, M., Driessen, A. J. M. & Konings, W. N. (1987). Protonmotive-force-driven leucine uptake in yeast plasma membrane vesicles. *FEBS Letters* **213**, 45–8.

Orlowski, J. H. & Barford, J. P. (1987). The mechanism of uptake of multiple sugars by *Saccharomyces cerevisiae* in batch culture under fully aerobic cultures. *Applied Microbiology and Biotechnology* **25**, 459–63.

Orpin, C. G. (1976). Studies on the rumen flagellate *Sphaeromonas communis*. *Journal of General Microbiology* **94**, 270–80.

Orpin, C. G. (1977a). The rumen flagellate *Piromonas communis*: its life-history and invasion of plant material in the rumen. *Journal of General Microbiology* **99**, 107–17.

Orpin, C. G. (1977b). The occurence of chitin in the cell walls of the rumen organisms *Neocallimastix frontalis*, *Piromonas communis* and *Sphaeromonas communis*. *Journal of General Microbiology* **99**, 215–18.

Orpin, C. G. (1993). Anaerobic fungi. In *Stress Tolerance of Fungi*, ed. D. H. Jennings, pp. 257–73. Marcel Dekker: New York.

Ortega, J. K. E., Keanini, R. G. & Manica, K. J. (1988a). Pressure probe technique to study transpiration in *Phycomyces*. *Plant Physiology* **87**, 11–14.

Ortega, J. K. E., Manica, K. J. & Keanini, R. G. (1988b). *Phycomyces*: turgor pressure behaviour during the light and avoidance growth responses. *Photochemistry and Photobiology* **48**, 697–704.

Ortega, M. D. & Rodríguez-Navarro, A. (1985). Potassium and rubidium effluxes in *Saccharomyces cerevisiae*. *Zeitschrift für Naturforschung* **40C**, 721–5.

Ortega, M. D. & Rodríguez-Navarro, A. (1986). Sodium ion transport in *Neurospora crassa*. *Physiologia Plantarum* **66**, 705–11.

Orth, A. B., Denny, M. & Tien, M. (1991). Overproduction of lignin-degrading enzymes by an isolate of *Phanerochaete chrysosporium*. *Applied and Environmental Microbiology* **57**, 2591–6.

Orthofer, R., Kubicek, C. P. & Röhr, M. (1979). Lipid levels and manganese deficiency in citric acid producing strains of *Aspergillus niger*. *FEMS Microbiology Letters* **5**, 403–6.

Oshima, Y. (1982). Regulatory circuits for gene expression: the metabolism of galactose and phosphate. In *The Molecular Biology of the Yeast Saccharomyces: Metabolism and Gene Expression*, ed. J. N. Strathern, E. W. Jones & J. R. Broach, pp. 159–80. Cold Spring Harbor Laboratory Press: Cold Spring Harbor, NY.

Oshima, Y. (1991). Impact of the Douglas–Hawthorne model as a paradigm for elucidating cellular regulatory mechanisms in fungi. *Genetics* **128**, 195–201.

Osmani, S.A., Marston, F. A. O., Selmes, I. P., Chapman, A. G. & Scrutton, M. C. (1981). Pyruvate carboxylase from *Aspergillus nidulans*. Regulatory properties. *European Journal of Biochemistry* **118**, 271–8.

Osmani, S.A. & Scrutton, M. C. (1983). The sub-cellular localization of pyruvate carboxylase and of some other enzymes in *Aspergillus nidulans*. *European Journal of Biochemistry* **133**, 551–60.

Osmani, S.A. & Scrutton, M. C. (1985). The subcellular localization and regulatory properties of pyruvate carboxylase from *Rhizopus arrhizus*. *European Journal of Biochemistry* **147**, 119–28.

Osothsilp, C. & Subden, R. C. (1986). Malate transport in *Schizosaccharomyces pombe*. *Journal of Bacteriology* **168**, 1439–43.

Otjen, L. & Blanchette, R. A. (1984). *Xylobolus frustulatus* decay of oak: patterns of a selective delignification and subsequent cellulose removal. *Applied and Environmental Microbiology* **47**, 670–6.

Otjen, L. & Blanchette, R. A. (1986a). A discussion of microstructural changes in wood during decomposition by white rot basidiomycetes. *Canadian Journal of Botany* **64**, 905–11.

Otjen, L. & Blanchette, R. A. (1986b). Selective delignification of birch wood (*Betula papyrifera*) by *Hirschioporus pargamenus* in the field and laboratory. *Holzforschung* **40**, 183–9.

Otjen, L., Blanchette, R. A. & Leatham, G. F. (1988). Lignin distribution in wood delignified by white-rot fungi: X-ray microanalysis of decayed wood treated with bromine. *Holzforschung* **42**, 281–8.

Otoguro, K., Awaya, J., Tanaka, H. & Omura, S. (1981). Saturated fatty acid-starved cells of *Saccharomyces cerevisiae* grown in the presence of cerulenin and oleic acid. *Journal of Biochemistry* **89**, 523–9.

Ottolenghi, P. (1967). The uptake of bovine serum albumin by a strain of *Saccharomyces* and its physico-pathological consequences. *Comptes rendus des travaux Carlsberg* **36**, 95–111.

Oura, E. (1977). Reaction products of yeast fermentation. *Process Biochemistry* **12**, 19–21.

Overman, S. A. & Romano, A. H. (1969). Pyruvate carboxylase of *Rhizopus nigricans* and its role in fumaric acid production. *Biochemical and Biophysical Research Communications* **37**, 457–63.

Paek, Y. L. & Weiss, R. L. (1989). Identification of an arginine carrier in the vacuolar membrane of *Neurospora crassa*. *Journal of Biological Chemistry* **264**, 7285–90.

Paietta, J. V. (1990). Molecular cloning and analysis of the *scon-2* negative regulatory gene of *Neurospora crassa*. *Molecular and Cellular Biology* **10**, 3630–7.

Paietta, J. V. (1992). Production of the CYS3 regulator, a bZIP DNA-binding protein, is sufficient to induce sulfur gene expression in *Neurospora crassa Molecular and Cellular Biology* **12**, 1568–77.

Paietta, J. V., Atkins, R. A., Lambowitz, A. M. & Marzluf, G. A. (1987). Molecular cloning and characterization of the *cys-3* regulatory gene in *Neurospora crassa. Molecular and Cellular Biology* **7**, 2506–11.

Pairunan, A. K., Robson, A. D. & Abbott, L. K. (1980). The effectiveness of vesicular–arbuscular mycorrhizas in increasing growth and phosphorus uptake by subterranean clover from phosphorus sources of different solubilities. *New Phytologist* **84**, 327–38.

Palacios, J. & Serrano, R. (1978). Proton permeability induced by polyene antibiotics. A plausible mechanism for their inhibition of maltose fermentation in yeast. *FEBS Letters* **91**, 198–201.

Pall, M. L. (1970). Amino acid transport in *Neurospora crassa*. III. Acidic amino acid transport. *Biochimica et Biophysica Acta* **211**, 513–20.

Pall, M. L. (1971). Amino acid transport in *Neurospora crassa*. IV. Properties and regulation of a methionine transport system. *Biochimica et Biophysica Acta* **233**, 201–14.

Pall, M. L. (1981). Adenosine 3′,5′-phosphate in fungi. *Microbiological Reviews* **45**, 462–80.

Pall, M. L. & Kelly, K. A. (1971). Specificity of transinhibition of amino acid transport in *Neurospora*. *Biochemical and Biophysical Research Communications* **42**, 940–7.

Palmer, J. G., Murmanis, L. & Highley, T. L. (1983). Visualization of hyphal sheath in wood-decay Hymenomycetes. II. White rotters. *Mycologia* **75**, 1005–10.

Palmer, J. M., Harvey, P. J. & Schoemaker, H. E. (1987). The role of peroxidases, radical cations and oxygen in the degradation of lignin. *Philosophical Transactions of the Royal Society* A **321**, 495–505.

Pan, S. S. & Nason, A. (1978). Purification and characterization of homogeneous assimilatory nicotinamide dinucleotide phosphate-nitrate reductase from *Neurospora crassa. Biochimica et Biophysica Acta* **523**, 297–313.

Panchal, C. J., Bast, L., Russell, I. & Stewart, G. G. (1988). Repression of xylose utilization by glucose in xylose-fermenting yeasts. *Canadian Journal of Microbiology* **34**, 1316–20.

Panchal, C. J., Peacock, L. & Stewart, G. G. (1982). Increased osmotolerance of genetically modified ethanol producing strains of *Saccharomyces* sp. *Biotechnology Letters* **4**, 639–44.

Panchal, C. J. & Stewart, G. G. (1981). Regulatory factors in alcohol fermentations. In *Current Developments in Yeast Research*, ed. G. G. Stewart & I. Russell, pp. 9–15. Pergamon Press: Canada.

Papendick, R. I. & Mulla, D. J. (1986). Basic principles of cell and tissue water relations. In *Water, Fungi and Plants,* ed. P. G. Ayres & L. Boddy, pp. 1–25. Cambridge University Press: Cambridge.

Pardo, J. P. & Slayman, C. W. (1988). The fluorescein isothiocyanate-binding site of the plasma-membrane H^+-ATPase of *Neurospora crassa. Journal of Biological Chemistry* **263**, 18664–8.

Pardo, J. P. & Slayman, C. W. (1989). Cysteine 532 and cysteine 545 are the *N*-ethylmaleimide-reactive residues of the *Neurospora* plasma membrane H^+-ATPase. *Journal of Biological Chemistry* **264**, 9373–9.

Park, D. & Robinson, P. M. (1966). Internal pressure of hyphal tips of fungi and its significance in morphogenesis. *Annals of Botany* N.S. **30**, 425–39.

Park, D. & Robinson, P. M. (1967). A fungal hormone controlling water distribution normally associated with cell ageing in fungi. *Symposium of the Society for Experimental Biology* **21**, 323–36.

Park, P. B. (1975). Biodeterioration in aircraft fuel systems. In *Microbial Aspects of the Biodeterioration of Materials*, ed. R. J. Gilbert & D. W. Lovelock, pp. 105–26. Academic Press: London.

Parkin, M. J. & Ross, I. S. (1986). The regulation of Mn^{2+} and Cu^{2+} uptake in cells of the yeast *Candida utilis* grown in continuous culture. *FEMS Microbiology Letters* **37**, 59–62.

Parkinson, S. M., Jones, R., Meharg, A. A., Wainwright, M. & Killham, K. (1991). The quantity and fate of carbon assimilated from $^{14}CO_2$ by *Fusarium oxysporium* grown under oligotrophic and near oligogrophic conditions. *Mycological Research* **95**, 1345–9.

Parkinson, S. M., Killham, K. & Wainwright, M. (1990). Assimilation of $^{14}CO_2$ by *Fusarium oxysporium* grown under oligotrophic conditions. *Mycological Research* **94**, 959–64.

Parkinson, S. M., Wainwright, M. & Killham, K. (1989). Observations on oligotrophic growth of fungi on silica gel. *Mycological Research* **93**, 529–34.

Passow, H. & Rothstein, A. (1960). The binding of mercury by the yeast cell in relation to changes in permeability. *Journal of General Physiology* **43**, 621–33.

Passow, H., Rothstein, A. & Loewenstein, B. (1959). An all-or-none response in the release of potassium by yeast cells treated with methylene blue and other basic redox dyes. *Journal of General Physiology* **43**, 97–107.

Paszczynska, A., Huynh, V.-B. & Crawford, R. (1985). Enzymatic activities of an extra-cellular or manganese-dependent peroxidase from *Phanerochaete chrysosporium*. *FEMS Microbiology Letters* **29**, 37–42.

Paszczynska, A., Huynh, V.-B. & Crawford, R. (1986). Comparison of ligninase-1 and peroxidase-M2 from the white-rot fungus *Phanerochaete chrysosporium*. *Archives of Biochemistry and Biophysics* **244**, 750–65.

Patching, J. W. & Rose, A. H. (1971). Cold osmotic shock in *Saccharomyces cerevisiae*. *Journal of Bacteriology* **108**, 451–9.

Pateman, J. A., Kinghorn, J. R. & Dunn, E. (1974). Regulatory aspects of L-glutamate transport in *Aspergillus nidulans*. *Journal of Bacteriology* **119**, 534–42.

Paton, F. M. & Jennings, D. H. (1988). The effect of sodium and potassium chloride and polyols on malate and glucose 6-phosphate dehyrogenases from the marine fungus *Dendryphiella salina*. *Transactions of the British Mycological Society* **91**, 205–15.

Paton, F. M. & Jennings, D. H. (1989). Evidence that *Thraustochytrium* is unable to synthesise lysine. *Mycological Research* **92**, 470–6.

Paton, W. H. N. & Budd, K. (1972). Zinc uptake in *Neocosmospora vasinfecta*. *Journal of General Microbiology* **72**, 173–4.

Payne, J. W. & Nisbet, T. M. (1981). Continuous monitoring of substrate uptake by microorganisms using fluorescamine: application to peptide transport by *Saccharomyces cerevisiae* and *Streptococcus faecalis*. *Journal of Applied Biochemistry* **3**, 447–58.

Payne, J. W. & Shallow, D. A. (1985). Studies on drug targeting in the pathogenic fungus *Candida albicans*: peptide transport mutants resistant to polyoxins, nikkomycins and bacilysin. *FEMS Microbiology Letters* **28**, 55–60.

Payton, M. A. & Roberts, C. F. (1976). Mutants of *Aspergillus nidulans* lacking pyruvate kinase. *FEBS Letters* **66**, 73–6.

Peinado, J. M., Barbero, A. & van Uden, N. (1987). Repression and inactivation by glucose of the maltose transport system of *Candida utilis*. *Applied Microbiology and Biotechnology* **26**, 154–7.

Peinado, J. M., Cameira-dos-Santos, P. J. & Loureiro-Dias, M. C. (1989). Regulation of glucose transport in *Candida utilis*. *Journal of General Microbiology* **135**, 195–201.

Peinado, J. M. & Loureiro-Dias, M. C. (1986). Reversible loss of affinity induced by glucose in the maltose-H$^+$ symport of *Saccharomyces cerevisiae*. *Biochimica et Biophysica Acta* **856**, 189–92.

Peleg, Y., Barak, A., Scrutton, M. C. & Goldberg, I. (1989). Malic acid accumulation by *Aspergillus flavus*. 3. C-13 NMR and isoenzyme analysis. *Applied Microbiology and Biotechnology* **30**, 176–83.

Peleg, Y., Rahamim, E., Kessel, M. & Goldberg, I. (1988*a*). Malic acid accumulation by *Aspergillus flavus*. 2. Crystals and hair-like processes formed by *A. flavus* in L-malic acid production medium. *Applied Microbiology and Biotechnology* **28**, 76–9.

Peleg, Y., Stieglitz, B. & Goldberg, I. (1988*b*). Malic acid accumulation by *Aspergillus flavus*. 1. Biochemical aspects of acid biosynthesis. *Applied Microbiology and Biotechnology* **28**, 69–75.

Peña, A. (1975). Studies on the mechanism of K$^+$ transport in yeast. *Archives of Biochemistry and Biophysics* **167**, 397–409.

Peña, A., Mora, M. A. & Carrasco, N. (1979). Uptake and effects of several cationic dyes on yeast. *Journal of Membrane Biology* **47**, 261–84.

Peña, A. & Ramirez, G. (1975). Interaction of ethidium bromide with the transport system for monovalent cations in yeast. *Journal of Membrane Biology* **23**, 369–84.

Peña, A. & Ramirez, G. (1991). An energy-dependent efflux system for potassium ions in yeast. *Biochimica et Biophysica Acta* **1068**, 237–44.

Pennincx, M. J., Jaspers, C. J. Legrain, M. J. (1983). The glutathione-dependent glyoxalase pathway in the yeast *Saccharomyces cerevisiae*. A vital defense line against methylglyoxal produced during glycerol catabolism. *Journal of Biological Chemistry* **258**, 6030–6.

Pepper, I. L. & Miller, R. H. (1978). Comparison of the oxidation of thiosulphate and elemental sulphur by two heterotrophic bacteria and *Thiobacillus thiooxidans*. *Soil Science* **126**, 9–14.

Perea, J. & Gancedo, C. (1978). Glucose transport in a glucose phosphate isomeraseless mutant of *Saccharomyces cerevisiae*. *Current Microbiology* **1**, 209–11.

Perl, M., Kearney, E. B. & Singer, T. P. (1976). Transport of riboflavin into yeast cells. *Journal of Biological Chemistry* **251**, 3221–8.

Perlin, D. S., Brown, C. L. & Haber, J. E. (1988). Membrane potential deficit in hygromycin B-resistant *pma1* mutants of *Saccharomyces cerevisiae*. *Journal of Biological Chemistry* **263**, 18118–22.

Perlin, D. S., Harris, S. L., Seto-Young, D. & Haber, J. E. (1989). Defective H$^+$-ATPase of hygromycin B-resistant *pma1* mutants from *Saccharomyces cerevisiae*. *Journal of Biological Chemistry* **264**, 21857–64.

Perlin, D. S., Kasamo, K., Brooker, R. J. & Slayman, C. W. (1984). Electrogenic H$^+$ translocation by the plasma membrane ATPase of *Neurospora*. Studies on plasma membrane vesicles and reconstituted enzyme. *Journal of Biological Chemistry* **259**, 7884–92.

Perlin, D. S., San Francisco, M. J. D., Slayman, C. W. & Rosen, B. P. (1986). H$^+$/ATP stoichiometry of proton pumps from *Neurospora crassa* and *Escherichia coli. Archives of Biochemistry and Biophysics* **248**, 53–61.

Perlman, D. (1978). Vitamins. In *Economic Microbiology*, vol. 2 *Primary Products of Metabolism*, ed. A. H. Rose, pp. 303–26. Academic Press: London.

Petrov, V. V. & Okorokov, L. A. (1990). Increase of the anion and proton permeability of *Saccharomyces carlsbergensis* plasmalemma by *n*-alcohols as a possible cause for de-energisation. *Yeast*, **6**, 311–18.

Pettrotta-Simpson, T. F., Talmadege, J. E. & Spencer, K. D. (1975). Specificity and genetics of *S*-adenosylmethionine transport in *Saccharomyces cerevisiae. Journal of Bacteriology* **123**, 516–22.

Pfyffer, G. E., Pfyffer, B. U. & Rast, D. M. (1986). The polyol pattern and chemosystematics of the fungi. *Sydowia* **39**, 160–201.

Pfyffer, G. E. & Rast, D. M. (1988). The polyol pattern of fungi as influenced by the carbohydrate nutrient source. *New Phytologist* **109**, 321–6.

Pfyffer, G. E. & Rast, D. M. (1989). Accumulation of acyclic polyols and trehalose as related to growth form and carbohydrate source in the dimorphic fungi *Mucor rouxii* and *Candida albicans. Mycopathologia* **105**, 25–33.

Philippi, F. (1893). Die Pilze Chiles soweit dieselben als Nahrungsmittel gebraucht werden. *Hedwigia* **32**, 115–18.

Pick, U., Bental, M., Chitlaru, E. & Weiss, M. (1990). Polyphosphate hydrolysis – a protective mechanism against alkaline stress. *FEBS Letters* **274**, 15–18.

Pickering, W. R. & Woods, R. A. (1972). The uptake and incorporation of purines by wild-type *Saccharomyces cerevisiae* and a mutant resistant 4-aminopyrazolo(3,4-d)pyrimidine. *Biochimica et Biophysica Acta* **264**, 45–58.

Pilkington, B. J. & Rose, A. H. (1989). Accumulation of sulphite by *Saccharomyces cerevisiae* and *Zygosaccharomyces bailii* as affected by phospholipid fatty-acyl unsaturation and chain length. *Journal of General Microbiology* **135**, 2423–8.

Pilz, F., Auling, G., Stephan, D., Rau, U. & Wagner, F. (1991). A high-affinity Zn^{2+} uptake system controls growth and biosynthesis of an extracellular, branched β-1,3-β-1,6-glucan in *Sclerotium rolfsii* ATCC 15205. *Experimental Mycology* **15**, 181–92.

Pinto, W. J. & Nes, W. R. (1983). Stereochemical specificity of sterols in *Saccharomyces cerevisiae. Journal of Biological Chemistry* **258**, 4472–6.

Piotrowska, M. (1980). Cross-pathway regulation of ornithine carbamoyl transferase in *Aspergillus nidulans. Journal of General Microbiology* **116**, 335–9.

Piotrowska, M., Stepién, P. P., Bartnik, E. & Zarkzewska, E. (1976). Basic and neutral amino acid transport in *Aspergillus nidulans. Journal of General Microbiology* **92**, 89–96.

Pirt, S. J. (1966). A theory of the mode of growth of fungi in the form of pellets in submerged culture. *Proceedings of the Royal Society* B **166**, 369–73.

Pirt, S. J. (1967). A kinetic study of the mode of growth of surface colonies of bacteria and fungi. *Journal of General Microbiology* **47**, 181–97.

Pitman, M. J. (1964). The effect of divalent cations on the uptake of salt by beetroot tissue. *Journal of Experimental Botany* **15**, 444–56.

Pitt, D. & Barnes, J. C. (1987). Hexose transport during calcium induced conidiation in *Penicillium notatum. Transactions of the British Mycological*

  Society **89**, 859–65.
Pitt, D. & Kaile, A. (1990). Transduction of the calcium signal with special reference to Ca^{2+}-induced conidiation in *Penicillium notatum*. In *Biochemistry of Cell Walls and Membranes in Fungi*, ed. P. J. Kuhn, A. P. J. Trinci, M. J. Jung, M. W. Goosey & L. G. Copping, pp. 283–98. Springer-Verlag: Berlin, Heidelberg, New York.
Pitt, D. & Ugalde, U. O. (1984). Calcium in fungi. *Plant, Cell and Environment* **7**, 467–75.
Pitt, J. I. (1975). Xerophilic fungi and the spoilage of foods of plant origin. In *Water Relations of Foods*, ed. R. B. Duckworth, pp. 273–307. Academic Press: London.
Pitt, J. I. & Hocking, A. D. (1977). Influence of solute and hydroge ion concentration on the water relations of some xerophilic fungi. *Journal of General Microbiology* **101**, 35–40.
Plassard, C., Moussain, D. & Salsac, C. L. (1984). Measure *in vitro* de l'activité nitrate réductase dans les thalles de *Hebeloma cylindrosporum* champignon basidiomycète. *Physiologie Végétale* **22**, 67–74.
Plassard, C., Scheromm, P., Moussain, D. & Salsac, L. (1991). Assimilation of mineral nitrogen and ion balance in the two partners of ectomycorrhizal symbiosis – data and hypothesis. *Experientia* **47**, 340–9.
Plesofsky-Vig, N. & Brambl, R. (1993). Heat shock proteins in fungi. In *Stress Tolerance of Fungi*, ed. D. H. Jennings, pp. 45–68. Marcel Dekker: New York.
Plesset, J., Palm, C. & McLaughlin, C. S. (1982). Induction of heat shock proteins and thermotolerance by ethanol in *Saccharomyces cerevisiae*. *Biochemical and Biophysical Research Communications* **108**, 1340–5.
Plunkett, B. E. (1956). The influence of factors of the aeration complex and light upon fruit-body form in pure cultures of an agaric and a polypore. *Annals of Botany* NS **20**, 563–86.
Plunkett, B. E. (1958). Translocation and pileus formation in *Polyporus brumalis*. *Annals of Botany* NS **22**, 237–49.
Poindexter, J. S. (1981). Oligotrophy: fast and famine existence. *Advances in Microbial Ecology* **5**, 63–89.
Polak, A. & Grenson, M. (1973). Evidence for a common transport system for cytosine, adenine and hypoxanthine in *Saccharomyces cerevisiae* and *Candida albicans*. *European Journal of Biochemistry* **36**, 276–82.
Polakis, E., Bartley, W. & Meek, G. A. (1965). Changes in the activities of respiratory enzymes during the aerobic growth of yeast. *Biochemical Journal* **97**, 298–302.
Ponta, J. & Broda, E. (1970). Mechanismen der Aufnahme von Zink durch Bäkerhefe. *Planta* **95**, 18–26.
Popp, J. L., Kalyanaraman, B. & Kirk, T. K. (1990). Lignin peroxidase oxidation of Mn^{2+} in the presence of veratryl alcohol, malonic or oxalic acid and oxygen. *Biochemistry* **29**, 10475–9.
Portillo, F. & Mazón, M. J. (1986). The *Saccharomyces cerevisiae* start mutant carrying the *cdc25* mutation is defective in activation of plasma membrane ATPase by glucose. *Journal of Bacteriology* **168**, 1254–7.
Portillo, F. & Serrano, R. (1989). Growth control strength and active site of yeast plasma membrane ATPase studies by site-directed mutagenesis. *European Journal of Biochemistry* **186**, 501–7.
Postma, E., Kuiper, A., Tomasouw, W. F., Scheffers, W. A. & van Dijken, J. P. (1989a). Competition for glucose between the yeasts *Saccharomyces*

cerevisiae and *Candida utilis*. *Applied and Environmental Microbiology* **55**, 3214–20.

Postma, E., Scheffers, W. A. & van Dijken, J. P. (1988). Adaptation of the kinetics of glucose transport to environmental conditions in the yeast *Candida utilis* CBS 621: a continuous-culture study. *Journal of General Microbiology* **134**, 1109–16.

Postma, E., Scheffers, W. A. & van Dijken, J. P. (1989b). Kinetics of growth and glucose transport in glucose-limited chemostat cultures of *Saccharomyces cerevisiae*. *Yeast* **5**, 159–65.

Postma, E., Verduyn, C., Scheffers, W. A. & van Dijken, J. P. (1989c). Enzymic analysis of the Crabtree effect in glucose-limited chemostat cultures of *Saccharomyces cerevisiae*. *Applied and Environmental Microbiology* **55**, 468–77.

Powers, P. A. & Pall, M. L. (1986). Cyclic AMP-dependent protein kinase of *Neurospora crassa*. *Biochemical and Biophysical Research Communications* **95**, 701–6.

Prasad, R. (1985). Lipids in the structure and function of yeast membrane. *Advances in Lipid Research* **21**, 187–242.

Prasad, R. & Rose, A. H. (1986). Involvement of lipids in solute transport in yeasts. *Yeasts* **2**, 205–20.

Prell, A., Páca, J. & Sigler, K. (1992). H^+ and K^+ fluxes and attendant processes in *Candida utilis* induced by inhibitory ethanol concentrations. *Folia Microbiologica* **37**, 39–42.

Prinz, R. & Weser, U. (1975). Naturally occurring Cu-thionein in *Saccharomyces cerevisiae*. *Journal of Physiological Chemistry* **356**, 767–76.

Prior, B. A., Kilian, S. G. & du Preez, J. C. (1989). Fermentation of D-xylose by the yeasts *Candida shehatae* and *Pichia stipitis*: prospects and problems. *Process Biochemistry* **24**, 21–32.

Prodouz, K. N. & Garrett, R. H. (1981). *Neurospora crassa* NAD(P)H-nitrite reductase. Studies of its composition and structure. *Molecular and General Genetics* **155**, 67–75.

Prokop, A., Ludvik, M. & Erikson, L. E. (1972). Growth models of cultures with two liquid phases. VIII. Experimental observations on droplet size and interfacial area. *Biotechnology and Bioengineering* **14**, 587–608.

Punja, Z. K. & Jenkins, S. F. (1984). Influence of medium composition on mycelial growth and oxalic acid production in *Sclerotium rolfsii*. *Mycologia* **76**, 947–50.

Purwin, C., Leidig, F. & Holzer, H. (1982). Cyclic AMP-dependent phosphorylation of fructose-1,6-bisphosphatase in yeast. *Biochemical and Biophysical Research Communications* **107**, 1482–9.

Rabanus, A. (1939). Über die Säureproduktion von Pilzen und deren Einfluss auf die Wirkung von Holzschutzmitteln. *Mitteilungen des Deutschen Forstvereins* **23**, 77–89.

Rabinowitz, J. C. & Snell, E. E. (1947). Vitamin B_6 group. Extraction procedures for the microbiological determination of vitamin B_6. *Analytical Chemistry* **19**, 277–80.

Radler, F. & Lang, E. (1982). Malatbildung bei Hefen. *Die Weinwissenschaft* **6**, 391–9.

Raeder, U. & Broda, P. (1984). Comparison of the lignin-degrading white rot fungi *Phanerochaete chrysosporium* and *Sporotrichum pulverulentum* at the DNA level. *Current Genetics* **8**, 499–506.

Raeder, U., Thompson, W. & Broda, P. (1989a). RFLP-based genetic map of

Phanerochaete chrysosporium ME 446: lignin peroxidase genes occur in clusters. *Molecular Microbiology* **3**, 911–18.

Raeder, U., Thompson, W. & Broda, P. (1989*b*). Genetic factors influencing lignin peroxidase activity in *Phanerochaete chrysosporium* ME 446. *Molecular Microbiology* **3**, 919–24.

Raguzzi, F., Lesuisse, E. & Crichton, R. R. (1988). Iron storage in *Saccharomyces cerevisiae*. *FEBS Letters* **231**, 253–8.

Ramasamy, K., Kelley, R. L. & Reddy, C. A. (1985). Lack of lignin degradation by glucose oxidase-negative mutants of *Phanerochaete chrysosporium*. *Biochemical and Biophysical Research Communications* **131**, 436–41.

Ramirez, J. A., Vacata, V., McCusker, J. H., Haber, J. E., Mortimer, R. K., Owne, W. G. & Lecar, H. (1989). ATP-sensitive K^+ channels in a plasma membrane H^+-ATPase mutant of the yeast *Saccharomyces cerevisiae*. *Proceedings of the National Academy of Sciences, USA* **96**, 7866–70.

Ramos, E. H., de Bongioanni, L. C., Claisse, M. L. & Stoppani, A. O. M. (1975). Energy requirements for the uptake of L-leucine by *Saccharomyces cerevisiae*. *Biochimica et Biophysica Acta* **394**, 470–81.

Ramos, E. H., de Bongioanni, L. C. & Stoppani, A. O. M. (1980). Kinetics of L-[^{14}C]leucine transport in *Saccharomyces cerevisiae*. Effect of energy coupling inhibitors. *Biochimica et Biophysica Acta* **599**, 214–31.

Ramos, J. & Cirillo, V. P. (1989). Role of cyclic AMP-dependent protein kinase in catabolite inactivation of the glucose and galactose transporters in *Saccharomyces cerevisiae*. *Journal of Bacteriology* **171**, 3545–8.

Ramos, J., Contreras, P. & Rodríguez-Navarro, A. (1985). A potassium transport mutant of *Saccharomyces cerevisiae*. *Archives of Microbiology* **143**, 88–93.

Ramos, J., Haro, R. & Rodríguez-Navarro, A. (1990). Regulation of potassium fluxes in *Saccharomyces cerevisiae*. *Biochimica et Biophysica Acta* **1029**, 211–17.

Ramos, J. & Rodríguez-Navarro, A. (1985). Rubidium transport in *Neurospora crassa*. *Biochimica et Biophysica Acta* **815**, 97–101.

Ramos, J. & Rodríguez-Navarro, A. (1986). Regulation and interconversion of the potassium transport systems of *Saccharomyces cerevisiae* as revealed by rubidium transport. *European Journal of Biochemistry* **154**, 307–11.

Ramos, J., Szkutnicka, K. & Cirillo, V. P. (1988). Relationship between low- and high-affinity glucose transport systems of *Saccharomyces cerevisiae*. *Journal of Bacteriology* **170**, 5375–7.

Rand, J. B. & Tatum, E. L. (1980*a*). Characterization and regulation of galactose transport in *Neurospora crassa*. *Journal of Bacteriology* **141**, 707–14.

Rand, J. B. & Tatum, E. L. (1980*b*). Fructose transport in *Neurospora crassa*. *Journal of Bacteriology* **142**, 763–7.

Rank, G. H. & Robertson, A. J. (1983). Protein and lipid composition of the yeast plasma membrane. In *Yeast Genetics. Fundamental and Applied Aspects*, ed. J. F. T. Spencer, D. M. Spencer & A. R. W. Smith, pp. 225–41. Springer-Verlag: Berlin, Heidelberg, New York.

Rao, U. S., Hennessey, J. P. Jr & Scarborough, G. A. (1991). Identification of the membrane-embedded regions of the *Neurospora crassa* plasma membrane H^+-ATPase. *Journal of Biological Chemistry* **266**, 14740–6.

Rasi-Caldogno, F., Pugliarello, M. C. & de Michaelis, M. I. (1987). The Ca^{2+}-transporting ATPase of plant plasma membrane catalyses an H^+/Ca^{2+} exchange. *Plant Physiology* **83**, 994–1000.

Rast, D. M., Horsch, M., Furter, R. & Gooday, G. W. (1991). A complex chitinolytic system in exponentially growing mycelium of *Mucor rouxii*: properties and function. *Journal of General Microbiology* **137**, 1797–810.

Rattray, J. B. M. (1988). Yeasts. In *Microbial Lipids*, vol. 1, ed. C. Ratledge & S. G. Wilkinson, pp. 555–679. Academic Press: London.

Rauser, W. E. (1990). Phytochelatins. *Annual Review of Biochemistry* **59**, 81–6.

Raven, J. A. (1974). Energetics of active phosphate influx in *Hydrodictyon africanum*. *Journal of Experimental Botany* **25**, 221–9.

Raven, J. A. (1987). The role of vacuoles. *New Phytologist* **106**, 357–422.

Rea, P. A. & Poole, R. J. (1985). Proton-translocating inorganic pyrophosphatase in red beet (*Beta vulgaris* L.) tonoplast vesicles. *Plant Physiology* **77**, 46–52.

Rea, P. A. & Poole, R. J. (1986). Chromatographic resolution of H^+-translocating pyrophosphate from H^+-translocating ATPase of higher plants. *Plant Physiology* **81**, 126–9.

Rea, P. A. & Sanders, D. (1987). Tonoplast energization: two H^+ pumps, one membrane. *Physiologia Plantarum* **71**, 131–41.

Read, D. J. & Stribley, D. P. (1975). Diffusion and translocation in some fungal culture systems. *Transactions of the British Mycological Society* **64**, 381–8.

Reed, R. J. (1984). Use and abuse of osmoterminology. *Plant, Cell and Environment* **7**, 165–70.

Reed, R. H., Chudek, J. A., Foster, R. & Gadd, G. M. (1987). Osmotic significance of glycerol accumulation in exponentially growing yeasts. *Applied and Environmental Microbiology* **53**, 2119–23.

Reese, E. T. & Mandels, M. (1984). Rolling with the times: production and application of *Trichoderma reesei* cellulase. *Annual Reports on Fermentation Processes* **7**, 1–20.

Reese, E. T., Siu, R. G. H. & Levinson, H. S. (1950). Biological degradation of soluble cellulose derivatives. *Journal of Bacteriology* **9**, 485–97.

Reese, R. N. & Wagner, G. J. (1987). Properties of tobacco (*Nicotiana tabacum*) cadmium-binding peptides. *Biochemical Journal* **241**, 641–7.

Reichert, U. & Forêt, M. (1977). Energy coupling in hypoxanthine transport of yeast. Potentiometric evidence for proton symport and potassium antiport. *FEBS Letters* **83**, 325–8.

Reichert, U., Schmidt, R. & Forêt, M. (1975). A possible mechanism of energy coupling in purine transport of *Saccharomyces cerevisiae*. *FEBS Letters* **52**, 100–2.

Reid, C. P. P., Crowley, D. E., Kim, H. J., Powell, P. E. & Szanislo, P. J. (1984). Utilization of iron by oat when supplied as ferrated synthetic chelate or as ferrated hydroxamate siderophores. *Journal of Plant Nutrition* **7**, 437–47.

Reid, I. D. (1979). The influence of nutrient balance on lignin degradation by the white-rot fungus *Phanerochaete chrysosporium*. *Canadian Journal of Botany* **57**, 2050–8.

Reid, I. D. & Seifert, K. A. (1982). Effect of an atmosphere of oxygen on growth, respiration, and lignin degradation by white-rot fungi. *Canadian Journal of Botany* **60**, 252–60.

Reig, J. A., Téllez-Iñon, M. T., Flawiá, M. M. & Torres, H. N. (1984). Activation of *Neurospora crassa* soluble adenylate cyclase by calmodulin. *Biochemical Journal* **221**, 541–3.

Reinert, W. R. & Marzluf, G. A. (1974). Regulation of sulfate metabolsim in *Neurospora crassa*: transport and accumulation of glucose 6-sulfate. *Biochemical Genetics* **12**, 97–108.

Reissig, J. L. & Kinney, S. G. (1983). Calcium as a branching signal in *Neurospora crassa. Journal of Bacteriology* **154**, 1397–1402.

Rendules, P. S. & Wolf, D. H. (1988). Proteinase function in yeast: biochemical and genetical approaches to a central mechanism of post-translational control in the eukaryotic cell. *FEMS Microbiology Reviews* **54**, 17–46.

Renosto, F., Martin, R. L. & Segel, I. H. (1991). Adenosine-5′-phosphosulfate kinase from *Penicillium chrysogenum*: ligand binding properties and the mechanism of substrate inhibition. *Archives of Biochemistry and Biophysics* **284**, 30–4.

Renosto, F., Martin, R. L., Wailes, L. M., Daley, L. A. & Segel, I. H. (1990). Regulation of inorganic sulfate activation in filamentous fungi. Allosteric inhibition of ATP sulfurylase by 3′-phosphoadenosine-5′-phosphosulfate. *Journal of Biological Chemistry* **265**, 10300–8.

Renosto, F., Ornitz, D. M., Peterson, D. & Segel, I. H. (1981). Nitrate reductase from *Penicillium chrysogenum. Journal of Biological Chemistry* **256**, 8616–25.

Renosto, F., Seubert, P. A. & Segel, I. H. (1984). Adenosine 5′-phosphosulfate kinase from *Penicillium chrysogenum*. Purification and kinetic characterization. *Journal of Biological Chemistry* **259**, 2113–23.

Reyes, E. & Ruiz-Herrera, J. (1972). Mechanism of maltose utilization by *Mucor rouxii. Biochimica et Biophysica Acta* **273**, 328–35.

Reyes, F. & Byrde, R. J. W. (1973). Partial purification and properties of a β-N-acetyl glucosaminidase from the fungus *Sclerotinia fructigena. Biochemical Journal* **131**, 381–8.

Reyes, F., Calatayud, J. & Martínez, M. J. (1988). Chitinolytic activity in the autolysis of *Aspergillus nidulans. FEMS Microbiology Letters* **49**, 239–43.

Reyes, F., Calatayud, J. & Martínez, M. J. (1989). Endochitinase from *Aspergillus nidulans* implicated in the autolysis of its cell wall. *FEMS Microbiology Letters* **60**, 119–24.

Reyes, F. & Lahoz, R. (1976). Liberation of protoplasts from mycelium of *Neurospora crassa* by means of enzymes from autolyzed cultures of this fungus. *Antonie van Leeuwenhoek* **42**, 457–60.

Riemersma, J. C. (1966). Effect of metal ions on the lysis of yeast cells by cationic dyes and surfactants. *Journal of Pharmacy and Pharmacology* **18**, 657–63.

Robbins, W. J. & Kavanagh, V. (1942). Vitamin deficiencies of the filamentous fungi. *Botanical Review* **8**, 411–71.

Roberts, K. R. & Marzluf, G. A. (1971). The specific interactions of chromate with the dual sulfate permease systems of *Neurospora crassa. Archives of Biochemistry and Biophysics* **142**, 651–9.

Robertson, N. F. (1959). Experimental control of hyphal branching and branch forms in hyphomycetous fungi. *Journal of the Linnean Society of London* **56**, 207–11.

Robertson, N. F. (1965). The fungal hypha. *Transactions of the British Mycological Society* **48**, 1–8.

Robertson, M. F. & Rizvi, S. R. H. (1968). Some observations on the water relations of the hyphae of *Neurospora crassa. Annals of Botany* N.S. **32**, 279–91.

Robinow, C. G. & Johnson, B. F. (1991). Yeast cytology: an overview. In *The Yeasts*, vol. 4 *Yeast Organelles*, ed. A. H. Rose & J. S. Harrison, pp. 7–120. Academic Press: London.

Robinson, J. H., Anthony, C. & Drabble, W. T. (1973a). The acidic amino-acid permease of *Aspergillus nidulans*. *Journal of General Microbiology* **79**, 53–63.

Robinson, J. H., Anthony, C. & Drabble, W. T. (1973*b*). Regulation of the acidic amino-acid permease of *Aspergillus nidulans*. *Journal of General Microbiology* **79**, 65–80.

Robinson, N. J. & Jackson, P. J. (1986). 'Metallothionein-like' metal complexes in angiosperms: their structure and function. *Physiologia Plantarum* **67**, 499–506.

Robinson, P. M. (1973). Oxygen-positive chemotropic factor in fungi. *New Phytologist* **72**, 1349–56.

Robson, G. D., Bell, S. D., Kuhn, P. J. & Trinci, A. P. J. (1987). Glucose and penicillin concentrations in agar medium below fungal colonies. *Journal of General Microbiology* **133**, 361–7.

Robson, G. D., Best, L. C., Wiebe, M. G. & Trinci, A. P. J. (1992). Choline transport in *Fusarium graminearum* A3/5. *FEMS Microbiology Letters* **92**, 247–52.

Robson, G. D., Wiebe, M. G. & Trinci, A. P. J. (1991). Low calcium concentrations induce increased branching in *Fusarium graminearum*. *Mycological Research* **95**, 561–5.

Rodríguez-Navarro, A. (1971). Inhibition by sodium and lithium in osmophilic yeasts. *Antonie van Leeuwenhoek* **37**, 225–31.

Rodríguez-Navarro, A., Blatt, M. R. & Slayman, C. L. (1986). A potassium-proton symport in *Neurospora crassa*. *Journal of General Physiology* **87**, 649–74.

Rodríguez-Navarro, A. & Ortega, M. D. (1982). The mechanism of sodium efflux in yeast. *FEBS Letters* **138**, 205–8.

Rodríguez-Navarro, A. & Ramos, J. (1984). Dual system for potassium transport in *Saccharomyces cerevisiae*. *Journal of Bacteriology* **159**, 940–5.

Rodríguez-Navarro, A. & Ramos, J. (1986). Two systems mediate rubidium uptake in *Neurospora crassa*: one exhibits the dual-uptake isotherm. *Biochimica et Biophysica Acta* **857**, 229–37.

Rodríguez-Navarro, A. & Sancho, E. D. (1979). Cation exchanges in yeast in the absence of magnesium. *Biochimica et Biophysica Acta* **552**, 322–30.

Rogers, P. J. & Stewart, P. R. (1974). Energetic efficiency and maintenance energy characteristics of *Saccharomyces cerevisiae* (wild type and petite) and *Candida parapsilosis* grown aerobically and micro-aerobically in continuous culture. *Archives of Microbiology* **99**, 25–46.

Rogers, T. O. & Lichstein, H. C. (1969*a*). Characterization of the biotin transport system in *Saccharomyces cerevisiae*. *Journal of Bacteriology* **100**, 557–64.

Rogers, T. O. & Lichstein, H. C. (1969*b*). Regulation of biotin transport in *Saccharomyces cerevisiae*. *Journal of Bacteriology* **100**, 565–72.

Romano, A. H. (1982). Facilitated diffusion of 6-deoxy-D-glucose in baker's yeast: evidence against phosphorylation-associated transport of glucose. *Journal of Bacteriology* **152**, 1295–7.

Romano, A. H., Bright, M. M. & Scott, W. E. (1967). Mechanism of fumaric acid accumulation by *Rhizopus nigricans*. *Journal of Bacteriology* **93**, 600–4.

Römheld, V. & Marschner, H. (1986). Mobilization of iron in the rhizosphere of different plant species. *Advances in Plant Nutrition* **2**, 155–204.

Roomans, G. M., Blasko, F. & Borst-Pauwels, G. W. F. H. (1977). Cotransport of phosphate and sodium by yeast. *Biochimica et Biophysica Acta* **467**, 65–71.

Roomans, G. M. & Borst-Pauwels, G. W. F. H. (1977). Interaction of phosphate with monovalent cation uptake in yeast. *Biochimica et Biophysica Acta* **470**, 84–91.

Roomans, G. M., Kuypers, G. A. J., Theuvenet, A. P. R. & Borst-Pauwels, G. W. F. H. (1979*a*). Kinetics of sulfate uptake by yeast. *Biochimica et Biophysica Acta* **551**, 197–206.

Roomans, G. M. & Seveus, L. A. (1976). Subcellular localization of diffusible ions in the yeast *Saccharomyces cerevisiae*: quantitative microprobe analysis of thin freeze-dried sections. *Journal of Cell Science* **21**, 119–27.

Roomans, G. M., Theuvenet, A. P. R., van den Berg, T. P. R. & Borst-Pauwels, G. W. F. H. (1979*b*). Kinetics of Ca^{2+} and Sr^{2+} uptake by yeast. Effects of pH, cations and phosphate. *Biochimica et Biophysica Acta* **551**, 187–96.

Roon, R. J., Even, H. L., Dunlop, P. & Larimore, F. (1975*a*). Methylamine and ammonia transport in *Saccharomyces cerevisiae*. *Journal of Bacteriology* **122**, 502–9.

Roon, R. J., Larimore, F. & Levy, J. S. (1975*b*). Inhibition of amino acid transport by ammonium ions in *Saccharomyces cerevisiae*. *Journal of Bacteriology* **124**, 325–31.

Roon, R. J., Levy, J. S. & Larimore, F. (1977*a*). Negative interactions between amino acids and methylamine/ammonia transport systems. *Journal of Biological Chemistry* **252**, 3599–604.

Roon, R. J., Meyer, G. M. & Larimore, F. (1977*b*). Evidence for a common component in kinetically distinct systems in *Saccharomyces cerevisiae*. *Molecular and General Genetics* **158**, 185–91.

Roos, W. (1989). Kinetic properties, nutrient-dependent regulation and energy coupling of amino-acid transport systems in *Penicillium cyclopium*. *Biochimica et Biophysica Acta* **978**, 119–33.

Rorison, I. H. (1960*a*). Some experimental aspects of the calcicole-calcifuge problem. I. The effects of competition and mineral nutritions upon seedling growth in the field. *Journal of Ecology* **48**, 585–99.

Rorison, I. J. (1960*b*). The calcicole-calcifuge problem. II. The effects of mineral nutrition on seedling growth in solution culture. *Journal of Ecology* **48**, 679–88.

Rosa, M. F. & Sá-Correia, I. (1991). *In vivo* activation by ethanol of plasma membrane ATPase of *Saccharomyces cerevisiae*. *Applied and Environmental Microbiology* **57**, 830–5.

Rosa, M. F. & Sá-Correia, I. (1992). Ethanol tolerance of plasma membrane ATPase in *Kluyveromyces marxianus* and *Saccharomyces cerevisiae*. *Enzyme and Microbial Technology* **14**, 23–7.

Rose, A. H. (1976). Chemical nature of membrane components. In *The Filamentous Fungus*, vol. 2 *Biosynthesis and Metabolism*, ed. J. E. Smith & D. R. Berry, pp. 308–27. Edward Arnold: London.

Rose, A. H. (1989). Transport and metabolism of sulphur dioxide in yeasts and filamentous fungi. In *Nitrogen, Phosphorus and Sulphur Utilization by Fungi*, ed. L. Boddy, R. Marchand & D. J. Read, pp. 59–70. Cambridge University Press: Cambridge.

Rose, D. (1975). Physical responses of yeast cells to osmotic shock. *Journal of Applied Bacteriology* **38**, 169–75.

Rose, M., Albig, W. & Entian, K. D. (1991). Glucose repression in *Saccharomyces cerevisiae* is directly associated with hexose phosphorylation by hexokinase-PI and hexokinase-PII. *European Journal of Biochemistry* **199**, 511–18.

Rosenberg, E. (1986). Microbial surfactants. *Critical Reviews in Biotechnology* 3, 109–32.

Rosing, J. & Slater, E. C. (1972). The values of $\Delta G^{0'}$ for the hydrolysis of ATP. *Biochimica et Biophysica Acta* 267, 275–90.

Ross, I. S. (1993). Membrane transport processes and response to exposure to heavy metals. In *Stress Tolerance of Fungi*, ed. D. H. Jennings, pp. 97–125. Marcel Dekker: New York.

Ross, I. S. & Parkin, M. J. (1989). Uptake of copper by *Candida utilis*. *Mycological Research* 93, 33–7.

Rossi, A. & Arst, H. N. Jr (1990). Mutants of *Aspergillus nidulans* able to grow at extremely acidic pH acidify the medium less than wild type when grown at more moderate pH. *FEMS Microbiology Letters* 66, 51–4.

Rothblatt, J. & Schekman, R. (1989). A hitchiker's guide to analysis of the secretory pathway in yeast. *Methods in Cell Biology* 32, 3–36.

Rothman, J. H., Yamashiro, C. T., Kane, P. M. & Stevens, T. H. (1989). Protein targeting to the yeast vacuole. *Trends in Biochemical Sciences* 14, 347–50.

Rothstein, A. (1954). Enzyme systems of the cell surface involved in the uptake of sugars by yeast. *Symposium of the Society for Experimental Biology* 8, 165–201.

Rothstein, A. (1963). Interactions of arsenate with the phosphate-transporting system of yeast. *Journal of General Physiology* 46, 1075–85.

Rothstein, A. (1974). Relationship of cation influxes and effluxes in yeast. *Journal of General Physiology* 64, 608–21.

Rothstein, A. & Bruce, M. (1958*a*). The potassium efflux and influx in yeast at different potassium concentrations. *Journal of Cell and Comparative Physiology* 51, 145–60.

Rothstein, A. & Bruce, M. (1958*b*). The efflux of potassium from yeast cells into a potassium-free medium. *Journal of Cellular and Comparative Physiology* 51, 439–55.

Rothstein, A. & Enns, L. H. (1946). The relationship of potassium to carbohydrate metabolism in baker's yeast. *Journal of Cellular and Comparative Physiology* 28, 231–52.

Rothstein, A., Frenkel, A. & Larrabee, C. (1948). The relationship of the cell surface to metabolism. III. Certain characteristics of the uranium complex with cell surface groups of yeast. *Journal of Cellular and Comparative Physiology* 28, 231–52.

Rothstein, A. & Hayes, A. D. (1956). The relationship of the cell surface to metabolism. XIII. The cation-binding properties of the yeast cell surface. *Archives of Biochemistry and Biophysics* 63, 87–99.

Rothstein, A., Hayes, A. D., Jennings, D. H. & Hooper, D. C. (1958). The active transport of Mg^{++} and Mn^{++} into the yeast cell. *Journal of General Physiology* 41, 585–94.

Roychowdhury, H. S. & Kapoor, M. (1988). Ethanol and carbon-source starvation enhance the accumulation of HSP80 in *Neurospora crassa*. *Canadian Journal of Microbiology* 34, 162–8.

Royt, P. W. (1981). Facilitated diffusion of 6-deoxy-D-glucose by the oxidative yeast *Kluyveromyces lactis*. *Archives of Microbiology* 130, 87–9.

Royt, P. W. (1983). Asymmetry of glucose transport in the yeast *Kluyveromyces lactis*. *Biochimica et Biophysica Acta* 728, 363–70.

Royt, P. W. & MacQuillan, A. M. (1976). Evidence for an inducible glucose transport system in *Kluyveromyces lactis*. *Biochimica et Biophysica Acta* 426, 302–16.

Royt, P. W. & Tombes, A. S. (1982). Transport of 2-deoxy-D-glucose by isolated protoplasts of the yeast *Kluyveromyces lactis*. *Canadian Journal of Microbiology* 28, 901–6.

Rudnick, G. (1986). ATP-driven H^+ pumping into intracellular organelles. *Annual Review of Physiology* 48, 403–13.

Rudolph, H. K., Antebi, A., Fink, G. R., Buckley, C. M., Dorman, T. E., Le Vitre, J., Davidow, L. S., Mao, J. & Moir, D. T. (1989). The yeast secretory pathway is perturbed by mutations in *PMR1*, a member of a Ca^{2+}-ATPase family. *Cell* 58, 133–45.

Ruel, K., Barnoud, F. & Eriksson, K.-E. (1981). Micromorphological and ultrastructural aspects of spruce wood degradation by wild-type *Sporotrichum pulverulentum* and its cellulase-less mutant *Cel44*. *Holzforschung* 35, 157–71.

Ruel, K. & Joseleau, J. P. (1991). Involvement of an extracellular glucan sheath during degradation of *Populus* wood by *Phanerochaete chrysosporium*. *Applied and Environmental Microbiology* 57, 374–84.

Ruml, T., Šilhánková, L. & Rauch, P. (1988). The irreversibility of thiamin transport in *Saccharomyces cerevisiae*. *Folia Microbiologica* 33, 372–6.

Ryan, F. J., Beadle, G. W. & Tatum, E. L. (1943). The tube method of measuring growth rate in *Neurospora*. *American Journal of Botany* 30, 784–99.

Ryan, H. (1967). Alcohol dehydrogenase activity and electron transport in living yeast. *Biochemical Journal* 105, 137–43.

Ryan, H., Ryan, J. P. & O'Connor, W. H. (1971). The effect of diffusible acids on potassium ion uptake by yeast. *Biochemical Journal* 125, 1081–5.

Ryan, J. P. & Ryan, H. (1972). The role of intracellular pH on the regulation of cation exchanges in yeast. *Biochemical Journal* 128, 138–46.

Rygiewicz, P. T., Bledsoe, C. S. & Zasoski, R. J. (1984a). Effects of ectomycorrhizal and solution pH on [^{15}N]ammonium uptake by coniferous seedlings. *Canadian Journal of Forest Research* 14, 885–92.

Rygiewicz, P. T., Bledsoe, C. S. & Zasoski, R. J. (1984b). Effects of ectomycorrhizal and solution pH on [^{15}N]nitrate uptake by coniferous seedlings. *Canadian Journal of Forest Research* 14, 893–9.

Rytka, J. (1975). Positive selection of general amino acid permease mutants in *Saccharomyces cerevisiae*. *Journal of Bacteriology* 121, 562–70.

Ryu, D. D. Y., Kim, C. & Mandels, M. (1984). Competitive adsorption of cellulase components and its significance in synergistic mechanisms. *Biotechnology and Bioengineering* 26, 488–96.

St Leger, R. J., Butt, T. M., Staples, R. C. & Roberts, D. W. (1990). Second messenger involvement in differentiation of the entomopathogenic fungus *Metarhizium anisopliae*. *Journal of General Microbiology* 136, 1779–89.

Salgueiro, P., Sá-Correia, I. & Novais, J. M. (1988). Ethanol-induced leakage in *Saccharomyces cerevisiae*: kinetics and relationship to yeast ethanol tolerance and alcohol fermentation productivity. *Applied and Environmental Microbiology* 54, 903–9.

Salmon, J. M. (1987). L-Malic acid permeates in resting cells of anaerobically grown *Saccharomyces cerevisiae*. *Biochimica et Biophysica Acta* 901, 30–4.

Saloheimo, M., Lehtovaara, P., Penttilä, M., Teeri, T. T., Stahlberg, J., Johansson, G., Pettersson, G., Claeyssens, M., Tomme, P. & Knowles, J. K. K. (1988). EG III, a new endoglucanase from *Trichoderma reesei*: the characterization of both gene and enzyme. *Gene* 63, 11–21.

Sand, F. E. M. J. (1973). Recent investigations on the microbiology of fruit juice concentrates. In *Technology of Fruit Juice Concentrates – Chemical Composition of Fruit Juices,* vol. 13. pp. 185–216. International Federation of Fruit Juice Producers, Scientific Technical Commission: Vienna.

Sanders, D. (1986). Generalized kinetic analysis of ion-driven cotransport systems. II. Random ligand binding as a simple explanation for non-Michaelian kinetics. *Journal of Membrane Biology* **90**, 67–87.

Sanders, D. (1988). Fungi. In *Solute Transport in Plant Cells and Tissues,* ed. D. A. Baker & J. L. Hall, pp. 106–65. Longman Scientific and Technical: Harlow.

Sanders, D. & Hansen, U.-P. (1981). Mechanism of Cl^- transport at the plasma membrane of *Chara corallina.* II. Transinhibition and the determination of H^+/Cl^- binding order from a reaction kinetic model. *Journal of Membrane Biology* **58**, 139–53.

Sanders, D., Hansen, U.-P. & Slayman, C. L. (1981). Role of the plasma membrane proton pump in pH regulation in non-animal cells. *Proceedings of the National Academy of Sciences, USA* **78**, 5903–7.

Sanders, D. & Slayman, C. L. (1982). Control of intracellular pH. Predominant role of oxidative metabolism, not proton transport, in the eukaryotic microorganism *Neurospora. Journal of General Physiology* **80**, 377–402.

Sanders, D., Slayman, C. L. & Pall, M. L. (1983). Stoichiometry of H^+/amino acid cotransport in *Neurospora crassa* revealed by current-voltage analysis. *Biochimica et Biophysica Acta* **735**, 67–76.

Sanders, F. E. T. & Tinker, P. B. (1973). Phosphate flow into mycorrhizal roots. *Pesticide Science* **4**, 385–95.

Sato, T., Ohsumi, Y. & Anraku, Y. (1984a). Substrate specificities of active transport systems for amino acids in vacuolar-membrane vesicles of *Saccharomyces cerevisiae. Journal of Biological Chemistry* **259**, 11505–8.

Sato, T., Ohsumi, Y. & Anraku, Y. (1984b). An arginine/histidine exchange transport system in vacuolar-membrane vesicles of *Saccharomyces cerevisiae. Journal of Biological Chemistry* **259**, 11509–11.

Scarborough, G. A. (1970a). Sugar transport in *Neurospora crassa. Journal of Biological Chemistry* **245**, 1694–8.

Scarborough, G. A. (1970b). Sugar transport in *Neurospora crassa.* II. A second glucose transport system. *Journal of Biological Chemistry* **245**, 3985–7.

Scarborough, G. A. (1975). Isolation and characterization of *Neurospora crassa* plasma membrane. *Journal of Biological Chemistry* **250**, 1106–11.

Scarborough, G. A. (1976). *Neurospora* plasma membrane ATPase is an electrogenic pump. *Proceedings of the National Academy of Sciences, USA* **73**, 1486–8.

Scarborough, G. A. (1977). Properties of the *Neurospora crassa* plasma membrane ATPase. *Archives of Biochemistry and Biophysics* **180**, 384–93.

Scarborough, G. A. (1980). Proton translocation catalyzed by the electrogenic ATPase in the plasma membrane of *Neurospora. Biochemistry* **19**, 2925–31.

Scarborough, G. A. (1988). Large-scale purification of plasma membrane H^+-ATPase from a cell wall-less mutant of *Neurospora. Methods in Eyzymology* **157**, 547–79.

Scarborough, G. A. & Addision, R. (1984). On the subunit composition of the *Neurospora* plasma membrane H^+-ATPase. *Journal of Biological Chemistry* **259**, 9109–14.

Scarborough, G. A. & Hennessy, J. P. Jr (1990). Identification of the major cytoplasmic regions of the *Neurospora crassa* plasma membrane H^+-ATPase using protein chemical techniques. *Journal of Biological Chemistry* **265**, 16145–9.

Scazzochio, C. & Arst, H. N. Jr (1989). Regulation of nitrate assimilation in *Aspergillus nidulans*. In *Molecular and Genetic Aspects of Nitrate Assimilation*, ed. J. L. Wray & J. K. Kinghorn, pp. 299–313. Oxford Science Publications: Oxford.

Schade, A. L. (1940). The nutrition of *Leptomitus*. *American Journal of Botany* 27, 376–84.

Scharff, T. G. & Maupin, W. C. (1960). Correlation of the metabolic effects of benzalkomium chloride with its membrane effects in yeast. *Biochemical Pharmacology* 5, 79–86.

Scheffers, W. A. (1961). On the inhibition of alcoholic fermentation in *Brettanomyces* yeast under anaerobic conditions. *Experientia* 17, 40–2.

Scheffers, W. A. (1966). Stimulation of fermentation in yeasts by acetoin and oxygen. *Nature, London* 210, 533–4.

Scheffers, W. A. (1987). Alcoholic fermentation. *Studies in Mycology* 30, 321–32.

Scheffers, W. A. & Wikén, T. O. (1969). The Custers effect (negative Pasteur effect) as a diagnostic criterion for the genus *Brettanomyces*. *Antonie van Leeuwenhoek* 35 (Suppl. Yeast Symposium) A31–A32.

Scheffey, C. (1988). Two approaches to construction of vibrating probes for electrical current measurement in solution. *Review of Scientific Instruments* 59, 787–92.

Scheiding, W., Thoma, M., Ross, A. & Schügerl, K. (1986). Modelling of the growth of *Trichoderma reesei* AM9414 on glucose and cellulose. *Journal of Biotechnology* 4, 101–18.

Schekman, R. (1985). Protein localization and membrane traffic in yeast. *Annual Review of Cell Biology* 1, 115–43.

Scheromm, P., Plassard, C. & Salsac, L. (1990a). Nitrate nutrition of maritime pine (*Pinus pinaster* Soland *in* Ait.) ectomycorrhizal with *Hebeloma cylindrosporium* Romagn. *New Phytologist* 114, 93–8.

Scheromm, P., Plassard, C. & Salsac, L. (1990b). Effect of nitrate and ammonium nutrition on the metabolism of the ectomycorrhizal basidiomycete *Hebeloma cylindrosporium* Romagn. *New Phytologist* 114, 227–34.

Scheromm, P., Plassard, C. & Salsac, L. (1990c). Nitrate reductase regulation in the mycorrhizal fungus *Hebeloma cylindrosporium* Romagn. cultured on nitrate or ammonium. *New Phytologist* 114, 441–7.

Scherrer, R., Louden, K. & Gerhardt, P. (1974). Porosity of the yeast cell wall and membrane. *Journal of Bacteriology* 118, 534–40.

Schlenk, F. (1970). The destructive effect of some proteins on the yeast cell membrane. *Biochimica Applicata* 17, 89–103.

Schlenk, F. & Dainko, J. L. (1965). Action of ribonuclease preparations on viable yeast cells and sphaeroplasts. *Journal of Bacteriology* 89, 428–36.

Schlenk, F., Dainko, J. L. & Svihla, G. (1970). The accumulation and intracellular destruction of biological sulfonium compounds in yeast. *Archives of Biochemistry and Biophysics* 140, 228–236.

Schloemer, R. H. & Garrett, R. H. (1974a). Nitrate transport system of *Neurospora crassa*. *Journal of Bacteriology* 118, 259–69.

Schloemer, R. H. & Garrett, R. H. (1974b). Uptake of nitrite by *Neurospora crassa*. *Journal of Bacteriology* 118, 270–4.

Schmidt, C. J., Whitten, B. K. & Nicholas, D. D. (1981). A proposed role for oxalic acid in non-enzymatic wood decay by brown rot fungi. *Proceedings of the American Wood-Preserving Association* 77, 157–64.

Schneider, H. (1989). Conversion of pentoses to ethanol and fungi. *Critical Reviews in Biotechnology* 9, 1–40.

Schneider, H., Lee, H., Barbosa, M. de F. S., Kubicek, C. P. & James, A. P. (1989). Physiological properties of a mutant of *Pachysolen tannophilus* deficient in NADPH-dependent D-xylose reductase. *Applied and Environmental Microbiology* **55**, 2877–81.

Schneider, R. P. & Wiley, R. P. (1971a). Kinetics characteristics of the two glucose transport systems in *Neurospora crassa. Journal of Bacteriology* **106**, 479–86.

Schneider, R. P. & Wiley, R. P. (1971b). Regulation of sugar transport in *Neurospora crassa. Journal of Bacteriology* **106**, 487–92.

Schoemaker, H. E. & Leisola, M. S. A. (1990). Degradation of lignin by *Phanerochaete chrysosporium. Journal of Biotechnology* **13**, 101–10.

Schreurs, W. J. A. & Harold, F. M. (1988). Transcellular proton current in *Achlya bisexualis* hyphae: relationship to polarized growth. *Proceedings of the National Academy of Sciences, USA* **85**, 1534–8.

Schriek, U. & Schwenn, J. D. (1986). Properties of the purified APS-kinase from *Escherichia coli* and *Saccharomyces cerevisiae. Archives of Microbiology* **145**, 32–8.

Schuddemat, J., de Boor, R., van Leeuwen, C. C. M., van den Broek, P. J. A. & van Steveninck, J. (1989a). Polyphosphate synthesis in yeast. *Biochimica et Biophysica Acta* **1010**, 191–8.

Schuddemat, J., van den Broek, P. J. A. & van Steveninck, J. (1986). Effect of xylose incubation on the glucose transport system in *Saccharomyces cerevisiae. Biochimica et Biophysica Acta* **861**, 489–93.

Schuddemat, J., van den Broek, P. J. A. & van Steveninck, J. (1988). The influence of ATP on sugar uptake mediated by the constitutive glucose carrier of *Saccharomyces cerevisiae. Biochimica et Biophysica Acta* **937**, 81–7.

Schuddemat, J., van Leeuwen, C. C. M., van den Broek, P. J. A. & van Steveninck, J. (1989b). Mechanism of transport-associated phosphorylation of 2-deoxy-D-glucose in the yeast *Kluyveromyces marxianus*: characterization of the phosphoryl source. *Biochimica et Biophysica Acta* **986**, 290–4.

Schuler, R. & Haselwandter, K. (1988). Hydroxamate siderophore production by ericoid mycorrhiza. *Journal of Plant Nutrition* **11**, 907–14.

Schultz, B. & Höfer, M. (1986). Utilization of lactose in non-respiring cells of the yeast *Debaryomyces polymorphus. Archives of Microbiology* **145**, 367–71.

Schuren, F. H. J. & Wessels, J. G. H. (1990). Two genes specifically expressed in fruiting dikaryons of *Schizophyllum commune*: homologies with a gene not regulated by the mating-type gene. *Gene* **90**, 199–205.

Schütte, K. H. (1956). Translocation in the fungi. *New Phytologist* **55**, 164–82.

Schwartz, H. & Radler, F. (1988). Formation of L(−)malate by *Saccharomyces cerevisiae* during fermentation. *Applied Microbiology and Biotechnology* **27**, 554–60.

Schweingruber, A. M., Dlugonski, J., Edenharter, E. & Schweingruber, M. E. (1991). Thiamine in *Schizosaccharomyces pombe*: dephosphorylation, intracellular pool, biosynthesis and transport. *Current Genetics* **19**, 249–54.

Schweingruber, M. E., Fluri, R., Maundrell, K., Schweingruber, A. M. & Dumeruth, E. (1986). Identification and characterization of thiamine repressible acid phosphatase in yeast. *Journal of Biological Chemistry* **261**, 15877–82.

Schweingruber, A. M., Frankhauser, H., Dlugonski, J., Steinmannloss, C. &

Schweingruber, M. E. (1992). Isolation and characterization of regulatory mutants of *Schizosaccharomyces pombe* involved in thiamine-regulated gene expression. *Genetics* **130**, 445–50.

Schwencke, J. (1991). Vacuoles, internal membrane systems and vesicles. In *The Yeasts*, vol. 4 *Yeast Organelles*, ed. A. H. Rose & J. S. Harrison, pp. 347–432. Academic Press: London.

Schwencke, J. & de Robichon-Szulmajster, H. (1976). The transport of S-adenosyl-L-methionine in isolated vacuoles and sphaeroplasts. *European Journal of Biochemistry* **65**, 49–60.

Schwenn, J. D., Krone, F. & Husmann, K. (1988). Yeast PAPS-reductase: properties and requirements of the purified enzyme. *Archives of Microbiology* **150**, 313–19.

Scott, A. I. & Beadling, L. C. (1974). Biosynthesis of patulin. Dehydrogenase and dioxygenase enzymes of *Penicillium patulum*. *Bioorganic Chemistry* **3**, 281–301.

Seaston, A., Inkson, C. & Eddy, A. A. (1973). The absorption of protons with specific amino acids and carbohydrates by yeast. *Biochemical Journal* **134**, 1031–4.

Sebald, W. & Hoppe, J. (1981). On the structure and genetics of the proteolipid of the ATP synthase complex. *Current Topics in Bioenergetics* **12**, 1–64.

Seeman, P. (1972). The membrane action of anaesthetics and tranquilizers. *Pharmacological Reviews* **24**, 583–655.

Selby, K. & Maitland, C. C. (1965). The fractionation of *Myrothecium verrucaria* cellulase by gel filtration. *Biochemical Journal* **94**, 578–83.

Sempuku, K. (1988). Photoinactivation of the thiamine transport system in *Saccharomyces cerevisiae* with azidobenzoyl derivatives of thiamine. *Biochimica et Biophysica Acta* **944**, 177–84.

Sempuku, K., Nishimura, H. & Iwashima, A. (1981). Photo-inactivation of the thiamine transport system in *Saccharomyces cerevisiae* with 4-azido-2-nitrobenzoyl thiamine. *Biochimica et Biophysica Acta* **645**, 226–8.

Serrano, R. (1978). Characterization of the plasma membrane ATPase of *Saccharomyces cerevisiae*. *Molecular and Cellular Biochemistry* **22**, 51–63.

Serrano, R. (1983). *In vivo* glucose activation of the yeast plasma membrane ATPase. *FEBS Letters* **156**, 11–14.

Serrano, R. (1984). Plasma membrane ATPase of fungi and plants as novel type of proton pumps. *Current Topics in Cellular Regulation* **23**, 87–126.

Serrano, R. (1985). *Plasma Membrane ATPase of Plants and Fungi*. CRC Press: Boca Raton, FL.

Serrano, R. (1988a). Structure and function of proton translocating ATPase in plasma membranes of plants and fungi. *Biochimica et Biophysica Acta* **947**, 1–28.

Serrano, R. (1988b). H$^+$-ATPase from plasma membranes of *Saccharomyces cerevisiae* and *Avena sativa* roots: purification and reconstitution. *Methods in Enzymology* **157**, 533–44.

Serrano, R. (1989). Structure and function of the plasma membrane ATPase. *Annual Review of Plant Physiology and Plant Molecular Biology* **49**, 61–94.

Serrano, R. & de la Fuente, G. (1974). Regulatory properties of the constitutive hexose transport in *Saccharomyces cerevisiae*. *Molecular and Cellular Biochemistry* **5**, 161–71.

Serrano, R., Kielland-Brandt, M. C. & Fink, G. R. (1986). Yeast plasma membrane ATPase is essential for growth and has homology with (Na$^+$+K$^+$)-, K$^+$- and Ca^{2+}-ATPases. *Nature, London* **319**, 689–93.

Serrano, R. & Portillo, F. (1990). Catalytic and regulatory sites of yeast plasma membrane H^+-ATPase studied by mutagenesis. *Biochimica et Biophysica Acta* **1018**, 195–9.

Severin, J., Langel, P. & Höfer, M. (1989). Analysis of the H^+/sugar symport in yeast under conditions of depolarized plasma membrane. *Journal of Bioenergetics and Biomembranes* **21**, 321–34.

Shane, B. & Snell, E. E. (1976). Transport and metabolism of vitamin B_6 in the yeast *Saccharomyces cerevisiae* 4228. *Journal of Biological Chemistry* **251**, 1042–51.

Shaw, G., Leake, J. R., Baker, A. J. M. & Read, D. J. (1990). The biology of mycorrhiza in the Ericaceae. 17. The role of mycorrhizal infection in the regulation of iron uptake by ericaceous plants. *New Phytologist* **115**, 251–8.

Sheard, J. & Farrar, J. F. (1987). Transport of sugar in *Phytophthora palmivora* (Butl.) Butl. *New Phytologist* **105**, 265–72.

Shere, S. M. & Jacobson, L. (1970). Mineral uptake in *Fusarium oxysporium* f. sp. *vasinfectum*. *Physiologia Plantarum* **23**, 51–62.

Sherwood, M. B. & Singer, E. D. (1944). Folic acid in cotton. *Journal of Biological Chemistry* **155**, 361–2.

Shieh, T. R. & Ware, J. H. (1968). Survey of microorganisms for the production of extracellular phytase. *Applied Microbiology* **16**, 1348–51.

Shieh, T. R., Wodzinski, R. J. & Ware, J. H. (1969). Regulation of the formation of acid phosphatases by inorganic phosphate in *Aspergillus ficuum*. *Journal of Bacteriology* **100**, 1161–5.

Shifrine, M. & Marr, A. G. (1963). The requirement of fatty acid by *Pityrosporium ovale*. *Journal of General Microbiology* **32**, 263–70.

Shih, C.-K., Wagner, R., Feinstein, S., Kanikennulat, C. & Neff, N. (1988). A dominant trifluoperazine resistant gene from *Saccharomyces cerevisiae* has homology with F_0F_1 ATP synthase and confers calcium-sensitive growth. *Molecular and Cellular Biology* **8**, 3094–103.

Shimada, M. (1979). Stereobiochemical approach to lignin biodegradation: possible significance of nonstereospecific oxidation catalyzed by laccase for lignin decomposition by white-rot fungi. In *Lignin Biodegradation: Microbiology, Chemistry and Potential Applications*, vol. I, ed. T. Kirk, T. Higuchi & H.-M. Chang, pp. 195–213. CRC Press: Boca Raton, FL.

Shimi, I. R. & Nour el Dein, M. S. (1962). Biosynthesis of itaconic acid by *Aspergillus terreus*. *Archiv für Mikrobiologie* **44**, 181–8.

Shoemaker, S. P. & Brown, R. D. Jr (1978). Characterization of endo-1,4-β-D-glucanases purified from *Trichoderma viride*. *Biochimica et Biophysica Acta* **523**, 147–61.

Shoemaker, S. P., Raymond, J. C. & Bruner, R. (1981). Cellulases: diversity among *Trichoderma* strains. In *Trends in Biology of Fermentation for Fuels and Chemicals*, ed. A. Hollaender, pp. 89–109. Plenum Press: New York.

Shoun, H., Kim, D.-H., Uchiyama, H. & Sugiyama, J. (1992). Denitrification by fungi. *FEMS Microbiology Letters* **94**, 277–82.

Shoun, H. & Tanimoto, T. (1991). Denitrification by the fungus *Fusarium oxysporium* and involvement of cytochrome P-450 in the respiratory nitrite reduction. *Journal of Biological Chemistry* **266**, 11078–82.

Shu, P. & Johnson, M. J. (1948). The interdependence of medium constituents in citric acid production by submerged fermentation. *Journal of Bacteriology* **56**, 577–85.

Siegenthaler, P. A., Belsky, M. M. & Goldstein, S. (1967). Phosphate uptake in a moderately marine fungus: a specific requirement for sodium. *Science* **155**, 93–4.

Siegenthaler, P. A., Belsky, M. M., Goldstein, S. & Menna, M. (1967). Phosphate uptake in an obligately marine fungus. II. Role of culture conditions, energy sources and inhibitors. *Journal of Bacteriology* **93**, 1281–8.

Sigler, K. & Höfer, M. (1991). Mechanism of acid extrusion in yeast. *Biochimica et Biophysica Acta* **1071**, 375–92.

Sills, A. M. & Stewart, G. G. (1982). Production of amylolytic enzymes by several yeast species. *Journal of the Institute of Brewing* **88**, 313–16.

Sims, A. P. & Barnett, J. A. (1978). The requirement of oxygen for the utilization of maltose, cellobiose and D-galactose by certain anaerobically fermenting yeasts (Kluyver effect). *Journal of General Microbiology* **106**, 277–88.

Sims, A. P. & Barnett, J. A. (1991). Levels of activity of enzymes involved in aerobic utilization of sugars by six yeast species: observations towards understanding the Kluyver effect. *FEMS Microbiology Letters* **77**, 295–8.

Sims, A. P. & Ferguson, A. R. (1972). The role of enzyme inactivation in the regulation of glutamine synthesis in yeast: *in vivo* studies using ^{15}N. In *6 Wissenschaftliche Konferenz der Gesellschaft Deutscher Naturforscher und Arti, Second International Symposium on Metabolic Interconversion of Enzymes, Rottach-Egern 1971*, ed. O. Wieland, E. Helmreich & H. Holzer, pp. 261–76. Springer-Verlag: Berlin, Heidelberg, New York.

Sims, A. P. & Ferguson, A. R. (1974). The regulation of glutamine metabolism in *Candida utilis*. Studies with ^{15}NH$_3$ to measure *in vivo* sites of glutamine synthesis. *Journal of General Microbiology* **80**, 143–58.

Sims, A. P. & Folkes, B. F. (1964). A kinetic study of the assimilation of [^{15}N]ammonia and the synthesis of amino acids in an exponentially growing culture of *Candida utilis*. *Proceedings of the Royal Society* B **159**, 479–502.

Sims, A. P., Stålbrand, H. & Barnett, J. A. (1991). The role of pyruvate decarboxylase in the Kluyver effect in the food yeast *Candida utilis*. *Yeast* **7**, 479–88.

Sims, A. P., Toone, J. & Box, V. (1974). The regulation of glutamine synthesis in the food yeast *Candida utilis*: the purification and subunit structure of glutamine synthetase and aspects of enzyme deactivation. *Journal of General Microbiology* **80**, 485–99.

Singh, D. P. & Lé John, H. B. (1975). Amino acid transport in a water-mould: the possible regulatory roles of calcium and N^6-(Δ^2-isopentyl)adenine. *Canadian Journal of Biochemistry* **53**, 975–88.

Singh, K. S. & Norton, R. S. (1991). Metabolic changes induced during adaptation of *Saccharomyces cerevisiae* to water stress. *Archives of Microbiology* **156**, 38–42.

Singh, M., Jayakumar, A. & Prasad, R. (1979a). The effect of altered ergosterol content on the transport of various amino acids in *Candida albicans*. *Biochimica et Biophysica Acta* **555**, 42–55.

Singh, M., Jayakumar, A. & Prasad, R. (1979b). Liquid composition and polyene antibiotic sensitivity in isolates of *Candida albicans*. *Microbios* **24**, 7–16.

Singleton, I., Wainwright, M. & Edyvean, R. G. D. (1990). Some factors influencing the adsorption of particulates by fungal mycelium. *Biorecovery* **1**, 271–89.

Sinha, U. (1969). Genetic control of uptake of amino acids in *Aspergillus nidulans*. *Genetics* **62**, 495–505.

Skinner, M. F. & Bowen, G. D. (1974). The uptake and translocation of phosphate by mycelial strands of mycorrhizas. *Soil Biology and Biochemistry* **6**, 53–6.

Slaughter, J. C. (1989). Sulphur compounds in fungi. In *Nitrogen, Phosphorus and Sulphur Utilization by Fungi*, ed. L. Boddy, R. Marchant & D. J. Read, pp. 91–105. Cambridge University Press: Cambridge.

Slayman, C. L. (1965). Electrical properties of *Neurospora crassa*. Effects of external cations in the intracellular potential. *Journal of General Microbiology* **49**, 69–92.

Slayman, C. L. (1970). Movement of ions and electrogenesis in micoorganisms. *American Zoologist* **10**, 377–92.

Slayman, C. L. (1973). Adenine nucleotide levels in *Neurospora* as influenced by conditions of growth and by other metabolic inhibitors. *Journal of Bacteriology* **114**, 752–66.

Slayman, C. L. (1977). Energetics and control of transport in *Neurospora*. In *Water Relations in Membrane Transport in Plants and Animals*, eds. A. H. Jungreis, T. K. Hodges, A. Kleinzeller & S. G. Schultz, pp. 69–87. Academic Press: New York.

Slayman, C. L. (1987). The plasma membrane ATPase of *Neurospora*: a proton-pumping electroenzyme. *Journal of Bioenergetics and Biomembranes* **19**, 1–20.

Slayman, C. L., Long, W. S. & Lu, C. Y.-H. (1973). The relationship between ATP and an electrogenic pump in the plasma membrane of *Neurospora crassa*. *Journal of Membrane Biology* **14**, 305–38.

Slayman, C. L. & Sanders, D. (1984). Electrical kinetics of proton pumping in *Neurospora*. In *Electronic Transport: Fundamental Principles and Physiological Implications*, ed. M. P. Blaustein and M. Lièberman, pp. 307–22. Raven Press: New York.

Slayman, C. L. & Slayman, C. W. (1968). Net uptake of potassium in *Neurospora*. Exchange for sodium and hydrogen ions. *Journal of General Physiology* **52**, 424–43.

Slayman, C. L. & Slayman, C. W. (1974). Depolarization of the plasma membrane of *Neurospora* during active transport of glucose: evidence for a proton-dependent cotransport system. *Proceedings of the National Academy of the Sciences, USA* **71**, 1935–9.

Slayman, C. W. & Slayman, C. L. (1970). Potassium transport in *Neurospora*. Evidence for a multisite carrier at high pH. *Journal of General Microbiology* **55**, 758–6.

Slayman, C. W. & Slayman, C. L. (1975). Energy coupling in the plasma membrane of *Neurospora*: ATP-dependent proton transport and proton-dependent cotransport. In *Molecular Aspects of Membrane Phenomena*, ed. H. R. Kaback, H. Neurath, G. K. Radda, R. Schwyzer & W. R. Wiley, pp. 233–48. Springer-Verlag: Berlin, Heidelberg, New York.

Slayman, C. W. & Tatum, E. L. (1964). Potassium transport in *Neurospora*. I. Intracellular sodium and potassium concentrations and cation requirements for growth. *Biochimica et Biophysica Acta* **88**, 578–92.

Slayman, C. W. & Tatum, E. L. (1965). Potassium transport in *Neurospora*. II. Measurement of steady-state potassium fluxes. *Biochimica et Biophysica Acta* **102**, 149–60.

Slininger, P. J., Bothast, R. J., Okos, M. R. & Ladisch, M. R. (1985). Comparative evaluation of ethanol production by xylose-fermenting yeasts presented with high xylose concentrations. *Biotechnology Letters* **7**, 431–6.

Slonimski, P. R. (1953). *La Formation des Enzymes Respiratoires chez la Levure*.

Masson et Cie: Paris.

Smith, F. A. (1966). Active phosphate uptake by *Nitella translucens*. *Biochimica et Biophysica Acta* **126**, 94–9.

Smith, F. A. (1972). A comparison of the uptake of nitrate, chloride and phosphate by excised beech mycorrhizas. *New Phytologist* **71**, 875–82.

Smith, R. & Scarborough, G. A. (1984). Large-scale isolation of the *Neurospora* plasma membrane H^+-ATPase. *Analytical Biochemistry* **138**, 156–63.

Solimene, R., Guerrini, A. M. & Donini, P. (1980). Levels of acid-soluble polyphosphate in growing cultures of *Saccharomyces cerevisiae*. *Journal of Bacteriology* **143**, 710–14.

Sollins, P., Cromack, J. Jr, Fugel, R. & Li, C. Y. (1981). Role of low-molecular-weight organic compounds in the inorganic nutrition of fungi and higher plants. In *The Fungal Community*, ed. D. T. Wicklow & G. C. Carroll, pp. 607–17. Marcel Dekker: New York.

Solomonson, L. P. & Barber, M. J. (1990). Assimilatory nitrate reductase: functional properties and regulation. *Annual Review of Plant Physiology and Plant Molecular Biology* **41**, 225–53.

Sols, A. & de la Fuente, G. (1961). Transport and hydrolysis in the utilizations of oligosaccharides by yeast. In *Membrane Transport and Metabolism*, ed. A. Kleinzeller & A. Koytk, pp. 361–77. Publishing House of the Czechoslovak Academy of Sciences: Prague.

Sophianopoulon, V. & Scazzocchio, C. (1989). The proline transport protein of *Aspergillus nidulans* is very similar to amino acid transporters of *Saccharomyces cerevisiae*. *Molecular Microbiology* **3**, 705–14.

Speir, T. W. & Ross, D. J. (1978). Soil phosphatase and sulphatase. In *Soil Enzymes*, ed. R. G. Burns, pp. 197–250. Academic Press: London.

Spence, K. D. (1971). Mutation of *Saccharomyces cerevisiae* preventing uptake of S-adenosylmethionine. *Journal of Bacteriology* **106**, 325–30.

Spencer-Martins, I. & van Uden, J. (1985a). Inactivation of active glucose transport in *Candida wikerhamii* is triggered by extra-cellular glucose. *FEMS Microbiology Letters* **28**, 277–9.

Spencer-Martins, I. & van Uden, J. (1985b). Catabolite interconversion of glucose transport systems in the yeast *Candida wikerhamii*. *Biochimica et Biophysica Acta* **812**, 168–72.

Spencer-Martins, I. & van Uden, J. (1986). Substrate and end product regulation of cellobiose uptake by *Candida wickerhamii*. *Applied Microbiology and Biotechnology* **25**, 272–6.

Spoerl, E., Williams, J. P. & Benedict, S. H. (1973). Increased sites of sugar transport in *Saccharomyces cerevisiae*. A result of sugar metabolism. *Biochimica et Biophysica Acta* **298**, 956–66.

Srebotnik, E., Messner, K. & Foisner, R. (1988). Penetrability of white rot-degraded pinewood by the lignin peroxidase of *Phanerochaete chrysosporium*. *Applied and Environmental Microbiology* **54**, 2608–18.

Sreenath, H. K., Chapman, T. W. & Jeffries, T. W. (1986). Ethanol production from D-xylose in batch fermentations with *Candida shehatae*: process variables. *Applied Microbiology and Biotechnology* **24**, 294–9.

Starling, A. P. & Ross, I. S. (1990). Uptake of manganese by *Penicillium notatum*. *Microbios* **63**, 93–100.

Starling, A. P. & Ross, I. S. (1991). Uptake of zinc by *Penicillium notatum*. *Mycological Research* **95**, 712–14.

Steffens, J. C. (1990). Heavy-metal binding peptides in plants. *Annual Review of Plant Physiology* **41**, 553–75.

Steinberg, R. A. (1948). The essentiality of calcium in the nutrition of fungi. *Science* **107**, 423.

Steinböck, F. A., Held, I., Choojun, S., Harmsen, H., Röhr, M., Kubicek-Pranz, E. M. & Kubicek, C. P. (1991). Regulatory aspects of carbohydrate metabolism in relation to citric acid accumulation by *Aspergillus niger*. *Acta Biotechnologica* **11**, 571–82.

Sternberg, D. & Mandels, G. R. (1982). β-Glucosidase induction and repression in the cellulolytic fungus *Trichoderma reesei*. *Experimental Mycology* **6**, 115–24.

Stevenson, R. M. & Lé John, H. B. (1978). Possible regulation of nucleoside transport and RNA synthesis in *Achlya*. *Canadian Journal of Biochemistry* **56**, 207–16.

Stewart, G. G., Garrison, I. F., Goring, T. E., Meleg, M., Pipasts, P. & Russell, I. (1976). Biochemical and genetical studies on yeast flocculation. *Kemia-Kemi* **3**, 465–79.

Stewart, G. G. & Goring, T. E. (1971). Effect of some monovalent and divalent cations on the flocculation of brewer's yeast stains. *Journal of the Institute of Brewing* **82**, 341–2.

Stewart, W. D. P. (1966). *Nitrogen Fixation in Plants*. The Athlone Press: London.

Stokes, P. M. & Lindsay, J. E. (1979). Copper tolerance and accumulation in *Penicillium ochro-chloron* isolated from copper-plating solution. *Mycologia* **71**, 796–806.

Stone, H. E. & Armentrout, V. N. (1985). Production of oxalic acid by *Sclerotium capivorum* during infection of onion. *Mycologia* **77**, 526–30.

Stone, J. E. & Scallan, A. M. (1965). Effects of component removal upon the porous structure of the cell wall of wood. *Journal of Polymer Science* **C11**, 13–25.

Straker, C. J. & Mitchell, D. T. (1986). The activity and characterization of acid phosphatases in endomycorrhizal fungi of the Ericaceae. *New Phytologist* **104**, 243–56.

Straker, C. J. & Mitchell, D. T. (1987). Kinetic characterization of a dual phosphate uptake system in the endomycorrhizal fungus of *Erica hispidula* L. *New Phytologist* **106**, 129–38.

Stratford, M. (1989). Yeast flocculation: calcium specificity. *Yeast* **5**, 487–96.

Stratford, M. (1992a). Yeast flocculation: a new perspective. *Advances in Microbial Physiology* **33**, 1–71.

Stratford, M. (1992b). Yeast flocculation - reconciliation of physiological and genetic viewpoints. *Yeast* **8**, 25–38.

Stratford, M., Morgan, P. & Rose, A. H. (1987). Sulphur dioxide resistance in *Saccharomyces cerevisiae* and *Saccharomyces ludwigii*. *Journal of General Microbiology* **133**, 2173–9.

Stratford, M. & Rose, A. H. (1986). Transport of sulphur dioxide by *Saccharomyces cerevisiae*. *Journal of General Microbiology* **132**, 1–6.

Strauss, J. & Kubicek, C. P. (1990). β-Glucosidase and cellulose formation by *Trichoderma reesei* mutant defective in constitutive β-glucosidase formation. *Journal of General Microbiology* **136**, 1321–6.

Streamer, M., Eriksson, K.-E. & Pettersson, B. (1975). Extracellular enzyme system utilized by the fungus *Sporotrichium pulverulentum* (*Chrysosporium lignorum*) for the breakdown of cellulose. Functional characterization of five *endo*-1,4-β-glucanases and one *exo*-1,4-β-glucanase. *European Journal of Biochemistry* **59**, 607–13.

Stroobant, P. & Scarborough, G. A. (1979). Active transport of calcium in *Neurospora* plasma membrane vesicles. *Proceedings of the National Academy of Sciences, USA* **76**, 3102–6.

Strullu, D. G., Harley, J. L., Gourret, J. P. & Garrec, J. P. (1983). A note of the relative phosphorous and calcium contents of metachromatic granules in *Fagus* mycorrhiza. *New Phytologist* **94**, 89–94.

Stump, R. F., Robinson, K. R., Harold, R. L. & Harold, F. M. (1980). Endogenous electrical currents in the water mold *Blastocladiella emersonii* during growth and sporulation. *Proceedings of the National Academy of Sciences, USA* **77**, 6673–7.

Subba Rao, P. V., Fritig, B., Vose, J. R. & Towers, G. H. N. (1971). An aromatic 3,4-oxygenase from *Tilletiopsis washingtoniensis* – oxidation of 3,4-dihydroxyphenylacetic acid to β-carboxymethylmuconolactone. *Phytochemistry* **10**, 51–6.

Subramanian, K. N., Weiss, R. L. & Davis, R. H. (1973). Use of external, biosynthetic and organellar arginine by *Neurospora*. *Journal of Bacteriology* **115**, 284–90.

Sugumaran, M. & Vaidyanathan, C. S. (1978). Metabolism of aromatic compounds. *Journal of the Indian Institute of Science* **60**, 57–123.

Sumadra, R. & Cooper, T. G. (1977). Allantoin transport in *Saccharomyces cerevisiae*. *Journal of Bacteriology* **131**, 839–47.

Sumadra, R., Gorski, M. & Cooper, T. G. (1976). Urea-transport-defective strains of *Saccharomyces cerevisiae*. *Journal of Bacteriology* **135**, 498–510.

Sumadra, R., Zacharski, C. A., Turoscy, V. & Cooper, T. G. (1978). Induction and inhibition of the allantoin permease in *Saccharomyces cerevisiae*. *Journal of Bacteriology* **135**, 498–510.

Suomalainen, H. & Oura, E. (1971). Yeast nutrition and solute uptake. In *The Yeasts*, vol. 2 *Physiology and Biochemistry of Yeasts*, ed. A. H. Rose & D. S. Harrison, pp. 3–74. Academic Press: London.

Suryanarayana, K., Thomas, D. de S. A. & Mutus, B. (1985). Calmodulin from the water mold *Achlya ambisexualis*: isolation and characterization. *Cell Biology International Reports* **9**, 389–400.

Sussman, M. R., Strickler, J. E., Hager, K. M. & Slayman, C. W. (1987). Location of dicyclohexyl carbodiimide-reactive glutamate residue in the *Neurospora crassa* plasma membrane H^+-ATPase. *Journal of Biological Chemistry* **262**, 4569–73.

Sutter, H.-P., Jones, E. B. G. & Wälchli, O. (1983). The mechanism of copper tolerance in *Poria placenta* (Fr.) Che. and *Poria vaillantii* (Pers.) Fr. *Material und Organismen* **18**, 241–62.

Sutton, C. D. & Gunary, D. (1969). Phosphate equilibria in soil. In *Ecological Aspects of the Mineral Nutrition of Plants*, ed. I. H. Rorison, pp. 127–34. Blackwell Scientific Publications: Oxford.

Sutton, D. D. & Lampen, J. O. (1962). Localization of sucrose and maltose fermenting systems in *Saccharomyces cerevisiae*. *Biochimica et Biophysica Acta* **56**, 303–12.

Svihla, G., Dainko, S. L. & Schlenk, F. (1963). Ultraviolet microscopy of purine compounds in the yeast vacuole. *Journal of Bacteriology* **85**, 399–409.

Svihla, G., Dainko, J. L. & Schlenk, F. (1969). Ultraviolet micrography of penetration of exogenous cytochrome *c* into the yeast cell. *Journal of Bacteriology* **100**, 498–504.

Svihla, G. & Schlenk, F. (1960). *S*-adenosyl-methionine in the vacuole of *Candida utilis*. *Journal of Bacteriology* **79**, 841–8.

Swamy, S., Uno, I. & Ishikawa, T. (1985). Regulation of cyclic AMP-dependent phosphorylation of cellular proteins by the incompatibility factors in *Coprinus cinereus*. *Journal of General and Applied Microbiology* **31**, 339–46.

Swenson, P. A. (1960). Leakage of phosphate compounds from ultraviolet-irradiated yeast cells. *Journal of Cellular and Comparative Physiology* **56**, 77–91.

Szaniszlo, P. J., Powell, P. E., Reid, C. P. P. & Cline, G. R. (1981). Production of hydroxamate siderophore iron chelators by ectomycorrhizal fungi. *Mycologia* **73**, 1158–74.

Sze, H. (1985). H$^+$-translocating ATPases: advances using membrane vesicles. *Annual Review of Plant Physiology* **36**, 175–208.

Szkutnicka, K., Tschopp, J. F., Andrews, L. & Cirillo, V. P. (1989). Sequence and structure of the yeast galactose transporter. *Journal of Bacteriology* **171**, 4486–93.

Tabak, H. H. & Cooke, W. B. (1968). Growth and metabolism of fungi in an atmosphere of nitrogen. *Mycologia* **60**, 115–40.

Taber, W. (1964). Sequential formation and accumulation of primary and secondary shunt metabolic products in *Claviceps purpurea*. *Applied Microbiology* **12**, 321–6.

Tachikawa, T., Miyakawa, T., Tsuchiya, E. & Fukui, S. (1987). A rapid and transient increase of cellular Ca^{2+} in response to mating pheromone in *Saccharomyces cerevisiae*. *Agricultural and Biological Chemistry* **51**, 1209–10.

Taghikhani, M., Lavi, L. W., Woost, P. G. & Griffin, C. C. (1984). Kinetics of D-glucose and 2-deoxy-D-glucose transport by *Rhodotorula glutinis*. *Biochimica et Biophysica Acta* **803**, 278–83.

Taiz, L. (1984). Plant cell extension: regulation of cell wall mechanical properties. *Annual Review of Plant Physiology* **35**, 585–657.

Taj-Aldeen, S. J. & Moore, D. (1982). The *ftr* cistron of *Coprinus cinereus* is the structural gene for a multifunctional transport carrier molecule. *Current Genetics* **5**, 209–13.

Takagi, M., Moriya, K. & Yano, K. (1980a). Induction of cytochrome P-450 in petroleum assimilating yeast. I. Selection of a strain and basic characteristics of cytochrome P-450 induction in the strain. *Cellular and Molecular Biology* **25**, 363–9.

Takagi, M., Moriya, K. & Yano, K. (1980b). Induction of cytochrome P-450 in petroleum-assimilating yeast. II. Comparison of protein synthesising activity in cells grown on glucose and *n*-tetradecane. *Cellular and Molecular Biology* **25**, 371–5.

Takahashi, T. & Sakaguchi, K. (1927). Acids formed by *Rhizopus* species. V. *Bulletin of the Agricultural Chemical Society of Japan* **3**, 59–62.

Takao, S. (1965). Organic acid production by basidiomycetes. *Applied Microbiology* **13**, 732–7.

Takebe, I. (1960). Choline sulfate as major soluble component of conidiospores of *Aspergillus niger*. *Journal of General and Applied Microbiology* **6**, 83–9

Takeuchi, Y., Schmid, J., Caldwell, J. H. & Harold, F. M. (1988). Transcellular ion currents and extension of *Neurospora crassa* hyphae. *Journal of Membrane Biology* **101**, 33–42.

Talbot, N. J., Coddington, A., Roberts, I. N. & Oliver, R. P. (1988). Diploid construction by protoplast fusion in *Fulvia frelva* (syn. *Cladosporium fulva*): genetic analysis of an imperfect plant pathogen. *Current Genetics* **14**, 567–72.

Tamai, Y., Toh-e, A. & Oshima, Y. (1985). Regulation of inorganic phosphate transport systems in *Saccharomyces cerevisiae*. *Journal of Bacteriology* **164**, 964–8.

Tanaka, A. & Fukui, S. (1989). Metabolism of *n*-alkanes. In *The Yeasts*, vol. 3 *Metabolism and Physiology of Yeasts*, ed. A. H. Rose & J. S. Harrison, pp. 261–88. Academic Press: London.

Tanaka, J.-I. & Fink, G. R. (1985). The histidine permease gene (*HIPI*) of *Saccharomyces cerevisiae*. *Gene* **38**, 205–14.

Tanner, W. & Beevers, H. (1990). Does transpiration have an essential function in long distance ion transport in plants? *Plant, Cell and Environment* **13**, 745–50.

Taylor, N. W. & Orton, W. L. (1978). Aromatic compounds and sugars in flocculation of *Saccharomyces cerevisiae*. *Journal of the Institute of Brewing* **84**, 113–14.

Techel, D., Gebauer, G., Kohler, W., Braumann, T., Jastorff, B. & Rensing, L. (1990). On the role of Ca^{2+}-calmodulin-dependent and cAMP-dependent phosphorylation in the circadian rhythm of *Neurospora crassa*. *Journal of Comparative Physiology* B **159**, 695–706.

Teeri, T. T. (1987). The cellulolytic enzyme system of *Trichoderma reesei*. Molecular cloning, characterization and expression of cellobiohydrolase genes. Ph.D. thesis, Technical Research Centre of Finland, Espoo.

Teeri, T. T., Jones, A., Kraulis, P., Rouvinen, J., Pentillä, M., Harkki, A., Nevalainen, H., Vanhanen, S., Saloheino, M. & Knowles, J. K. C. (1990). Engineering *Trichoderma* and its cellulases. In Trichoderma reesei *Cellulases: Biochemistry, Genetics, Physiology and Applications*, ed. C. P. Kubicek, D. E. Eveleigh, H. Esterbaur, W. Steiner & E. M. Kubicek-Prans, pp. 156–67. Royal Society of Chemistry: Cambridge.

Teh, J. S. (1975). Glucose transport and its inhibition by short-chain *n*-alkanes in *Cladosporium resinae*. *Journal of Bacteriology* **122**, 832–40.

Téllez-Iñón, M. T., Ulloa, R. M., Glikin, G. C. & Torres, F. M. (1985). Characterization of *Neurospora crassa* cyclic AMP phosphodiesterase activated by calmodulin. *Biochemical Journal* **232**, 425–30.

Tempest, D. W. & Neijssel, O. M. (1980). Growth yield values in relation to respiration. In *Diversity of Bacterial Respiratory Systems*, vol. 1, ed. C. J. Knowles, pp. 1–31. CRC Press: Boca Raton, FL.

Tempest, D. W. & Neijssel, O. M. (1984). The status of Y_{ATP} and maintenance energy as biologically interpretable phenomena. *Annual Review of Microbiology* **38**, 459–86.

Thain, J. F. & Girvin, D. (1987). Translocation through established mycelium of *Neurospora crassa* on a nutrient-free substrate. *Transactions of the British Mycological Society* **89**, 45–9.

Thatcher, F. S. (1939). Osmotic and permeability relations in the nutrition of fungus parasites. *American Journal of Botany* **26**, 449–58.

Theodorou, C. (1971). The phytase activity of the mycorrhizal fungus *Rhizopogon roseolus*. *Soil Biology and Biochemistry* **10**, 33–7.

Theodorou, M. K., Lowe, S. E. & Trinci, A. P. J. (1992). Anaerobic fungi and the rumen ecosystem. In *The Fungal Community: Its Organisation and Role in the Ecosystem*, ed. G. C. Carroll & D. T. Wicklow, pp. 43–72. Marcel Dekker: New York.

Theuvenet, A. P. R. & Borst-Pauwels, G. W. F. H. (1976*a*). Surface charge and the kinetics of two-site mediated ion-translocation. The effects of UO_2^{2+} and La^{2+} upon the Rb^+-uptake into yeast cells. *Bioelectrochemistry and*

Bioenergetics **3**, 230–40.

Theuvenet, A. P. R. & Borst-Pauwels, G. W. F. H. (1976b). Kinetics of ion translocation across charged membranes mediated by a two-site transport mechanism. Effects of polyvalent cautions upon rubidium uptake into yeast cells. *Biochimica et Biophysica Acta* **426**, 745–56.

Theuvenet, A. P. R. & Borst-Pauwels, G. W. F. H. (1976c). The influence of surface charge on the kinetics of ion-translocation across biological membranes. *Journal of Theoretical Biology* **57**, 313–29.

Theuvenet, A. P. R. & Borst-Pauwels, G. W. F. H. (1983). Effect of surface potential on Rb^+ in yeast. The effect of pH. *Biochimica et Biophysica Acta* **734**, 62–9.

Theuvenet, A. P. R., Roomans, G. M. & Borst-Pauwels, G. W. F. H. (1977). Intracellular pH and the kinetics of Rb^+ uptake by yeast non carrier versus mobile carrier-mediated uptake. *Biochimica et Biophysica Acta* **469**, 272–80.

Theuvenet, A. P. R., van de Wijngaard, W. H. M., van de Rijke, J. W. & Borst-Pauwels, G. W. F. H. (1984). Application of 9-amino-acridine as a probe of the surface potential experienced by cation transporters in the plasma membrane of yeast cells. *Biochimica et Biophysica Acta* **775**, 161–8.

Thevelein, J. M. (1984). Regulation of trehalose mobilization in fungi. *Microbiological Reviews* **48**, 42–59.

Thevelein, J. M. (1991). Fermentable sugars and intracellular acidification as specific activators of the RAS adenylate cyclase signalling pathway in yeast – the relationship to nutrient-induced cell cycle control. *Molecular Microbiology* **5**, 1301–8.

Thieken, A. & Winkelmann, G. (1992). Rhizoferrin – a complexone type of siderophore of the Mucorales and Entomophthorales (Zygomycetes). *FEMS Microbiology Letters* **94**, 37–41.

Thomas, D., Jacquemin, I. & Surdin-Kerjan, Y. (1992). MET4, a leucine zipper protein, and centromere-binding factor-1⁻ are both required for transcriptional activation of sulfur metabolism in *Saccharomyces cerevisiae*. *Molecular and Cellular Biology* **12**, 1719–27.

Thomas, D. S., Hossack, J. A. & Rose, A. H. (1978). Plasma-membrane lipid composition and ethanol tolerance in *Saccharomyces cerevisiae*. *Archives of Microbiology* **117**, 239–45.

Thomas, D. S. & Rose, A. H. (1979). Inhibitory effect of ethanol on growth and solute accumulation by *Saccharomyces cerevisiae* as affected by plasma membrane lipid composition. *Archives of Microbiology* **122**, 49–55.

Thompson, W., Brownlee, C. & Jennings, D. H. & Mortimer, A. M. (1987). Localized cold-induced inhibition of translocation in mycelia and strands of *Serpula lacrimans*. *Journal of Experimental Botany* **38**, 889–99.

Thompson, W., Eamus, D. & Jennings, D. H. (1985). Water flux through mycelium of *Serpula lacrimans*. *Transactions of the British Mycological Society* **84**, 601–8.

Thomson, B. D., Clarkson, D. T. & Brain, P. (1990). Kinetics of phosphorus uptake by the germ-tubes of the vesicular–arbuscular mycorrhizal fungus *Gigaspora marginata*. *New Phytologist* **116**, 647–53.

Thornton, J. D., Galpin, M. F. J. & Jennings, D. H. (1976). The bursting of hyphal tips of *Dendryphiella salina* and the relevance of the phenomenon to repression of glucose transport. *Journal of General Microbiology* **96**, 145–53.

Tien, M., Kersten, P. J. & Kirk, T. K. (1987). Selection and improvement of lignin-degrading microorganisms: potential strategy based on lignin model-amino acid adducts. *Applied and Environmental Microbiology* **53**, 242–5.

Tien, M. & Kirk, T. K. (1983). Lignin-degrading enzyme from the hymenomycete *Phanerochaete chrysosporium* Burds. *Science* **221**, 661–3.

Tien, M. & Kirk, T. K. (1984). Lignin-degrading enzyme from *Phanerochaete chrysosporium*: purification, characterization and catalytic properties of a unique H_2O_2-requiring oxygenase. *Proceedings of the National Academy of Sciences of the USA* **81**, 2280–4.

Tien, M., Kirk, T. K., Bull, C. & Fee, J. A. (1986). Steady-state and transient-state kinetic studies on the oxidation of 3,4-dimethoxybenzyl alcohol catalyzed by the ligninase of *Phanaerochaete chrysosporium* Burds. *Journal of Biological Chemistry* **261**, 1687–93.

Tien, M. & Myer, S. B. (1990). Selection and characterization of mutants of *Phanerochaete chrysosporium* exhibiting ligninolytic activity under nutrient-rich conditions. *Applied and Environmental Microbiology* **56**, 2540–4.

Tijssen, J. P. F., Beekes, H. W. & van Steveninck, J. (1981). Localization of polyphosphate at the outside of the yeast plasma membrane. *Biochimica et Biophysica Acta* **649**, 529–32.

Tijssen, J. P. F., Dubbelman, T. M. A. R. & van Steveninck, J. (1983). Isolation and characterization of polyphosphates from the yeast cell surface. *Biochimica et Biophysica Acta* **760**, 143–8.

Tijssen, J. P. F., van den Broek, P. J. A. & van Steveninck, J. (1984). The involvement of high-energy phosphate in 2-deoxy-D-glucose transport in *Kluyveromyces marxianus*. *Biochimica et Biophysica Acta* **778**, 87–93.

Tijssen, J. P. F. & van Steveninck, J. (1984). Detection of a yeast polyphosphate fraction localized outside the plasma membrane by the method of phosphorus-31 nuclear magnetic resonance. *Biochemical and Biophysical Research Communications* **119**, 447–51.

Tijssen, J. P. F., van Steveninck, J. & de Bruijn, W. C. (1985). Cytochemical staining of a yeast polyphosphate fraction, localised outside the plasma membrane. *Protoplasma* **125**, 124–8.

Tinker, P. B. (1969). The transport of ions in the soil around plant roots. In *Ecological Aspects of the Mineral Nutrition of Plants*, ed. I. H. Rorison, pp. 135–47. Blackwell Scientific Publications: Oxford.

Tinker, P. B. (1975). Effects of vesicular–arbuscular mycorrhizas on higher plants. *Symposium of the Society for Experimental Biology* **29**, 325–49.

Tohoyama, H., Tomoyasu, T., Inouhe, M., Joho, M. & Murayama, T. (1992). The gene for cadmium metallothionein from cadmium-resistant yeast appears to be identical to *CUP1* in a copper-resistant strain. *Current Genetics* **21**, 275–380.

Toivola, A., Yarrow, D., van den Bosch, E., van Dijken, J. P. & Scheffers, W. A. (1984). Alcoholic fermentation of D-xylose by yeasts. *Applied and Environmental Microbiology* **47**, 1221–3.

Tomlinson, N., Campbell, J. J. R. & Trussel, P. C. (1950). The influence of zinc, iron, copper and manganese on the production of citric acid by *Aspergillus niger*. *Journal of Bacteriology* **59**, 217–27.

Tomsett, A. B. (1989). The genetics and biochemistry of nitrate assimilation in ascomycete fungi. In *Nitrogen, Phosphorus and Sulphur Utilization by Fungi*, eds. L. Boddy, R. Marchant & D. J. Read, pp. 33–57. Cambridge University Press: Cambridge.

Tomsett, A. B. (1993). Genetics and molecular biology of metal tolerance in fungi. In *Stress Tolerance of Fungi*, ed. D. H. Jennings, pp. 69–95. Marcel Dekker: New York.

Tomsett, A. B. & Cove, D. J. (1979). Deletion mapping of the *niiA*, *niaD* gene region of *Aspergillus nidulans*. *Genetic Research* **34**, 19–32.

Tomsett, A. B. & Garrett, R. H. (1980). The isolation and characterization of mutants defective in nitrate assimilation in *Neurospora crassa*. *Genetics* **95**, 649–60.

Tomsett, A. B. & Garrett, R. H. (1981). Biochemical analysis of mutants defective in nitrate assimilation in *Neurospora crassa*: evidence for autogenous control by nitrate reductase. *Molecular and General Genetics* **184**, 183–90.

Tomsett, A. B. & Thurman, D. A. (1988). Molecular biology of metal tolerances in plants. *Plant, Cell and Environment* **11**, 383–94.

Torii, K. & Bandurski, R. D. (1967). Yeast sulfate reducing system. III. An intermediate in the reduction of 3′-PAPS to sulfite. *Biochimica et Biophysica Acta* **136**, 286–95.

Tortora, P., Birtel, M., Lenz, A. G. & Holzer, H. (1981). Glucose-dependent metabolic interconversion of fructose-1,6-bisphosphatase in yeast. *Biochemical and Biophysical Research Communications* **100**, 688–95.

Townsend, B. B. (1954). Morphology and development of fungal rhizomorphs. *Transactions of the British Mycological Society* **37**, 222–33.

Tresner, H. D. & Hayes, J. A. (1971). Sodium chloride tolerance of terrestrial fungi. *Applied Microbiology* **22**, 210–13.

Trevithick, J. R. & Metzenberg, R. L. (1966). Genetic alteration of pore size and other properties of the *Neurospora* cell wall. *Journal of Bacteriology* **92**, 1016–21.

Trinci, A. P. J. (1970). Kinetics of the growth of mycelial pellets of *Aspergillus nidulans*. *Archiv für Mikrobiologie* **88**, 135–45.

Trinci, A. P. J. (1971). Influence of the width of the peripheral growth zone on the radial growth rate of fungal colonies on solid media. *Journal of General Microbiology* **67**, 325–44.

Trinci, A. P. J. (1978). The duplication cycle and vegetative development in moulds. In *The Filamentous Fungi*, vol. 3 *Developmental Mycology*, ed. J. E. Smith & D. R. Berry, pp. 132–63. Edward Arnold: London.

Trinci, A. P. J. (1992). Myco-protein: a twenty-year overnight success story. *Mycological Research* **96**, 1–13.

Trinci, A. P. J. & Collinge, A. J. (1973). Structure and plugging of septa of wild type and spreading colonial mutants of *Neurospora crassa*. *Archiv für Mikrobiologie* **91**, 355–64.

Trinci, A. P. J. & Collinge, A. J. (1974). Occlusion of the septal pores of damaged hyphae of *Neurospora crassa* by hexagonal crystals. *Protoplasma* **80**, 57–67.

Trinci, A. P. J. & Righelato, R. C. (1970). Changes in constituents and ultrastructure during autolysis of glucose-starved *Penicillium chrysogenum*. *Journal of General Microbiology* **60**, 239–50.

Trinci, A. P. J. & Thurston, C. F. (1976). Transition to the non-growing states in eukaryotic micro-organisms. *Symposium of the Society for General Microbiology* **26**, 55–80.

Trivedi, A., Khare, S., Singhal, G. S. & Prasad, R. (1982). Effect of phosphatidylcholine and phosphatidylethanolamine enrichment on the structure and function of the yeast cell membrane. *Biochimica et Biophysica Acta* **692**, 202–9.

Literature cited   583

Trivedi, A., Singhal, G. S. & Prasad, R. (1983). Effect of phosphatidylserine enrichment on amino acid transport in yeast. *Biochimica et Biophysica Acta* **729**, 85–9.

Trüper, H. G. & Galinski, E. A. (1986). Concentrated brines as habitats for microorganisms. *Experientia* **42**, 1182–7.

Turian, G. (1979a). Cytochemical gradients and mitochondrial exclusions in the apices of vegetative hyphae. *Experientia* **35**, 1164–6.

Turian, G. (1979b). Polarity of elongation growth generated and sustained by anisotropic distribution of the protons ejected from mitochondria into the cytosol of hyphal apices (*Neurospora* model)? *Archives des Sciences, Genève* **32**, 251–4.

Turian, G. (1980). Germination of conidia (*Neurospora, Trichoderma*): evidence of an electrogenic sink of protons mitochondrially-extruded into emerging tubes. *Bulletin de la Société botanique suisse* **90**, 203–12.

Turoscy, V. & Cooper, T. G. (1979). Allantoate transport in *Saccharomyces cerevisiae*. *Journal of Bacteriology* **140**, 971–9.

Tweedie, J. W. & Segel, I. H. (1970). Specificity of transport processes for sulfur, selenium and molybdenum anions by filamentous fungi. *Biochimica et Biophysica Acta* **196**, 95–106.

Uchida, E., Ohsumi, Y. & Anraku, Y. (1985). Purification and properties of H^+-translocating Mg^{2+}-adenosine triphosphatase from vacuolar membranes of *Saccharomyces cerevisiae*. *Journal of Biological Chemistry* **260**, 1090–5.

Uchida, E., Ohsumi, Y. & Anraku, Y. (1988). Characterization and function of the catalytic subunit of a H^+-translocating adenosine triphosphate from vacuolar membranes of *Saccharomyces cerevisiae*. *Journal of Biological Chemistry* **263**, 45–51.

Ueda, Y. & Oshima, Y. (1975). A constitutive mutation *phoT* of the repressible acid phosphatase synthesis with inability to transport inorganic phosphate in *Saccharomyces cerevisiae*. *Molecular and General Genetics* **136**, 255–9.

Ulaszewski, S., Hilger, F. & Goffeau, A. (1989). Cyclic AMP controls the plasma membrane H^+-ATPase activity from *Saccharomyces cerevisiae*. *FEBS Letters* **245**, 131–6.

Ullrich, W. R. (1983). Uptake and reduction of nitrate: algae and fungi. In *Encyclopaedia of Plant Physiology*, vol. 15 *Inorganic Plant Nutrition*, ed. A. Läuchli & R. L. Bieleski, pp. 376–97. Springer-Verlag: Berlin, Heidelberg, New York.

Ullrich, W. R. (1987). Nitrate and ammonium uptake in green algae and higher plants: mechanism and relationship with nitrate metabolism. In *Inorganic Nitrogen Metabolism*, ed. W. R. Ulrich, P. J. Aparicio, P. J. Syrett & F. Castillo, pp. 32–8. Springer-Verlag: Berlin, Heidelberg, New York.

Ulmer, D. C., Leisola, M. S. A. & Fiechter, A. (1984). Possible induction of the ligninolytic system of *Phanerochaete chrysosporium*. *Journal of Biotechnology* **1**, 13–24.

Ulmer, D. C., Leisola, M., Puhakka, J. & Fiechter, A. (1983). *Phanerochaete chrysosporium*: growth pattern and lignin degradation. *European Journal of Applied Microbiology and Biotechnology* **18**, 153–7.

Ulmer, D., Leisola, M. S. A., Schmidt, B. H. & Fiechter, A. (1983). Rapid degradation of isolated lignins by *Phanerochaete chrysosporium*. *Applied and Environmental Microbiology* **45**, 1795–801.

Umemoto, A., Yoshihisa, F., Hirata, R. & Anraku, Y. (1990). Roles of the *VMA3* gene product, subunit c of the vacuolar membrane H^+-ATPase on vacuolar acidification and protein transport. A study with

VMA3-disrupted mutants of *Saccharomyces cerevisiae*. *Journal of Biological Chemistry* **265**, 18447–53.

Unezawa, C. & Kishi, T. (1989). Vitamin metabolism. In *The Yeasts*, vol. 3 *Metabolism and Physiology of Yeasts*, ed. A. H. Rose & J. S. Harrison, pp. 457–88. Academic Press: London.

Unkles, S. E. (1989). Fungal biotechnology and the nitrate assimilation pathway. In *Molecular Aspects of Nitrate Assimilation*, ed. J. L. Wray & J. R. Kinghorn, pp. 341–63. Oxford Science Publications: Oxford.

Unkles, S. E., Campbell, E. I., Carrez, D., Grieve, C., Contreras, R., Fiers, W., van den Hondel, C. A. M. J. J. & Kinghorn, J. R. (1989). Transformation of *Aspergillus niger* using homologous *niaD* gene. *Gene* **78**, 157–66.

Unkles, S. E., Hawker, K. L., Grieve, C., Campbell, E. I., Montague, P. & Kinghorn, J. R. (1990). *crnA* encodes a nitrate transporter in *Aspergillus nidulans*. *Proceedings of the National Academy of Sciences, USA* **88**, 204–8.

Urbanus, J. F. L. M., van den Ende, H. & Koch, B. (1978). Calcium oxalate crystals in the wall of *Mucor mucedo*. *Mycologia* **70**, 829–42.

Ushio, K. & Nakata, Y. (1989). Isolation and characterization of mutants defective in salt tolerance in *Zygosaccharomyces rouxii*. *Journal of Fermentation and Bioengineering* **68**, 165–9.

Ushio, K., Sawatani, M. & Nakata, Y. (1991). Effect of altered sterol composition on the salt tolerance of *Zygosaccharomyces rouxii*. *Journal of Fermentation and Biochemistry* **71**, 390–6.

Ussing, H. H. (1947). Interpretation of the exchange of radio-sodium in isolated muscle. *Nature, London* **160**, 262–3.

Utter, M. F., Duell, E. A. & Bernofsky, C. (1968). Alterations in the respiratory enzymes of mitochondria of growing and resting yeast. In *Aspects of Yeast Metabolism*, ed. A. K. Mills & H. A. Krebs, pp. 197–215. Blackwell Scientific Publications: Oxford.

Vaheri, M., Leisola, M. & Kauppinen, V. (1978). Transglycosylation products of the cellulose system of *Trichoderma reesei*. *Biotechnology Letters* **1**, 41–6.

Vaheri, M. P., Vaheri, M. E. O. & Kauppinen, V. S. (1979). Formation and release of cellulolytic enzymes during growth of *Trichoderma reesei* on cellulose and glycerol. *European Journal of Applied Microbiology and Biotechnology* **8**, 73–80.

Vallejo, C. G. & Serrano, R. (1989). Physiology of mutants with reduced expression of plasma membrane H^+-ATPase. *Yeast* **5**, 307–19.

van Brunt, J., Caldwell, J. H. & Harold, F. M. (1982). Circulation of potassium across the plasma membrane of *Blastocladiella emersonii*: K^+ channel. *Journal of Bacteriology* **150**, 1449–61.

van de Mortel, J. B. J., Mulders, D., Korthout, H., Theuvenet, A. P. R. & Borst-Pauwels, G. W. F. H. (1988). Transient hyperpolarization of yeast by glucose and ethanol. *Biochimica et Biophysica Acta* **936**, 421–8.

van de Mortel, J. B. J., Theuvenet, A. P. R. & Borst-Pauwels, G. W. F. H. (1990). A putative K^+-selective channel in the plasma membrane of yeast that is blocked by micromolar concentrations of external divalent cations and is insensitive to tetraethyl-ammonium. *Biochimica et Biophysica Acta* **1026**, 220–4.

van den Broek, P. J. A., Christianse, K. & van Steveninck, J. (1982). The energetics of D-fucose transport in *Saccharomyces fragilis*. The influence of the protonmotive force on sugar accumulation. *Biochimica et Biophysica Acta* **692**, 231–7.

van den Broek, P. J. A., de Bruijne, A. W. & van Steveninck, J. (1987). The role of ATP in the control of H^+-galactoside symport in the yeast *Kluyveromyces marxianus*. *Biochemical Journal* **242**, 729–34.

van den Broek, P. J. A., Haasnoot, C. J. P., van Leeuwen, C. C. M. & van Steveninck, J. (1982). The influence of uncouplers on facilitated diffusion of sorbose in *Saccharomyces cerevisiae*. *Biochimica et Biophysica Acta* **689**, 429–36.

van den Broek, P. J. A., Schuddemat, J., van Leeuwen, C. C. M. & van Steveninck, J. (1986). Characterization of 2-deoxyglucose and 6-deoxyglucose transport in *Kluyveromyces marxianus*: evidence for two different transport mechanisms. *Biochimica et Biophysica Acta* **860**, 626–31.

van den Broek, P. J. A. & van Steveninck, J. (1980). Kinetic analysis of simultaneous occurring proton-sorbose symport and passive sorbose transport in *Saccharomyces fragilis*. *Biochimica et Biophysica Acta* **602**, 419–32.

van den Broek, P. J. A. & van Steveninck, J. (1981). Kinetic analysis of 2-deoxy-D-glucose uptake in *Saccharomyces fragilis*. *Biochimica et Biophysica Acta* **649**, 305–9.

van den Broek, P. J. A. & van Steveninck, J. (1982). Kinetic analysis of H^+/methyl-β-D-thiogalactoside symport in *Saccharomyces fragilis*. *Biochimica et Biophysica Acta* **693**, 213–20.

van der Walt, J. P. (1970). Criteria and methods used in classification. In *The Yeasts: A Taxonomic Study*, ed. J. Lodder, pp. 34–113. North-Holland: Amsterdam.

van Dijken, J. P., Otto, R. & Harder, W. (1976). Growth of *Hansenula polymorpha* in a methanol-limited chemostat. Physiological response due to involvement of methanol oxidase as a key enzyme in methanol metabolism. *Archives of Microbiology* **111**, 137–46.

van Dijken, J. P. & Scheffers, W. A. (1986). Redox balances in the metabolism of sugars by yeast. *FEMS Microbiology Reviews* **32**, 199–224.

van Laere, A. (1989). Trehalose, reserve and/or stress metabolite? *FEMS Microbiology Review* **63**, 201–10.

van Langenhove, H., Wuyts, E. & Schamp, N. (1986). Elimination of hydrogen sulphide from odorous air by a wood bark filter. *Water Research* **20**, 1471–6.

van Leeuwen, C. C. M., Postma, E., van den Broek, P. J. A. & van Steveninck, J. (1991). Proton-motive force-driven D-galactose transport in plasma membrane vesicles from the yeast *Kluyveromyces marxianus*. *Journal of Biological Chemistry* **266**, 12146–51.

van Steveninck, J. (1968). Transport and transport-associated phosphorylation of 2-deoxy-D-glucose by yeast. *Biochimica et Biophysica Acta* **163**, 386–94.

van Steveninck, J. (1969). The mechanism of transmembrane glucose transport in yeast: evidence for phosphorylation associated with transport. *Archives of Biochemistry and Biophysics* **130**, 244–52.

van Steveninck, J. (1970). The transport mechanism of α-methylglucoside in yeast: evidence for transport-associated phosphorylation. *Biochimica et Biophysica Acta* **203**, 376–84.

van Steveninck, J. (1972). Transport and transport-associated phosphorylation of galactose in *Saccharomyces cerevisiae*. *Biochimica et Biophysica Acta* **274**, 575–83.

van Steveninck, J. & Rothstein, A. (1965). Sugar transport and metal binding in yeast. *Journal of General Microbiology* **49**, 235–46.

van Tuinen, D., Perez, R. O., Marmé, D. & Turian, G. (1984). Calcium, calmodulin-dependent protein phosphorylation in *Neurospora crassa*.

FEBS Letters **176**, 317–20.

van Uden, N. (1967). Transport-limited fermentation and growth of *Saccharomyces cerevisiae* and its competitive inhibition. *Archiv für Mikrobiologie* **58**, 155–68.

van Uden, J. (1969). Kinetics of nutrient limited growth. *Annual Review of Microbiology* **23**, 473–86.

van Urk, H., Postma, E., Scheffers, W. A. & van Dijken, J. P. (1989). Glucose transport in Crabtree-positive and Crabtree-negative yeasts. *Journal of General Microbiology* **135**, 2399–406.

van Zyl, P. J., Kilian, S. G. & Prior, B. A. (1990). The role of an active transport mechanism in glycerol accumulation during osmoregulation by *Zygosaccharomyces rouxii*. *Applied Microbiology and Biotechnology* **34**, 231–6.

van Zyl, P. J. & Prior, B. A. (1990). Water relations of polyol accumulation by *Zygosaccharomyces rouxii* in continuous culture. *Applied Microbiology and Biotechnology* **33**, 12–17.

van Zyl, P. J., Prior, B. A. & Kilian, S. G. (1991). Regulation of glycerol metabolism in *Zygosaccharomyces rouxii* in response to osmotic stress. *Applied Microbiology and Biotechnology* **36**, 369–74.

Väre, H. (1990). Aluminium polyphosphate in the ectomycorrhizal fungus *Suillus variegatus* (Fr.) O. Kuze. as revealed by energy dispersive spectroscopy. *New Phytologist* **116**, 663–8.

Väre, H. & Markkola, A. M. (1990). Aluminium tolerance of the ectomycorrhizal fungus *Suillus variegatus*. *Symbiosis* **9**, 83–6.

Vaughn, L. E. & Davis, R. H. (1981). Purification of vacuoles from *Neurospora crassa*. *Molecular and Cellular Biology* **1**, 797–806.

Veenhuis, M. & Harder, W. (1989). Microbodies. In *The Yeasts*, vol. 4 *Yeast Organelles*, ed. A. H. Rose & J. S. Harrison, pp. 601–53. Academic Press: London.

Veenhuis, M., van Dijken, J. P. & Harder, W. (1983). The significance of peroxisomes in the metabolism of one-carbon compounds in yeasts. *Advances in Microbial Physiology* **24**, 1–82.

Vega, J. M. (1976). A reduced pyridine nucleotides–diaphorase activity associated with the assimilatory nitrite reductase complex from *Neurospora crassa*. *Archives of Microbiology* **109**, 237–42.

Veness, R. G. & Evans, C. S. (1989). The role of hydrogen peroxide in the degradation of crystalline cellulose by basidiomycete fungi. *Journal of General Microbiology* **135**, 2799–806.

Venkatadri, R. & Irvine, R. L. (1990). Effect of agitation on ligninase activity and ligninase production by *Phanerochaete chrysosporium*. *Applied and Environmental Microbiology* **56**, 2684–91.

Verduyn, C., Frank, J., van Dijken, J. P. & Scheffers, W. A. (1985a). Multiple forms of xylose reductase in *Pachysolen tannophilus* CBS 4044. *FEMS Microbiology Letters* **30**, 313–17.

Verduyn, C., van Kleef, R., Frank, J., Schreuder, H., van Dijken, J. P. & Scheffers, W. A. (1985b). Properties of the NAD(P)H-dependent xylose reductase from the xylose-fermenting yeast *Pichia stipitis*. *Biochemical Journal* **226**, 669–77.

Verduyn, C., van Wijngaaden, C. P., Scheffers, W. A. van Dijken, J. P. (1991). Hydrogen peroxide as an electron acceptor for mitochondrial respiration in the yeast *Hansenula polymorpha*. *Yeast* **7**, 137–46.

Verhoff, F. H. & Spradlin, J. F. (1976). Mass and energy balance analysis of metabolic pathways applied to citric acid production by *Aspergillus niger*. *Biotechnology and Bioengineering* **18**, 425–33.

Verma, R. S., Spencer-Martins, I. & van Uden, N. (1987). Role of *de novo* protein synthesis in the interconversion of glucose transport systems in the yeast *Pichia ohmeri*. *Biochimica et Biophysica Acta* **900**, 139–44.

Vézina, L.-P., Margolis, H. A., McAfee, B. J. & Delaney, S. (1989). Changes in the activity of enzymes involved with primary nitrogen metabolism due to ectomycorrhizal symbiosis in jack pine seedlings. *Physiologia Plantarum* **75**, 55–62.

Vidal, M., Buckley, A. M., Hilger, F. & Gaber, R. F. (1990). Direct selection for mutants with increased K^+ transport in *Saccharomyces cerevisiae*. *Genetics* **125**, 313–21.

Villalobo, A., Boutry, M. & Gouffeau, A. (1981). Electrogenic proton translocation coupled to ATP hydrolysis by the plasma membrane Mg^{2+}-dependent ATPase of yeast in reconstituted proteoliposomes. *Journal of Biological Chemistry* **256**, 12081–7.

Viswanatha, T., Bayer, E. & Wilchek, M. (1975). Reversibility of the affinity-labelled biotin transport system in yeast cells. *Biochimica et Biophysica Acta* **410**, 152–6.

Vogel, K. & Hinnen, A. (1990). The yeast phosphatase system. *Molecular Microbiology* **4**, 2013–17.

Vojtek, A. B. & Fraenkel, D. G. (1990). Phosphorylation of yeast hexokinases. *European Journal of Biochemistry* **190**, 371–5.

Volc, J., Sedmera, P. & Musilek, V. (1978). Glucose-2-oxidase activity and accumulation of D-arabino-2-hexulose in cultures of the basidiomycete *Oudemansiella mucida*. *Folia Microbiologica* **23**, 292–8.

von Hedenström, M. & Höfer, M. (1979). The effect of nystatin on active transport in *Rhodotorula glutinis* (*gracilis*) is restricted to the plasma membrane. *Biochimica et Biophysica Acta* **555**, 169–74.

von Richter, A. A. (1912). Über einen osmophilen Organisms den Hefepilz *Zygosaccharomyces mellis acidi* sp. n. *Mykologisches Zentralblatt* **1**, 67–76.

Wada, Y., Ohsumi, Y., Tanifuji, M., Kasai, M. & Anraku, Y. (1987). Vacuolar ion channel in the yeast *Saccharomyces cerevisiae*. *Journal of Biological Chemistry* **36**, 17260–3.

Wagner, W. W. & Davidson, R. W. (1954). Heartrots in living trees. *Botanical Reviews* **29**, 61–134.

Wainwright, M. (1978). Sulphur oxidising microorganisms on vegetation and in soils exposed to atmospheric pollution. *Environmental Pollution* **17**, 167–74.

Wainwright, M. (1984). Sulfur oxidation in soils. *Advances in Agronomy* **37**, 349–96.

Wainwright, M. (1988). Metabolic diversity of fungi in relation to growth and mineral cycling in soil. *Transactions of the British Mycological Society* **90**, 159–70.

Wainwright, M. (1989). Inorganic sulphur oxidation by fungi. In *Nitrogen, Phosphorus and Sulphur Utilization by Fungi*, ed. L. Boddy, R. Marchant & D. J. Read, pp. 71–89. Cambridge University Press: Cambridge.

Wainwright, M. (1993). Oligotrophic growth of fungi – stress or natural state? In *Stress Tolerance of Fungi*, ed. D. H. Jennings, pp. 127–44. Marcel Dekker: New York.

Wainwright, M. & Grayston, S. J. (1988). Fungal growth and stimulation by thiosulphate under oligocarbotrophic conditions. *Transactions of the British Mycological Society* **91**, 149–56.

Wainwright, M., Grayston, S. J. & de Jong, P. (1986). Absorption of insoluble compounds by mycelium of the fungus *Mucor flavus*. *Enzyme and Microbial Technology* **8**, 601–6.

Wainwright, M., Singleton, I. & Edyvean, R. G. J. (1990). Magnetite adsorption as a means of making fungal biomass susceptible to a magnetic field. *Biorecovery* **2**, 37–53.

Waites, M. J. & Quayle, J. R. (1980). Dihydroxyacetone: a product of xylulose 5-phosphate-dependent fixation of formaldehyde by methanol-grown *Candida boidinii*. *Journal of General Microbiology* **118**, 321–7.

Walker, L. P. & Wilson, D. B. (1991). Enzymic hydrolysis of cellulose – an overview. *Bioresource Technology* **36**, 3–14.

Walker, N. A. & Pitman, M. G. (1976). Measurement of fluxes across membranes. In *Encyclopedia of Plant Physiology*, vol. 2 *Transport in Plants*, Part A *Cells*, ed. U. Lüttge & M. G. Pitman, pp. 93–126. Springer-Verlag, Berlin, Heidelberg, New York.

Wang, S.-Y., Moriyama, Y., Mandel, M., Hulmes, J. D., Pan, Y.-C. E., Danho, W., Nelson, H. & Nelson, J. (1988). Cloning of cDNA encodes a 32-kDa protein. An accessory polypeptide of the proton ATPase from chromaffin granules. *Journal of Biological Chemistry* **263**, 17638–42.

Wang, Y., Leigh, R. A., Kaestner, K. H. & Sze, H. (1986). Electrogenic H^+-pumping pyrophosphatase in tonoplast vesicles of oat roots. *Plant Physology* **81**, 497–502.

Wariishi, H., Dunford, H. B., MacDonald, I. D. & Gold, M. H. (1989). Manganese peroxidase from the lignum-degrading basidiomycete *Phanerochaete chrysosporium*. Transient state kinetics and reaction mechanism. *Journal of Biological Chemistry* **264**, 3335–40.

Wariishi, H. & Gold, M. H. (1989). Lignin peroxidase compound. III. Formation, inactivation and conversion to the native enzyme. *FEBS Letters* **243**, 165–8.

Warncke, J. & Slayman, C. L. (1980). Metabolic modulation of stoichiometry in a proton pump. *Biochimica et Biophysica Acta* **591**, 224–33.

Warth, A. D. (1977). Mechanism of resistance of *Saccharomyces bailii* to benzoic, sorbic and other weak acids used as food preservatives. *Journal of Applied Bacteriology* **43**, 215–30.

Warth, A. D. (1989). Transport of benzoic and propanoic acids by *Zygosaccharomyces bailii*. *Journal of General Microbiology* **135**, 1383–90.

Warth, A. D. (1991). Mechanism of action of benzoic acid on *Zygosaccharomyces bailii* – effects on glycolytic metabolite levels, energy production and intracellular pH. *Applied and Environmental Microbiology* **57**, 3410–14.

Watanabe, T. (1968a). Cellulolytic enzymes. IV. Fractionation of a cellulolytic enzyme system. *Journal of Fermentation Technology* **46**, 299–302.

Watanabe, T. (1968b). Cellulolytic enzymes. V. Some properties of cellulase fraction I. *Journal of Fermentation Technology* **46**, 303–7.

Watanabe, Y., Shirmizu, M. & Tamai, Y. (1991). Molecular cloning and sequencing of plasma membrane H^+-ATPase gene from the salt tolerant yeast *Zygosaccharomyces rouxii*. *Journal of Biochemistry* **110**, 237–40.

Watanabe, Y. & Takakuwa, M. (1984). Effect of sodium chloride on lipid composition of *Saccharomyces rouxii*. *Agricultural and Biological Chemistry* **48**, 2415–22.

Watanabe, Y. & Takakuwa, M. (1987). A salt tolerance of yeast. 2. Change of liquid composition of *Zygosaccharomyces rouxii* after transfer to high sodium chloride culture medium. *Journal of Fermentation Technology* **65**,

365–70.

Watanabe, Y. & Takakuwa, M. (1988). A salt tolerance of yeast. 3. Depressions of synthesis of linoleyl residues in *Zygosaccharomyces rouxii* cells by sodium chloride. *Journal of Fermentation Technology* **66**, 461–6.

Watanabe, Y. & Tamai, Y. (1992). Inhibition of cell growth in *Zygosaccharomyces rouxii* by protonionophore and plasma membrane ATPase inhibitor in the presence of a high concentration of sodium chloride. *Bioscience, Biotechnology and Biochemistry* **56**, 342–343.

Watanabe, Y., Yamaguchi, M., Sanemitu, Y. & Tamai, Y. (1991). Characterization of plasma membrane proton-ATPase from salt tolerant *Zygosaccharomyces rouxii* cells. *Yeast* **7**, 599–606.

Watson, K. (1980). Thermal adaptation in yeast: correlation of substrate transport with membrane liquid composition in psychrophilic and thermotolerant strains. *Biochemical Society Transactions* **6**, 293–6.

Watson, K. & Cavicchioli, R. (1983). Acquisition of ethanol tolerance in yeast cells by heat shock. *Biotechnology Letters* **5**, 683–8.

Webb, S. R. & Lee, H. (1990). Regulation of D-xylose utilization by hexoses in pentose-fermenting yeasts. *Biotechnology Advances* **8**, 685–98.

Webster, J., Procter, M. C. F., Davey, R. A. & Duller, G. A. (1988). Measurement of the electrical charge on some basidiospores and an assessment of two possible mechanisms of ballistospore propulsion. *Transactions of the British Mycological Society* **91**, 193–203.

Wedzicha, B. L. C. (1984). *Chemistry of Sulphur Dioxide in Food.* Elsevier Applied Science: London.

Weete, J. D. (1980). *Lipid Biochemistry of Fungi and Other Organisms.* Plenum Press: New York.

Weimberg, R. (1975). Polyphosphate levels in nongrowing cells of *Saccharomyces mellis* as determined by magnesium iron and the phenomenon of 'überkompensation'. *Journal of Bacteriology* **121**, 1122–330.

Weimberg, R. & Orton, W. L. (1965). Synthesis and breakdown of the polyphosphate fraction and acid phosphomonoesterase of *Saccharomyces mellis* and their locations in the cell. *Journal of Bacteriology* **89**, 740–7.

Weiss, R. L. (1973). Intracellular localization of arginine and ornithine pools in *Neurospora*. *Journal of Biological Chemistry* **248**, 5409–13.

Wells, J. M. & Boddy, L. (1990). Wood decay and phosphorus and fungal biomass allocation in mycelial cord systems. *New Phytologist* **116**, 285–95.

Wells, J. M., Hughes, C. & Boddy, L. (1990). The fate of soil-derived phosphorus in mycelial cord systems of *Phanerochaete velutina* and *Phallus impudicus*. *New Phytologist* **114**, 595–606.

Wésolovski-Louvel, M., Goffrini, P., Ferrero, I. & Fukuhara, I. (1992). Glucose transport in the yeast *Kluyveromyces lactis*. I. Properties of an inducible low-affinity glucose transporter gene. *Molecular and General Genetics* **233**, 89–96.

Wessels, J. G. H. (1986). Cell wall synthesis in apical hyphal growth in fungi. *International Review of Cytology* **104**, 37–79.

Wessels, J. G. H. (1988). A steady-state model for apical wall growth in fungi. *Acta Botanica Neerlandica* **37**, 3–16.

Wessels, J. G. H. (1991). Fungal growth and development: a molecular perspective. In *Frontiers in Mycology*, ed. D. L. Hawkesworth, pp. 27–48. CAB International: Wallingford.

Wessels, J. G. H. & Sietsma, J. H. (1981). Fungal cell walls: a survey. In *Encyclopaedia of Plant Physiology*, vol. 13B *Plant Carbohydrates*, Part II *Extracellular Carbohydrates*, ed. W. Tanner & F. A. Loewus, pp. 352–94.

Springer-Verlag: Berlin, Heidelberg, New York.

Westermark, U. & Eriksson, K.-E. (1974a). Carbohydrate-dependent enzymic quinone reduction during lignin degradation. *Acta Chimica Scandinavica* **28B**, 204–8.

Westermark, U. & Eriksson, K.-E. (1974b). Cellobiose: quinone oxidoreductase a new wood-degrading enzyme from white-rot fungi. *Acta Chimica Scandinavica* **28B**, 209–14.

Westermark, U. & Eriksson, K.-E. (1975). Purification and properties of cellobiose: quinone oxidoreductase of *Sporotrichium pulverulentum*. *Acta Chimica Scandinavica* **29B**, 419–24.

Wethered, J. M. & Jennings, D. H. (1985). Major solutes contributing to solute potential of *Thraustochytrium aureum* and *T. roseum* after growth in media of different salinities. *Transactions of the British Mycological Society* **85**, 439–46.

Wethered, J. M., Metcalf, E. C. & Jennings, D. H. (1985). Carbohydrate metabolism in the fungus *Dendryphiella salina*. VIII. The contribution of polyols and ions to the mycelial solute potential in relation to the external osmoticum. *New Phytologist* **101**, 631–49.

Whelan, W. L., Gocke, E. & Manney, T. R. (1979). The *Can1* locus of *Saccharomyces cerevisiae*: fine structure analysis and forward mutation rates. *Genetics* **91**, 35–51.

White, C. & Gadd, G. M. (1986). Uptake and cellular distribution of copper, cobalt and cadmium in strains of *Saccharomyces cerevisiae* cultured on elevated concentrations of these metals. *FEMS Microbiology Ecology* **38**, 277–84.

White, C. & Gadd, G. M. (1987a). The uptake and cellular distribution of zinc in *Saccharomyces cerevisiae*. *Journal of General Microbiology* **133**, 727–37.

White, C. & Gadd, G. M. (1987b). Inhibition of H^+ efflux and K^+ efflux in yeast by heavy metals. *Toxicity Assessment* **2**, 437–48.

White, M. J., Lopes, J. M. & Henry, S. A. (1991). Inositol metabolism in yeasts. *Advances in Microbial Physiology* **32**, 1–51.

Whitney, K. D. & Arnott, H. J. (1986). Morphology and development of calcium oxalate deposites in *Gilbertella persicaria* (Mucorales). *Mycologia* **78**, 42–51.

Whitney, K. D. & Arnott, H. J. (1987). Calcium oxalate crystal morphology and development in *Agaricus bisporus*. *Mycologia* **79**, 180–7.

Whitney, K. D. & Arnott, H. J. (1988). The effect of calcium on mycelial growth and calcium oxalate crystal formation in *Gilbertella persicaria* (Mucorales). *Mycologia* **80**, 707–15.

Wiame, J. M., Grenson, M. & Arst, H. J. Jr (1985). Nitrogen catabolite repressions in yeasts and filamentous fungi. *Advances in Microbial Physiology* **26**, 1–87.

Wiemken, A. (1990). Trehalose in yeast, stress protectant rather than reserve carbohydrate. *Antonie van Leeuwenhoek* **58**, 209–17.

Wiemken, A., Matile, P. & Moor, H. (1970). Vacuolar dynamics in synchronously budding yeast. *Archiv für Mikrobiologie* **70**, 89–103.

Wiemken, A. & Nurse, P. (1973). Isolation and characterization of the amino acid pools located within the cytoplasm and vacuoles of *Candida utilis*. *Planta* **109**, 293–306.

Wieringa, R. T. (1966). Solid media with elemental sulphur for detection of sulphur oxidising microbes. *Antonie van Leeuwenhoek* **32**, 183–6.

Wijsman, M. R., van Dijken, J. P., van Kleeff, B. H. A. & Scheffers, W. A.

(1984). Inhibition of fermentation and growth in batch cultures of the yeast *Brettanomyces intermedius* upon a shift from aerobic to anaerobic conditions (Custers effect). *Antonie van Leeuwenhoek* **50**, 183–92.

Wikén, T., Scheffers, W. A. & Verhaar, A. J. M. (1961). On the existence of a negative Pasteur effect in yeasts classified in the genus *Brettanomyces* Kufferath et Van Laer. *Antonie van Leeuwenhoek* **27**, 401–33.

Wilcox, W. W. (1970). Anatomical changes in wood cell walls attacked by fungi and bacteria. *Botanical Review* **36**, 1–27.

Wilcoxson, R. D. & Subbarayadu, S. (1968). Translocation to and accumulation of phosphorus-32 in sclerotia. *Canadian Journal of Botany* **46**, 85–8.

Wilkins, D. A. (1991). The influence of sheathing (ecto-)mycorrhizas of trees on the uptake and toxicity of metals. *Agriculture, Ecosystems and Environment* **35**, 245–60.

Wilkins, P. O. Cirillo, V. P. (1965). Sorbose counterflow as a measure of intracellular glucose in baker's yeast. *Journal of Bacterialogy* **90**, 1605–10.

Williams, R. J., Eakin, R. E. & Snell, E. E. (1940). Relationship of inositol, thiamin, biotin, pantothenic acid and vitamin B6 to the growth of yeasts. *Journal of the American Chemical Society* **62**, 1204–7.

Williamson, J. R. & Monck, J. R. (1989). Hormone effects on cellular Ca^{2+} fluxes. *Annual Review of Physiology* **51**, 107–24.

Wills, C. (1990). Regulation of sugar and ethanol metabolism in *Saccharomyces cerevisiae*. *Critical Reviews in Biochemistry and Molecular Biology* **25**, 245–80.

Winkelmann, G. (1979). Surface ion polymers in hydroxy acids. A model of iron supply in sideramine-free fungi. *Archives of Microbiology* **121**, 43–51.

Winkelmann, G. (1991). *Handbook of Microbial Iron Chelates*. CRC Press: Boca Raton, FL.

Winkelmann, G. (1992). Structures and functions of fungal siderophores containing hydroxamate and complexone type iron binding ligands. *Mycological Research* **96**, 529–34.

Winkelmann, G. & Braun, V. (1981). Stereo-selective recognition of ferrichrome by fungi and bacteria. *FEMS Microbiology Letters* **11**, 237–41.

Winkelmann, G. & Huschka, H.-G. (1987). Molecular recognition and transport of siderophores in fungi. In *Iron Transport in Microbes, Plants and Animals*, ed. G. Winkelmann, D. van der Helm & J. B. Neilands, pp. 317–36. VCH: Weinheim.

Winkelmann, G., van der Helm, D. & Neilands, J. B. (1987). *Iron Transport in Microbes, Plants and Animals*. VCH: Weinheim.

Winkelmann, G. & Zähner, H. (1973). Stoffurechsel produkte von Mikroorganismen. 115. Mitteilung. Eisenaufnahme bei *Neurospora crassa*. 1. Zur Specifität des Eisentransportes. *Archives of Microbiology* **88**, 49–60.

Winskill, N. (1983). Tricarboxylic acid activity in relation to itaconic acid biosynthesis by *Aspergillus terreus*. *Journal of General Microbiology* **129**, 2877–83.

Wolf, D. H. (1980). Control of metabolism in yeast and other lower eukaryotes through action of proteinases. *Advances in Microbial Physiology* **21**, 267–338.

Wolf, D. H. (1986). Cellular control in the eukaryotic cell through action of proteinases: the yeast *Saccharomyces cerevisiae* as a model organism. *Microbiological Sciences* **3**, 107–16.

Wolf, F. T. (1955). Nutrition and metabolism of the tobacco wilt *Fusarium*. *Bulletin of the Torrey Botanical Club* **82**, 343–54.

Wolf, H. J. & Hansen, R. S. (1979). Isolation and characterization of methane-utilizing yeasts. *Journal of General Microbiology* **114**, 187–94.

Wolfinbarger, L. Jr (1980a). Transport and utilization of amino acids by fungi. In *Microorganisms and Nitrogen Sources*, ed. J. W. Payne, pp. 63–87. John Wiley: Chichester.

Wolfinbarger, L. Jr (1980b). Transport and utilization of peptides by fungi. *Microorganisms and Nitrogen Sources*, ed. J. W. Payne, pp. 281–300. John Wiley: Chichester.

Wolfinbarger, L. Jr & de Busk, A. G. (1971). Molecular transport. I. *In vivo* studies of transport mutants of *Neurospora crassa* with altered amino acid competition mutants. *Archives of Biochemistry and Biophysics* **144**, 503–11.

Wolfinbarger, L. Jr, Jervis, H. H. & de Busk, A. G. (1971). Active transport of L-aspartic acid in *Neurospora crassa*. *Biochimica et Biophysica Acta* **249**, 63–8.

Wolfinbarger, L. Jr & Kay, W. W. (1973). Transport of C_4-dicarboxylic acids in *Neurospora crassa*. *Biochimica et Biophysica Acta* **307**, 243–57.

Wolfinbarger, L. Jr & Marzluf, G. A. (1974). Peptide utilization by amino acid auxotrophs of *Neurospora crassa*. *Journal of Bacteriology* **119**, 371–8.

Wolfinbarger, L. Jr & Marzluf, G. A. (1975a). Size restriction on utilization of peptides by amino acid auxotrophs of *Neurospora crassa*. *Journal of Bacteriology* **122**, 949–56.

Wolfinbarger, L. Jr & Marzluf, G. A. (1975b). Specificity and regulation of peptide transport in *Neurospora crassa*. *Archives of Biochemistry and Biophysics* **171**, 637–44.

Wolfinbarger, L. Jr & Marzluf, G. A. (1976). Characterization of a mutant of *Neurospora crassa* sensitive to L-tyrosine. *Journal of General Microbiology* **93**, 189–93.

Wood, H. G. & Clark, J. E. (1988). Biological aspects of inorganic polyphosphates. *Annual Review of Biochemistry* **57**, 235–60.

Wood, T. M. (1969). The cellulose of *Fusarium solani*. Resolution of the enzyme complex. *Biochemical Journal* **115**, 457–64.

Wood, T. M. (1989). The mechanism of cellulose degradation by enzymes from aerobic and anaerobic fungi. In *Enzyme Systems for Lignocellulose Degradation*, ed. M. P. Coughlan, pp. 17–35. Elsevier Applied Science: London.

Wood, T. M. & McCrae, S. I. (1978). The mechanism of cellulose action with particular reference to the C_1 component. In *Proceedings of the Symposium on the Bioconversion of Cellulosic Substance into Energy, Chemicals and Microbial Protein*, ed. T. K. Ghose, pp. 114–41. IIT: New Delhi.

Wood, T. M., McCrae, S. I., Wilson, C. A., Bhat, K. M. & Gow, L. A. (1988). Aerobic and anaerobic fungal cellulases with special reference to their mode of attachment on crystalline cellulose. In *Biochemistry and Genetics of Cellulose Degradation*, ed. J.-P. Aubert, P. Beguin & J. Millet, pp. 31–52. Academic Press: London.

Wood, T. M., Wilson, C. A., McCrae, S. I. & Joblin, K. N. (1986). A highly active extracellular cellulose from the anaerobic rumen fungus *Neocallimastix frontalis*. *FEMS Microbiology Letters* **34**, 37–40.

Wooton, J. C. (1983). Re-assessment of ammonium iron affinities of NADP-specific glutamate dehydrogenases. *Biochemical Journal* **209**, 527–31.

Wösten, H. A. B., Moukha, S. M., Sietsma, J. H. & Wessels, J. G. H. (1991). Localization of growth and secretion of proteins in *Aspergillus niger*.

Journal of General Microbiology **137**, 2017–23.

Wray, J. L. & Kinghorn, J. R. (eds.) (1989). *Molecular and Genetic Aspects of Nitrate Assimilation*. Oxford Science Publications: Oxford.

Wright, J. L. C. & Vining, L. C. (1973). Carboxylic acids. In *Handbook of Microbiology*, vol. III *Microbial Products*, ed. A. I. Laskin & H. Lechevalier, pp. 27–50. CRC Press: Cleveland, OH.

Wyn Jones, R. G., Brady, C. J. & Spiers, J. (1979). Ionic and osmotic relations in plant cells. In *Recent Advances in the Biochemistry of Cereals*, ed. D. L. Laidman & R. G. Wyn Jones, pp. 61–103. Academic Press: London.

Wyn Jones, R. G. & Pollard, A. (1983). Proteins, enzymes and inorganic ions. In *Encyclopaedia of Plant Physiology*, vol. 15 *Inorganic Plant Nutrition*, ed. A. Läuchli & R. L. Bieleski, pp. 528–62. Springer-Verlag: Berlin, Heidelberg, New York.

Yadan, J. C., Gonneau, M., Sarthou, P. & Le Goffic, F. (1984). Sensitivity of nikkomycin Z in *Candida albicans*, role of peptide permeases. *Journal of Bacteriology* **160**, 884–8.

Yagi, T. (1988). Intracellular levels of glycerol necessary for initiation of growth under salt-stressed conditions in a salt-tolerant yeast *Zygosaccharomyces rouxii*. *FEMS Microbiology Letters* **49**, 25–30.

Yagi, T. (1991). Effects of increases and decreases in the external salinity on the intracellular glycerol and inorganic iron content in the salt-tolerant yeast *Zygosaccharomyces rouxii*. *Microbios* **68**, 109–18.

Yagi, T., Nogami, A. & Nishi, T. (1992). Salt tolerance and glycerol accumulation of a respiration-deficient mutant isolated from the petite-mutant, salt-tolerant yeast *Zygosaccharomyces rouxii*. *FEMS Microbiology Letters* **92**, 289–94.

Yagi, T. L Tada, K. (1988). Isolation and characterization of salt-sensitive mutants of salt-tolerant yeast *Zygosaccharomyces rouxii*. *FEMS Microbiology Letters* **49**, 317–21.

Yamamoto, L. A. & Segel, I. H. (1966). The inorganic sulphate transport system of *Penicillium chrysogenum*. *Archives of Biochemistry and Biophysics* **114**, 523–38.

Yamashiro, C. T., Kane, P. M., Wolczyk, D. F., Preston, R. A. & Stevens, T. H. (1990). Role of vacuolar acidification in protein sorting and zymogen activation: a genetic analysis of the yeast vacuolar proton-translocating ATPase. *Molecular and Cellular Biology* **10**, 3737–49.

Yanagita, Y., Abdel-Ghany, M. A., Raden, D., Nelson, N. & Racker, E. (1987). Polypeptide-dependent protein kinase from baker's yeast. *Proceedings of the National Academy of Sciences, USA* **84**, 925–9.

Yang, C. Y.-D. & Mitchell, J. E. (1965). Certain effects on reproduction in *Phthium* sp. *Phytopathology* **55**, 1127–31.

Yang, H. H., Effland, M. J. & Kirk, T. K. (1980). Factors influencing fungal degradation of lignin in a representative lignocellulosic thermomechanical pulp. *Biotechnology and Bioengineering* **22**, 65–77.

Yano, T., Kodama, T. & Yamada, K. (1961). Fundamental studies on the aerobic fermentation. VIII. Oxygen transfer within a mould pellet. *Agricultural and Biological Chemistry* **25**, 580–4.

Yarlett, N., Orpin, C. G., Munn, E. A., Yallett, N. C. & Greenwood, C. A. (1986). Hydrogenosomes in the rumen fungus *Neocallimastix patriciarum*. *Biochemical Journal* **236**, 729–39.

Yazdi, M. T., Woodward, J. R. & Radford, A. (1990). Cellulase production in

594 Literature cited

Neurospora crassa: the enzymes of the complex and their regulation. *Enzyme and Microbial Technology* **12**, 116–19.

Yoshimoto, A. & Sato, R. (1968a). Studies on yeast sulfite reductase. I. Purification and characterization. *Biochimica et Biophysica Acta* **153**, 555–75.

Yoshimoto, A. & Sato, R. (1968b). Studies on yeast sulfite reductase. II. Partial purification and properties of incomplete sulfite reductases. *Biochimica et Biophysica Acta* **153**, 576–88.

Yoshimoto, A. & Sato, R. (1970). Studies on yeast sulfite reductase. III. Further characterization. *Biochimica et Biophysica Acta* **220**, 190–205.

Youatt, J. (1993). Calcium and microorganisms. *Critical Reviews in Microbiology* **19**, 83–97.

Youatt, J., Gow, N. A. R. & Gooday, G. W. (1988). Bioelectric and biosynthetic aspects of cell polarity in *Allomyces macrogynus*. *Protoplasma* **146**, 118–26.

Young, N. & Ashford, A. E. (1992). Changes during development in the permeability of sclerotia of *Sclerotina minor* to an apoplastic tracer. *Protoplasma* **167**, 205–14.

Zadrazil, F. & Reininger, P. (eds.) (1988). *Treatment of Lignocellulases with White Rot Fungi*. Elsevier Applied Science: London.

Zauner, E. & Dellweg, H. (1983). Purification and properties of the assimilatory nitrate reductase from the yeast *Hansenula anomala*. *European Journal of Applied Microbiology and Biotechnology* **17**, 90–5.

Zehender, C. & Böck, A. (1964). Wachstumsund Ernährungs-bedingungen des Abwasserpilzes *Leptomitus lacteus*. *Zentralblatt für Bakteriologie und Parasitenkunde* Abt. II, **117**, 399–411.

Zerez, C. R., Weiss, R. L., Franklin, C. & Bowman, B. J. (1986). The properties of arginine transport in vacuolar membrane vesicles of *Neurospora crassa*. *Journal of Biological Chemistry* **261**, 8877–82.

Zimmermann, U. (1978). Physics of turgor and osmoregulation. *Annual Review of Plant Physiology* **29**, 121–48.

Zimmermann, U. & Steudle, E. (1978). Physical aspects of water relations of cells. *Advances in Botanical Research* **6**, 45–117.

Zlotnik, H., Fernandez, M. P., Bowers, B. & Cabib, E. (1984). *Saccharomyces cerevisiae* mannoproteins form an external wall layer that determines wall porosity. *Journal of Bacteriology* **159**, 1018–26.

Zmijewski, J. M. & MacQuillan, A. M. (1975). Dual effects of glucose on dicarboxylase acid transport in *Kluyveromyces lactis*. *Canadian Journal of Microbiology* **21**, 473–80.

Index

α-factor 47, 228, 249
A2318 347
Absidia glauca 299
Absidia spp. 197
acetate metabolism 94, 382–4
acetoin 103
acetyl-ornithine transaminase 236
acetyl CoA-dependent pyruvate carboxylase 390
acetylglutamate kinase 236
acetylglutamate synthase 236
acetylglutamyl-phosphate reductase 236
acetylornithine-glutamate acetyl transferase 236
acetylornithine transaminase 236
Achlya ambisexualis 347, 350
Achlya bisexualis
  apical wall properties 408–9
  hyphal branching 349
  plama membrane H^+-ATPase 12
  plasmolysis 434
  transhyphal current 35–6
  turgor potential 405
Achlya spp.
  amino acid transport 35, 213–15
  2-deoxy-D-glucose/sugars transport 118, 215
  nucleoside transport 215
  transhyphal current 34–8
acid phophatase
  presence in the vacuole 68
  regulation by pH of the medium 395
  repression by thiamine 310
  secretion 47
acid protease 40
aconitase 193, 390

Acremonium 299
acridine orange 8
acyl CoA oxidase 193
acyl-CoA synthetase 193
adenine as a growth factor 307
adenosine 5′-phosphosulphate kinase (APS kinase) 295
adenosylhomocysteinase 297
adenylate cyclase 170, 351
adhesion to surfaces 57
affinity constant (K_T) 25–31
Agaricales requirement for thiamine 308
Agaricus bisporus
  calcium oxalate in hyphal walls 357
  calmodulin 350
  copper 361
  metallothionein 361
  polyphosphate content of spores 277
  protein utilisation 249–50
aicraft fuel 190
alanine : 2-oxoglutarate aminotransferase 241
albumin 114–15
alcohol dehydrogenase 22, 68, 108, 110, 194, 383, 440
alcohol oxidase 186–8
aldehyde dehydrogenase 194
alkaline phosphatase 68
alkane utilisation
  chain length 190
  emulsifying agents 192
  gluconeogenesis 191
  glyoxylate cycle 191
  hair-like structures on the wall 192
  incorporation into cells 191–3
  oxidation 191, 193–4

alkane utilisation *continued*
  single cell protein 190
all-or-none process 328
Allium cepa (onion) 285, 454
Allomyces arbuscula 347
Allomyces javanicus 88
Allomyces javanicus var. *macrogynus* 196
Allomyces macrogynus
  rhizoids 38
  trans-hyphal current 34–5
Allomyces spp. 307
Alternaria alternata 420
Alternaria tenuis 299
aluminium
  acid rain 375
  forest decline 375
  mycorrhizal fungi
    nitrogen nutrition 375
    phosphorus 376
    polyphosphate granules 376
    tolerance 376
  potassium efflux 330
  siderophores 369
Amanita muscaria
  CEC of stipe 57
  siderophore 372
  sulphur oxidation 299
Amanita pantherina 57
Amanita sp. 374
Ambrosiozyma spp. 197
ω-amidase 238–9
amino acids
  cytoplasmic and vacuolar 67, 83–4
  metabolism 234–42, 247–8
  osmotic role 415–17
  sulphur-containing, requirement for 289, 295–6
  transport
    Achlya 213–15
    Aspergillus nidulans 215–17
    Neurospora crassa 217–20
    Penicillium 221–3
    tonoplast transport 82–4
p-aminobenzoic acid 306
ammonia
  basidiospore formation, inhibition 240
  inability to grow on 198
  incorporation into carbon skeletons 235–8, 240–2, 247–8
  mycorrhizal fungi, assimilation in 243–8
  release into the medium 249
  scavenging system 240
  transport 199–200, 208, 319
amylase 40–1, 47

anaerobic fungi 108–9, 158
anaerobiosis 92
anisomycin 311
antimycin 16, 130
antiport 2–3
Apium graveolens (celery) 454
Aqualinderella fermentans 92
Araispora 91
arginase 236, 241
arginosuccinate lyase 236
arginosuccinate synthetase 236
Armillaria mellea
  nitrate utilisation 196
  phosphorus pools 456–3
  translocation
    bidirectional 462–3
    dry matter flux 462
    vessel hyphae 463
Arrhenius plot 333
arsenate transport 262
aryl sulphatase 289
aryl-β-glucosidase 40
Ascobolus denuda 196
Ascobolus leveillei 196
Ascochyta pisi 196
ascorbic acid 61
Ashbya gossypii 314
asparagine synthetase 241
aspartate : 2-oxoglutarate aminotransferase 241
aspartate carbamoyltransferase 236
Aspergillus aculeatus 393
Aspergillus amstelodami 206
Aspergillus ficuum 255
Aspergillus flavus
  calcium malate crystals 390
  growth on silica gel 396
  organic acid production 393
  sulphur oxidation 299
Aspergillus fumigatus 299, 370
Aspergillus nidulans
  acetate metabolism 383
  amino acid transport 215–17
  apical protoplasts 37–8
  bicarbonate and growth 395–6
  calcium, calmodulin-dependent protein kinases 347
  carbohydrate metabolism 90–1
  carbon dioxide fixation 395–6
  choline-*O*-sulphate, sulphur storage 296
  denitrification, inability to bring about 198
  general amino acid control 233
  genes
    acuA, acuD, acuC, acuE 383

acuG 90, 383
areA 201–3
cnx 203
crnA 199, 202–3
facB 383
frA, galD, galE, glcB, manA, mnr455 90
niaD, niiA, nirA 202–3
pkiA, pppA 90
prnB 217
pycA 383
sbA, sorB 90
glucose, galactose and fructose transport 118
glutamine, signal for metabolic control 230
mitochondrial functioning 385
mutants
  nap3 215
  pacC-5 253–4
  resistance to acid media 395
nitrate
  reductase 201–4
  transport 199
  utilisation 103
nitrogen catabolite control 230–1
nitrogen utilisation 201–5, 230–4
NMR studies 89
osmotic shock 430
pentose phosphate pathway and related enzymes, increase in 93, 104
pH of medium, regulatory role 395
polyols 24, 421
protoplasts, potassium : sodium ratio 37–8
pyruvate carboxylase 386
solutes, internal concentrations 414
sulphate transport 290–1
transhyphal current 34
wall, ions in 418
Aspergillus niger
  calmodulin 350
  citric acid synthesis 385–9
  dioxygenase 182
  growth on silica gel 396
  manganese transport 338
  mitochondrial functioning 385
  nitrate assimilation genes 202
  organic acid synthesis 393
  oxalate biosynthetic pathways 358
  oxalate in culture 357
  pellets 32
  plasmolysis 434
  polyphosphate content of mycelium and spores 276–7
  protein secretion 41

solubilisation of soil phosphorus 252
sulphur oxidation 299
Aspergillus ochreaceus 369, 393
Aspergillus oryzae 202, 393
Aspergillus phoenicus 157
Aspergillus repens 396
Aspergillus restrictus 420, 423
Aspergillus sojae 393
Aspergillus spp.
  calcium and sporulation 347
  citric acid producers 393
  gluconic acid producers 393
  growth factors not required 308
  malic acid producers 393
  nitrate utilisation 196
  phytate breakdown 255
  tolerance of low water potentials 420, 423
Aspergillus terreus 388–90
Aspergillus wentii
  growth and turgor potential 400–1, 407–8
  organic acid synthesis 393
  osmotic, turgor and water potentials 404
ATP sulphurylase (sulphate adenylyl transferase) 294
Aureobasidium pullulans
  cadmium and copper transport 345
  dioxygenase 182
  sulphur oxidation 299
Auricularia auricularis
autolysis
  conditions 447–8
  effect on morphology 448
  enzymes 448
  translocation of products 448
auxin 19
auxotrophic 305
azaserine 242
azide 5, 6, 75, 146, 223, 227, 289, 329

bacilysin 225, 228
Basidiobolus ranarum 197
Basidiobolus variarum 34
basidiomycete fungi
  fruit body 240–1, 317, 464–7
  mycelium, two types 165
  radiocaesium content 317
  spore discharge 398
  see also ectomycorrhiza of forest trees
bean (*Phaseolus vulgaris*) 252
beech (*Fagus sylvatica*) mycorrhiza
  ammonia and nitrate uptake 243–6
  cytoplasmic orthophosphate concentration 272
  fluxes of nutrients 31, 243, 245

beech (*Fagus sylvatica*) mycorrhiza *continued*
  fungal sheath
    nitrogen, proportion in 246
    phosphorus, proportion in 285
  nitrogen metabolism 243–5, 248
  partners 245
  phosphate
    pools 285
    uptake 282–4
benzalkonium chloride 329
benzoic acid 380–1
Betula papyrifera 374
Betula sp. 374
bicarbonate pump 396
biochemical pH-stat 394–5
biotin (vitamin B₇)
  function 306
  regulation of cell concentration 312
  transport 311–12
bivalent metals *see* multivalent metals
Bjerkandera adusta 171
Blastocladia spp. 92
Blastocladiella emersonii
  calcium channels 357
  calmodulin 350
  gamma particles 250
  inorganic nitrogen, inability to grow on 196
  rhizoids 39
  transhyphal current 34
Blastocladiella pringsheimii 88
Blastocladiella spp.
  lactic acid as fermentation end-product 92
  sulphur amino acids as growth factors 307
Boletus edulis 372
Botryotinia convoluta 196
Botrytis allii 196
Botrytis cinerea
  nitrate utilisation 196
  oxalate in culture 357
  plasmolysis 434
Boyle–van't Hoff relationship 426–7
Brettanomyces intermedius 73, 102–3
Brettanomyces lambicus 205
Brettanomyces naardenensis 104
Brettanomyces spp.
  glucose utilisation 102–3
  nitrate utilisation 197
bromophenol blue 205
brown-rot fungi 158–60
buffering capacity, intracellular 395
Bullera spp. 197
2,3-butanedione 9–10
butyric acid, acidification of hyphae 335

cadmium
  metallothioneins 360–4
  toxicity 337–8, 345
  transport 336–9, 341
caesium
  basidiomycete fruit bodies 317
  Chernobyl 317
  transport 322–3
calcium
  amino acid transport 215, 222
  ATPase 355
  ATP-powered exchange 352–5
  calmodulin 350
  channels 356–7
  cytoplasmic concentration 73, 77
  ethanol tolerance 116
  flocculation 53–4
  mating pheromone signal transduction in yeasts 349
  oxalate
    ecological significance 357
    importance 358–9
    production in culture 357
    walls 357–8
  plasma membrane
    efflux 352–5
    electrochemical driving force 352–4
    influx 344–5, 347, 355–7
  signalling system 349
  tonoplast transport 77, 85
Calluna vulgaris 374
calmidazolium 329
calmodulin antagonists 329
Candida albicans
  ethanol, effects on 111–12, 116
  nikkomycin-resistant mutants 225
  peptide utilisation 223–5
  plasma membrane
    H⁺-ATPase 13
    lipid composition 60
  polyoxin-resistant mutants 225
Candida boidinii 185
Candida flareri 314
Candida glabrata 49, 362–4
Candida intermedia 118
Candida lipolytica 386
Candida nitrophila
  nitrate reductase 204, 206
  nitrogen metabolism 204–5, 241–2
Candida pintolopesii 102
Candida shehatae
  ethanol and proton diffusion 113
  monosaccharide transport 118
  xylose fermentation 104–6

Index

599

Candida sphaerica (=*Kluyveromyces marxianus*)
Candida spp.
  amino acid transport 209
  methanol fermentation 104–6
  nitrate utilisation 197
  thiamine, lack of growth depression 312
Candida tenuis 104
Candida tropicalis
  alkane uptake 191–2
  ethanol on morphological transitions 111–12
  dioxygenases 182
  plasma membrane H^+-ATPase 13
  phosphate transport 267–8
  vacuoles, isolation of 74
Candida utilis
  amino acid pools 69
  ammonia metabolism 241–2
  copper transport 338
  Crabtree negative 29, 102
  ethanol and proton diffusion 113
  glucose transport 29–30, 32–3, 119
  glutamate synthesis 241–2
  GOGAT, absence of 241
  growth in continuous culture 29–30, 32–3, 119
  invertase 143
  iron accumulation 346
  L-lactic and citric acid transport 378
  maltose utilisation 147
  manganese transport 338
  mitochondria 319–20
  nitrate utilisation 199, 205
  pentose phosphate enzymes, increased activity with nitrate 103–4
  phosphate
    cytoplasmic concentration 73, 272
    transport 267
  polyphosphates 272
  potassium and oxidative phosphorylation 319–20
  potassium and sodium concentrations in cytoplasm and vacuole 70
  potassium/rubidium and growth 319–20
  selenate-resistant mutants 301
  sulphite transport 301
  transhydrogenation, absence of 103
  vacuole
    isolation 67–71
    transport of *S*-adenosyl-L-methionine 82
  xylose isomerase, absence of 104
  zinc transport 338, 345–6

Candida wickerhamii
  cellobiose utilisation 143
  ethanol and proton diffusion 113
  glucose transport 119
carbamoyl-phosphate synthetase 241
carbamoyl phosphate synthetase A 235–6
carbon polymer utilisation, complexity of study 148–9
carbonyl cyanide *m*-chlorophenylhydrazone (CCCP) 64, 301, 329, 338–9, 439
carbonyl cyanide *p*-trifluoromethoxyphenylhydrazone (FCCP) 315, 339, 355
cardinal osmotic potentials 422–3
carnitine acetyl transferase 193
carry-over 305
catabolite inactivation 99, 132, 147
catalase 180, 187–8, 193
cation exchange capacity (CEC) 56–7
CCCP *see* carbonyl cyanide *m*-chlorophenylhydrazone
CDP-diacylglycerol synthase 316
cell deactivation 109
cellobiase (also β-glucosidase) 143–5, 150–8, 160, 179–80
cellobiose oxidase 179–80
cellobiose : quinone oxidoreductase 179–80
cellodextrins 152, 155
cellulases
  assay 152–3
  substrates 152
cellulose utilisation
  adaptive nature 155–6
  amorphogenesis 154, 158
  brown-rot fungi
    Fe(II) / hydrogen peroxide 158–9
    oxalic acid 159–60
  (C_1–C_2) concept of cellulolytic activity 154–5
  enzymes
    adsorption and desorption 155–6
    assay 152–3
    induction 156–8
    mass transfer 156
    multiplicity of forms 153–4
    product inhibition 156
    purification difficulties 151–3
    solubilisation of crystalline cellulose 151
    synergism 155
    types 150–1
  kinetic analysis 155
  rumen fungi 158
  Trichoderma reesii 149–58
Cenococcum geophilum 372, 376

Cephalosporium acremonium 196
Cephalosporium sp. 396
Cephalothecium roseum 196
Ceratocystis fimbriata 349
Ceratocystis ulmi 350
Ceratostomella fimbriata 197
Ceratostomella ulmi 197
Cercospora apii 196
cerulenin 61
cetrimide 329
cetylpyridinium chloride 329
Chaetomium cochlioides 196
Chaetomium convolutum 196
Chaetomium fresenii 369
Chaetomium globosum 196
Chaetomium spp. 308, 347
CHAPS 168
Chara 168
chemiosmosis 2–3
chemolithotrophy 298
chemotropism 38–39
chitinase 68
chloride
    concentration in cytoplasm and vacuole 70
    transport 421
p-chloromercuribenzoate 301
Choanephora cucurbitarum 197
choline
    alteration of plasma membrane lipids 60
    inositol metabolism, interrrelation 314–16
choline kinase 316
choline-*O*-sulphate
    sulphur storage 296
    transport 221, 288, 291–2
chromaffin granules 78
chromate 291
Chrysosporium fastidium
    osmotic, turgor and water potentials 404
    osmotic shock 430
    solutes, internal concentrations 416, 418
    tolerance of low water potentials 420,
        423–4
Chrysosporium lignosum (=*Sporotrichum
    pulverulentum*)
Chytridium spp. 88, 92, 196
cis-aconitate decarboxylase 390
Citeromyces spp. 197
citrate synthase 68, 193
citric acid cycle 100, 108, 238, 358, 382–92
Cladosporium cladosporoides 420
Cladosporium herbarum 396
Cladosporium resinae 119, 190
Claviceps purpurea 276
Clitocybe albirhiza 57

Clostridium acetobutylicum 312
cobalt
    binding in *Saccharomyces cerevisiae* 127
    vacuoles 80
    vitamin B_{12} 336
Cokeromyces recurratus 369
Colleotrichum lagenarium 196
Colleotrichum lindemuthianum 196
Collybia tuberosa 196
Collybia velutipes 196, 466
colony
    ageing 447
    autolytic enzymes 447–8
    translocation of autolytic products 447–8
    vacuolation 447
compatible solutes
    compounds produced by microorganisms
        444
    energy dissipation 444–6
    haloprotection 443
compound 48/80 329
concanavalin A 7, 55, 74
continuous culture 27–33, 319, 345–6
copper
    metallothioneins 360–5
    potassium efflux 329
    transport 339, 345, 346
    toxicity 331, 364–7, 374
Coprinus cinereus
    ammonia metabolism 240–1
    cAMP-dependent protein kinase 350
    fruit body development 240–1
    glucose transport 119
    protein
        inclusions 250
        utilisation 249–50
    transhyphal current 34
    urea cycle 240
Coprinus lagopus 350
Coprinus sp. 88
coprogens 307, 368–70
coprophilous fungi 240, 420, 422
Cordyceps militaris 196
Coriolus versicolor 171, 182
Cortinarius ahaii 57
Cortinarius glaucopus 57
cotransport stoichiometry 15-16
Crabtree effect 101–3
Cronartium ribicola 198
Cryptococcus laurentii 48–9, 143
Cryptococcus spp. 197
Cunninghamella blakesleeana 365
Cunninghamella echinulata 358
Cunninghamella elegans 369

Cunninghamella spp. 255
current–voltage analysis 333
Custers effect 101–3
cyanide 18, 119, 353
cyanocobalamin (vitamin B_{12}) 306
cyclic AMP (cAMP) 18, 170
cyclic AMP phosphorylation 19, 100
cyclic AMP-dependent protein kinase 19
cycloheximide 116, 119–21, 129, 147, 169, 222, 266, 311
Cylindrocephalum sp. 182
γ-cystathionase 297
γ-cystathionine synthase 297
cysteine synthase 297
cysteinylglycine dipeptidase 297
β-cystathionase 297
β-cystathionine synthase 297
cytochrome $aa-3$ 442
cytochrome $b-557$ 204
cytochrome c peroxidase 190
cytochrome oxidase 312
cytochrome $P-450$ 193–4, 198
cytoplasm
    amino acids 67–71, 235–7
    calcium concentrations 73–7
    pH 16–17
    phosphate concentrations 265, 269, 272
    potassium, sodium and chloride concentrations 70–1

Dasyscypha capita 357
DCCD *see* N, N'-dicyclohexylcarbodiamide
DEAE-dextran 46, 67, 75, 80, 84, 351
deaminase 220
Debaryomyces hansenii
    alkaline ATPase activity 441
    dioxygenase 118
    glycerol 419, 429, 431, 433, 437, 444–5
    osmotic shock 428–33
    plasma membrane H^+-ATPase 13, 438
    potassium absorption and organic acids 394
    sodium/glycerol co-transporter 433
    sodium chloride
        in vitro activity of NAD^+-dependent glycerol 3-phosphate dehydrogenase 437
        lag phase 440
        tolerance mutants 443
    sodium/proton antiport 439
    tolerance of low water potentials 420, 423
    wall volume 435
Debaryomyces spp., ability to utilise nitrite but not nitrate 196

Debaryomyces subglobosus 182
Dekkera intermedia 145, 205
Dekkera spp.
    nitrate utilisation
Dendryphiella salina
    cytoplasmic pH 73
    glucose transport 118
    mannitol, incorporation into glycogen 130–1
    membrane potential 440
    3-*O*-methyl glucose, effect on metabolism 130–1
    nitrate
        increased flux of carbon through the pentose phosphate pathway 93
        utilisation 196
    organic acid concentrations 394
    pellets 448
    plasma membrane H^+-ATPase 13, 37
    polyol metabolism 24, 94–5, 418–21, 429, 444–6
    potassium efflux and bivalent cations 321
    potassium, sodium and chloride concentrations in cytoplasm and vacuole 70
    potassium : sodium ratio in hyphal apex 38
    potassium selectivity of wall 56
    radiotracer compartmental analysis 71–2
    sodium chloride and *in vitro* activity of malic dehydrogenase and glucose 6-phosphatase 437
    solutes, internal concentrations 416
    L-sorbose, effect on metabolism 131
    sulphate, protoplasmic concentration 296
    wall
        ions 418
        volume 435
denitrification 198
2-deoxy-D-glucose (2DG) 16, 119–39, 301
6-deoxy-D-glucose (6DG) 119, 122, 125, 129, 131, 138
Dermocystidium sp. 268
DES *see* diethylstilboestrol
desthiobiotin 307
detector peptides 224
2DG *see* 2-deoxy-D-glucose
6DG *see* 6-deoxy-D-glucose
3,4-diaminopyridine 37
3,5-di-*tert*-butyl-4-hydroxybenzilidinematononitrile (SF 6847) 38
dichlorophenolindophenol 205
diethylstilboestrol (DES) 329, 338–9
dihydroxyacetone kinase 188

dihydroxyacetone synthase 188–9
dihydroxymaleic acid 176
dimethyl sulphide 296
2,4-dinitrophenol 82, 119, 121–2, 129, 147,
    211, 264, 267, 293, 301, 308, 315, 329,
    338–9, 381, 433
Dio-9 329, 347
dioxygenase 181, 194
disaccharide utilisation
    chromogenic aglycon 141–2
    external hydrolysis 140–8
    non-metabolised analogues 142, 146, 148
    transport 145–8
dithiothreitol 176
DNP *see* 2,4-dinitrophenol
Dothidella quercus 196
downshock 426–9

ectomycorrhiza of forest trees
    aluminium tolerance 375–6
    calcium oxalate 358
    metal tolerance 374
    nitrogen metabolism 242–8
    nutrient absorption, enhancement of 31,
        243, 245
    protein utilisation 250
    sheath as a barrier to movement 246
    siderophores 372–4
    solubilisation of insoluble soil phosphorus
        253
    surface reducing activity 376
    symbiocalcicole 376
Eeniella spp. 103
EGTA 349
electrical double layer 52
electrophysiological studies 4, 12, 332–5
electroporation 73
endo-(sarco-)plasmic reticulum pump 355
endo-(1→4)-β-D-glucan-4-glucanohydrolase
    150–1, 160, 178–9
endoglucanases 151–5, 178–80
Endogone sp. 270
Endomyces magnusii 118, 275
Endomycopsis sp.182
endoplasmic reticulum 67
Endothia parasitica 197
energy dissipation 239
enolase 110
enoyl-CoA hydratase 193
enzyme proteolysis 100
enzyme secretion 39–41
Epicoccum nigrum 299
1,2-epoxypropane 53

Eremascus spp. 433
Eremothecium asbyii 312
ergosterol 60, 65, 74, 114–15, 307
ethanol
    accumulation 109–11
    denaturation of proteins 109–10
    extracellular acidification 113
    fermentation end-product 28, 87–92
    maltose transport, prevention of
        inactivation 147
    membrane transport, effect on 112–13
    plasma membrane
        H⁺-ATPase 116
        passive distribution across 110–11
    stress proteins 111, 116–17
    tolerance
        lipids role of 109, 113–15
        membranes, locus of action of ethanol
            109, 112–17
        mutants 109
        xylose utilisation, inhibition 106
ethanolamine 60
ethidium bromide 329–30
L-ethionyl-L-alanine 228
Eucalyptus marginata 358
Eucalyptus pilularis 283, 376
Eucryphia cordifolia 184
Eurotium amstelodami 404
Eurotium spp. 420, 423
exchange diffusion 2
exo-(1→4)-β-glucanase 178–80
exo-β-(1→3)-glucanase, presence in the
    vacuole 68

facultative anaerobe 97
FAD-linked glycerol-3-phosphate
    dehydrogenase 193
fatty acids
    alkane utilisation 191–4
    ethanol tolerance 115
    hydrogen peroxide production 178
    thiamine transport, inhibition 308
FCCP *see* carbonyl cyanide *p*-
    trifluoromethoxyphenylhydrazone
fermentation
    ability of yeasts for 101
    extracellular products of 28
ferredoxin 108
ferrichromes 367–72
ferricrocin 370
ferrioxamine 370
ferrirubrin 370
Filobasidiella neoformans var. *bacillispora* 49
Filobasidium spp. 197

flocculation
 calcium-bridging hypothesis 53–4
 'lectin'-like hypothesis 54–5
fluorecamine 224
fluorescein isothiocyanate (FITC) 10
fluorescence imaging 73
fluorescent peptides 224
fluorescent probes 16
flux 25, 31–3
Fomes lignosus 171
Fomes officinalis 307
formaldehyde dehydrogenase 187–9
formate dehydrogenase 187–8
β-D-fructofuranosidase *see* invertase
fructose 1,6-bisphosphatase 90–1, 100
fructose 1,6-bisphosphate aldolase 90–1,
 110
Fulvia fulva 202
fumaric acid synthesis 390
fura-2 73, 351
fusarinines 369
Fusarium culmorum 420
Fusarium episphaeria 299
Fusarium equisesti
 growth and turgor potential 402, 407–8
 osmotic, turgor and water potentials 404
Fusarium graminearum
 calcium and hyphal branching 349
 choline transport 316
 growth in fermenter 31
Fusarium lini 87
Fusarium oxysporium
 denitrification 198
 dioxygenase 182
 growth in absence of exogenous carbon
 source 397
 nitrate assimilation genes 202
 potassium absorption and charge balance
 394–5
 sulphur oxidation 299
Fusarium oxysporium f. *nicotianae* 88
Fusarium roseum 371
Fusarium solani 252, 396
Fusarium sp. 396
Fusarium spp.
 growth factors not required 308
 nitrate utilisation 196
 silica gel, growth on 397
Fusarium tricinctum 299
futile cycle
 glutamine 238–9, 245
 glycerol 245
 mannitol 245
 trehalose 245–246

α-D-galactopyranosidase *see* melibiase
galactokinase 90–1
galactose utilisation 90–1, 107, 117–18
Geophyllum trabeum (=*Lenzites trabea*)
Geosmithia argillacea 299
Geosmithia emersonii 299
Gibberella fujikuroi 202
Gibbs free energy changes, calcium efflux
 354
Gibertella persicaria 358
Gliocladium roseum 396
Glomerella cingulata 196
Glomus sp. 373
(1→4)-β-glucan cellobiohydrolase 150–1
β-glucanase 150, 157
β-D-glucoside glucohydrolase (*also* β-D-
 glucoside) *see* cellobiase
gluconeogenesis 191
glucosamine 119, 123
glucose oxidase 178, 180
glucose 6-phosphatase 90–1
glucose 6-phosphate dehydrogenase 68, 90–
 91, 93, 437
glucose 6-phosphate isomerase
glucose repression 97–101, 106, 180, 194,
 378–80
glucose 6-sulphate utilisation 288
glucose utilisation
 ability to regenerate NADP from NADPH
 anaerobically 103–6
 fermentation 91–107
 gluconeogenesis 191
 glycolysis in fungi other than yeasts 89–92
 lactic acid as end-product of fermentation
 91–2
 NMR studies 89–91
 non-utilising fungi 91
 pentose phosphate pathway
 contribution to the dissimilation of
 glucose 87–93, 440
 evidence 92–3
 generation of NADPH 92–3
 randomisation of radiolabel 89
 stoichiometric conversion to ethanol 87–9
 transhydrogenation 103–5
 yeasts 96–104
α, α'-glucosidase 1-glucohydrolase *see*
 trehalase
glutamate decarboxylase 241
glutamate dehydrogenase (GDH)
 control by nitrogen and carbon circuits
 237–8
 NAD 231, 235–8, 240–1, 248
 NADP 235–9, 240–2, 247–9

glutamate kinase 236
glutamine (amide): 2-oxoglutarate
    aminotransferase (GOGAT) 238–9,
    240, 247–8
glutamine
    cycle 238–9
    signal for metabolic control 230–1
glutamine synthetase (GS) 238, 241–2, 247–8
glutamine transaminase 238–9
L-glutamine : D-fructose 6-phosphate
    amidotransferase 91
glutamyl-phosphate reductase 236
γ-glutamyl cysteine synthetase 297
γ-glutamyl transpeptidase 297
glutathione
    metal complexes 364
    sulphur storage 296–7
glutathione synthetase 297
glyceraldehyde-phosphate dehydrogenase
    90–1
glycerol-3-phosphate acyltransferase 193
glycerol
    compatible solute 443–6
    cycle 445
    loss 429–31
    osmotic response 416–17, 419, 440
    synthesis stimulated by sodium chloride
    440
glycerol kinase 90–1
glycerol phosphate dehydrogenase 90–1
α-glycerophosphatase 68
β-D-glycosidase *see* cellobiase
glycolysis 87–92, 95, 108
glyoxalate bypass 382–3
glyoxylate cycle 191
glyoxysomes 188, 382–3
Golgi apparatus 67
growth factors
    culture media 304
    definitions 303
    niche, in relation to 303–4
    requirement, establishing 305–8
    roles 306
    transport 308–16
Gymnoascus setosus 196
Gymnosporangium juniperi-virginianae 198
Gyromitra gigas 57

Hanseniaspora uvarum 49
Hansenula anomola
    L-malic acid transport 379
    nitrate reductase 204, 206
Hansenula polymorpha
    catalase-negative mutant 190

ethanol and proton diffusion 113
methanol utilisation 186–90
Hansenula spp. 186, 197
heartwood 185
heavy metals
    classification 359–61
    binding 359–65
    mechanisms of tolerance 364–7, 374
Hebeloma crustuliniforme
    growth in ammonia and nitrate 245–6
    nitrate reductase 245–6
    surface reducing activity 376
Hebeloma cylindrosporum 245
Hebeloma sp. 247
Helminthosporium spp. 196
heterologous proteins, secretion 48
hexadecane 192
hexokinase 90–1, 110, 125–6, 130, 142
hexose monophosphate shunt *see* pentose
    phosphate pathway
Hilaria jamesii 373
homocysteine methyltransferase 297
homocysteine synthase 297
homoserine acetyltransferase 297
hydraulic conductivity 403, 406
hydrocarbon utilisation *see* alkane utilisation

Hydrodictyon africanum 271
hydrogen peroxide
    cellulose degradation 158–9, 178–80
    lignin breakdown 171–8, 181–3
    methanol utilisation 186–90
hydrogen sulphide 251–2
hydrogenase 108
hydrogenosomes 108
hydroquinone 61
3-hydroxyacyl-CoA dehydrogenase 193
hydroxylamine hydrochloride 61
hydroxymethylpyrimidine kinase 310
Hygrophorus pudorinus 57
Hymenoscyphus ericae
    copper and zinc tolerance 374
    protein utilisation 250
    siderophore 372
    sulphur oxidation 299
    surface reducing activity 376
hyperosmotic shock 426
hyphal
    apex 39–41
        bursting 399–400
        lytic enzymes 398
        plasicity/rigidity 399
    branching 41–2
    extension 33–9

Hypholoma fasciculare 299
Hypogymnia physodes 268
hypoosmotic shock 426
Hysterangium crassum 358

illicit transport 224
2,3-imino-squaline 61
immunogold labelling
enzyme secretion 41
in lignin breakdown 175
indo-1 73
indole-3-acetic acid (auxin) 19–21
inducer exclusion 233
Inonotus tomentosus 233
inositol
growth factor 307
metabolism 314–6
polyols 94, 419
inositol phosphates 215
inositol-1-phosphate synthase 316
myo-inositol hexaphosphatase 255
invertase (β-D-fructofuranosidase)
fructosyl transferase ability 157
presence in the vacuole 68
properties of wall enzyme 40–1,143
secretion 47
iodoacetamide 301
iodoacetate 127, 308, 381
ion channels 38, 332, 356–7
ion selective electrodes 16, 85, 351
iron
accumulation in *Candida utilis*
siderophores 367–74
storage 369–70
vacuoles 80
Irpex lacteus 149
isocitrate lyase 193
isocitrate dehydrogenase 437
itaconic acid synthesis 388–90

K_S *see* saturation constant
K_T *see* affinity constant
3-ketoacyl-CoA thiolase
Kloekeria brevis 308
Kluyver effect 102
Kluyveromyces fragilis 115
Kluyveromyces lactis
DEAE dextran penetration of wall 46
glucose transport 120
killer polypeptide 49
lactose transport 147–8
succinate transport 379

Kluyveromyces marxianus
kynurenine hydroxylase 61
L-lactic acid transport 378
lactose transport 148
L-malic acid transport 378
maltose negative strain 146
α-methyl glucoside transport 146
monosaccharide transport 131–3
polyphosphate
content 279
in wall 131–2, 274–5
protein secretion 49
Kluyveromyces spp. 197, 311
kynurenine hydroxylase 61

Laccaria bicolor 245
Laccaria laccata 245
Laccaria proxima 374
laccase 179–83
Lactarius deliciosus 57
Lactarius hepaticus 245
Lactarius hibbardae 374
Lactarius rufus 245, 374
Lactarius sp. 248
Lactarius subdulcis 243–6
lactic acid 91–2, 109
lactic dehydrogenase 108
lactonase 179
Lagenidium giganteum 347
Lambertella corni-maris 196
Lentinus edodes 171
Lentinus tegrinus 196, 466
Lenzites trabea 160, 197
Leptomitus lacteus 91, 196
lethicin 307
Leucosporidium spp. 197
Leucosypha sp. 243
lichens 195, 268
lignin
acetovanillone-oxidising activity 69
breakdown (delignification)
activation by lignin, veratryl alcohol,
veratraldehyde *p*-coumaric acid,
sinapic acid 170
brown-rot fungi 160
cyclic AMP levels 170
Cα-Cβ cleavage 171
carbon, sulphur and nitrogen limitation,
requirement for 169–170
cellulose breakdown in parallel 161–2,
180
dearomatisation of benzene nuclei 181
demethylation of aromatic methoxyl
groups 160

lignin *continued*
  breakdown (delignification) *continued*
    enzymes involved *see* cellulose
      utilisation, ligninase, manganese
      peroxidase
    hydrogen peroxide, source of 178
    inactivation 168
    liquid culture 163–6
    oxygen 161, 166, 170–2, 179–83
    pH of wood 180–1
    radical cations, production of 172–3
    selective 183–4
    ^{14}C-labelled 167
  carbon source for growth 167–8
  energy content 167–8
  isolation 162
  structure 161–2
  synthesis 161–2
ligninase (lignin peroxidase)
  liquid culture
    enhancement of production by
      polypropylene glycol or polyethylene
      glycol 168
    inactivation by agitation 168
  location in glucan sheath 167
  mechanism of action 171–5
  molecular mass 171
  pH optimum 171
  primary growth, absence from 169–70
  radical cations 172–3
  reactions catalysed 173
  repression by glutamate and other nitrogen
    compounds 169–70
  secondary metabolite 169
  veratryl alcohol 173–6
  wood, location on surface 175
lignocellulose breakdown
  complexities of study 160–1
  restrictive nature of term 160
  see also cellulose utilisation, *Phanerochaete
    chrysosporium*, white-rot fungi
Lineweaver-Burk plot 200, 211, 293
linoleic acid 115
linolenic acid 115
lipids
  mitochondrial membrane 64–5
  plasma membrane 58–65
  tonoplast 59, 65
  wall 59
liposomes 11
long-chain alcohol dehydrogenase 193
long-chain aldehyde dehydrogenase 193
lower fungi, spore release 398
Lycoperdon pyriforme 57

lyotropic effects 318, 443
lysine as growth factor 307
lysosome 249

Macrochytridium spp. 92
Macrosporium sarcinaeforme 196
magnesium
  ethanol tolerance 116
  hyphal tip bursting 399
  transport
    plasma membrane 340–1
    tonoplast 85
malate dehydrogenase 108, 193, 437
malate synthase 193
malic acid synthesis 390–2
malic enzyme 108
maltose transport 145–7
manganese
  citric acid synthesis 387–8
  hyphal tip bursting 399
  ligninolytic agent 176–7
  peroxidase 176–7
  production of Mn(III) 176–7, 181
  transport 338–9, 346
  vacuole 80, 85
mannitol
  compatible solute 444
  cycle 445
  symmetrical with respect to radiolabel 89
  utilisation 88, 94
mannitol dehydrogenase 91
mannitol kinase 90–1
mannitol 1-phosphate dehydrogenase 90–1,
    93
mannitol 1-phosphate phosphatase 90–1
mannokinase 91
α-mannosidase 68
Marasmius fulvobulbilosus 196
Marasmius spp. 196
marine fungi 45, 436
mathematical modelling 26
matric forces 415
matric potential 422
Melampsora lini 289
melibiase 143
mercury 329
metabolic end-products 109–11
metallothioneins
  classification 360
  metal tolerance, requirement for 364
  production 364
  properties 360–4
Metchnikowia reukauffii 13, 118
methanol oxidase 178

methanol utilisation
  carbon assimilation, process of 189
  filamentous fungi 186
  formaldehyde fixation 186–9
  production of energy 186, 189
  single cell protein 190
  xylulose monophosphate pathway 186–8
Metharzium anisopliae 350
methionine as growth factor 307
methyl viologen 205
methylene blue 22, 328–9
miconazole 329
microbodies 188, 250
Mindeniella spp. 92
mitochondria
  absence in anaerobic fungi 108
  alkane assimilation 193–4
  citric acid cycle 383–91
  F_0F_1 ATPases 6–7, 77
  oxidation of NADH 189
  polyphosphate 275
  potassium and phosphorylation sites 319–20
  proton production 37
molybdate 289–90, 293
molybdenum, cofactor for nitrate reductase 202–5
Monascus bisporus 420, 423
Monilia sp. 299
Monilinia fruticola 197
monooxidase (mixed function oxidase) 193
Monod equation 26–7, 33
monosaccharide utilisation *see* glucose utilisation
monosaccharide transport
  kinetic approach 140
  Kluyveromyces marxianus 131–3
  Neurospora crassa 137–9
  other fungi 118–23
  Rhodotorula glutinis 133–7
  Saccharomyces cerevisiae 117, 124–31
  species and strain differences 139–40
Morchella esculenta 57
Mortierella isabellina 299
Mortierella rhizogena 197
Mössbauer spectroscopy 370
mucilage 45
Mucor flavus
  binding of particulates 56
  growth on silica gel 396
  inability to assimilate nitrate 197
  sulphur oxidation 299
Mucor hiemalis
  growth and turgor potential 400–1, 407–8

  inability to assimilate nitrate 197
  osmotic, turgor and water potentials 404
Mucor mucedo
  calcium oxalate in walls 358
  rhizoferrin 369
  transhyphal current 34
Mucor plumbeus 358
Mucor pursillus 198
Mucor racemosus 273, 276
Mucor rouxii 396, 399
Mucor spp.
  lactic acid as fermentation end-product 92
  growth factors not required 308
  inability to utilise nitrate 197
  phytate breakdown 255
multivalent metals
  binding to wall 337
  complex formation 359–64, 365–72
  mycorrhiza 372–6
  surface potential 266, 322–4, 342–3
  trace requirements, difficulty of demonstrating 336
  transport
    energy dependence 341
    interactions with phosphate 342
    less specific systems 340, 342–3
    membrane potential as driving force 344–345
    specific systems 336–40
    tolerance, in relation to 365–6
Mutinus caninus 271
Myceliophthora thermophila 299
Mycobacterium smegmalis 312
Mycocandida riboflavina 312
mycorrhiza
  ericoid
    metal tolerance 374
    phytate, release of phosphate from 286
    protein utilisation 250
    siderophores 372–3
  v-a
    organic phosphorus, inability to utilise 286–7
    polyphosphate 285
    soil, increased exploitation of 285
    translocation 454–6
  see beech mycorrhiza *and* ectomycorrhiza of forest trees
Mycotypha africana 369
myristate as growth factor 307
Myrothecium verrucaria 149

N,N'-dicyclohexylcarbodiimide (DCCD) 6, 10, 77, 308, 329

N, N', N''-triacetyl fusarinine C, hydrolysis of
   ornithine ester bond 370
N-(ethoxycarbonyl)-2-ethoxy-1,2-
   dihydroquinoline 9–10
N-ethylmaleimide 9–10, 119
NAD⁺-linked isocitrate dehydrogenase 193
NAD⁺-linked glycerol-3-phosphate
   dehydrogenase 193, 437, 440
NAD malate dehydrogenase 391
NADH-linked sideramine reductase 371
NADH / NADPH-cytochrome *c* reductase
   (NCR) 19–21
NADH (NADPH) ferredoxin
   oxidoreductase 108
NADP-linked isocitrate dehydrogenase 103,
   104, 193
NADPH-cytochrome P-450 (cytochrome *c*)
   reductase 194
Neocallimastix frontalis 108–9, 158
Neocosmospora vasinfecta
   cadmium transport 366
   chloride transport 421
   glucose transport 121
   metal tolerance 366
   nitrate utilisation 196
   osmotic potential 416
   osmotic shock 429
   solutes, internal concentrations 414–16
   zinc transport 338
Neurospora crassa
   acetate metabolism 383
   adenyl cyclase 217–21
   amino acid transport 199–200
   ammonia/methylamine transport 199 200
   calcium
      cytoplasmic concentration 349–54
      hyphal branching 349
   calmodulin 350
   chromate 289
   citrate, impermeability to 380
   copper 361
   cyclic AMP
      phosphodiesterase 350
      plasma membrane H⁺-ATPase,
         regulation 18
   cyclic AMP-dependent protein kinase 350
   deaminase, extracellular 220
   dioxygenase 182
   fructose transport 139
   galactose transport 138
   general amino acid control 233
   general amino acid transport system 217
   genes
      am 235

cpc1 234
cys-3, cys-13, cys-14 289
nit-1 203
nit-2 202, 220
nit-3 202
nit-4 201
nit-6 202
nit-7, nit-8 203
nmr 231
nuc-1⁺, nuc-2⁺ 259, 261
pcon 261
pho-2⁺, pho-3⁺ 256, 261
pho-4⁺ 256, 266
preg⁺ 259, 261
scon-1, scon-2 289
van⁺ 256
vma-1, vma-2 76
glucose transport 37, 137–8
glucose 6-sulphate transport 288
glutamate dehydrogenase 231, 235–9
glutamine, signal for metabolic control
   230–1
glutamine synthetase 238
GOGAT 238–9
growth tube 447
hyphal tip bursting 399
large spherical cells 332
metal tolerance 361–6
metallothionein 361
mitochondria
   ATPase 6–7
   functioning 385
   proton production 37
   role in nitrogen metabolism 235–6
molybdenum 202–5
mutants
   ATP-sulphurylase 291
   bat 217
   inl 448
   mtr 217
   non-spreading 399
   nuc-1, nuc-2 254
   OS-1 227
   pcon, preg 260
   sl 7
NADH-linked sideramine reductase
   370
nitrate
   reductase 196, 201–5
   transport 199
nitrite
   reductase 204
   transport 199
nitrogen catabolite repression 231–9

nitrogen utilisation
  favoured sources 230
  fine control of metabolism 234–9
  regulation 230–9
nucleic acids growth on 254
peptide utilisation 225–9
phosphate transport 37, 256–61
phosphorus family of proteins 259
plasma membrane H$^+$-ATPase
  electrophysiological studies 4–7, 333–4
  four state reaction kinetic model 5
  isolation 7–9
  molecular structure and transmembrane
    topography 9–11
plasmolysis 434
polyol concentration 419, 421
polyphosphate metabolism 272–80
potassium/rubidium transport 332–5
radiotracer compartment analysis 71–2
siderophores 369–71
sodium transport 335
solutes, internal concentrations 414–15
stress proteins 116–17
succinate transport 377–80
sulphate transport 289–91
sulphur, scavenging by methionine
    transport system 217, 219
tonoplast
  arginine transport 82–3
  H$^+$-ATPase 74–9
transhyphal current
translocation 451–2
tungstate 289
turgor potential 404
vacuoles
  isolation 74
  potassium and sodium concentrations
    70
  putrecine 80
  role in nitrogen metabolism 235–6
  vanadate transport 256
  wall porosity 39
Neurospora spp. 255
niacin *see* nicotinamide
nickel
  toxicity 374
  transport 366
  urease 336
nicotinamide (niacin) 306
nicotinic acid (vitamin B$_3$) 306
nigericin 83
nikkomycins 225, 228
Nile blue 329
Nitella sp. 6

Nitella translucens 271
nitrate
  nitrous oxide evolution 198
  reductase
    absence of 198
    mycorrhizal fungi 246
    properties 204
    regulation 201–4
  transport 199
  vacuolar pH inhibitor 6, 75
nitrite
  nitrous oxide evolution 198
  reductase 204
  transport 199
nitrogen
  catabolite control (or regulation or
    repression) 201, 208, 230–5
  denitrification 198
  favoured sources 195–8
  fine control of metabolism 234–9
  fixation, absence of 195, 396
  inorganic nitrogen
    inability to grow on 196–8
    utilisation 200–5, 237–8
  regulation 229–48
  translocation 455, 457
p-nitrophenylacetate esterase 68
nitrous oxide, anaerobic evolution of 198
Nothofagus donbeyi 184
nuclear magnetic resonance (NMR)
  carbon metabolism 89–91
  cytoplasmic orthophosphate
    concentrations 269, 272
  detection of radical cations 175
  determination of cytoplasmic and vacuolar
    pH 16, 21, 72–3, 207, 280
  phosphate transport 267
  polyphosphate metabolism 272, 275, 282
nuclease 255
nucleic acids, breakdown 252
nystatin 64, 74, 82, 329, 338

Oidiodendron griseum 372
oleic acid 114–5
oligocarbonotrophy 396
oligomycin 6, 75
oligotrophy 396–7
3-*O*-methyl glucose (3OMG) 119, 120, 452,
    457
one-carbon compounds *see* methanol
    utilisation
Ophiobolus (Gaumannomyces) graminis
    196
Ophiobolus miyabeanus 196

organic acids
  accumulation in the vacuole 86
  biosynthesis 384–92
  carboxylation/decarboxylation 394–5
  cation absorption 393–4
  charge balance 392–5
  enzyme levels, regulation of
  'long' biosynthetic path 385
  oligotrophy 395–7
  osmotic potential 377
  pH
    biochemical pH-stat 394–5
    external, prevention of fall of 247
    internal 392–5
    phosphorus solubilisation 251–3
    protons, loss of 377
    range of acids produced by fungi 392
    'short' biosynthetic path 385
    transport 86, 377–82
ornithine : 2-oxo-acid aminotransferase 241
ornithine acetyl transferase 241
ornithine carbamoyl-transferase 235–6, 241
ornithine decarboxylase 236
ornithine transaminase 236
osmo-acclimation 426
osmoduric 421
osmophile 421
osmotic potential *see* water relations
osmotic regulation (osmoregulation) 426
osmotic shock 215
osmotolerant 421
osmotophilic 421
osmotrophic 421
Ostracoblabe implexa 91
overflow metabolism 384–5
overshoot 311
oxalacetase 386
oxaloacetate hydrolase 386
oxalate
  biosynthetic pathways 358
  calcium 357–9
  cellulose degradation 159
  ecological significance 358
  lignin breakdown 177, 180–1
  pectin degrading enzymes 359
  pH regulation 359
  solubilisation of phosphate 253
  structural role 359
oxalate decarboxylase 386
oxonol *V* 8
oxygen
  chemotropism 42–3
  lignin breakdown 161, 166, 170–2,
    179–83

potassium and yield value 319
  requirement for growth 92

Pachysolen spp. 197
Pachysolen tannophilus 104–6
palmitic acid 114, 307
palo podrido 184
pantothenic acid (vitamin B₅)
Panus tigrinus 171
parasitic fungi xiv
Pasteur effect 97–9, 101–2
Paxillus involutus
  solubilisation of insoluble soil phosphorus
    253
  sulphur oxidation 299
  zinc tolerance of *Betula* 374
Paxillus sp. 374
Paxillus tinctorius 376
pellet
  anaerobic conditions 31–2
  autolysis 448
Penicillium bilaii 252
Penicillium chrysogenum
  amino acid transport 221–3
  APS kinase 295
  choline-*O*-sulphate transport 221, 288
  general amino acid transport system 221
  growth on silica gel 396
  methylamine transport 200
  nitrate
    assimilation 202–4
    reductase 204–6
  osmotic shock 430
  PAPS 295
  pellets 448
  solutes, internal concentrations 417–18
  sulphate transport 290-1
  sulphite transport 300
Penicillium cyclopium 221–3
Penicillium digitatum 88
Penicillium expansum 198
Penicillium notatum
  calmodulin 350
  glucose transport 121
  growth on silica gel 396
  manganese transport 339
  sulphate transport 290–2
  zinc transport 339
Penicillium ochro-chloron 367
Penicillium patulum 182
Penicillium pinetorum 299
Penicillium rubrum 396
Penicillium sp. 370
Penicillium spinulosum 182

Penicillium spp.
  calcium requirement for sporulation 347
  coprogen production 267
  growth factors not required 308
  methanol utilisation 185
  nitrate utilisation 196
  phytate breakdown 255
  tolerance of low water potentials 420, 423
Penicillium urticae 183
pentose phosphate pathway *see* glucose
  utilisation
pentose utilisation *see* xylose utilisation
peptide utilisation 223–9
Perenospora spp. 58–9
peripheral growth zone 38
periplasm 142, 309
peroxidase 167, 180–9
peroxisomes 187–9
Phallus impudicus
  growth and turgor potential 411–13
  osmotic, turgor and water potentials 411
  phosphate transport 269
  phytate breakdown 255
Phanerochaete chrysosporium
  acetovanillone-oxidising activity 169
  cellulose degradation 178–80
  fruiting structures 165
  β-(1 → 3)-(1 → 6)-D-glucan sheath round
    hyphae 167
  glucose level of medium, regulation 170
  glucose transport 170
  lignocellulose breakdown
    choice for studies 165
    review of observations 165–83
  media for growth 166–7
  molecular studies 165–7
  mutants
    lacking glucose oxidase 178
    lip 176
    PSBL-1 170
  NAD, NADH, NADP and NADPH
    levels in the medium 176
  protein content of cultures 168–9
  strain ME446, similarity with
    Sporotrichum pulverulentum
    Novobranova 165
Phanerochaete sp. 271
Phanerochaete vellutina 299, 411
Phellinus noxius 404
phenol oxidases 181–3
phenols 180–3
phenosafranine 181–3, 329
phenylglyoxyl 9–10
Phlebia radiata 171

Phlychtochytrium 436
Phoma apiicola 196
Phoma betae 196
phosphatases 253–5
phosphate
  aluminium 376
  cytoplasmic concentration 265, 269, 272
  interactions with multivalent metals
  organic
    breakdown 253–5
    inositol phosphates 255
    phosphomonoesters 253–4
    phosphodiesters 254
    phosphatases 253–5
  transport
    plasma membrane 255–69
    phosphate-binding proteins 267–8
    tonoplast 85
phosphatidylcholine 60, 114–15, 307
phosphatidylethanolamine 60
phosphatidylethanolamine methyltransferase
  316
phosphatidylinositol synthase 316
phosphatidylserine 61, 114–15
phosphatidylserine carboxylase 61
phosphatidylserine synthase 316
phosphite 268
3′-phosphoadenosine-5′-phosphosulphate
  (PAPS) 294–6
6-phosphogluconate dehydrogenase 90–1, 93
phosphodiesterase 170, 395
phosphoenolpyruvate carboxykinase 247
phosphoenolpyruvate carboxylase 395
phosphofructokinase 90–1, 93, 110, 387, 440
phosphoglucoisomerase 90–1, 93
phosphoglucomutase 90–1, 110
phosphoglycerate kinase 110
phosphoglycerate mutase 110
phosphoinositidase C 61
phosphoketoepimerase 90–1
phosphoketoisomerase 90–1
phospholipid methyltransferase 316
phospholipids
  breakdown 254
  phosphate deficiency 280
  plasma membrane 74
  tonoplast 74
phosphomannans 254
phosphomannose isomerase 90–1
phosphomannose mutase 90–1
phosphomethylpyrimidine kinase 310
phosphoriboisomerase 90–1
phosphorus
  aluminium 375–6

phosphorus *continued*
compartmentation in the protoplasm 269–82
family of proteins 259
soil
  calcium/aluminium/iron phosphates 251
  soluble phosphate 251
  see solubilisation of soil phosphorus
translocation
utilisation
Phycomyces blakesleeanus
calcium
  growth requirement 347
  vacuoles 85
calmodulin 350
ferritin 369
inability to utlise nitrate 197
sporangiophores
  avoidance response 409
  calcium location 85
  hydraulic conductivity 406
  light response 409
  turgor potential 409
  wall 409
phylloplane 396
phytate 255
Phytophthora bahamensis 198
Phytophthora cactorum 88, 347
Phytophthora cinnamomi
growth and turgor potential 402, 404, 407–8
osmotic shock 430
solutes, internal concentrations 417–18
tolerance of low water potentials 420
Phytophthora epistomum 307
Phytophthora fragariae 198, 347
Phytophthora infestans 198
Phytophthora megasperma f. sp.*glycinea* 13
Phytophthora palmivora
calcium requirement for morphogenesis 347
glucose transport 121
phosphate/phosphite transport 268
polyphosphates 272
Phytophthora sp. 198, 399
Picea excelsa 248
Picea sitchensis 243–4
Picea sp. 374
Pichia guillermondii 143
Pichia heedii 121
Pichia miso 431
Pichia ohmeri 121
Pichia segobiensis 104
Pichia spp., nitrate utilisation 197

Pichia stipitis (presumptive telomorph of *Candida shehatae*) 104–6, 122
Pilobolus longipes 122
Pilobolus sp. 307
Pinus contorta 244–5
Pinus pinaster 244–5
Pinus radiata 358, 462
Pinus rigida 375
Pinus sp. 374
Pinus spp. 462
Pinus strobus 242, 244
Piromonas communis 158
Pisolithus sp. 373
Pisolithus tinctorius
aluminium tolerance 375
nitrogen metabolism 243–5, 247–8
siderophores 372
surface reducing activity 376
sulphur oxidation 299
Pisonia grandis 376
Pityrosporium ovale 307
plasma membrane
Ca^{2+}-ATPase 355
determination of surface area 31
fluidity 60, 63, 302
H^{+}-ATPase
  properties 4–19
  proton pump 4
  transmembrane structure 13–14
  potassium transport 333–4
  lipid and sterol content and functioning 58–65
phosphoryl groups 323
redox systems 19–24
sodium exclusion 438–40
surface area 31
surface potential 266, 322–4, 376
tolerance to ethanol and sulphur dioxide 112–17
wall, addressed to 434–5
plasmolysis 434
Pleurage curvicolla 196
Pleurotus ostreatus 171, 196
poly-L-lysine 46
polycations 46, 65
polyene antibiotics 61
polyethylene glycol 399, 401, 408–9, 433
polyols
compatible solutes 443–6
constancy of total concentration 418–21
metabolism 93–5
osmotic response 95, 415–21
proton-consuming and proton-producing reactions 24, 446

taxonomic distribution 95–6
polyoxins 225
polyphosphate
  acid-soluble 273–8
  acid-insoluble 273–8
  aluminium 376
  complexes with other polymers 274
  cycle 278–9
  kinase 280
  metabolism 275–82
  mitochondria 275
  nucleus 275
  Oomycetes, presence in 272
  role
    alkaline stress protection 282
    cytoplasmic orthophosphate
      concentration 281–2
    energy dissipation 282
    phosphagen hypothesis 281
    phosphorus reservoir 280–1
  sugar transport, involvement in
    129–32
  translocation 285, 455, 466
  vacuole 81, 275
  v–a mycorrhiza 285, 455
  wall 274–5
  zinc depletion 345–6
Polyporus brumalis 466
Polystictus versicolor 171, 182
Poria vaillantii
  copper tolerance 367
  growth factor requirements 307, 308
Postia (Poria) placenta 159, 367
potassium
  concentration in cytoplasm and vacuole
    70–1
  efflux 327–30
  growth rate 318–20
  lyotropic effects 318
  organic acids, balance 393–4
  oxidative phosphorylation 318–19
  replacement by ammonium/rubidium
    319
  ribosome functioning 318
  surface potential 322–4
  transport 320–35
pressure probe 405, 408
proline oxidase 236
protein
  breakdown 387
  sorting 79
  utilisation 249–50
  wall 55
proteinase A (acidic endoproteinase) 68

proteinases
  regulation role in
    release of nitrogen in starvation 249
    utilisation of extracellular proteins
      249–50
  vacuole 68
protons
  chemotropism 42–3
  economy 3
  pump 4, 79–81
  stoichiometry of a transport system
    15–16
prototrophic 305
Pseudotsuga menziesii 245
Pseudotsuga taxifolia 289
Psilocybe cubensis 466
Puccinia spp. 198, 295
Puccinia graminis 198, 295, 304
Puccinia graminis tritici 289
purine and pyrimidine transport 212, 215
putrecine in vacuoles 80
pyrazole 21
pyridoxine (also pyridoxal; vitamin B_6)
  function 306
  transport 312–13
pyrimidine-specific carbamoyl-phosphate
  synthetase 236
Pyronema confluens 196
pyrophosphatase 68, 79–81
pyrroline 5-carboxylate dehydrogenase 236
pyrroline 5-carboxylate reductase 236
pyruvate carboxylase 108, 358, 386, 390, 392
pyruvate decarboxylase 108, 110
pyruvate kinase 90–1
Pythiogeton spp. 92
Pythiomorpha gonapodyoides 196
Pythium debaryanum 196
Pythium graminicola 347
Pythium intermedium 196
Pythium irregulare 196
Pythium spp. 58-59
Pythium ultimum 252, 420

radical cations 172–3
raffinose 141, 145
Ramaria largentii 57
random ligand binding 268
ras-adenylate cyclase signalling pathway 18
redox dyes 22–4
reflection coefficient 403
Rhipidium spp. 92
Rhizoctonia solani
  ineffective in solubilisation of insoluble soil
    phosphorus 252

Rhizoctonia solani continued
  nitrate utilisation 196
  tolerance of low water potentials 420
rhizoferrin 367
rhizoids 39
rhizomorphs
  translocation 460–6
  vessel hyphae 463–5
Rhizophlyctis rosea 197
Rhizopogon luteolus 358, 462
Rhizopogon roseolus 299
Rhizopogon sp. 374
Rhizopus arrhizus 385, 390
Rhizopus javanicus 195
Rhizopus microsporus var. *rhizopodiformis*
  369
Rhizopus nigricans 197, 452–3
Rhizopus oryzae 197
Rhizopus spp.
  fumaric acid synthesis 390
  growth factors not required 308
  lactic acid as fermentation end-product 92
  phytate breakdown 255
  translocation 413
Rhizopus stolonifer 365
rhizosphere 373, 396
Rhododendron ponticum 374
Rhodosporidium spp. 197
*Rhodosporidium toruloides see Rhodotorula
  glutinis*
rhodotorucine A
Rhodotorula 93
Rhodotorula glutinis (= *Rhodosporidium
  toruloides*)
  calcium, transient uptake 349
  Crabtree negative 102
  glucosamine transport 134
  monosaccharide transport 133–7
  nitrate reductase 204–6
  nystatin-resistant strain 64
  pentitol transport 137
Rhodotorula gracilis (= *Rhodotorula glutinis*)
Rhodotorula minuta 369
Rhodotorula mucilaginosa 182
Rhodotorula pilimanae 369, 371
Rhodotorula sp. 299
rhodotorulic acid 369
riboflavin (vitamin B2)
  microorganisms producing high
    concentrations 314
  transport 313
ribonuclease 39–40, 68, 255
ribulose bisphosphate carboxylase/oxygenase
  397

rubidium
  growth rate 319
  replacement of potassium 319
  transport 320–5
  translocation 452
rumen fungi 108–9, 158
Russula brevipes 51
Russula spp. 248
Russula xerampelina 57
rust fungi, plasmolysis of haustoria 434

S-adenosyl-L-methionine, tonoplast
    transport 82
S-adenosylmethionine demethylase 297
S-adenosylmethionine synthetase 297
S-formylglutathione hydrolase 187
Saccharomyces bayanus
Saccharomyces bisporus var. *mellis* 276
Saccharomyces carlsbergensis
  adaptation to arsenate 262
  calcium, magnesium, manganese and zinc
    transport at the tonoplast 85–6
  electrophoretic properties 52
  ethanol, effect on transport 116
  polyphosphate incorporation into wall
    274, 276
  potassium concentrations in cytoplasm
    and vacuole 70
Saccharomyces cerevisiae
  α-factor 47, 228, 249
  alkanes, inability to use 191
  amino acids
    anabolic uptake systems 211
    catabolic uptake systems 207–11
    ions, movement of other 207–9
    pools 63
    shock excretion 213
    transport across plasma membrane 60–3
    vacuolar content 212
  ammonia transport 206, 319
  ammonia/calcium/magnesium/sodium-
    loaded cells 330
  arsenate transport 262
  ATP synthesis via fermentation and
    respiration 96–9
  biotin transport 311–12
  bud growth and abscission 49–50
  cadmium 361–2
  calcium
    ATPase 355
    calmodulin protein-dependent kinases
      350
    cytoplasmic concentration 351
    influx 344–5, 355–57

calmodulin 350
catabolite inactivation 99, 147
cell wall
    charged groups 50-3
    electrophoretic properties 52
    porosity 45-50
    proteins 55
    shrinkage 433-5
choline, transport and metabolism 314-16
cobalt 366
copper
    resistance 361-2, 366
    transport 339, 366
Crabtree positive 29, 102
cyclic AMP-dependent protein kinases
    350
cytoplasmic pH 394-5
2-deoxy-D-glucose, effect on phosphorus
    metabolism 131
disaccharide utilisation 140-8
ergosterol synthesis requires molecular
    oxygen 97, 307
ethanol
    fermentation 28, 392
    formation 87-9
    tolerance 109-17
facultative anaerobe 97
flocculation
    calcium-bridging hypothesis 53-4
    'lectin-like' hypothesis 54-5
fructose transport 117-31
galactose
    transport 117-18
    utilisation 96-7
general amino acid control 233-4
general amino acid transport system 63,
    207
genes
    CAN1 315
    control of glucose repression 101
    CUP1 361-2
    CTR 315
    ENA1 438-9
    ENA2 438
    FUR4 211
    GAL2 127
    gdhCR (URE2) 231
    HIP1 315
    HXT1 127
    HXT2 120, 127
    ITR1, ITR2 314-16
    MAL1, MAL2, MAL3, MAL4, MAL6
        146
    MGL1, MGL2, MGL3, MGL4 145

ORF2 127
PHO2 266
PHO3 309
PHO48 266
PMA1 17
PMR1(SSC1) 355
PMR2 355
SNF3 127
TRK1, TRK2 325
URE2 231
VMA1 76-7, 85
VMA2 76
VMA3 76-7, 85
VMA4 79
glucose utilisation 87-9, 96-101
glucose/2-deoxy-D-glucose transport
    facilitated diffusion 29, 124, 129
    kinetics 125, 129
    metabolic state, effect of 125-7
    genes 127-8
    glucose 6-phosphate as regulatory agent
        128
    growth 27-30, 32-3, 125
    kinases, involvement in 125
    phosphorylation, as part of the process
        129-30
    sensing role 128
    vesicles 129
glucose repression 97-101
glutamine, signal for metabolic control
    230-1
glutathione, sulphur storage 296-7
glycerol 430-1, 433
GOGAT, absence of 238
hydroxymethylpyrimidine transport 311
inositol transport and metabolism 314-16
invertase 68
malic acid synthesis 392
maltose transport 100, 145-7
mating aggregates 53
melibiase 143
metal tolerance 361-2, 366
metallothioneins 361-2
methylamine transport 306
α-methyl glucoside transport 145-6
mitochondria 19, 385-6
monovalent cation transport
    bivalent cations, importance for 320-1
    genetic studies 324-5
    kinetic studies 320-4
    physiological studies 326-7
    potassium efflux 327-30
    sodium efflux 330-2
    surface potential 322-4

multivalent metal transport 339–45
mutants
   chr 293
   cls 85
   cpc1 234
   decreased ATPase actvity 330
   gap 207
   hxt2 127
   mep-1, mep-2 200
   mgl2 145
   mmn9 47
   Pet⁻ 85
   potassium transport 324–5
   riboflavin-requiring 313
   salt tolerance 442–3
   sel 293
   snf3 127
   trk1 325
   vma4 79
nickel transport 366
nitrogen utilisation
   favoured sources 230
   fine control of metabolism 234–8
   inducer exclusion 233
   rapid adaptation 231
   regulation 229–38
   starvation 249
organic acid transport 377–82
osmometer, perfect 427
osmotic potential 405
osmotic shock 426–8, 430–2
overshoot 311
Pasteur effect 97–9
peptidases, no evidence for extracellular 228
peptide utilisation 228–9, 249
phosphate
   cytoplasmic concentration 265
   transport 261–7
phosphate-binding proteins 267–8
phospholipid metabolism 60–1
plasma membrane
   gross chemical composition 74
   H⁺-ATPase 12–19, 327
   lipid composition and transport 59–63
   phosphatidylinositol, as main charged lipid 316
   potassium channels 332
   redox systems 21–3
polyphosphate
   content 273, 275–6
   cycle 278–80
   mitochondria 275
   wall 51

vacuole 81
potassium
   channels 332
   organic acids 394–5
   phosphate 262, 266
   transport 252, 320–32
protein degradation 248–9
proteinases 249
purine and pyrimidine transport 211–13
ras-adenylate cyclase signalling pathway 18
resistance to bacilysin, L-ethionyl-L-alanine, nikkomycin 225
respiratory enzymes, repression by glucose 225
riboflavin transport 313–14
secretion of homologous and heterologous proteins 47–50
sodium
   efflux 330–2
   glycolysis, inhibition 440
   phosphate transport 261, 265–6
   transport 320–32
   trehalose synthesis 442
strains
   ade2 79
   301N 361–2
   metal tolerant 361
   NCYC 366 62
   poky f 17–18
   respiratory deficient 331
   WT 293
   Y185 62
sterols 59–60
stress proteins 116–17
succinate synthesis 391–2
sulphate transport 292–4
sulphur fluxes 296–7
surface potential 266, 322–4, 376
thiamine, depression of growth 310
thiamine/pyrithiamine transport 308–11
tonoplast
   basic amino acid transport 83–84, 237
   calcium transport 84–85
   gross chemical composition 74
   H⁺-ATPase 75–9
   monovalent cation channels 86
   pyrophosphatase 79–81
total monovalent cation concentrations in cytoplasm and vacuole 70
transhydrogenation, absence of 103
trehalose 441–2, 444–6
trehalose transport 147–8
turgor potential 405

unsaturated fatty acid synthesis requires molecular oxygen 97
urea transport 207
vacuoles
  amino acids 83–4
  biogenesis, defects in 442
  calcium 73, 84–5
  cobalt 80
  enzymes 68
  iron 80
  nitrogen metabolism, role in 234–7
  polyphosphate 275
  zinc 80
vitamin B_6 transport 312–13
wall shrinkage 433–5
water potentials, tolerance of low 420, 423
Saccharomyces fragilis (= *Kluyveromyces marxianus*)
Saccharomyces pastorianus 86
Saccharomyces sake 114–15
Saccharomyces spp. 197
Saccharomyces unisporus 113
Saccharomyces uvarum
  ethanol tolerance 115
  pyridoxine/pyridoxal/pyridoxamine transport 312–13
  potassium transport gene
Saccharomycodes ludwigii 301
salicyl hydroxamic acid (SHAM)-sensitive oxidase 92
Salix viminalis 283
salting out 318
Saprolegnia delica 88
Saprolegnia ferax
  calcium, growth requirement 247
  growth and turgor potential 400–1, 409
  osmotic potential 400–7
  stretch-activated ion channels 38
  turgor potential 400–1, 405–8
Sapromyces spp. 92
saprotrophic fungi xiv, 304
Sarcodon imbricatus 57
saturation constant (K_S) 26
Schizophyllum commune
  dioxygenase 182
  fruiting genes, *Sc1 Sc2 Sc4* 57–8
  hyphal tip bursting 399
  transhyphal current 34
Schizosaccharomyces maldevoran 242
Schizosaccharomyces pombe
  Ca^{2+}-ATPase 355
  cadmium 362
  calcium uptake 347
  copper 362

DEAE-dextran penetration of walls 46
genes
  cta3 355
  sod2 331, 439
glucose and 6-deoxy-D-glucose transport 123
GOGAT 241
L-malic acid transport 379
metallothioneins 362–363
myo-inositol transport 316
plasma membrane H^+-ATPase 12–13
sodium efflux 439
thiamine transport 310–11
Schizosaccharomyces spp. 197, 312
Schizosaccharomyces versitalis 242
Schwanniomyces occidentalis 49
Scleroderma flavidum 374
Scleroderma spp. 374
sclerotia 467
Sclerotinia minor 196
Sclerotinia sclerotiorum 196, 357, 434
Sclerotium bataticola 196
Sclerotium cepivorum 359
Sclerotium rolfsii
  calcium oxalate in hyphal walls 357
  oxalate biosynthesis 357–9
  pectin-degrading enzymes 359
  solubilisation of insoluble soil phosphorus 252
  zinc transport 339
SCN *see* thiocyanate
Scopulariopsis brevicaulis 196
secondary metabolism, shunting to 384–5
selective delignification 183–5
selenate 293
Septoria nodorum 196, 202
serine acetyltransferase 297
Serpula lacrymans
  calcium oxalate in hyphal walls 358
  droplets 406, 458–9
  growth and turgor potential 411–13
  hydraulic conductivity 406
  osmotic and turgor potentials 404
  tolerance of low water potentials 420
  translocation
    direct monitoring of radioactivity 456–8
    effect of low temperature 459–60
    mechanism 459–60
    phloem translocation, similarity to 460
    trehalose as major translocate 459
    velocity 456–8
    water content of substratum 459
S-formylglutathione dehydrogenase 187

S-formylglutathione hydrolase 187
siderophores
  aluminium 369
  coprogens 369
  ferrichromes 369
  fusarinines 368
  gallium 369
  iron
    ferric 367
    release 370–1
    storage 370
  mycorrhiza 372–4
  rhizoferrin 369
  rhodotorulic acid 368–9
  terminology 367
  transport 369, 371–2
  types
    complexerone 369
    hydroxamate 367–9
single cell protein 190
site-directed mutagenesis 14
sodium
  active extrusion 438–40
  chloride and fungal growth 421, 435–8
  concentration in cytoplasm and vacuole
    70
  efflux 330–5
  glucose symporter 2
  glycerol co-transporter 433
  lyotropic effects 318
  micro-element question 317
  phosphate transport requirement 261, 265–
    6
  transport 320–35, 438–9
sodium dodecylsulphate (SDS) 274
soil fungi, tolerance of low water potentials
    420, 423
solubilisation of insoluble soil phosphorus
  effect on plant growth 252
  microbial mechanisms 251–2
  mycorrhizal fungi 253
  nitrogen, role of 253
  organic acids 251–3
  organic phosphorus 253–5
solubilisation of metal ions 252
solute potential *see* osmotic potential
sophorolipids 192
sophorose 157
sorbitol (glucitol) by-pass 95
sorbitol dehydrogenase 90–1
Sordaria fimicola 196
spatial aspects of nutrient aquisition 41–3
Sphaerobolus stellatus 196
Sphaeromonas communis 158

Sphaeropsis malorum 196
Sphaerostilbe repens 349
spore release 398
Sporidiobolus spp. 196
Sporobolomyces roseus 205
Sporobolomyces spp. 182, 197
Sporodina grandis 197
Sporopachydermia spp. 195
Sporotrichum pulverulentum see
    Phanerochaete chrysosporium
Stecherinum fimbriatum 271
Stemphylium sp. 452–3
stereology 31
Stereum gauspatum 88
Sterigmatomyces spp. 197
sterols in the plasma membrane 59, 64–5, 74
stress proteins 111, 116–17
stretch-activated ion channels 38
Suillus bovinus
  rhizomorphs
    translocation 462
    water movement 463
  sulphur oxidation 299
Suillus brevipes 372
Suillus granulatus 376
Suillus spp. 374
Suillus tomentosus 57
Suillus variegatus 376
suloctidil 329
sulphatases 388
sulphate
  protoplasmic concentration 296
  transport 289–94
sulphite reductase 295
sulphorhodamine G 467
sulphotransferases 295
sulphur compounds
  inorganic, oxidation 298–300
  metabolism 294–7, 364
  metallothioneins 360–5
  storage 296–7
  translocation 455
  volatile 288
sulphur dioxide
  beverage/food preservative 288
  equilibria in solution 300
  tolerance 300–2
  transport 300–1
superoxide dismutase 92
surface potential
  monovalent cation transport 322–4
  phosphate transport 266, 376
  phosphoryl groups 323
  zeta potential 324

symbiocalcicoles 376
symport 2-3

Talaromyces emersonii 149
techoic acids 254
telurite 293
tetrapentyl ammonium chloride 408
tetraphenylphosphonium (TPP$^+$) 133-4, 211, 326
Thelephora terrestris 283
thermocouple psychrometry 400, 404-5, 411-13
thermotolerance 117
thiamine (vitamin B$_1$)
  analogues 309
  binding protein 309-10
  function 306
  photoinactivation of the transport system 309
  requirement 305
  transport 308-11
thiamine-phosphate pyrophosphorylase 310
thiocyanate (SCN) 6, 75, 135
Thraustochytrium aureum
  ammonia, inability to grow on 196
  lysine as growth factor 307
  plasma membrane H$^+$-ATPase 12
  sodium exclusion 438
  sodium and growth 436
  solutes, internal concentrations 417
Thraustochytrium sp. 268
Tilletiopsis washingtonensis 182
thiophenol 224
toluylene blue 329
tonoplast
  elasticity 74
  H$^+$-ATPase
    genes 76-9
    properties 6-7, 75
    structure 75-9
  potential difference across 81
  pyrophosphatase 79-81
  resistance to
    nystatin 74
    polycations 75
  sensitivity to pH 75
  solute transfer across 81-6
  sterol content 74
Torulopsis bombicola 192
Torulopsis candida 123
Torulopsis glabrata 145
Torulopsis spp. 185
trace metals 337
training for metal tolerance 365

Trametes cingulata 171
transhyphal currents 33-9
transaldolase 90-1
transglycosylation 157
transhydrogenation 103
transinhibition 217, 222, 290-1
transketolase 90-1, 189
translocation
  basidiomycete fruit bodies 466-7
  bidirectional 462-3
  evaporational flow 466-7
  flux 455, 459, 463
  mechanism
    diffusion 449, 453-4
    contractile systems 450, 455-6
    solution flow 450, 452, 456-60
  moulds 449-54
  Petri-dish culture 449-51
  rhizomorphs
    bidirectional translocation 462-3
    dry matter flux 462
    septal pores 463-4
    vessel hyphae 463-5
    water movement 456-66
  Serpula lacrymans 456-60
  transpiration 466-7
  turgor gradients 460-1
  v-a mycorrhiza 454-6
  velocity 457
transport
  activation by phophatases 253-4
  active 1-4
  charge balance 134-5, 209, 264
  random ligand binding to carrier 268
  transinhibition 219, 222, 290-1
  stoichiometry 4-5, 15-16, 220
  surface potential 266, 322-4, 376
trehalase (α, α'-glucoside 1-glucohydrolase) 40, 68, 145
trehalose
  compatible solute 442-6
  cycle 445
  stress metabolite 440-1, 446
  transport 147-8
  water stress plating hypersensitivity 441-2
triacetylfusarinine 370
tricarboxylic acid cycle *see* citric acid cycle
Trichoderma harzianum 34, 299
Trichoderma koningii 149
Trichoderma lignorum 196, 350
Trichoderma reesii
  cellulose utilisation 145, 149-58
  cellulases, nomenclature of 154
  strain QM6a 149

Trichoderma sp. 299
Trichoderma spp. 185, 299
Trichoderma viride
  DNA breakdown 254
  sulphur oxidation 299
  translocation 452–3
Tricholoma saponaceum 57
Trichophyton interdigitale 347
Trichophyton megninii, histidine as growth
    factor 307
Trichosporon spp. 196
trifluoperazine 329
triose phosphate isomerase 90–1, 110
triphenyl tin 329
triphenylphosphonium (TMP$^+$) 133–4
trisaccharide utilisation *see* disaccharide
    utilisation
trivalent metals *see* multivalent metals
Tsuga heterophila 245
tungstate 289
tunicamycin 309
turgor
  maintenance 426
  potential
    anomolous values 412–413
    determination 400
    growth 400–3, 408–9, 411–13
    values 401–2, 408–9, 411
  regulation 426
  yield threshold 403
Tween 115, 168
tyrosinase 181–3

UDP-galactose pyrophosphorylase 90–1
UDP-glucose epimerase 90–1
UDP-glucose pyrophosphorylase 90–1
uniport 2–3
upshock 426–9
uranyl ions 51, 127
urea amidohydrolase 236
urea cycle 240
urea transport 207–8
urease 236, 241
Ustilago maydis
  glucose transport 123
  nitrate assimilation genes 202
Ustilago sphaerogena 267

Vaccinium macrocarpon 374
vacuolar membranes *see* tonoplast
vacuole
  amino acids within 235–7
  determination of composition 67–73
  green plant 66

  isolation 71, 74–5
  large algal coenocyte 66
  lysosome 249
  polyphosphate 8
  processing of vacuolar enzymes 67
  proteinases 68
  withdrawal of sap for analysis 66
vanadate 6–7, 75, 334
Varia granulosa 182
veratryl alcohol
  involvement in ligninase redox cycle 176
  oxidation by ligninase 173–5
  oxidising agent 175
  wood, ability to penetrate 175
veratryl aldehyde 173
Verticillium atro-album 196
vessel hyphae 463–5
vibrating micro-electrode (vibrating probe)
    33
Viets effect 324
vitamin B$_1$ *see* thiamine
vitamin B$_2$ *see* riboflavin
vitamin B$_3$ *see* nicotinic acid
vitamin B$_4$ *see* pantothenic acid
vitamin B$_5$ *see* pyridoxine
vitamin B$_7$ *see* biotin
vitamin B$_{12}$ *see* cyanocobalamin
vitamins 303, 306
volumetric elastic modulus 409–10
volumetric hydraulic conductance 403
Volvariella volvacea 249–50

wall
  absorption of particulates 56
  alkanes, adherence 191–2
  calcium oxalate 357–9
  cation exchange capacity 56–7
  charged polymers 45
  enzymes 44
  expansion 398
  elastic extensibility 406
  electrophoretic properties 52
  flocculation 53–5
  hairs 192
  hydrophobicity 56–8
  ion binding 53–6, 336, 418
  lipids within 59
  lytic enzymes 398–9
  phosphatases 47, 253–4
  physical properties, alteration 399
  polymers, plasicity/rigidity 399
  polyphosphate 44
  porosity 39–40, 45–50

relative volume 44, 435
shrinkage 434–5
wall yielding coefficient (extensibility) 403, 406
Wallemia sebe 418, 423
water potential *see* water relations
water relations
  basidiospore discharge 389
  compatible solutes 443–6
  conditions far from equilibrium 411–15
  osmo-acclimation 426
  osmoregulation 426
  osmotic potential
    cardinal 422–3
    determination 400, 415
    maximum 422–3
    minimum 422–3
    optimum 422–3
    values 404–5, 415–18
  osmotic shock 426
  spore release, lower fungi 389
  tolerance of microorganisms to low water potentials 420
  turgor maintenance and regulation 426
  potential
    determination 400, 408
    growth 400–3
    values 404–5
  volume regulation 426
  wall elasticity 410, 425–6
  water potential 399, 403–6, 411, 416–17
water stress plating hypersensitivity 441–2
wheat (*Triticum aestivum*) 252
Wickerhamiella spp. 197
Williopsis mrakii 49
willow (*Salix viminalis*) 383
wood decay fungi 159–85, 420, 422
Woronin bodies 250

X-polyldipeptidyl aminopeptidase 68
X-ray microanalysis 38, 71, 329
Xeromyces bisporus
  growth and turgor potential 402, 404, 408
  osmotic potential 418
  tolerance to low water potentials 423–4
xerophilic 421
xerotolerant 421
Xylaria mali 196
xylanase 153
xylose isomerase 104
xylose reductase (NADPH) 104–6

xylose utilisation
  biotechnological significance 106–7
  ethanol, inhibition by 106–7
  fungi capable of 88
  glucose repression 106–7
  pathways 106–7
xylulose monophosphate pathway 186–7

Yarrowia lypolitica (= *Candida lypolitica*) 386
yeasts
  alkane utilisation 191–4
  amines, one carbon, utilisation but inability to grow on 189–90
  bottom (lager) 143
  Crabtree effect 102–3
  Custers effect 101–3
  disaccharide utilisation 140–8
  ethanol tolerance 109–17
  facultatively fermentative 101
  glucose and glycoside utilisation 95–104, 141
  galactose and galactoside utilisation 107, 141
  growth factors 305, 307–16
  Kluyver effect 102, 107
  methanol utilisation 186–90
  monosaccharide utilisation 95–109
  non-fermentative 101
  obligately aerobic 93
  obligately fermentative 101
  organic acid utilisation 377–82
  Pasteur effect 97–99, 101
  phytate, inability to break down 255
  RNA, inability to break down 254
  single cell protein 190
  sulphur dioxide tolerance 300–2
  top (ale) 143
  xylose utilisation 104–7
yield threshold 403
yield value 319

zeta potential 324
zinc
  accumulation during growth 345–6
  potassium efflux 344
  toxicity 336, 374
  translocation 455
  transport 336–46
Zygorrhynchus moelleri 197, 299
Zygorrhynchus spp. 308
Zygorrhynchus vuieminii 299

Zygosaccharomyces bailii
  benzoic acid transport 381
  permeability to other organic acids 381–2
Zygosaccharomyces rouxii
  glycerol 429–31, 433
  osmotic shock 430–3
  plasma membrane
    H$^+$-ATPase 12
    lipid composition 60

tolerance to sodium chloride
  adaptation to 431
  in vitro activity of isocitrate
    dehydrogenase 437
  mutants 64, 442
sodium extrusion 438–9
tolerance of low water potentials 420, 423
Zygosaccharomycodes bailii
  sulphur dioxide transport 301–2